*Continued on inside back cover*

# BRIEF CALCULUS
## WITH APPLICATIONS

### SECOND EDITION

# BRIEF CALCULUS
# WITH APPLICATIONS

## SECOND EDITION

ROLAND E. LARSON

ROBERT P. HOSTETLER
with the assistance of David E. Heyd

The Pennsylvania State University
The Behrend College

D. C. HEATH AND COMPANY
Lexington, Massachusetts   Toronto

International Standard Book Number: 0-669-12060-X

Library of Congress Catalog Card Number: 86-82096

# Preface

*Brief Calculus with Applications, Second Edition* is designed for use in a beginning calculus course for students in business, economics, management, and the social and life sciences. In writing this edition, we were guided by two primary objectives formed over many years of teaching calculus. For the student, our objective was to write in a precise and readable manner, with the basic concepts, techniques, and applications clearly defined and demonstrated. For the instructor, our objective was to create a comprehensive teaching instrument that employs proven pedagogical techniques, thus freeing the instructor to make the most efficient use of classroom time.

As full-time calculus instructors, we have found the following features to be valuable aids to the teaching and learning of calculus.

## FEATURES

**Introductory Examples**   Each section in Chapters 1–10 begins with a one-page motivational example designed to show the applicability of the material in the section. These examples are to be read for enjoyment and not for complete mastery, since many of the concepts and techniques involved are discussed in the material that follows.

**Prerequisites**   Chapter 0 is designed as a quick review of the algebra needed to study calculus. An instructor may elect to skip Chapter 0 and begin the course with Chapter 1, and in such cases, Chapter 0 can serve as an algebraic reference for students.

**Examples**   The text contains over 600 titled examples that have been chosen to illustrate specific concepts or problem-solving techniques.

**Exercises**   Many new exercises were added to the Second Edition and the text now contains over 4,500 exercises. The answers to the odd-numbered exercises are given in the back of the text.

**Applications**  There are over 500 applications taken from a variety of fields, with a special emphasis on applications in business and economics. An index of applications is given on the front inside covers of the text.

**Graphics**  There are over 1,700 figures and graphs in the text. Of these, approximately 450 are in the actual exercise sets.

**Theorems and Definitions**  Special care has been taken to state the theorems and definitions simply, without sacrificing accuracy. The theorems and definitions are set off by a color screen to emphasize their importance.

**Section Topics**  Each section begins with a list of the major topics that are covered in the section.

**Remarks**  The text contains special instructional notes in the form of ''Remarks.'' These appear after definitions, theorems, or examples and are designed to give additional insight, help avoid common errors, or describe generalizations.

**Calculators**  Special emphasis is given to the use of hand calculators in sections dealing with limits, numerical integration, Taylor polynomials, and Newton's Method. In addition, many of the exercise sets contain problems identified by the symbol ▦ as calculator exercises.

**Study Aids**  For students who need algebraic help, the *Study and Solutions Guide* by Dianna L. Zook contains step-by-step solutions to the odd-numbered exercises. Answers to the even-numbered exercises are given in the *Instructors Guide* by Ann R. Kraus.

## CHANGES IN THE SECOND EDITION

In writing this edition, we were guided by suggestions given by instructors and students who used the first edition.

- Several new sections were added to the text: Section 2.8 (Related Rates), Section 3.8 (Differentials), Section 4.5 (The Definite Integral as the Limit of a Sum), Section 6.7 (Random Variables and Probability), Section 6.8 (Expected Value, Standard Deviation, and Median), Section 7.3 (First-Order Linear Differential Equations), and Section 9.3 ($p$-Series and the Ratio Test). Also, the first section in which surfaces in space are introduced was expanded to two sections— Section 8.1 (The Three-Dimensional Coordinate System) and Section 8.2 (Surfaces in Space).
- Many of the existing sections in the first edition were rewritten. For instance, the presentation of limits and continuity in Sections 1.5 and 1.6 is essentially new.
- Chapter summaries and review exercises were added to the ends of the chapters (beginning with Chapter 1).
- Reference tables were added to the back of the text.
- Computer software is now available to supplement the text. The software consists

of exploratory and directed activities that reinforce the learning and enhance the teaching of calculus. The programs feature a carefully designed user interface and are structured as reusable tools that a student will be able to use in other mathematics courses.

## ACKNOWLEDGMENTS

We would like to thank the many people who have helped us at various stages of this project. Their encouragement, criticisms, and suggestions have been invaluable to us. Special thanks go to the reviewers of the first and second editions: Miriam E. Connellan, Marquette University; Bruce H. Edwards, University of Florida; Roger A. Engle, Clarion University of Pennsylvania; William C. Huffman, Loyola University of Chicago; James A. Kurre, Pennsylvania State University; Norbert Lerner, State University of New York at Cortland; Earl H. McKinney, Ball State University; Eldon L. Miller, University of Mississippi; Maurice L. Monahan, South Dakota State University; Stephen B. Rodi, Austin Community College; DeWitt L. Sumners, Florida State University; Jonathan Wilkin, Northern Virginia Community College; Melvin R. Woodard, Indiana University of Pennsylvania; and Robert A. Yawin, Springfield Technical Community College.

We would like to thank our publisher, D. C. Heath and Company, and we are especially appreciative of the assistance given by the following people: Mary Lu Walsh, mathematics acquisitions editor; Anne Marie Jones, developmental editor; Cathy Cantin, senior production editor; Sally Steele, designer; Mike O'Dea, production coordinator; and Martha Shethar, photo researcher.

Several other people worked on this project with us and we appreciate their help. David E. Heyd assisted us in writing the text. Dianna L. Zook wrote the *Student Solutions Guide*. Ann R. Kraus wrote the *Instructors Guide*. Linda L. Matta proofread the galleys. Timothy R. Larson prepared the art and proofread the galleys. Linda M. Bollinger typed part of the manuscript and proofread the galleys. Nancy K. Stout typed part of the manuscript. Helen Medley proofread the manuscript.

On a personal level, we are grateful to our children for their support during the past several years, and to our wives, Deanna Gilbert Larson and Eloise Hostetler, for their love, patience, and understanding.

If you have suggestions for improving this text, please feel free to write to us. Over the past 15 years we have received many useful comments from both instructors and students and we value these very much.

*Roland E. Larson*
*Robert P. Hostetler*

# Contents

# BRIEF CALCULUS
## WITH APPLICATIONS

SECOND EDITION

# A Precalculus Review

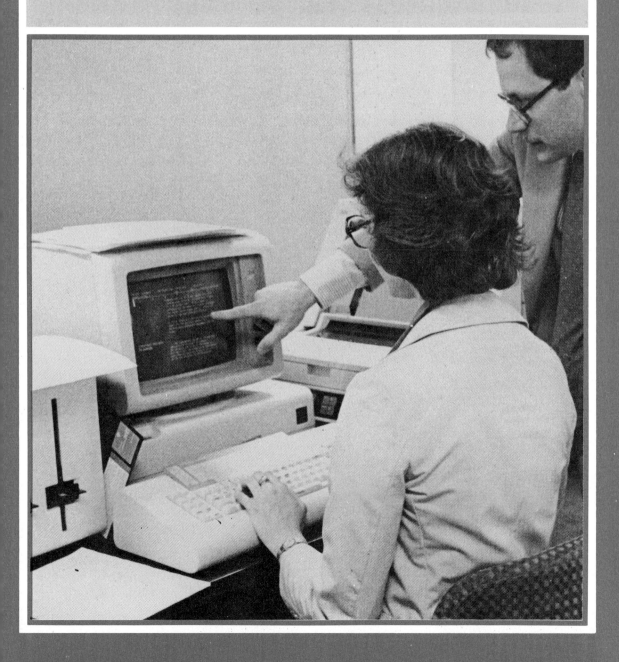

# The Real Line and Order

To represent the real numbers, we use a coordinate system called the **real line** (or $x$-axis), as shown in Figure 0.1. The **positive direction** (to the right) is denoted by an arrowhead and indicates the direction of increasing values of $x$. The real number corresponding to a particular point on the real line is called the **coordinate** of the point. As shown in Figure 0.1, it is customary to label those points whose coordinates are integers.

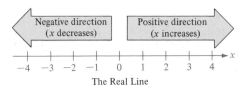

The Real Line

FIGURE 0.1

The point on the real line corresponding to zero is called the **origin.** Numbers to the right of the origin are **positive,** and numbers to the left of the origin are **negative.** We use the term **nonnegative** to describe a number that is either positive or zero.

The importance of the real line is that it provides us with a conceptually perfect picture of the real numbers. That is, each point on the real line corresponds to one and only one real number, and each real number corresponds to one and only one point on the real line. This type of relationship is called a **one-to-one correspondence,** as indicated in Figure 0.2.

Each of the four points in Figure 0.2 corresponds to a real number that can be expressed as the ratio of two integers. (Note that $1.85 = \frac{37}{20}$ and $-2.6 = -\frac{13}{5}$.) We

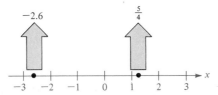

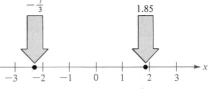

Every point on the real line corresponds to a real number.

Every real number corresponds to a point on the real line.

FIGURE 0.2

call such numbers **rational.** Rational numbers have either terminating or infinite repeating decimal representations.

| Terminating decimals | Infinite repeating decimals |
|---|---|
| $\dfrac{2}{5} = 0.4$ | $\dfrac{1}{3} = 0.333\ldots = 0.\overline{3}$* |
| $\dfrac{7}{8} = 0.875$ | $\dfrac{12}{7} = 1.714285714285\ldots = 1.\overline{714285}$ |

Real numbers that are not rational are called **irrational,** and they cannot be represented as the ratio of two integers (or as terminating or infinite repeating decimals). To represent an irrational number, we usually resort to a decimal approximation. Some irrational numbers occur so frequently in applications that mathematicians have invented special symbols to represent them. For example, the symbols $\sqrt{2}$, $\pi$, and $e$ represent irrational numbers whose decimal approximations are as follows:

$$\sqrt{2} \approx 1.4142135623$$

$$\pi \approx 3.1415926535$$

$$e \approx 2.7182818284$$

▩ **Remark:** We use $\approx$ to mean *approximately equal to.* Remember that even though we cannot represent irrational numbers *exactly* as terminating decimals, they can be represented *exactly* by points on the real line, as shown in Figure 0.3.

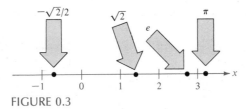

FIGURE 0.3

---

*The bar indicates which digits repeat.

## Order and Inequalities

One important property of the real numbers is that they are **ordered:** 0 is less than 1, $-3$ is less than $-2.5$, $\pi$ is less than $\frac{22}{7}$, and so on. We can visualize this property on the real line by observing that $a$ is less than $b$ if and only if $a$ lies to the left of $b$. Symbolically, we denote "$a$ is less than $b$" by the inequality

$$a < b$$

For example, the inequality $\frac{3}{4} < 1$ follows from the fact that $\frac{3}{4}$ lies to the left of 1 on the real line, as shown in Figure 0.4.

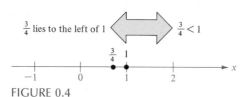

FIGURE 0.4

When three real numbers $a$, $x$, and $b$ are ordered such that $a < x$ and $x < b$, we say that $x$ is **between** $a$ and $b$ and write

$$a < x < b \qquad \text{$x$ is between $a$ and $b$}$$

The set of *all* real numbers between $a$ and $b$ is called the **open interval** between $a$ and $b$ and is denoted by $(a, b)$. An interval of the form $(a, b)$ does not contain the "endpoints" $a$ and $b$. Intervals that include their endpoints are called **closed** and are denoted by $[a, b]$. Intervals of the form $[a, b)$ and $(a, b]$ are called **half-open intervals.** Table 0.1 pictures the nine types of intervals on the real line.

TABLE 0.1   Intervals on the real line

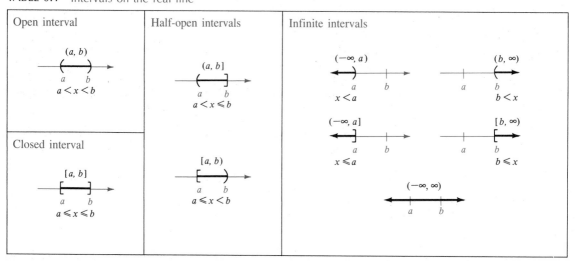

**Remark:** Note that a square bracket is used to denote "less than or equal to" ($\leq$). Furthermore, we use the symbols $\infty$ and $-\infty$ to denote positive and negative infinity. These symbols do not denote real numbers; they merely enable us to describe unbounded conditions more concisely. For instance, the interval $[b, \infty)$ is unbounded to the right since it includes *all* real numbers that are greater than $b$.

**EXAMPLE 1**

**Intervals on the Real Line**

Describe the intervals on the real line that correspond to the temperature ranges (in degrees Fahrenheit) for water in the following two states:
(a) liquid                                    (b) gas

**SOLUTION**

(a) Since water is in a liquid state at temperatures that are greater than $32°$ and less than $212°$, we have the open interval $(32, 212)$. If we let $x$ represent the temperature of water, this interval consists of all $x$ such that

$$32 < x < 212$$

as shown in Figure 0.5.

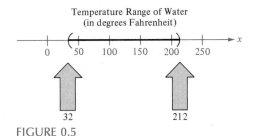

Temperature Range of Water
(in degrees Fahrenheit)

FIGURE 0.5

(b) Since water is in a gaseous state (steam) at temperatures that are greater than or equal to $212°$, we have the interval $[212, \infty)$. This interval consists of all temperatures $x$ such that

$$212 \leq x$$

as shown in Figure 0.6.

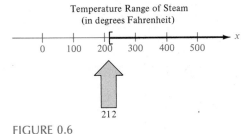

Temperature Range of Steam
(in degrees Fahrenheit)

FIGURE 0.6

In calculus we are frequently required to ''solve inequalities'' involving variable expressions such as $3x - 4 < 5$. We say $x = a$ is a **solution** to an inequality if the inequality is true when $a$ is substituted for $x$. The set of all values of $x$ that satisfy an inequality is called the **solution set** of the inequality. The following properties are useful for solving inequalities. (Similar properties are obtained if $<$ is replaced by $\leq$ and $>$ is replaced by $\geq$.)

**Properties of inequalities**

1. Transitive Property: $a < b$ and $b < c \Rightarrow a < c$
2. Adding inequalities: $a < b$ and $c < d \Rightarrow a + c < b + d$
3. Multiplying by a (positive) constant: $a < b \Rightarrow ac < bc$
4. Multiplying by a (negative) constant: $a < b \Rightarrow ac > bc$
5. Subtracting a constant: $a < b \Rightarrow a - c < b - c$
6. Adding a constant: $a < b \Rightarrow a + c < b + c$

■ **Remark:** Note that we *reverse the inequality* when we multiply by a negative number. For example, if $x < 3$, then $-4x > -12$. This principle also applies to division. Thus, if $-2x > 4$, then $x < -2$.

**EXAMPLE 2**

**Solving an Inequality**

Find the solution set of the inequality $3x - 4 < 5$.

**SOLUTION**

To find the solution set, we attempt to rewrite the inequality in such a way that $x$ appears alone on either the right or the left side. To do this, we use appropriate properties of inequalities as follows. First, we can add 4 to both sides of the inequality to obtain

$$3x - 4 < 5$$

$$3x - 4 + 4 < 5 + 4$$

$$3x < 9$$

Now, multiplying both sides of this inequality by $\frac{1}{3}$ gives us

$$\frac{1}{3}(3x) < \frac{1}{3}(9)$$

$$x < 3$$

Thus, the solution set is given by the interval $(-\infty, 3)$, as shown in Figure 0.7.

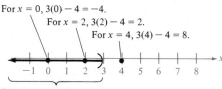

For $x = 0, 3(0) - 4 = -4$.
For $x = 2, 3(2) - 4 = 2$.
For $x = 4, 3(4) - 4 = 8$.

Solution set for $3x - 4 < 5$.
FIGURE 0.7

■ **Remark:** In Example 2, all five inequalities listed as steps in the solution have the same solution set, and we call them **equivalent.**

Once you have solved an inequality, it is a good idea to check some $x$-values in your solution interval to see if they satisfy the original inequality. You might also check some values outside your solution interval to verify that they do not satisfy the inequality. For example, in Figure 0.7 we see that when $x = 0$ or $x = 2$ the inequality is satisfied, but when $x = 4$ the inequality is not satisfied.

**EXAMPLE 3**

**Solving a Double Inequality**

Find the interval corresponding to the set of $x$-values that satisfy the inequality

$$-3 \leq 2 - 5x \leq 12$$

Note that this double inequality means that $-3 \leq 2 - 5x$ *and* $2 - 5x \leq 12$.

**SOLUTION**

Although two inequalities are involved in this problem, we can work with both simultaneously. We begin by subtracting 2 from all three expressions to obtain

$$-3 - 2 \leq 2 - 5x - 2 \leq 12 - 2$$
$$-5 \leq -5x \leq 10$$

Now, we divide all three expressions by $-5$ (making sure to reverse both inequalities) to obtain

$$\frac{-5}{-5.} \geq \frac{-5x}{-5} \geq \frac{10}{-5}$$
$$1 \geq x \geq -2$$

Thus, the interval representing the solution is $[-2, 1]$, as shown in Figure 0.8.

FIGURE 0.8

The inequalities in Examples 2 and 3 involve first-degree polynomials. To solve inequalities involving polynomials of higher degree, we use a result from algebra—a polynomial can change signs *only* at its zeros (the values that make the polynomial zero). Between two consecutive zeros a polynomial must be entirely positive or entirely negative. This means that when the real zeros of a polynomial are put in order, they divide the real line into **test intervals** in which the polynomial has no sign changes. For example, the polynomial

$$x^2 - x - 6 = (x - 3)(x + 2)$$

can change signs only at $x = -2$ and $x = 3$. This means that to determine the sign of the polynomial in the intervals $(-\infty, -2)$, $(-2, 3)$, and $(3, \infty)$, we need to test only *one value* from each interval. This procedure is demonstrated in the next two examples.

**EXAMPLE 4**

**Solving a Second-Degree Polynomial Inequality**

Find the solution set for the inequality $x^2 < x + 6$.

**SOLUTION**

We begin by grouping all terms to one side of the inequality, to obtain a polynomial. Then, through factoring or some other means, we find the zeros of the polynomial.

$$x^2 < x + 6 \qquad \text{Given}$$

$$x^2 - x - 6 < 0 \qquad \text{Polynomial form}$$

$$(x - 3)(x + 2) < 0 \qquad \text{Factor}$$

Thus, the polynomial has $x = -2$ and $x = 3$ as its zeros, and we can solve the inequality by testing the sign of the polynomial in each of the following intervals:

$$x < -2, \qquad -2 < x < 3, \qquad x > 3$$

To test an interval, we choose a representative number in the interval and compute the sign of each factor. For example, for any $x < -2$, both of the factors $(x - 3)$ and $(x + 2)$ are negative. Consequently, the product (of two negative numbers) is positive and the inequality is *not* satisfied in the interval $x < -2$. We suggest the testing format shown in Figure 0.9.

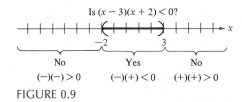

FIGURE 0.9

Since the inequality is satisfied only by the center test interval, we conclude that the solution set is given by the interval

$$-2 < x < 3 \qquad \text{Solution set}$$

**EXAMPLE 5**

**Solving a Second-Degree Polynomial Inequality**

Find the solution set for the inequality $2x^2 + 5x - 12 \geq 0$.

**SOLUTION**

We begin by factoring the polynomial to obtain

$$2x^2 + 5x - 12 \geq 0 \qquad \text{Given}$$

$$(2x - 3)(x + 4) \geq 0 \qquad \text{Factored form}$$

Thus the zeros of the polynomial are $x = \frac{3}{2}$ and $x = -4$, which implies that the test intervals are

$$x < -4, \qquad -4 < x < \frac{3}{2}, \qquad x > \frac{3}{2}$$

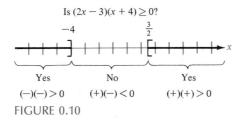

Is $(2x - 3)(x + 4) \geq 0$?

| Yes | No | Yes |
|-----|-----|-----|
| $(-)(-) > 0$ | $(+)(-) < 0$ | $(+)(+) > 0$ |

FIGURE 0.10

By testing the sign of $(2x - 3)(x + 4)$ in each of these intervals (see Figure 0.10), we see that the solution set is given by the intervals

$$x \leq -4 \quad \text{and} \quad x \geq \frac{3}{2} \qquad \text{Solution set}$$

Using the notation for the **union of two sets,** we can denote the solution set as $(-\infty, -4] \cup [\frac{3}{2}, \infty)$.

Inequalities are frequently used to describe conditions that occur in business and science. For instance, the inequality

$$118 \leq W \leq 138$$

describes the recommended weight $W$ for a woman whose height is 5 feet 6 inches. In Example 6, we show how an inequality can be used to describe the production level of a manufacturing plant.

**EXAMPLE 6**

**A Business Application**

In addition to fixed overhead costs of $500 per day, the cost of producing $x$ units of a certain item is $2.50 per unit. During the month of August, the total cost of production varied from a high of $1,325 to a low of $1,200 per day. Find the high and low *production levels* during the month.

**SOLUTION**

Since it costs $2.50 to produce 1 unit, it will cost $2.5x$ to produce $x$ units. Furthermore, since the fixed cost per day is $500, the total daily cost of producing $x$ units is

$$C = 2.5x + 500$$

Now since the cost ranged between $1,200 and $1,325, we can write

$$1200 \leq 2.5x + 500 \leq 1325$$

$$1200 - 500 \leq 2.5x + 500 - 500 \leq 1325 - 500$$

$$700 \leq 2.5x \leq 825$$

$$\frac{700}{2.5} \leq \frac{2.5x}{2.5} \leq \frac{825}{2.5}$$

$$280 \leq x \leq 330$$

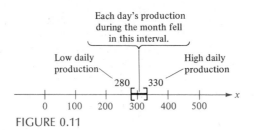

FIGURE 0.11

Thus, we see that the daily production levels during the month varied between a low of 280 units and a high of 330 units, as pictured in Figure 0.11.

## SECTION EXERCISES 0.1

In Exercises 1–10, determine whether the real number is rational or irrational.

1. $0.7$

2. $-3678$

3. $\dfrac{3\pi}{2}$

4. $3\sqrt{2} - 1$

5. $4.34\overline{51451}$

6. $\dfrac{22}{7}$

7. $\sqrt[3]{64}$

8. $0.8\overline{1778177}$

9. $\sqrt[3]{60}$

10. $2e$

In Exercises 11–14, determine whether the given value of $x$ satisfies the inequality.

11. $5x - 12 > 0$
   (a) $x = 3$        (b) $x = -3$
   (c) $x = \dfrac{5}{2}$       (d) $x = \dfrac{3}{2}$

12. $x + 1 < \dfrac{2x}{3}$
   (a) $x = 0$        (b) $x = 4$
   (c) $x = -4$      (d) $x = -3$

13. $0 < \dfrac{x - 2}{4} < 2$
   (a) $x = 4$        (b) $x = 10$
   (c) $x = 0$        (d) $x = \dfrac{7}{2}$

14. $-1 < \dfrac{3 - x}{2} \le 1$
   (a) $x = 0$        (b) $x = \sqrt{5}$
   (c) $x = 1$        (d) $x = 5$

In Exercises 15 and 16, complete the given table by filling in the appropriate interval notation, inequality, and graph.

15.

| Interval notation | Inequality | Graph |
|---|---|---|
|  |  | (graph from $-3$ to $0$) |
| $(-\infty, -4]$ |  |  |
|  | $3 \le x \le \dfrac{11}{2}$ |  |
| $(-1, 7)$ |  |  |

16.

| Interval notation | Inequality | Graph |
|---|---|---|
|  |  | (graph from $98$ to $102$) |
|  | $10 < x$ |  |
| $(\sqrt{2}, 8]$ |  |  |
|  | $\dfrac{1}{3} < x \le \dfrac{22}{7}$ |  |

In Exercises 17–34, solve the inequality and graph the solution on the real number line.

**17.** $x - 5 \geq 7$  **18.** $2x > 3$

**19.** $4x + 1 < 2x$  **20.** $2x + 7 < 3$

**21.** $2x - 1 \geq 0$  **22.** $3x + 1 \geq 2x + 2$

**23.** $4 - 2x < 3x - 1$  **24.** $x - 4 \leq 2x + 1$

**25.** $-4 < 2x - 3 < 4$  **26.** $0 \leq x + 3 < 5$

**27.** $\dfrac{3}{4} > x + 1 > \dfrac{1}{4}$  **28.** $-1 < -\dfrac{x}{3} < 1$

**29.** $\dfrac{x}{2} + \dfrac{x}{3} > 5$  **30.** $\dfrac{x}{2} - \dfrac{x}{3} > 5$

**31.** $x^2 \leq 3 - 2x$  **32.** $x^2 - x \leq 0$

**33.** $x^2 + x - 1 \leq 5$  **34.** $2x^2 + 1 < 9x - 3$

**35.** $P$ dollars is invested at a (simple) interest rate of $r$. After $t$ years the balance in the account is given by

$$A = P + Prt$$

In order for an investment of $1,000 to grow to *more than* $1,250 in two years, what must the interest rate be?

**⊞ \* 36.** A doughnut shop at a shopping mall sells a dozen doughnuts for $2.95. Beyond the fixed costs (for rent, utilities, and insurance) of $150 per day, it costs $1.45 for enough materials (flour, sugar, etc.) and labor to produce each dozen doughnuts. If the daily profit *varies between* $50 and $200, between what levels (in dozens) do the daily sales vary?

**⊞ 37.** The revenue for selling $x$ units of a product is

$$R = 115.95x$$

and the cost of producing $x$ units is

$$C = 95x + 750$$

In order to obtain a profit, the revenue must be *greater than* the cost. For what values of $x$ will this product return a profit?

**⊞ 38.** A utility company has a fleet of vans for which the annual operating cost per van is

$$C = 0.32m + 2300$$

where $m$ is the number of miles traveled by a van in a year. What number of miles will yield an annual operating cost per van that is *less than* $10,000?

**⊞ 39.** A square region is to have an area of *at least* 500 square meters. What must the length of the sides of the region be?

**⊞ 40.** An isosceles right triangle is to have an area of *at least* 32 square feet. What must the length of each side be?

In Exercises 41 and 42, determine which of the two given real numbers is greater.

**⊞ 41.** (a) $\pi$ or $\dfrac{355}{113}$  (b) $\pi$ or $\dfrac{22}{7}$

**⊞ 42.** (a) $\dfrac{224}{151}$ or $\dfrac{144}{97}$  (b) $\dfrac{73}{81}$ or $\dfrac{6427}{7132}$

In Exercises 43 and 44, determine whether each statement is true or false, given $a < b$.

**43.** (a) $-2a < -2b$  (b) $a + 2 < b + 2$

(c) $6a < 6b$  (d) $\dfrac{1}{a} < \dfrac{1}{b}$

**44.** (a) $a - 4 < b - 4$  (b) $4 - a < 4 - b$

(c) $-3b < -3a$  (d) $\dfrac{a}{4} < \dfrac{b}{4}$

---

\*The symbol ⊞ indicates that a calculator may be helpful in solving the exercise.

# Absolute Value and Distance on the Real Line

The absolute value of a real number $a$ is denoted by $|a|$ and is defined as follows.

**Definition of absolute value**

The **absolute value** of a real number $a$ is

$$|a| = \begin{cases} a, & \text{if } a \geq 0 \\ -a, & \text{if } a < 0 \end{cases}$$

At first glance it may appear from this definition that an absolute value can be negative, but that is not possible. For example, let $a = -3$. Then, since $-3 < 0$, we have

$$|a| = |-3|$$
$$= -(-3) = 3$$

Similarly,

$$|2 - 7| = |-5|$$
$$= -(-5) = 5$$

and

$$|7 - 2| = |5| = 5$$

In working with absolute values, the following properties are useful.

| Properties of absolute value | 1. Multiplication: $\|ab\| = \|a\|\|b\|$ |
| --- | --- |
| | 2. Division: $\left\|\dfrac{a}{b}\right\| = \dfrac{\|a\|}{\|b\|}, \; b \neq 0$ |
| | 3. Power: $\|a^n\| = \|a\|^n$ |
| | 4. Square root: $\sqrt{a^2} = \|a\|$ |

Be sure you understand the fourth property in this list. A common error in algebra is to imagine that by squaring a number and then taking the square root, we come back to the original number. But this is true only if the original number is nonnegative. For instance, if $a = 2$, then

$$\sqrt{2^2} = \sqrt{4} = 2$$

but if $a = -2$, then

$$\sqrt{(-2)^2} = \sqrt{4} = 2$$

The reason for this is that (by definition) the square root symbol $\sqrt{\phantom{x}}$ denotes only the nonnegative root.

### Distance on the Real Line

Given two distinct points on the real line, we will make use of each of the distances pictured in Figure 0.12:

1. The **directed distance from $a$ to $b$** denoted by $b - a$
2. The **directed distance from $b$ to $a$** denoted by $a - b$
3. The **distance between $a$ and $b$** denoted by $\|a - b\|$ or $\|b - a\|$

In Figure 0.12, note that since $b$ is to the right of $a$, the directed distance from $a$ to $b$ (moving to the right) is positive. Moreover, since $a$ is to the left of $b$, the directed distance from $b$ to $a$ (moving to the left) is negative. The distance *between* two points on the real line can never be negative.

The following relationship between distance and square roots is useful. In Section 1.1, we will see that the "square root" version of the distance between two points on the real line generalizes nicely to give the distance between two points in the plane.

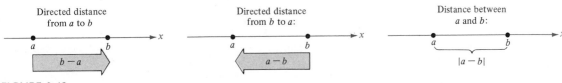

FIGURE 0.12

The distance $d$ between points $x_1$ and $x_2$ on the real line is given by
$$d = |x_2 - x_1| = \sqrt{(x_2 - x_1)^2}$$

Note that the order of subtraction with $x_1$ and $x_2$ does not matter since
$$|x_2 - x_1| = |x_1 - x_2| \quad \text{and} \quad (x_2 - x_1)^2 = (x_1 - x_2)^2$$

**EXAMPLE 1**

**Finding Distance on the Real Line**

Determine the distance between $-3$ and $4$ on the real line. What is the directed distance from $-3$ to $4$? From $4$ to $-3$?

**SOLUTION**

The distance between $-3$ and $4$ is given by
$$|4 - (-3)| = |7| = 7 \quad \text{or} \quad |-3 - 4| = |-7| = 7$$
The directed distance from $-3$ to $4$ is
$$4 - (-3) = 7$$
The directed distance from $4$ to $-3$ is
$$-3 - 4 = -7$$
(See Figure 0.13.)

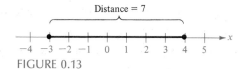

FIGURE 0.13

Absolute values are useful for defining intervals on the real line, as demonstrated in the following two examples.

**EXAMPLE 2**

**Using Absolute Value to Define an Interval on the Real Line**

Find the interval on the real line that contains all numbers that lie no more than 2 units from 3.

**SOLUTION**

Let $x$ be any point in this interval. Then we wish to find all $x$ such that the distance between $x$ and 3 is less than or equal to 2. This implies that
$$|x - 3| \leq 2$$
Requiring the absolute value of $x - 3$ to be less than or equal to 2 means that $x - 3$ must lie between $-2$ and 2, and hence we write
$$-2 \leq x - 3 \leq 2$$

Solving this pair of inequalities, we have

$$-2 + 3 \le x - 3 + 3 \le 2 + 3$$

$$1 \le x \le 5 \qquad \text{Solution set}$$

Therefore, the interval is [1, 5], as shown in Figure 0.14.

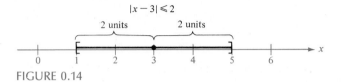

FIGURE 0.14

**EXAMPLE 3**

**Solving an Inequality That Results in Two Intervals**

Find the intervals on the real line that contain all numbers that lie more than 3 units from $-2$.

**SOLUTION**

Since the distance between $x$ and $-2$ is given by $|x - (-2)|$, we write

$$|x - (-2)| > 3$$

$$|x + 2| > 3$$

Thus, we are hunting for $x$-values such that the absolute value of $x + 2$ is greater than 3. This can happen in two ways:

1. If $x + 2$ is positive, then $|x + 2| = x + 2$, and we have

$$x + 2 > 3 \implies 1 < x$$

2. If $x + 2$ is negative, then $|x + 2| = -(x + 2)$, and we have

$$-x - 2 > 3 \implies x < -5$$

Therefore, $x$ can lie in either of the intervals $(-\infty, -5)$ or $(1, \infty)$ as shown in Figure 0.15, and the solution set is

$$(-\infty, -5) \cup (1, \infty) \qquad \text{Solution set}$$

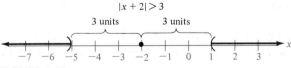

FIGURE 0.15

Examples 2 and 3 suggest the following general results for absolute value and inequalities.

**Intervals defined
by absolute value**

Single interval, $|x - a| \leq d$:

$$a - d \leq x \leq a + d$$

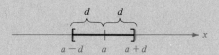

Two intervals, $|x - a| \geq d$:

$$x \leq a - d \quad \text{or} \quad a + d \leq x$$

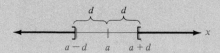

■■■ **Remark:** Be sure you see that inequalities of the form $|x - a| \geq d$ have solution sets consisting of two intervals. To describe the two intervals without using absolute values, we must use *two* separate inequalities. For instance, we would write

$$x \leq 0 \quad \text{or} \quad 4 \leq x \qquad \text{Correct}$$

and not

$$4 \leq x \leq 0 \qquad \text{Incorrect}$$

## EXAMPLE 4

### A Business Application

A large manufacturer hired a quality control firm to determine the reliability of a certain product. Using statistical methods, the firm determined that the manufacturer could expect $0.35\% \pm 0.17\%$ of the units to be defective. If the manufacturer offers a money-back guarantee on this product, how much should be budgeted to cover the refunds on 100,000 units? (Assume that the retail price is \$8.95.)

## SOLUTION

If we let $r$ represent the percentage of defective units (written in decimal form), we know that $r$ will differ from 0.0035 by at most 0.0017. Using inequalities, we can write

$$0.0035 - 0.0017 \leq r \leq 0.0035 + 0.0017$$

$$0.0018 \leq r \leq 0.0052$$

Now, letting $x$ be the number of defective units out of 100,000, we know that $x = 100,000r$ and we have

$$0.0018(100,000) \leq 100,000r \leq 0.0052(100,000)$$

$$180 \leq x \leq 520$$

Finally, letting $C$ be the cost of refunds, we have $C = 8.95x$, and it follows that the total cost of refunds for 100,000 units should fall within the interval given by

$$180(8.95) \le 8.95x \le 520(8.95)$$

$$\$1{,}611 \le C \le \$4{,}654$$

Figure 0.16 illustrates these results graphically.

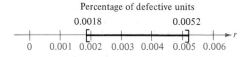

Percentage of defective units

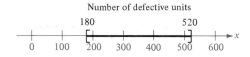

Number of defective units

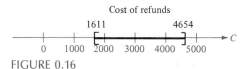

Cost of refunds

FIGURE 0.16

In Example 4 we concluded that the manufacturer should expect to spend between $1,611 and $4,654 for refunds. Of course, the safer budget figure for refunds would be the higher of these estimates. However, from a statistical point of view, the most representative estimate would be the average of these two extremes. Graphically, this average is the **midpoint** of the interval that has these two numbers as its endpoints, as shown in Figure 0.17.

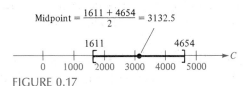

$$\text{Midpoint} = \frac{1611 + 4654}{2} = 3132.5$$

FIGURE 0.17

| Midpoint of an interval | The **midpoint** of the interval with endpoints $a$ and $b$ is found by taking the average of the endpoints: |
|---|---|

$$\text{Midpoint} = \frac{a + b}{2}$$

## EXAMPLE 5

### Finding the Midpoint of an Interval

Find the midpoint of each of the following intervals:

(a) $[-5, 7]$     (b) $(-12, -1)$     (c) $(3, 21]$

**SOLUTION**

For each interval we find the midpoint by taking the average of the two endpoints as follows:

(a) Midpoint $= \dfrac{-5 + 7}{2} = \dfrac{2}{2} = 1$

(b) Midpoint $= \dfrac{-12 + (-1)}{2} = -\dfrac{13}{2}$

(c) Midpoint $= \dfrac{3 + 21}{2} = \dfrac{24}{2} = 12$

## SECTION EXERCISES 0.2

In Exercises 1–8, find (a) the directed distance from $a$ to $b$, (b) the directed distance from $b$ to $a$, and (c) the distance between $a$ and $b$.

**1.**

**2.**

**3.**

**4.**

**5.** $a = 126$, $b = 75$

**6.** $a = -126$, $b = -75$

**7.** $a = 9.34$, $b = -5.65$

**8.** $a = \dfrac{16}{5}$, $b = \dfrac{112}{75}$

In Exercises 9–16, find the midpoint of the given interval.

**9.**

**10.**

**11.**

**12.**

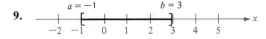

**13.** $[7, 21]$     **14.** $[8.6, 11.4]$

**15.** $[-6.85, 9.35]$     **16.** $[-4.6, -1.3]$

In Exercises 17–30, solve the inequality and graph the solution on the real line.

**17.** $|x| < 5$     **18.** $|2x| < 6$

**19.** $\left|\dfrac{x}{2}\right| > 3$     **20.** $|5x| > 10$

**21.** $|x + 2| < 5$     **22.** $|3x + 1| \geq 4$

**23.** $\left|\dfrac{x - 3}{2}\right| \geq 5$     **24.** $|2x + 1| < 5$

**25.** $|10 - x| > 4$     **26.** $|25 - x| \geq 20$

**27.** $|9 - 2x| < 1$     **28.** $\left|1 - \dfrac{2x}{3}\right| < 1$

**29.** $|x - a| \leq b$     **30.** $|2x - a| \geq b$

In Exercises 31–42, use absolute values to describe each interval (or pair of intervals) on the real line.

**31.**

**32.**

**33.**

**34.**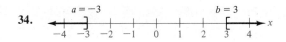

$a = -3$          $b = 3$

**35.**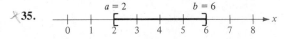

$a = 2$          $b = 6$

**36.**

$a = -7$          $b = -1$

**37.**

$a = 0$          $b = 4$

**38.**

$a = 20$          $b = 24$

**39.** All numbers *less than* 2 units from 4.
**40.** All numbers *more than* 6 units from 3.
**41.** $y$ is *at most* 2 units from $a$.
**42.** $y$ is *less than* $h$ units from $c$.

**43.** The heights, $h$, of two-thirds of the members of a certain population satisfy the inequality

$$\left| \frac{h - 68.5}{2.7} \right| \leq 1$$

where $h$ is measured in inches. Determine the interval on the real line in which these heights lie.

**44.** To determine whether a coin is fair, an experimenter tosses it 100 times and records the number of heads, $x$. Statistical theory states that the coin should be declared unfair if

$$\left| \frac{x - 50}{5} \right| \geq 1.645$$

For what values of $x$ will the coin be declared unfair?

**45.** The estimated daily production, $x$, at a refinery is given by

$$|x - 2,250,000| < 125,000$$

where $x$ is measured in barrels of oil. Determine the high and low production levels.

**46.** A stock market analyst predicts that over the next year the price, $p$, of a certain stock will not change from its current price of $\$33\frac{1}{8}$ by more than $\$2$. Use absolute values to write this prediction as an inequality.

In Exercises 47–50, use the definition of absolute value to prove each property.

**47.** $|ab| = |a||b|$

**48.** $|a - b| = |b - a|$

**49.** $\left| \dfrac{a}{b} \right| = \dfrac{|a|}{|b|}, \; b \neq 0$

**50.** $\left| \dfrac{1}{a} \right| = \dfrac{1}{|a|}, \; a \neq 0$

# Exponents and Radicals

Much of the work in this text involves algebraic expressions such as

$$2x^3 + 3, \qquad \sqrt{x^2 + 5x - 2}, \qquad (x + 3)(2x - 5)^{-1}$$

It is essential that you be familiar with the algebraic methods of simplifying and evaluating such expressions, and we devote this and the next two sections to reviewing some of the basic results and techniques of algebra. We begin by summarizing the basic properties of exponents.

**Properties of exponents**

1. Whole number exponents: $x^n = \underbrace{x \cdot x \cdot x \cdots x}_{n \text{ factors}}$

2. Zero exponent: $x^0 = 1, \ x \neq 0$

3. Negative exponents: $x^{-n} = \dfrac{1}{x^n}, \ x \neq 0$

4. Radicals (principal $n$th root*): $\sqrt[n]{x} = a \Longrightarrow x = a^n$

5. Rational exponents $(1/n)$: $x^{1/n} = \sqrt[n]{x}$

6. Rational exponents $(m/n)$: $x^{m/n} = (x^{1/n})^m = (\sqrt[n]{x})^m$ or
$x^{m/n} = (x^m)^{1/n} = \sqrt[n]{x^m}$

7. Special convention (positive square root): $\sqrt[2]{x} = \sqrt{x}$

---

*If $n$ is even, then the principal $n$th root is defined to be positive. For example, $\sqrt{4} = +2$ and $\sqrt[4]{81} = +3$.

**EXAMPLE 1**

**Evaluating Expressions with Exponents**

Evaluate each expression for the given value of $x$.

| *Expression* | *x-value* |
|---|---|
| (a) $y = -2x^2$ | $x = 4$ |
| (b) $y = 3x^{-3}$ | $x = -1$ |
| (c) $y = (-x)^2$ | $x = \dfrac{1}{2}$ |
| (d) $y = 2x^{1/2}$ | $x = 4$ |
| (e) $y = (2x)^{1/2}$ | $x = 4$ |
| (f) $y = \sqrt[3]{x^2}$ | $x = 8$ |
| (g) $y = (x^2 - 9)^{-3/2}$ | $x = 5$ |
| (h) $y = \sqrt{x}$ | $x = 3$ |

**SOLUTION**

(a) $y = -2(4^2) = -2(16) = -32$

(b) $y = 3(-1)^{-3} = \dfrac{3}{(-1)^3} = \dfrac{3}{-1} = -3$

(c) $y = \left(-\dfrac{1}{2}\right)^2 = \dfrac{1}{4}$

(d) $y = 2\sqrt{4} = 2(2) = 4$

(e) $y = \sqrt{2(4)} = \sqrt{2}\sqrt{4} = 2\sqrt{2} \approx 2.828$      Calculator

(f) $y = 8^{2/3} = (8^{1/3})^2 = 2^2 = 4$

(g) $y = (5^2 - 9)^{-3/2} = (16)^{-3/2} = \dfrac{1}{(16)^{3/2}} = \dfrac{1}{(\sqrt{16})^3} = \dfrac{1}{4^3} = \dfrac{1}{64}$

(h) $y = \sqrt{3} \approx 1.732$      Calculator

**Remark:** In parts (b) and (g) of Example 1, we converted to positive exponent form before evaluating the expression. This is a good practice to follow.

In calculus, the form of an algebraic expression is often critical and you must be familiar with the following operations with exponents. We use these operations to expand or simplify an expression.

**Operations with exponents**

1. Multiplying like bases:

$$x^n x^m = x^{n+m}$$  Add exponents

2. Dividing like bases:

$$\frac{x^n}{x^m} = x^{n-m}$$  Subtract exponents

3. Removing parentheses:

$$(xy)^n = x^n y^n$$

$$\left(\frac{x}{y}\right)^n = \frac{x^n}{y^n}$$

$$(x^n)^m = x^{nm}$$

4. Special conventions:

$$-x^n = -(x^n), \quad -x^n \neq (-x)^n$$

$$cx^n = c(x^n), \quad cx^n \neq (cx)^n$$

$$x^{n^m} = x^{(n^m)}, \quad x^{n^m} \neq (x^n)^m$$

---

**EXAMPLE 2**

**Simplifying Expressions with Exponents**

Simplify each of the following expressions:

(a) $2x^2(x^3)$

(b) $(3x)^2 \sqrt[3]{x}$

(c) $\dfrac{3x^2}{(\sqrt{x})^3}$

(d) $\dfrac{5x^4}{(x^2)^3}$

(e) $x^{-1}(2x^2)$

(f) $\dfrac{-\sqrt{x}}{5x^{-1}}$

**SOLUTION**

(a) $2x^2(x^3) = 2x^{2+3} = 2x^5$

(b) $(3x)^2 \sqrt[3]{x} = 9x^2 x^{1/3} = 9x^{2+(1/3)} = 9x^{7/3}$

(c) $\dfrac{3x^2}{(\sqrt{x})^3} = \dfrac{3x^2}{(x^{1/2})^3} = 3\left(\dfrac{x^2}{x^{3/2}}\right) = 3x^{2-(3/2)} = 3x^{1/2}$

(d) $\dfrac{5x^4}{(x^2)^3} = \dfrac{5x^4}{x^6} = \dfrac{5}{x^{6-4}} = \dfrac{5}{x^2}$

(e) $x^{-1}(2x^2) = 2x^{-1}x^2 = 2x^{2-1} = 2x$

(f) $\dfrac{-\sqrt{x}}{5x^{-1}} = -\dfrac{1}{5}\left(\dfrac{x^{1/2}}{x^{-1}}\right) = -\dfrac{1}{5}x^{(1/2)+1} = -\dfrac{1}{5}x^{3/2}$

---

Note in Example 2 that one characteristic of simplified expressions is the absence of negative exponents. Another characteristic of simplified expressions is that we usually prefer to write sums and differences in *factored form*. To do this, we can use the **distributive property:**

$$abx^n + acx^{n+m} = ax^n(b + cx^m)$$

Study the next example carefully to be sure that you understand the concepts involved in the factoring process.

**EXAMPLE 3**

**Factoring Sums and Differences of Exponential Expressions**

Simplify the following expressions by factoring:
(a) $2x^2 - x^3$
(b) $2x^3 + x^2$
(c) $2x^{1/2} + 4x^{5/2}$
(d) $2x^{-1/2} + 3x^{5/2}$
(e) $3(x + 1)^{1/2}(2x - 3)^{5/2} - 6(x + 1)^{3/2}(2x - 3)^{3/2}$
(f) $(x + 1)^{-1/2}(2x - 3)^{5/2} + 3(x + 1)^{1/2}(2x - 3)^{3/2}$

**SOLUTION**

(a) $2x^2 - x^3 = x^2(2 - x)$
(b) $2x^3 + x^2 = x^2(2x + 1)$
(c) $2x^{1/2} + 4x^{5/2} = 2x^{1/2}(1 + 2x^2)$

(d) $2x^{-1/2} + 3x^{5/2} = x^{-1/2}(2 + 3x^3) = \dfrac{2 + 3x^3}{\sqrt{x}}$

(e) $3(x + 1)^{1/2}(2x - 3)^{5/2} - 6(x + 1)^{3/2}(2x - 3)^{3/2}$
$$= 3(x + 1)^{1/2}(2x - 3)^{3/2}[(2x - 3) - 2(x + 1)]$$
$$= 3(x + 1)^{1/2}(2x - 3)^{3/2}(2x - 3 - 2x - 2)$$
$$= 3(x + 1)^{1/2}(2x - 3)^{3/2}(-5)$$
$$= -15(x + 1)^{1/2}(2x - 3)^{3/2}$$

(f) $(x + 1)^{-1/2}(2x - 3)^{5/2} + 3(x + 1)^{1/2}(2x - 3)^{3/2}$
$$= (x + 1)^{-1/2}(2x - 3)^{3/2}[(2x - 3) + 3(x + 1)]$$
$$= (x + 1)^{-1/2}(2x - 3)^{3/2}(2x - 3 + 3x + 3)$$
$$= (x + 1)^{-1/2}(2x - 3)^{3/2}(5x)$$
$$= \dfrac{5x(2x - 3)^{3/2}}{(x + 1)^{1/2}}$$

Be sure you see that in Example 3 we subtract exponents when factoring—even if the exponents are negative. For example, in part (d), we subtracted the exponent

$-\frac{1}{2}$ as follows:

$$\left(\frac{1}{2}\right) - \left(-\frac{1}{2}\right) \quad \left(\frac{5}{2}\right) - \left(-\frac{1}{2}\right)$$

$$2x^{-1/2} + 3x^{5/2} = x^{-1/2}(2x^0 + 3x^3) = x^{-1/2}(2 + 3x^3) = \frac{2 + 3x^3}{\sqrt{x}}$$

**EXAMPLE 4**

**Factors Involving Quotients**

Simplify the following expressions by factoring:

(a) $\dfrac{3x^2 + x^4}{2x}$

(b) $\dfrac{\sqrt{x} + x^{3/2}}{x}$

(c) $\dfrac{3}{5}(x + 1)^{5/3} + \dfrac{3}{4}(x + 1)^{8/3}$

(d) $\dfrac{3(x + 2)^2(x - 1)^3 - 3(x + 2)^3(x - 1)^2}{[(x - 1)^3]^2}$

**SOLUTION**

(a) $\dfrac{3x^2 + x^4}{2x} = \dfrac{x^2(3 + x^2)}{2x} = \dfrac{x^{2-1}(3 + x^2)}{2} = \dfrac{x(3 + x^2)}{2}$

(b) $\dfrac{\sqrt{x} + x^{3/2}}{x} = \dfrac{x^{1/2}(1 + x)}{x} = \dfrac{1 + x}{x^{1-(1/2)}} = \dfrac{1 + x}{\sqrt{x}}$

(c) $\dfrac{3}{5}(x + 1)^{5/3} + \dfrac{3}{4}(x + 1)^{8/3} = \dfrac{12}{20}(x + 1)^{5/3} + \dfrac{15}{20}(x + 1)^{8/3}$

$$= \dfrac{3}{20}(x + 1)^{5/3}[4 + 5(x + 1)]$$

$$= \dfrac{3}{20}(x + 1)^{5/3}(4 + 5x + 5)$$

$$= \dfrac{3}{20}(x + 1)^{5/3}(5x + 9)$$

(d) $\dfrac{3(x + 2)^2(x - 1)^3 - 3(x + 2)^3(x - 1)^2}{[(x - 1)^3]^2}$

$$= \dfrac{3(x + 2)^2(x - 1)^2[(x - 1) - (x + 2)]}{(x - 1)^6}$$

$$= \dfrac{3(x + 2)^2(x - 1 - x - 2)}{(x - 1)^{6-2}}$$

$$= \dfrac{-9(x + 2)^2}{(x - 1)^4}$$

### The Domain of an Algebraic Expression

When working with algebraic expressions involving $x$, we face the potential difficulty of substituting a value of $x$ for which the expression is not defined (does not produce a real number). For example, the expression

$$\sqrt{2x + 3}$$

is *not defined* when $x = -2$ since

$$\sqrt{2(-2) + 3} = \sqrt{-4 + 3} = \sqrt{-1} \neq real\ number$$

The set of all values for which an expression is defined is called its **domain.** Thus, the domain of $\sqrt{2x + 3}$ is the set of all values of $x$ such that $\sqrt{2x + 3}$ is a real number. In order for $\sqrt{2x + 3}$ to represent a real number, it is necessary that

$$2x + 3 \geq 0$$

$$2x \geq -3$$

$$x \geq -\frac{3}{2}$$

In other words, $\sqrt{2x + 3}$ is defined only for those values of $x$ that lie in the interval $[-\frac{3}{2}, \infty)$, as shown in Figure 0.18.

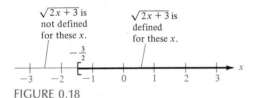

FIGURE 0.18

---

**EXAMPLE 5**

**Finding the Domain of an Algebraic Expression**

Find the intervals on which the following expressions are defined:

(a) $\sqrt{3x - 2}$

(b) $\dfrac{1}{\sqrt{3x - 2}}$

(c) $\sqrt[3]{9x + 1}$

(d) $\sqrt{2 + x} + \sqrt{3 - x}$

**SOLUTION**

(a) The domain of $\sqrt{3x - 2}$ consists of all $x$ such that

$$3x - 2 \geq 0$$

But this implies that $x \geq \frac{2}{3}$. Therefore, the domain is

$$\left[\frac{2}{3}, \infty\right)$$

(b) The domain of $1/\sqrt{3x - 2}$ is the same as the domain of the expression in part (a), *except* that this expression is not defined when $3x - 2 = 0$. Since this occurs when $x = \frac{2}{3}$, we conclude that the domain is $(\frac{2}{3}, \infty)$.

(c) Since $\sqrt[3]{9x + 1}$ is defined for *all* real numbers, we conclude that its domain is $(-\infty, \infty)$.

(d) The sum

$$\sqrt{2 + x} + \sqrt{3 - x}$$

is defined for all $x$ that are in the domain of *both* radical expressions. Since the domain of $\sqrt{2 + x}$ is $x \geq -2$ and the domain of $\sqrt{3 - x}$ is $x \leq 3$, we conclude that the domain of the sum is the intersection of the intervals

$$[-2, \infty) \quad \text{and} \quad (-\infty, 3]$$

Thus, the domain of $\sqrt{2 + x} + \sqrt{3 - x}$ is

$$[-2, 3]$$

as shown in Figure 0.19.

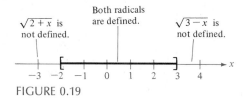

FIGURE 0.19

## SECTION EXERCISES 0.3

In Exercises 1–20, evaluate the given expression for the indicated value of $x$.

| Expression | x-value | Expression | x-value |
|---|---|---|---|
| **1.** $-3x^3$ | $x = 2$ | **2.** $\dfrac{x^2}{2}$ | $x = 6$ |
| **3.** $4x^{-3}$ | $x = 2$ | **4.** $7x^{-2}$ | $x = 4$ |
| **5.** $\dfrac{1 + x^{-1}}{x^{-1}}$ | $x = 2$ | **6.** $x - 4x^{-2}$ | $x = 3$ |
| **7.** $3x^2 - 4x^3$ | $x = -2$ | **8.** $5(-x)^3$ | $x = 3$ |
| **9.** $6x^0 - (6x)^0$ | $x = 10$ | **10.** $\dfrac{1}{(-x)^{-3}}$ | $x = 4$ |
| **11.** $\sqrt[3]{x^2}$ | $x = 27$ | **12.** $\sqrt{x^3}$ | $x = \dfrac{1}{9}$ |
| **13.** $x^{-1/2}$ | $x = 4$ | **14.** $x^{-3/4}$ | $x = 16$ |
| **15.** $x^{-2/5}$ | $x = -32$ | **16.** $(x^{2/3})^3$ | $x = 10$ |
| **17.** $500x^{60}$ | $x = 1.01$ | **18.** $\dfrac{10{,}000}{x^{120}}$ | $x = 1.075$ |
| **19.** $\sqrt[3]{x}$ | $x = -154$ | **20.** $\sqrt[6]{x}$ | $x = 325$ |

In Exercises 21–34, simplify the given expression.

**21.** $5x^4(x^2)$  
**22.** $(8x^4)(2x^3)$  
**23.** $6y^2(2y^4)^2$  
**24.** $z^{-3}(3z^4)$  
**25.** $10(x^2)^2$  
**26.** $(4x^3)^2$  
**27.** $\dfrac{7x^2}{x^{-3}}$  
**28.** $\dfrac{r^4}{r^6}$  
**29.** $\dfrac{12(x + y)^3}{9(x + y)}$  
**30.** $\left(\dfrac{12s^2}{9s}\right)^3$  
**31.** $\dfrac{3x\sqrt{x}}{x^{1/2}}$  
**32.** $(\sqrt[3]{x^2})^3$  
**33.** $\left(\dfrac{\sqrt{2}\sqrt{x^3}}{\sqrt{x}}\right)^4$  
**34.** $(2x^2yz^5)^0$

In Exercises 35–40, simplify by removing all possible factors from the radical.

**35.** (a) $\sqrt{8}$  
(b) $\sqrt{18}$  
**36.** (a) $\sqrt[3]{\dfrac{16}{27}}$  
(b) $\sqrt[3]{\dfrac{24}{125}}$  
**37.** (a) $\sqrt[3]{16x^5}$  
(b) $\sqrt[4]{32x^4z^5}$

**38.** (a) $\sqrt[4]{(3x^2y^3)^4}$      (b) $\sqrt[3]{54x^7}$

**39.** (a) $\sqrt{75x^2y^{-4}}$      (b) $\sqrt[3]{5(x-y)^3}$

**40.** (a) $\sqrt[5]{96b^6c^3}$      (b) $\sqrt{72(x+1)^4}$

In Exercises 41–56, insert the required factor (or factors) in the parentheses.

**41.** $y^4 - 4y^2 = y^2(y+2)(\phantom{xxxx})$

**42.** $(x+y)z^2 - (x+y) = (x+y)(z+1)(\phantom{xxxx})$

**43.** $\dfrac{3}{4}x + \dfrac{1}{2} = \dfrac{1}{4}(\phantom{xxxx})$

**44.** $\dfrac{2}{3}x^2 + \dfrac{1}{3}x + 5 = \dfrac{2}{3}(\phantom{xxxx})$

**45.** $\sqrt{x} + x\sqrt{x} = \sqrt{x}(\phantom{xxxx})$

**46.** $x\sqrt{y+1} + \sqrt{y+1} = \sqrt{y+1}(\phantom{xxxx})$

**47.** $x^2(x^3-1)^4 = (\phantom{xxxx})(x^3-1)^4(3x^2)$

**48.** $x(1-2x^2)^3 = (\phantom{xxxx})(1-2x^2)^3(-4x)$

**49.** $5x\sqrt{1+x^2} = (\phantom{xxxx})\sqrt{1+x^2}(2x)$

**50.** $\dfrac{1}{\sqrt{x}(1+\sqrt{x})^2} = (\phantom{xxxx})\dfrac{1}{(1+\sqrt{x})^2}\left(\dfrac{1}{2\sqrt{x}}\right)$

**51.** $3x^{1/2} + 4x^{3/2} = x^{1/2}(\phantom{xxxx})$

**52.** $5x^{1/3} - 4x^{4/3} = x^{1/3}(\phantom{xxxx})$

**53.** $3x^{-1/2} + 4x^{3/2} = x^{-1/2}(\phantom{xxxx})$

**54.** $5x^{-1/3} - 4x^{5/3} = x^{-1/3}(\phantom{xxxx})$

**55.** $\dfrac{1}{2}x(x+1)^{-1/2} + (x+1)^{1/2} = \dfrac{1}{2}(x+1)^{-1/2}(\phantom{xxxx})$

**56.** $\dfrac{3}{2}x(x+1)^{1/2} + (x+1)^{3/2} = \dfrac{3}{2}(x+1)^{1/2}(\phantom{xxxx})$

In Exercises 57–66, find the domain of the given expression.

**57.** $\sqrt{x-1}$      **58.** $\sqrt{5-2x}$

**59.** $\sqrt{x^2+3}$      **60.** $\sqrt[5]{1-x}$

**61.** $\dfrac{1}{\sqrt[3]{x-1}}$      **62.** $\dfrac{1}{\sqrt{x+4}}$

**63.** $\dfrac{1}{\sqrt[4]{2x-6}}$      **64.** $\dfrac{\sqrt{x-1}}{x+1}$

**65.** $\sqrt{x-1} + \sqrt{5-x}$      **66.** $\dfrac{1}{\sqrt{2x+3}} + \sqrt{6-4x}$

# Factoring Polynomials

The Fundamental Theorem of Algebra states that every $n$th-degree polynomial

$$a_n x^n + a_{n-1} x^{n-1} + \cdots + a_1 x + a_0$$

has precisely $n$ **zeros*** (or equivalently, the polynomial equation $a_n x^n + a_{n-1} x^{n-1} + \cdots + a_1 x + a_0 = 0$ has precisely $n$ **roots**). The problem of finding the zeros of a polynomial is equivalent to the problem of factoring the polynomial into linear factors. For example, the factorization

$$x^3 - 3x + 2 = (x - 1)(x - 1)(x + 2)$$

implies that $x = 1$ is a (repeated) zero of $x^3 - 3x + 2$ and $x = -2$ is a zero of $x^3 - 3x + 2$. We can check this by substitution. That is, if we let $x = 1$, then

$$(1)^3 - 3(1) + 2 = 1 - 3 + 2 = 0$$

and if we let $x = -2$, then

$$(-2)^3 - 3(-2) + 2 = -8 + 6 + 2 = 0$$

We summarize these steps as follows:

| | |
|---|---|
| $x^3 - 3x + 2 = 0$ | Set polynomial equal to zero |
| $(x - 1)(x - 1)(x + 2) = 0$ | Linear factors |
| $x - 1 = 0 \qquad x + 2 = 0$ | Set factors equal to zero |
| $x = 1 \qquad\qquad x = -2$ | Real zeros |

Similarly, the factorization

---

*The zeros may be repeated or imaginary.

$$x^5 - x = x(x - 1)(x + 1)(x^2 + 1)$$

implies that $x = 0$, $1$, and $-1$ are the three real zeros of $x^5 - x$. [The factor $(x^2 + 1)$ yields the two imaginary zeros $x = \pm\sqrt{-1}$. However, we are not concerned with imaginary zeros in this text.]

Factorization is *not* a simple problem, and there are a wide variety of techniques that we use to factor polynomials. Many of these techniques depend upon recognition of special products, and we suggest that you study the following summary carefully.

## Special Products and Factorization Techniques

Quadratic Formula

$$ax^2 + bx + c = 0 \implies x = \frac{-b \pm \sqrt{b^2 - 4ac}}{2a}$$

Example

$$x^2 + 3x - 1 = 0 \implies x = \frac{-3 \pm \sqrt{13}}{2}$$

Special products

$$x^2 - a^2 = (x - a)(x + a)$$
$$x^3 - a^3 = (x - a)(x^2 + ax + a^2)$$
$$x^3 + a^3 = (x + a)(x^2 - ax + a^2)$$
$$x^4 - a^4 = (x - a)(x + a)(x^2 + a^2)$$

Examples

$$x^2 - 9 = (x - 3)(x + 3)$$
$$x^3 - 8 = (x - 2)(x^2 + 2x + 4)$$
$$x^3 + 64 = (x + 4)(x^2 - 4x + 16)$$
$$x^4 - 16 = (x - 2)(x + 2)(x^2 + 4)$$

Binomial Theorem*

$$(x + a)^2 = x^2 + 2ax + a^2$$
$$(x - a)^2 = x^2 - 2ax + a^2$$
$$(x + a)^3 = x^3 + 3ax^2 + 3a^2x + a^3$$
$$(x - a)^3 = x^3 - 3ax^2 + 3a^2x - a^3$$
$$(x + a)^4 = x^4 + 4ax^3 + 6a^2x^2 + 4a^3x + a^4$$
$$(x - a)^4 = x^4 - 4ax^3 + 6a^2x^2 - 4a^3x + a^4$$

Examples

$$(x + 3)^2 = x^2 + 6x + 9$$
$$(x^2 - 5)^2 = x^4 - 10x^2 + 25$$
$$(x + 2)^3 = x^3 + 6x^2 + 12x + 8$$
$$(x - 1)^3 = x^3 - 3x^2 + 3x - 1$$
$$(x + 2)^4 = x^4 + 8x^3 + 24x^2 + 32x + 16$$
$$(x - 4)^4 = x^4 - 16x^3 + 96x^2 - 256x + 256$$

$$(x + a)^n = x^n + nax^{n-1} + \frac{n(n - 1)}{2!}a^2x^{n-2} + \frac{n(n - 1)(n - 2)}{3!}a^3x^{n-3} + \cdots + na^{n-1}x + a^n$$

$$(x - a)^n = x^n - nax^{n-1} + \frac{n(n - 1)}{2!}a^2x^{n-2} - \frac{n(n - 1)(n - 2)}{3!}a^3x^{n-3} + \cdots \pm na^{n-1}x \mp a^n$$

Factoring by grouping

$$acx^3 + adx^2 + bcx + bd = ax^2(cx + d) + b(cx + d)$$
$$= (ax^2 + b)(cx + d)$$

Example

$$3x^3 - 2x^2 - 6x + 4 = x^2(3x - 2) - 2(3x - 2)$$
$$= (x^2 - 2)(3x - 2)$$

---

*The *factorial* symbol ! is defined as follows: $0! = 1$, $1! = 1$, $2! = 2 \cdot 1 = 2$, $3! = 3 \cdot 2 \cdot 1 = 6$, $4! = 4 \cdot 3 \cdot 2 \cdot 1 = 24$, and so on.

When the Quadratic Formula is used, the term inside the square root (the **discriminant**) can be used to classify the zeros of $ax^2 + bx + c$ into the following three cases:

1. Two (distinct) real zeros: $b^2 - 4ac > 0$
2. One (repeated) real zero: $b^2 - 4ac = 0$
3. Two imaginary zeros:     $b^2 - 4ac < 0$

In the case of two distinct real zeros, we can graphically represent the zeros as the endpoints of an interval. The Quadratic Formula gives the midpoint of this interval as well as the distance to the endpoints, as shown in Figure 0.20.

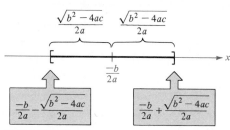

FIGURE 0.20

**EXAMPLE 1**

**Applying the Quadratic Formula**

Use the Quadratic Formula to find all real zeros of the following:
(a) $4x^2 + 6x + 1$                      (b) $x^2 + 6x + 9$
(c) $2x^2 - 6x + 5$

**SOLUTION**

(a) Using $a = 4$, $b = 6$, and $c = 1$, we have

$$x = \frac{-b \pm \sqrt{b^2 - 4ac}}{2a} = \frac{-6 \pm \sqrt{36 - 16}}{8}$$

$$= \frac{-6 \pm \sqrt{20}}{8}$$

$$= \frac{-6 \pm 2\sqrt{5}}{8}$$

$$= \frac{\cancel{2}(-3 \pm \sqrt{5})}{\cancel{2}(4)}$$

$$= \frac{-3 \pm \sqrt{5}}{4}$$

Thus, there are two real zeros:

$$x = \frac{-3 - \sqrt{5}}{4} \approx -1.309 \quad \text{and} \quad x = \frac{-3 + \sqrt{5}}{4} \approx -0.191$$

(b) In this case $a = 1$, $b = 6$, and $c = 9$, and the Quadratic Formula yields

$$x = \frac{-b \pm \sqrt{b^2 - 4ac}}{2a} = \frac{-6 \pm \sqrt{36 - 36}}{2} = -\frac{6}{2} = -3$$

Thus, there is one real zero: $x = -3$.

(c) For this quadratic equation $a = 2$, $b = -6$, and $c = 5$. Thus,

$$x = \frac{-b \pm \sqrt{b^2 - 4ac}}{2a} = \frac{6 \pm \sqrt{36 - 40}}{4} = \frac{6 \pm \sqrt{-4}}{4}$$

Since $\sqrt{-4}$ is imaginary, there are no real zeros.

In Example 1 the zeros in part (a) are irrational, and the zeros in part (c) are imaginary. In both of these cases we say that the quadratic is **irreducible** since it cannot be factored into linear factors with rational coefficients. The following example shows how to find the zeros associated with *reducible* quadratics. Remember that you can always use the Quadratic Formula to find the zeros of a second-degree polynomial. However, as you become more familiar with some of the simpler cases, you will soon learn to recognize quadratics whose zeros are simple rational numbers. In the following example, we use factoring to find the zeros of each quadratic, and we suggest that you try using the Quadratic Formula to obtain the same zeros.

**EXAMPLE 2**

**Factoring Quadratics**

Find the zeros of the following quadratic polynomials:
(a) $x^2 - 5x + 6$                (b) $x^2 - 5x - 6$
(c) $2x^2 + 5x - 3$

**SOLUTION**

(a) Since

$$x^2 - 5x + 6 = (x - 2)(x - 3)$$

the zeros are $x = 2$ and $x = 3$.

(b) Since

$$x^2 - 5x - 6 = (x + 1)(x - 6)$$

the zeros are $x = -1$ and $x = 6$.

(c) Since

$$2x^2 + 5x - 3 = (2x - 1)(x + 3)$$

the zeros are $x = \frac{1}{2}$ and $x = -3$.

In calculus we will encounter several expressions of the form

$$\sqrt{ax^2 + bx + c}$$

To find the domain of this expression, we first note that it will be defined at precisely those $x$-values for which

$$ax^2 + bx + c \geq 0$$

Thus, the problem of finding the interval (or intervals) for which this expression is defined is equivalent to solving a quadratic inequality, as demonstrated in Section 0.1.

**EXAMPLE 3**

**Finding the Domain of a Radical Expression**

Find the domain of each of the following expressions:

(a) $\sqrt{x^2 - 3x + 2}$     (b) $\sqrt{-x^2 + 3x + 4}$     (c) $\sqrt{x^2 - 2x + 2}$

**SOLUTION**

(a) Since

$$x^2 - 3x + 2 = (x - 1)(x - 2)$$

we know that the zeros of the quadratic are $x = 1$ and $x = 2$, and we must test the sign of the quadratic in the three intervals $(-\infty, 1)$, $(1, 2)$, and $(2, \infty)$, as shown in Figure 0.21. After testing each of these intervals, we see that the quadratic is negative in the center interval and positive in the outer two intervals. Moreover, since the quadratic is zero when $x = 1$ and $x = 2$, we conclude that the domain of $\sqrt{x^2 - 3x + 2}$ is

$$(-\infty, 1] \quad \text{and} \quad [2, \infty) \qquad \text{Domain}$$

(b) Since

$$-x^2 + 3x + 4 = (-x + 4)(x + 1)$$

we know that the zeros of the quadratic are $x = -1$ and $x = 4$. By checking points inside and outside the interval $[-1, 4]$, we see that the radical is defined only inside this interval, as shown in Figure 0.22.

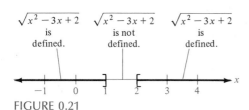

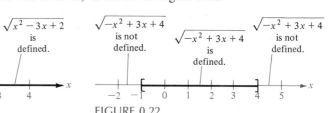

FIGURE 0.21

FIGURE 0.22

(c) Since $x^2 - 2x + 2 = 0$ has no real roots, we observe that the quadratic is positive when $x = 0$ and conclude that the radical is defined for all values of $x$. (See Figure 0.23.)

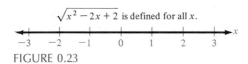

FIGURE 0.23

## Factoring Polynomials of Degree Three or More

It is often very difficult to find the zeros of polynomials of degree three or more. However, if one of the zeros of a polynomial is known, then we can use that zero to reduce the degree of the polynomial. For example, if we know that $x = 2$ is a zero of $x^3 - 4x^2 + 5x - 2$, then we know that $(x - 2)$ is a factor, and we can use long division to reduce the polynomial as follows:

$$
\begin{array}{r}
x^2 - 2x + 1 \\
x - 2 \overline{) x^3 - 4x^2 + 5x - 2} \\
\underline{x^3 - 2x^2} \\
-2x^2 + 5x \\
\underline{-2x^2 + 4x} \\
x - 2 \\
\underline{x - 2}
\end{array}
$$

Thus, we know that

$$x^3 - 4x^2 + 5x - 2 = (x - 2)(x^2 - 2x + 1)$$

and finally, by factoring the quadratic expression, we have

$$x^3 - 4x^2 + 5x - 2 = (x - 2)(x - 1)(x - 1)$$

■■■ **Remark:** Note that even if the quadratic expression had not happened to factor, we would still be "home free" since we could always apply the Quadratic Formula.

As an alternative to long division, many people prefer to use **synthetic division** to reduce the degree of a polynomial. We outline this procedure as follows.

**Synthetic division for a cubic polynomial**

Given $x = x_1$ is a zero of $ax^3 + bx^2 + cx + d$,

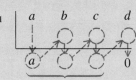

Coefficients for quadratic factor

Vertical pattern: *Add terms.*
Diagonal pattern: *Multiply by the given zero, $x_1$.*

For example, performing synthetic division on the polynomial

$$x^3 - 4x^2 + 5x - 2$$

using the given zero, $x = 2$, we have

$$
\begin{array}{r|rrrr}
2 & 1 & -4 & 5 & -2 \\
  &   & 2 & -4 & 2 \\
\hline
  & 1 & -2 & 1 & 0
\end{array}
$$

$$(x - 2)(x^2 - 2x + 1) = x^3 - 4x^2 + 5x - 2$$

When you use synthetic division you must remember to take *all* coefficients into account—*even if some of them are zero*. For instance, if we know that $x = -2$ is a zero of $x^3 + 3x + 14$, we apply synthetic division as follows:

$$
\begin{array}{r|rrrr}
-2 & 1 & 0 & 3 & 14 \\
   &   & -2 & 4 & -14 \\
\hline
   & 1 & -2 & 7 & 0
\end{array}
$$

$$(x + 2)(x^2 - 2x + 7) = x^3 + 3x + 14$$

### The Rational Zero Theorem

There is a systematic way to find the *rational* zeros of a polynomial. We can use the **Rational Zero Theorem** (also called the Rational Root Theorem).

| **Rational Zero Theorem** | If a polynomial $$a_n x^n + a_{n-1} x^{n-1} + \cdots + a_1 x + a_0$$ has integer coefficients, then every *rational* zero is of the form $x = p/q$, where $p$ is a factor of $a_0$ and $q$ is a factor of $a_n$. |
| --- | --- |

To use the Rational Zero Theorem we make a list of all possible rational zeros. This list will consist of ratios of the form

$$\frac{p}{q} = \frac{\text{factor of constant term}}{\text{factor of leading coefficient}}$$

For example, the possible rational zeros of the polynomial

$$2x^4 - 7x^3 + 5x^2 - 7x + 3$$

are $x = \pm 1$, $\pm \frac{1}{2}$, $\pm 3$, and $\pm \frac{3}{2}$. By testing we find the two rational zeros to be $\frac{1}{2}$ and 3. Note that to test for possible zeros, we simply evaluate the polynomial for the given test values. For instance, $x = 1$ is not a zero since

$$2(1)^4 - 7(1)^3 + 5(1)^2 - 7(1) + 3 = -4$$

but $x = 3$ is a zero of the polynomial since

$$2(3)^4 - 7(3)^3 + 5(3)^2 - 7(3) + 3 = 0$$

**EXAMPLE 4**

**Using the Rational Zero Theorem**

Find all real zeros of $2x^3 + 3x^2 - 8x + 3$.

**SOLUTION**

$$\boxed{2}x^3 + 3x^2 - 8x + \boxed{3}$$

Factors of constant term: $\pm 1$, $\pm 3 \leftarrow$

$\rightarrow$ Factors of leading coefficient: $\pm 1$, $\pm 2$

The possible rational zeros are the factors of the constant term divided by the factors of the leading coefficient:

$$1, \quad -1, \quad 3, \quad -3, \quad \frac{1}{2}, \quad -\frac{1}{2}, \quad \frac{3}{2}, \quad -\frac{3}{2}$$

By testing these possible zeros we see that $x = 1$ works since

$$2(1)^3 + 3(1)^2 - 8(1) + 3 = 2 + 3 - 8 + 3 = 0$$

Now by synthetic division we have

$$
\begin{array}{r|rrrr}
1 & 2 & 3 & -8 & 3 \\
  &   & 2 & 5 & -3 \\
\hline
  & 2 & 5 & -3 & 0
\end{array}
$$

$$(x - 1)(2x^2 + 5x - 3) = 2x^3 + 3x^2 - 8x + 3$$

Finally, by factoring the quadratic

$$2x^2 + 5x - 3 = (2x - 1)(x + 3)$$

we have

$$2x^3 + 3x^2 - 8x + 3 = (x - 1)(2x - 1)(x + 3)$$

and we conclude that the zeros are

$$x = 1, \quad x = \frac{1}{2}, \quad x = -3$$

---

**SECTION EXERCISES 0.4**

In Exercises 1–10, use the Quadratic Formula to find all real zeros of the given second-degree polynomial.

1. $6x^2 - x - 1$
2. $8x^2 - 2x - 1$
3. $4x^2 - 12x + 9$
4. $9x^2 + 12x + 4$
5. $y^2 + 4y + 1$
6. $x^2 + 6x - 1$
7. $3x^2 - 2x - 2$
8. $2s^2 - 7s + 3$
9. $2s^2 - 7s + 4$
10. $2s^2 - 7s + 5$

In Exercises 11–14, use the Quadratic Formula to find all real roots of the given equation.

11. $x^2 - 2x + 3 = 0$
12. $x^2 + 1 = x(3 - x)$
13. $x + 1 = \dfrac{3}{x}$
14. $(z + 1)^2 = 2z^2$

In Exercises 15–30, write the second-degree polynomial as the product of two linear factors.

**15.** $x^2 - 4x + 4$

**16.** $x^2 + 10x + 25$

**17.** $4x^2 + 4x + 1$

**18.** $9x^2 - 12x + 4$

**19.** $x^2 + x - 2$

**20.** $2x^2 - x - 1$

**21.** $3x^2 - 5x + 2$

**22.** $x^2 - xy - 2y^2$

**23.** $x^2 - 4xy + 4y^2$

**24.** $a^2b^2 - 2abc + c^2$

**25.** $2y^2 + 21yz - 36z^2$

**26.** $60x^2 - 61xy + 15y^2$

**27.** $16y^2 - 9$

**28.** $z^2 - 100$

**29.** $(x - 1)^2 - 4$

**30.** $25 - (z + 5)^2$

In Exercises 31–44, completely factor the given polynomial.

**31.** $81 - y^4$

**32.** $x^4 - 16$

**33.** $x^3 - 8$

**34.** $x^3 - 27$

**35.** $y^3 + 64$

**36.** $z^3 + 125$

**37.** $x^3 - 27$

**38.** $(x - a)^3 + b^3$

**39.** $x^3 - 4x^2 - x + 4$

**40.** $x^3 - x^2 - x + 1$

**41.** $2x^3 - 3x^2 + 4x - 6$

**42.** $x^3 - 5x^2 - 5x + 25$

**43.** $2x^3 - 4x^2 - x + 2$

**44.** $x^3 - 7x^2 - 4x + 28$

In Exercises 45–68, find all real roots of the given equation.

**45.** $x^2 - 5x = 0$

**46.** $2x^2 - 3x = 0$

**47.** $x^2 - 9 = 0$

**48.** $x^2 - 25 = 0$

**49.** $x^2 - 3 = 0$

**50.** $x^2 - 8 = 0$

**51.** $(x - 3)^2 - 9 = 0$

**52.** $(x + 1)^2 - 8 = 0$

**53.** $x^2 + x - 2 = 0$

**54.** $x^2 + 5x + 6 = 0$

**55.** $x^2 - 5x + 6 = 0$

**56.** $x^2 + x - 20 = 0$

**57.** $2x^2 - x - 1 = 0$

**58.** $3x^2 - 5x + 2 = 0$

**59.** $(x - 5)(x + 3) = 33$

**60.** $(x - 1)(x + 2) = 4$

**61.** $x^3 + 64 = 0$

**62.** $x^3 - 216 = 0$

**63.** $x^4 - 16 = 0$

**64.** $x^4 - 625 = 0$

**65.** $x^3 - x^2 - 4x + 4 = 0$

**66.** $2x^3 + x^2 + 6x + 3 = 0$

**67.** $(x + 2)^2(x - 1) + (x + 2)(x - 1)^2 = 0$

**68.** $(x + 5)(x - 2) + (x + 5)^2 = 0$

In Exercises 69–76, find the interval (or intervals) on which the given expression is defined.

**69.** $\sqrt{x^2 - 7x + 12}$

**70.** $\sqrt{x^2 - 4}$

**71.** $\sqrt{4 - x^2}$

**72.** $\sqrt{144 - 9x^2}$

**73.** $\sqrt{12 - x - x^2}$

**74.** $\sqrt{x^2 + 4}$

**75.** $\sqrt{x^2 - 3x + 3}$

**76.** $\sqrt{-x^2 + 2x - 2}$

In Exercises 77–82, use synthetic division to complete the indicated factorization.

**77.** $x^3 + 8 = (x + 2)(\phantom{xxxx})$

**78.** $x^3 - 2x^2 - x + 2 = (x + 1)(\phantom{xxxx})$

**79.** $2x^3 - x^2 - 2x + 1 = (x - 1)(\phantom{xxxx})$

**80.** $x^4 - 16x^3 + 96x^2 - 256x + 256 = (x - 4)(\phantom{xxxx})$

**81.** $x^4 + 2x^3 - 6x^2 - 18x - 27 = (x - 3)(\phantom{xxxx})$

**82.** $x^5 - 243 = (x - 3)(\phantom{xxxx})$

In Exercises 83–94, use the Rational Zero Theorem as an aid in finding all real roots of the given equation.

**83.** $x^3 - x^2 - x + 1 = 0$

**84.** $x^3 - x^2 - 4x + 4 = 0$

**85.** $x^3 - 6x^2 + 11x - 6 = 0$

**86.** $x^3 + 2x^2 - 5x - 6 = 0$

**87.** $4x^3 - 4x^2 - x + 1 = 0$

**88.** $18x^3 - 9x^2 - 8x + 4 = 0$

**89.** $x^3 - 3x^2 - 3x - 4 = 0$

**90.** $4x^3 - 6x^2 + 2x + 3 = 0$

**91.** $z^3 + 8z^2 + 11z - 2 = 0$

**92.** $3y^3 + 11y^2 - y - 1 = 0$

**93.** $x^4 - 13x^2 + 36 = 0$

**94.** $x^4 - x^3 - 7x^2 + x + 6 = 0$

**95.** Use the Binomial Formula

$$(x + a)^5 = x^5 + 5x^4a + 10x^3a^2 + 10x^2a^3 + 5xa^4 + a^5$$

to factor

$$x^5 - 10x^4 + 40x^3 - 80x^2 + 80x - 32$$

# Fractions and Rationalization

In the final section of this chapter we review operations involving fractional expressions such as

$$\frac{2}{x}, \qquad \frac{x^2 + 2x - 4}{x + 6}, \qquad \frac{1}{\sqrt{x^2 + 1}}$$

The first two expressions have polynomials as both numerator and denominator and are called **rational expressions.** A rational expression is **proper** if the degree of the numerator is less than the degree of the denominator. For example, $x/(x^2 + 1)$ is proper. If the degree of the numerator is greater than or equal to the degree of the denominator, then the rational expression is **improper.** For example, $x^2/(x^2 + 1)$ is improper.

To add, subtract, multiply, or divide rational (or other fractional) expressions, we use the following familiar rules for operating with fractions.

**Operations with fractions**

1. Add fractions (find a common denominator): $\dfrac{a}{b} + \dfrac{c}{d} = \dfrac{a}{b}\left(\dfrac{d}{d}\right) + \dfrac{c}{d}\left(\dfrac{b}{b}\right) = \dfrac{ad}{bd} + \dfrac{bc}{bd} = \dfrac{ad + bc}{bd}$

2. Subtract fractions (find a common denominator): $\dfrac{a}{b} - \dfrac{c}{d} = \dfrac{a}{b}\left(\dfrac{d}{d}\right) - \dfrac{c}{d}\left(\dfrac{b}{b}\right) = \dfrac{ad}{bd} - \dfrac{bc}{bd} = \dfrac{ad - bc}{bd}$

3. Multiply fractions: $\left(\dfrac{a}{b}\right)\left(\dfrac{c}{d}\right) = \dfrac{ac}{bd}$

4. Divide fractions (invert and multiply): $\dfrac{a/b}{c/d} = \left(\dfrac{a}{b}\right)\left(\dfrac{d}{c}\right) = \dfrac{ad}{bc}, \ \dfrac{a/b}{c} = \dfrac{a/b}{c/1} = \left(\dfrac{a}{b}\right)\left(\dfrac{1}{c}\right) = \dfrac{a}{bc}$

5. Cancel: $\dfrac{\cancel{a}b}{\cancel{a}c} = \dfrac{b}{c}, \quad \dfrac{ab + ac}{ad} = \dfrac{\cancel{a}(b + c)}{\cancel{a}d} = \dfrac{b + c}{d}$

**EXAMPLE 1**

**Adding and Subtracting Rational Expressions**

Perform the indicated operations and simplify:

(a) $x + \dfrac{1}{x}$

(b) $\dfrac{1}{x+1} - \dfrac{2}{2x-1}$

**SOLUTION**

(a) $x + \dfrac{1}{x} = \dfrac{x^2}{x} + \dfrac{1}{x} = \dfrac{x^2 + 1}{x}$

(b) $\dfrac{1}{x+1} - \dfrac{2}{2x-1} = \dfrac{(2x-1)}{(x+1)(2x-1)} - \dfrac{2(x+1)}{(x+1)(2x-1)}$

$= \dfrac{2x - 1 - 2x - 2}{2x^2 + x - 1}$

$= \dfrac{-3}{2x^2 + x - 1}$

In adding (or subtracting) fractions whose denominators have no common factors, it is convenient to use the following "cross multiplication" pattern:

$$\frac{a}{b} + \frac{c}{d} = \frac{a \times c}{(b) \times (d)} = \frac{ad + bc}{bd}$$

For instance, in part (b) of Example 1 we could have used cross multiplication as follows:

$$\frac{1}{x+1} - \frac{2}{2x-1} = \frac{(2x-1) - 2(x+1)}{(x+1)(2x-1)}$$

$$= \frac{2x - 1 - 2x - 2}{(x+1)(2x-1)}$$

$$= \frac{-3}{2x^2 + x - 1}$$

In Example 1 the denominators of the rational expressions have no common factors. When the denominators do have common factors, it is best to find the least common denominator before adding or subtracting. For instance, to add $1/x$ and $2/x^2$, we recognize the least common denominator to be $x^2$ and write

$$\frac{1}{x} + \frac{2}{x^2} = \frac{x}{x^2} + \frac{2}{x^2} = \frac{x+2}{x^2}$$

This is further demonstrated in Example 2.

**EXAMPLE 2**

**Adding and Subtracting Rational Expressions**

Perform the indicated operation and simplify:

(a) $\dfrac{x}{x^2 - 1} + \dfrac{3}{x + 1}$

(b) $\dfrac{1}{2(x^2 + 2x)} - \dfrac{1}{4x}$

**SOLUTION**

(a) Since $x^2 - 1 = (x + 1)(x - 1)$, the least common denominator is $x^2 - 1$, and we have

$$\frac{x}{x^2 - 1} + \frac{3}{x + 1} = \frac{x}{(x - 1)(x + 1)} + \frac{3}{x + 1}$$

$$= \frac{x}{(x - 1)(x + 1)} + \frac{3(x - 1)}{(x - 1)(x + 1)}$$

$$= \frac{x + 3x - 3}{(x - 1)(x + 1)}$$

$$= \frac{4x - 3}{x^2 - 1}$$

(b) In this case, the least common denominator is $4x(x + 2)$.

$$\frac{1}{2(x^2 + 2x)} - \frac{1}{4x} = \frac{1}{2x(x + 2)} - \frac{1}{2(2x)}$$

$$= \frac{2}{2(2x)(x + 2)} - \frac{x + 2}{2(2x)(x + 2)}$$

$$= \frac{2 - x - 2}{4x(x + 2)}$$

$$= \frac{-\cancel{x}}{4\cancel{x}(x + 2)} \qquad\qquad \text{Cancel } x$$

$$= \frac{-1}{4(x + 2)}$$

To add more than two fractions we must find a denominator that is common to all the fractions. For instance, to add $\frac{1}{2}$, $\frac{1}{3}$, and $\frac{1}{5}$, we use a (least) common denominator of 30 and write

$$\frac{1}{2} + \frac{1}{3} + \frac{1}{5} = \frac{15}{30} + \frac{10}{30} + \frac{6}{30} = \frac{31}{30}$$

To add more than two rational expressions, we use a similar procedure, as demonstrated in Example 3.

EXAMPLE 3

**Adding More Than Two Rational Expressions**

Perform the indicated addition of rational expressions:

(a) $\dfrac{A}{x+2} + \dfrac{B}{x-3} + \dfrac{C}{x+4}$

(b) $\dfrac{A}{x+2} + \dfrac{B}{(x+2)^2} + \dfrac{C}{x-1}$

SOLUTION

(a) The least common denominator is $(x+2)(x-3)(x+4)$, and we have

$$\dfrac{A}{x+2} + \dfrac{B}{x-3} + \dfrac{C}{x+4}$$

$$= \dfrac{A(x-3)(x+4) + B(x+2)(x+4) + C(x+2)(x-3)}{(x+2)(x-3)(x+4)}$$

$$= \dfrac{A(x^2 + x - 12) + B(x^2 + 6x + 8) + C(x^2 - x - 6)}{(x+2)(x-3)(x+4)}$$

$$= \dfrac{Ax^2 + Bx^2 + Cx^2 + Ax + 6Bx - Cx - 12A + 8B - 6C}{(x+2)(x-3)(x+4)}$$

$$= \dfrac{(A + B + C)x^2 + (A + 6B - C)x + (-12A + 8B - 6C)}{(x+2)(x-3)(x+4)}$$

(b) Here the least common denominator is $(x+2)^2(x-1)$.

$$\dfrac{A}{x+2} + \dfrac{B}{(x+2)^2} + \dfrac{C}{x-1}$$

$$= \dfrac{A(x+2)(x-1) + B(x-1) + C(x+2)^2}{(x+2)^2(x-1)}$$

$$= \dfrac{A(x^2 + x - 2) + B(x-1) + C(x^2 + 4x + 4)}{(x+2)^2(x-1)}$$

$$= \dfrac{Ax^2 + Cx^2 + Ax + Bx + 4Cx - 2A - B + 4C}{(x+2)^2(x-1)}$$

$$= \dfrac{(A + C)x^2 + (A + B + 4C)x + (-2A - B + 4C)}{(x+2)^2(x-1)}$$

In calculus we use an operation (called differentiation) that tends to produce "messy" expressions when applied to fractional expressions. This is especially true if the fractional expression involves radicals. When differentiation is used, it is important to be able to simplify these messy expressions so that we can obtain more manageable forms. All of the expressions in Examples 4 and 5 are the result of differentiation. In each case note how much *simpler* the simplified form is than the original form.

**EXAMPLE 4**

**Simplifying an Expression Involving Radicals**

(a) $\dfrac{\sqrt{x+1} - \dfrac{x}{2\sqrt{x+1}}}{x+1}$

(b) $\left(\dfrac{1}{x + \sqrt{x^2+1}}\right)\left(1 + \dfrac{2x}{2\sqrt{x^2+1}}\right)$

**SOLUTION**

(a) $\dfrac{\sqrt{x+1} - \dfrac{x}{2\sqrt{x+1}}}{x+1} = \dfrac{\dfrac{2(x+1)}{2\sqrt{x+1}} - \dfrac{x}{2\sqrt{x+1}}}{x+1}$

$= \dfrac{\dfrac{2x+2-x}{2\sqrt{x+1}}}{\dfrac{x+1}{1}}$

$= \dfrac{x+2}{2\sqrt{x+1}}\left(\dfrac{1}{x+1}\right)$

$= \dfrac{x+2}{2(x+1)^{3/2}}$

(b) $\left(\dfrac{1}{x + \sqrt{x^2+1}}\right)\left(1 + \dfrac{2x}{2\sqrt{x^2+1}}\right) = \left(\dfrac{1}{x + \sqrt{x^2+1}}\right)\left(1 + \dfrac{x}{\sqrt{x^2+1}}\right)$

$= \left(\dfrac{1}{x + \sqrt{x^2+1}}\right)\left(\dfrac{\sqrt{x^2+1}}{\sqrt{x^2+1}} + \dfrac{x}{\sqrt{x^2+1}}\right)$

$= \left(\dfrac{1}{x + \sqrt{x^2+1}}\right)\left(\dfrac{x + \sqrt{x^2+1}}{\sqrt{x^2+1}}\right)$

$= \dfrac{1}{\sqrt{x^2+1}}$

**EXAMPLE 5**

**Simplifying an Expression Involving Radicals**

$\dfrac{-x\left[\dfrac{2x}{2\sqrt{x^2+1}}\right] + \sqrt{x^2+1}}{x^2} + \left(\dfrac{1}{x + \sqrt{x^2+1}}\right)\left(1 + \dfrac{2x}{2\sqrt{x^2+1}}\right)$

**SOLUTION**

From Example 4 we already know that the second part of this sum simplifies to $1/\sqrt{x^2+1}$. Try filling in the steps to show that the first part simplifies as follows:

$\dfrac{-x\left[\dfrac{2x}{2\sqrt{x^2+1}}\right] + \sqrt{x^2+1}}{x^2} = \dfrac{1}{x^2\sqrt{x^2+1}}$

Therefore, the sum is

$$\frac{-x\left[\dfrac{2x}{2\sqrt{x^2+1}}\right]+\sqrt{x^2+1}}{x^2}+\left(\frac{1}{x+\sqrt{x^2+1}}\right)\left(1+\frac{2x}{2\sqrt{x^2+1}}\right)$$

$$=\frac{1}{x^2\sqrt{x^2+1}}+\frac{1}{\sqrt{x^2+1}}$$

$$=\frac{1}{x^2\sqrt{x^2+1}}+\frac{x^2}{x^2\sqrt{x^2+1}}$$

$$=\frac{x^2+1}{x^2\sqrt{x^2+1}}$$

$$=\frac{\sqrt{x^2+1}}{x^2}$$

## Rationalization Techniques

In working with quotients involving radicals it is often convenient to move the radical expression from the denominator to the numerator, or vice versa. For example, we can move $\sqrt{2}$ from the denominator to the numerator in the following quotient by multiplying by $\sqrt{2}/\sqrt{2}$.

| *Radical in denominator* | | *Rationalize the denominator* | | *Radical in numerator* |
|---|---|---|---|---|
| $\dfrac{1}{\sqrt{2}}$ | $\Longrightarrow$ | $\dfrac{1}{\sqrt{2}}\left(\dfrac{\sqrt{2}}{\sqrt{2}}\right)$ | $\Longrightarrow$ | $\dfrac{\sqrt{2}}{2}$ |

We call this process **rationalizing the denominator.** Similarly, if the radical is moved from the numerator to the denominator, the process is called **rationalizing the numerator.** The three most common rationalizing techniques are given in the following list.

---

**Rationalizing techniques**

To rationalize the numerator or denominator of an expression, use the following:

1. If numerator or denominator is $\sqrt{a}$, multiply by $\dfrac{\sqrt{a}}{\sqrt{a}}$.

2. If numerator or denominator is $\sqrt{a}-\sqrt{b}$, multiply by $\dfrac{\sqrt{a}+\sqrt{b}}{\sqrt{a}+\sqrt{b}}$.

3. If numerator or denominator is $\sqrt{a}+\sqrt{b}$, multiply by $\dfrac{\sqrt{a}-\sqrt{b}}{\sqrt{a}-\sqrt{b}}$.

Note that the success of the second and third techniques stems from the following elimination of radicals:

$$(\sqrt{a} - \sqrt{b})(\sqrt{a} + \sqrt{b}) = (\sqrt{a})^2 + \sqrt{a}\sqrt{b} - \sqrt{a}\sqrt{b} - (\sqrt{b})^2$$

$$= a - b$$

**EXAMPLE 6**

**Rationalizing Single-Term Denominators and Numerators**

(a) $\dfrac{3}{\sqrt{12}}$

(b) $\dfrac{\sqrt{x+1}}{2}$

**SOLUTION**

(a) $\dfrac{3}{\sqrt{12}} = \dfrac{3}{2\sqrt{3}} = \dfrac{3}{2\sqrt{3}}\left(\dfrac{\sqrt{3}}{\sqrt{3}}\right) = \dfrac{3\sqrt{3}}{2(3)} = \dfrac{\sqrt{3}}{2}$

(b) $\dfrac{\sqrt{x+1}}{2} = \dfrac{\sqrt{x+1}}{2}\left(\dfrac{\sqrt{x+1}}{\sqrt{x+1}}\right) = \dfrac{x+1}{2\sqrt{x+1}}$

**EXAMPLE 7**

**Rationalizing Double-Term Denominators and Numerators**

(a) $\dfrac{1}{\sqrt{5} + \sqrt{2}}$

(b) $\dfrac{1}{\sqrt{x} - \sqrt{x+1}}$

(c) $\dfrac{x - \sqrt{x^2 - x}}{3x}$

**SOLUTION**

(a) $\dfrac{1}{\sqrt{5} + \sqrt{2}} = \dfrac{1}{\sqrt{5} + \sqrt{2}}\left(\dfrac{\sqrt{5} - \sqrt{2}}{\sqrt{5} - \sqrt{2}}\right)$

$= \dfrac{\sqrt{5} - \sqrt{2}}{5 - 2} = \dfrac{\sqrt{5} - \sqrt{2}}{3}$

(b) $\dfrac{1}{\sqrt{x} - \sqrt{x+1}} = \dfrac{1}{\sqrt{x} - \sqrt{x+1}}\left(\dfrac{\sqrt{x} + \sqrt{x+1}}{\sqrt{x} + \sqrt{x+1}}\right)$

$= \dfrac{\sqrt{x} + \sqrt{x+1}}{x - (x+1)} = -\sqrt{x} - \sqrt{x+1}$

(c) $\dfrac{x - \sqrt{x^2 - x}}{3x} = \dfrac{x - \sqrt{x^2 - x}}{3x}\left(\dfrac{x + \sqrt{x^2 - x}}{x + \sqrt{x^2 - x}}\right)$

$= \dfrac{x^2 - (x^2 - x)}{3x(x + \sqrt{x^2 - x})} = \dfrac{x}{3x(x + \sqrt{x^2 - x})}$

$= \dfrac{1}{3(x + \sqrt{x^2 - x})}$

In Exercises 1–30, perform the indicated operations and simplify your answer.

**1.** $\dfrac{5}{x-1} + \dfrac{x}{x-1}$

**2.** $\dfrac{2x-1}{x+3} + \dfrac{1-x}{x+3}$

**3.** $\dfrac{2x}{x^2+2} - \dfrac{1-3x}{x^2+2}$

**4.** $\dfrac{5x+10}{2x-1} - \dfrac{2x+10}{2x-1}$

**5.** $\dfrac{4}{x} - \dfrac{3}{x^2}$

**6.** $\dfrac{5}{x-1} + \dfrac{3}{x}$

**7.** $\dfrac{2}{x+2} - \dfrac{1}{x-2}$

**8.** $\dfrac{x}{x^2+x-2} - \dfrac{1}{x+2}$

**9.** $\dfrac{5}{x-3} + \dfrac{3}{3-x}$

**10.** $\dfrac{x}{2-x} + \dfrac{2}{x-2}$

**11.** $\dfrac{1}{x^2-x-2} - \dfrac{x}{x^2-5x+6}$

**12.** $\dfrac{x-1}{x^2+5x+4} + \dfrac{2}{x^2-x-2} + \dfrac{10}{x^2+2x-8}$

**13.** $\dfrac{A}{x-6} + \dfrac{B}{x+3}$

**14.** $\dfrac{A}{x+1} + \dfrac{B}{(x+1)^2} + \dfrac{C}{x-2}$

**15.** $\dfrac{A}{x-5} + \dfrac{B}{x+5} + \dfrac{C}{(x+5)^2}$

**16.** $\dfrac{Ax+B}{x^2+2} + \dfrac{C}{x-4}$

**17.** $-\dfrac{1}{x} + \dfrac{2}{x^2+1}$

**18.** $\dfrac{2}{x+1} + \dfrac{1-x}{x^2-2x+3}$

**19.** $\dfrac{-x}{(x+1)^{3/2}} + \dfrac{2}{(x+1)^{1/2}}$

**20.** $2\sqrt{x}(x-2) + \dfrac{(x-2)^2}{2\sqrt{x}}$

**21.** $\dfrac{2-t}{2\sqrt{1+t}} - \sqrt{1+t}$

**22.** $-\dfrac{\sqrt{x^2+1}}{x^2} + \dfrac{1}{\sqrt{x^2+1}}$

**23.** $\left(2x\sqrt{x^2+1} - \dfrac{x^3}{\sqrt{x^2+1}}\right) \div (x^2+1)$

**24.** $\left(\sqrt{x^3+1} - \dfrac{3x^3}{2\sqrt{x^3+1}}\right) \div (x^3+1)$

**25.** $\left(\dfrac{1}{(x+\Delta x)^2} - \dfrac{1}{x^2}\right) \div \Delta x$

**26.** $\left(\dfrac{x+\Delta x}{x+\Delta x+1} - \dfrac{x}{x+1}\right) \div \Delta x$

**27.** $\dfrac{(x^2+2)^{1/2} - x^2(x^2+2)^{-1/2}}{x^2}$

**28.** $\dfrac{x(x+1)^{-1/2} - (x+1)^{1/2}}{x^2}$

**29.** $\dfrac{\dfrac{\sqrt{x+1}}{\sqrt{x}} - \dfrac{\sqrt{x}}{\sqrt{x+1}}}{2(x+1)}$

**30.** $\dfrac{\dfrac{2x^2}{3(x^2-1)^{2/3}} - (x^2-1)^{1/3}}{x^2}$

In Exercises 31–54, rationalize the numerator or denominator and simplify.

**31.** $\dfrac{3}{\sqrt{27}}$

**32.** $\dfrac{5}{\sqrt{10}}$

**33.** $\dfrac{\sqrt{2}}{3}$

**34.** $\dfrac{\sqrt{26}}{2}$

**35.** $\dfrac{x}{\sqrt{x-4}}$

**36.** $\dfrac{4y}{\sqrt{y+8}}$

**37.** $\dfrac{\sqrt{y^3}}{6y}$

**38.** $\dfrac{x\sqrt{x^2+4}}{3}$

**39.** $\dfrac{49(x-3)}{\sqrt{x^2-9}}$

**40.** $\dfrac{10(x+2)}{\sqrt{x^2-x-6}}$

**41.** $\dfrac{5}{\sqrt{14}-2}$

**42.** $\dfrac{13}{6+\sqrt{10}}$

**43.** $\dfrac{2x}{5-\sqrt{3}}$

**44.** $\dfrac{x}{\sqrt{2}+\sqrt{3}}$

**45.** $\dfrac{1}{\sqrt{6}+\sqrt{5}}$

**46.** $\dfrac{\sqrt{15}+3}{12}$

**47.** $\dfrac{\sqrt{3}-\sqrt{2}}{x}$

**48.** $\dfrac{x-1}{\sqrt{x}+x}$

**49.** $\dfrac{2x-\sqrt{4x-1}}{2x-1}$

**50.** $\dfrac{10}{\sqrt{x}+\sqrt{x+5}}$

**51.** $\dfrac{x+1}{\sqrt{x^2-2}-\sqrt{x}}$

**52.** $\dfrac{x-1}{\sqrt{3}-\sqrt{4x-x^2}}$

**53.** $\dfrac{8x}{\sqrt{17x-1}}$

**54.** $\dfrac{1-x^2}{\sqrt{x}-\sqrt{x^3}}$

In Exercises 1–4, place the appropriate inequality sign ($<$ or $>$) between the two real numbers.

**1.** (a) $\dfrac{3}{2} \; \square \; 7$  (b) $\pi \; \square \; -6$

**2.** (a) $-4 \; \square \; -8$  (b) $\dfrac{5}{6} \; \square \; \dfrac{2}{3}$

**3.** (a) $-\dfrac{3}{7} \; \square \; -\dfrac{8}{7}$  (b) $\dfrac{3}{4} \; \square \; \dfrac{11}{16}$

**4.** (a) $-\dfrac{5}{4} \; \square \; 6$  (b) $-1.75 \; \square \; -2.5$

In Exercises 5–10, sketch the solution set of the given inequality.

**5.** $3 - 2x \le 0$  **6.** $5x + 7 > 32$

**7.** $|x - 2| \le 3$  **8.** $|3x - 2| \le 0$

**9.** $4 < (x + 3)^2$  **10.** $\dfrac{1}{|x|} < 1$

In Exercises 11 and 12, find the midpoint of the given interval.

**11.** $\left[\dfrac{7}{8}, \dfrac{5}{2}\right]$  **12.** $\left[-1, \dfrac{3}{2}\right]$

In Exercises 13–16, use inequality notation to describe the given expression.

**13.** $x$ is nonnegative.

**14.** $y$ is *greater than* 5 and *less than or equal to* 12.

**15.** The area $A$ is *no more than* 12 square meters.

**16.** The probability, $p$, of rain is *at least* 0.75.

In Exercises 17–20, use absolute value notation to describe the given expression.

**17.** The distance between $x$ and 5 is *no more than* 3.

**18.** The distance between $x$ and $-10$ is *at least* 6.

**19.** $y$ is *at least* 2 units from $a$.

**20.** $y$ is *within* $c$ units of $L$.

In Exercises 21–30, simplify the given expression.

**21.** $(10xy)^2(3y^3)$  **22.** $x^4(\sqrt{x})^3$

**23.** $\dfrac{4x^4y^3}{(3xy)^4}$  **24.** $(5x^2y^4z^4)^3(5x^2y^4z^4)^{-3}$

**25.** $(4a^{-2}b^3)^{-3}$  **26.** $\left(\dfrac{10x^{-2}y^3}{x^4}\right)^0$

**27.** $(\sqrt[3]{(x-y)^2})^6$  **28.** $\sqrt{36(x+1)^5}$

**29.** $\sqrt[4]{(3x^2y^3)^2}$  **30.** $\sqrt{5x^2y}\sqrt{3y}$

In Exercises 31–44, completely factor the given polynomial.

**31.** $2x^3 - 2x^2 - 4x$  **32.** $2ay^3 - 7ay^2 - 15ay$

**33.** $63rs^2 - 7r^3$  **34.** $80 - 5z^2$

**35.** $4x^3y - 4x^2y^2 + xy^3$

**36.** $150x + 120x^2 + 24x^3$

**37.** $6x^4 - 48xy^3$  **38.** $27x^2y - x^2y^4$

**39.** $(x^2 + 2y^2)^2 - 9x^2y^2$

**40.** $(x^2 + y^2)^2 - 4x^2y^2$

**41.** $4x^2(2x - 1) + 2x(2x - 1)^2$

**42.** $9x(3x - 9)^2 + (3x - 9)^3$

**43.** $xy + 2y^2 + 2xz + 4yz$

**44.** $12x - 24y - 6xy + 12y^2$

In Exercises 45–50, insert the missing factor or factors.

**45.** $\dfrac{3}{4}x^2 - \dfrac{5}{6}x + 4 = \dfrac{1}{12}(\underline{\phantom{xxx}})$

**46.** $\dfrac{2}{3}x^4 - \dfrac{3}{8}x^3 + \dfrac{5}{6}x^2 = \dfrac{x^2}{24}(\underline{\phantom{xxx}})$

**47.** $x^3 - 1 = (x - 1)(\underline{\phantom{xxx}})$

**48.** $x^6 - y^6 = (x - y)(x + y)(\underline{\phantom{xx}})(\underline{\phantom{xx}})$

**49.** $x^4 - 2x^2y^2 + y^4 = (x + y)^2(\underline{\phantom{xx}})^2$

**50.** $z^4 - 5z^3 + 8z - 40 = (z - 5)(\underline{\phantom{xx}})(\underline{\phantom{xx}})$

In Exercises 51–58, use the Quadratic Formula to find the real roots (if any) of the given polynomial equation.

**51.** $16x^2 + 8x - 3 = 0$  **52.** $x^2 - 2x - 2 = 0$

**53.** $x^2 + 8x - 4 = 0$  **54.** $(y - 5)^2 = 2y$

**55.** $16t^2 + 4t + 3 = 0$  **56.** $49x^2 - 28x + 4 = 0$

**57.** $5.1x^2 - 1.7x - 3.2 = 0$

**58.** $422x^2 - 506x - 347 = 0$

In Exercises 59–62, use synthetic division to simplify the given rational expression.

**59.** $\dfrac{3x^3 - 17x^2 + 15x - 25}{x - 5}$

**60.** $\dfrac{-x^3 + 75x - 250}{x + 10}$

**61.** $\dfrac{5 - 3x + 2x^2 - x^3}{x + 1}$

**62.** $\dfrac{5x^3 - 6x^2 + 8}{x - 4}$

In Exercises 63–66, use the Rational Zero Theorem as an aid in finding all real zeros of the given polynomial.

**63.** $2x^3 - 9x^2 - 6x + 5$

**64.** $3x^3 - 16x^2 - 13x + 6$

**65.** $x^4 - 4x^3 + 3x^2 + 2x$

**66.** $9x^4 - 10x^2 + 1$

In Exercises 67–70, perform the indicated operations and simplify.

**67.** $x - 1 + \dfrac{1}{x + 2} + \dfrac{1}{x - 1}$

**68.** $\dfrac{A}{x - 2} + \dfrac{B}{(x - 2)^2} + \dfrac{C}{x + 2}$

**69.** Rationalize the numerator:

$$\frac{\sqrt{x + \Delta x} - \sqrt{x}}{\Delta x}$$

**70.** Perform the indicated operations, then rationalize the numerator:

$$\left( \frac{1}{\sqrt{x + \Delta x}} - \frac{1}{\sqrt{x}} \right) \div \Delta x$$

**71.** The daily cost of producing $x$ units of a certain product is

$$C = 800 + 0.04x + 0.0002x^2, \quad 0 \le x$$

If the cost is \$1,680, how many units are produced?

**72.** The weekly cost of operating a certain manufacturing plant is

$$C = 0.5x^2 + 15x + 5000, \quad 0 \le x$$

where $x$ is the number of units produced. If, for a certain week, the cost was \$11,500, how many units were produced?

# CHAPTER 1

# Functions, Graphs, and Limits

# The Cartesian Plane and the Distance Formula

## INTRODUCTORY EXAMPLE
## Prime Interest Rate

The prime interest rate is the annual rate of interest that commercial banks charge their most creditworthy business customers. The prime rate is usually reserved for large loans that are based solely on the good reputation of the borrower. Smaller loans to individuals usually carry interest rates that are higher than the prime lending rate. One of the main factors that determine the prime rate is the current rate of inflation. It stands to reason that the interest paid on a loan must exceed the inflation rate in order for the lender to make money.

Table 1.1 shows the prime rate for selected years between 1960 and 1984. These data are plotted on a **rectangular coordinate system** with appropriate units (see Figure 1.1). Note that each plotted point relates two numbers: one number corresponding to the year (the horizontal axis) and one number corresponding to the prime rate (the vertical axis).

TABLE 1.1

| Year | 1960 | 1965 | 1970 | 1973 | 1974 | 1975 | 1976 | 1977 | 1978 | 1979 | 1980 | 1981 | 1982 | 1983 | 1984 |
|---|---|---|---|---|---|---|---|---|---|---|---|---|---|---|---|
| Prime rate (%) | 4.82 | 4.54 | 7.91 | 8.03 | 10.81 | 7.86 | 6.84 | 6.83 | 9.06 | 12.67 | 15.26 | 18.87 | 14.86 | 10.79 | 11.51 |

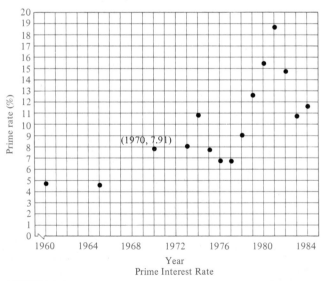

FIGURE 1.1

Just as we can represent real numbers by points on the real line, we can represent ordered pairs of real numbers by points in a plane called the **rectangular coordinate system,** or the **Cartesian plane,** after the French mathematician René Descartes (1596–1650).

The Cartesian plane is formed by using two real lines intersecting at right angles, as shown in Figure 1.2. The horizontal real line is called the ***x*-axis,** and the vertical real line is called the ***y*-axis.** The point of intersection of these two axes is called the **origin,** and the lines divide the plane into four parts called **quadrants.**

Each point in the plane corresponds to an **ordered pair** $(x, y)$ of real numbers $x$ and $y$, called **coordinates** of the point. The ***x*-coordinate** represents the directed distance from the $y$-axis to the point, and the ***y*-coordinate** represents the directed distance from the $x$-axis to the point, as shown in Figure 1.3.

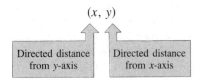

$(x, y)$

| Directed distance from $y$-axis | Directed distance from $x$-axis |

■ **Remark:** It is customary to use the notation $(x, y)$ to denote both a point in the plane and an open interval on the real line. This should cause no confusion because the nature of a specific problem will show which we are talking about.

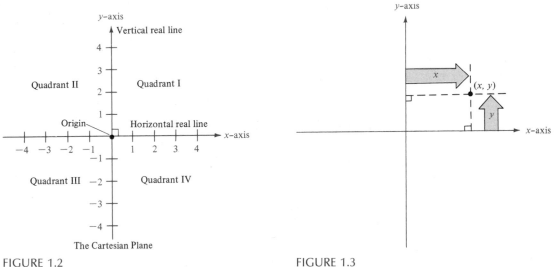

FIGURE 1.2                    FIGURE 1.3

## EXAMPLE 1

**Plotting Points in the Cartesian Plane**

Plot the points $(-1, 2)$, $(3, 4)$, $(0, 0)$, $(3, 0)$, and $(-2, -3)$ in the Cartesian Plane.

## SOLUTION

To plot the point

$$(-1, 2)$$

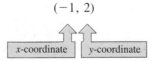

we sketch a vertical line through $-1$ on the $x$-axis and a horizontal line through $2$ on the $y$-axis. The intersection of these two lines is the point $(-1, 2)$. The other five points can be plotted in a similar way, and are shown in Figure 1.4.

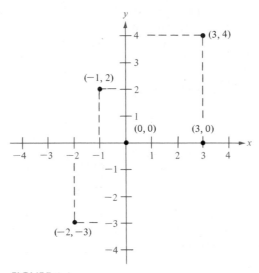

FIGURE 1.4

The beauty of a rectangular coordinate system is that it allows us to visualize the relationships between the variables $x$ and $y$. It would be hard to overestimate the importance of Descartes's coordinatization of the plane. Today, his ideas are in common use in virtually every scientific and business-related field. In Example 2, notice how much your intuition is enhanced by the use of a graphical presentation.

## EXAMPLE 2

**An Application of a Rectangular Coordinate System**

The numbers of U.S. veterans with service in Vietnam who were discharged between 1966 and 1979 are given in Table 1.2. Plot these points on a rectangular coordinate system.

TABLE 1.2

| Year | 1966 | 1967 | 1968 | 1969 | 1970 | 1971 | 1972 | 1973 | 1974 | 1975 | 1976 | 1977 | 1978 | 1979 |
|------|------|------|------|------|------|------|------|------|------|------|------|------|------|------|
| Number (in 1000s) | 56 | 144 | 321 | 485 | 560 | 504 | 300 | 179 | 135 | 86 | 60 | 64 | 78 | 68 |

**SOLUTION**   The points are shown in Figure 1.5. Note that the break in the *x*-axis indicates that we have omitted the numbers between 0 and 1966.

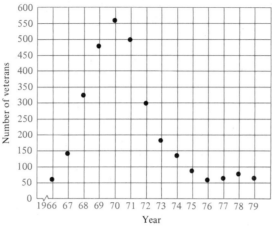

FIGURE 1.5

### The Distance Formula

We have seen how to determine the distance between two points $x_1$ and $x_2$ on the real line (Section 0.2). We now turn our attention to the slightly more difficult problem of finding the distance between two points in the plane. Recall from the Pythagorean Theorem that, for a right triangle with hypotenuse *c* and sides *a* and *b*, we have the relationship $a^2 + b^2 = c^2$, as shown in Figure 1.6. (The converse is also true. That is, if $a^2 + b^2 = c^2$, then the triangle is a right triangle.)

Suppose we wish to determine the distance *d* between two points $(x_1, y_1)$ and $(x_2, y_2)$ in the plane. With these two points, a right triangle can be formed, as shown in Figure 1.7. By finding the distance between $y_1$ and $y_2$, we see that the length of

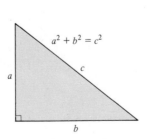

Pythagorean Theorem

FIGURE 1.6

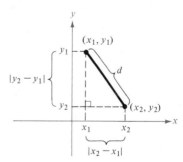

Distance Between Two Points

FIGURE 1.7

the vertical side of the triangle is $|y_2 - y_1|$. Similarly, the length of the horizontal side is $|x_2 - x_1|$. By the Pythagorean Theorem, we then have

$$d^2 = |x_2 - x_1|^2 + |y_2 - y_1|^2$$

or

$$d = \sqrt{|x_2 - x_1|^2 + |y_2 - y_1|^2}$$

Replacing $|x_2 - x_1|^2$ and $|y_2 - y_1|^2$ by the equivalent expressions $(x_2 - x_1)^2$ and $(y_2 - y_1)^2$ allows us to write

$$d = \sqrt{(x_2 - x_1)^2 + (y_2 - y_1)^2}$$

We choose the positive square root for $d$ because the distance *between* two points is not a directed distance. Since the two differences $(x_2 - x_1)$ and $(y_2 - y_1)$ appear as squared terms in the Distance Formula, it doesn't matter in which order we subtract the $x$- and $y$-coordinates. In other words, an equivalent version of the Distance Formula is

$$d = \sqrt{(x_1 - x_2)^2 + (y_1 - y_2)^2}$$

We have therefore established the following result.

**The Distance Formula**

The distance $d$ between the points $(x_1, y_1)$ and $(x_2, y_2)$ is

$$d = \sqrt{(x_2 - x_1)^2 + (y_2 - y_1)^2}$$

**EXAMPLE 3**

**Finding the Distance Between Two Points in the Plane**

Find the distance between the points $(-2, 1)$ and $(3, 4)$.

**SOLUTION**

Letting $(x_1, y_1) = (-2, 1)$ and $(x_2, y_2) = (3, 4)$, we apply the Distance Formula to obtain

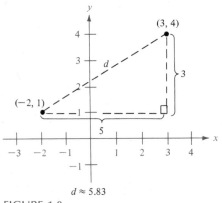

$$d = \sqrt{(x_2 - x_1)^2 + (y_2 - y_1)^2}$$
$$= \sqrt{[3 - (-2)]^2 + (4 - 1)^2}$$
$$= \sqrt{(5)^2 + (3)^2}$$
$$= \sqrt{25 + 9}$$
$$= \sqrt{34} \approx 5.83 \quad \text{(See Figure 1.8.)}$$

$d \approx 5.83$

FIGURE 1.8

**Remark:** In Example 3 note that we use the symbol $\approx$ to mean approximately equal to.

**EXAMPLE 4**

**An Application of the Distance Formula**

Use the Distance Formula to show that the points $(2, 1)$, $(4, 0)$, and $(5, 7)$ are the vertices of a right triangle.

**SOLUTION**

The three points are plotted in Figure 1.9. Using the Distance Formula, we find the lengths of the three sides to be

$$d_1 = \sqrt{(5 - 2)^2 + (7 - 1)^2} = \sqrt{9 + 36} = \sqrt{45}$$
$$d_2 = \sqrt{(4 - 2)^2 + (0 - 1)^2} = \sqrt{4 + 1} = \sqrt{5}$$
$$d_3 = \sqrt{(5 - 4)^2 + (7 - 0)^2} = \sqrt{1 + 49} = \sqrt{50}$$

Since

$$d_1{}^2 + d_2{}^2 = 45 + 5 = 50 = d_3{}^2$$

we can apply the Pythagorean Theorem to conclude that the triangle must be a right triangle.

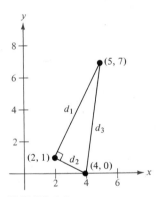

FIGURE 1.9

The figures provided in Examples 3 and 4 were not really essential to the solution. *Nevertheless,* we strongly recommend that you develop the habit of including sketches with your problem solutions, even if they are not specifically required. Throughout many years of teaching calculus, we have found that students who do well with the technical aspects of calculus are very often the same students who have a good grasp of the visual aspects of the subject.

**EXAMPLE 5**

**An Application of the Distance Formula**

In a football game, a quarterback throws a pass from the 5-yard line, 20 yards from the sideline. The pass is caught by a wide receiver on the 45-yard line, 50 yards from the same sideline, as shown in Figure 1.10. How long was the pass?

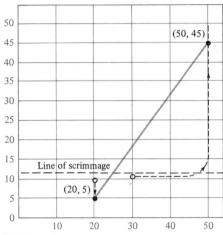

**FIGURE 1.10**

**SOLUTION**

Using the Distance Formula we find the distance to be

$$d = \sqrt{(50 - 20)^2 + (45 - 5)^2}$$
$$= \sqrt{900 + 1600}$$
$$= \sqrt{2500}$$
$$= 50 \text{ yards}$$

**Remark:** Note in Example 5 that the scale along the goal line does not normally appear on a football field. However, when we use coordinate geometry to solve real-life problems, we are free to place the coordinate system any way that is convenient to the solution of the problem.

The next example illustrates an important type of problem in coordinate (or analytic) geometry. In the problem, we are given a description of a point in the plane and are asked to find the coordinates of the point. To do this, we assign variables to the unknown coordinate (or coordinates) and then use the given description to form an equation involving the unknown values. Usually, to determine the unknown values we need to form the same number of equations as there are unknowns.

**EXAMPLE 6**

**Finding Points at a Specified Distance from a Given Point**

Find $x$ so that the distance between $(x, 3)$ and $(2, -1)$ is 5.

**SOLUTION**

Using the Distance Formula we have

$$d = \sqrt{(x - 2)^2 + (3 + 1)^2} = 5$$

By squaring both sides we have

$$(x^2 - 4x + 4) + 16 = 25$$
$$x^2 - 4x - 5 = 0$$
$$(x - 5)(x + 1) = 0$$

Therefore, $x = 5$ or $x = -1$, and we conclude that both of the points $(5, 3)$ and $(-1, 3)$ lie 5 units from the point $(2, -1)$, as shown in Figure 1.11.

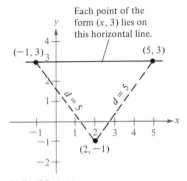

FIGURE 1.11

## Midpoint Formula

We now look at a formula for finding the **midpoint** of the line segment joining two points in the plane.

**The Midpoint Formula**

The midpoint of the line segment joining the points $(x_1, y_1)$ and $(x_2, y_2)$ is

$$\left( \frac{x_1 + x_2}{2}, \frac{y_1 + y_2}{2} \right)$$

**Proof**

This formula is just what we should expect. To find the midpoint of a line segment, we simply find the "average" values of the respective coordinates of the two endpoints. To prove this, we use Figure 1.12 and show that

$$d_1 = d_2 \quad \text{and} \quad d_1 + d_2 = d_3$$

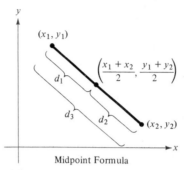

Midpoint Formula

**FIGURE 1.12**

Using the Distance Formula we obtain

$$d_1 = \sqrt{\left(\frac{x_1 + x_2}{2} - x_1\right)^2 + \left(\frac{y_1 + y_2}{2} - y_1\right)^2} = \frac{1}{2}\sqrt{(x_2 - x_1)^2 + (y_2 - y_1)^2}$$

$$d_2 = \sqrt{\left(x_2 - \frac{x_1 + x_2}{2}\right)^2 + \left(y_2 - \frac{y_1 + y_2}{2}\right)^2} = \frac{1}{2}\sqrt{(x_2 - x_1)^2 + (y_2 - y_1)^2}$$

$$d_3 = \sqrt{(x_2 - x_1)^2 + (y_2 - y_1)^2}$$

Thus, it follows that $d_1 = d_2$ and $d_1 + d_2 = d_3$.

**EXAMPLE 7**

**The Midpoint of a Line Segment**

Find the midpoint of the line segment joining the points $(-5, -3)$ and $(9, 3)$, as shown in Figure 1.13.

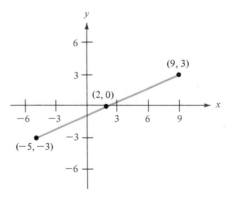

**FIGURE 1.13**

**SOLUTION**

We let $(x_1, y_1) = (-5, -3)$ and $(x_2, y_2) = (9, 3)$

Then, using the Midpoint Formula we have

$$\text{Midpoint} = \left(\frac{x_1 + x_2}{2}, \frac{y_1 + y_2}{2}\right) = \left(\frac{-5 + 9}{2}, \frac{-3 + 3}{2}\right) = (2, 0)$$

Although we developed the Midpoint Formula in the context of geometry, it can be applied to problems that don't appear to be geometrical in nature. For instance, in Example 8 we use the Midpoint Formula to estimate retail sales.

**EXAMPLE 8**

**An Application of the Midpoint Formula**

A business had annual retail sales of $240,000 in 1980 and $312,000 in 1986. Find the sales for 1983, assuming that the annual increase in sales followed a *linear* pattern.

**SOLUTION**

To make the computations simpler we let $t = 0$ represent the year 1980 and $t = 6$ represent the year 1986. Similarly, if we measure the retail sales in 1000s of dollars, then the sales for 1980 and 1986 are represented by

$$(0, 240) \quad \text{and} \quad (6, 312)$$

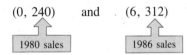
1980 sales      1986 sales

Since 1983 is midway between 1980 and 1986 and since the growth pattern is linear, we can use the Midpoint Formula to find the 1983 sales as follows:

$$\text{Midpoint} = \left(\frac{0 + 6}{2}, \frac{240 + 312}{2}\right) = (3, 276)$$

1983 sales

Thus, the 1983 sales were $276,000, as indicated in Figure 1.14.

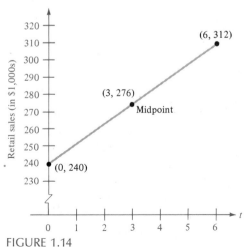

FIGURE 1.14

In Exercises 1–4, (a) find the length of each side of the right triangle, and (b) show that these lengths satisfy the Pythagorean Theorem.

**1.**

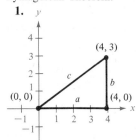

**2.**

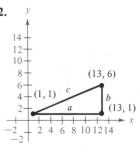

**3.**

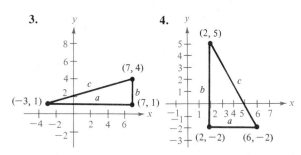

**4.**

In Exercises 5–12, (a) plot the points, (b) find the distance between the points, and (c) find the midpoint of the line segment joining the points.

**5.** (2, 1), (4, 5)

**6.** (−3, 2), (3, −2)

**7.** $\left(\frac{1}{2}, 1\right), \left(-\frac{3}{2}, -5\right)$

**8.** $\left(\frac{2}{3}, -\frac{1}{3}\right), \left(\frac{5}{6}, 1\right)$

**9.** (2, 2), (4, 14)

**10.** (−3, 7), (1, −1)

**11.** $(1, \sqrt{3}), (-1, 1)$

**12.** $(-2, 0), (0, \sqrt{2})$

In Exercises 13–16, show that the given points form the vertices of the indicated figure. (A rhombus is a polygon whose sides are all of the same length.)

| Vertices | Figure |
|---|---|
| **13.** (4, 0), (2, 1), (−1, −5) | Right triangle |
| **14.** (1, −3), (3, 2), (−2, 4) | Isosceles triangle |
| **15.** (0, 0), (1, 2), (2, 1), (3, 3) | Rhombus |
| **16.** (0, 1), (3, 7), (4, 4), (1, −2) | Parallelogram |

In Exercises 17–20, use the Distance Formula to determine whether the given points are collinear (lie on the same line). (Hint: Use the model shown in Example 4 and determine whether $d_1 + d_2 = d_3$.)

**17.** (0, −4), (2, 0), (3, 2)

**18.** (0, 4), (7, −6), (−5, 11)

**19.** (−2, 1), (−1, 0), (2, −2)

**20.** (−1, 1), (3, 3), (5, 5)

In Exercises 21 and 22, find $x$ so that the distance between the points is 5.

**21.** (0, 0), (x, −4)

**22.** (2, −1), (x, 2)

In Exercises 23 and 24, find $y$ so that the distance between the points is 8.

**23.** (0, 0), (3, y)

**24.** (5, 1), (5, y)

In Exercises 25 and 26, find the relationship between $x$ and $y$ so that $(x, y)$ is equidistant from the two given points.

**25.** (4, −1), (−2, 3)

**26.** $\left(3, \frac{5}{2}\right), (-7, -1)$

**27.** Use the Midpoint Formula successively to find the three points that divide the line segment joining $(x_1, y_1)$ and $(x_2, y_2)$ into four equal parts.

**28.** Show that $(\frac{1}{3}[2x_1 + x_2], \frac{1}{3}[2y_1 + y_2])$ is one of the points of trisection of the line segment joining $(x_1, y_1)$ and $(x_2, y_2)$. Also, find the midpoint of

$$\left(\frac{1}{3}[2x_1 + x_2], \frac{1}{3}[2y_1 + y_2]\right) \quad \text{and} \quad (x_2, y_2)$$

to find the second point of trisection of the line segment.

**29.** Use the result of Exercise 27 to find the points that divide the line segment joining the given points into four equal parts.

(a) (1, −2), (4, −1)     (b) (−2, −3), (0, 0)

**30.** Use the result of Exercise 28 to find the points of trisection of the line segment joining the given points.

(a) (1, −2), (4, 1)     (b) (−2, −3), (0, 0)

In Exercises 31 and 32, plot the points given in the accompanying table on a rectangular coordinate system. (Use Example 2 of this section as a model.)

**31.** The following table gives the number of subscribers (in millions) to cable TV:

| Year | 1976 | 1977 | 1978 | 1979 |
|---|---|---|---|---|
| Subscribers | 11.0 | 12.2 | 13.4 | 15.0 |

| Year | 1980 | 1981 | 1982 | 1983 |
|---|---|---|---|---|
| Subscribers | 17.5 | 21.5 | 25.4 | 29.4 |

**32.** The following table gives the daily production of crude oil in millions of barrels by OPEC:

| Year | 1974 | 1975 | 1976 | 1977 | 1978 |
|------|------|------|------|------|------|
| Production | 17.7 | 16.0 | 18.6 | 19.2 | 18.5 |

| Year | 1979 | 1980 | 1981 | 1982 | 1983 |
|------|------|------|------|------|------|
| Production | 21.1 | 19.0 | 15.8 | 11.7 | 10.3 |

**33.** The base and height of the trusses for the roof of a house are 32 feet and 5 feet, respectively.
   (a) Find the distance from the eaves to the peak of the roof.
   (b) Use the result of part (a) to find the number of square feet of roofing required for the house if the length of the house is 40 feet.

**34.** A guy wire is stretched from a broadcasting tower at a point 200 feet above the ground to an anchor 125 feet from the base. How long is the wire?

In Exercises 35 and 36, use the Midpoint Formula to estimate the sales of a company for 1983, given the sales in 1980 and 1986. Assume that the annual increase in sales followed a linear pattern.

**35.**

| Year | 1980 | 1986 |
|------|------|------|
| Sales | $520,000 | $740,000 |

**36.**

| Year | 1980 | 1986 |
|------|------|------|
| Sales | $4,200,000 | $5,650,000 |

In Exercises 37 and 38, use Figure 1.15 showing retail sales in billions of dollars.

**37.** Approximate retail sales for
   (a) June 1983
   (b) January 1984
   (c) January 1985
   (d) June 1985

**38.** Approximate the percentage increase or decrease in retail sales from
   (a) February 1983 to May 1985
   (b) February 1984 to April 1984

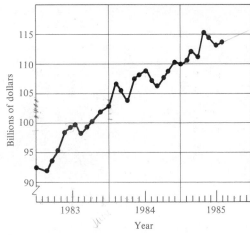

FIGURE 1.15

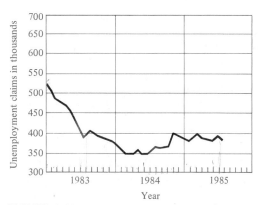

FIGURE 1.16

In Exercises 39 and 40, use Figure 1.16 showing unemployment claims in thousands.

**39.** Approximate the number of claims for
   (a) January 1983          (b) June 1983
   (c) June 1984             (d) June 1985

**40.** Approximate the percentage increase or decrease in unemployment claims from
   (a) January 1983 to June 1983
   (b) June 1984 to November 1984

**41.** A line segment has $(x_1, y_1)$ as one endpoint and $(x_m, y_m)$ as its midpoint. Find $(x_2, y_2)$, the other endpoint of the line segment.

**42.** Assume that the retail sales shown in Figure 1.15 are increasing in a linear pattern. Use the result of Exercise 41 and the retail sales for January 1984 and January 1985 to predict the sales for January 1986.

# Graphs of Equations

## INTRODUCTORY EXAMPLE
## A Temperature Model

During a normal day the air temperature drops to a low during the night and rises to a high during the day. For example, Table 1.3 gives the temperature (in degrees Fahrenheit) for a typical 24-hour period.

One of the goals of applied mathematics is to find equations that describe real-world phenomena. We call such equations **mathematical models.** A model that gives a reasonably accurate description of the temperature pattern indicated in Table 1.3 is the equation*

$$y = 0.00026157t^5 - 0.01306453t^4 + 0.18365236t^3$$
$$- 0.34077620t^2 - 3.79407878t$$
$$+ 60.11462451$$

where $t = 1$ corresponds to 1 A.M. and $t = 13$ corresponds to 1 P.M. For instance, when $t = 1$, the temperature given by the model is

$$y \approx 56.15°$$

and when $t = 10$, the temperature given by the model is

$$y \approx 67.26°$$

Figure 1.17 shows the graph of this model. Note the comparison between the temperatures given by the model and the actual temperatures.

TABLE 1.3

| Time (A.M.) | 1 | 2 | 3 | 4 | 5 | 6 |
|---|---|---|---|---|---|---|
| Temp. (F°) | 55 | 53 | 50 | 50 | 49 | 50 |

| Time (A.M.) | 7 | 8 | 9 | 10 | 11 | 12 |
|---|---|---|---|---|---|---|
| Temp. (F°) | 51 | 56 | 60 | 65 | 73 | 78 |

| Time (P.M.) | 1 | 2 | 3 | 4 | 5 | 6 |
|---|---|---|---|---|---|---|
| Temp. (F°) | 85 | 86 | 84 | 81 | 80 | 70 |

| Time (P.M.) | 7 | 8 | 9 | 10 | 11 | 12 |
|---|---|---|---|---|---|---|
| Temp. (F°) | 68 | 64 | 59 | 58 | 57 | 57 |

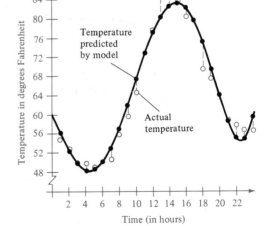

FIGURE 1.17

---

*This equation was generated by a computer using a process called least squares regression analysis.

- **The Graph of an Equation**
- **Intercepts of a Graph**
- **Circles**
- **Points of Intersection**
- **Mathematical Models**

In Section 1.1 we used a coordinate system to graphically represent the relationship between two quantities. There, the graphical picture consisted of a collection of points in the plane (see Example 2 in Section 1.1).

Frequently the relationship between two quantities is expressed in the form of an equation. For instance, degrees on the Fahrenheit scale are related to degrees on the Celsius scale by the equation $F = \frac{9}{5}C + 32$. In this section we introduce a basic procedure for determining the geometric picture associated with such an equation.

## The Graph of an Equation

Consider the equation

$$3x + y = 7$$

If $x = 2$ and $y = 1$, the equation is satisfied, and we call the point $(2, 1)$ a **solution point** of the equation. Of course, there are other solution points, such as $(1, 4)$ and $(0, 7)$. We can make up a **table of values** for $x$ and $y$ by choosing arbitrary values for $x$ and determining the corresponding values for $y$. To determine the values for $y$, it is convenient to write the equation in the form

$$y = 7 - 3x$$

TABLE 1.4

| $x$ | 0 | 1 | 2 | 3 | 4 |
|---|---|---|---|---|---|
| $y$ | 7 | 4 | 1 | $-2$ | $-5$ |

Thus, $(0, 7)$, $(1, 4)$, $(2, 1)$, $(3, -2)$, and $(4, -5)$ are all solution points of the equation $3x + y = 7$, as shown in Table 1.4. There are actually an infinite number of solution points for the equation, and the set of all such points is called the **graph** of the equation, as shown in Figure 1.18.

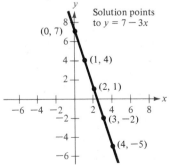

FIGURE 1.18

**Remark:** Even though we refer to the sketch shown in Figure 1.18 as the graph of $y = 7 - 3x$, it actually represents only a *portion* of the graph. The entire graph is a line that would extend off the page.

| **Definition of the graph of an equation in two variables** | The **graph of an equation** involving two variables $x$ and $y$ is the set of all points in the plane that are solution points to the equation. |

**EXAMPLE 1**

**Sketching the Graph of an Equation**

Sketch the graph of the equation $y = x^2 - 2$.

**SOLUTION**

First we make a table of values (Table 1.5) by choosing several convenient values of $x$ and calculating the corresponding values of $y = x^2 - 2$.

TABLE 1.5

| $x$ | $-2$ | $-1$ | $0$ | $1$ | $2$ | $3$ |
|---|---|---|---|---|---|---|
| $y$ | $2$ | $-1$ | $-2$ | $-1$ | $2$ | $7$ |

Next we plot these points in the plane, as in Figure 1.19. Finally, we connect the points by a *smooth curve,* as shown in Figure 1.20.

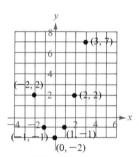

FIGURE 1.19

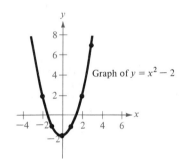

Graph of $y = x^2 - 2$

FIGURE 1.20

**Remark:** The graph shown in Example 1 is called a **parabola.** The graph of any second-degree equation of the form

$$y = ax^2 + bx + c, \quad a \neq 0$$

has a similar shape. If $a > 0$, the parabola opens upward, as in Figure 1.20, and if $a < 0$, the parabola opens downward.

We call the curve-sketching method demonstrated in Example 1 the **point-plotting method.** Its basic features are given in the following summary.

<table>
<tr><td>**The point-plotting method of graphing**</td><td>1. Make up a table of several solution points of the equation.<br>2. Plot these points in the plane.<br>3. Connect the points with a smooth curve.</td></tr>
</table>

Steps 1 and 2 of the point-plotting method are usually easy, but step 3 can be the source of some major difficulties. For instance, how would you connect the four points in Figure 1.21? Without additional points or further information about the equation, any one of the three graphs in Figure 1.22 would be reasonable.

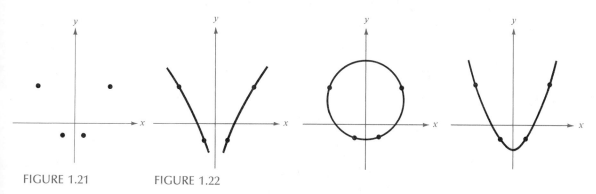

FIGURE 1.21        FIGURE 1.22

Obviously, with too few solution points we could badly misrepresent the graph of a given equation. How many points should be plotted? For a straight-line graph two points are sufficient. For more complicated graphs we need many more points. More sophisticated graphing techniques are discussed in later chapters. In the meantime we suggest that you plot enough points so as to reveal the essential behavior of the graph. The more solution points you plot, the more accurate your graph will be. (A programmable calculator is very useful for determining the solution points.)

## Intercepts of a Graph

Two solution points that are often easy to determine are those having zero as either the $x$- or $y$-coordinate. These points are called **intercepts** because they are the points at which the graph intersects the $x$- or $y$-axis.

<table>
<tr><td>**Definition of intercepts**</td><td>1. The point $(a, 0)$ is called an **$x$-intercept** of the graph of an equation if it is a solution point of the equation.<br>2. The point $(0, b)$ is called a **$y$-intercept** of the graph of an equation if it is a solution point of the equation.</td></tr>
</table>

**■■■ Remark:** Some texts denote the *x*-intercept as the *x*-coordinate of the point $(a, 0)$ rather than the point itself. Unless it is necessary to make a distinction, we will use the term *intercept* to mean either the point or the coordinate.

Of course, it is possible for a graph to have no intercepts, or several. For instance, consider the four graphs in Figure 1.23.

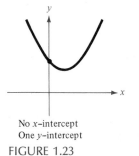

No *x*-intercept
One *y*-intercept

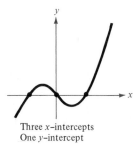

Three *x*-intercepts
One *y*-intercept

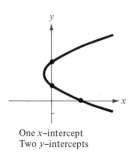

One *x*-intercept
Two *y*-intercepts

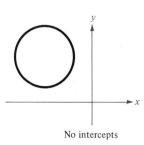

No intercepts

FIGURE 1.23

---

**Finding intercepts**

1. To find *x*-intercepts, let *y* be zero and solve the equation for *x*.
2. To find *y*-intercepts, let *x* be zero and solve the equation for *y*.

---

**EXAMPLE 2**

**Finding *x*- and *y*-Intercepts**

Find the *x*- and *y*-intercepts for the graphs of the following equations:
(a) $y = x^3 - 4x$             (b) $y^2 - 3 = x$

**SOLUTION**

(a) Let $y = 0$. Then $0 = x(x^2 - 4)$ has solutions $x = 0$ and $x = \pm 2$.

$$x\text{-intercepts: } (0, 0), (2, 0), (-2, 0)$$

Let $x = 0$. Then $y = 0$.

$$y\text{-intercept: } (0, 0)$$

(See Figure 1.24.)

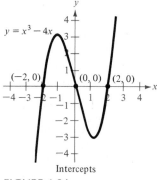

FIGURE 1.24

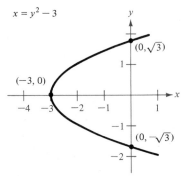

FIGURE 1.25

(b) Let $y = 0$. Then $-3 = x$.

$$x\text{-intercept: } (-3, 0)$$

Let $x = 0$. Then $y^2 - 3 = 0$ has solutions $y = \pm\sqrt{3}$.

$$y\text{-intercepts: } (0, \sqrt{3}), (0, -\sqrt{3})$$

(See Figure 1.25.)

## Circles

Although useful, the point-plotting method of sketching can be tedious. For common types of equations, it is simpler to learn to recognize the general nature of the graph from the equation's form. For example, in the next section, we will learn to recognize equations whose graphs are straight lines. In this section, we consider the graph of another easily recognized graph—that of a **circle.**

By referring to the circle in Figure 1.26, we see that a point $(x, y)$ is on the circle if and only if its distance from the center $(h, k)$ is $r$. Expressing this relationship in terms of the Distance Formula, we have

$$\sqrt{(x - h)^2 + (y - k)^2} = r$$

as the condition that the coordinates of $(x, y)$ must satisfy. By squaring both sides of this equation, we obtain the **standard form of the equation of a circle.**

| **Standard form of the equation of a circle** | The point $(x, y)$ lies on the circle of radius $r$ and center $(h, k)$ if and only if $$(x - h)^2 + (y - k)^2 = r^2$$ |
| --- | --- |

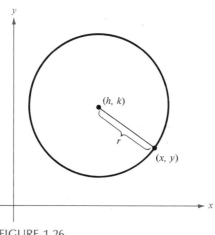

FIGURE 1.26

■■■ **Remark:** Note that the phrase "if and only if" is a way of stating two implications in one sentence. One implication states that "if the point $(x, y)$ lies on the circle, then it satisfies the given equation." The other implication is the *converse*, which says that "if the point $(x, y)$ satisfies the given equation, then it lies on the circle."

From this result we see that the standard form of the equation of a circle with its center at the origin is simply

$$x^2 + y^2 = r^2$$

**EXAMPLE 3**

**Finding the Equation of a Circle**

The point $(3, 4)$ lies on a circle whose center is at $(-1, 2)$, as shown in Figure 1.27. Find the standard form of the equation for this circle.

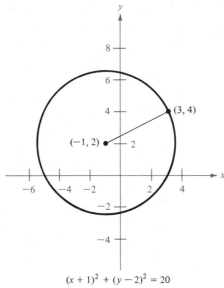

$(x + 1)^2 + (y - 2)^2 = 20$

FIGURE 1.27

**SOLUTION**

The radius of the circle is the distance between $(-1, 2)$ and $(3, 4)$. Thus,

$$r = \sqrt{[3 - (-1)]^2 + (4 - 2)^2} = \sqrt{16 + 4} = \sqrt{20}$$

Therefore, the standard form of the equation for this circle is

$$[x - (-1)]^2 + (y - 2)^2 = (\sqrt{20})^2$$

$$(x + 1)^2 + (y - 2)^2 = 20 \qquad \text{Standard form}$$

■■■■■■■■■

If we remove the parentheses in the standard equation of Example 3 we obtain

$$(x + 1)^2 + (y - 2)^2 = 20 \qquad \text{Standard form}$$

$$x^2 + 2x + 1 + y^2 - 4y + 4 = 20$$

$$x^2 + y^2 + 2x - 4y - 15 = 0 \qquad \text{General form}$$

The third equation is written in the **general form of the equation of a circle:**

$$Ax^2 + Ay^2 + Dx + Ey + F = 0, \quad A \neq 0$$

The general form of the equation of a circle is less useful than the equivalent standard form. For instance, it is not immediately apparent from the general equation of the circle in Example 3 that the center is at $(-1, 2)$ and the radius is $\sqrt{20}$. Before graphing the equation of a circle, it is best to write the equation in standard form. This can be accomplished by **completing the square,** as demonstrated in the following example.

**EXAMPLE 4**

**Completing the Square**

Sketch the graph of the circle whose general equation is

$$4x^2 + 4y^2 + 20x - 16y + 37 = 0$$

**SOLUTION**

To complete the square we will first divide by 4 so that the coefficients of $x^2$ and $y^2$ are both 1.

$$4x^2 + 4y^2 + 20x - 16y + 37 = 0 \qquad \text{General form}$$

$$x^2 + y^2 + 5x - 4y + \frac{37}{4} = 0 \qquad \text{Divide by 4}$$

$$(x^2 + 5x + \quad) + (y^2 - 4y + \quad) = -\frac{37}{4} \qquad \text{Group terms}$$

$$\left(x^2 + 5x + \frac{25}{4}\right) + (y^2 - 4y + 4) = -\frac{37}{4} + \frac{25}{4} + 4$$

$\qquad\quad$ (Half)$^2$ $\qquad\qquad$ (Half)$^2$ $\qquad\qquad\qquad$ Complete square

$$\left(x + \frac{5}{2}\right)^2 + (y - 2)^2 = 1 \qquad \text{Standard form}$$

Note that we complete the square by adding the square of half the coefficient of $x$ and the square of half the coefficient of $y$ to both sides of the equation. Therefore, the circle is centered at $\left(-\frac{5}{2}, 2\right)$, and its radius is 1, as shown in Figure 1.28.

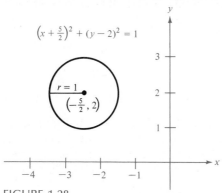

FIGURE 1.28

The general equation $Ax^2 + Ay^2 + Dx + Ey + F = 0$ may not always represent a circle. In fact, such an equation will have no solution points if the procedure of completing the square yields the impossible result

$$(x - h)^2 + (y - k)^2 = \text{negative number}$$

At the time analytic geometry was being developed, the two major branches of mathematics—algebra and geometry—were largely independent of each other. Circles belonged to geometry and equations belonged to algebra. The coordination of the points on a circle and the solutions to an equation belongs to what is now called analytic geometry. It is important that you develop skill in using analytic geometry to move back and forth between algebra and geometry. For instance, in Example 3 we are given a geometric description of a circle and are asked to find an algebraic equation for the circle. Thus, we are moving from geometry to algebra. Similarly, in Example 4 we are given an algebraic equation and asked to sketch a geometric picture. In this case we are moving from algebra to geometry. These two examples illustrate the two most common types of problems in analytic geometry.

1. Given a graph, find its equation.
2. Given an equation, find its graph.

### Points of Intersection

Since each point of a graph is a solution point of its corresponding equation, a **point of intersection** of two graphs is simply a solution point that satisfies both of the corresponding equations. Moreover, the points of intersection of two graphs can be found by solving the corresponding equations simultaneously.

**EXAMPLE 5**

**Finding Points of Intersection**

Find all points of intersection of the graphs of

$$x - y = 1 \quad \text{and} \quad x^2 - y = 3$$

**SOLUTION**

It is helpful to begin by making a sketch for each equation on the *same* coordinate plane, as shown in Figure 1.29. From the figure it appears that the two graphs have two points of intersection. To find these two points, we proceed as follows.

$$y = x^2 - 3 \quad \text{and} \quad y = x - 1 \qquad \text{Solve each equation for } y$$

$$x^2 - 3 = x - 1 \qquad \text{Equate } y\text{-values}$$

$$x^2 - x - 2 = 0$$

$$(x - 2)(x + 1) = 0 \qquad \text{Factor}$$

Therefore, $x = 2$ or $x = -1$. The corresponding values of $y$ are obtained by substituting $x = 2$ and $x = -1$ into either of the original equations. For instance, if we choose the equation $y = x - 1$, then the values of $y$ are 1 and $-2$, respectively. Therefore, the two points of intersection are $(2, 1)$ and $(-1, -2)$.

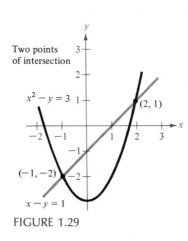

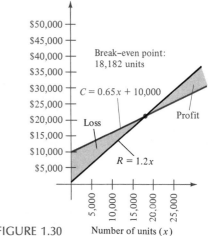

FIGURE 1.29

FIGURE 1.30    Number of units ($x$)

Many applications involve finding the point of intersection of two graphs. A common one from business is called **break-even analysis.** The marketing of a new product typically requires a substantial investment. When enough units have been sold so that the total revenue has offset the total cost we say that the sale of the product has reached the **break-even point.** We denote the **total cost** of producing $x$ units of a product by $C$, and the **total revenue** received from selling $x$ units of the product by $R$. Thus, we can find the break-even point by setting the cost $C$ equal to the revenue $R$ and solving for $x$.

**EXAMPLE 6**

**An Application: Break-Even Analysis**

A small home business is set up with an investment of $10,000 for equipment. The business manufactures a product at a cost of $0.65 per unit. If the product sells for $1.20, how many units must be sold before the business breaks even?

**SOLUTION**

The cost of selling $x$ units is

$$C = 0.65x + 10,000$$

and the revenue obtained by selling $x$ units is

$$R = 1.2x$$

Since the break-even point occurs when $R = C$, we have

$$R = C$$

$$1.2x = 0.65x + 10,000$$

$$0.55x = 10,000$$

$$x = \frac{10,000}{0.55} \approx 18,182 \text{ units}$$

Note in Figure 1.30 that sales less than the break-even point correspond to an overall loss, whereas sales greater than the break-even point correspond to a profit.

### Mathematical Models

In this text we will be concerned primarily with the use of equations as **mathematical models** of real-world phenomena. In developing a mathematical model to represent actual data, we strive for two (often conflicting) goals—accuracy and simplicity. That is, we would like the model to be simple enough to be workable and at the same time accurate enough to produce meaningful results. The next example describes a typical mathematical model.

**EXAMPLE 7**

### A Mathematical Model

The median income (between 1955 and 1980) for married couples in the United States is given in Table 1.6. A mathematical model for these data is given by

$$y = 0.0332t^2 - 0.1292t + 5.0536$$

where $y$ represents the median income in 1000s of dollars and $t$ represents the year, with $t = 0$ corresponding to 1955. Using a graph, compare the data with the model and use the model to predict the median income for 1985.

TABLE 1.6

| Year | 1955 | 1960 | 1965 | 1970 | 1975 | 1980 |
|---|---|---|---|---|---|---|
| Income (in 1000s) | 4.6 | 5.9 | 7.3 | 10.5 | 14.9 | 23.1 |

**SOLUTION**

Table 1.7 and Figure 1.31 compare the model's values with the actual values.

TABLE 1.7

| $t$ | 0 | 5 | 10 | 15 | 20 | 25 |
|---|---|---|---|---|---|---|
| $y$ | 5.1 | 5.2 | 7.1 | 10.6 | 15.8 | 22.6 |
| Actual income | 4.6 | 5.9 | 7.3 | 10.5 | 14.9 | 23.1 |

To predict the median income for 1985 we let $t = 30$ and calculate $y$ as follows:

$$y = 0.0332(30)^2 - 0.1292(30) + 5.0536 \approx 31.1$$

Thus, we estimate the 1985 median income to be $31,100.

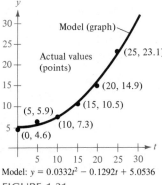

Model: $y = 0.0332t^2 - 0.1292t + 5.0536$

FIGURE 1.31

## SECTION EXERCISES 1.2

In Exercises 1–4, determine whether the points are solution points for the given equation.

**1.** $2x - y - 3 = 0$

  (a) $(1, 2)$      (b) $(1, -1)$      (c) $(4, 5)$

**2.** $x^2 + y^2 = 4$

  (a) $(1, -\sqrt{3})$    (b) $\left(\frac{1}{2}, -1\right)$      (c) $\left(\frac{3}{2}, \frac{7}{2}\right)$

**3.** $x^2y - x^2 + 4y = 0$

  (a) $\left(1, \frac{1}{5}\right)$      (b) $\left(2, \frac{1}{2}\right)$      (c) $(-1, -2)$

**4.** $x^2 - xy + 4y = 3$

  (a) $(0, 2)$      (b) $\left(-2, -\frac{1}{6}\right)$      (c) $(3, -6)$

In Exercises 5–14, find the intercepts of the graph of the given equation.

**5.** $2x - y - 3 = 0$      **6.** $y = (x - 1)(x - 3)$

**7.** $y = x^2 + x - 2$      **8.** $y^2 = x^3 - 4x$

**9.** $y = x^2\sqrt{9 - x^2}$      **10.** $xy = 4$

**11.** $y = \dfrac{x - 1}{x - 2}$      **12.** $y = \dfrac{x^2 + 3x}{(3x + 1)^2}$

**13.** $x^2y - x^2 + 4y = 0$      **14.** $y = x^2 + 1$

In Exercises 15–20, match the given equation with the correct graph. [Graphs are labeled (a), (b), (c), (d), (e), and (f).]

**15.** $y = x - 2$      **16.** $y = -\dfrac{1}{2}x + 2$

**17.** $y = x^2 + 2x$      **18.** $y = \sqrt{9 - x^2}$

**19.** $y = |x| - 2$      **20.** $y = x^3 - x$

(a)

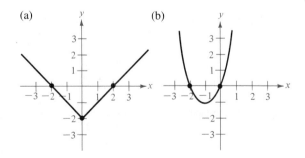

(b)

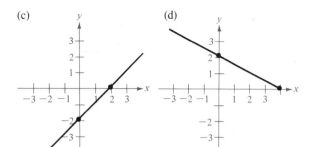

(c)      (d)

(e)      (f)

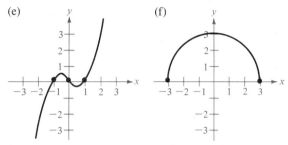

In Exercises 21–39, sketch the graph of the given equation, and plot the intercepts.

**21.** $y = x$      **22.** $y = x + 3$

**23.** $y = -3x + 2$      **24.** $y = 2x - 3$

**25.** $y = 1 - x^2$      **26.** $y = x^2 + 3$

**27.** $y = x^3 + 2$      **28.** $y = 1 - x^3$

**29.** $y = (x + 2)^2$      **30.** $y = (x - 1)^2$

**31.** $y = \sqrt{x}$      **32.** $y = \sqrt[3]{x}$

**33.** $y = -\sqrt{x - 3}$      **34.** $y = -\sqrt[3]{x}$

**35.** $y = |x - 2|$      **36.** $y = -|x - 2|$

**37.** $y = \dfrac{1}{x}$      **38.** $y = \dfrac{1}{x^2 + 1}$

**39.** $x = y^2 - 4$

**40.** (a) Sketch the graph of $y = 3x^4 - 4x^3$ by completing the accompanying table and plotting the resulting solution points.

| $x$ | $-1$ | $0$ | $1$ | $2$ |
|---|---|---|---|---|
| $y$ | | | | |

  (b) Find additional solution points for the equation by completing the accompanying table. Use these points to refine the graph of part (a).

| $x$ | $-0.75$ | $-0.50$ | $-0.25$ | $0.25$ | $0.50$ | $0.75$ | $1.33$ |
|---|---|---|---|---|---|---|---|
| $y$ | | | | | | | |

In Exercises 41–48, write the general form of the equation of the specified circle.

**41.** Center: $(0, 0)$; radius: 3

**42.** Center: $(0, 0)$; radius: 5

**43.** Center: $(2, -1)$; radius: 4

**44.** Center: $(-4, 3)$; radius: $\dfrac{5}{8}$

**45.** Center: $(-1, 2)$; solution point: $(0, 0)$

**46.** Center: $(3, -2)$; solution point: $(-1, 1)$

**47.** Endpoints of a diameter: $(0, 0)$, $(6, 8)$

**48.** Endpoints of a diameter: $(-4, -1)$, $(4, 1)$

In Exercises 49–56, use the process of completing the square to write the equation of the circle in standard form. Sketch the graph of the circle.

**49.** $x^2 + y^2 - 2x + 6y + 6 = 0$

**50.** $x^2 + y^2 - 2x + 6y - 15 = 0$

**51.** $x^2 + y^2 - 2x + 6y + 10 = 0$

**52.** $3x^2 + 3y^2 - 6y - 1 = 0$

**53.** $2x^2 + 2y^2 - 2x - 2y - 3 = 0$

**54.** $4x^2 + 4y^2 - 4x + 2y - 1 = 0$

**55.** $16x^2 + 16y^2 + 16x + 40y - 7 = 0$

**56.** $x^2 + y^2 - 4x + 2y + 3 = 0$

In Exercises 57–66, find the points of intersection (if any) of the graphs of the equations and check your results.

**57.** $x + y = 2$, $2x - y = 1$

**58.** $2x - 3y = 13$, $5x + 3y = 1$

**59.** $x + y = 7$, $3x - 2y = 11$

**60.** $x^2 + y^2 = 25$, $2x + y = 10$

**61.** $x^2 + y^2 = 5$, $x - y = 1$

**62.** $x^2 + y = 4$, $2x - y = 1$

**63.** $y = x^3$, $y = x$

**64.** $y = \sqrt{x}$, $y = x$

**65.** $y = x^4 - 2x^2 + 1$, $y = 1 - x^2$

**66.** $y = x^3 - 2x^2 + x - 1$, $y = -x^2 + 3x - 1$

**67.** A person setting up a part-time business makes an initial investment of $5,000. The unit cost of the product is $21.60, and the selling price is $34.10.
  (a) Find equations for the total cost $C$ and total revenue $R$ for $x$ units.
  (b) Find the break-even point by finding the point of intersection of the cost and revenue equations of part (a).

**68.** A certain car model costs $14,500 with a gasoline engine and $15,450 with a diesel engine. The number of miles per gallon of fuel for cars with these two engines is 22 and 31, respectively. Assume that the price of both types of fuel is $1.389/gal.

(a) Show that the cost $C_g$ of driving the gasoline-powered car $x$ miles is

$$C_g = 14,500 + \frac{1.389x}{22}$$

and the cost $C_d$ of driving the diesel model $x$ miles is

$$C_d = 15,450 + \frac{1.389x}{31}$$

(b) Find the break-even point; that is, find the mileage at which the diesel-powered car becomes more economical than the gasoline-powered car.

In Exercises 69 and 70, find the sales necessary to break even ($R = C$) for the given cost $C$ of $x$ units and the given revenue $R$ obtained from selling $x$ units. (Round your answer *up* to the nearest whole unit.)

**69.** $C = 8650x + 250,000$, $R = 9950x$

**70.** $C = 5.5\sqrt{x} + 10,000$, $R = 3.29x$

**71.** The following table gives the consumer price index (CPI) for the years 1970–1983. In the base year of 1967, CPI = 100.

| Year | 1969 | 1970 | 1971 | 1972 | 1973 |
|------|------|------|------|------|------|
| CPI  | 109.8 | 116.3 | 121.3 | 125.3 | 133.1 |

| Year | 1974 | 1975 | 1976 | 1977 | 1978 |
|------|------|------|------|------|------|
| CPI  | 147.7 | 161.2 | 170.5 | 181.5 | 195.3 |

| Year | 1979 | 1980 | 1981 | 1982 | 1983 |
|------|------|------|------|------|------|
| CPI  | 217.7 | 247.0 | 272.3 | 288.6 | 297.4 |

A mathematical model for the CPI during this period is

$$y = 0.79t^2 + 4.68t + 114.68$$

where $y$ represents the CPI and $t$ represents the year, with $t = 0$ corresponding to 1970.
  (a) Use a graph to compare the CPI with the model.
  (b) Use the model to predict the CPI for 1985.

**72.** From the model of Exercise 71, we obtain the model

$$V = \frac{100}{0.79t^2 + 4.68t + 114.68}$$

where $V$ represents the purchasing power of the dollar (in terms of constant 1967 dollars) and $t$ represents the year, with $t = 0$ corresponding to 1970. Use this model to complete the following table:

| $t$ | 0 | 2 | 4 | 6 | 8 | 10 | 12 | 14 |
|-----|---|---|---|---|---|----|----|----|
| $V$ |   |   |   |   |   |    |    |    |

**73.** The average number of acres per farm in the United States is given in the following table:

| Year | 1950 | 1960 | 1965 | 1970 | 1975 | 1980 | 1984 |
|------|------|------|------|------|------|------|------|
| Acres | 213 | 297 | 340 | 374 | 391 | 427 | 437 |

A mathematical model for these data is given by

$$y = -0.09t^2 + 9.65t + 212.48$$

where $y$ represents the average number of acres per farm and $t$ represents the year, with $t = 0$ corresponding to 1950.

(a) Use a graph to compare the actual number of acres per farm with that given by the model.

(b) Use the model to predict the average number of acres per farm in the United States in 1990.

**74.** The farm population in the United States as a percentage of the total population is given in the following table:

| Year | 1950 | 1955 | 1960 | 1965 | 1970 | 1975 | 1980 | 1983 |
|------|------|------|------|------|------|------|------|------|
| % | 15.3 | 11.6 | 8.7 | 6.4 | 4.8 | 4.2 | 2.7 | 2.5 |

A mathematical model for these data is given by

$$y = \frac{500}{14 + 5t}$$

where $y$ represents the percentage and $t$ represents the year, with $t = 0$ corresponding to 1950.

(a) Use a graph to compare the actual percentage with that given by the model.

(b) Use the model to predict the farm population as a percentage of the total population in 1990.

# Lines in the Plane; Slope

## INTRODUCTORY EXAMPLE
## Straight-line Depreciation

Most business expenses can be applied as tax deductions to the year in which the expense occurs. An important exception to this is the cost of property with a useful life of more than one year. For example, the Internal Revenue Service does not allow businesses to deduct the entire cost of buildings, cars, furniture, or equipment in one year. Such costs must be spread out over the useful life of the property. This procedure is called **depreciation.** If the *same amount* is depreciated each year, the procedure is called **straight-line depreciation.**

Suppose an item that has an initial cost of $C$ has a salvage value of $S$ after $n$ years. The total depreciation that can be claimed as a business expense during the $n$ years is

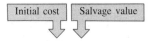

Total depreciation $= C - S$

With the straight-line method of depreciation, the amount of depreciation that can be claimed each year is

$$D = \frac{C - S}{n}$$

Table 1.8 and Figure 1.32 illustrate the depreciation of a $12,000 machine over an 8-year period. The salvage value at the end of the 8 years is $2,000. This means that $1,250 is depreciated each year. The nondepreciated value $y$ of the machine after $t$ years is given by the **linear equation**

$$y = -1250t + 12,000$$

Since the nondepreciated value *decreases* by 1250 each year, we say that the line shown in Figure 1.32 has a negative **slope** of $-1250$.

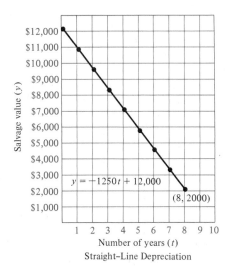

FIGURE 1.32

TABLE 1.8

| $t$ | 0 | 1 | 2 | 3 | 4 | 5 | 6 | 7 | 8 |
|---|---|---|---|---|---|---|---|---|---|
| $y$ ($) | 12,000 | 10,750 | 9,500 | 8,250 | 7,000 | 5,750 | 4,500 | 3,250 | 2,000 |

■ **The Slope of a Line**
■ **Equations of Lines**
■ **Linear Interpolation**
■ **Parallel and Perpendicular Lines**

The simplest mathematical model for relating two variables is the **linear equation** $y = mx + b$. By letting $x = 0$ we see that the line* given by this equation crosses the $y$-axis at $y = b$. In other words, its $y$-intercept is $(0, b)$. The steepness or slope of the line is given by $m$.

$$y = mx + b$$

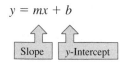

By the **slope** of a line we mean the number of units the line rises (or falls) vertically for each unit of horizontal change from left to right, as shown in Figure 1.33.

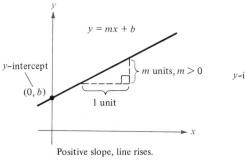

Positive slope, line rises.                    Negative slope, line falls.

FIGURE 1.33

| **The slope-intercept form of the equation of a line** | The graph of the equation $y = mx + b$ is a line whose slope is $m$ and $y$-intercept is $(0, b)$. |
|---|---|

Once we have determined the slope and the $y$-intercept of a line it is a relatively simple matter to sketch its graph.

―――――――

*Note that we use the term *line* to mean *straight line*.

**EXAMPLE 1**

**Graphing a Linear Equation**

Sketch the graphs of the following linear equations:

(a) $y = 2x + 1$                     (b) $y = 2$

(c) $x + y = 2$

**SOLUTION**

(a) Since $b = 1$, the $y$-intercept is $(0, 1)$. Moreover, since the slope is $m = 2$ this line *rises* 2 units for each unit the line moves to the right, as shown in Figure 1.34.

(b) By writing this equation in the form $y = (0)x + 2$, we see that the $y$-intercept is $(0, 2)$ and the slope is zero. A zero slope implies that the line is horizontal— that is, it doesn't rise *or* fall, as shown in Figure 1.35.

(c) By writing the equation in slope-intercept form,

$$x + y = 2 \qquad \text{Given equation}$$

$$y = -x + 2 \qquad \text{Slope-intercept form}$$

we see that the $y$-intercept is $(0, 2)$. Moreover, since the slope is $m = -1$, this line *falls* 1 unit for each unit the line moves to the right, as shown in Figure 1.36.

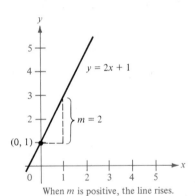

When *m* is positive, the line rises.

FIGURE 1.34

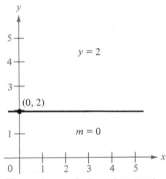

When *m* is zero, the line is horizontal.

FIGURE 1.35

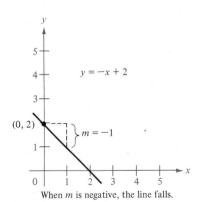

When *m* is negative, the line falls.

FIGURE 1.36

Note that in Example 1 we gave no examples of vertical lines. Vertical lines have equations of the form

$$x = a \qquad \text{Vertical line}$$

Since such an equation cannot be written in the form $y = mx + b$, we say that the slope of a vertical line is undefined, as indicated in Figure 1.37.

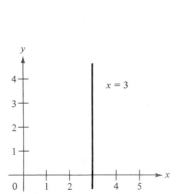

When the line is vertical, the slope is undefined.

FIGURE 1.37

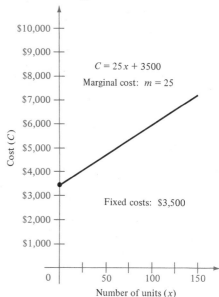

FIGURE 1.38

## EXAMPLE 2

### A Business Application

A manufacturing company determines that the total cost in dollars of producing $x$ units of a certain product is

$$C = 25x + 3500$$

Describe the practical significance of the $y$-intercept and slope of the line given by this equation.

## SOLUTION

The $y$-intercept $(0, 3500)$ tells us that the cost of producing zero units is $3,500. We call these the **fixed costs** of production, and they include costs such as product development and rent that must be paid regardless of the number of units produced. The slope of $m = 25$ tells us that the cost of producing each unit is $25.00, as shown in Figure 1.38. In the next chapter, we will see that economists refer to this unit cost as the **marginal cost.** Thus, if the production increases by one unit, then the "margin" or extra amount of cost is $25.00. For instance, the cost of producing 100 units is

$$C = 25(100) + 3500 = \$6,000$$

and the cost of producing 101 units is $25 more:

$$C = 25(101) + 3500 = \$6,025$$

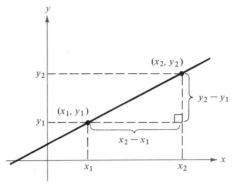

FIGURE 1.39

Now that we can determine the slope of a line from its equation, let us suppose that the equation is not given. How then can we determine the slope? For instance, suppose we want to find the slope of the line passing through the two points $(x_1, y_1)$ and $(x_2, y_2)$ shown in Figure 1.39. As we move from left to right along this line a change of $(y_2 - y_1)$ units in the vertical direction corresponds to a change of $(x_2 - x_1)$ units in the horizontal direction. We denote these two changes by the symbols

$$\Delta y = y_2 - y_1 = \text{the change in } y$$

and

$$\Delta x = x_2 - x_1 = \text{the change in } x$$

($\Delta$ is the Greek capital letter delta, and the symbols $\Delta y$ and $\Delta x$ are read "delta $y$" and "delta $x$.") We use the ratio of $\Delta y$ to $\Delta x$ to calculate the slope of the line and write

$$\text{Slope} = \frac{\Delta y}{\Delta x} = \frac{y_2 - y_1}{x_2 - x_1}$$

**The slope of a line passing through two points**

The **slope** $m$ of the line passing through the points $(x_1, y_1)$ and $(x_2, y_2)$ is

$$m = \frac{\Delta y}{\Delta x} = \frac{y_2 - y_1}{x_2 - x_1}$$

where $x_1 \neq x_2$.

When this formula is used for slope, the *order of subtraction* is important. Given two points on a line, you are free to label either one of them as $(x_1, y_1)$ and the other as $(x_2, y_2)$. However, once you have done this, you must form the numerator and denominator using the same order of subtraction:

$$m = \frac{y_2 - y_1}{x_2 - x_1} \qquad m = \frac{y_1 - y_2}{x_1 - x_2} \qquad m = \frac{y_2 - y_1}{x_1 - x_2}$$

$$\underbrace{\phantom{m = \frac{y_2 - y_1}{x_2 - x_1}}}_{\text{Correct}} \qquad \underbrace{\phantom{m = \frac{y_1 - y_2}{x_1 - x_2}}}_{\text{Correct}} \qquad \underbrace{\phantom{m = \frac{y_2 - y_1}{x_1 - x_2}}}_{\text{Incorrect}}$$

**EXAMPLE 3**

**Finding the Slope of a Line Determined by Two Points**

Find the slopes of the lines passing through the following pairs of points:
(a) $(-2, 0)$ and $(3, 1)$          (b) $(-1, 2)$ and $(2, 2)$
(c) $(0, 4)$ and $(1, -1)$

**SOLUTION**

(a) Letting $(x_1, y_1) = (-2, 0)$ and $(x_2, y_2) = (3, 1)$, we obtain a slope of

$$m = \frac{y_2 - y_1}{x_2 - x_1} \quad \Longleftarrow \boxed{\text{Difference in } y\text{-values}}$$
$$\quad\quad\quad \Longleftarrow \boxed{\text{Difference in } x\text{-values}}$$

$$= \frac{1 - 0}{3 - (-2)}$$

$$= \frac{1}{5}$$

(b) The slope of the line passing through $(-1, 2)$ and $(2, 2)$ is

$$m = \frac{2 - 2}{2 - (-1)} = \frac{0}{3} = 0$$

(c) The slope of the line passing through $(0, 4)$ and $(1, -1)$ is

$$m = \frac{-1 - 4}{1 - 0} = \frac{-5}{1} = -5$$

See Figure 1.40.

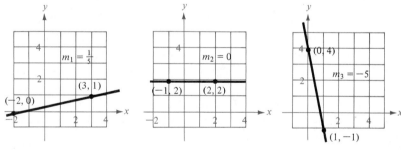

If $m$ is positive, the line rises.

If $m$ is zero, the line is horizontal.

If $m$ is negative, the line falls.

FIGURE 1.40

## Finding an Equation for a Line

If we know the slope of a line and its $y$-intercept then we can find an equation of the line using the slope-intercept form, $y = mx + b$. But suppose we know the slope and some other point on the line (a point other than the $y$-intercept). Then how can we find an equation for the line? The formula for the slope of the line passing through two points gives us the answer. For if $(x_1, y_1)$ is a point lying on a line of slope $m$ and $(x, y)$ is *any other* point on the line, then

$$\frac{y - y_1}{x - x_1} = m$$

This equation, involving two variables $x$ and $y$, can be rewritten in the form

$$y - y_1 = m(x - x_1)$$

which we call the **point-slope form** of the equation of a line.

| **Point-slope form of the equation of a line** | The equation of the line with slope $m$ passing through the point $(x_1, y_1)$ is given by $$y - y_1 = m(x - x_1)$$ |
|---|---|

The point-slope form is the most useful for *finding* the equation of a line. You should memorize this formula; we will use it repeatedly throughout the text.

**EXAMPLE 4**

**The Point-Slope Form of the Equation of a Line**

Find the equation of the line that has a slope of 3 and passes through the point $(1, -2)$, as shown in Figure 1.41.

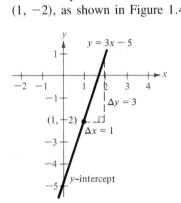

FIGURE 1.41

**SOLUTION**

Using the point-slope form we have

$$y - y_1 = m(x - x_1)$$
$$y - (-2) = 3(x - 1)$$
$$y + 2 = 3x - 3$$
$$y = 3x - 5$$

The point-slope form can be used to find the equation of a line passing through two points $(x_1, y_1)$ and $(x_2, y_2)$. To do this, we use two steps. We first find the slope of the line

$$m = \frac{y_2 - y_1}{x_2 - x_1}$$

and then use the point-slope form to obtain the equation

$$y - y_1 = \frac{y_2 - y_1}{x_2 - x_1}(x - x_1)$$

This is sometimes called the **two-point form** of the equation of a line. We demonstrate its use in the next example.

**EXAMPLE 5**

**An Application**

Total U.S. sales (including inventories) during the first two quarters of 1978 were 539.9 and 560.2 billion dollars, respectively. Assuming a *linear growth pattern,* predict the total sales during the fourth quarter of 1978.

**SOLUTION**

Referring to Figure 1.42, let $(1, 539.9)$ and $(2, 560.2)$ be two points on the line representing total U.S. sales, where $x$ represents the quarter and $y$ represents the sales in billions of dollars. The slope of the line passing through these two points is

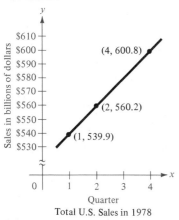

FIGURE 1.42

$$m = \frac{y_2 - y_1}{x_2 - x_1}$$

$$= \frac{560.2 - 539.9}{2 - 1}$$

$$= 20.3$$

and by the point-slope form, the equation of the line is

$$y - y_1 = m(x - x_1)$$

$$y - 539.9 = 20.3(x - 1)$$

$$y = 20.3(x - 1) + 539.9$$

$$y = 20.3x + 519.6$$

Now, using this linear model, we estimate the fourth-quarter sales ($x = 4$) to be

$$y = (20.3)(4) + 519.6 = 600.8 \text{ billion}$$

(In this particular case the estimate proves to be quite good. The actual fourth-quarter sales in 1978 were 600.5 billion dollars.)

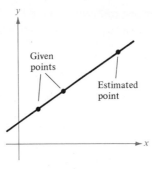 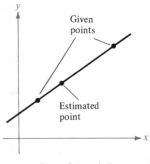

FIGURE 1.43　　　Linear Extrapolation　　　　Linear Interpolation

The prediction method illustrated in Example 5 is called **linear extrapolation.** Note in Figure 1.43 that an extrapolated point does not lie between the given points. When the estimated point lies between two given points, we call the procedure **linear interpolation.**

Since the slope of a vertical line is not defined, its equation cannot be written in the slope-intercept form. However, every line has an equation that can be written in the **general form**

$$Ax + By + C = 0$$

where $A$ and $B$ are not both zero. For instance, the vertical line given by $x = a$ can be represented by the general form $x - a = 0$. We summarize the five most common forms of equations of lines in the following list.

| **Equations of lines** | |
|---|---|
| | 1. General form: $\quad Ax + By + C = 0$ |
| | 2. Vertical line: $\quad x = a$ |
| | 3. Horizontal line: $\quad y = b$ |
| | 4. Slope-intercept form: $y = mx + b$ |
| | 5. Point-slope form: $\quad y - y_1 = m(x - x_1)$ |

## Parallel Lines and Perpendicular Lines

The slope of a line is a convenient tool for determining if two lines are parallel or perpendicular.

**Parallel lines and perpendicular lines**

1. Two distinct nonvertical lines are parallel if and only if their slopes are equal. That is,

$$m_1 = m_2$$

2. Two nonvertical lines are perpendicular if and only if their slopes are related by the equation

$$m_1 = -\frac{1}{m_2}$$

**EXAMPLE 6**

**Finding Parallel and Perpendicular Lines**

Find the equation of the line that passes through the point $(2, -1)$ and is
(a) parallel to the line $2x - 3y = 5$
(b) perpendicular to the line $2x - 3y = 5$

**SOLUTION**

Writing the equation $2x - 3y = 5$ in the slope-intercept form, we have

$$3y = 2x - 5$$

$$y = \frac{2}{3}x - \frac{5}{3}$$

Therefore, the given line has a slope of $m = \frac{2}{3}$, as shown in Figure 1.44.

(a) Any line parallel to the given line $2x - 3y = 5$ must also have a slope of $\frac{2}{3}$. Thus, the line through $(2, -1)$ that is parallel to the line $2x - 3y = 5$ has an equation of the form

$$y - (-1) = \frac{2}{3}(x - 2)$$

$$3(y + 1) = 2(x - 2)$$

$$3y + 3 = 2x - 4$$

$$-2x + 3y = -7$$

or

$$2x - 3y = 7$$

(Note the similarity to the original equation $2x - 3y = 5$.)

(b) Any line perpendicular to the line $2x - 3y = 5$ must have a slope of $-\frac{3}{2}$. Therefore, the line through $(2, -1)$ that is perpendicular to the line $2x - 3y = 5$ has the equation

$$y - (-1) = -\frac{3}{2}(x - 2)$$

$$2(y + 1) = -3(x - 2)$$

or

$$3x + 2y = 4$$

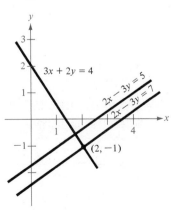

FIGURE 1.44

In Exercises 1–6, estimate the slope of the given line from its graph.

**1.**

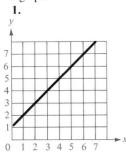

**2.**

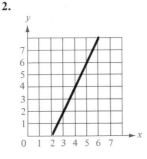

**3.**

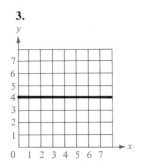

**4.**

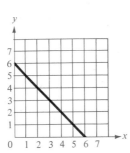

**5.**

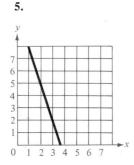

**6.**

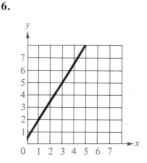

In Exercises 7–14, plot the points and find the slope of the line passing through the given points.

**7.** $(3, -4)$, $(5, 2)$

**8.** $(-2, 1)$, $(4, -3)$

**9.** $\left(\dfrac{1}{2}, 2\right)$, $(6, 2)$

**10.** $\left(-\dfrac{3}{2}, -5\right)$, $\left(\dfrac{5}{6}, 4\right)$

**11.** $(-6, -1)$, $(-6, -4)$

**12.** $(2, 1)$, $(2, 5)$

**13.** $(1, 2)$, $(-2, 2)$

**14.** $\left(\dfrac{7}{8}, \dfrac{3}{4}\right)$, $\left(\dfrac{5}{4}, -\dfrac{1}{4}\right)$

In Exercises 15–22, use the given point on the line and the slope of the line to find three additional points that the line passes through. (The solution is not unique.)

| | Point | Slope | | Point | Slope |
|---|---|---|---|---|---|
| **15.** | $(2, 1)$ | $m = 0$ | **16.** | $(-3, 4)$ | $m$ undefined |
| **17.** | $(5, -6)$ | $m = 1$ | **18.** | $(-2, -2)$ | $m = 2$ |
| **19.** | $(1, 7)$ | $m = -3$ | **20.** | $(10, -6)$ | $m = -1$ |
| **21.** | $(-8, 1)$ | $m$ undefined | **22.** | $(-3, -1)$ | $m = 0$ |

In Exercises 23–28, find the slope and $y$-intercept (if possible) of the line specified by the given equation.

**23.** $x + 5y = 20$

**24.** $2x + y = 40$

**25.** $4x - 3y = 18$

**26.** $6x - 5y = 15$

**27.** $x = 4$

**28.** $y = -1$

In Exercises 29–36, find an equation for the line that passes through the given points, and sketch the graph of the line.

**29.** $(2, 1)$, $(0, -3)$

**30.** $(-3, -4)$, $(1, 4)$

**31.** $(0, 0)$, $(-1, 3)$

**32.** $(-3, 6)$, $(1, 2)$

**33.** $(2, 3)$, $(2, -2)$

**34.** $(6, 1)$, $(10, 1)$

**35.** $(1, -2)$, $(3, -2)$

**36.** $\left(\dfrac{7}{8}, \dfrac{3}{4}\right)$, $\left(\dfrac{5}{4}, -\dfrac{1}{4}\right)$

In Exercises 37–46, find an equation of the line that passes through the given point and has the indicated slope. Sketch the line.

| | Point | Slope | | Point | Slope |
|---|---|---|---|---|---|
| **37.** | $(0, 3)$ | $m = \dfrac{3}{4}$ | **38.** | $(-1, 2)$ | $m$ is undefined |
| **39.** | $(0, 0)$ | $m = \dfrac{2}{3}$ | **40.** | $(-1, -4)$ | $m = \dfrac{1}{4}$ |
| **41.** | $(0, 5)$ | $m = -2$ | **42.** | $(-2, 4)$ | $m = -\dfrac{3}{5}$ |
| **43.** | $(0, 2)$ | $m = 4$ | **44.** | $\left(0, -\dfrac{2}{3}\right)$ | $m = \dfrac{1}{6}$ |
| **45.** | $\left(0, \dfrac{2}{3}\right)$ | $m = \dfrac{3}{4}$ | **46.** | $(0, 4)$ | $m = 0$ |

**47.** Find an equation of the vertical line with $x$-intercept at 3.

**48.** Find an equation of the horizontal line with $y$-intercept at $-5$.

In Exercises 49–54, write an equation of the line through the given point (a) parallel to the given line and (b) perpendicular to the given line.

| | Point | Line | | Point | Line |
|---|---|---|---|---|---|
| **49.** | $(-3, 2)$ | $x + y = 7$ | **50.** | $(2, 1)$ | $4x - 2y = 3$ |
| **51.** | $(-6, 4)$ | $3x + 4y = 7$ | **52.** | $\left(\dfrac{7}{8}, \dfrac{3}{4}\right)$ | $5x + 3y = 0$ |
| **53.** | $(-1, 0)$ | $y = -3$ | **54.** | $(2, 5)$ | $x = 4$ |

**55.** Find the equation of the line giving the relationship between the temperature in degrees Celsius $C$ and degrees Fahrenheit $F$. Use the fact that water freezes at 0° Celsius (32° Fahrenheit) and boils at 100° Celsius (212° Fahrenheit).

**56.** Use the result of Exercise 55 to complete the following table.

| $C$ |  | $-10°$ | $10°$ |  |  | $177°$ |
|---|---|---|---|---|---|---|
| $F$ | 0° |  |  | 68° | 90° |  |

**57.** A company reimburses its sales representatives $95 per day for lodging and meals plus 25¢ per mile driven. Write a linear equation giving the daily cost $C$ to the company in terms of $x$, the number of miles driven.

**58.** A manufacturing company pays its assembly line workers $9.50 per hour *plus* an additional piece-work rate of $0.75 per unit produced. Find a linear equation for the hourly wages $W$ in terms of the number of units produced per hour $t$.

**59.** A small business purchases a piece of equipment for $875. After 5 years the equipment will be obsolete and have no value. Write a linear equation giving the value $y$ of the equipment during the five years it will be used. (Let $t$ represent the time in years.)

**60.** A company constructs a warehouse for $825,000. It has an estimated useful life of 25 years, after which its value is expected to be $75,000. Use straight-line depreciation to write a linear equation giving the value $y$ of the warehouse during its 25 years of useful life. (Let $t$ represent the time in years.)

**61.** A real estate office handles an apartment complex with 50 units. When the rent is $380 per month, all 50 units are occupied. However, when the rent is $425, the average number of occupied units drops to 47. Assume that the relationship between the monthly rent $p$ and the demand $x$ is linear. (Note: Here we use the term *demand* to refer to the number of occupied units.)

   (a) Write a linear equation giving the quantity demanded $x$ in terms of the rent $p$.

   (b) (Linear Extrapolation) Use this equation to predict the number of units occupied if the rent is raised to $455.

   (c) (Linear Interpolation) Predict the number of units occupied if the rent is lowered to $395.

**62.** Personal income (in billions of dollars) in the United States between 1980 and 1983 is given in the following table:

| Year | 1980 | 1981 | 1982 | 1983 |
|---|---|---|---|---|
| $t$ | 0 | 1 | 2 | 3 |
| Income $y$ | 2165 | 2429 | 2585 | 2744 |

   (a) Assume that the relationship between $y$ and $t$ is linear and write an equation for the line passing through (0, 2165) and (3, 2744).

   (b) (Linear Interpolation) Use this equation to estimate personal income in 1981 and 1982. Compare the estimate with the actual amount.

   (c) (Linear Extrapolation) Predict personal income in 1985.

   (d) What information is given by the slope of the line in part (a)?

**63.** A contractor purchases a piece of equipment for $26,500. The equipment requires an average expenditure of $5.25 per hour for fuel and maintenance, and the operator is paid $9.50 per hour.

   (a) Write a linear equation giving the total cost $C$ of operating this equipment $t$ hours.

   (b) Given that customers are charged $25 per hour of machine use, write an equation for the revenue $R$ derived from $t$ hours of use.

   (c) Use the formula for profit ($P = R - C$) to write an equation for the profit derived from $t$ hours of use.

   (d) Find the number of hours the equipment must be operated for the contractor to break even.

**64.** Rework Exercise 63 assuming that the hourly wage for the operator is raised to $17.50 and the hourly rate that customers are charged is raised to $40.

# Functions

Two types of mathematical models that economists use to analyze a market are supply and demand functions:

$$p = S(x) \qquad \text{Supply function}$$

and $\quad p = D(x) \qquad$ Demand function

where $x$ is the number of units produced (or sold) and $p$ is the price per unit. In each case, we say that the price $p$ is a **function** of the production level $x$.

For example, suppose that the supply function for the monthly production of hard winter wheat in the United States is given by the *linear* model

$$p = S(x) = \frac{1}{7}x + 2$$

where $x$ is measured in millions of bushels and $p$ is the price per bushel. From Figure 1.45, we can see that if the price per bushel were $2.00, then (according to this model) the production of wheat would come to a stand-still, since no one would be willing to grow and sell wheat for such a low price. On the other hand, if the price per bushel were $3.00, then enough farmers would be willing to grow wheat to increase the production level to 7 million bushels per month. This supply function is typical in that as the price per unit increases, more and more farmers become willing to grow wheat and the monthly production increases.

On the other hand, the typical demand function tends to show a decrease in the quantity demanded with each increase in price. For instance, suppose that the demand function for hard winter wheat in the United States is given by the model

$$p = D(x) = -\frac{4}{11}x + 9$$

From Figure 1.45, we can see that increases in the quantity demanded $x$ correspond to decreases in the price $p$.

In an idealized situation, with no other factors present to influence the market, the production level should stabilize at the point of intersection of the graphs of the supply and demand functions. We call this point the **equilibrium point.**

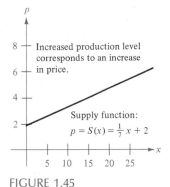

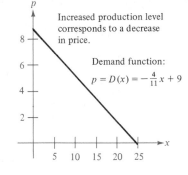

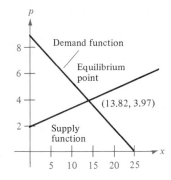

FIGURE 1.45

In many common relationships between two variables the value of one of the variables depends on the value of the other. For example, the sales tax on an item depends on its selling price, the distance an object moves in a given amount of time depends on its speed, the pressure of a gas in a closed container depends on its temperature, and the area of a circle depends on its radius.

Consider the relationship between the area of a circle and its radius. This relationship can be expressed by the equation

$$A = \pi r^2$$

Considering the radius of a circle to be positive, we have within the set of positive numbers a free choice for the value of $r$. The value of $A$ then depends on our choice of $r$. Thus, we refer to $A$ as the **dependent variable** and $r$ as the **independent variable.**

In calculus we are primarily interested in relationships such that to every value of the independent variable there corresponds *one and only one* value of the dependent variable. We call this type of correspondence a **function.**

| | |
|---|---|
| **Definition of a function** | A **function** is a relationship between two variables such that to each value of the independent variable there corresponds exactly one value of the dependent variable.<br><br>The **domain** of the function is the set of all values of the independent variable for which the function is defined. The **range** of the function is the set of all values taken on by the dependent variable. |

■■ **Remark:** If to each value of the dependent variable there corresponds exactly one value of the independent variable, then the function is called **one-to-one.** Geometrically, this means that a function is one-to-one if every horizontal line intersects the graph of the function at most once.

Although functions can be described by various means, we often specify functions by formulas, or equations.

**EXAMPLE 1**
■■■■■

**Determining Functional Relationships from Equations**

Which of the following equations define $y$ as a function of $x$?

(a) $x + y = 1$               (b) $x^2 + y^2 = 1$

(c) $x^2 + y = 1$             (d) $x + y^2 = 1$

**SOLUTION**

To determine whether or not an equation defines a function, it is helpful to isolate the dependent variable on the left-hand side. Thus, to see if the equation $x + y = 1$ defines $y$ as a function of $x$, we write $y = 1 - x$.

In this form, we can see that for any given value of $x$, there is precisely one value of $y$. Therefore, $y$ is a function of $x$. Table 1.9 summarizes the results for the four given equations. Note that those equations that assign two values ($\pm$) to the dependent variable for a given value of the independent variable do not define functions. For instance, in part (b) if $x = 0$, then the equation

$$y = \pm\sqrt{1 - x^2}$$

indicates that $y = +1$ or $y = -1$.

TABLE 1.9

| Original equation | Rewritten equation with $y$ on left | Is $y$ a function of $x$? |
|---|---|---|
| (a) $x + y = 1$ | $y = 1 - x$ | yes |
| (b) $x^2 + y^2 = 1$ | $y = \pm\sqrt{1 - x^2}$ | no (two values of $y$ for some values of $x$) |
| (c) $x^2 + y = 1$ | $y = 1 - x^2$ | yes |
| (d) $x + y^2 = 1$ | $y = \pm\sqrt{1 - x}$ | no |

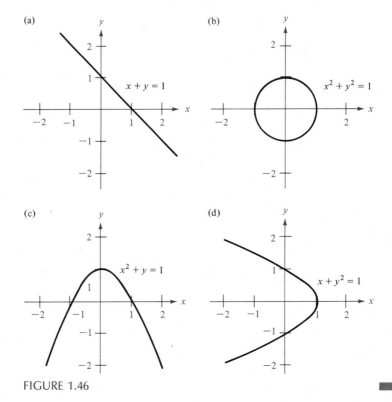

FIGURE 1.46

In sketching the graph of a function, the standard convention is to let the horizontal axis represent the independent variable (usually we use $x$). When this convention is used, the test described in Example 1 has a nice graphical interpretation called the **vertical line test.** This test states that if every vertical line intersects the graph of an equation at most once, then the equation defines $y$ as a function of $x$. (Compare this test to the definition of a function, which states that for a given value of $x$ *in the domain* there corresponds exactly one value of $y$.) For instance, in Figure 1.46 we can see that the graphs in parts (a) and (c) pass the vertical line test, but those shown in parts (b) and (d) do not.

The domain of a function may be explicitly described along with the function, or it may be *implied* by an equation used to define the function. (The implied domain is the set of all real numbers for which the equation is defined.) For example, the function given by

$$y = \frac{1}{x^2 - 4}$$

has an implied domain which consists of all real $x$ other than $x = \pm 2$. These two values are excluded from the domain because division by zero is undefined.

Another common type of implied domain is that used to avoid even roots of negative numbers. For example, the function given by

$$y = \sqrt{x}$$

has an implied domain of $x \geq 0$ or $[0, \infty)$.

**EXAMPLE 2**

**Finding the Domain and Range of a Function**

Determine the domain and range for the function of $x$ given by

$$y = \sqrt{x - 1}$$

**SOLUTION**

Since $\sqrt{x - 1}$ is not defined for $x - 1 < 0$ (that is, for $x < 1$), we conclude that the domain of the function is the interval

$$x \geq 1 \qquad \text{or} \qquad [1, \infty) \qquad \text{Domain}$$

To find the range, we observe that $y = \sqrt{x - 1}$ is never negative. Moreover, as $x$ takes on the various values in the domain, $y$ takes on all nonnegative values, and we find the range to be

$$y \geq 0 \qquad \text{or} \qquad [0, \infty) \qquad \text{Range}$$

The graph of the function, as shown in Figure 1.47, lends further support to our conclusions.

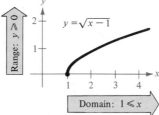

FIGURE 1.47

Sometimes a function is defined using more than one equation. This procedure is demonstrated in Example 3.

**EXAMPLE 3**

**A Function Defined by More Than One Equation**

Determine the domain and range for the function of $x$ given by

$$y = \begin{cases} \sqrt{x-1}, & \text{if } x \geq 1 \\ 1-x, & \text{if } x < 1 \end{cases}$$

**SOLUTION**

Since the function is defined for $x \geq 1$ and $x < 1$, the domain of the function is the entire set of real numbers. On the portion of the domain for which $x \geq 1$, the function behaves as in Example 2. For $x < 1$, the value of $1 - x$ is positive, and therefore the range of the function is $y \geq 0$ or $[0, \infty)$, as shown in Figure 1.48.

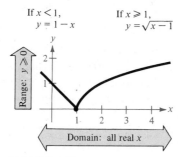

FIGURE 1.48

## Functional Notation

When using an equation to define a function, we generally isolate the dependent variable on the left side of the equation. For instance, writing the equation $x + 2y = 1$ in the form

$$y = \frac{1-x}{2}$$

indicates that $y$ is the dependent variable. In **functional notation,** this equation has the form

$$f(x) = \frac{1-x}{2}$$

This notation has the advantage of clearly identifying the dependent variable as $f(x)$ while at the same time providing a name "$f$" for the function. The symbol $f(x)$ is read "$f$ of $x$." To denote the value of the dependent variable when $x = a$, we use the symbol $f(a)$. For example, the value of $f$ when $x = 3$ is

$$f(3) = \frac{1-(3)}{2} = \frac{-2}{2} = -1$$

Similarly,

$$f(0) = \frac{1 - (0)}{2} = \frac{1}{2}$$

$$f(-2) = \frac{1 - (-2)}{2} = \frac{3}{2}$$

The values $f(3)$, $f(0)$, and $f(-2)$ are called **functional values,** and they lie in the range of $f$. This means that the values $f(3)$, $f(0)$, and $f(-2)$ are $y$-values, and thus the points $(3, f(3))$, $(0, f(0))$, and $(-2, f(-2))$ lie on the graph of $f$.

**EXAMPLE 4**

**Evaluating a Function**

Find the value of the function given by

$$f(x) = 2x^2 - 4x + 1$$

when $x$ is $-1$, 0, and 2. Is $f$ one-to-one?

**SOLUTION**

When $x = -1$, the value of $f$ is given by

$$f(-1) = 2(-1)^2 - 4(-1) + 1 = 2 + 4 + 1 = 7$$

When $x = 0$, the value of $f$ is given by

$$f(0) = 2(0)^2 - 4(0) + 1 = 0 - 0 + 1 = 1$$

When $x = 2$, the value of $f$ is given by

$$f(2) = 2(2)^2 - 4(2) + 1 = 8 - 8 + 1 = 1$$

Since two different values of $x$ yield the same value for $f(x)$, the function is *not* one-to-one, as shown in Figure 1.49.

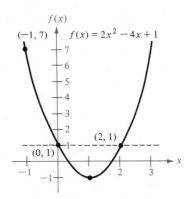

FIGURE 1.49

■■ **Remark:** Graphically, we can see that the function $y = f(x)$ is not one-to-one if a horizontal line intersects the graph of the function more than once. For instance, in Example 4, the line $y = 1$ intersects the graph of the function twice, and hence the function is not one-to-one.

Example 4 suggests that the role of the variable $x$ in the equation

$$f(x) = 2x^2 - 4x + 1$$

is simply that of a placeholder. The same function $f$ can be properly described by using parentheses instead of $x$. Thus, $f$ could be defined by the equation

$$f(\quad) = 2(\quad)^2 - 4(\quad) + 1$$

To evaluate $f(-2)$, we simply place $-2$ in each set of parentheses:

$$f(-2) = 2(-2)^2 - 4(-2) + 1 = 2(4) + 8 + 1 = 17$$

**EXAMPLE 5**

**Evaluating a Function**

For the function $f$ given by $f(x) = x^2 - 4x + 7$, evaluate

(a) $f(x + \Delta x)$ 

(b) $\dfrac{f(x + \Delta x) - f(x)}{\Delta x}$

**SOLUTION**

We begin by writing the equation for $f$ in the form

$$f(\quad) = (\quad)^2 - 4(\quad) + 7$$

(a) To evaluate $f$ at $x + \Delta x$, we substitute $x + \Delta x$ into each set of parentheses as follows:

$$f(x + \Delta x) = (x + \Delta x)^2 - 4(x + \Delta x) + 7 = x^2 + 2x\Delta x + (\Delta x)^2 - 4x - 4\Delta x + 7$$

(b) Using the result of part (a), we have

$$\frac{f(x + \Delta x) - f(x)}{\Delta x} = \frac{[(x + \Delta x)^2 - 4(x + \Delta x) + 7] - [x^2 - 4x + 7]}{\Delta x}$$

$$= \frac{x^2 + 2x\Delta x + (\Delta x)^2 - 4x - 4\Delta x + 7 - x^2 + 4x - 7}{\Delta x}$$

$$= \frac{2x\Delta x + (\Delta x)^2 - 4\Delta x}{\Delta x} = 2x + \Delta x - 4$$

■ **Remark:** The ratio in part (b) of Example 5 is called a *difference quotient* and has a special significance in calculus.

Although we generally use $f$ as a convenient function name and $x$ as the independent variable, we can use other symbols. For instance, the following equations all define the same function:

$$f(x) = x^2 - 4x + 7$$

$$f(t) = t^2 - 4t + 7$$

$$g(s) = s^2 - 4s + 7$$

## Combinations of Functions

Two functions can be combined in various ways to create new functions. For example, if

$$f(x) = 2x - 3 \quad \text{and} \quad g(x) = x^2 + 1$$

we can form the functions

$$f(x) + g(x) = (2x - 3) + (x^2 + 1) = x^2 + 2x - 2 \qquad \text{Sum}$$

$$f(x) - g(x) = (2x - 3) - (x^2 + 1) = -x^2 + 2x - 4 \qquad \text{Difference}$$

$$f(x)g(x) = (2x - 3)(x^2 + 1) = 2x^3 - 3x^2 + 2x - 3 \qquad \text{Product}$$

$$\frac{f(x)}{g(x)} = \frac{2x - 3}{x^2 + 1} \qquad \text{Quotient}$$

We can combine two functions in yet another way called the **composition** of two functions.

| | |
|---|---|
| **Definition of composite function** | Let $f$ and $g$ be functions such that the range of $g$ is in the domain of $f$. Then the function whose values are given by $f(g(x))$ is called the **composite** of $f$ with $g$. |

The composite of $f$ with $g$ may not be equal to the composite of $g$ with $f$. This is illustrated in the following example.

**EXAMPLE 6**

**Forming Composite Functions**

Given $f(x) = 2x - 3$ and $g(x) = x^2 + 1$, find

(a) $f(g(x))$                                 (b) $g(f(x))$

**SOLUTION**

(a) Since

$$f(\blacksquare) = 2(\blacksquare) - 3$$

we have

$$f(g(x)) = 2(g(x)) - 3$$
$$= 2(x^2 + 1) - 3$$
$$= 2x^2 - 1$$

(b) Since

$$g(\blacksquare) = (\blacksquare)^2 + 1$$

we have

$$g(f(x)) = (f(x))^2 + 1$$
$$= (2x - 3)^2 + 1$$
$$= 4x^2 - 12x + 10$$

Note in Example 6 that $f(g(x)) \neq g(f(x))$, and in general this is the case. An important instance in which these two composite functions are equal occurs when

$$f(g(x)) = x = g(f(x))$$

We call such functions **inverses** of each other, as stated in the following definition.

| | |
|---|---|
| **Definition of inverse functions** | Two functions $f$ and $g$ are **inverses** of each other if $$f(g(x)) = x, \quad \text{for each } x \text{ in the domain of } g$$ and $\qquad\qquad g(f(x)) = x, \quad \text{for each } x \text{ in the domain of } f$ We denote $g$ by $f^{-1}$ (read "$f$ inverse"). |

■■■ **Remark:** For inverse functions $f$ and $g$, the range of $g$ must be equal to the domain of $f$, and vice versa.

Don't be confused by the "exponential notation" for inverse functions. Whenever we write $f^{-1}(x)$, we will *always* be referring to the inverse of the function $f$ and not to the reciprocal of $f$.

**EXAMPLE 7**
■■■■■■■

**Demonstrating Properties of Inverse Functions**

Show that the following functions are inverses of each other:

$$f(x) = 2x^3 - 1$$

and

$$g(x) = \sqrt[3]{\frac{x+1}{2}}$$

**SOLUTION**

First note that both composite functions exist, since the domain and range of both $f$ and $g$ consist of the set of all real numbers. The composite of $f$ and $g$ is given by

$$f(g(x)) = 2\left(\sqrt[3]{\frac{x+1}{2}}\right)^3 - 1$$

$$= 2\left(\frac{x+1}{2}\right) - 1 = x + 1 - 1 = x$$

The composite of $g$ with $f$ is given by

$$g(f(x)) = \sqrt[3]{\frac{(2x^3 - 1) + 1}{2}}$$

$$= \sqrt[3]{\frac{2x^3}{2}} = \sqrt[3]{x^3} = x$$

Since $f(g(x)) = x = g(f(x))$, we know that $f$ and $g$ are inverses of each other.

The graphs of the functions $f$ and $f^{-1}$ are mirror images of each other (with respect to the line $y = x$), as shown in Figure 1.50. This reflective property of inverse functions is formally stated as follows.

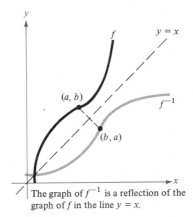

The graph of $f^{-1}$ is a reflection of the graph of $f$ in the line $y = x$.

FIGURE 1.50

| **Reflective property of inverse functions** | The graph of $f$ contains the point $(a, b)$ if and only if the graph of $f^{-1}$ contains the point $(b, a)$. |

This geometric property suggests an *algebraic* procedure for finding the inverse of a function. Since $(a, b)$ lies on the graph of $f$ if and only if $(b, a)$ lies on the graph of $f^{-1}$, we can find the inverse of a function by algebraically interchanging the roles of $x$ and $y$. This procedure is demonstrated in the next example.

**EXAMPLE 8**

**Finding the Inverse of a Function**

Find the inverse of the function given by $f(x) = \sqrt{2x - 3}$.

**SOLUTION**

Substituting $y$ for $f(x)$, we have $y = \sqrt{2x - 3}$. Now to find the inverse function, we simply solve for $x$ in terms of $y$. Since $y$ is nonnegative, squaring both sides gives an equivalent equation:

$$\sqrt{2x - 3} = y \qquad \text{Given } y \text{ as a function of } x$$
$$2x - 3 = y^2$$
$$2x = y^2 + 3$$
$$x = \frac{y^2 + 3}{2} \qquad \text{Solve for } x \text{ as a function of } y$$

Thus, the inverse function has the form

$$f^{-1}(\blacksquare) = \frac{(\blacksquare)^2 + 3}{2}$$

Using $x$ as the independent variable, we write

$$f^{-1}(x) = \frac{x^2 + 3}{2}, \quad x \geq 0$$

Note from the graphs of these two functions in Figure 1.51 that the domain of $f^{-1}$ coincides with the range of $f$.

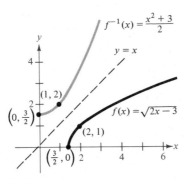

FIGURE 1.51

Not all functions possess an inverse. In fact, for a function $f$ to have an inverse it is necessary that $f$ be one-to-one. The next example is a case in point.

**EXAMPLE 9**

**A Function That Has No Inverse**

Find the inverse (if it exists) of the function given by

$$f(x) = x^2 - 1$$

(Assume the domain of $f$ is the set of all real numbers.)

**SOLUTION**

First we note that

$$f(2) = (2)^2 - 1 = 3$$

and $\qquad f(-2) = (-2)^2 - 1 = 3$

Thus, $f$ is not one-to-one, and it has no inverse (see Figure 1.52). This same conclusion can be obtained by substituting $y$ for $f(x)$ and solving for $x$ as follows:

$$x^2 - 1 = y$$

$$x^2 = y + 1$$

$$x = \pm\sqrt{y + 1}$$

This last equation does not define $x$ as a function of $y$, and thus $f$ has no inverse.

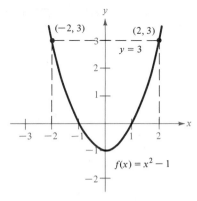

$$f(x) = x^2 - 1$$

$f$ is not one-to-one
and has no inverse.

FIGURE 1.52

## SECTION EXERCISES 1.4

1. Given $f(x) = 2x - 3$, find
   (a) $f(0)$      (b) $f(-3)$
   (c) $f(x - 1)$      (d) $f(1 + \Delta x)$
2. Given $f(x) = x^2 - 2x + 2$, find
   (a) $f\left(\dfrac{1}{2}\right)$      (b) $f(-1)$

   (c) $f(c)$      (d) $f(x + \Delta x)$
3. Given $f(x) = x^2$, find
   (a) $f(-2)$      (b) $f(6)$
   (c) $f(c)$      (d) $f(x + \Delta x)$

4. Given $f(x) = \dfrac{1}{x}$, find

   (a) $f(2)$      (b) $f\left(\dfrac{1}{4}\right)$

   (c) $f(x + \Delta x)$      (d) $f(x + \Delta x) - f(x)$
5. Given $f(x) = \dfrac{|x|}{x}$, find

   (a) $f(2)$      (b) $f(-2)$
   (c) $f(x^2)$      (d) $f(x - 1)$
6. Given $f(x) = |x| + 4$, find
   (a) $f(2)$      (b) $f(-2)$
   (c) $f(x^2)$      (d) $f(x + \Delta x) - f(x)$
7. Given $f(x) = 3x - 1$, find and simplify

   $$\frac{f(x + \Delta x) - f(x)}{\Delta x}$$

8. Given $f(x) = 5 - 4x$, find and simplify

   $$\frac{f(x + \Delta x) - f(x)}{\Delta x}$$

9. Given $f(x) = x^2 - x + 1$, find and simplify

   $$\frac{f(2 + \Delta x) - f(2)}{\Delta x}$$

10. Given $f(x) = x^3$, find and simplify

   $$\frac{f(x + \Delta x) - f(x)}{\Delta x}$$

11. Given $f(x) = x^3 - x$, find and simplify

   $$\frac{f(x) - f(1)}{x - 1}$$

12. Given $f(x) = 1/\sqrt{x - 1}$, find and simplify

   $$\frac{f(x) - f(2)}{x - 2}$$

In Exercises 13–22, find the domain and range of the given function. (Use interval notation.)

13. $f(x) = 4 - 2x$

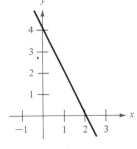

14. $f(x) = x^3$

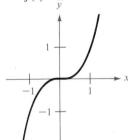

**15.** $f(x) = x^2$

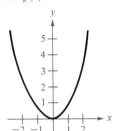

**16.** $f(x) = 4 - x^2$

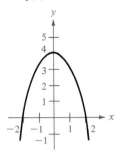

In Exercises 23–32, use the vertical line test to determine whether $y$ is a function of $x$.

**23.** $y = x^2$

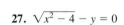

**24.** $y = x^3 - 1$

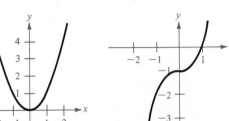

**17.** $f(x) = \sqrt{x - 1}$

**18.** $f(x) = \sqrt{1 - x}$

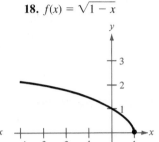

**25.** $x - y^2 = 0$

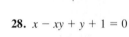

**26.** $x^2 + y^2 = 9$

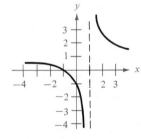

**19.** $f(x) = \sqrt{9 - x^2}$

**20.** $f(x) = \dfrac{1}{|x|}$

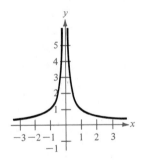

**27.** $\sqrt{x^2 - 4} - y = 0$

**28.** $x - xy + y + 1 = 0$

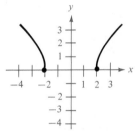

**29.** $x^2 = xy - 1$

**30.** $x = |y|$

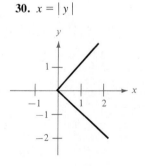

**21.** $f(x) = |x - 2|$

**22.** $f(x) = \dfrac{|x|}{x}$

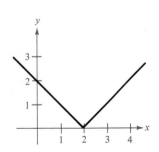

**31.** $x^2 - 4y^2 + 4 = 0$

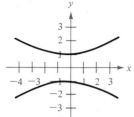

**32.** $y = |x - 3| + |x - 1|$

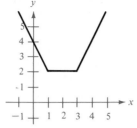

In Exercises 33–40, identify the equations that determine $y$ as a function of $x$.

**33.** $x^2 + y^2 = 4$

**34.** $3x - 2y + 5 = 0$

**35.** $2x + 3y = 4$

**36.** $x + y^2 = 4$

**37.** $x^2 + y = 4$

**38.** $x^2 + y^2 - 2x - 4y + 1 = 0$

**39.** $y^2 = x^2 - 1$

**40.** $x^2y - x^2 + 4y = 0$

In Exercises 41–48, find (a) $f(x) + g(x)$, (b) $f(x) \cdot g(x)$, (c) $f(x)/g(x)$, and if defined, find (d) $f(g(x))$ and (e) $g(f(x))$.

**41.** $f(x) = x + 1$; $g(x) = x - 1$

**42.** $f(x) = 2x - 5$; $g(x) = 1 - x$

**43.** $f(x) = x^2$; $g(x) = 1 - x$

**44.** $f(x) = 2x - 5$; $g(x) = 5$

**45.** $f(x) = x^2 + 5$; $g(x) = \sqrt{1 - x}$

**46.** $f(x) = \sqrt{x^2 - 4}$; $g(x) = \dfrac{x^2}{x^2 + 1}$

**47.** $f(x) = \dfrac{1}{x}$; $g(x) = \dfrac{1}{x^2}$

**48.** $f(x) = \dfrac{x}{x + 1}$; $g(x) = x^3$

In Exercises 49–54, (a) show that $f$ and $g$ are inverse functions by showing that $f(g(x)) = x$ and $g(f(x)) = x$, and (b) graph $f$ and $g$ on the same set of coordinate axes.

**49.** $f(x) = x^3$; $g(x) = \sqrt[3]{x}$

**50.** $f(x) = \dfrac{1}{x}$; $g(x) = \dfrac{1}{x}$

**51.** $f(x) = 5x + 1$; $g(x) = \dfrac{x - 1}{5}$

**52.** $f(x) = 9 - x^2$, $x \geq 0$; $g(x) = \sqrt{9 - x}$, $x \leq 9$

**53.** $f(x) = \sqrt{x - 4}$; $g(x) = x^2 + 4$, $x \geq 4$

**54.** $f(x) = 1 - x^3$; $g(x) = \sqrt[3]{1 - x}$

In Exercises 55–64, find the inverse of $f$. Then graph both $f$ and $f^{-1}$ on the same set of axes.

**55.** $f(x) = 2x - 3$

**56.** $f(x) = 6 - 3x$

**57.** $f(x) = x^5$

**58.** $f(x) = x^3 + 1$

**59.** $f(x) = \sqrt{x}$

**60.** $f(x) = x^2$, $x \geq 0$

**61.** $f(x) = \sqrt{4 - x^2}$, $0 \leq x \leq 2$

**62.** $f(x) = \sqrt{x^2 - 4}$, $x \geq 2$

**63.** $f(x) = x^{2/3}$, $x \geq 0$

**64.** $f(x) = x^{3/5}$

In Exercises 65–70, determine whether the given function is one-to-one, and if so, find its inverse.

**65.** $f(x) = ax + b$

**66.** $f(x) = \sqrt{x - 2}$

**67.** $f(x) = x^2$

**68.** $f(x) = x^4$

**69.** $f(x) = |x - 2|$

**70.** $f(x) = 3$

**71.** Use the graph of $f(x) = \sqrt{x}$ below to sketch the graph of each of the following.

(a) $y = \sqrt{x} + 2$

(b) $y = -\sqrt{x}$

(c) $y = \sqrt{x - 2}$

(d) $y = \sqrt{x + 3}$

(e) $y = \sqrt{x - 4}$

(f) $y = 2\sqrt{x}$

**72.** Use the accompanying graph of $f(x) = \sqrt[3]{x}$ to sketch the graph of each of the following.

(a) $y = \sqrt[3]{x} - 1$

(b) $y = \sqrt[3]{x} + 1$

(c) $y = \sqrt[3]{x - 1}$

(d) $y = -\sqrt[3]{x} - 2$

(e) $y = \dfrac{\sqrt[3]{x}}{2}$

(f) $y = \sqrt[3]{x + 1} - 1$

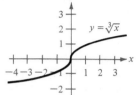

**73.** Use the given graph to write formulas for the functions whose graphs are shown in parts (a)–(d).

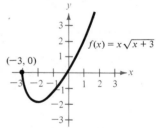

$f(x) = x\sqrt{x + 3}$

(a)

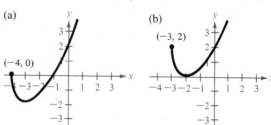

(b)

(c)

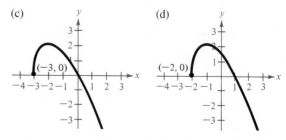

(d)

**74.** Use the given graph to write formulas for the functions whose graphs are shown in parts (a) through (d).

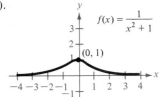
$$f(x) = \frac{1}{x^2 + 1}$$
(0, 1)

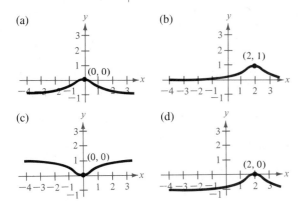

(a)

(0, 0)

(b)

(2, 1)

(c)

(0, 0)

(d)

(2, 0)

**75.** The rectangle shown in Figure 1.53 has a perimeter of 100 feet. Express the area $A$ of the rectangle as a function of $x$.

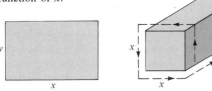

FIGURE 1.53    FIGURE 1.54

**76.** The rectangular package with square cross sections shown in Figure 1.54 has a combined length and girth (perimeter of a cross section) of 108 inches. Express the volume $V$ as a function of $x$.

**77.** Express the value $V$ of a farm having $500,000 worth of buildings, livestock, and equipment in terms of the number of acres $x$ on the farm if each acre is valued at $1,750.

**78.** The inventor of a new game believes that the variable cost for producing the game is $0.95 per unit, and the fixed costs are $6,000.
(a) Write the total cost $C$ as a function of the number of games sold $x$.
(b) Find a formula for the average cost per unit $\bar{C} = C/x$.
(c) The selling price for each game is $1.69. How many units must be sold before the average cost per unit falls below the selling price?

**79.** The demand function for a particular commodity is given by
$$p = \frac{14.75}{1 + 0.01x}, \quad x \geq 0$$
where $p$ is the price per unit and $x$ is the number of units sold.
(a) Find $x$ as a function of $p$.
(b) Use the result of part (a) to find the number of units sold when the price is $10.

**80.** The demand function for a particular product is given by
$$p = 289 - \frac{(x + 5)^2}{10,000}, \quad 0 \leq x \leq 1695$$
where $p$ is the price per unit and $x$ is the number of units sold.
(a) Find $x$ as a function of $p$.
(b) Use the result of part (a) to find the number of units sold when the price is $189.

**81.** A power station is on one side of a river that is one-half mile wide. A factory is 3 miles downstream on the other side of the river, as shown in Figure 1.55. It costs $10/ft to run the power lines on land and $15/ft to run them underwater. Write the cost $C$ of running the line from the power station to the factory as a function of $x$.

Power station

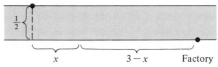

$\frac{1}{2}$

$x$    $3 - x$    Factory

FIGURE 1.55

**82.** A radio manufacturer charges $90 per unit for units that cost $60 to produce. To encourage large orders from distributors, the manufacturer will reduce the price by $0.01 per unit for each unit in excess of 100. (For example, an order of 200 units would have a price of $89 per unit.) This price reduction is discontinued when the price per unit drops to $75.
(a) Write the price per unit $p$ as a function of the order size $x$.
(b) Write the profit $P$ as a function of the order size $x$. [Note: $P = R - C = px - 60x$]

**83.** Assume that the amount of money deposited in a bank is proportional to the square of the interest rate the bank pays on the money. That is, $d = kr^2$, where $d$ is the total deposit, $r$ is the interest rate, and $k$ is the proportionality constant. Assuming the bank can reinvest the money for a return of 18%, write the bank's profit $P$ as a function of the interest rate $r$.

# Limits

### INTRODUCTORY EXAMPLE
### Compound Interest

There are many ways of computing the interest earned in a savings account. The most common of these methods involves *compound interest*—interest that is periodically added to the principal. Each time interest is added to the principal, the interest is said to be compounded. The result of compounding is that, starting with the second compounding, the account earns interest on interest in addition to earning interest on the principal. Compound interest rates are given as annual rates, no matter how many times the interest is compounded per year.

As an example of the effect of various compounding periods, consider an account paying 12% (annual percentage rate) on an initial deposit of $10,000. If $A(x)$ represents the balance after *one* year and $x$ represents the compounding period, then the formula for the balance is

$$A(x) = 10,000(1 + 0.12x)^{1/x}$$

Table 1.10 shows the balance after one year for various compounding periods. From this table, it appears that from an investor's point of view, it is better to have smaller compounding periods. For instance, one compounding period (simple interest) produces a balance of $11,200.00, but quarterly compounding produces a balance of $11,255.09—an increase of $55.09, as shown in Figure 1.56.

However, even though the balance increases as the compounding period approaches zero, it appears from the table that we cannot obtain a balance that is larger than $11,274.97. Mathematically, we describe this phenomenon by saying that the **limit** of the balance as $x$ approaches zero is $11,274.97,* and we write

$$\lim_{x \to 0} A(x) = 11,274.97$$

---

*This number is rounded to the nearest cent. The actual limit is $10,000e^{0.12} \approx 11,274.96852$.

TABLE 1.10 Balance After One Year

| Compounding period | Balance ($) |
|---|---|
| year: $x = 1$ | 11,200.00 |
| quarter: $x = \dfrac{1}{4}$ | 11,255.09 |
| month: $x = \dfrac{1}{12}$ | 11,268.25 |
| hour: $x = \dfrac{1}{8,760}$ | 11,274.96 |
| minute: $x = \dfrac{1}{525,600}$ | 11,274.97 |
| second: $x = \dfrac{1}{31,536,000}$ | 11,274.97 |

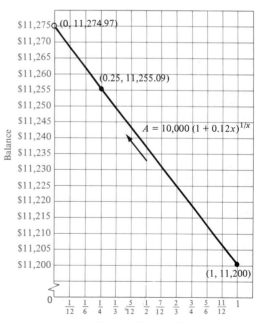

Compounding period in years

FIGURE 1.56

- **The Limit of a Function**
- **Properties of Limits**
- **Evaluation of Limits**
- **One-Sided Limits**
- **Unbounded Behavior**

The notion of a limit is fundamental to the study of calculus, and it is important for you to acquire a good working knowledge of limits before continuing your study of calculus.

Just what do we mean by the term "limit"? In everyday language we refer to the speed limit, a wrestler's weight limit, the limit of one's endurance, the limits of modern technology, or stretching a spring to its limits. These terms all suggest that a limit is a type of bound, which on some occasions may not be reached but on other occasions may be reached or even exceeded.

Consider an ideal spring that will break only if a weight of 10 or more pounds is attached to it. Suppose we want to determine how far the spring will stretch without breaking. We could conduct an experiment by increasing the weight attached to the spring and measuring the spring length $s$ at each weight, as shown in Figure 1.57. As the attached weight nears 10 pounds, we would need to use smaller and smaller increments so as not to reach the 10-pound maximum. By recording the successive spring lengths, we should be able to determine the value $L$ that $s$ approaches as the weight $w$ approaches 10 pounds. Symbolically, we write

$$s \rightarrow L \qquad \text{as} \qquad w \rightarrow 10 \quad (w < 10)$$

and we say that $L$ is the **limit** of the length $s$.

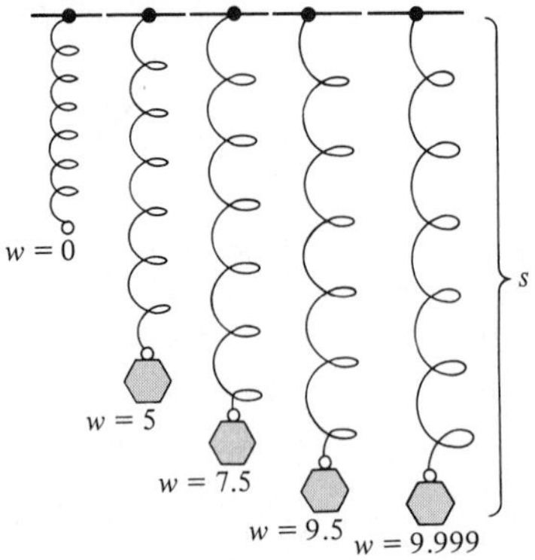

What is the limit of $s$ as $w$ approaches 10 lb?
FIGURE 1.57

## The Limit of a Function

A mathematical limit is much like the limit of a spring. For instance, suppose that we wanted to find the limit of the function given by

$$f(x) = \frac{x^3 - 1}{x - 1}$$

as $x$ approaches 1. Note that we are not interested in determining the value of $f(1)$ because $f$ is not defined at $x = 1$. What we seek, however, is the value that $f(x)$ approaches as $x$ approaches 1. To perform this experiment, we could program the function into a calculator and evaluate $f(x)$ for several values of $x$ near 1. In approaching 1 from the left, we could use the values

$$x = 0.5,\ 0.75,\ 0.9,\ 0.99,\ 0.999$$

and from the right, we could use

$$x = 1.5,\ 1.25,\ 1.1,\ 1.01,\ 1.001$$

Table 1.11 gives the values of $f(x)$ that correspond to these 10 different values of $x$.

TABLE 1.11

| | $x$ approaches 1 from the left | | | | | | $x$ approaches 1 from the right | | | | |
|---|---|---|---|---|---|---|---|---|---|---|---|
| $x$ | 0.5 | 0.75 | 0.9 | 0.99 | 0.999 | 1 | 1.001 | 1.01 | 1.1 | 1.25 | 1.5 |
| $f(x)$ | 1.750 | 2.313 | 2.710 | 2.970 | 2.997 | ? | 3.003 | 3.030 | 3.310 | 3.813 | 4.750 |
| | $f(x)$ approaches 3 | | | | | | $f(x)$ approaches 3 | | | | |

From Table 1.11, it seems reasonable to guess that as $x$ approaches 1, the limit of $f(x)$ is 3. We denote this limit by

$$f(x) \to 3 \qquad \text{as} \qquad x \to 1$$

or by the equation

$$\lim_{x \to 1} f(x) = 3$$

Even though $f(x)$ is undefined when $x = 1$, it appears from Table 1.11 that we can force $f(x)$ to be arbitrarily close to the value 3 by choosing values of $x$ sufficiently close to 1. This suggests the following definition.

**Definition of limit**  If $f(x)$ becomes arbitrarily close to a single number $L$ as $x$ approaches $c$ from either side, then we write

$$\lim_{x \to c} f(x) = L$$

and say that the **limit** of $f(x)$ as $x$ approaches $c$ is $L$.

■■■ **Remark:** The phrase "$x$ approaches $c$" means that no matter how close $x$ comes to the value $c$, there is another value of $x$ (different from $c$) in the domain of $f$ that is even closer to $c$.

Inherent in the definition of a limit is the assumption that a function cannot approach two different limits at the same time. This means that *if the limit of a function exists, it is unique.* Moreover, throughout this text, whenever we write

$$\lim_{x \to c} f(x) = L$$

we imply two statements—the limit exists *and* the limit is $L$.

### Properties of Limits

It is *incorrect* to think of a limit as a number that the values of a function get close to but can never reach. Many times the limit of $f(x)$ as $x$ approaches $c$ is simply $f(c)$. Whenever the limit of $f(x)$ as $x$ approaches $c$ is

$$\lim_{x \to c} f(x) = f(c)$$

Direct substitution

we say that the limit can be evaluated by **direct substitution.** It is important that you learn to recognize the types of functions that have this property. Some basic functions that have this property are given in the following list.

---

**Properties of limits**    Let $b$ and $c$ be real numbers, and let $n$ be a positive integer.

1. $\lim\limits_{x \to c} b = b$ 
2. $\lim\limits_{x \to c} x = c$
3. $\lim\limits_{x \to c} x^n = c^n$ 
4. $\lim\limits_{x \to c} \sqrt[n]{x} = \sqrt[n]{c}$

---

■■■ **Remark:** In Property 4, if $n$ is even we require $c > 0$.

**EXAMPLE 1**

**Evaluating a Limit**

Find the following limits:

(a) $\lim\limits_{x \to 2} x^2$ 
(b) $\lim\limits_{x \to 2} 3$

**SOLUTION**

(a) Using the properties of limits, we can evaluate this limit by direct substitution. That is,

$$\lim_{x \to 2} x^2 = 2^2 = 4$$

(b) In this case, the constant function $f(x) = 3$ is unaffected by the value of $x$ and the limit is simply

$$\lim_{x \to 2} 3 = 3$$

By combining the properties of limits with the following rules for operating with limits, we can find limits for a wide variety of algebraic functions.

---

| **Operations with limits** | Let $b$ and $c$ be real numbers and let $n$ be a positive integer. If the limits of $f(x)$ and $g(x)$ exist as $x$ approaches $c$, then the following operations are valid. |
|---|---|

1. Constant multiple: $\lim_{x \to c} [bf(x)] = b[\lim_{x \to c} f(x)]$

2. Addition: $\lim_{x \to c} [f(x) \pm g(x)] = \lim_{x \to c} f(x) \pm \lim_{x \to c} g(x)$

3. Multiplication: $\lim_{x \to c} [f(x)g(x)] = [\lim_{x \to c} f(x)][\lim_{x \to c} g(x)]$

4. Division: If $\lim_{x \to c} g(x) \neq 0$, $\lim_{x \to c} \dfrac{f(x)}{g(x)} = \dfrac{\lim_{x \to c} f(x)}{\lim_{x \to c} g(x)}$

5. Power: $\lim_{x \to c} [f(x)]^n = [\lim_{x \to c} f(x)]^n$

6. Radical*: $\lim_{x \to c} \sqrt[n]{f(x)} = \sqrt[n]{\lim_{x \to c} f(x)}$

**EXAMPLE 2**

**Evaluating a Limit**

Find the limit:

$$\lim_{x \to 2} (x^2 + 2x - 3)$$

**SOLUTION**

$$\lim_{x \to 2} (x^2 + 2x - 3) = \lim_{x \to 2} x^2 + \lim_{x \to 2} 2x - \lim_{x \to 2} 3$$

$$= 4 + 4 - 3 = 5$$

Example 2 is an illustration of the following important result. That is, the limit of a polynomial can be evaluated by direct substitution.

| **The limit of a polynomial function** | If $p$ is a polynomial function and $c$ is any real number, then $$\lim_{x \to c} p(x) = p(c)$$ |
|---|---|

### Techniques for Evaluating Limits

Now that we know that the limit of a polynomial function can be found by direct substitution, we look at some techniques for reducing other limits to this form. These techniques are based on the following important theorem. Basically, the theorem tells us that if two functions agree at all but a single point $c$, then they must have identical limit behavior at $x = c$.

---

*If $n$ is even and $\lim_{x \to c} f(x) = L$, then we require $L > 0$.

To apply the Replacement Theorem, we use a result from algebra that states that for a polynomial function $p$, $p(c) = 0$ if and only if $(x - c)$ is a factor of $p(x)$. For instance, to find the limit

$$\lim_{x \to 1} \frac{x^3 - 1}{x - 1}$$

we note that both the numerator and the denominator have a value of zero when $x = 1$. This implies that $(x - 1)$ is a factor of both the numerator and the denominator and we can cancel this common factor to obtain

$$\frac{x^3 - 1}{x - 1} = \frac{(x - 1)(x^2 + x + 1)}{(x - 1)}$$

$$= x^2 + x + 1, \quad x \neq 1$$

Thus, the rational function $(x^3 - 1)/(x - 1)$ and the polynomial function $x^2 + x + 1$ agree for all values of $x$ other than $x = 1$, and we can apply the Replacement Theorem to obtain

$$\lim_{x \to 1} \frac{x^3 - 1}{x - 1} = \lim_{x \to 1} (x^2 + x + 1)$$

$$= (1^2 + 1 + 1)$$

$$= 3$$

Figure 1.58 illustrates this result graphically. Note that the two graphs are identical except that the graph of $g$ contains the point $(1, 3)$ whereas this point is missing on the graph of $f$. (We denote this missing point by an open dot.)

This **cancellation technique** for evaluating a limit is further demonstrated in the following example.

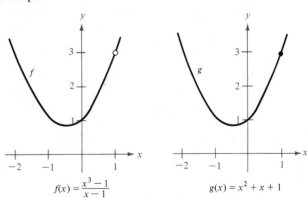

FIGURE 1.58

$$f(x) = \frac{x^3 - 1}{x - 1}$$

$$g(x) = x^2 + x + 1$$

**EXAMPLE 3**

**Cancellation Technique**

Find the limit:

$$\lim_{x \to -3} \frac{x^2 + x - 6}{x + 3}$$

**SOLUTION**

Direct substitution fails to give the limit since both the numerator and the denominator are zero when $x = -3$.

$$\lim_{x \to -3} \frac{x^2 + x - 6}{x + 3} \qquad \begin{array}{l} \lim_{x \to -3} (x^2 + x - 6) = 0 \\ \lim_{x \to -3} (x + 3) = 0 \end{array}$$

However, since the limits of both the numerator and the denominator are zero, we know that they have a *common factor* of $(x + 3)$. Thus, for all $x \neq -3$, we can cancel this factor to obtain

$$\frac{x^2 + x - 6}{x + 3} = \frac{(x + 3)(x - 2)}{x + 3} = x - 2$$

Since the polynomial $x - 2$ agrees with the original rational function for every $x$ other than $x = -3$, we can apply the Replacement Theorem to conclude that

$$\lim_{x \to -3} \frac{x^2 + x - 6}{x + 3} = \lim_{x \to -3} (x - 2) = -5$$

This result is shown graphically in Figure 1.59.

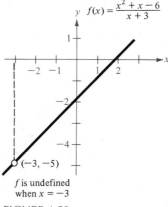

$y$ \quad $f(x) = \dfrac{x^2 + x - 6}{x + 3}$

$(-3, -5)$

$f$ is undefined
when $x = -3$

FIGURE 1.59

**EXAMPLE 4**

**Evaluating Limits**

Find the following limits:

(a) $\displaystyle \lim_{x \to -2} \frac{x^2 + x - 2}{x - 2}$

(b) $\displaystyle \lim_{x \to 0} \frac{(x + 2)^2 - 4}{x}$

**SOLUTION**

(a) Since the limit of the denominator is not zero, we can use the limit of a quotient (Property 4 of Operations with Limits) to obtain

$$\lim_{x \to -2} \frac{x^2 + x - 2}{x - 2} = \frac{4 - 2 - 2}{-2 - 2} = \frac{0}{-4} = 0$$

(b) Since the limits of both the numerator and the denominator are zero, we use cancellation to obtain

$$\lim_{x \to 0} \frac{x^2 + 4x + 4 - 4}{x} = \lim_{x \to 0} \frac{x(x + 4)}{x} = \lim_{x \to 0} (x + 4) = 4$$

In Example 3 and Example 4(b), direct substitution produced the meaningless fractional form 0/0. We call such an expression an **indeterminate form** since we cannot (from the form alone) determine the limit. When you are trying to evaluate a limit and encounter this form, remember that you must change the fraction so that the new denominator does not have zero as its limit. One way to do this is to *cancel like factors* (as in Example 3). A second way is to *rationalize the numerator* as shown in the following example.

**EXAMPLE 5**

**Evaluating a Limit by Rationalizing the Numerator**

Find the limit:

$$\lim_{x \to 0} \frac{\sqrt{x + 1} - 1}{x}$$

**SOLUTION**

By direct substitution we get the indeterminate form 0/0:

$$\lim_{x \to 0} \frac{\sqrt{x + 1} - 1}{x} \qquad \Longleftarrow \quad \lim_{x \to 0} (\sqrt{x + 1} - 1) = 0$$
$$\Longleftarrow \quad \lim_{x \to 0} x = 0$$

In this case, we change the form of the fraction by rationalizing the numerator:

$$\frac{\sqrt{x + 1} - 1}{x} = \left( \frac{\sqrt{x + 1} - 1}{x} \right) \left( \frac{\sqrt{x + 1} + 1}{\sqrt{x + 1} + 1} \right) = \frac{(x + 1) - 1}{x(\sqrt{x + 1} + 1)}$$

$$= \frac{x}{x(\sqrt{x + 1} + 1)} = \frac{1}{\sqrt{x + 1} + 1}$$

Therefore,

$$\lim_{x \to 0} \frac{\sqrt{x+1}-1}{x} = \lim_{x \to 0} \frac{1}{\sqrt{x+1}+1} = \frac{1}{1+1} = \frac{1}{2}$$

Table 1.12 reinforces our conclusion that this limit is $\frac{1}{2}$.

TABLE 1.12

| | x approaches 0 from the left | | | | x approaches 0 from the right | | | |
|---|---|---|---|---|---|---|---|---|
| x | −0.25 | −0.1 | −0.01 | −0.001 | 0 | 0.001 | 0.01 | 0.1 | 0.25 |
| f(x) | 0.5359 | 0.5132 | 0.5013 | 0.5001 | ? | 0.4999 | 0.4988 | 0.4881 | 0.4721 |

| | f(x) approaches $\frac{1}{2}$ | | | f(x) approaches $\frac{1}{2}$ | |
|---|---|---|---|---|---|

**Remark:** For a review of the algebraic aspects of rationalization, see Section 0.5.

## One-Sided Limits

In our original example of a spring, we could test for the limit of the length of the spring only by using weights that are less than 10 pounds. However, the definition of the limit

$$\lim_{x \to c} f(x) = L$$

requires that we let $x$ approach $c$ from the right as well as from the left. To denote these two possible approaches to $c$, we use the symbols

$$x \to c^-$$

read "$x$ approaches $c$ from the left," and

$$x \to c^+$$

read "$x$ approaches $c$ from the right." Because of the importance of the "limits" obtained from these one-sided approaches, we give each a special name.

**One-sided limits**

1. $\lim_{x \to c^-} f(x)$ is called the **limit from the left.**

2. $\lim_{x \to c^+} f(x)$ is called the **limit from the right.**

**EXAMPLE 6**

**Evaluating One-Sided Limits**

Find the limit as $x \to 1$ from the left and the limit as $x \to 1$ from the right for the function given by

$$f(x) = \frac{|2x-2|}{x-1}$$

**SOLUTION**    From the graph of $f$, as shown in Figure 1.60, we see that $f(x) = -2$ for all $x < 1$. Therefore, the limit from the left is

$$\lim_{x \to 1^-} \frac{|2x - 2|}{x - 1} = -2$$

Moreover, since $f(x) = 2$ for all $x > 1$, the limit from the right is

$$\lim_{x \to 1^+} \frac{|2x - 2|}{x - 1} = 2$$

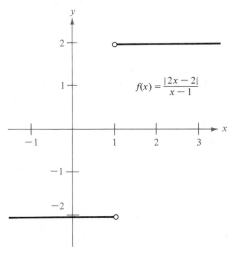

FIGURE 1.60

In Example 6 note that the function approaches different limits from the left and from the right. In this case, we say that the limit of $f(x)$ as $x \to 1$ does not exist. In order for the limit of a function to exist as $x \to c$, it must be true that *both* one-sided limits exist and are equal.

**Existence of a limit**    If $f$ is a function and $c$ and $L$ are real numbers, then

$$\lim_{x \to c} f(x) = L$$

if and only if

$$\lim_{x \to c^-} f(x) = L \quad \text{and} \quad \lim_{x \to c^+} f(x) = L$$

One-sided limits of polynomial functions can also be evaluated by direct substitution, and we use this property of polynomial functions in the following example.

**EXAMPLE 7**    **Evaluating One-Sided Limits**

For the function $f(x) = x^2 + 3$, find the following limits:

(a) $\lim_{x \to 1^-} f(x)$           (b) $\lim_{x \to 1^+} f(x)$           (c) $\lim_{x \to 1} f(x)$

**SOLUTION**

(a) Since $f$ is a polynomial function, we can use direct substitution to determine that

$$\lim_{x \to 1^-} (x^2 + 3) = 1^2 + 3 = 4$$

(b) Similarly, the limit from the right is given by

$$\lim_{x \to 1^+} (x^2 + 3) = 1^2 + 3 = 4$$

(c) Since $f(x)$ approaches the same limit from the left and from the right, we conclude that

$$\lim_{x \to 1} (x^2 + 3) = 4$$

**EXAMPLE 8**

**Evaluating One-Sided Limits**

Find the limit of $f(x)$ as $x$ approaches 1:

$$f(x) = \begin{cases} 4 - x, & \text{for } x < 1 \\ 4x - x^2, & \text{for } x > 1 \end{cases}$$

**SOLUTION**

Remember we are concerned about the value of $f$ near $x = 1$ rather than at $x = 1$. Thus, for $x < 1$, $f(x)$ is given by the polynomial $4 - x$, and we can use direct substitution to obtain

$$\lim_{x \to 1^-} f(x) = \lim_{x \to 1^-} (4 - x) = 4 - 1 = 3$$

For $x > 1$, $f(x)$ is given by the polynomial function $4x - x^2$, and we can use direct substitution to obtain

$$\lim_{x \to 1^+} f(x) = \lim_{x \to 1^+} (4x - x^2) = 4 - 1 = 3$$

Since the one-sided limits both exist and are equal to 3, we have

$$\lim_{x \to 1} f(x) = 3$$

A sketch further illustrates our result (see Figure 1.61).

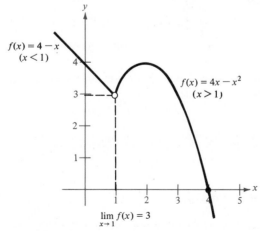

FIGURE 1.61

In the next example, we use one-sided limits to show that the limit of the "postage stamp" (or step) function does not exist at a point where the functional values abruptly change.

**EXAMPLE 9**

**Limits That Differ from the Right and the Left**

Assume the cost of sending first-class mail is 22¢ for the first ounce and 18¢ for each additional ounce. Letting $x$ represent the weight of a letter and $f(x)$ the cost of postage, we have

$$f(x) = 22 + 18n, \quad n < x \le n + 1 \qquad (n = 0, 1, 2, 3, \ldots)$$

Show that the limit of $f(x)$ as $x \to 2$ does not exist.

**SOLUTION**

The graph of $f$ is shown in Figure 1.62. Observe that

$$f(x) = \begin{cases} 40, & 1 < x \le 2 \\ 58, & 2 < x \le 3 \end{cases}$$

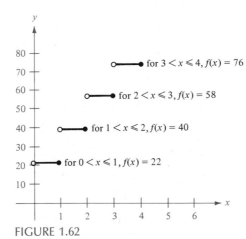

FIGURE 1.62

Thus the limit of $f(x)$ as $x$ approaches 2 from the left is

$$\lim_{x \to 2^-} f(x) = 40$$

whereas the limit of $f(x)$ as $x$ approaches 2 from the right is

$$\lim_{x \to 2^+} f(x) = 58$$

Since these one-sided limits are not equal, the limit of $f(x)$ as $x \to 2$ *does not exist.*

**Unbounded Behavior**

In Example 9 we looked at a limit that failed to exist at a point because the limits from the right and left differ. We close this section by looking at another important way in which a limit can fail to exist. It may happen that, as $x \rightarrow c$, $f(x)$ will increase (or decrease) without bound. Our last example describes such a situation.

**EXAMPLE 10**

**An Unbounded Function**

Evaluate (if possible) the limit

$$\lim_{x \to 2} \frac{3}{x - 2}$$

**SOLUTION**

From Figure 1.63 and Table 1.13, we can see that $f(x)$ *decreases without bound* as $x$ approaches 2 from the left and $f(x)$ *increases without bound* as $x$ approaches 2 from the right. Symbolically, we can write this as

$$\lim_{x \to 2^-} \frac{3}{x - 2} = -\infty$$

and

$$\lim_{x \to 2^+} \frac{3}{x - 2} = \infty$$

Since $f$ is unbounded as $x$ approaches 2, we conclude that the limit *does not exist*.

TABLE 1.13

| | x approaches 2 from the left | | | | | | x approaches 2 from the right | | | |
|---|---|---|---|---|---|---|---|---|---|---|
| x | 1 | 1.5 | 1.9 | 1.99 | 1.999 | 2 | 2.001 | 2.01 | 2.1 | 2.5 | 3.0 |
| f(x) | −3 | −6 | −30 | −300 | −3000 | ? | 3000 | 300 | 30 | 6 | 3 |
| | f(x) decreases without bound | | | | | | f(x) increases without bound | | | |

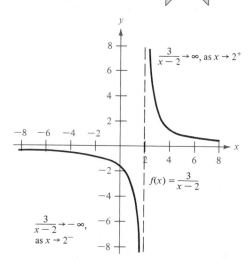

FIGURE 1.63

**Remark:** The equals sign in the statement

$$\lim_{x \to c^+} f(x) = \infty$$

does not mean that the limit exists! On the contrary, it tells us how the limit *fails to exist* by denoting the unbounded behavior of $f(x)$ as $x$ approaches $c$. (Remember that the symbol $\infty$ does *not* denote a number but is used to describe the situation in which the value of the function is becoming large without bound.)

## SECTION EXERCISES 1.5

In Exercises 1–4, use the graph to visually determine the indicated limit.

**1.**

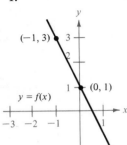

(a) $\lim_{x \to 0} f(x)$

(b) $\lim_{x \to -1} f(x)$

**2.**

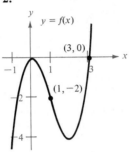

(a) $\lim_{x \to 1} f(x)$

(b) $\lim_{x \to 3} f(x)$

**3.**

(a) $\lim_{x \to 0} g(x)$

(b) $\lim_{x \to -1} g(x)$

**4.**

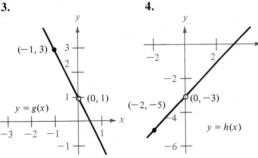

(a) $\lim_{x \to -2} h(x)$

(b) $\lim_{x \to 0} h(x)$

In Exercises 5–10, use the graph to visually determine the indicated limit, if it exists.

(a) $\lim_{x \to c^+} f(x)$     (b) $\lim_{x \to c^-} f(x)$     (c) $\lim_{x \to c} f(x)$

**5.**

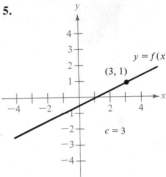

**6.**

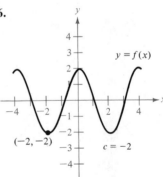

**7.**

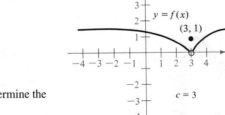

**8.**

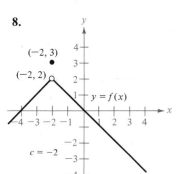

**9.**

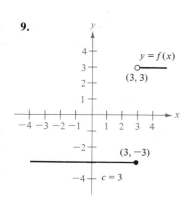

**10.**

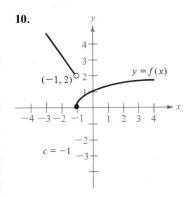

In Exercises 11–20, find the indicated limit.

**11.** $\lim_{x \to 2} x^2$

**12.** $\lim_{x \to -3} (3x + 2)$

**13.** $\lim_{x \to 0} (2x - 1)$

**14.** $\lim_{x \to 1} (1 - x^2)$

**15.** $\lim_{x \to 2} (-x^2 + x - 2)$

**16.** $\lim_{x \to 1} (3x^3 - 2x^2 + 4)$

**17.** $\lim_{x \to 3} \sqrt{x + 1}$

**18.** $\lim_{x \to 4} \sqrt[3]{x + 4}$

**19.** $\lim_{x \to 2} \frac{1}{x}$

**20.** $\lim_{x \to -3} \frac{2}{x + 2}$

In Exercises 21 and 22, use the given limits to find

(a) $\lim_{x \to c} [f(x) + g(x)]$    (b) $\lim_{x \to c} [f(x)g(x)]$

(c) $\lim_{x \to c} \frac{f(x)}{g(x)}$

**21.** $\lim_{x \to c} f(x) = 2, \ \lim_{x \to c} g(x) = 3$

**22.** $\lim_{x \to c} f(x) = \frac{3}{2}, \ \lim_{x \to c} g(x) = \frac{1}{2}$

In Exercises 23–54, find the limit (if it exists).

**23.** $\lim_{x \to -1} \frac{x^2 - 1}{x + 1}$

**24.** $\lim_{x \to -1} \frac{2x^2 - x - 3}{x + 1}$

**25.** $\lim_{x \to 3} \frac{x - 3}{x^2 - 9}$

**26.** $\lim_{x \to 2} \frac{2 - x}{x^2 - 4}$

**27.** $\lim_{t \to 5} \frac{t - 5}{t^2 - 25}$

**28.** $\lim_{t \to 1} \frac{t^2 + t - 2}{t^2 - 1}$

**29.** $\lim_{x \to -2} \frac{x^3 + 8}{x + 2}$

**30.** $\lim_{x \to -1} \frac{x^3 - 1}{x - 1}$

**31.** $\lim_{x \to 1^-} \frac{x}{x^2 - x}$

**32.** $\lim_{x \to 1^+} \frac{5}{1 - x}$

**33.** $\lim_{x \to -2^-} \frac{1}{x + 2}$

**34.** $\lim_{x \to 0^-} \frac{x + 1}{x}$

**35.** $\lim_{x \to 5} \frac{2}{(x - 5)^2}$

**36.** $\lim_{x \to 3^+} \frac{x}{x^2 - 2x - 3}$

**37.** $\lim_{x \to 0} \frac{|x|}{x}$

**38.** $\lim_{x \to 2} \frac{|x - 2|}{x - 2}$

**39.** $\lim_{x \to 3} f(x)$, where $f(x) = \begin{cases} \dfrac{x + 2}{2}, & x \le 3 \\ \dfrac{12 - 2x}{3}, & x > 3 \end{cases}$

**40.** $\lim_{s \to 1} f(s)$, where $f(s) = \begin{cases} s, & s \le 1 \\ 1 - s, & s > 1 \end{cases}$

**41.** $\lim_{x \to 1} f(x)$, where $f(x) = \begin{cases} x^3 + 1, & x < 1 \\ x + 1, & x \ge 1 \end{cases}$

**42.** $\lim_{x \to 2} f(x)$, where $f(x) = \begin{cases} x^2 - 4x + 6, & x < 2 \\ -x^2 + 4x - 2, & x \ge 2 \end{cases}$

**43.** $\lim_{\Delta x \to 0} \frac{2(x + \Delta x) - 2x}{\Delta x}$

**44.** $\lim_{\Delta x \to 0} \frac{5(x + \Delta x) - 3 - (5x - 3)}{\Delta x}$

**45.** $\lim_{\Delta t \to 0} \frac{(t + \Delta t)^2 + 3 - (t^2 + 3)}{\Delta t}$

**46.** $\lim_{\Delta x \to 0} \frac{(1 + \Delta x)^3 - 1}{\Delta x}$

**47.** $\lim_{\Delta x \to 0} \frac{(x + \Delta x)^2 - 2(x + \Delta x) + 1 - (x^2 - 2x + 1)}{\Delta x}$

**48.** $\lim\limits_{\Delta x \to 0} \dfrac{(x + \Delta x)^3 - x^3}{\Delta x}$

**49.** $\lim\limits_{x \to 0} \dfrac{\sqrt{3 + x} - \sqrt{3}}{x}$

**50.** $\lim\limits_{x \to 0} \dfrac{\sqrt{2 + x} - \sqrt{2}}{x}$

**51.** $\lim\limits_{x \to 0} \dfrac{[1/(2 + x)] - (1/2)}{x}$

**52.** $\lim\limits_{x \to 0} \dfrac{[1/(x + 4)] - (1/4)}{x}$

**53.** $\lim\limits_{x \to 4} \dfrac{\sqrt{x} - 2}{x - 4}$

**54.** $\lim\limits_{x \to 1} \dfrac{(1/\sqrt{x}) - 1}{x - 1}$

In Exercises 55–62, complete a table to estimate the given limit.

**55.** $\lim\limits_{x \to 2} (5x + 4)$

**56.** $\lim\limits_{x \to 2} \dfrac{x - 2}{x^2 - x - 2}$

**57.** $\lim\limits_{x \to 2} \dfrac{x - 2}{x^2 - 4}$

**58.** $\lim\limits_{x \to 0} \dfrac{\sqrt{x + 3} - \sqrt{3}}{x}$

**59.** $\lim\limits_{x \to 0} \dfrac{\sqrt{x + 2} - \sqrt{2}}{x}$

**60.** $\lim\limits_{x \to 2^-} \dfrac{2 - x}{\sqrt{4 - x^2}}$

**61.** $\lim\limits_{x \to 0} \dfrac{[1/(2 + x)] - (1/2)}{2x}$

**62.** $\lim\limits_{x \to 2} \dfrac{x^5 - 32}{x - 2}$

**63.** Find $\lim\limits_{x \to 0} f(x)$, given

$$4 - x^2 \le f(x) \le 4 + x^2, \text{ for all } x$$

**64.** The following limit will be introduced in Section 5.1:

$$\lim\limits_{x \to 0} (1 + x)^{1/x} = e \approx 2.718$$

Show the reasonableness of this limit by completing the following table.

| $x$ | $-0.1$ | $-0.01$ | $-0.001$ | 0.001 | 0.01 | 0.1 |
|---|---|---|---|---|---|---|
| $(1 + x)^{1/x}$ | | | | | | |

# SECTION 1.6

# Continuity

## INTRODUCTORY EXAMPLE
### Continuously Compounded Interest

In the Introductory Example to Section 1.5 we saw that the balance in a savings account tends to increase as the number of compoundings per year increases. For instance, monthly compounding produces a higher balance than does quarterly compounding. Figure 1.64 shows the balance in an account paying 12% compounded quarterly on an initial deposit of $10,000. Note that a *full* quarter must pass before interest is added to the account. For any time $t$ such that $0 \leq t < \frac{1}{4}$, the balance is $10,000. Then, at the end of the first quarter, the balance jumps by $300, and we say that the graph is **discontinuous** at the point $(\frac{1}{4}, 10,300)$. Other discontinuities occur at the end of the second, third, and fourth quarters.

In Chapter 5 we will see that the maximum balance that can be obtained by increasing the number of compoundings per year is given by a procedure called *continuous compounding*. Continuous compounding is often compared to compounding every second. However, the distinction between these two types of compounding is comparable to the difference between a digital watch and a watch with a second hand. The digital watch registers abrupt changes from one second to the next, whereas the watch with the second hand continuously registers changes in time. Figure 1.65 shows the balance in an account that is compounded continuously. Note that the graph has no jumps or breaks. In mathematical terms, we say that the graph in Figure 1.65 is **continuous.**

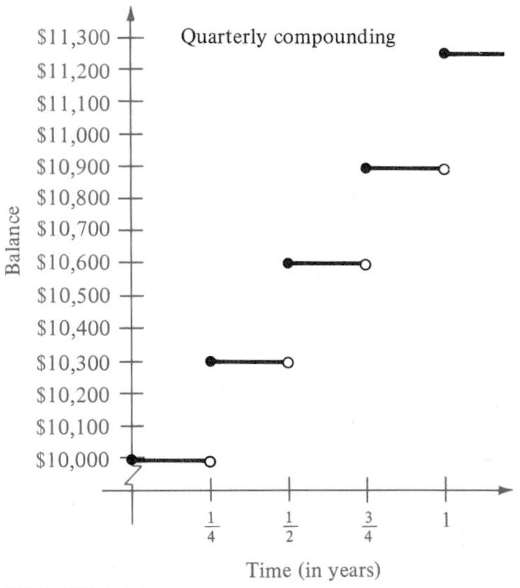

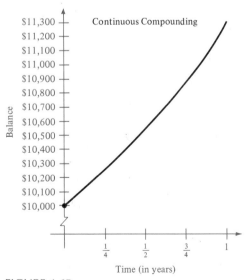

FIGURE 1.64

FIGURE 1.65

■ **Continuity**
■ **Discontinuity**
■ **Properties of Continuous Functions**

In mathematics the term "continuous" has much the same meaning as it does in everyday usage. To say that a function is continuous at $x = c$ means that there is no interruption in the graph of $f$ at $c$. Its graph is unbroken at $c$, and there are no holes, jumps, or gaps. Roughly speaking, we say that a function is continuous (on an interval) if its graph (on the interval) can be traced with pencil and paper without lifting the pencil from the paper. As simple as this concept may seem initially, its precise definition eluded mathematicians for many years. In fact, it was not until the early 1800s that a careful definition was finally developed.

Before looking at this definition, let's consider the function whose graph is shown in Figure 1.66. This figure identifies three values of $x$ at which the graph of $f$ is not continuous. At all other points ($x \neq c_1, c_2, c_3$) in the interval $(a, b)$, the graph of $f$ is uninterrupted, and we say it is **continuous** at such points. Where the graph is **discontinuous,** we observe the following:

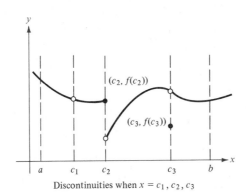

Discontinuities when $x = c_1, c_2, c_3$

FIGURE 1.66

1. At $x = c_1$, $f(c_1)$ is not defined.
2. At $x = c_2$, $\lim_{x \to c_2} f(x)$ does not exist.
3. At $x = c_3$, $f(c_3) \neq \lim_{x \to c_3} f(x)$.

Thus, it appears that continuity at $x = c$ is destroyed by any one of these conditions. This brings us to the following definition.

| **Definition of continuity** | ***Continuity at a Point:*** A function is said to be **continuous** at $c$ if the following three conditions are met: |
| --- | --- |

    1. $f(c)$ is defined.    2. $\lim_{x \to c} f(x)$ exists.    3. $\lim_{x \to c} f(x) = f(c)$

***Continuity on an Open Interval:*** A function is said to be **continuous on an interval** $(a, b)$ if it is continuous at each point in the interval.

A function is said to be **discontinuous** at $c$ if it is defined on an open interval containing $c$ (except possibly at $c$) and $f$ is not continuous at $c$.

In Section 1.5 we studied several types of functions that meet the three conditions for continuity. Specifically, if *direct substitution* can be used to evaluate the limit of a function at a point $c$, then the function is continuous at $c$. Two types of functions that have this property are polynomial functions and rational functions.

**Continuity of polynomial and rational functions**

1. A polynomial function is continuous at every real number.
2. A rational function is continuous at every real number *in its domain*.

**EXAMPLE 1**

**Continuity of a Polynomial Function**

Discuss the continuity of the following functions:
(a) $f(x) = x^2 - 2x + 3$  (b) $f(x) = x^3 - x$
(c) $f(x) = x^4 - 3x^2 + 2$

**SOLUTION**

Each of these functions is a *polynomial* function. Therefore each one is continuous on the entire real line, as shown in Figure 1.67.

(a)

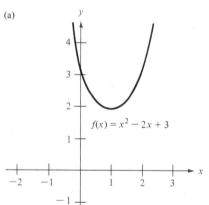

$f(x) = x^2 - 2x + 3$

Continuous on $(-\infty, \infty)$

(b)

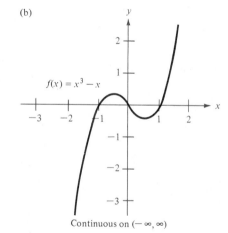

$f(x) = x^3 - x$

Continuous on $(-\infty, \infty)$

(c) $f(x) = x^4 - 3x^2 + 2$

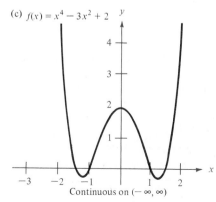

Continuous on $(-\infty, \infty)$

FIGURE 1.67

Polynomial functions are one of the most important types of functions used in calculus. Be sure you see from this example that the graph of a polynomial function is continuous and has no holes, breaks, or jumps on the entire real line.

On the other hand, the graph of a rational function can have holes or jumps, as illustrated in the following example.

**EXAMPLE 2**

**Continuity of a Rational Function**

Discuss the continuity of the following functions:

(a) $f(x) = \dfrac{1}{x}$ 　　　　(b) $f(x) = \dfrac{x^2 - 1}{x - 1}$ 　　　　(c) $f(x) = \dfrac{1}{x^2 + 1}$

**SOLUTION**

Each of these functions is rational and is therefore continuous at every point in its domain. (See Figure 1.68.)

(a) The domain of $f(x) = 1/x$ consists of all real numbers other than $x = 0$. Therefore, this rational function is continuous on the intervals $(-\infty, 0)$ and $(0, \infty)$.

(b) The domain of $f(x) = (x^2 - 1)/(x - 1)$ consists of all real numbers other than $x = 1$. Therefore, this rational function is continuous on the intervals $(-\infty, 1)$ and $(1, \infty)$.

(c) The domain of $f(x) = 1/(x^2 + 1)$ is the entire real line. Therefore, this rational function is continuous on the entire real line.

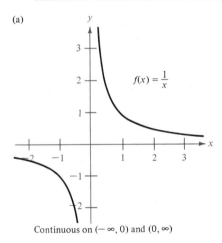

(a)

$f(x) = \dfrac{1}{x}$

Continuous on $(-\infty, 0)$ and $(0, \infty)$

(b)

$(1, 2)$

$f(x) = \dfrac{x^2 - 1}{x - 1}$

Continuous on $(-\infty, 1)$ and $(1, \infty)$

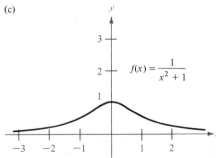

(c)

$f(x) = \dfrac{1}{x^2 + 1}$

Continuous on $(-\infty, \infty)$

FIGURE 1.68

Discontinuities fall into two categories: **removable** and **nonremovable.** A discontinuity at $x = c$ is called removable if $f$ can be made continuous by defining $f$ at $x = c$ so as to "plug up the hole" in the graph of $f$. For instance, in Figure 1.68(b) the discontinuity at $x = 1$ can be removed by defining $f(1) = 2$. The discontinuity at $x = 0$ in Figure 1.68(a) is nonremovable since there is no value we can assign to $f(0)$ to make the graph continuous. The particular type of nonremovable discontinuity shown in Figure 1.68(a) is called an **infinite discontinuity.**

### Continuity on a Closed Interval

The intervals discussed in Examples 1 and 2 are open. To discuss continuity on a closed interval, we use the concept of one-sided limits, as defined in the previous section.

| | |
|---|---|
| **Definition of continuity on a closed interval** | Let $f$ be defined on a closed interval $[a, b]$. If $f$ is continuous on the open interval $(a, b)$ and $$\lim_{x \to a^+} f(x) = f(a) \qquad \text{and} \qquad \lim_{x \to b^-} f(x) = f(b)$$ then $f$ is said to be **continuous on $[a, b]$.** |

■■ **Remark:** If $f$ is continuous on $[a, b]$, we say it is **continuous from the right** at $a$ and **continuous from the left** at $b$.

**EXAMPLE 3**

**Continuity at the Endpoint of an Interval**

Discuss the continuity of

$$f(x) = \sqrt{3 - x}$$

**SOLUTION**

First, $f$ is defined for all $x \leq 3$. Furthermore, $f$ is continuous from the left at $x = 3$ because

$$\lim_{x \to 3^-} f(x) = \lim_{x \to 3^-} \sqrt{3 - x} = 0 = f(3)$$

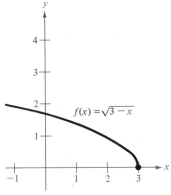

$f(x) = \sqrt{3 - x}$

FIGURE 1.69

For all $x < 3$, the function $f$ satisfies the three requirements for continuity. Thus, we conclude that $f$ is continuous on the interval $(-\infty, 3]$, as shown in Figure 1.69.

## EXAMPLE 4

**Continuity on a Closed Interval**

Discuss the continuity of

$$g(x) = \begin{cases} 5 - x, & -1 \leq x \leq 2 \\ x^2 - 1, & 2 < x \leq 3 \end{cases}$$

**SOLUTION**

The polynomial functions $5 - x$ and $x^2 - 1$ are continuous on the intervals $[-1, 2)$ and $(2, 3]$, respectively. Thus, to conclude that $g$ is continuous on the entire interval $[-1, 3]$, we need only check the behavior of $g$ when $x = 2$. Taking the one-sided limits when $x = 2$, we see that

$$\lim_{x \to 2^-} g(x) = \lim_{x \to 2^-} (5 - x) = 3$$

and

$$\lim_{x \to 2^+} g(x) = \lim_{x \to 2^+} (x^2 - 1) = 3$$

Since these two limits are equal, we have

$$\lim_{x \to 2} g(x) = g(2) = 3$$

Thus, $g$ is continuous at $x = 2$, and consequently it is continuous on the entire interval $[-1, 3]$. The graph of $g$ is shown in Figure 1.70.

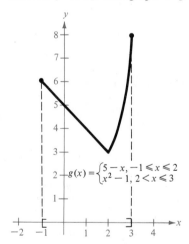

FIGURE 1.70

In the context of one-sided limits and one-sided continuity it is interesting to look at the **greatest integer function.** This function is denoted by

$$[\![x]\!] = \text{greatest integer} \leq x$$

For instance,

$$[\![-2.1]\!] = \text{greatest integer} \leq -2.1 = -3$$

$$[\![-2]\!] = \text{greatest integer} \leq -2 = -2$$

$$[\![1.5]\!] = \text{greatest integer} \leq 1.5 = 1$$

In the graph of the greatest integer function (Figure 1.71), note how the graph jumps up one unit at each integer and thus is discontinuous at each integer. Each of these discontinuities is nonremovable.

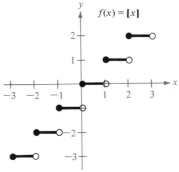

FIGURE 1.71      Greatest Integer Function

In applications the domain of the greatest integer function is often restricted to nonnegative values of $x$. In such cases this function serves the purpose of **truncating** the decimal portion of $x$. For example, $[\![1.345]\!]$ is truncated to 1, and $[\![3.57]\!]$ is truncated to 3.

**EXAMPLE 5**

**An Application of the Greatest Integer Function**

A bookbinding company produces 10,000 books in an 8-hour shift. The fixed costs *per shift* amount to $5,000, and the unit cost *per book* is $3. Using the greatest integer function, we can write the cost of producing $x$ books as

$$C(x) = 5000\left(1 + \left[\!\left[\frac{x-1}{10,000}\right]\!\right]\right) + 3x$$

Sketch the graph of this cost function and discuss its continuity.

**SOLUTION**

Note that during the first 8-hour shift

$$\left[\!\left[\frac{x-1}{10,000}\right]\!\right] = 0, \quad 1 \le x \le 10,000$$

and we have

$$C(x) = 5000\left(1 + \left[\!\left[\frac{x-1}{10,000}\right]\!\right]\right) + 3x = 5000 + 3x$$

During the second 8-hour shift,

$$\left[\!\left[\frac{x-1}{10,000}\right]\!\right] = 1, \quad 10,001 \le x \le 20,000$$

and we have

$$C(x) = 5000\left(1 + \left[\!\left[\frac{x-1}{10,000}\right]\!\right]\right) + 3x = 10,000 + 3x$$

The graph of $C$ is given in Figure 1.72. Note that the graph has discontinuities at $x = 10{,}000$, $20{,}000$, and so forth.

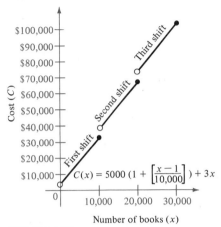

FIGURE 1.72

Because continuity is defined in terms of limits, it is not surprising to learn that continuous functions and limits possess many of the same properties. We summarize these in the following list.

| **Properties of continuous functions** | Let $f$ and $g$ be continuous at $c$ and let $k$ be a real number. Then the following functions are also continuous at $c$: |
|---|---|

1. Sum or difference: $f \pm g$          2. Constant multiple: $kf$
3. Product: $fg$                          4. Quotient: $f/g$, $\quad g(c) \neq 0$
5. Composition: $f(g(x))$, $\quad$ where $f$ is continuous at $g(c)$

## SECTION EXERCISES 1.6

In Exercises 1–6, find the discontinuities, if any, for the given function.

**1.** $f(x) = -\dfrac{x^3}{2}$     **2.** $f(x) = \dfrac{x^2 - 1}{x}$     **3.** $f(x) = \dfrac{x^2 - 1}{x + 1}$     **4.** $f(x) = \dfrac{1}{x^2 - 4}$

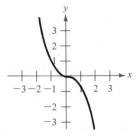 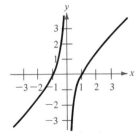 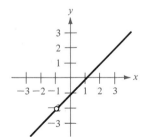 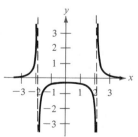

**5.** $f(x) = \begin{cases} x, & x < 1 \\ 2, & x = 1 \\ 2x - 1, & x > 1 \end{cases}$  **6.** $f(x) = \dfrac{[\![x]\!]}{2} + x$

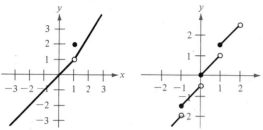

In Exercises 7–26, find the discontinuities (if any) for the given function. Which of the discontinuities are removable?

**7.** $f(x) = x^2 - 2x + 1$

**8.** $f(x) = \dfrac{1}{x^2 + 1}$

**9.** $f(x) = \dfrac{1}{x - 1}$

**10.** $f(x) = \dfrac{x}{x^2 - 1}$

**11.** $f(x) = \dfrac{x}{x^2 + 1}$

**12.** $f(x) = \dfrac{x - 3}{x^2 - 9}$

**13.** $f(x) = \dfrac{x + 2}{x^2 - 3x - 10}$

**14.** $f(x) = \dfrac{x - 1}{x^2 + x - 2}$

**15.** $f(x) = \begin{cases} x, & x \le 1 \\ x^2, & x > 1 \end{cases}$

**16.** $f(x) = \begin{cases} -2x + 3, & x < 1 \\ x^2, & x \ge 1 \end{cases}$

**17.** $f(x) = \begin{cases} \dfrac{x}{2} + 1, & x \le 2 \\ 3 - x, & x > 2 \end{cases}$

**18.** $f(x) = \begin{cases} -2x, & x \le 2 \\ x^2 - 4x + 1, & x > 2 \end{cases}$

**19.** $f(x) = \begin{cases} 3 + x, & x \le 2 \\ x^2 + 1, & x > 2 \end{cases}$

**20.** $f(x) = \begin{cases} |x - 2| + 3, & x < 0 \\ x + 5, & x \ge 0 \end{cases}$

**21.** $f(x) = \dfrac{|x + 2|}{x + 2}$  **22.** $f(x) = \dfrac{|4 - x|}{4 - x}$

**23.** $f(x) = [\![x - 1]\!]$  **24.** $f(x) = x - [\![x]\!]$

**25.** $h(x) = f(g(x)),\ f(x) = \dfrac{1}{\sqrt{x}},\ g(x) = x - 1,\ x > 1$

**26.** $h(x) = f(g(x)),\ f(x) = \dfrac{1}{x - 1},\ g(x) = x^2 + 5$

In Exercises 27–30, sketch the graph of the given function to determine any points of discontinuity.

**27.** $f(x) = \dfrac{x^2 - 16}{x - 4}$  **28.** $g(x) = \dfrac{2x^2 + x}{x}$

**29.** $f(x) = \dfrac{x^3 + x}{x}$  **30.** $h(x) = \dfrac{x^4 - 1}{x^2 - 1}$

**31.** Determine the constant $a$ so that the following function is continuous:
$$f(x) = \begin{cases} x^3, & x \le 2 \\ ax^2, & x > 2 \end{cases}$$

**32.** Determine the constants $a$ and $b$ so that the following function is continuous:
$$f(x) = \begin{cases} 2, & x \le -1 \\ ax + b, & -1 < x < 3 \\ -2, & x \ge 3 \end{cases}$$

**33.** A union contract guarantees a 9% increase each year for five years. For a salary of $38,500, the salary $S$ is
$$S = 38{,}500(1.09)^{[\![t]\!]}, \quad 0 \le t < 6$$
where $t = 0$ corresponds to 1985. Sketch a graph of this function and discuss its continuity.

**34.** A dial-direct long distance call costs $1.52 for the first two minutes and $0.86 for each additional minute or fraction thereof. Use the greatest integer function to write the cost $C$ of a call in terms of the time $t$ (in minutes). Sketch a graph of this function and discuss its continuity.

**35.** The number of units in inventory in a company is
$$N = 25\left(2\left[\!\!\left[\dfrac{t + 2}{2}\right]\!\!\right] - t\right), \quad t \ge 0$$
where $t$ is the time in months. Sketch the graph of this function and discuss its continuity. How often does this company replenish its inventory?

**36.** A deposit of $5,000 is made in a savings plan that pays 12% compounded semiannually. The amount in the account after $t$ years is given by the function
$$A = 5000(1.06)^{[\![2t]\!]}, \quad t \ge 0$$
Graph this function and discuss its continuity.

**37.** The cost (in millions of dollars) of removing $x$ percent of the pollutants being emitted from the smokestack of a certain factory is given by
$$C = \dfrac{2x}{100 - x}, \quad 0 \le x < 100$$
Graph this function and discuss its continuity. What is the appropriate domain of the function?

# CHAPTER 1 SUMMARY

## Important Terms

Rectangular coordinate system
Cartesian plane
$x$-axis, $y$-axis
Origin
Quadrant
Ordered pair
$x$-coordinate, $y$-coordinate
Solution point of an equation
Graph of an equation
Parabola
$x$-intercept, $y$-intercept
Circle
Point of intersection
Mathematical model
Linear equation
Slope of a line
Linear interpolation
Linear extrapolation
Dependent variable
Independent variable
Function
Domain of a function
Range of a function

One-to-one function
Functional notation
Functional value
Composition of two functions
Inverse function
Limit of a function
$x$ approaches $c$
Indeterminate form
One-sided limit
Limit from the left
Limit from the right
Decreases without bound
Increases without bound
Continuous at a point
Continuous on an open interval
Discontinuous at a point
Removable discontinuity
Nonremovable discontinuity
Continuous on a closed interval
Continuous from the right
Continuous from the left
Greatest integer function

## Important Techniques

Plotting a point $(x, y)$
Finding the distance between two points
Sketching the graph of an equation
Finding $x$- and $y$-intercepts
Completing the square
Finding points of intersection
Sketching the graph of a line
Finding the equation of a line

Finding the domain of a function
Finding the range of a function
Using the vertical line test
Evaluating a function
Finding the inverse of a function
Finding the limit of a function
Determining the continuity of a function

## Important Formulas and Theorems

Distance Formula:

$$d = \sqrt{(x_2 - x_1)^2 + (y_2 - y_1)^2}$$

Midpoint Formula:

$$\text{Midpoint} = \left( \frac{x_1 + x_2}{2}, \frac{y_1 + y_2}{2} \right)$$

Standard form of equation of circle:

$$(x - h)^2 + (y - k)^2 = r^2$$

General form of equation of circle:

$$Ax^2 + Ay^2 + Dx + Ey + F = 0$$

Equation of line (slope-intercept form):

$$y = mx + b$$

Slope of line:

$$m = \frac{y_2 - y_1}{x_2 - x_1}$$

Change in $x$:

$$\Delta x = x_2 - x_1$$

Change in $y$:

$$\Delta y = y_2 - y_1$$

Equation of line (point-slope form):

$$y - y_1 = m(x - x_1)$$

Equation of line (general form):

$$Ax + By + C = 0$$

Equation of vertical line:

$$x = a$$

Equation of horizontal line:

$$y = b$$

Slopes of parallel lines:

$$m_1 = m_2$$

Slopes of perpendicular lines:

$$m_1 = -\frac{1}{m_2}$$

Composite of $f$ with $g$:

$$f(g(x))$$

Inverse functions:

$$f(f^{-1}(x)) = x = f^{-1}(f(x))$$

# REVIEW EXERCISES FOR CHAPTER 1

In Exercises 1–10, find (a) the distance between the two points, (b) the coordinates of the midpoint of the line segment between the two points, and (c) the general form of the equation of the line through the points.

1. $(0, 0)$, $(6, 0)$
2. $(0, 0)$, $(0, 10)$
3. $(-2, -1)$, $(2, 2)$
4. $(-1, 4)$, $(2, 0)$
5. $(2, 1)$, $(14, 6)$
6. $(-2, 2)$, $(3, -10)$
7. $(-1, 0)$, $(6, 2)$
8. $(1, 6)$, $(4, 2)$
9. $\left(\dfrac{1}{3}, \dfrac{4}{3}\right)$, $\left(\dfrac{2}{3}, \dfrac{1}{6}\right)$
10. $(0.64, 0.45)$, $(1.32, 4.68)$

In Exercises 11 and 12, determine the value of $t$ so that the points are on the same line.

11. $(-2, 5)$, $(0, t)$, $(1, 1)$
12. $(-6, 1)$, $(1, t)$, $(10, 5)$

In Exercises 13 and 14, find the general form of the line through the specified point and satisfying the given characteristic.

13. Point: $(-2, 4)$
    (a) Slope is $\dfrac{7}{16}$
    (b) Parallel to the line $5x - 3y = 3$
    (c) Passes through the origin
    (d) Parallel to the $y$-axis

14. Point: $(1, 3)$
    (a) Slope is $-\dfrac{2}{3}$
    (b) Perpendicular to the line $x + y = 0$
    (c) Passes through the point $(2, 4)$
    (d) Parallel to the $x$-axis

In Exercises 15–18, sketch the graph of the equation of the line.

15. $4x - 2y = 6$
16. $0.02x + 0.15y = 0.25$
17. $-\dfrac{1}{3}x + \dfrac{5}{6}y = 1$
18. $51x + 17y = 102$

In Exercises 19–22, determine the radius and center of the given circle and sketch its graph.

19. $x^2 + y^2 + 6x - 2y + 1 = 0$
20. $4x^2 + 4y^2 - 4x + 8y = 11$
21. $x^2 + y^2 + 6x - 2y + 10 = 0$
22. $x^2 + y^2 - 6x + 8y = 0$

In Exercises 23 and 24, find the general form of the equation of the circle with specified center and radius. Then determine if each of the points is inside, outside, or on the circle.

23. Center: $(1, 2)$; radius: 3
    (a) $(1, 5)$        (b) $(0, 0)$
    (c) $(-2, 1)$       (d) $(0, 4)$
24. Center: $(2, 1)$; radius: 2
    (a) $(1, 1)$        (b) $(4, 2)$
    (c) $(0, 1)$        (d) $(3, 1)$

In Exercises 25 and 26, find the point(s) of intersection of the graphs of the given equations.

25. $3x - 4y = 8$, $x + y = 5$
26. $x - y - 1 = 0$, $y - x^2 = -7$

In Exercises 27–30, sketch the graph of the given equation and use the vertical line test to determine if the equation expresses $y$ as a function of $x$.

27. $x - 2y = 0$
28. $y - x^3 + 1 = 0$
29. $x^2 + y^2 = 16$
30. $y = \dfrac{4}{x^2 + 1}$

In Exercises 31 and 32, use $f$ and $g$ to find the following:
(a) $f(x) + g(x)$        (b) $f(x) - g(x)$
(c) $f(x)g(x)$           (d) $\dfrac{f(x)}{g(x)}$
(e) $f(g(x))$            (f) $g(f(x))$

31. $f(x) = 1 - x^2$, $g(x) = 2x + 1$
32. $f(x) = 2x - 3$, $g(x) = \sqrt{x + 1}$

33. The sum of two positive numbers is 500. Let one of the numbers be $x$ and express the product $P$ of the two numbers as a function of $x$.

34. The product of two positive numbers is 120. Let one of the numbers be $x$ and express the sum of the two numbers as a function of $x$.

35. When a wholesaler sold a certain product at $25 per unit, sales were 800 units per week. However, after a price increase of $5 per unit, the average number of units sold dropped to 775 units per week. Write the quantity demanded $x$ as a linear function of the price $p$.

36. A small business uses a van for making deliveries. The cost per hour of fuel for the van is

$$C = \frac{v^2}{600}$$

where $v$ is the average speed for the trip in miles per hour. The driver is paid $5 per hour. Write the total cost $T$ for making a 210-mile trip as a function of $v$.

In Exercises 37–52, find the given limit (if it exists).

**37.** $\lim\limits_{x \to 2} (5x - 3)$

**38.** $\lim\limits_{x \to 2} (2x + 9)$

**39.** $\lim\limits_{x \to 2} (5x - 3)(2x + 9)$

**40.** $\lim\limits_{x \to 2} \dfrac{5x - 3}{2x + 9}$

**41.** $\lim\limits_{t \to 3} \dfrac{t^2 + 1}{t}$

**42.** $\lim\limits_{t \to 1} \dfrac{t + 1}{t - 2}$

**43.** $\lim\limits_{t \to 0} \dfrac{t^2 + 1}{t}$

**44.** $\lim\limits_{t \to 2} \dfrac{t + 1}{t - 2}$

**45.** $\lim\limits_{x \to -2} \dfrac{x + 2}{x^2 - 4}$

**46.** $\lim\limits_{x \to 3^-} \dfrac{x^2 - 9}{x - 3}$

**47.** $\lim\limits_{x \to 0^+} \left(x - \dfrac{1}{x}\right)$

**48.** $\lim\limits_{x \to (1/2)} \dfrac{2x - 1}{6x - 3}$

**49.** $\lim\limits_{x \to 0} \dfrac{[1/(x + 1)] - 1}{x}$

**50.** $\lim\limits_{s \to 0} \dfrac{(1/\sqrt{1 + s}) - 1}{s}$

**51.** $\lim\limits_{\Delta x \to 0} \dfrac{(x + \Delta x)^3 - (x + \Delta x) - (x^3 - x)}{\Delta x}$

**52.** $\lim\limits_{\Delta x \to 0} \dfrac{1 - (x + \Delta x)^2 - (1 - x^2)}{\Delta x}$

In Exercises 53 and 54, estimate the given limit by completing the following table.

| $x$ | 1.1 | 1.01 | 1.001 | 1.0001 |
|---|---|---|---|---|
| $f(x)$ | | | | |

**53.** $\lim\limits_{x \to 1^+} \dfrac{\sqrt{2x + 1} - \sqrt{3}}{x - 1}$

**54.** $\lim\limits_{x \to 1^+} \dfrac{1 - \sqrt[3]{x}}{x - 1}$

In Exercises 55–60, determine if the given limit statement is true or false.

**55.** $\lim\limits_{x \to 0} \dfrac{|x|}{x} = 1$

**56.** $\lim\limits_{x \to 0} x^3 = 0$

**57.** $\lim\limits_{x \to 2} f(x) = 3$, $f(x) = \begin{cases} 3, & x \le 2 \\ 0, & x > 2 \end{cases}$

**58.** $\lim\limits_{x \to 3} f(x) = 1$, $f(x) = \begin{cases} x - 2, & x \le 3 \\ -x^2 + 8x - 14, & x > 3 \end{cases}$

**59.** $\lim\limits_{x \to 0} \sqrt{x} = 0$

**60.** $\lim\limits_{x \to 0} \sqrt[3]{x} = 0$

In Exercises 61–68, determine the points of discontinuity (if any) of the given function.

**61.** $f(x) = [\![ x + 3 ]\!]$

**62.** $f(x) = \dfrac{3x^2 - x - 2}{x - 1}$

**63.** $f(x) = \begin{cases} \dfrac{3x^2 - x - 2}{x - 1}, & x \ne 1 \\ 0, & x = 1 \end{cases}$

**64.** $f(x) = \begin{cases} 5 - x, & x \le 2 \\ 2x - 3, & x > 2 \end{cases}$

**65.** $f(x) = \dfrac{1}{(x - 2)^2}$

**66.** $f(x) = \dfrac{x + 2}{x}$

**67.** $f(x) = \dfrac{3}{x + 1}$

**68.** $f(x) = \dfrac{x + 1}{2x + 2}$

**69.** Determine the value of $c$ so that the following function is continuous:
$$f(x) = \begin{cases} x + 3, & x \le 2 \\ cx + 6, & x > 2 \end{cases}$$

**70.** Determine the values of $b$ and $c$ so that the following function is continuous:
$$f(x) = \begin{cases} x + 1, & 1 < x < 3 \\ x^2 + bx + c, & |x - 2| \ge 1 \end{cases}$$

# Differentiation

# The Derivative and the Slope of a Curve

## INTRODUCTORY EXAMPLE
### Energy Required for Walking and Running

Humans use different types of motions to walk and to run. When a person walks (at a speed of up to 2.5 meters per second), each foot is on the ground for slightly more than half the time. There are times when both feet are on the ground and at no time are both feet off the ground. In contrast, when a person is running each foot is on the ground for less than half the time. Thus, there are times when a runner has both feet off the ground.

Not only are the motions required for walking and running quite different; the energy patterns for these two types of motions are also quite different. (See Figure 2.1.) For a runner, the energy corresponding to a certain speed is given by the *linear* model

$$E = 250x + 100, \quad 2.5 \le x$$

where $x$ is the speed measured in meters per second and $E$ is the energy measured in watts. The slope of the graph of this function is *constant*—that is, it is the same at each point on the graph. In calculus, we call the function that gives this slope the **derivative** and denote it by

$$\frac{dE}{dx} = 250$$

For a person walking, the energy corresponding to a certain speed is given by the *quadratic* model

$$E = 155x^2 - 65x + 160, \quad 0.5 \le x \le 2.5$$

The slope of the graph of this function is *not constant*— that is, as the speed increases, the slope of the **tangent line to the graph** changes. For a given speed $x$, the slope of the graph of this function is given by the derivative

$$\frac{dE}{dx} = 310x - 65$$

Walking

Running

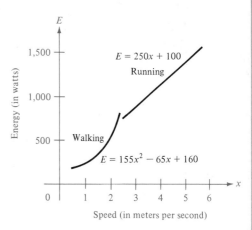

FIGURE 2.1

- **Tangent Line to a Curve**
- **Slope of a Curve**
- **Derivative of a Function**
- **Four-Step Process for Finding Derivatives**

In Section 1.3, we saw how the slope of a line can be used to describe the rate at which the line rises or falls. For a line, this rate (or slope) is the same at every point on the line. For graphs (or curves) other than lines, the rate at which the graph rises or falls changes from point to point. For instance, in Figure 2.2, the graph is rising more quickly at the point $(x_1, y_1)$ than it is at the point $(x_2, y_2)$. Then the graph levels off at the point $(x_3, y_3)$ and is falling at the point $(x_4, y_4)$.

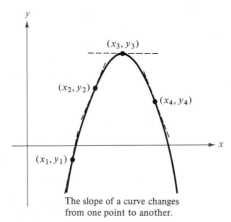

The slope of a curve changes from one point to another.

FIGURE 2.2

To determine the rate at which a curve rises or falls at a point, we use the **tangent line** to the curve at the point. For instance, in Figure 2.3 the tangent line to the graph of $f$ at $P$ is the line that best approximates the graph of $f$ at that point. The problem of finding the slope of a curve at a point thus becomes one of finding the slope of the line tangent to the curve at that point.

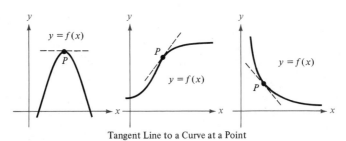

Tangent Line to a Curve at a Point

FIGURE 2.3

**EXAMPLE 1**

**Approximating the Slope of a Curve at a Point**

Use the graph in Figure 2.4 to approximate the slope of the graph of

$$f(x) = x^2$$

at the point $(1, 1)$.

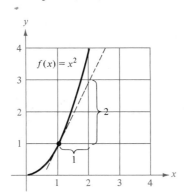

FIGURE 2.4

**SOLUTION**

From the graph of $f(x) = x^2$, we see that the tangent line at $(1, 1)$ rises approximately 2 units for each unit change in $x$. Thus, the *slope of the tangent line* at $(1, 1)$ is given by

$$\text{Slope} = \frac{\text{change in } y}{\text{change in } x} \approx \frac{2}{1} = 2$$

Finally, we conclude that the *slope of the curve* at $(1, 1)$ is approximately 2.

■■■ **Remark:** When visually approximating the slope of a curve, note that the scales on the horizontal and vertical axes may differ. When this happens (as it frequently does in applications), the slope of the tangent line is distorted, and you must be careful to account for the scale differences.

**EXAMPLE 2**

**Approximating the Slope of a Curve at a Point**

Figure 2.5 graphically depicts the average daily temperature (in degrees Fahrenheit) in Duluth, Minnesota. Estimate the slope of this curve at the indicated point and give a physical interpretation of the result.

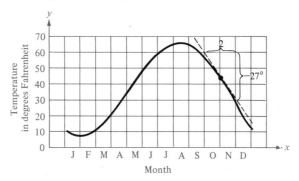

FIGURE 2.5

**SOLUTION**

From the graph in Figure 2.5, we see that the tangent line at the given point falls approximately 27 units for each 2-unit change in $x$. Thus, we estimate the slope at the given point to be

$$\text{Slope} = \frac{\text{change in } y}{\text{change in } x} \approx \frac{-27}{2} = -13.5 \text{ (degrees per month)}$$

This means that we can expect the average November temperature to be roughly 13.5 degrees *lower* than the average October temperature.

### Finding the Slope of a Curve with the Limit Process

In Examples 1 and 2 we approximated the slope of a curve at a point by making a careful graph and then "eyeballing" the tangent line at the point of tangency. A more precise method of approximating tangent lines makes use of a line through the point of tangency and a second point on the curve, as shown in Figure 2.6. We call such a line a **secant line.** (This use of the word *secant* comes from the Latin *secare,* meaning "to cut.") If $(x, f(x))$ is the point of tangency and $(x + \Delta x, f(x + \Delta x))$ is a second point on the graph of $f$, then the slope of the secant line through these two points is given by

$$m_{\sec} = \frac{f(x + \Delta x) - f(x)}{x + \Delta x - x}$$

$$= \frac{f(x + \Delta x) - f(x)}{\Delta x} \qquad \text{Slope of secant line}$$

This formula is called a **difference quotient.** The denominator is called the **change in $x$,** and the numerator is called the **change in $y$,** denoted by

$$\Delta y = f(x + \Delta x) - f(x) \qquad \text{Change in } y$$

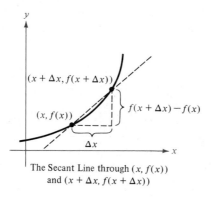

The Secant Line through $(x, f(x))$
and $(x + \Delta x, f(x + \Delta x))$

FIGURE 2.6

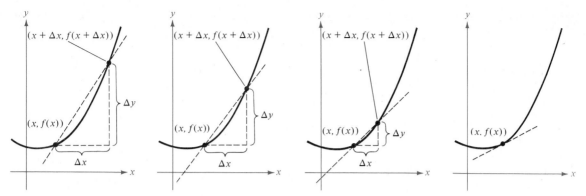

As Δx approaches 0, the secant lines approach the tangent line.

FIGURE 2.7

The beauty of this procedure is that we can obtain better and better approximations to the slope of the tangent line by choosing the second point closer and closer to the point of tangency, as shown in Figure 2.7.

Finally, by resorting to the limit process we can determine the *exact* slope of the tangent line at $(x, f(x))$ to be

$$m = \lim_{\Delta x \to 0} \frac{f(x + \Delta x) - f(x)}{\Delta x}$$   Slope of tangent line

We use this limit as our formal definition of the slope of a curve at a point.

| **Definition of the slope of a curve** | The slope $m$ of the graph of $y = f(x)$ at $(x, f(x))$ is equal to the slope of its tangent line at $(x, f(x))$, and it is determined by the formula $$m = \lim_{\Delta x \to 0} \frac{f(x + \Delta x) - f(x)}{\Delta x}$$ provided this limit exists. |
|---|---|

■■■ **Remark:** Note that $\Delta x$ is used as a variable to represent the change in $x$ in this definition. Other variables may be used. For instance, this definition is sometimes written as

$$m = \lim_{h \to 0} \frac{f(x + h) - f(x)}{h}$$

To calculate the slope of the tangent line to a curve by its limit definition we follow a general four-step process.

**Four-step process**

To find the slope of the graph of $y = f(x)$ at the point $(x, f(x))$, use the following four steps.

1. Evaluate $f$ at $(x + \Delta x)$: $\qquad\qquad f(x + \Delta x)$

2. Subtract $f(x)$ to find change in $y$: $\qquad f(x + \Delta x) - f(x)$

3. Divide by $\Delta x$ to form difference quotient: $\quad \dfrac{f(x + \Delta x) - f(x)}{\Delta x}$

4. Let $\Delta x \to 0$ to obtain slope: $\qquad\quad \displaystyle\lim_{\Delta x \to 0} \dfrac{f(x + \Delta x) - f(x)}{\Delta x}$

**EXAMPLE 3**

**Finding the Slope by the Four-Step Process**

Find a formula for the slope of the line tangent to the graph of

$$f(x) = 2x - 3$$

at any point $(x, y)$.

**SOLUTION**

By the four-step process,

1. $\qquad f(x + \Delta x) = 2(x + \Delta x) - 3$

2. $\qquad f(x + \Delta x) - f(x) = (2x + 2\Delta x - 3) - (2x - 3) = 2\Delta x$

3. $\qquad \dfrac{f(x + \Delta x) - f(x)}{\Delta x} = \dfrac{2\Delta x}{\Delta x} = 2$

4. $\displaystyle\lim_{\Delta x \to 0} \dfrac{f(x + \Delta x) - f(x)}{\Delta x} = \lim_{\Delta x \to 0} 2 = 2$

Therefore, the slope at any point $(x, y)$ is

$$m = 2$$

as shown in Figure 2.8.

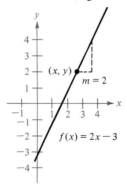

**FIGURE 2.8**

In Example 3 the slope of the graph of $f$ is constant. This is not surprising since the graph of $f$ is a line. Of course, not all graphs have a constant slope, as shown in the next example.

**EXAMPLE 4**

**Finding the Slope by the Four-Step Process**

Find a formula for the slope of the graph of

$$f(x) = x^2 + 1$$

What is the slope at $(-2, 5)$, at $(0, 1)$, and at $(2, 5)$?

**SOLUTION**

By the four-step process,

1. $$f(x + \Delta x) = (x + \Delta x)^2 + 1$$

2. $$f(x + \Delta x) - f(x) = x^2 + 2x(\Delta x) + (\Delta x)^2 + 1 - x^2 - 1$$
$$= 2x\Delta x + (\Delta x)^2$$

3. $$\frac{f(x + \Delta x) - f(x)}{\Delta x} = \frac{\Delta x(2x + \Delta x)}{\Delta x} = 2x + \Delta x$$

4. $$\lim_{\Delta x \to 0} \frac{f(x + \Delta x) - f(x)}{\Delta x} = \lim_{\Delta x \to 0} (2x + \Delta x) = 2x$$

Therefore, at any point $(x, y)$ on the graph of $f(x) = x^2 + 1$, the slope is given by the formula

$$m = 2x$$

To find the slope at a particular point, we substitute the *x-coordinate* of the point into this formula as follows:

At $(-2, 5)$, the slope is $m = 2(-2) = -4$.

At $(0, 1)$, the slope is $m = 2(0) = 0$.

At $(2, 5)$, the slope is $m = 2(2) = 4$.

Note in Figure 2.9 that the graph of $f$ has a negative slope at $(-2, 5)$, a zero slope at $(0, 1)$, and a positive slope at $(2, 5)$.

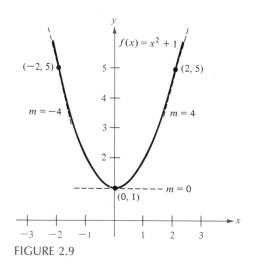

FIGURE 2.9

**EXAMPLE 5**

**Finding a Tangent Line by the Four-Step Process**

Find the equation of the line tangent to the graph of

$$f(x) = \sqrt{x}$$

at the point (4, 2).

**SOLUTION**

By the four-step process,

1. $$f(x + \Delta x) = \sqrt{x + \Delta x}$$

2. $$f(x + \Delta x) - f(x) = \sqrt{x + \Delta x} - \sqrt{x}$$

3. $$\frac{f(x + \Delta x) - f(x)}{\Delta x} = \frac{\sqrt{x + \Delta x} - \sqrt{x}}{\Delta x}$$

4. $$\lim_{\Delta x \to 0} \frac{f(x + \Delta x) - f(x)}{\Delta x} = \lim_{\Delta x \to 0} \frac{\sqrt{x + \Delta x} - \sqrt{x}}{\Delta x}$$

To find this limit, we rationalize the numerator, as follows:

$$\frac{\sqrt{x + \Delta x} - \sqrt{x}}{\Delta x} = \left( \frac{\sqrt{x + \Delta x} - \sqrt{x}}{\Delta x} \right) \left( \frac{\sqrt{x + \Delta x} + \sqrt{x}}{\sqrt{x + \Delta x} + \sqrt{x}} \right)$$

$$= \frac{(x + \Delta x) - x}{\Delta x(\sqrt{x + \Delta x} + \sqrt{x})}$$

$$= \frac{\Delta x}{\Delta x(\sqrt{x + \Delta x} + \sqrt{x})} = \frac{1}{\sqrt{x + \Delta x} + \sqrt{x}}$$

Now taking the limit we find the slope at any point $(x, y)$ to be given by the formula

$$m = \lim_{\Delta x \to 0} \frac{1}{\sqrt{x + \Delta x} + \sqrt{x}} = \frac{1}{2\sqrt{x}}$$

Using this formula, we find the slope of the tangent line at (4, 2) to be

$$m = \frac{1}{2\sqrt{4}} = \frac{1}{4}$$

as shown in Figure 2.10. Finally, using the point-slope form, we have for the equation of the tangent line at (4, 2)

$$y - 2 = \frac{1}{4}(x - 4)$$

$$y = \frac{1}{4}x + 1$$

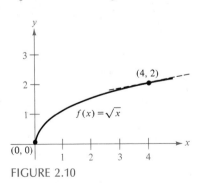

FIGURE 2.10

## The Derivative of a Function

We have now arrived at a crucial point in our study of calculus, for the limit

$$\lim_{\Delta x \to 0} \frac{f(x + \Delta x) - f(x)}{\Delta x}$$

is used to define one of the two fundamental quantities of calculus—namely, the **derivative.**

---

**Definition of the derivative**

The **derivative of $f$ at $x$** is given by

$$f'(x) = \lim_{\Delta x \to 0} \frac{f(x + \Delta x) - f(x)}{\Delta x}$$

provided the limit exists.

---

A function is said to be **differentiable** at $x$ if its derivative exists at $x$, and the process of finding the derivative is called **differentiation.**

In addition to $f'(x)$ (read "$f$ prime of $x$"), various other notations are used to denote the derivative. The most commonly used notations are

$$\frac{dy}{dx}, \quad y', \quad \frac{d}{dx}[f(x)], \quad D_x(y)$$

The notation $dy/dx$ is read as the "derivative of $y$ with respect to $x$," and using limit notation, we write

$$\frac{dy}{dx} = \lim_{\Delta x \to 0} \frac{\Delta y}{\Delta x}$$

Thus, we have

$$\frac{dy}{dx} = \lim_{\Delta x \to 0} \frac{\Delta y}{\Delta x} = \lim_{\Delta x \to 0} \frac{f(x + \Delta x) - f(x)}{\Delta x} = f'(x)$$

Since the derivative $f'(x)$ and the slope of the tangent line to the graph of $f$ are both defined by the limit

$$\lim_{\Delta x \to 0} \frac{f(x + \Delta x) - f(x)}{\Delta x}$$

we can use the four-step process to determine $f'(x)$.

**EXAMPLE 6**

**Finding the Derivative by the Four-Step Process**

Find the derivative of

$$f(x) = x^3 + 2x$$

---

## SOLUTION

1. $$f(x + \Delta x) = (x + \Delta x)^3 + 2(x + \Delta x)$$

2. $$f(x + \Delta x) - f(x) = x^3 + 3x^2\Delta x + 3x(\Delta x)^2 + (\Delta x)^3 + 2x + 2\Delta x - (x^3 + 2x$$
$$= 3x^2\Delta x + 3x(\Delta x)^2 + (\Delta x)^3 + 2\Delta x$$

3. $$\frac{f(x + \Delta x) - f(x)}{\Delta x} = \frac{\Delta x[3x^2 + 3x\Delta x + (\Delta x)^2 + 2]}{\Delta x}$$

4. $$\lim_{\Delta x \to 0} \frac{f(x + \Delta x) - f(x)}{\Delta x} = \lim_{\Delta x \to 0} [3x^2 + 3x\Delta x + (\Delta x)^2 + 2]$$
$$= 3x^2 + 2$$

Therefore, the derivative of $f$ is

$$f'(x) = 3x^2 + 2$$

---

## EXAMPLE 7

**Finding the Derivative by the Four-Step Process**

Determine the derivative of $y$ with respect to $t$ for the function given by

$$y = \frac{2}{t}$$

## SOLUTION

Considering $y = f(t)$, we have

1. $$f(t + \Delta t) = \frac{2}{t + \Delta t}$$

2. $$f(t + \Delta t) - f(t) = \frac{2}{t + \Delta t} - \frac{2}{t}$$
$$= \frac{2t - 2t - 2\Delta t}{t(t + \Delta t)}$$
$$= \frac{-2\Delta t}{t(t + \Delta t)}$$

3. $$\frac{f(t + \Delta t) - f(t)}{\Delta t} = \frac{-2\Delta t}{\Delta t[t(t + \Delta t)]}$$

4. $$\lim_{\Delta t \to 0} \frac{f(t + \Delta t) - f(t)}{\Delta t} = \lim_{\Delta t \to 0} \frac{-2}{t(t + \Delta t)} = -\frac{2}{t^2}$$

Therefore, the derivative of $y$ with respect to $t$ is

$$\frac{dy}{dt} = -\frac{2}{t^2}$$

---

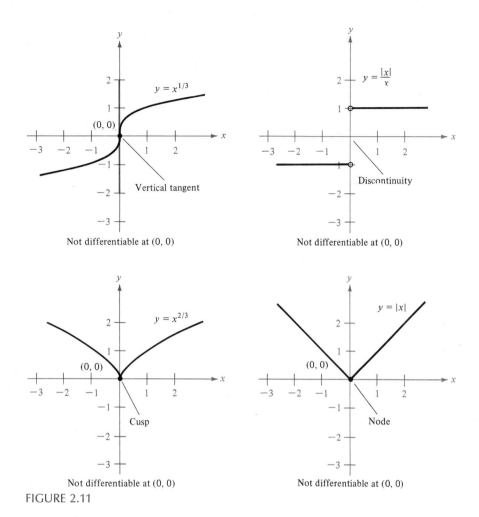

FIGURE 2.11

Not every function is differentiable. Figure 2.11 shows some common types of problems that destroy differentiability—vertical tangent lines, discontinuities, and sharp turns in a graph. Each of the functions shown in Figure 2.11 is differentiable at every value of $x$ *except $x = 0$.*

From Figure 2.11, we can see that continuity is not a strong enough condition to guarantee differentiability. (All but one of the functions are continuous at $(0, 0)$, yet none is differentiable there.) On the other hand, if a function is differentiable at a point then it must be continuous. We state this important result in the following theorem.

**Differentiability implies continuity**

If a function is differentiable at $x = c$, then it is continuous at $x = c$.

## SECTION EXERCISES 2.1

In Exercises 1–4, trace the given curve on another piece of paper and sketch the tangent line at each of the points $(x_1, y_1)$ and $(x_2, y_2)$. Determine whether the slope of each tangent line is positive, negative, or zero.

**1.**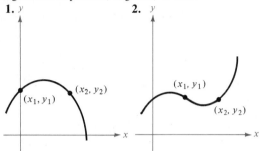

**2.**

**3.**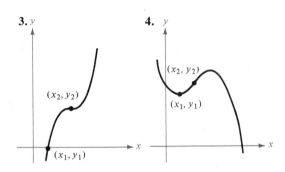

**4.**

In Exercises 5–10, estimate the slope of the curve at the point $(x, y)$.

**5.**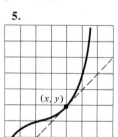

**6.**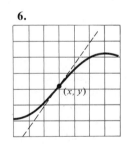

**7.**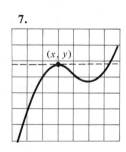

**8.**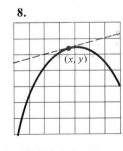

**9.**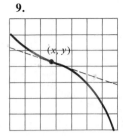

**10.**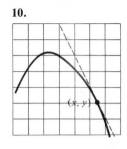

In Exercises 11–22, use the four-step process to find the derivative. (For the first four exercises, you should know what the derivative will be prior to using the four-step process.)

**11.** $f(x) = 3$       **12.** $f(x) = -4$

**13.** $f(x) = -5x + 3$       **14.** $f(x) = \dfrac{1}{2}x$

**15.** $f(x) = x^2$       **16.** $f(x) = 1 - x^2$

**17.** $f(x) = 2x^2 + x - 1$       **18.** $f(x) = x - x^2$

**19.** $h(t) = \sqrt{t - 1}$       **20.** $g(s) = \dfrac{1}{s - 1}$

**21.** $f(t) = t^3 - 12t$       **22.** $f(t) = t^3 + t^2$

In Exercises 23–30, find an equation of the tangent line to the graph of $f$ at the indicated point. Then verify your answer by sketching both the graph of $f$ and the tangent line.

| Function | Point of Tangency |
|---|---|
| **23.** $f(x) = 6 - 2x$ | $(2, 2)$ |
| **24.** $f(x) = 6$ | $(-2, 6)$ |
| **25.** $f(x) = x^2 - 2$ | $(2, 2)$ |
| **26.** $f(x) = x^2 + 2x + 1$ | $(-3, 4)$ |
| **27.** $f(x) = x^3$ | $(2, 8)$ |
| **28.** $f(x) = x^3$ | $(-2, -8)$ |
| **29.** $f(x) = \sqrt{x + 1}$ | $(3, 2)$ |
| **30.** $f(x) = \dfrac{1}{x + 1}$ | $(0, 1)$ |

In Exercises 31 and 32, find an equation of the tangent line(s) to the curve parallel to the given line.

**31.** Curve: $y = x^3$; line: $3x - y + 1 = 0$

**32.** Curve: $y = \dfrac{1}{\sqrt{x}}$; line: $x + 2y - 6 = 0$

In Exercises 33 and 34, find the general form of the equations of the two tangent lines to the curve that pass through the specified point.

**33.** Curve $y = 4x - x^2$; point: (2, 5)

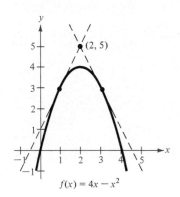

$f(x) = 4x - x^2$

**34.** Curve: $y = x^2$; point: (1, −3)

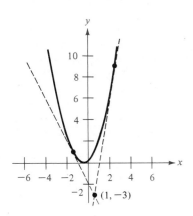

(1, −3)

In Exercises 35–38, determine where the function is not differentiable.

**35.** $f(x) = |x + 3|$

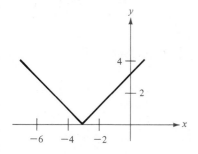

**36.** $f(x) = \dfrac{2x}{x - 1}$

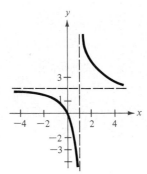

**37.** $f(x) = \begin{cases} 4 - x^2, & 0 < x \\ x^2 - 4, & x \le 0 \end{cases}$

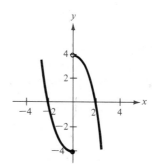

**38.** $f(x) = (x - 3)^{2/3}$

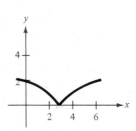

# Some Rules for Differentiation

## INTRODUCTORY EXAMPLE
### Home Mortgage Payments

When a loan is repaid in monthly installments, the size of the monthly payment is a function of the total number of years in the loan period. For example, if a home mortgage of $75,000 at 16% is taken for 20 years the monthly payment is

$1,043.44    20-year payment

If the same mortgage is taken for 30 years the monthly payment drops to

$1,008.57    30-year payment

Table 2.1 shows the monthly payment for this loan for loan periods ranging from 5 years to 60 years. Figure 2.12 graphically shows the relationship between the monthly payment and the number of years in the loan period.

For the graph shown in Figure 2.12, we can interpret the **slope** to be the rate at which the monthly payment is changing for each year that is added to the loan payment. (For this particular graph the slope is always negative, which indicates that the monthly payment decreases as the number of years in the loan period increases.) For instance, if the loan period is changed from 10 to 11 years, the monthly payment will drop by approximately $51. However, if the loan period is changed from 25 to 26 years, the monthly payment will only drop by approximately $3. Note that the graph levels off as the number of years increases. For the homeowner this means that extending a mortgage period beyond 20 years is a relatively inefficient way of reducing the monthly payment.

TABLE 2.1

| Number of years | Monthly payment (dollars) | Slope | Total interest (dollars) |
|---|---|---|---|
| 5 | 1,823.85 | −238.83 | 34,431.00 |
| 10 | 1,256.35 | −51.19 | 75,762.00 |
| 15 | 1,101.53 | −17.78 | 123,275.40 |
| 20 | 1,043.44 | −7.20 | 175,425.60 |
| 25 | 1,019.17 | −3.10 | 230,751.00 |
| 30 | 1,008.57 | −1.37 | 288,085.20 |
| 35 | 1,003.85 | −0.61 | 346,617.00 |
| 40 | 1,001.74 | −0.28 | 405,835.20 |
| 45 | 1,000.78 | −0.12 | 465,421.20 |
| 50 | 1,000.35 | −0.06 | 525,210.00 |
| 55 | 1,000.16 | −0.03 | 585,105.60 |
| 60 | 1,000.07 | −0.01 | 645,050.40 |

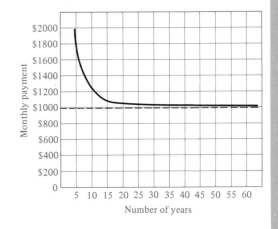

FIGURE 2.12

■ **The Constant Rule**
■ **The (Simple) Power Rule**
■ **The Constant Multiple Rule**
■ **The Sum Rule**

In Section 2.1 we found derivatives by the limit definition of the derivative. This procedure is rather tedious even for simple functions, and fortunately there are rules that greatly simplify the differentiation process. These rules permit us to calculate the derivative without the *direct* use of limits.

---

**Constant Rule**

The derivative of a constant function is zero:

$$\frac{d}{dx}[c] = 0, \quad c \text{ is a constant}$$

---

Proof

Let $f(x) = c$. Then by the limit definition,

$$f'(x) = \lim_{\Delta x \to 0} \frac{f(x + \Delta x) - f(x)}{\Delta x}$$

$$= \lim_{\Delta x \to 0} \frac{c - c}{\Delta x} = \lim_{\Delta x \to 0} 0 = 0$$

Therefore,

$$\frac{d}{dx}[c] = 0$$

---

■ **Remark:** Note in Figure 2.13 that the Constant Rule is equivalent to saying that the slope of a horizontal line is zero.

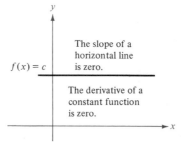

$f(x) = c$

The slope of a horizontal line is zero.

The derivative of a constant function is zero.

FIGURE 2.13

---

**EXAMPLE 1**

**Finding the Derivative of a Constant Function**

(a) $\dfrac{d}{dx}[7] = 0$

(b) If $f(x) = 0$, then $f'(x) = 0$.

(c) If $y = 2$, then $\dfrac{dy}{dx} = 0$.

(d) If $g(t) = -\dfrac{3}{2}$, then $g'(t) = 0$.

Before deriving the next rule we review the procedure for expanding a binomial. Recall that

$$(x + \Delta x)^2 = x^2 + 2x\Delta x + (\Delta x)^2$$

and $$(x + \Delta x)^3 = x^3 + 3x^2\Delta x + 3x(\Delta x)^2 + (\Delta x)^3$$

and the general binomial expansion for a positive integer $n$ is

$$(x + \Delta x)^n = x^n + nx^{n-1}\Delta x + \underbrace{\frac{n(n-1)x^{n-2}}{2}(\Delta x)^2 + \cdots + (\Delta x)^n}$$

$(\Delta x)^2$ is a factor of these terms

We use the binomial expansion in proving a special case of the following rule.

**(Simple) Power Rule**

$\dfrac{d}{dx}[x^n] = nx^{n-1}$,  $n$ is any real number

Proof

Although the Power Rule is true for any real number $n$, the binomial expansion applies only when $n$ is a positive integer. Thus, for the time being, we give a proof for the case where $n$ is a positive integer. Sections 2.4 and 2.7 contain proofs for negative integers and arbitrary rationals, respectively. Letting $f(x) = x^n$, we have

$$f'(x) = \lim_{\Delta x \to 0} \frac{f(x + \Delta x) - f(x)}{\Delta x} = \lim_{\Delta x \to 0} \frac{(x + \Delta x)^n - x^n}{\Delta x}$$

Now applying the binomial expansion rule we obtain

$$f'(x) = \lim_{\Delta x \to 0} \frac{x^n + nx^{n-1}\Delta x + [n(n-1)x^{n-2}/2](\Delta x)^2 + \cdots + (\Delta x)^n - x^n}{\Delta x}$$

$$= \lim_{\Delta x \to 0} \left[ nx^{n-1} + \frac{n(n-1)x^{n-2}}{2}\Delta x + \cdots + (\Delta x)^{n-1} \right]$$

$$= nx^{n-1} + 0 + \cdots + 0$$

$$= nx^{n-1}$$

**Remark:** The case of the Power Rule for $n = 1$ is worth memorizing as a separate rule. That is,

$$\frac{d}{dx}[x] = 1$$

**EXAMPLE 2**

**Applying the Power Rule**

Find the derivatives of the following functions:

(a) $f(x) = x^3$  (b) $y = \dfrac{1}{x^2}$

**SOLUTION**

(a) By the Power Rule we have

$$f(x) = x^3 \qquad \text{Function}$$

$$f'(x) = 3x^2 \qquad \text{Derivative}$$

(b) By the Power Rule we have

$$y = \frac{1}{x^2} = x^{-2} \qquad \text{Function}$$

$$\frac{dy}{dx} = (-2)x^{-3} = -\frac{2}{x^3} \qquad \text{Derivative}$$

In part (b) of Example 2 note that *before* differentiating we rewrite $1/x^2$ as $x^{-2}$. Rewriting is the first step in *many* differentiation problems.

| Given: | Rewrite: | Differentiate: | Simplify: |
|--------|----------|----------------|-----------|
| $y = \dfrac{1}{x^2}$ | $y = x^{-2}$ | $\dfrac{dy}{dx} = (-2)x^{-3}$ | $\dfrac{dy}{dx} = -\dfrac{2}{x^3}$ |

In proving the following differentiation rule, we use the following property of limits:

$$\lim_{x \to a} cg(x) = c[\lim_{x \to a} g(x)]$$

Since the derivative is defined as a limit it is not surprising that many of the properties of limits produce corresponding properties for derivatives.

**Constant Multiple Rule**

If $f$ is a differentiable function of $x$ and $c$ is a real number, then

$$\frac{d}{dx}[cf(x)] = cf'(x), \quad c \text{ is a constant}$$

**Proof**   Applying the definition of the derivative, we have

$$\frac{d}{dx}[cf(x)] = \lim_{\Delta x \to 0} \frac{cf(x + \Delta x) - cf(x)}{\Delta x}$$

$$= \lim_{\Delta x \to 0} c\left[\frac{f(x + \Delta x) - f(x)}{\Delta x}\right]$$

$$= c\left[\lim_{\Delta x \to 0} \frac{f(x + \Delta x) - f(x)}{\Delta x}\right]$$

$$= cf'(x)$$

Informally, the Constant Multiple Rule states that constants can be factored out of the differentiation process.

$$\frac{d}{dx}[cf(x)] = c\frac{d}{dx}[\bigcirc f(x)] = cf'(x)$$

The usefulness of this rule is often overlooked, especially when the constant appears in the denominator, as follows:

$$\frac{d}{dx}\left[\frac{f(x)}{c}\right] = \frac{d}{dx}\left[\frac{1}{c}f(x)\right] = \frac{1}{c}\left[\frac{d}{dx}[\bigcirc f(x)]\right] = \frac{1}{c}f'(x)$$

To use the Constant Multiple Rule to best advantage be on the lookout for constants that can be factored out *before* differentiating.

**EXAMPLE 3**   **Applying the Power Rule and the Constant Multiple Rule**
Differentiate the following functions:

(a) $y = 2x^{1/2}$

(b) $f(t) = \dfrac{4t^2}{5}$

**SOLUTION**   (a) By the Constant Multiple Rule and the Power Rule we have

$$\frac{dy}{dx} = \frac{d}{dx}[2x^{1/2}] = \overbrace{2\frac{d}{dx}[x^{1/2}]}^{\substack{\text{Constant}\\ \text{Multiple Rule}}} = 2\overbrace{\left(\frac{1}{2}x^{-1/2}\right)}^{\substack{\text{Power}\\ \text{Rule}}} = x^{-1/2} = \frac{1}{\sqrt{x}}$$

(b) We begin by rewriting $f(t)$ as

$$f(t) = \frac{4t^2}{5} = \frac{4}{5}t^2$$

Then by the Constant Multiple and Power Rules we have

$$f'(t) = \frac{d}{dt}\left[\frac{4}{5}t^2\right] = \frac{4}{5}\left[\frac{d}{dt}(t^2)\right] = \frac{4}{5}(2t) = \frac{8}{5}t$$

In practice it is helpful to combine the Constant Multiple and Power Rules into one rule. The combination rule is

$$\frac{d}{dx}[cx^n] = cnx^{n-1}, \quad n \text{ is a real number, } c \text{ is a constant}$$

For instance, in part (b) of Example 3 we can apply this combination rule to obtain

$$\frac{d}{dx}\left[\frac{4}{5}t^2\right] = \left(\frac{4}{5}\right)(2)t = \frac{8}{5}t$$

The three functions in the next example are very simple, yet errors are frequently made in differentiating a constant multiple of the first power of $x$. Keep in mind that

$$\frac{d}{dx}[cx] = c, \quad c \text{ is a constant}$$

**EXAMPLE 4**

**Applying the Constant Multiple Rule**

| *Function* | *Derivative* |
|---|---|
| (a) $y = -\dfrac{3x}{2}$ | $y' = -\dfrac{3}{2}$ |
| (b) $y = 3\pi x$ | $y' = 3\pi$ |
| (c) $y = -\dfrac{x}{2}$ | $y' = -\dfrac{1}{2}$ |

The next rule is one that you might expect to be true, and you may have used it without thinking about it. For instance, if you were to find the derivative of $y = 3x + 2x^3$, you would probably write

$$y' = 3 + 6x^2$$

without questioning your answer. The validity of differentiating a sum term by term is given in the following rule.

**Sum Rule**

The derivative of the sum of two differentiable functions is the sum of their derivatives.

$$\frac{d}{dx}[f(x) + g(x)] = f'(x) + g'(x)$$

Proof

Let $h(x) = f(x) + g(x)$. Then the derivative of $h$ is

$$h'(x) = \lim_{\Delta x \to 0} \frac{h(x + \Delta x) - h(x)}{\Delta x}$$

$$= \lim_{\Delta x \to 0} \frac{f(x + \Delta x) + g(x + \Delta x) - f(x) - g(x)}{\Delta x}$$

$$= \lim_{\Delta x \to 0} \frac{f(x + \Delta x) - f(x) + g(x + \Delta x) - g(x)}{\Delta x}$$

$$= \lim_{\Delta x \to 0} \left[ \frac{f(x + \Delta x) - f(x)}{\Delta x} + \frac{g(x + \Delta x) - g(x)}{\Delta x} \right]$$

$$= \lim_{\Delta x \to 0} \frac{f(x + \Delta x) - f(x)}{\Delta x} + \lim_{\Delta x \to 0} \frac{g(x + \Delta x) - g(x)}{\Delta x}$$

$$= f'(x) + g'(x)$$

Thus,

$$\frac{d}{dx}[f(x) + g(x)] = f'(x) + g'(x)$$

By a similar procedure it can be shown that the derivative of a difference is the difference of the derivatives. That is,

$$\frac{d}{dx}[f(x) - g(x)] = f'(x) - g'(x)$$

Furthermore, either rule can be extended to the derivative of the sum or difference of any finite number of functions. For instance, if

$$y = f(x) + g(x) - h(x) - k(x)$$

then

$$y' = f'(x) + g'(x) - h'(x) - k'(x)$$

With the four differentiation rules given so far in this section we can now differentiate *any* polynomial function. This is illustrated in the next two examples.

**EXAMPLE 5**

**Applying the Sum Rule**

Find the slope of the graph of the polynomial function

$$f(x) = x^3 - 4x + 2$$

at the point $(1, -1)$.

**SOLUTION**

The derivative of $f(x)$ is

$$f'(x) = 3x^2 - 4$$

Therefore, the slope of the graph of $f$ at $(1, -1)$ is given by

$$\text{Slope} = f'(1) = 3(1)^2 - 4 = -1$$

as shown in Figure 2.14.

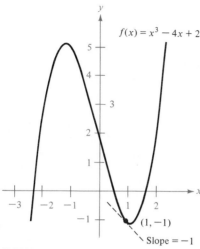

$f(x) = x^3 - 4x + 2$

$(1, -1)$

Slope $= -1$

FIGURE 2.14

■■■ **Remark:** Example 5 illustrates the use of the derivative for determining the shape of the graph of a function. A rough sketch of the graph of $f(x) = x^3 - 4x + 2$ might lead you to think that the point $(1, -1)$ is a minimum point of the graph. However, after finding the slope at this point to be $-1$, we can conclude that the graph continues to drop to the right of the point before it turns upward. (We will look in detail at the problem of finding local minimum and maximum points of a graph in Section 3.2.)

**EXAMPLE 6**

**Applying the Sum Rule**

Find the equation of the tangent line to the graph of

$$g(x) = -\frac{x^4}{2} + 3x^3 - 2x$$

at the point $(-1, -\frac{3}{2})$.

**SOLUTION**

The derivative of $g$ is given by

$$g'(x) = -\frac{4}{2}x^3 + 9x^2 - 2$$

$$= -2x^3 + 9x^2 - 2$$

which implies that the slope of the graph of $g$ at the point $(-1, -\frac{3}{2})$ is given by

$$\text{Slope} = g'(-1) = -2(-1)^3 + 9(-1)^2 - 2$$

$$= 2 + 9 - 2 = 9$$

Thus, using the point-slope form, the equation of the tangent line at $(-1, -\frac{3}{2})$ is

$$y + \frac{3}{2} = 9(x + 1)$$

$$y = 9x + \frac{15}{2}$$

Parentheses play an important role in the use of the Constant Multiple Rule and Power Rule. In the following example, be sure you understand the mathematical conventions involving the use of parentheses.

**EXAMPLE 7**  **Using Parentheses When Differentiating**

| Function | Rewrite | Differentiate | Simplify |
|---|---|---|---|
| (a) $y = \dfrac{5}{2x^3}$ | $y = \dfrac{5}{2}(x^{-3})$ | $y' = \dfrac{5}{2}(-3x^{-4})$ | $y' = -\dfrac{15}{2x^4}$ |
| (b) $y = \dfrac{5}{(2x)^3}$ | $y = \dfrac{5}{8}(x^{-3})$ | $y' = \dfrac{5}{8}(-3x^{-4})$ | $y' = -\dfrac{15}{8x^4}$ |
| (c) $y = \dfrac{7}{3x^{-2}}$ | $y = \dfrac{7}{3}(x^2)$ | $y' = \dfrac{7}{3}(2x)$ | $y' = \dfrac{14x}{3}$ |
| (d) $y = \dfrac{7}{(3x)^{-2}}$ | $y = 63(x^2)$ | $y' = 63(2x)$ | $y' = 126x$ |

When differentiating functions involving radicals, we rewrite the function in terms of rational exponents, as shown in the next example.

**EXAMPLE 8**  **Differentiating a Function Involving a Radical**

| Function | Rewrite | Differentiate | Simplify |
|---|---|---|---|
| (a) $y = \sqrt{x}$ | $y = x^{1/2}$ | $y' = \left(\dfrac{1}{2}\right)x^{-1/2}$ | $y' = \dfrac{1}{2\sqrt{x}}$ |
| (b) $y = \dfrac{1}{2\sqrt[3]{x^2}}$ | $y = \dfrac{1}{2}x^{-2/3}$ | $y' = \dfrac{1}{2}\left(-\dfrac{2}{3}\right)x^{-5/3}$ | $y' = -\dfrac{1}{3x^{5/3}}$ |
| (c) $y = \sqrt{2x}$ | $y = \sqrt{2}x^{1/2}$ | $y' = \dfrac{\sqrt{2}}{2}x^{-1/2}$ | $y' = \dfrac{1}{\sqrt{2x}}$ |

Now that you are able to find derivatives without the four-step process, try looking back at the examples in Section 2.1. See if you can obtain the same derivative (or slope) using differentiation rules.

In Exercises 1–4, find the slope of the tangent line to $y = x^n$ at the point $(1, 1)$.

**1.** (a) $y = x^2$

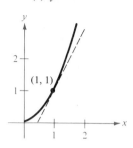

(b) $y = x^{1/2}$

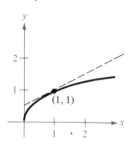

**2.** (a) $y = x^{3/2}$

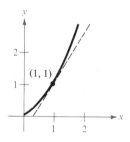

(b) $y = x^3$

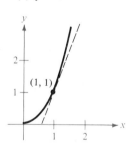

**3.** (a) $y = x^{-1}$

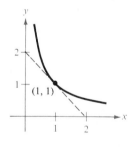

(b) $y = x^{-3/2}$

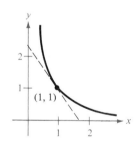

**4.** (a) $y = x^{-1/2}$

(b) $y = x^{-2}$

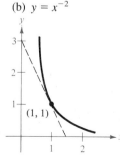

In Exercises 5–20, use the differentiation rules in this section to find the derivative of the given function.

**5.** $y = 3$

**6.** $f(x) = -2$

**7.** $f(x) = x + 1$

**8.** $g(x) = 3x - 1$

**9.** $g(x) = x^2 + 4$

**10.** $y = t^2 + 2t - 3$

**11.** $f(t) = -2t^2 + 3t - 6$

**12.** $y = x^3 - 9$

**13.** $s(t) = t^3 - 2t + 4$

**14.** $f(x) = 2x^3 - x^2 + 3x - 1$

**15.** $y = 4t^{3/4}$

**16.** $h(x) = x^{5/2}$

**17.** $f(x) = 4\sqrt{x}$

**18.** $g(x) = 5\sqrt[3]{x} + 2$

**19.** $y = 4x^{-2} + 2x^2$

**20.** $s = 4t^{-1} + 1$

In Exercises 21–26, complete the table using Example 7 as a model.

| Function | Rewrite | Derivative | Simplify |
|---|---|---|---|
| **21.** $y = \dfrac{1}{3x^3}$ | | | |
| **22.** $y = \dfrac{2}{3x^2}$ | | | |
| **23.** $y = \dfrac{1}{(3x)^3}$ | | | |
| **24.** $y = \dfrac{\pi}{(3x)^2}$ | | | |
| **25.** $y = \dfrac{\sqrt{x}}{x}$ | | | |
| **26.** $y = \dfrac{4}{x^{-3}}$ | | | |

In Exercises 27–32, find the value of the derivative of the function at the indicated point.

| Function | Point |
|---|---|
| **27.** $f(x) = \dfrac{1}{x}$ | $(1, 1)$ |
| **28.** $f(x) = -\dfrac{1}{2} + \dfrac{7}{5}x^3$ | $\left(0, -\dfrac{1}{2}\right)$ |
| **29.** $f(t) = 3 - \dfrac{3}{5t}$ | $\left(\dfrac{3}{5}, 2\right)$ |
| **30.** $y = 3x\left(x^2 - \dfrac{2}{x}\right)$ | $(2, 18)$ |
| **31.** $y = (2x + 1)^2$ | $(0, 1)$ |
| **32.** $f(x) = 3(5 - x)^2$ | $(5, 0)$ |

In Exercises 33–44, find $f'(x)$.

**33.** $f(x) = x^2 - \dfrac{4}{x}$

**34.** $f(x) = x^2 - 3x - 3x^{-2} + 5x^{-3}$

**35.** $f(x) = x^3 - 3x - \dfrac{2}{x^4}$  **36.** $f(x) = x^2 + 4x + \dfrac{1}{x}$

**37.** $f(x) = \dfrac{x^3 - 3x^2 + 4}{x^2}$  **38.** $f(x) = \dfrac{2x^2 - 3x + 1}{x}$

**39.** $f(x) = x(x^2 + 1)$  **40.** $f(x) = (x^2 + 2x)(x + 1)$

**41.** $f(x) = x^{4/5}$  **42.** $f(x) = x^{1/3} - 1$

**43.** $f(x) = \sqrt[3]{x} + \sqrt[5]{x}$  **44.** $f(x) = \dfrac{1}{\sqrt[3]{x^2}}$

In Exercises 45 and 46, find an equation of the tangent line to the given function at the indicated point.

| Function | Point |
|---|---|
| **45.** $y = x^4 - 3x^2 + 2$ | $(1, 0)$ |
| **46.** $y = x^3 + x$ | $(-1, -2)$ |

In Exercises 47–50, determine the point(s) (if any) at which the given function has a horizontal tangent line.

**47.** $y = x^4 - 2x^2 + 2$  **48.** $y = \dfrac{1}{2}x^2 + 5x$

**49.** $f(x) = x^3 + x$  **50.** $y = x^2 + 1$

**51.** If $h(x) = f(x) + C$, use the Constant Rule and the Sum Rule to show that $h'(x) = f'(x)$.

**52.** Given the functions $f(x) = x^2$, $g(x) = x^2 + 2$, and $h(x) = x^2 - 1$,
 (a) Sketch the graphs of $f$, $g$, and $h$ on the same set of axes.
 (b) Sketch the tangent lines on the graphs of part (a) if $x = 1$.
 (c) Find $f'(1)$, $g'(1)$, and $h'(1)$, the slope of the tangent lines of part (b). (See Exercise 51.)

**53.** Use the Constant Rule, the Constant Multiple Rule, and the Sum Rule to find $h'(1)$ given that $f(x) = x^3$ and $f'(1) = 3$, where $h(x)$ is defined as follows:
 (a) $h(x) = f(x) - 2$  (b) $h(x) = 2f(x)$

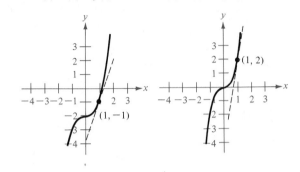

 (c) $h(x) = -f(x)$  (d) $h(x) = 4 - f(x)$

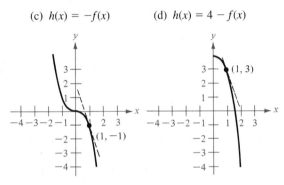

**54.** Show that the graphs of the equations $y = x$ and $y = 1/x$ have tangent lines that are perpendicular to each other at their points of intersection.

# Rates of Change: Velocity and Marginals

## INTRODUCTORY EXAMPLE
### The Rate of Inflation

The most commonly used measure of the rate of inflation in the United States is the Consumer Price Index (CPI). This index is published by the United States Bureau of Labor Statistics and is a statistical measure of the change in price of a fixed amount of selected goods and services. The CPI actually consists of several indices—food, rent, fuel, apparel, medical, and so on. Table 2.2 lists the CPI for all items from 1967 to 1984.

The numbers in the index represent prices relative to the base year of 1967. That is, each number in the CPI represents the year's price for goods and services that cost $100.00 in 1967. For example, from Table 2.2, we see that goods and services costing $100.00 in 1967 would have cost approximately $116.30 in 1970 and $247.00 in 1980.

To measure the rate of inflation in a given year, we take the **average rate of change** in the CPI and divide by the CPI for January of that year. For example, the average rate of change in the CPI for 1979 was given by

$$\frac{\text{Average rate}}{\text{of change}} = \frac{\text{CPI}(1980) - \text{CPI}(1979)}{1980 - 1979} = 29.3$$

This implies that the rate of inflation during 1979 was

$$\text{Rate of inflation} = \frac{29.3}{217.7} = 0.135 = 13.5\%$$

Figure 2.15 graphically depicts the CPI for 1950–1984.

TABLE 2.2

| Year | CPI (all items) | Inflation rate (%) |
|------|-----------------|--------------------|
| 1967 | 100.0 | 4.2 |
| 1968 | 104.2 | 5.4 |
| 1969 | 109.8 | 5.9 |
| 1970 | 116.3 | 4.3 |
| 1971 | 121.3 | 3.3 |
| 1972 | 125.3 | 6.2 |
| 1973 | 133.1 | 11.0 |
| 1974 | 147.7 | 9.1 |
| 1975 | 161.2 | 5.8 |
| 1976 | 170.5 | 6.5 |
| 1977 | 181.5 | 7.7 |
| 1978 | 195.3 | 11.5 |
| 1979 | 217.7 | 13.5 |
| 1980 | 247.0 | 10.2 |
| 1981 | 272.3 | 6.0 |
| 1982 | 288.6 | 3.0 |
| 1983 | 297.4 | 4.5 |
| 1984 | 310.7 | — |

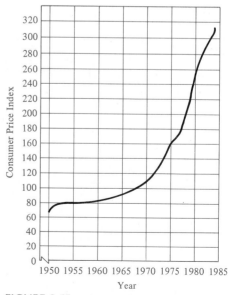

FIGURE 2.15

- **Average Rate of Change**
- **Instantaneous Rate of Change**
- **The Derivative as a Rate of Change**
- **Marginals**

We have already seen how the derivative is used to determine the slope of a curve. We now consider another interpretation—namely, as a way of determining the rate of change in one variable with respect to another. There are numerous applications of the notion of rate of change. A few examples are population growth rates, unemployment rates, production rates, and rate of water flow. Although rates of change often involve change with respect to time, we can actually investigate the rate of change with respect to any related variable.

## Average Rate of Change

When determining the rate of change of one variable with respect to another we must be careful to distinguish between *average* and *instantaneous* rates of change. The distinction between these two rates of change is comparable to the distinction between the slope of the secant line through two points on a curve and the slope of the tangent line at one point on a curve.

| **Definition of average rate of change** | If $y = f(x)$, then the **average rate of change** of $y$ with respect to $x$ on the interval $[x, x + \Delta x]$ is given by $$\text{Average rate of change} = \frac{\Delta y}{\Delta x}$$ |
|---|---|

**EXAMPLE 1**

### Finding the Average Rate of Change over an Interval

A drug is administered to a patient. The drug concentration in the patient's bloodstream is monitored over 10-minute intervals for two hours. Find the average rates of change (in milligrams per minute) over the following intervals for the concentrations in Table 2.3:

(a) [0, 10]    (b) [0, 20]    (c) [100, 110]

TABLE 2.3

| $t$ (min) | 0 | 10 | 20 | 30 | 40 | 50 | 60 | 70 | 80 | 90 | 100 | 110 | 120 |
|---|---|---|---|---|---|---|---|---|---|---|---|---|---|
| $C$ (mg) | 0 | 2 | 17 | 37 | 55 | 73 | 89 | 103 | 111 | 113 | 113 | 103 | 68 |

**SOLUTION**

(a) For the interval [0, 10], the average rate of change is

$$\frac{\Delta C}{\Delta t} = \frac{2 - 0}{10 - 0} = \frac{2}{10} = 0.2 \text{ mg/min}$$

(b) For the interval [0, 20], the average rate of change is

$$\frac{\Delta C}{\Delta t} = \frac{17 - 0}{20 - 0} = \frac{17}{20} = 0.85 \text{ mg/min}$$

(c) For the interval [100, 110], the average rate of change is

$$\frac{\Delta C}{\Delta t} = \frac{103 - 113}{110 - 100} = \frac{-10}{10} = -1 \text{ mg/min}$$

Note in Example 1 that the average rate of change is positive when the concentration increases and negative when the concentration decreases, as shown in Figure 2.16.

**EXAMPLE 2**

**Finding the Average Rate of Change**

If a free-falling object is dropped from a height of 100 feet, its height $h$ at time $t$ is given by

$$h = -16t^2 + 100$$

where $h$ is measured in feet and $t$ is measured in seconds. Find the average rate of change of the height over the following intervals. (See Figure 2.17.)
(a) [1, 2]  (b) [1, 1.5]  (c) [1, 1.1]

**SOLUTION**

Using the equation $h = -16t^2 + 100$, we determine the heights at $t = 1, 1.1, 1.5,$ and 2, as shown in Table 2.4.

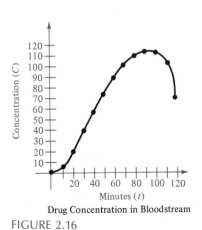

FIGURE 2.16

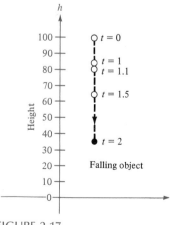

FIGURE 2.17

**TABLE 2.4**

| $t$ (sec) | 1 | 1.1 | 1.5 | 2 |
|-----------|-----|-------|-----|-----|
| $h$ (ft)  | 84  | 80.64 | 64  | 36  |

(a) For the interval [1, 2], the object falls from a height of 84 feet to a height of 36 feet. Thus, the average rate of change is

$$\frac{\Delta h}{\Delta t} = \frac{84 - 36}{1 - 2} = \frac{48}{-1} = -48 \text{ ft/sec}$$

(b) For the interval [1, 1.5], the average rate of change is

$$\frac{\Delta h}{\Delta t} = \frac{84 - 64}{1 - 1.5} = \frac{20}{-0.5} = -40 \text{ ft/sec}$$

(c) For the interval [1, 1.1], the average rate of change is

$$\frac{\Delta h}{\Delta t} = \frac{84 - 80.64}{1 - 1.1} = \frac{3.36}{-0.1} = -33.6 \text{ ft/sec}$$

### Instantaneous Rate of Change

Suppose that in Example 2 we wanted to find the rate of change in $h$ at the instant $t = 1$ second. We call this the **instantaneous rate of change** of the height when $t = 1$. Just as we approximated the slope of the tangent line by the slope of the secant line, we can approximate the instantaneous rate of change at $t = 1$ by calculating the average rate of change over a small interval [1, 1 + $\Delta t$], as shown in Table 2.5.

**TABLE 2.5**

$\Delta t$ approaches 0 →

| $\Delta t$ | 1 | 0.5 | 0.1 | 0.01 | 0.001 | 0.0001 | | 0 |
|------------|-----|-----|-------|--------|---------|----------|---|-----|
| $\dfrac{\Delta h}{\Delta t}$ | $-48$ | $-40$ | $-33.6$ | $-32.16$ | $-32.016$ | $-32.0016$ | | $-32$ |

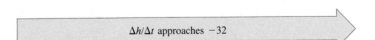

$\Delta h / \Delta t$ approaches $-32$ →

From Table 2.5 it seems reasonable to conclude that the instantaneous rate of change of the height when $t = 1$ is $-32$ ft/sec. We will verify this conclusion after presenting the following definition.

| **Definition of instantaneous rate of change** | The **instantaneous rate of change** of $y$ at $x$ is the limit of the average rate of change on the interval $[x, x + \Delta x]$: |
|---|---|

$$\lim_{\Delta x \to 0} \frac{\Delta y}{\Delta x} = \lim_{\Delta x \to 0} \frac{f(x + \Delta x) - f(x)}{\Delta x}$$

The limit in the preceding definition is the same as the limit in the definition of the derivative of $f$ at $x$. This is an important observation, and we have arrived at a second major interpretation of the derivative—as an *instantaneous rate of change in one variable with respect to another*. We make the following summary statement concerning the derivative and its interpretations.

| **Interpretations of the derivative** | Let the function given by $y = f(x)$ be differentiable at $x$. Then its derivative |
|---|---|

$$\frac{dy}{dx} = f'(x) = \lim_{\Delta x \to 0} \frac{f(x + \Delta x) - f(x)}{\Delta x}$$

denotes both of the following:

1. The *slope* of the graph of $f$ at $x$.
2. The *instantaneous rate of change* in $y$ with respect to $x$.

**Remark:** In future work with derivatives we will use "rate of change" to mean "instantaneous rate of change."

Now let us return to the problem of finding the rate of change in the height of the falling object in Example 2 when $t = 1$. Considering the derivative as a rate of change, we see that if

$$h = -16t^2 + 100$$

then

$$\frac{dh}{dt} = -32t$$

and we conclude that the rate of change in $h$ when $t = 1$ is

$$\frac{dh}{dt} = -32 \text{ ft/sec}$$

We call this particular rate of change the **velocity** of the falling body when $t = 1$. Note that negative velocity indicates that the object's height is decreasing. The use of the derivative to compute velocity is important in a wide variety of problems in physics and engineering.

**EXAMPLE 3**

**Using the Derivative to Find Velocity**

At time $t = 0$, a diver jumps from a diving board that is 32 feet high. The position of the diver is given by

$$h(t) = -16t^2 + 16t + 32$$

where $h(t)$ is measured in feet and $t$ is measured in seconds. (See Figure 2.18.)

---

(a) When does the diver hit the water?
(b) What is the diver's velocity at impact?

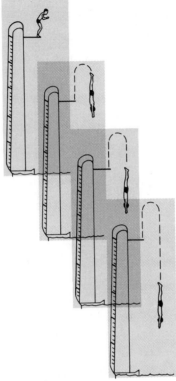

FIGURE 2.18

**SOLUTION**

(a) To find the time at which the diver hits the water, we let $h(t) = 0$ and solve for $t$:

$$-16t^2 + 16t + 32 = 0$$

$$-16(t^2 - t - 2) = 0$$

$$-16(t + 1)(t - 2) = 0$$

$$t = -1, 2$$

The solution $t = -1$ doesn't make sense, so we conclude that the diver hits the water when $t = 2$ seconds.

(b) The velocity at time $t$ is given by the derivative

$$h'(t) = -32t + 16$$

Therefore, the velocity at time $t = 2$ is

$$h'(2) = -32(2) + 16 = -64 + 16 = -48 \text{ ft/sec}$$

## Marginals

Another important use of rates of change is in the field of economics. Economists refer to **marginal profit, marginal revenue,** and **marginal cost** as the rates of change of the profit, revenue, and cost with respect to the number of units produced or sold. An equation that relates these three quantities is

$$P = R - C$$

where

$$P = \text{total profit}$$

$$R = \text{total revenue}$$

$$C = \text{total cost}$$

Differentiating each of these gives the *marginals,* a term used in economics to denote derivatives:

$$\frac{dP}{dx} = \text{marginal profit}$$

$$\frac{dR}{dx} = \text{marginal revenue}$$

$$\frac{dC}{dx} = \text{marginal cost}$$

In many problems in business and economics the number of units produced or sold is restricted to positive integer values, as shown in Figure 2.19. (Of course, it could happen that a sale involves half or quarter units, but it is hard to conceive of a sale involving $\sqrt{2}$ units.) The variable that denotes such units is called a **discrete variable.** To analyze a function of a discrete variable $x$, we temporarily assume that $x$ is able to take on any real value in a given interval, as shown in Figure 2.20. Then we use the methods of calculus to find the $x$-value that corresponds to the marginal revenue, maximum profit, least cost, or whatever is called for. Finally, we round the solution off to the nearest sensible $x$-value—cents, dollars, units, or days, depending on the context of the problem.

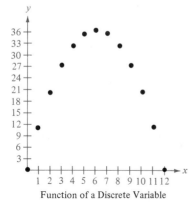

FIGURE 2.19     Function of a Discrete Variable

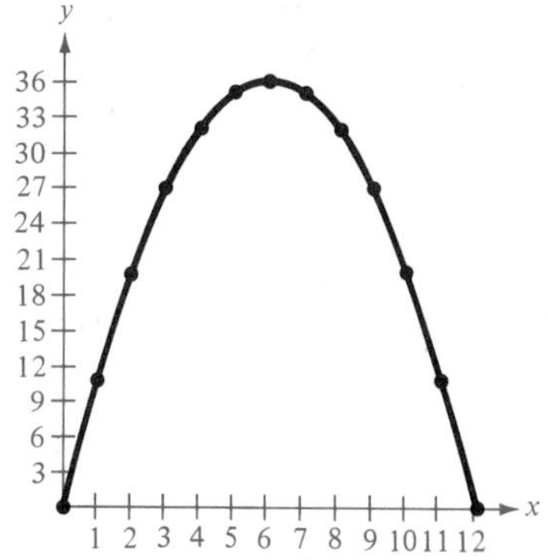

FIGURE 2.20     Function of a Continuous Variable

**EXAMPLE 4**

**Finding the Marginal Profit**

A manufacturer determines that the profit derived from selling $x$ units of a certain item is given by

$$P = 0.0002x^3 + 10x$$

(a) Find the marginal profit for a production level of 50 units.
(b) Compare this to the actual gain in profit obtained by increasing the production level from 50 to 51 units.

**SOLUTION**

(a) Since

$$P = 0.0002x^3 + 10x$$

the marginal profit is given by the derivative

$$\frac{dP}{dx} = 0.0006x^2 + 10$$

When $x = 50$, the marginal profit is

$$\frac{dP}{dx} = (0.0006)(50)^2 + 10 = 1.5 + 10 = \$11.50 \text{ per unit}$$

(b) For $x = 50$ the actual profit is

$$P = (0.0002)(50)^3 + 10(50) = 25 + 500 = \$525.00$$

and for $x = 51$ the actual profit is

$$P = (0.0002)(51)^3 + 10(51) = 26.53 + 510 = \$536.53$$

Thus the additional profit obtained by increasing the production level from 50 to 51 units is

$$536.53 - 525.00 = \$11.53$$

Note that the actual profit increase of \$11.53 (when $x$ increases from 50 to 51) can be approximated by the marginal profit of \$11.50 per unit (when $x = 50$), as shown in Figure 2.21.

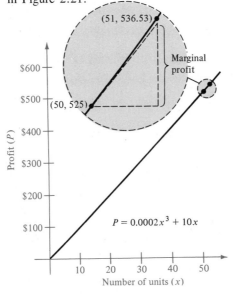

FIGURE 2.21

**Remark:** The reason the marginal profit gives a good approximation of the actual change in profit is that the graph of $P$ is nearly straight over the interval $50 \le x \le 51$. We will discuss the validity of this approximation method later. (See Section 3.8, Introductory Example.)

The profit function in Example 4 is quite unusual in that the profit continues to increase as long as the number of units sold increases. In practice it is more common to encounter situations in which sales can be increased only by lowering the price per item. Such reductions in price will ultimately cause the profit to decline.

We define the number of units $x$ that consumers are willing to purchase at a given price per unit $p$ by the **demand function**

$$p = f(x) \qquad \text{Demand function}$$

The total revenue $R$ is then related to the price per unit and the quantity demanded (or sold) by the equation

$$R = xp$$

**EXAMPLE 5**

**Finding the Marginal Revenue**

A fast-food restaurant has determined that the monthly demand for their hamburgers is given by

$$p = \frac{60,000 - x}{20,000}$$

Figure 2.22 shows that as the price decreases, the demand increases. Table 2.6 shows the demand for hamburgers at various prices.

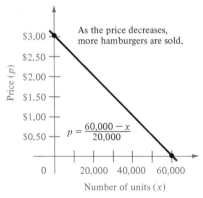

FIGURE 2.22

TABLE 2.6

| $x$ | 60,000 | 50,000 | 40,000 | 30,000 | 20,000 | 10,000 | 0 |
|---|---|---|---|---|---|---|---|
| $p$ | 0 | $0.50 | $1.00 | $1.50 | $2.00 | $2.50 | $3.00 |

Find the increase in revenue per hamburger for monthly sales of 20,000 hamburgers. In other words, find the marginal revenue when $x = 20,000$.

**SOLUTION**

Since the demand is given by

$$p = \frac{60,000 - x}{20,000}$$

the revenue is given by $R = xp$, and we have

$$R = xp = x\left(\frac{60,000 - x}{20,000}\right) = \frac{1}{20,000}(60,000x - x^2)$$

By differentiating, we find the marginal revenue to be

$$\frac{dR}{dx} = \frac{1}{20,000}(60,000 - 2x)$$

Finally, when $x = 20,000$, the marginal revenue is

$$\frac{dR}{dx} = \frac{1}{20,000}[60,000 - 2(20,000)] = \frac{20,000}{20,000} = \$1 \text{ per unit}$$

■■■ **Remark:** Writing a demand function in the form $p = f(x)$ is a convention often used in economics. From a consumer's point of view, it might seem more reasonable to think that the quantity demanded is a function of the price. However, mathematically the two points of view are equivalent, since a typical demand function is one-to-one and therefore possesses an inverse. For instance, in Example 5, we could have written the demand as a function of the price as follows:

$$x = 60,000 - 20,000p$$

**EXAMPLE 6**

**Finding the Marginal Profit**

Suppose that in Example 5 the cost $C$ of producing $x$ hamburgers is

$$C = 5000 + 0.56x, \quad 0 \le x \le 50,000$$

Find the profit *and* the marginal profit for the following production levels:
(a) $x = 20,000$  (b) $x = 24,400$  (c) $x = 30,000$

**SOLUTION**

From Example 5 we know that the total revenue from selling $x$ units is

$$R = \frac{1}{20,000}(60,000x - x^2)$$

Since the total profit is given by $P = R - C$, we have

$$P = \frac{1}{20,000}(60,000x - x^2) - 5000 - 0.56x$$

$$= 3x - \frac{x^2}{20,000} - 5000 - 0.56x$$

$$= 2.44x - \frac{x^2}{20,000} - 5000$$

Thus, the marginal profit is given by

$$\frac{dP}{dx} = 2.44 - \frac{x}{10,000}$$

(a) When $x = 20,000$ the profit and marginal profit are

$$P = (2.44)(20,000) - \frac{(20,000)^2}{20,000} - 5000 = \$23,800.00$$

$$\frac{dP}{dx} = 2.44 - \frac{20,000}{10,000} = \$0.44 \text{ per unit}$$

(b) When $x = 24,400$ the profit and marginal profit are

$$P = (2.44)(24,400) - \frac{(24,400)^2}{20,000} - 5000 = \$24,768.00$$

$$\frac{dP}{dx} = 2.44 - \frac{24,400}{10,000} = 0$$

(c) When $x = 30,000$ the profit and marginal profit are

$$P = (2.44)(30,000) - \frac{(30,000)^2}{20,000} - 5000 = \$23,200.00$$

$$\frac{dP}{dx} = 2.44 - \frac{30,000}{10,000} = -\$0.56 \text{ per unit}$$

**Remark:** In Example 6 we can see that when more than 24,400 hamburgers are sold the marginal profit is negative. This means that increasing production beyond this point will *reduce* rather than increase profit, as shown in Figure 2.23.

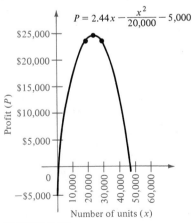

$$P = 2.44x - \frac{x^2}{20,000} - 5,000$$

FIGURE 2.23

1. The average hourly wage paid to U.S. workers for selected years from 1970 to 1983 is shown in the following table:

| Year | 1970 | 1975 | 1980 | 1982 | 1983 |
|------|------|------|------|------|------|
| Wage | $3.23 | $4.53 | $6.66 | $7.68 | $8.02 |

Find the average rate of change in hourly wages per year for the following periods of time:
(a) 1970 to 1975          (b) 1975 to 1980
(c) 1980 to 1982          (d) 1982 to 1983

2. Personal savings in billions of dollars in the United States from 1980 to 1983 is shown in the following table:

| Year | 1980 | 1981 | 1982 | 1983 |
|------|------|------|------|------|
| Savings | $6.0 | $6.7 | $6.2 | $5.0 |

Find the average rate of change in personal savings per year for the following periods of time:
(a) 1980 to 1981          (b) 1980 to 1982
(c) 1980 to 1983          (d) 1981 to 1983

In Exercises 3–12, sketch the graph of the function and find its average rate of change over the indicated interval. Compare this to the instantaneous rates of change at the endpoints of the interval.

|        *Function*        |     *Interval*     |
|--------------------------|--------------------|
| 3. $f(t) = 2t + 7$       | $[1, 2]$           |
| 4. $f(t) = 3t - 1$       | $\left[0, \dfrac{1}{3}\right]$ |
| 5. $h(x) = 1 - x^2$      | $[0, 1]$           |
| 6. $h(x) = x^2 - 4$      | $[-2, 2]$          |
| 7. $f(t) = t^2 - 3$      | $[2, 2.1]$         |
| 8. $f(x) = x^2 - 6x - 1$ | $[-1, 3]$          |
| 9. $f(x) = \dfrac{1}{x}$ | $[1, 4]$           |
| 10. $f(x) = x^{-1/2}$    | $[1, 4]$           |
| 11. $g(x) = 4\sqrt{x}$   | $[1, 9]$           |
| 12. $g(x) = x^3 - 1$     | $[-1, 1]$          |

13. Suppose the effectiveness $E$ (on a scale from 0 to 1) of a pain-killing drug $t$ hours after entering the bloodstream is given by

$$E(t) = \frac{1}{27}(9t + 3t^2 - t^3), \quad 0 \le t \le 4.5$$

Find the average rate of change of $E$ over the indicated interval and compare this to the instantaneous rates of change at the endpoints of the interval.
(a) $[0, 1]$              (b) $[1, 2]$
(c) $[2, 3]$              (d) $[3, 4]$

14. At 0° Celsius, the wind-chill-corrected temperature is given by

$$T(w) = 33 - 1.43(10\sqrt{w} - w + 10.45)$$

where $T$ is measured in degrees Celsius and the wind speed $w$ is measured in meters per second. Find the instantaneous rate at which $T$ is changing when the wind speed is
(a) 4 m/sec              (b) 9 m/sec

15. The height $s$ at time $t$ of a silver dollar dropped from the World Trade Center is given by

$$s = -16t^2 + 1350$$

where $s$ is measured in feet and $t$ is measured in seconds.
(a) Find the average velocity on the interval $[1, 2]$.
(b) Find the instantaneous velocity when $t = 1$ and $t = 2$.
(c) How long will it take the dollar to hit the ground?
(d) Find the velocity of the dollar when it hits the ground.

16. Suppose the height of an object fired straight up from ground level is given by

$$h(t) = 200t - 16t^2$$

where $h$ is measured in feet and $t$ is measured in seconds.
(a) How fast is the object moving as it leaves the ground (when $t = 0$)?
(b) How fast is the object moving after 3 seconds?

17. Suppose the position of an accelerating car is given by

$$s(t) = 10t^{3/2}, \quad 0 \le t \le 10$$

where $s$ is measured in feet and $t$ is measured in seconds. Find the velocity of the car when
(a) $t = 0$              (b) $t = 1$
(c) $t = 4$              (d) $t = 9$

18. Derive the equation for the velocity of an object whose position is given by

$$s(t) = t^2 + 2t$$

where $s$ is measured in meters and $t$ is measured in seconds.

In Exercises 19–22, use the cost function to find the marginal cost for producing $x$ units.

**19.** $C = 4500 + 1.47x$

**20.** $C = 104{,}000 + 7200x$

**21.** $C = 55{,}000 + 470x - \dfrac{1}{4}x^2$, $0 \le x \le 940$

**22.** $C = 100(9 + 3\sqrt{x})$

In Exercises 23–26, use the revenue function to find the marginal revenue for selling $x$ units.

**23.** $R = 50x - \dfrac{1}{2}x^2$

**24.** $R = 30x - x^2$

**25.** $R = -6x^3 + 8x^2 + 200x$

**26.** $R = 50(20x - x^{3/2})$

In Exercises 27–30, use the profit function to find the marginal profit for selling $x$ units.

**27.** $P = -2x^2 + 72x - 145$

**28.** $P = -\dfrac{1}{4}x^2 + 2000x - 1{,}250{,}000$

**29.** $P = -\dfrac{1}{4000}x^2 + 12.2x - 25{,}000$

**30.** $P = -\dfrac{1}{2}x^3 + 30x^2 - 164.25x - 1000$

**31.** The revenue from producing $x$ units of a product is

$$R = 12x - 0.001x^2$$

(a) Find the additional revenue if production is increased from 5000 to 5001 units.

(b) Find the marginal revenue when 5000 units are produced.

**32.** The revenue from renting $x$ apartments is

$$R = 25(900 + 32x - x^2)$$

(a) Find the additional revenue if the number of rentals is increased from 14 to 15.

(b) Find the marginal revenue when 14 apartments are rented.

**33.** If the average fuel cost is $1.35 per gallon, then the annual fuel cost of driving a car 15,000 mi/yr is approximately

$$C = \frac{20{,}250}{x}$$

where $x$ is the number of miles per gallon.

(a) Complete the following table and use it to sketch the graph of the cost function:

| $x$ | 10 | 15 | 20 | 25 | 30 | 35 | 40 |
|---|---|---|---|---|---|---|---|
| $C$ | | | | | | | |

(b) Complete the following table for the marginal cost:

| $x$ | 10 | 15 | 20 | 25 | 30 | 35 | 40 |
|---|---|---|---|---|---|---|---|
| $\dfrac{dC}{dx}$ | | | | | | | |

(c) Find the change in annual fuel cost when $x$ increases from 10 to 11 and compare this with the marginal cost when $x = 10$.

(d) Find the change in annual fuel cost when $x$ increases from 30 to 31 and compare this with the marginal cost when $x = 30$.

**34.** If $Q$ is the order size when the inventory is replenished, then the annual inventory cost for a certain manufacturer is given by

$$C = \frac{1{,}008{,}000}{Q} + \frac{63Q}{10}$$

(a) Complete the following table for the marginal cost:

| $Q$ | 300 | 350 | 400 | 450 | 500 |
|---|---|---|---|---|---|
| $\dfrac{dC}{dQ}$ | | | | | |

(b) Find the change in annual cost when $Q$ is increased from 350 to 351 and compare this with the marginal cost when $Q = 350$.

**35.** The profit for producing $x$ units of a product is given by

$$P = 2400 - 403.4x + 32x^2 - 0.664x^3$$

where $10 \le x \le 25$. Find the marginal profit when

(a) $x = 10$      (b) $x = 20$

(c) $x = 23$      (d) $x = 25$

**36.** The profit for producing $x$ units of a product is given by

$$P = 36{,}000 + 2048\sqrt{x} - \frac{1}{8x^2}$$

where $150 \le x \le 275$. Find the marginal profit when

(a) $x = 150$      (b) $x = 175$

(c) $x = 200$      (d) $x = 225$

(e) $x = 250$      (f) $x = 275$

**37.** The monthly demand function and the cost function for $x$ quarts of oil at a local service station are given, respectively, by

$$p = \frac{1100 - x}{400}$$

and

$$C = 65 + 1.25x$$

(a) Find the monthly revenue as a function of $x$.
(b) Find the profit as a function of $x$.
(c) Complete the following table for marginal revenue, marginal profit, and profit:

| $x$ | 200 | 250 | 300 | 350 | 400 |
|---|---|---|---|---|---|
| $\dfrac{dR}{dx}$ | | | | | |
| $\dfrac{dP}{dx}$ | | | | | |
| $P$ | | | | | |

**38.** The demand function and the cost function for $x$ units of a product are given, respectively, by

$$p = 25(20 - \sqrt{x})$$

and

$$C = 42.50 + 1.25x$$

(a) Find the revenue as a function of $x$.
(b) Find the profit as a function of $x$.
(c) Complete the following table for marginal revenue, marginal profit, and profit.

| $x$ | 50 | 100 | 150 | 200 |
|---|---|---|---|---|
| $\dfrac{dR}{dx}$ | | | | |
| $\dfrac{dP}{dx}$ | | | | |
| $P$ | | | | |

**39.** Since 1790 the center of population of the United States has been gradually moving westward. Use Figure 2.24 to estimate the rate (in miles per year) at which the center of population was moving *westward* during the period
(a) from 1790 to 1900    (b) from 1900 to 1970

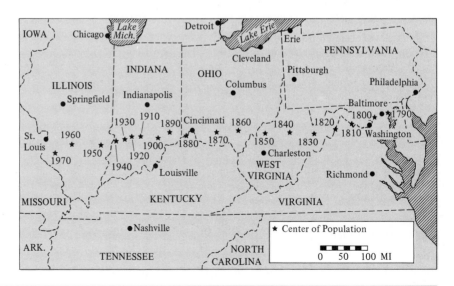

FIGURE 2.24

# The Product and Quotient Rules

## INTRODUCTORY EXAMPLE
### The Solubility of Oxygen in Water

When air is in contact with undersaturated water, some of the oxygen in the air is absorbed into the water. The amount of oxygen absorbed depends on the temperature, salinity, and pressure of the water. Cold water absorbs more oxygen than warm water; fresh water absorbs more oxygen than salt water; and water at low altitudes absorbs more oxygen than water at high altitudes. In 1964, Montgomery, Thom, and Cockburn* published the results of their experiments measuring the solubility of oxygen in pure water at sea level. From their measurements, they constructed a mathematical model (see Figure 2.25) for solubility:

$$S = \frac{468}{31.6 + t}, \quad 0 \le t$$

where $S$ is measured in milligrams of oxygen per liter of water and $t$ is the temperature in degrees Celsius. In this section, you will learn how to use the **Quotient Rule** to differentiate rational functions such as this one. By this rule, the rate of change of the solubility of oxygen in water with respect to the temperature is given by

$$\frac{dS}{dt} = -\frac{468}{(31.6 + t)^2}$$

Note that the fact that this rate of change is always negative $(0 \le t)$ means that as the temperature increases the solubility decreases. This coincides with our assertion that cold water absorbs more oxygen than warm water.

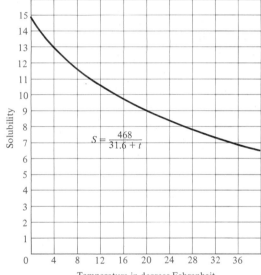

Temperature in degrees Fahrenheit

Solubility of Oxygen in Pure Water at Sea Level

FIGURE 2.25

---

*Data from H. A. C. Montgomery, N. S. Thom, and A. Cockburn, Determination of dissolved oxygen by the Winkler method and the solubility of oxygen in pure water at sea level, *J. Appl. Chem.* **14** (1964), 180–196.

■ **The Product Rule**
■ **The Quotient Rule**

In Section 2.2 we saw that the derivative of a sum or difference of two functions is simply the sum or difference of their derivatives. The rules for the derivative of a product or quotient of two functions are not so simple, and you may find the results surprising. For each rule, we list a verbal statement of the rule as well as a formula. We have found that memorizing the verbal statements is more helpful than trying to memorize the formula associated with each rule.

**Product Rule**

The derivative of the product of two differentiable functions is equal to the first function times the derivative of the second, plus the second times the derivative of the first.

$$\frac{d}{dx}[f(x)g(x)] = f(x)g'(x) + g(x)f'(x)$$

Proof

Some mathematical proofs, such as the proof of the Sum Rule in Section 2.2, are straightforward. Others involve clever steps that may appear to you to be unmotivated. This proof involves such a step—adding and subtracting the same quantity. (This step is shown in color.) To begin, let

$$F(x) = f(x)g(x)$$

Then the derivative of $F$ is given by

$$F'(x) = \lim_{\Delta x \to 0} \frac{F(x + \Delta x) - F(x)}{\Delta x}$$

$$= \lim_{\Delta x \to 0} \frac{f(x + \Delta x)g(x + \Delta x) - f(x)g(x)}{\Delta x}$$

$$= \lim_{\Delta x \to 0} \frac{f(x + \Delta x)g(x + \Delta x) - f(x + \Delta x)g(x) + f(x + \Delta x)g(x) - f(x)g(x)}{\Delta x}$$

$$= \lim_{\Delta x \to 0} \left[ f(x + \Delta x)\frac{g(x + \Delta x) - g(x)}{\Delta x} + g(x)\frac{f(x + \Delta x) - f(x)}{\Delta x} \right]$$

$$= \underbrace{\lim_{\Delta x \to 0} f(x + \Delta x)\frac{g(x + \Delta x) - g(x)}{\Delta x}}_{f(x)g'(x)} + \underbrace{\lim_{\Delta x \to 0} g(x)\frac{f(x + \Delta x) - f(x)}{\Delta x}}_{g(x)f'(x)}$$

The first of these two limits is given by

$$\left[ \lim_{\Delta x \to 0} f(x + \Delta x) \right]\left[ \lim_{\Delta x \to 0} \frac{g(x + \Delta x) - g(x)}{\Delta x} \right] = f(x)g'(x)$$

Similarly, the second limit is $g(x)f'(x)$, and we have

$$F'(x) = f(x)g'(x) + g(x)f'(x)$$

## EXAMPLE 1

**Finding the Derivative of a Product**

Find the derivative of

$$y = (3x - 2x^2)(5 + 4x)$$

**SOLUTION**

By the Product Rule,

$$\frac{dy}{dx} = \overbrace{(3x - 2x^2)}^{\text{(first)}} \overbrace{\frac{d}{dx}[5 + 4x]}^{\left(\substack{\text{derivative} \\ \text{of second}}\right)} + \overbrace{(5 + 4x)}^{\text{(second)}} \overbrace{\frac{d}{dx}[3x - 2x^2]}^{\left(\substack{\text{derivative} \\ \text{of first}}\right)}$$

$$= (3x - 2x^2)(4) + (5 + 4x)(3 - 4x)$$

$$= (12x - 8x^2) + (15 - 8x - 16x^2)$$

$$= 15 + 4x - 24x^2$$

■ **Remark:** In Example 1 we could have bypassed the Product Rule by first multiplying $(3x - 2x^2)$ by $(5 + 4x)$ to get $15x + 2x^2 - 8x^3$ and then differentiating this result. After we introduce the Chain Rule in Section 2.5 you will see that the Product Rule cannot always be bypassed by first multiplying the given factors.

Be sure you see that, in general, the derivative of the product of two functions is *not equal to* the product of the derivatives of the two functions. This is evident in Example 1, since

$$\frac{d}{dx}[(3x - 2x^2)(5 + 4x)] = 15 + 4x - 24x^2$$

whereas

$$\left(\frac{d}{dx}[3x - 2x^2]\right)\left(\frac{d}{dx}[5 + 4x]\right) = (3 - 4x)(4) = 12 - 16x$$

## EXAMPLE 2

**Finding the Derivative of a Product**

Find the derivative of

$$f(x) = \left(\frac{1}{x} + 1\right)(x - 1)$$

**SOLUTION**

We begin by rewriting $f(x)$ as

$$f(x) = (x^{-1} + 1)(x - 1)$$

Then using the Product Rule we have

$$f'(x) = (x^{-1} + 1)\frac{d}{dx}[x - 1] + (x - 1)\frac{d}{dx}[x^{-1} + 1]$$

$$= (x^{-1} + 1)(1) + (x - 1)(-x^{-2})$$

$$= \frac{1}{x} + 1 - \frac{x - 1}{x^2} = \frac{x + x^2 - x + 1}{x^2} = \frac{x^2 + 1}{x^2}$$

We now have two differentiation rules dealing with products—the Constant Multiple Rule and the Product Rule.* The difference between these two rules is that the Constant Multiple Rule deals with the product of a constant and a variable,

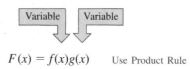

$$F(x) = cf(x) \qquad \text{Use Constant Multiple Rule}$$

whereas the Product Rule deals with the product of two variable quantities,

$$F(x) = f(x)g(x) \qquad \text{Use Product Rule}$$

We compare the use of these two rules in the next example.

**EXAMPLE 3**

**Comparing the Product Rule and Constant Multiple Rule**

Find the derivatives of the following:

(a) $y = 2xg(x)$ (b) $y = 2g(x)$

**SOLUTION**

(a) By the Product Rule we have

$$\frac{dy}{dx} = (2x)\frac{d}{dx}[g(x)] + g(x)\frac{d}{dx}[2x]$$

$$= 2xg'(x) + g(x)(2)$$

$$= 2[xg'(x) + g(x)]$$

(b) By the Constant Multiple Rule we have

$$\frac{dy}{dx} = 2g'(x)$$

The Product Rule can be extended to cover products involving more than two factors. For example, if $f$, $g$, and $h$ are differentiable functions of $x$, then

$$\frac{d}{dx}[f(x)g(x)h(x)] = f'(x)g(x)h(x) + g'(x)f(x)h(x) + h'(x)f(x)g(x)$$

---

*Actually we can consider the Constant Multiple Rule to be a special case of the Product Rule by writing

$$\frac{d}{dx}[cf(x)] = c\frac{d}{dx}[f(x)] + f(x)\frac{d}{dx}[c] = cf'(x) + f(x)(0) = cf'(x)$$

## The Quotient Rule

In Section 2.2 we saw that through the use of the Constant Rule, Power Rule, Constant Multiple Rule, and Sum Rule we are able to differentiate *any* polynomial function. By combining these rules with the following important rule, we are now able to differentiate *any* rational function.

**Quotient Rule**

The derivative of the quotient of two differentiable functions is equal to the denominator times the derivative of the numerator minus the numerator times the derivative of the denominator, divided by the square of the denominator.

$$\frac{d}{dx}\left[\frac{f(x)}{g(x)}\right] = \frac{g(x)f'(x) - f(x)g'(x)}{[g(x)]^2}, \quad g(x) \neq 0$$

**Proof**

As in the proof of the Product Rule, a key step in this proof involves adding and subtracting the same quantity. We let

$$F(x) = \frac{f(x)}{g(x)}$$

Then the derivative of $F$ is given by

$$F'(x) = \lim_{\Delta x \to 0} \frac{F(x + \Delta x) - F(x)}{\Delta x}$$

$$= \lim_{\Delta x \to 0} \frac{\dfrac{f(x + \Delta x)}{g(x + \Delta x)} - \dfrac{f(x)}{g(x)}}{\Delta x}$$

$$= \lim_{\Delta x \to 0} \frac{g(x)f(x + \Delta x) - f(x)g(x + \Delta x)}{\Delta x g(x)g(x + \Delta x)}$$

$$= \lim_{\Delta x \to 0} \frac{g(x)f(x + \Delta x) - f(x)g(x) + f(x)g(x) - f(x)g(x + \Delta x)}{\Delta x g(x)g(x + \Delta x)}$$

$$= \frac{\displaystyle\lim_{\Delta x \to 0} \frac{g(x)[f(x + \Delta x) - f(x)]}{\Delta x} - \lim_{\Delta x \to 0} \frac{f(x)[g(x + \Delta x) - g(x)]}{\Delta x}}{\displaystyle\lim_{\Delta x \to 0} [g(x)g(x + \Delta x)]}$$

$$= \frac{g(x)\left[\displaystyle\lim_{\Delta x \to 0} \dfrac{f(x + \Delta x) - f(x)}{\Delta x}\right] - f(x)\left[\displaystyle\lim_{\Delta x \to 0} \dfrac{g(x + \Delta x) - g(x)}{\Delta x}\right]}{\displaystyle\lim_{\Delta x \to 0} [g(x)g(x + \Delta x)]}$$

$$= \frac{g(x)f'(x) - f(x)g'(x)}{[g(x)]^2}$$

As suggested previously, you should memorize the verbal statement of the Quotient Rule. The following form may assist you in memorizing the rule:

$$\frac{d}{dx}[\text{quotient}] = \frac{(\text{denominator})\dfrac{d}{dx}[\text{numerator}] - (\text{numerator})\dfrac{d}{dx}[\text{denominator}]}{(\text{denominator})^2}$$

This form certainly points out that the derivative of the quotient of two functions is *not* given by the quotient of their respective derivatives.

**EXAMPLE 4**

**Finding the Derivative of a Quotient**

Find the derivative of

$$y = \frac{2x^2 - 4x + 3}{2 - 3x}$$

**SOLUTION**

By the Quotient Rule,

$$y' = \frac{(2 - 3x)\dfrac{d}{dx}[2x^2 - 4x + 3] - (2x^2 - 4x + 3)\dfrac{d}{dx}[2 - 3x]}{(2 - 3x)^2}$$

$$= \frac{(2 - 3x)(4x - 4) - (2x^2 - 4x + 3)(-3)}{(2 - 3x)^2}$$

$$= \frac{-12x^2 + 20x - 8 + (6x^2 - 12x + 9)}{(2 - 3x)^2}$$

$$= \frac{-6x^2 + 8x + 1}{(2 - 3x)^2}$$

**EXAMPLE 5**

**Finding the Derivative of a Quotient**

In the introductory example to this section we saw that the solubility of oxygen in pure water at sea level is given by

$$S = \frac{468}{31.6 + t}, \quad 0 \le t$$

where $S$ is measured in milligrams of oxygen per liter of water and $t$ is the temperature in degrees Celsius. Use the Quotient Rule to find the rate of change of $S$ with respect to $t$.

**SOLUTION**

By the Quotient Rule,

$$\frac{dS}{dt} = \frac{(31.6 + t)(0) - (468)(1)}{(31.6 + t)^2}$$

$$= -\frac{468}{(31.6 + t)^2}$$

**Remark:** Note the use of parentheses in Examples 4 and 5. A liberal use of parentheses is a good practice with all types of differentiation problems. For instance, with the Quotient Rule it is a good idea to enclose all factors and derivatives in parentheses and to pay special attention to the subtraction required in the numerator.

When we introduced differentiation rules in Section 2.2 we emphasized the need for rewriting *before* differentiating. The next example illustrates this point with the Quotient Rule.

**EXAMPLE 6**

**Rewriting before Differentiating**

Find the derivative of

$$y = \frac{3 - (1/x)}{x + 5}$$

**SOLUTION**

First we rewrite $y$ as

$$y = \frac{(3x - 1)/x}{x + 5}$$

$$= \frac{3x - 1}{x(x + 5)}$$

$$= \frac{3x - 1}{x^2 + 5x}$$

Now by the Quotient Rule we have

$$\frac{dy}{dx} = \frac{(x^2 + 5x)(3) - (3x - 1)(2x + 5)}{(x^2 + 5x)^2}$$

$$= \frac{(3x^2 + 15x) - (6x^2 + 13x - 5)}{(x^2 + 5x)^2}$$

$$= \frac{-3x^2 + 2x + 5}{(x^2 + 5x)^2}$$

Not every quotient needs to be differentiated by the Quotient Rule. For instance, the quotients in the next example can be considered as products of a constant times a function of $x$. In such cases the Constant Multiple Rule is more convenient than the Quotient Rule.

## EXAMPLE 7

### Rewriting before Differentiating

| Given function | Rewrite | Differentiate | Simplify |
|---|---|---|---|
| (a) $y = \dfrac{x^2 + 3x}{6}$ | $y = \dfrac{1}{6}(x^2 + 3x)$ | $y' = \dfrac{1}{6}(2x + 3)$ | $y' = \dfrac{2x + 3}{6}$ |
| (b) $y = \dfrac{5x^4}{8}$ | $y = \dfrac{5}{8}x^4$ | $y' = \dfrac{5}{8}(4x^3)$ | $y' = \dfrac{5}{2}x^3$ |
| (c) $y = \dfrac{-3(3x - 2x^2)}{7x}$ | $y = -\dfrac{3}{7}(3 - 2x)$ | $y' = -\dfrac{3}{7}(-2)$ | $y' = \dfrac{6}{7}$ |
| (d) $y = \dfrac{9}{5x^2}$ | $y = \dfrac{9}{5}(x^{-2})$ | $y' = \dfrac{9}{5}(-2x^{-3})$ | $y' = -\dfrac{18}{5x^3}$ |

When we introduced the Power Rule,

$$\frac{d}{dx}[x^n] = nx^{n-1}$$

in Section 2.2 we stated that the rule is valid for any real number $n$. However, we proved only the case where $n$ is a positive integer. We now prove this rule for the case where $n$ is a negative integer.

## EXAMPLE 8

### Proof of a Special Case of the Power Rule

Use the Quotient Rule to prove the Power Rule for the case where $n$ is a negative integer.

**SOLUTION**

If $n$ is a negative integer, then there exists a positive integer $k$ such that $n = -k$. By letting

$$x^n = x^{-k} = \frac{1}{x^k}$$

we can use the Quotient Rule to obtain

$$\frac{d}{dx}[x^n] = \frac{d}{dx}\left[\frac{1}{x^k}\right] = \frac{x^k(0) - (1)(kx^{k-1})}{(x^k)^2}$$

$$= \frac{0 - kx^{k-1}}{x^{2k}}$$

$$= -kx^{-k-1}$$

$$= nx^{n-1}$$

Therefore, if $n$ is a negative integer, we also have

$$\frac{d}{dx}[x^n] = nx^{n-1}$$

EXAMPLE 9

**Combining the Product and Quotient Rules**

Differentiate

$$y = \frac{(1 - 2x)(3x + 2)}{5x - 4}$$

**SOLUTION**

Here we have a product within a quotient, and we could first multiply the factors in the numerator, then apply the Quotient Rule. However, to show how the Product Rule can be used within the Quotient Rule, we proceed with the equation as written.

$$y' = \frac{(5x - 4)\dfrac{d}{dx}[(1 - 2x)(3x + 2)] - (1 - 2x)(3x + 2)\dfrac{d}{dx}[5x - 4]}{(5x - 4)^2}$$

$$= \frac{(5x - 4)[(1 - 2x)(3) + (3x + 2)(-2)] - (1 - 2x)(3x + 2)(5)}{(5x - 4)^2}$$

$$= \frac{(5x - 4)(-12x - 1) - (1 - 2x)(15x + 10)}{(5x - 4)^2}$$

$$= \frac{(-60x^2 + 43x + 4) - (-30x^2 - 5x + 10)}{(5x - 4)^2} = \frac{-30x^2 + 48x - 6}{(5x - 4)^2}$$

**EXAMPLE 10**

**Rate of Change of Blood Pressure**

As blood moves from the heart through the major arteries out to the capillaries and back through the veins, the systolic pressure continuously drops. (See Figure 2.26.) Suppose that this pressure is given by

$$P = \frac{25t^2 + 125}{t^2 + 1}, \quad 0 \le t \le 10$$

where $P$ is measured in millimeters of mercury and $t$ is measured in seconds. At what rate is the pressure changing 5 seconds after blood leaves the heart?

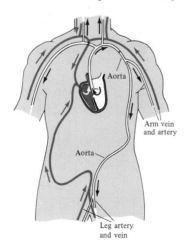

FIGURE 2.26

Aorta

Arm vein and artery

Aorta

Leg artery and vein

**SOLUTION**     By the Quotient Rule we have

$$\frac{dP}{dt} = \frac{(t^2 + 1)(50t) - (25t^2 + 125)(2t)}{(t^2 + 1)^2}$$

$$= \frac{50t^3 + 50t - 50t^3 - 250t}{(t^2 + 1)^2}$$

$$= -\frac{200t}{(t^2 + 1)^2}$$

When $t = 5$, the rate of change is

$$\frac{dP}{dt} = -\frac{200(5)}{26^2} \approx -1.48 \text{ mm/sec}$$

Therefore, the pressure is *dropping* at the rate of 1.48 mm/sec when $t = 5$ seconds.

In the examples in this section much of the work in obtaining the final form of the answer occurs *after* the differentiation is complete. Direct application of differentiation rules often yields answers that are not in simplest form, as shown in Table 2.7. Two characteristics of the simplest form of an algebraic expression are (1) the absence of negative exponents and (2) the combining of like terms.

TABLE 2.7

|  | $f'(x)$ after differentiating | $f'(x)$ after simplifying |
|---|---|---|
| Example 1 | $(3x - 2x^2)(4) + (5 + 4x)(3 - 4x)$ | $15 + 4x - 24x^2$ |
| Example 2 | $(x^{-1} + 1)(1) + (x - 1)(-x^{-2})$ | $\dfrac{x^2 + 1}{x^2}$ |
| Example 4 | $\dfrac{(2 - 3x)(4x - 4) - (2x^2 - 4x + 3)(-3)}{(2 - 3x)^2}$ | $\dfrac{-6x^2 + 8x + 1}{(2 - 3x)^2}$ |
| Example 9 | $\dfrac{(5x - 4)[(1 - 2x)(3) + (3x + 2)(-2)] - (1 - 2x)(3x + 2)(5)}{(5x - 4)^2}$ | $\dfrac{-30x^2 + 48x - 6}{(5x - 4)^2}$ |

## SECTION EXERCISES 2.4

In Exercises 1–10, find the value of the derivative of the function at the indicated point.

*Function* · *Point*

1. $f(x) = x^2(3x^3 - 1)$ · $(1, 2)$

2. $f(x) = (x^2 + 1)(2x + 5)$ · $(-1, 6)$

3. $f(x) = \frac{1}{3}(2x^3 - 4)$ · $\left(0, -\frac{4}{3}\right)$

4. $f(x) = \dfrac{5 - 6x^2}{7}$ · $\left(1, -\frac{1}{7}\right)$

5. $g(x) = (x^2 + 3x - 1)(x + 3)$ · $(-2, -3)$

*Function* · *Point*

6. $g(x) = (x^2 - 2x + 1)(x^3 - 1)$ · $(1, 0)$

7. $h(x) = \dfrac{x}{x - 5}$ · $(6, 6)$

8. $h(x) = \dfrac{x^2}{x + 3}$ · $\left(-1, \frac{1}{2}\right)$

9. $f(t) = \dfrac{t^2 + 1}{4t}$ · $\left(1, \frac{1}{2}\right)$

10. $f(x) = \dfrac{x + 1}{x - 1}$ · $(2, 3)$

In Exercises 11–16, complete the table without using the Quotient Rule.

| Function | Rewrite | Derivative | Simplify |
|---|---|---|---|

**11.** $y = \dfrac{x^2 + 2x}{x}$

**12.** $y = \dfrac{4x^{3/2}}{x}$

**13.** $y = \dfrac{7}{3x^3}$

**14.** $y = \dfrac{4}{5x^2}$

**15.** $y = \dfrac{3x^2 - 5}{7}$

**16.** $y = \dfrac{x^2 - 4}{x + 2}$

In Exercises 17–40, differentiate the given function.

**17.** $f(x) = (x^3 - 3x)(2x^2 + 3x + 5)$
**18.** $f(x) = (x - 1)(x^2 - 3x + 2)$
**19.** $h(t) = (3t + 4)(t^4 - 5t^2 + 2)$
**20.** $g(t) = (t^5 - 1)(4t^2 - 7t - 3)$
**21.** $h(s) = (s^3 - 2)^2$      **22.** $h(x) = (x^2 - 1)^2$
**23.** $f(x) = \sqrt[3]{x}(\sqrt{x} + 3)$      **24.** $f(x) = \sqrt[3]{x}(x + 1)$
**25.** $f(x) = \dfrac{3x - 2}{2x - 3}$      **26.** $f(x) = \dfrac{x^3 + 3x + 2}{x^2 - 1}$
**27.** $f(x) = \dfrac{3 - 2x - x^2}{x^2 - 1}$      **28.** $f(x) = x^3\left(\dfrac{x - 2}{x + 1}\right)$
**29.** $f(x) = (x^5 - 3x)\left(\dfrac{1}{x^2}\right)$      **30.** $f(x) = x\left(1 - \dfrac{2}{x + 1}\right)$
**31.** $h(t) = \dfrac{t + 1}{t^2 + 2t + 2}$      **32.** $g(s) = \dfrac{5}{s^2 + 2s + 2}$
**33.** $f(x) = \dfrac{x + 1}{\sqrt{x}}$      **34.** $f(x) = \dfrac{\sqrt{x}}{x + 1}$

**35.** $g(x) = \left(\dfrac{x + 1}{x + 2}\right)(2x - 5)$

**36.** $f(x) = \left(\dfrac{x^2 - x - 3}{x^2 + 1}\right)(x^2 + x + 1)$

**37.** $f(x) = (3x^3 + 4x)(x - 5)(x + 1)$
**38.** $f(x) = (x^2 - x)(x^2 + 1)(x^2 + x + 1)$

**39.** $f(x) = \dfrac{x^2 + c^2}{x^2 - c^2}$, $c$ is a constant

**40.** $f(x) = \dfrac{c^2 - x^2}{c^2 + x^2}$

In Exercises 41–44, find an equation of the tangent line to the graph of the given function at the indicated point.

| Function | Point |
|---|---|

**41.** $f(x) = \dfrac{x}{x - 1}$      (2, 2)

**42.** $f(x) = (x - 1)(x^2 - 2)$      (0, 2)
**43.** $f(x) = (x^3 - 3x + 1)(x + 2)$      (1, −3)

**44.** $f(x) = \dfrac{x - 1}{x + 1}$      $\left(2, \dfrac{1}{3}\right)$

In Exercises 45 and 46, determine the point(s), if any, where the graph of the function has a horizontal tangent.

**45.** $f(x) = \dfrac{x^2}{x - 1}$      **46.** $f(x) = \dfrac{x^2}{x^2 + 1}$

In Exercises 47 and 48, use the demand function to find the rate of change in the demand $x$ for the given price $p$.

**47.** $x = 150\left(1 - \dfrac{p}{p + 1}\right)$, $p = \$4$

**48.** $x = 300 - p - \dfrac{2p}{p + 1}$, $p = \$3$

**49.** The function
$$f(t) = \dfrac{t^2 - t + 1}{t^2 + 1}$$
measures the percentage of the normal level of oxygen in a pond, where $t$ is the time in weeks after organic waste is dumped into the pond. Find the rate of change of $f$ with respect to $t$ when
(a) $t = 0.5$      (b) $t = 2$      (c) $t = 8$

**50.** Suppose that the temperature $T$ of food placed in a refrigerator is given by the equation
$$T = 10\left(\dfrac{4t^2 + 16t + 75}{t^2 + 4t + 10}\right)$$
where $t$ is the time in hours. What is the initial temperature of the food? Find the rate of change of $T$ with respect to $t$ when
(a) $t = 1$      (b) $t = 3$      (c) $t = 5$      (d) $t = 10$

**51.** A population of 500 bacteria is introduced into a culture. The number of bacteria in the population at any given time is given by
$$P = 500\left(1 + \dfrac{4t}{50 + t^2}\right)$$
where $t$ is measured in hours. Find the rate of change of the population when $t = 2$.

**52.** The percentage $P$ of defective parts produced by a new employee $t$ days after the employee starts work is given by
$$P = \dfrac{t + 1750}{50(t + 2)}$$
Find the rate of change of $P$ when
(a) $t = 1$   (b) $t = 10$

# The Chain Rule

A set of gears is constructed as in Figure 2.27 so that the second and third gears are on the same axle. As the first axle revolves, it drives the second axle, which in turn drives the third axle. Let

$y$ = Revolutions per minute for first axle

$u$ = Revolutions per minute for second axle

$x$ = Revolutions per minute for third axle

Since the circumference of the second gear is 3 times that of the first, the first axle must make 3 revolutions to turn the second axle once. That is,

$$y = 3u \qquad \frac{dy}{du} = 3$$

Similarly, the second gear must make 2 revolutions to turn the third axle once, and we have

$$u = 2x \qquad \frac{du}{dx} = 2$$

Combining these two results, we know that the first axle must make 6 revolutions to turn the third axle once. That is,

$$y = 6x \qquad \frac{dy}{dx} = 6$$

In other words, the rate of change of $y$ with respect to $x$ is the product of the rate of change of $y$ with respect to $u$ and the rate of change of $u$ with respect to $x$.

$$\frac{dy}{dx} = \left(\frac{dy}{du}\right)\left(\frac{du}{dx}\right) = (3)(2) = 6$$

We call this result the **Chain Rule.**

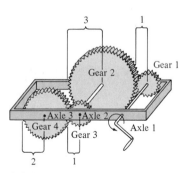

Axle 1: $y$ revolutions per minute
Axle 2: $u$ revolutions per minute
Axle 3: $x$ revolutions per minute

FIGURE 2.27

■ **The Chain Rule**
■ **The General Power Rule**
■ **Algebraic Simplification Techniques**

We have yet to discuss one of the most powerful rules in differential calculus—the **Chain Rule.** This differentiation rule deals with composite functions and adds a surprising versatility to the rules we discussed in Sections 2.2 and 2.4. For example, compare the following functions. Those on the left can be differentiated without the Chain Rule, and those on the right are best done with the Chain Rule.

| *Without the Chain Rule* | *With the Chain Rule* |
|---|---|
| $y = x^2 + 1$ | $y = \sqrt{x^2 + 1}$ |
| $y = x + 1$ | $y = (x + 1)^{-1/2}$ |
| $y = 3x + 2$ | $y = (3x + 2)^5$ |

Basically the Chain Rule says that if

$$y \text{ changes } \frac{dy}{du} \text{ times as fast as } u$$

and

$$u \text{ changes } \frac{du}{dx} \text{ times as fast as } x$$

then

$$y \text{ changes } \frac{dy}{du}\frac{du}{dx} \text{ times as fast as } x$$

**The Chain Rule**

If $y = f(u)$ is a differentiable function of $u$ and $u = g(x)$ is a differentiable function of $x$, then $y = f(g(x))$ is a differentiable function of $x$, and

$$\frac{dy}{dx} = \frac{dy}{du}\frac{du}{dx}$$

or, equivalently,

$$\frac{d}{dx}[f(g(x))] = f'(u)g'(x)$$

In applying the Chain Rule it is helpful to think of the composite function $y = f(g(x))$ as having two parts—an *inside* and an *outside*—as follows.

outside

$$y = f(g(x)) = f(u)$$

$u = g(x)$
inside

Example 1 illustrates two cases.

**EXAMPLE 1**

**Decomposition of a Composite Function**

| $y = f(g(x))$ | $u = g(x)$ (inside) | $y = f(u)$ (outside) |
|---|---|---|
| (a) $y = \dfrac{1}{x+1}$ | $u = x + 1$ | $y = \dfrac{1}{u}$ |
| (b) $y = \sqrt{3x^2 - x + 1}$ | $u = 3x^2 - x + 1$ | $y = \sqrt{u}$ |

**EXAMPLE 2**

**Using the Chain Rule**

Find the derivative of

$$y = (x^2 + 1)^3$$

**SOLUTION**

To apply the Chain Rule we identify the inside function $u$ as follows:

$$y = \underbrace{(x^2 + 1)}_{u}{}^3 = u^3$$

Then by the Chain Rule we have

$$\frac{dy}{dx} = \underbrace{3(x^2 + 1)^2}_{\frac{dy}{du}}\underbrace{(2x)}_{\frac{du}{dx}} = 6x(x^2 + 1)^2$$

The function in Example 2 is an instance of one of the most common types of composite functions—that is, functions of the form

$$y = [u(x)]^n$$

The rule for differentiating such *power functions* is called the **General Power Rule,** and it is a special case of the Chain Rule.

---

**General Power Rule**

If $y = [u(x)]^n$, where $u$ is a differentiable function of $x$ and $n$ is a real number, then

$$\frac{dy}{dx} = n[u(x)]^{n-1}\frac{du}{dx}$$

or, equivalently,

$$\frac{d}{dx}[u^n] = nu^{n-1}u'$$

---

Proof

Since $y = u^n$, we apply the Chain Rule to obtain

$$\frac{dy}{dx} = \left(\frac{dy}{du}\right)\left(\frac{du}{dx}\right) = \frac{d}{du}[u^n]\frac{du}{dx}$$

Now, by the (simple) Power Rule in Section 2.2, we have

$$\frac{d}{du}[u^n] = nu^{n-1}$$

and it follows that

$$\frac{dy}{dx} = nu^{n-1}\frac{du}{dx}$$

**EXAMPLE 3**

**Using the General Power Rule**

Find the derivative of

$$f(x) = (3x - 2x^2)^3$$

**SOLUTION**

If we let $u = 3x - 2x^2$, then

$$f(x) = (3x - 2x^2)^3 = u^3$$

and by the General Power Rule, the derivative is

$$f'(x) = \overbrace{3}^{n}\overbrace{(3x - 2x^2)^2}^{u^{n-1}}\overbrace{\frac{d}{dx}[3x - 2x^2]}^{u'}$$

$$= 3(3x - 2x^2)^2(3 - 4x)$$

$$= (9 - 12x)(3x - 2x^2)^2$$

**EXAMPLE 4**

**Using the General Power Rule**

Find the derivative of

$$y = \sqrt[3]{(x^2 + 2)^2}$$

**SOLUTION**

We begin by rewriting the function as

$$y = (x^2 + 2)^{2/3}$$

Then, by the General Power Rule (with $u = x^2 + 2$), we have

$$y' = \overbrace{\frac{2}{3}}^{n}\overbrace{(x^2 + 2)^{-1/3}}^{u^{n-1}}\overbrace{(2x)}^{u'}$$

$$= \frac{4x}{3\sqrt[3]{x^2 + 2}}$$

The derivative of a quotient can sometimes be found more easily with the General Power Rule than with the Quotient Rule, especially when the numerator is a constant. Compare the solution given in the next example with that given in Example 5 in the previous section.

**EXAMPLE 5**

**Applying the General Power Rule to Quotients**

Find the derivative of

$$S = \frac{468}{31.6 + t}, \quad 0 \le t$$

**SOLUTION**

Rewriting this function in the form

$$S = 468(31.6 + t)^{-1}$$

we can apply the General Power Rule to obtain

$$\frac{dS}{dt} = (-1)(468)(31.6 + t)^{-2}(1)$$

$$= -\frac{468}{(31.6 + t)^2}$$

**Simplifying Derivatives**

Throughout this chapter we have emphasized writing derivatives in simplified form. The reason for this, as we show in the next chapter, is that many applications involving the derivative require simplified forms. The next three examples illustrate some useful simplification techniques for the "raw derivatives" of functions involving products, quotients, and composites. (You may want to review some additional examples of simplification in Section 0.5.)

**EXAMPLE 6**

**Simplification by Factoring Least Powers**

Differentiate

$$f(x) = x^2\sqrt{1 - x^2}$$

**SOLUTION**

$$f(x) = x^2(1 - x^2)^{1/2} \qquad \text{Rewrite}$$

$$f'(x) = x^2\frac{d}{dx}[(1 - x^2)^{1/2}] + (1 - x^2)^{1/2}\frac{d}{dx}[x^2] \qquad \text{Product Rule}$$

$$= x^2\left[\frac{1}{2}(1 - x^2)^{-1/2}(-2x)\right] + (1 - x^2)^{1/2}(2x) \qquad \text{Power Rule}$$

$$= -x^3(1 - x^2)^{-1/2} + 2x(1 - x^2)^{1/2}$$

$$= x(1 - x^2)^{-1/2}[-x^2(1) + 2(1 - x^2)] \qquad \text{Factor}$$

$$= x(1 - x^2)^{-1/2}(2 - 3x^2)$$

$$= \frac{x(2 - 3x^2)}{\sqrt{1 - x^2}} \qquad \text{Simplify}$$

In Example 6 note that we subtract exponents when factoring. Thus, when $(1 - x^2)^{-1/2}$ is factored out of $(1 - x^2)^{1/2}$, the *remaining* factor has an exponent of $\frac{1}{2} - (-\frac{1}{2}) = 1$. Thus, $(1 - x^2)^{1/2} = (1 - x^2)^{-1/2}(1 - x^2)^1$. The next example further demonstrates this procedure.

**EXAMPLE 7**

**Simplifying the Derivative of a Quotient**

Differentiate

$$f(x) = \frac{x}{\sqrt[3]{x^2 + 4}}$$

**SOLUTION**

$$f(x) = \frac{x}{(x^2 + 4)^{1/3}} \qquad \text{Rewrite}$$

$$f'(x) = \frac{(x^2 + 4)^{1/3}(1) - x(1/3)(x^2 + 4)^{-2/3}(2x)}{(x^2 + 4)^{2/3}} \qquad \text{Quotient Rule}$$

$$= \frac{1}{3}(x^2 + 4)^{-2/3}\left[\frac{3(x^2 + 4) - (2x^2)(1)}{(x^2 + 4)^{2/3}}\right] \qquad \text{Factor}$$

$$= \frac{1}{3}(x^2 + 4)^{-2/3}\left[\frac{x^2 + 12}{(x^2 + 4)^{2/3}}\right]$$

$$= \frac{x^2 + 12}{3(x^2 + 4)^{4/3}} \qquad \text{Simplify}$$

**EXAMPLE 8**

**Differentiating a Quotient Raised to a Power**

Find the derivative of

$$y = \left(\frac{3x - 1}{x^2 + 3}\right)^2$$

**SOLUTION**

$$\frac{dy}{dx} = 2\overbrace{\left(\frac{3x - 1}{x^2 + 3}\right)}^{u^{n-1}}\overbrace{\frac{d}{dx}\left[\frac{3x - 1}{x^2 + 3}\right]}^{u'} \quad \overset{n}{}$$

$$= \left[\frac{2(3x - 1)}{x^2 + 3}\right]\left[\frac{(x^2 + 3)(3) - (3x - 1)(2x)}{(x^2 + 3)^2}\right]$$

$$= \frac{2(3x - 1)(3x^2 + 9 - 6x^2 + 2x)}{(x^2 + 3)^3}$$

$$= \frac{2(3x - 1)(-3x^2 + 2x + 9)}{(x^2 + 3)^3}$$

Try finding $y'$ using the Quotient Rule on

$$y = \frac{(3x - 1)^2}{(x^2 + 3)^2}$$

and compare the results.

**EXAMPLE 9**

**An Application of the Chain Rule**

A sum of \$1,000 is deposited in an account with an interest rate of $r$ percent compounded quarterly. At the end of 5 years the balance in the account is given by

$$A = 1000\left(1 + \frac{r}{400}\right)^{20}$$

Find the rate of change of $A$ with respect to $r$ for the following rates:
(a) $r = 8\%$        (b) $r = 10\%$        (c) $r = 12\%$

**SOLUTION**

By the General Power Rule, the rate of change of $A$ with respect to $r$ is

$$\frac{dA}{dr} = (20)(1000)\left(1 + \frac{r}{400}\right)^{19}\left(\frac{1}{400}\right)$$

$$= 50\left(1 + \frac{r}{400}\right)^{19}$$

(a) When $r = 8$, the rate of change of $A$ with respect to $r$ is

$$\frac{dA}{dr} = 50\left(1 + \frac{8}{400}\right)^{19}$$

$$\approx \$72.84 \text{ per percentage point}$$

(b) When $r = 10$, the rate of change of $A$ with respect to $r$ is

$$\frac{dA}{dr} = 50\left(1 + \frac{10}{400}\right)^{19}$$

$$\approx \$79.93 \text{ per percentage point}$$

(c) When $r = 12$, the rate of change of $A$ with respect to $r$ is

$$\frac{dA}{dr} = 50\left(1 + \frac{12}{400}\right)^{19}$$

$$\approx \$87.68 \text{ per percentage point}$$

■■■ **Remark:** In Figure 2.28 note that the rate of change of $A$ with respect to $r$ increases as $r$ increases.

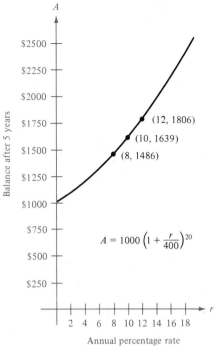

FIGURE 2.28

We have now discussed all the rules needed to differentiate *any* algebraic function. For convenience we list these rules here. We urge you to memorize them, as this will greatly increase your efficiency in differentiating.

**Summary of differentiation rules**

Let $u$ and $v$ be differentiable functions of $x$, and let $y$ be a differentiable function of $u$.

1. Constant Rule:

$$\frac{d}{dx}[c] = 0, \quad c \text{ is a constant}$$

2. Constant Multiple Rule:

$$\frac{d}{dx}[cu] = c\frac{du}{dx}, \quad c \text{ is a constant}$$

3. Sum Rule:

$$\frac{d}{dx}[u \pm v] = \frac{du}{dx} \pm \frac{dv}{dx}$$

4. Product Rule:

$$\frac{d}{dx}[uv] = u\frac{dv}{dx} + v\frac{du}{dx}$$

5. Quotient Rule:

$$\frac{d}{dx}\left[\frac{u}{v}\right] = \frac{v\dfrac{du}{dx} - u\dfrac{dv}{dx}}{v^2}$$

6. Power Rules:

$$\frac{d}{dx}[x^n] = nx^{n-1}$$

$$\frac{d}{dx}[u^n] = nu^{n-1}\frac{du}{dx}$$

7. Chain Rule:

$$\frac{dy}{dx} = \frac{dy}{du}\frac{du}{dx}$$

## SECTION EXERCISES 2.5

In Exercises 1–8, complete the table using Example 1 as a model.

| $y = f(g(x))$ | $u = g(x)$ | $y = f(u)$ |
| --- | --- | --- |

**1.** $y = (6x - 5)^4$
**2.** $y = (x^2 - 2x + 3)^3$
**3.** $y = (4 - x^2)^{-1}$
**4.** $y = (x^2 + 1)^{4/3}$
**5.** $y = \sqrt{x^2 - 1}$
**6.** $y = \sqrt{9 - x^2}$
**7.** $y = \dfrac{1}{3x + 1}$

**8.** $y = \dfrac{1}{\sqrt{x + 1}}$

In Exercises 9–54, find the derivative of the function.

**9.** $y = (2x - 7)^3$
**10.** $y = (3x^2 + 1)^4$
**11.** $f(x) = 2(x^2 - 1)^3$
**12.** $g(x) = 3(9x - 4)^4$
**13.** $g(x) = (4 - 2x)^3$
**14.** $h(t) = (1 - x^2)^4$
**15.** $h(x) = (6x - x^3)^2$
**16.** $f(x) = (4x - x^2)^3$
**17.** $f(x) = (x^2 - 9)^{2/3}$
**18.** $f(t) = (9t + 2)^{2/3}$
**19.** $f(t) = \sqrt{t + 1}$
**20.** $g(x) = \sqrt{2x + 3}$
**21.** $s(t) = \sqrt{t^2 + 2t - 1}$
**22.** $y = \sqrt[3]{3x^3 + 4x}$
**23.** $y = \sqrt[3]{9x^2 + 4}$
**24.** $g(x) = \sqrt{x^2 - 2x + 1}$
**25.** $y = 2\sqrt{4 - x^2}$
**26.** $f(x) = -3\sqrt[4]{2 - 9x}$
**27.** $f(x) = (25 + x^2)^{-1/2}$
**28.** $g(x) = (6 + 3x^2)^{-2/3}$
**29.** $h(x) = (4 - x^3)^{-4/3}$
**30.** $f(x) = (4 - 3x)^{-5/2}$
**31.** $y = \dfrac{1}{x - 2}$
**32.** $s(t) = \dfrac{1}{t^2 + 3t - 1}$
**33.** $f(t) = \left(\dfrac{1}{t - 3}\right)^2$
**34.** $y = -\dfrac{4}{(t + 2)^2}$
**35.** $f(x) = \dfrac{3}{x^3 - 4}$
**36.** $f(x) = \dfrac{1}{(x^2 - 3x)^2}$
**37.** $y = \dfrac{1}{\sqrt{x + 2}}$
**38.** $g(t) = \dfrac{1}{t^2 - 2}$
**39.** $g(x) = \dfrac{3}{\sqrt[3]{x^3 - 1}}$
**40.** $s(x) = \dfrac{1}{\sqrt{x^2 - 3x + 4}}$
**41.** $y = x\sqrt{2x + 3}$
**42.** $y = t\sqrt{t + 1}$

**43.** $y = t^2\sqrt{t - 2}$
**44.** $y = \dfrac{x}{\sqrt{25 + x^2}}$
**45.** $f(x) = x^2(x - 2)^4$
**46.** $f(x) = x(3x - 9)^3$
**47.** $f(t) = (t^2 - 9)\sqrt{t + 2}$
**48.** $y = \sqrt{x}(x - 2)^2$
**49.** $f(x) = \sqrt{\dfrac{x + 1}{x}}$
**50.** $y = \sqrt{\dfrac{2x}{x + 1}}$
**51.** $g(t) = \dfrac{3t^2}{\sqrt{t^2 + 2t - 1}}$
**52.** $f(x) = \dfrac{x + 1}{2x - 3}$
**53.** $f(x) = \dfrac{\sqrt{x^2 + 1}}{x}$
**54.** $g(x) = \sqrt{x - 1} + \sqrt{x + 1}$

In Exercises 55 and 56, find an equation of the tangent line to the graph of $f$ at the given point.

| Function | Point |
| --- | --- |
| **55.** $f(x) = \sqrt{3x^2 - 2}$ | $(3, 5)$ |
| **56.** $f(x) = x\sqrt{x^2 + 5}$ | $(2, 6)$ |

**57.** A sum of \$1,000 is deposited in an account with an interest rate of $r$ percent compounded monthly. At the end of 5 years the balance in the account is given by

$$A = 1000\left(1 + \frac{r}{1200}\right)^{60}$$

Find the rate of change of $A$ with respect to $r$ for the following rates:

(a) $r = 8\%$    (b) $r = 10\%$    (c) $r = 12\%$
(Compare these results with those of Example 9 to see the effect of increasing the frequency of compounding.)

**58.** Find the rate of change in the demand $x$ of a product if the price is \$2 and the demand function is given by

$$x = 25\left(10 - \frac{p}{\sqrt{p^2 + 1}}\right)$$

**59.** Find the rate of change in the supply $x$ of a product if the price is $150 and the supply function is given by

$$x = 10\sqrt{(p - 100)^2 + 25}, \quad p \geq \$100$$

**60.** A certain automobile depreciates according to the formula

$$V = \frac{7500}{1 + 0.4t + 0.1t^2}$$

where $t$ is measured in years and $t = 0$ represents the time of purchase (in years). At what rate is the value of the car changing

(a) 1 year after purchase

(b) 2 years after purchase

**61.** An assembly plant purchases small electric motors to install in one of their products. The plant management estimates that the cost per motor over the next five years is given by

$$C = 4(1.52t + 10)^{3/2}$$

where $t$ is the time in years. Complete the following table for the rate of change of the cost for each of the next five years.

| $t$ | 1 | 2 | 3 | 4 | 5 |
|-----|---|---|---|---|---|
| $\dfrac{dC}{dt}$ | | | | | |

**62.** The number $N$ of bacteria in a certain culture after $t$ days is given by

$$N = 400\left[1 - \frac{3}{(t^2 + 2)^2}\right]$$

Complete the following table showing the rate of change in the number of bacteria.

| $t$ | 0 | 1 | 2 | 3 | 4 |
|-----|---|---|---|---|---|
| $\dfrac{dN}{dt}$ | | | | | |

# Higher-Order Derivatives

## INTRODUCTORY EXAMPLE
### Acceleration of a Falling Object

If a rock is dropped from a height of 100 feet, its height is given by the position function

$$s(t) = -16t^2 + 100 \qquad \text{Position function}$$

where the height $s$ is measured in feet and the time $t$ is measured in seconds with $t = 0$ corresponding to the instant the rock is dropped. (See Figure 2.29.) We know from Section 2.3 that the velocity of the rock is given by the derivative

$$v(t) = s'(t) = -32t \qquad \text{Velocity function}$$

In Table 2.8 we can see that the velocity of the rock changes as the rock falls. For instance, at time $t = 1$, the velocity is $-32$ ft/sec, but at time $t = 2$, the speed is $-64$ ft/sec. (Note that a negative velocity indicates that the rock is falling, rather than rising.)

In calculus we use the term **acceleration** to describe changes in velocity. Moreover, just as the derivative of the position function gives the velocity, the derivative of the velocity function gives the acceleration. Thus, we have

$$a(t) = v'(t) = -32 \qquad \text{Acceleration function}$$

For free-falling objects, we see that the acceleration is a constant $-32$ ft/sec$^2$, and we call this the **acceleration due to gravity.**

Note that the acceleration function is actually the derivative of a derivative. That is, we can obtain the acceleration function by differentiating the position function *twice*. We call the acceleration function the **second derivative** of the position function, and we denote this using a double prime notation as follows:

$$a(t) = s''(t) \qquad \text{Second derivative}$$

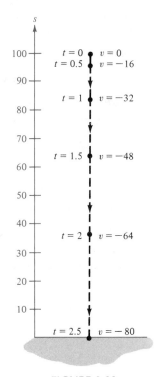

FIGURE 2.29

TABLE 2.8

| Time (sec) | 0 | 0.5 | 1 | 1.5 | 2 | 2.5 |
|---|---|---|---|---|---|---|
| Height (ft) | 100 | 96 | 84 | 64 | 36 | 0 |
| Velocity (ft/sec) | 0 | −16 | −32 | −48 | −64 | −80 |

■ **Second, Third, and Higher-Order Derivatives**
■ **Acceleration**

Since the derivative of a function is itself a function, we may consider finding its derivative. If this is done the result is again a function that may be able to be differentiated. By repeating this process, we obtain what are called **higher-order derivatives.** For instance, the derivative of $f'$ is called the **second derivative** of $f$ and is denoted by $f''$.

$$\frac{d}{dx}[f'(x)] = f''(x) \qquad \text{Second derivative}$$

Similarly, we define the **third derivative** of $f$ as the derivative of $f''$ and denote it by $f'''$.

$$\frac{d}{dx}[f''(x)] = f'''(x) \qquad \text{Third derivative}$$

So long as each successive derivative is differentiable we can continue in this manner to obtain derivatives of higher orders.

■ **Remark:** In the context of higher-order derivatives, we often refer to the derivative of $f$ as the **first derivative.**

**EXAMPLE 1**

**Finding Higher-Order Derivatives**

Find the first five derivatives of

$$f(x) = 2x^4 - 3x^2$$

**SOLUTION**

By successive differentiation we have

$$f'(x) = 8x^3 - 6x \qquad \text{First derivative}$$

$$f''(x) = 24x^2 - 6 \qquad \text{Second derivative}$$

$$f'''(x) = 48x \qquad \text{Third derivative}$$

$$f^{(4)}(x) = 48 \qquad \text{Fourth derivative}$$

$$f^{(5)}(x) = 0 \qquad \text{Fifth derivative}$$

In Example 1 note that we drop the prime notation after the third derivative and use parentheses around the order of the derivative. There are several commonly used notations for higher-order derivatives, as indicated in the following list.

**Notation for higher-order derivatives**

1. First derivative: $\quad y', \quad f'(x), \quad \dfrac{dy}{dx}, \quad \dfrac{d}{dx}[f(x)], \quad D_x(y)$

2. Second derivative: $\quad y'', \quad f''(x), \quad \dfrac{d^2y}{dx^2}, \quad \dfrac{d^2}{dx^2}[f(x)], \quad D_x^2(y)$

3. Third derivative: $\quad y''', \quad f'''(x), \quad \dfrac{d^3y}{dx^3}, \quad \dfrac{d^3}{dx^3}[f(x)], \quad D_x^3(y)$

4. Fourth derivative: $\quad y^{(4)}, \quad f^{(4)}(x), \quad \dfrac{d^4y}{dx^4}, \quad \dfrac{d^4}{dx^4}[f(x)], \quad D_x^4(y)$

$\qquad\qquad\qquad\qquad \vdots$

5. $n$th derivative: $\quad y^{(n)}, \quad f^{(n)}(x), \quad \dfrac{d^ny}{dx^n}, \quad \dfrac{d^n}{dx^n}[f(x)], \quad D_x^n(y)$

---

**EXAMPLE 2**

**Finding Higher-Order Derivatives**

Find the value of $g'''(2)$ for the function given by

$$g(t) = -t^4 + 2t^3 + t + 4$$

**SOLUTION**

We begin by differentiating three times to obtain

$$g'(t) = -4t^3 + 6t^2 + 1$$
$$g''(t) = -12t^2 + 12t$$
$$g'''(t) = -24t + 12$$

Then, we evaluate the third derivative of $g$ at $t = 2$.

$$g'''(2) = -24(2) + 12 = -36$$

In Examples 1 and 2 we found higher-order derivatives of *polynomial* functions. Note that with each successive differentiation, the degree of the polynomial drops by one. Eventually, polynomial higher-order derivatives degenerate to a constant function. To be exact, the $n$th-order derivative of an $n$th-degree polynomial function given by

$$f(x) = a_nx^n + a_{n-1}x^{n-1} + \cdots + a_1 + a_0$$

is the constant function

$$f^{(n)}(x) = n!a_n, \quad n! = 1 \cdot 2 \cdot 3 \cdots n$$

(Each derivative of order higher than $n$ is the zero function.) Polynomial functions are the *only* functions with this characteristic. For other functions, successive differentiation will never produce a constant function. Finding higher-order derivatives for nonpolynomial functions can be cumbersome. However, sometimes a pattern will develop, as illustrated in the next example.

**EXAMPLE 3**

**Finding Higher-Order Derivatives**

Find the first four derivatives of

$$y = \frac{1}{x}$$

**SOLUTION**

We begin by rewriting the function as $y = x^{-1}$. Then using the Power Rule repeatedly, we obtain

$$y' = (-1)x^{-2} = -\frac{1}{x^2}$$

$$y'' = (-1)(-2)x^{-3} = \frac{2}{x^3}$$

$$y''' = (-1)(-2)(-3)x^{-4} = -\frac{6}{x^4}$$

$$y^{(4)} = (-1)(-2)(-3)(-4)x^{-5} = \frac{24}{x^5}$$

**Remark:** Note that in Example 3 the pattern following by the higher-order derivatives is

$$y^{(n)} = (-1)^n \frac{n!}{x^{n+1}}$$

In Chapter 3 we will use the second derivative of a function to determine the concavity of its graph. For the time being we concentrate on what is perhaps the most common application of a higher-order derivative—that of finding the acceleration of an object moving along a straight path.

## Acceleration

In Section 2.3 we saw that the velocity of an object moving along a straight path is given by the derivative of its position function. In other words, the rate of change of the position with respect to time is defined to be the velocity. In a similar way, we define the rate of change of the velocity with respect to time to be the **acceleration.** Thus, we have

$$s = f(t) \qquad \text{Position function}$$

$$\frac{ds}{dt} = f'(t) \qquad \text{Velocity function}$$

$$\frac{d^2s}{dt^2} = f''(t) \qquad \text{Acceleration function}$$

To find the position, velocity, or acceleration at a particular time $t$, we substitute the given value of $t$ into the appropriate function.

**EXAMPLE 4**

**Finding the Acceleration of a Moving Object**

A ball is thrown into the air from the top of a cliff, as shown in Figure 2.30. The height of the ball is given by the position function

$$s = -16t^2 + 48t + 160$$

where $s$ is measured in feet and $t$ is measured in seconds. Find the acceleration of the ball.

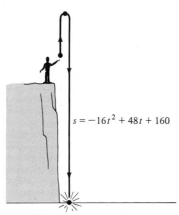

$$s = -16t^2 + 48t + 160$$

FIGURE 2.30

**SOLUTION**

To find the acceleration we differentiate the position function twice as follows.

$$s = -16t^2 + 48t + 160 \qquad \text{Position function (ft)}$$

$$\frac{ds}{dt} = -32t + 48 \qquad \text{Velocity (ft/sec)}$$

$$\frac{d^2s}{dt^2} = -32 \qquad \text{Acceleration (ft/sec}^2\text{)}$$

Thus, for this particular object, the acceleration is the same,

$$-32 \text{ ft/sec}^2$$

for any time $t$.

▬▬ **Remark:** The abbreviation ft/sec² is read "feet per second squared" or, more formally, "feet per second per second."

In Example 4 the constant acceleration of the ball is due to the **gravitation** (or gravity) of the earth. The acceleration due to gravity varies somewhat from point to point on the earth, but $-32$ ft/sec$^2$ is often used as an approximation. Note that the negative value of this acceleration indicates that the ball is being pulled *down*—toward the earth.

Although the acceleration exerted on a falling object is relatively constant near the surface of the earth, it varies greatly throughout our solar system. Large planets exert a much greater gravitational pull than do small planets or satellites.

**EXAMPLE 5**

**The Acceleration of a Free-Falling Object on the Moon**

An astronaut standing on the surface of the moon throws a rock into the air, as shown in Figure 2.31. The height of the rock is given by

$$s = -\frac{27}{10}t^2 + 27t + 6$$

where $s$ is measured in feet and $t$ is measured in seconds. Find the acceleration of the rock.

FIGURE 2.31

**SOLUTION**

$$s = -\frac{27}{10}t^2 + 27t + 6 \qquad \text{Position function}$$

$$\frac{ds}{dt} = -\frac{27}{5}t + 27 \qquad \text{Velocity}$$

$$\frac{d^2s}{dt^2} = -\frac{27}{5} \qquad \text{Acceleration}$$

Thus, the acceleration at any time $t$ is approximately $-5.4$ ft/sec$^2$ (about one-sixth of that exerted by the earth).

In the free-falling problems discussed in Examples 4 and 5, the fact that the acceleration is constant does *not* imply that the velocity is constant. Rather, in both problems the object thrown into the air has an initial positive velocity (upward), then as the gravitational force acts on the object it slows down. Finally, it loses its upward motion, reaches a maximum height, and begins to fall. As the object falls, the gravitational force is acting in the direction of motion and thus the object continues to pick up speed until it hits the ground. Table 2.9 illustrates this pattern by listing the height, velocity, and acceleration of the object in Example 5 at 1-second intervals.

TABLE 2.9

| Time $t$ | 0 | 1 | 2 | 3 | 4 | 5 | 6 | 7 | 8 | 9 | 10 |
|---|---|---|---|---|---|---|---|---|---|---|---|
| Height $s(t)$ | 6 | 30.3 | 49.2 | 62.7 | 70.8 | 73.5 | 70.8 | 62.7 | 49.2 | 30.3 | 6 |
| Velocity $s'(t)$ | 27 | 21.6 | 16.2 | 10.8 | 5.4 | 0 | −5.4 | −10.8 | −16.2 | −21.6 | −27 |
| Acceleration $s''(t)$ | −5.4 | −5.4 | −5.4 | −5.4 | −5.4 | −5.4 | −5.4 | −5.4 | −5.4 | −5.4 | −5.4 |

We have already noted that velocity can be considered as positive or negative. (Normally we use a coordinate system so that positive velocity denotes motion upward or to the right and negative velocity denotes motion downward or to the left.) When we wish to talk about the magnitude of the velocity, without indicating the direction of motion, we use the term **speed.**

In the next example we look at the motion of an accelerating automobile. In this case, the acceleration is not constant.

**EXAMPLE 6**

**Finding the Acceleration of a Moving Object**

Suppose that the *velocity* (in feet per second) of an automobile starting from rest is given by

$$v = \frac{80t}{t + 5}$$

Find the velocity and acceleration of the automobile when $t = 0, 5, 10, \ldots, 60$ seconds.

**SOLUTION**

The position of the car is shown in Figure 2.32. Since acceleration represents the rate of change of the velocity, we determine the acceleration at time $t$ to be

$$\text{Acceleration} = \frac{dv}{dt} = \frac{(t + 5)(80) - (80t)(1)}{(t + 5)^2}$$

$$= \frac{80t + 400 - 80t}{(t + 5)^2}$$

$$= \frac{400}{(t + 5)^2} \text{ ft/sec}^2$$

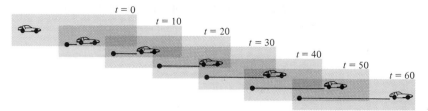

FIGURE 2.32

Table 2.10 compares the velocity and acceleration of the automobile at 5-second intervals during its first minute of travel.

TABLE 2.10

| $t$ | 0 | 5 | 10 | 15 | 20 | 25 | 30 | 35 | 40 | 45 | 50 | 55 | 60 |
|---|---|---|---|---|---|---|---|---|---|---|---|---|---|
| $v$ | 0 | 40 | 53.3 | 60.0 | 64.0 | 66.7 | 68.6 | 70.0 | 71.1 | 72.0 | 72.7 | 73.3 | 73.8 |
| $\dfrac{dv}{dt}$ | 16 | 4 | 1.78 | 1.00 | 0.64 | 0.44 | 0.33 | 0.25 | 0.20 | 0.16 | 0.13 | 0.11 | 0.09 |

Note from Table 2.10 that the acceleration approaches zero as the velocity levels off. This observation agrees with our experience—when riding in an accelerating car, we do not feel the velocity, but we feel the acceleration. In other words, we feel changes in velocity.

## SECTION EXERCISES 2.6

In Exercises 1–16, find the second derivative of the given function.

**1.** $f(x) = 5 - 4x$

**2.** $f(x) = 3x - 1$

**3.** $f(x) = x^2 - 2x + 1$

**4.** $f(x) = 3x^2 + 4x$

**5.** $g(t) = \dfrac{1}{3}t^3 - 4t^2 + 2t - 7$

**6.** $h(t) = 5t^4 + 10t^2 + 3$

**7.** $g(t) = t^{-1/3}$

**8.** $f(t) = \dfrac{3}{4t^2}$

**9.** $f(x) = 4(x^2 - 1)^2$

**10.** $f(x) = 3(2 - x^2)^3$

**11.** $f(x) = x\sqrt[3]{x}$

**12.** $f(x) = \sqrt{9 - x^2}$

**13.** $f(x) = \dfrac{x + 1}{x - 1}$

**14.** $g(t) = -\dfrac{4}{(t + 2)^2}$

**15.** $f(x) = (x - 1)(x^2 - 3x + 2)$

**16.** $h(s) = (s^2 - 2s + 1)(s^3 - 1)$

In Exercises 17–24, find the third derivative of the given function.

**17.** $f(x) = x^5 - 3x^4$

**18.** $f(x) = x^4 - 2x^3$

**19.** $f(x) = 2x(x - 1)^2$

**20.** $f(x) = (x - 1)^2$

**21.** $f(x) = \dfrac{3}{(4x)^2}$

**22.** $f(x) = \dfrac{1}{x}$

**23.** $f(x) = \sqrt{4 - x}$

**24.** $s(t) = \sqrt{2t + 3}$

In Exercises 25–30, find the indicated derivative.

**25.** Given $f'(x) = 2x^2$, find $f''(x)$.

**26.** Given $f''(x) = 20x^3 - 36x^2$, find $f'''(x)$.

**27.** Given $f''(x) = (2x - 2)/x$, find $f'''(x)$.

**28.** Given $f'''(x) = 2\sqrt{x - 1}$, find $f^{(4)}(x)$.

**29.** Given $f^{(4)}(x) = 2x + 1$, find $f^{(6)}(x)$.

**30.** Given $f(x) = x^3 - 2x$, find $f''(x)$.

In Exercises 31–38, find the second derivative and solve the equation $f''(x) = 0$.

**31.** $f(x) = x^3 - 9x^2 + 27x - 27$

**32.** $f(x) = 3x^3 - 9x + 1$

**33.** $f(x) = (x + 1)(x - 2)(x - 5)$

**34.** $f(x) = 3x^4 - 18x^2$

**35.** $f(x) = x^4 - 8x^3 + 18x^2 - 16x + 2$

**36.** $f(x) = x\sqrt{4 - x^2}$

**37.** $f(x) = \dfrac{x}{x^2 + 3}$

**38.** $f(x) = \dfrac{x}{x^2 + 1}$

**39.** A ball is thrown upward from ground level, and its height at any time is given by

$$s(t) = -16t^2 + 48t$$

(a) Find expressions for the velocity and acceleration of the ball.
(b) Find the time when the ball is at its highest point by finding the time when the velocity is zero.
(c) Find the height at the time given in part (b).

**40.** A coin is dropped from the top of the Washington Monument (a height of 550 feet), and its height above ground is given by

$$s(t) = -16t^2 + 550$$

(a) Find expressions for the velocity and acceleration of the coin.
(b) Find the time when the coin reaches ground level ($s = 0$).
(c) Use the time found in part (b) to find the velocity of the coin at the time of impact with the ground.

**41.** The velocity of an automobile starting from rest is given by

$$\frac{ds}{dt} = \frac{90t}{t + 10} \ \text{ft/sec}$$

Complete the following table showing the velocity and acceleration at 10-second intervals during the first minute of travel.

| $t$ | 0 | 10 | 20 | 30 | 40 | 50 | 60 |
|---|---|---|---|---|---|---|---|
| $\dfrac{ds}{dt}$ | | | | | | | |
| $\dfrac{d^2s}{dt^2}$ | | | | | | | |

**42.** A car is traveling at a rate of 66 ft/sec (about 45 mi/hr) when the brakes are applied. The position function for the car is

$$s(t) = -8.25t^2 + 66t$$

where $s$ is measured in feet and $t$ is measured in seconds. Use this function to complete the following table.

| $t$ | 0 | 1 | 2 | 3 | 4 |
|---|---|---|---|---|---|
| $s(t)$ | | | | | |
| $v(t)$ | | | | | |
| $a(t)$ | | | | | |

# Implicit Differentiation

## INTRODUCTORY EXAMPLE
## The Whispering Gallery

An interesting phenomenon in acoustics is the *whispering gallery*. A whispering gallery contains two special points that are acoustically close together even though they may be physically many feet apart, as shown in Figure 2.33. If the gallery is constructed perfectly, then nearly all of the sound originating at one of the points bounces off the gallery wall and ceiling in such a way that it is directed toward the other point. Thus, even a low whisper made by a person at one of the points can be heard by a person at the other point.

The base of the gallery is an ellipse, and each of the two points is called a *focus* of this ellipse. The dome of the gallery is shaped like half of an egg—that is, cross sections containing both focus points are semiellipses having the same shape as (half of) the base. As sound leaves either one of the focus points, it is reflected off the dome in such a way that it makes equal angles (coming and going) with the tangent to the elliptical cross section at the point of contact, as shown in Figure 2.34. By determining

the slope of the tangent line at a point on an ellipse, we can prove that the sound from one focus point will travel to the other focus point.

In this section, we will study a procedure called **implicit differentiation** that can be used to show that for an ellipse given by the equation

$$\frac{x^2}{a^2} + \frac{y^2}{b^2} = 1$$

the slope of the tangent line at the point $(x, y)$ is

$$\frac{dy}{dx} = -\frac{xb^2}{ya^2}$$

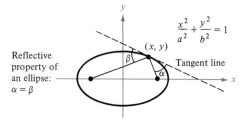

Reflective property of an ellipse: $\alpha = \beta$

FIGURE 2.34

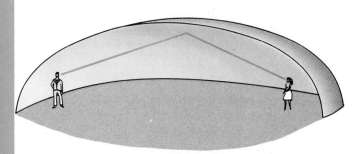

Whispering Gallery

FIGURE 2.33

■ **Implicit and Explicit Functions**
■ **Implicit Differentiation**

Up to this point our functions involving two variables have generally been expressed in the **explicit form** $y = f(x)$. That is, one of the two variables has been explicitly given in terms of the other. For example,

$$y = 3x - 5, \qquad s = -16t^2 + 20t, \qquad u = 3w - w^2$$

are all written in explicit form, and we say that $y$, $s$, and $u$ are functions of $x$, $t$, and $w$, respectively.

However, many functions are not given explicitly and are only implied by a given equation. For instance, the function

$$y = \frac{1}{x}$$

is defined **implicitly** by the equation

$$xy = 1$$

Suppose that you were asked to find $dy/dx$ in this equation. As it turns out this is a relatively simple task, and you would probably begin by solving the equation for $y$.

| *Implicit form* | *Explicit form* | *Derivative* |
|---|---|---|
| $xy = 1$ | $y = \dfrac{1}{x} = x^{-1}$ | $\dfrac{dy}{dx} = -x^{-2} = -\dfrac{1}{x^2}$ |

This procedure works well whenever we can easily solve for the function explicitly. However, we cannot use this procedure in cases in which we are unable to solve for $y$ as a function of $x$. For instance, how would we find $dy/dx$ in the equation

$$x^2 - 2y^3 + 4y = 2$$

where it is very difficult to express $y$ as a function of $x$ explicitly? To do this we use a procedure called **implicit differentiation** in which we assume $y$ is a differentiable function of $x$.

To understand how to find $dy/dx$ implicitly, you must realize that the differentiation is taking place *with respect to x*. This means that when we differentiate terms involving $x$ alone we can differentiate as usual. *But* when we differentiate terms involving $y$ we must apply the Chain Rule, because we are assuming that $y$ is defined implicitly as a function of $x$. Study the next example carefully. Note in particular how the Chain Rule is used to introduce the $dy/dx$ factors.

**EXAMPLE 1**
■■■■■■■■

**Applying the Chain Rule**

Differentiate the following with respect to $x$:

(a) $3x^2$     (b) $2y^3$     (c) $x + 3y$     (d) $xy^2$

**SOLUTION**

(a) The only variable in this expression is $x$. Therefore, to differentiate *with respect to x* we use the Simple Power Rule combined with the Constant Multiple Rule and obtain

$$\frac{d}{dx}[3x^2] = 6x$$

(b) This case is different. The variable in the expression is $y$, and yet we are asked to differentiate with respect to $x$. To do this we assume that $y$ is a differentiable function of $x$ and use the Chain Rule.

$$\frac{d}{dx}[2y^3] = \overset{c}{2}\ \overset{n}{(3)}\ \overset{u^{n-1}}{y^2}\ \overset{u'}{\frac{dy}{dx}} \qquad \text{Chain Rule}$$

$$= 6y^2\frac{dy}{dx}$$

(c) This expression involves both $x$ and $y$. By the Sum Rule and the Constant Multiple Rule we have

$$\frac{d}{dx}[x + 3y] = 1 + 3\frac{dy}{dx}$$

(d) By the Product Rule and the Chain Rule we have

$$\frac{d}{dx}[xy^2] = x\frac{d}{dx}[y^2] + y^2\frac{d}{dx}[x] \qquad \text{Product Rule}$$

$$= x\left(2y\frac{dy}{dx}\right) + y^2(1) \qquad \text{Chain Rule}$$

$$= 2xy\frac{dy}{dx} + y^2$$

For equations involving $x$ and $y$ we suggest the following procedure for finding $dy/dx$ implicitly.

**Implicit differentiation**

Given an equation involving $x$ and $y$ and assuming $y$ is a differentiable function of $x$, we can find $dy/dx$ as follows:

1. Differentiate both sides of the equation *with respect to x*.
2. Collect all terms involving $dy/dx$ on the left side of the equation and move all other terms to the right side of the equation.
3. Factor $dy/dx$ out of the left side of the equation.
4. Solve for $dy/dx$ by dividing both sides of the equation by the left-hand factor that does not contain $dy/dx$.

**EXAMPLE 2**

**Implicit Differentiation**

Find $dy/dx$ given that $x$ and $y$ are related by the equation

$$y^3 + y^2 - 5y - x^2 = -4$$

**SOLUTION**

1. Differentiating both sides of the equation with respect to $x$, we have

$$\frac{d}{dx}[y^3 + y^2 - 5y - x^2] = \frac{d}{dx}[-4]$$

$$\frac{d}{dx}[y^3] + \frac{d}{dx}[y^2] - \frac{d}{dx}[5y] - \frac{d}{dx}[x^2] = \frac{d}{dx}[-4]$$

$$3y^2\frac{dy}{dx} + 2y\frac{dy}{dx} - 5\frac{dy}{dx} - 2x = 0$$

2. Collecting the $dy/dx$ terms on the left, we have

$$3y^2\frac{dy}{dx} + 2y\frac{dy}{dx} - 5\frac{dy}{dx} = 2x$$

3. Factoring $dy/dx$ out of the left side, we have

$$\frac{dy}{dx}(3y^2 + 2y - 5) = 2x$$

4. Finally, we divide by $(3y^2 + 2y - 5)$ to obtain

$$\frac{dy}{dx} = \frac{2x}{3y^2 + 2y - 5}$$

**Remark:** From Example 2 note that implicit differentiation can produce an expression for $dy/dx$ that contains both $x$ and $y$.

To see how we can use an *implicit derivative* consider the graph of the equation $y^3 + y^2 - 5y - x^2 = -4$ shown in Figure 2.35. The derivative found in Example 2 gives us a formula for the slope of the tangent line at a point on this graph. The slopes at several points on the graph are as follows:

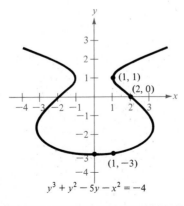

$y^3 + y^2 - 5y - x^2 = -4$

FIGURE 2.35

| Point on graph | Slope of graph |
|:---:|:---:|
| $(2, 0)$ | $-\dfrac{4}{5}$ |
| $(1, -3)$ | $\dfrac{1}{8}$ |
| $x = 0$ | $0$ |
| $(1, 1)$ | undefined |

**EXAMPLE 3**

**Finding the Slope of a Curve Implicitly**

Determine the slope of the tangent line to the ellipse given by

$$x^2 + 4y^2 = 4$$

at the point $(\sqrt{2}, -1/\sqrt{2})$, as shown in Figure 2.36.

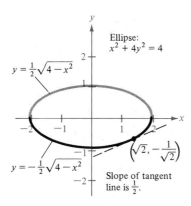

FIGURE 2.36

**SOLUTION**

1. Implicit differentiation of both sides of the equation $x^2 + 4y^2 = 4$ with respect to $x$ yields

$$\frac{d}{dx}[x^2 + 4y^2] = \frac{d}{dx}[4]$$

$$2x + 8y\left(\frac{dy}{dx}\right) = 0$$

2. Collecting terms involving $dy/dx$ gives us

$$8y\left(\frac{dy}{dx}\right) = -2x$$

3. Solving for $dy/dx$ we have

$$\frac{dy}{dx} = \frac{-2x}{8y} = -\frac{x}{4y}$$

Therefore at $(\sqrt{2}, -1/\sqrt{2})$ the slope is

$$\frac{dy}{dx} = \frac{-\sqrt{2}}{-4/\sqrt{2}} = \frac{1}{2}$$

**Remark:** To see the benefit of implicit differentiation, we suggest you do Example 3 using the explicit function

$$y = -\frac{1}{2}\sqrt{4 - x^2}$$

The graph of this function is the lower half of the ellipse.

**EXAMPLE 4**

**Finding the Slope of a Curve Implicitly**

Determine the slope of the graph of

$$3(x^2 + y^2)^2 = 100xy$$

at the point $(3, 1)$.

**SOLUTION**

1.
$$\frac{d}{dx}[3(x^2 + y^2)^2] = \frac{d}{dx}[100xy]$$

$$3(2)(x^2 + y^2)\left(2x + 2y\frac{dy}{dx}\right) = 100\left[x\frac{dy}{dx} + y(1)\right]$$

2.   $$12y(x^2 + y^2)\frac{dy}{dx} - 100x\frac{dy}{dx} = 100y - 12x(x^2 + y^2)$$

3.   $$[12y(x^2 + y^2) - 100x]\frac{dy}{dx} = 100y - 12x(x^2 + y^2)$$

4.
$$\frac{dy}{dx} = \frac{100y - 12x(x^2 + y^2)}{-100x + 12y(x^2 + y^2)}$$

$$= \frac{25y - 3x(x^2 + y^2)}{-25x + 3y(x^2 + y^2)}$$

At the point $(3, 1)$ the slope of the graph is

$$\frac{dy}{dx} = \frac{25(1) - 3(3)(3^2 + 1^2)}{-25(3) + 3(1)(3^2 + 1^2)}$$

$$= \frac{25 - 90}{-75 + 30}$$

$$= \frac{-65}{-45} = \frac{13}{9}$$

as shown in Figure 2.37. This graph is called a **lemniscate.**

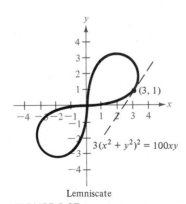

Lemniscate

**FIGURE 2.37**

**EXAMPLE 5**

**An Application of Implicit Differentiation**

The demand function for a certain commodity is given by

$$p = \frac{3,000,000}{x^3 + 10,000x + 1,000,000}$$

where the price $p$ is measured in dollars and the demand $x$ is measured in 1000s of units, as shown in Figure 2.38. Find the rate of change of the demand with respect to the price when $x = 100$. (That is, find $dx/dp$ when $x = 100$.)

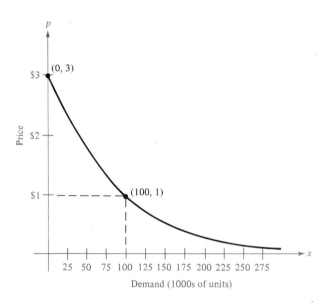

**FIGURE 2.38**

Demand (1000s of units)

**SOLUTION**

We begin by rewriting the equation in the form

$$x^3 + 10,000x + 1,000,000 = \frac{3,000,000}{p}$$

Differentiating both sides with respect to $p$, we have

$$3x^2\frac{dx}{dp} + 10,000\frac{dx}{dp} = -\frac{3,000,000}{p^2}$$

$$\frac{dx}{dp}(3x^2 + 10,000) = -\frac{3,000,000}{p^2}$$

$$\frac{dx}{dp} = -\frac{3,000,000}{p^2(3x^2 + 10,000)}$$

Now when $x = 100$ the price is

$$p = \frac{3,000,000}{100^3 + 10,000(100) + 1,000,000} = 1$$

Substituting the values $x = 100$ and $p = 1$ into the formula for $dx/dp$, we find the rate of change of the demand with respect to the price to be

$$\frac{dx}{dp} = -\frac{3,000,000}{1^2[3(100^2) + 10,000]} = -75$$

This means that when $x = 100$, the demand is changing at the rate of 75 thousand units per dollar drop in the price.

In Example 5 the derivative $dx/dp$ does not represent the slope of the graph of the demand function. Since the demand function is given by $p = f(x)$, the slope of the graph is given by $dp/dx$. These two derivatives are related as follows:

$$\frac{dx}{dp} = \frac{1}{dp/dx}$$

This formula gives us an alternative method of solving Example 5. That is, we could solve for $dp/dx$, evaluate the result at $x = 100$, and then form the reciprocal to obtain $dx/dp$.

In Section 2.2 we listed the Simple Power Rule as

$$\frac{d}{dx}[x^n] = nx^{n-1}$$

Although we stated that this rule holds when $n$ is any real number, we have given proofs only for the cases when $n$ is a positive integer (Section 2.2) and when $n$ is a negative integer (Section 2.4). Using implicit differentiation we can now prove the validity of the Power Rule when $n$ is any rational number.

**EXAMPLE 6**

**Proof of the Power Rule for Rational Exponents**

Let
$$y = x^n$$

where $n$ is the rational number $p/q$ ($p$ and $q$ are integers). By raising both sides of this equation to the $q$th power, we have

$$y = x^n \quad \Longrightarrow \quad y^q = x^{nq}$$

Since $q$ and $nq$ are both integers, we implicitly differentiate both sides of the equation with respect to $x$ to obtain

$$qy^{q-1}\frac{dy}{dx} = nqx^{nq-1}$$

$$q(x^n)^{q-1}\frac{dy}{dx} = nqx^{nq-1}$$

$$x^{nq-n}\frac{dy}{dx} = nx^{nq-1}$$

$$\frac{dy}{dx} = nx^{n-1}$$

# SECTION EXERCISES 2.7

In Exercises 1–6, find $dy/dx$ by implicit differentiation and find the slope of the tangent line at the indicated point on the graph.

**1.** $3x^2 - 2y + 5 = 0$

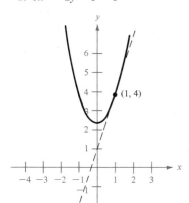

**2.** $4x + 3y^2 - 15 = 0$

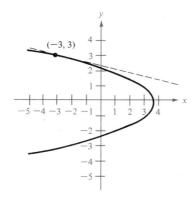

**3.** $x - y^2 = 0$

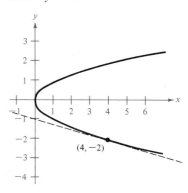

**4.** $x^2 + y^2 = 4$

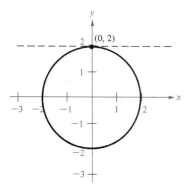

**5.** $xy = 4$

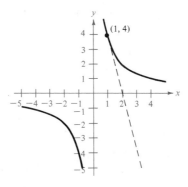

**6.** $x^2 - y^3 = 0$

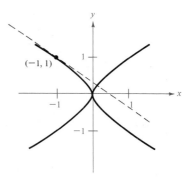

In Exercises 7–22, find $dy/dx$ by implicit differentiation and evaluate the derivative at the indicated point.

| Equation | Point | Equation | Point |
|---|---|---|---|
| **7.** $x^2 + y^2 = 25$ | $(3, 4)$ | **15.** $x^3y^3 - y = x$ | $(0, 0)$ |
| **8.** $x^2 - y^2 = 16$ | $(4, 0)$ | **16.** $x^3 + y^3 = 2xy$ | $(1, 1)$ |
| **9.** $y + xy = 4$ | $(-5, -1)$ | **17.** $x^{1/2} + y^{1/2} = 9$ | $(16, 25)$ |
| **10.** $4xy + x^2 = 5$ | $(1, 1)$ | **18.** $\sqrt{xy} = x - 2y$ | $(4, 1)$ |
| **11.** $x^2 - y^3 = 3$ | $(2, 1)$ | **19.** $x^{2/3} + y^{2/3} = 5$ | $(8, 1)$ |
| **12.** $x^3 + y^3 = 8$ | $(0, 2)$ | **20.** $(x + y)^3 = x^3 + y^3$ | $(-1, 1)$ |
| **13.** $x^3 - xy + y^2 = 4$ | $(0, -2)$ | **21.** $x^3 - 2x^2y + 3xy^2 = 38$ | $(2, 3)$ |
| **14.** $x^2y + y^2x = -2$ | $(2, -1)$ | **22.** $y^2 = \dfrac{x^2 - 9}{x^2 + 9}$ | $(3, 0)$ |

In Exercises 23–26, find $dy/dx$ implicitly and explicitly (the explicit functions are shown on the graph) and show that the two results are equivalent. Find the slope of the tangent line at the point given on the graph.

**23.** $x^2 + y^2 = 25$

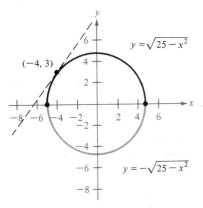

$y = \sqrt{25 - x^2}$

$(-4, 3)$

$y = -\sqrt{25 - x^2}$

**24.** $x - y^2 - 1 = 0$

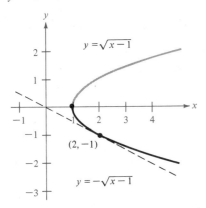

$y = \sqrt{x-1}$

$(2, -1)$

$y = -\sqrt{x-1}$

**25.** $9x^2 + 16y^2 = 144$

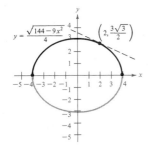

$y = \frac{\sqrt{144 - 9x^2}}{4}$

$\left(2, \frac{3\sqrt{3}}{2}\right)$

**26.** $4y^2 - x^2 = 7$

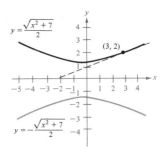

$y = \frac{\sqrt{x^2 + 7}}{2}$

$(3, 2)$

$y = -\frac{\sqrt{x^2 + 7}}{2}$

In Exercises 27 and 28, find equations for the tangent lines to the given circle at the indicated points.

| Function | Points |
|---|---|
| **27.** $x^2 + y^2 = 169$ | $(5, 12)$ and $(-12, 5)$ |
| **28.** $x^2 + y^2 = 9$ | $(0, 3)$ and $(2, \sqrt{5})$ |

In Exercises 29 and 30, the demand function for a certain commodity is given. Use implicit differentiation to find $dx/dp$.

**29.** $p = \dfrac{200 - x}{2x}, \quad 0 < x \le 200$

**30.** $p = \sqrt{\dfrac{500 - x}{2x}}, \quad 0 < x \le 500$

In Exercises 31 and 32, show that the tangent lines are perpendicular at the points of intersection of the given graphs.

**31.** $2x^2 + y^2 = 6, \ y^2 = 4x$

**32.** $y^2 = x^3, \ 2x^2 + 3y^2 = 5$

**33.** The speed $S$ of blood that is $r$ centimeters from the center of an artery is given by

$$S = C(R^2 - r^2)$$

where $C$ is a constant, $R$ is the radius of the artery, and $S$ is measured in centimeters per second. Suppose a drug is administered and the artery begins dilating at the rate of $dR/dt$. At a constant distance $r$, find the rate at which $S$ changes with respect to $t$ for $C = 1.76 \times 10^5$, $R = 1.2 \times 10^{-2}$, and $dR/dt = 10^{-5}$.

# Related Rates

A residential section of a city obtains its water from a large cylindrical storage tank, as shown in Figure 2.39. The tank has a radius of 20 feet and a height of 40 feet and therefore has a capacity of

$$\text{Volume} = \pi r^2 h = \pi (20^2)(40) \approx 50{,}265.5 \text{ cubic feet}$$

Although water is entering the tank at a constant rate, the rate at which the residents use the water varies. For example, on a normal working day the peak usage occurs at about 7 A.M. To permit monitoring of the water supply, the tank is equipped with a flotation device that automatically records the height of water in the tank. The height of the water in the tank during a 12-hour period beginning at midnight is shown in Figure 2.40. From this graph we can see that the rate of change of the height at 7 A.M. is approximately −4 ft/hr. That is, at 7 A.M., the rate of change of the height is

$$\frac{dh}{dt} = -4 \text{ ft/hr}$$

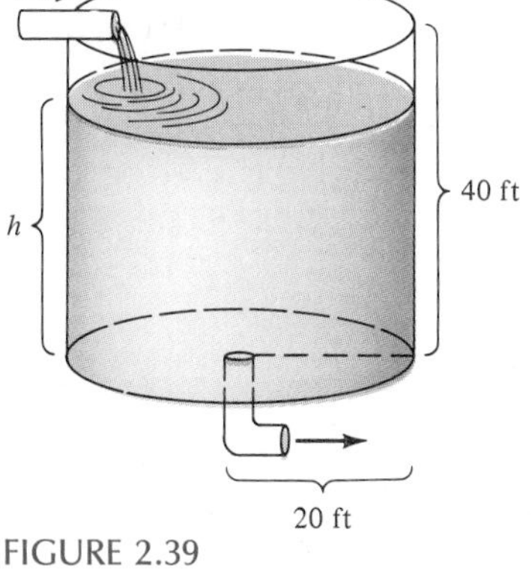

Since the volume of water in the tank at any time is related to the height of the water by the equation

$$V = 400\pi h$$

the rate of change of the volume is related to the rate of change of the height by the equation

$$\frac{dV}{dt} = 400\pi \frac{dh}{dt}$$

which means that the rate of change of the volume of the tank at 7 A.M. is

$$\frac{dV}{dt} = 400\pi(-4) = -5026.5 \text{ ft}^3/\text{hr}$$

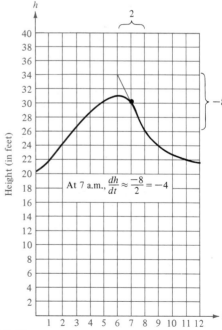

FIGURE 2.39

FIGURE 2.40

Time (in hours)

### ▆ Related Rate Problems

In this section we consider problems involving two (or more) variables that are changing with respect to time. If two such variables are related to each other, then their rates of change with respect to time are also related. For instance, if $x$ and $y$ are related by the equation

$$y = 2x$$

then their rates of change are related by the equation

$$\frac{dy}{dt} = 2\frac{dx}{dt}$$

**EXAMPLE 1**
▬▬▬▬▬▬▬

**Two Rates That Are Related**

Suppose that $x$ and $y$ are both differentiable functions of $t$ and are related by the equation $y = x^2 + 3$. Find $dy/dt$ when $x = 1$, given that $dx/dt = 2$ when $x = 1$.

**SOLUTION**

Using the Chain Rule we can differentiate both sides of the equation *with respect to t.*

$$y = x^2 + 3 \qquad \text{Given equation}$$

$$\frac{d}{dt}[y] = \frac{d}{dt}[x^2 + 3] \qquad \text{Differentiate with respect to } t$$

$$\frac{dy}{dt} = 2x\frac{dx}{dt} + 0 \qquad \text{Chain Rule}$$

$$= 2x\frac{dx}{dt}$$

Now when $x = 1$ and $dx/dt = 2$ we have

$$\frac{dy}{dt} = 2(1)(2) = 4$$

█████████████

In Example 1 we are *given* the following mathematical model.

Given equation: $y = x^2 + 3$

Given rate: $\dfrac{dx}{dt} = 2$ when $x = 1$

Find: $\dfrac{dy}{dt}$ when $x = 1$

Now let's look at an example in which we must *create* the model from a verbal description.

**EXAMPLE 2**

### An Application Involving Related Rates

A pebble is dropped into a calm pool of water, causing ripples in the form of concentric circles, as shown in Figure 2.41. The radius $r$ of the outer ripple is increasing at a constant rate of 1 ft/sec. When this radius is 4 feet, at what rate is the total area $A$ of the disturbed water changing?

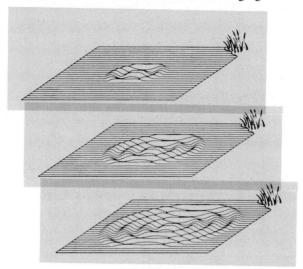

FIGURE 2.41

**SOLUTION**

The variables $r$ and $A$ are related by the equation for the area of a circle, $A = \pi r^2$. To solve this problem, we must remember that the rate of change of the radius $r$ is given by its derivative $dr/dt$. Thus, the problem can be summarized by the following model.

$$\text{Given equation: } A = \pi r^2$$

$$\text{Given rate: } \frac{dr}{dt} = 1 \text{ when } r = 4$$

$$\text{Find: } \frac{dA}{dt} \text{ when } r = 4$$

With this information we proceed as in Example 1.

$$\frac{d}{dt}[A] = \frac{d}{dt}[\pi r^2]$$

$$\frac{dA}{dt} = 2\pi r \frac{dr}{dt}$$

Thus, when $r = 4$ we have

$$\frac{dA}{dt} = 2\pi(4)(1) = 8\pi \text{ ft}^2/\text{sec}$$

The solution shown in Example 2 illustrates the following four solution steps for related-rate problems.

**Procedure for solving related-rate problems**

1. Assign symbols to all *given* quantities and quantities *to be determined*. Make a sketch and label the quantities if feasible.
2. Write an equation involving the variables whose rates of change either are given or are to be determined.
3. Using the Chain Rule, implicitly differentiate both sides of the equation *with respect to time t.*
4. Substitute into the resulting equation all known values for the variables and their rates of change. Then solve for the required rate of change.

■■■ **Remark:** Be sure you notice the order of steps 3 and 4. We do not substitute the known values for the variables until after we have differentiated.

Table 2.11 shows the mathematical models for some common rates of change used in the first step of the solution to related rate problems.

TABLE 2.11

| Verbal statement | Mathematical model |
|---|---|
| The velocity of a car after traveling one hour is 50 mi/hr. | $x$ = distance traveled <br> $\dfrac{dx}{dt} = 50$ when $t = 1$ |
| Water is being pumped into a swimming pool at the rate of 10 ft³/min. | $V$ = volume of water in pool <br> $\dfrac{dV}{dt} = 10$ ft³/min |
| A population of bacteria is increasing at the rate of 2000 per hour. | $x$ = number in population <br> $\dfrac{dx}{dt} = 2000$ |

**EXAMPLE 3**

**An Inflating Balloon**

Air is being pumped into a spherical balloon at the rate of 4.5 in³/min, as shown in Figure 2.42. Find the rate of change of the radius when the radius is 2 inches.

**SOLUTION**

1. We let $V$ be the volume of the balloon and let $r$ be its radius. Since the volume is increasing at the rate of 4.5 in³/min, we know that at time $t$ the rate of change of the volume is $dV/dt = \frac{9}{2}$. Thus the problem can be stated as follows:

$$\text{Given: } \frac{dV}{dt} = \frac{9}{2} \qquad \text{Constant rate}$$

$$\text{Find: } \frac{dr}{dt} \text{ when } r = 2$$

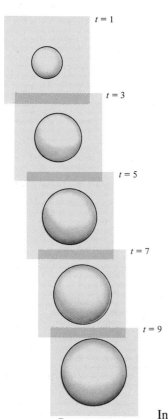

$t = 1$

$t = 3$

$t = 5$

$t = 7$

$t = 9$

Expanding Balloon

FIGURE 2.42

2. To find the rate of change of the radius we must find an equation that relates the radius $r$ to the volume $V$. This is given by the formula for the volume of a sphere.

$$V = \frac{4}{3}\pi r^3$$

3. Now we implicitly differentiate *with respect to* $t$ to obtain

$$\frac{dV}{dt} = 4\pi r^2 \left(\frac{dr}{dt}\right)$$

$$\frac{dr}{dt} = \frac{1}{4\pi r^2}\left(\frac{dV}{dt}\right)$$

4. Finally, when $r = 2$, the rate of change of the radius is

$$\frac{dr}{dt} = \frac{1}{16\pi}\left(\frac{9}{2}\right) \approx 0.09 \text{ in/min}$$

In Example 3 note that the volume is increasing at a *constant* rate but the radius is increasing at a *variable* rate. Thus, when we say that two rates are related we do not mean that they are proportional. In this particular case the radius is growing more and more slowly as $t$ increases. This is illustrated in Figure 2.42 and Table 2.12.

TABLE 2.12

| $t$ | 1 | 3 | 5 | 7 | 9 | 11 |
|---|---|---|---|---|---|---|
| $V = 4.5t$ | 4.5 | 13.5 | 22.5 | 31.5 | 40.5 | 49.5 |
| $r = \sqrt[3]{\dfrac{3V}{4\pi}}$ | 1.02 | 1.48 | 1.75 | 1.96 | 2.13 | 2.28 |
| $\dfrac{dr}{dt}$ | 0.34 | 0.16 | 0.12 | 0.09 | 0.08 | 0.07 |

## EXAMPLE 4

### The Velocity of an Airplane Tracked by Radar

An airplane is flying at an elevation of 6 miles on a flight path that is directly over a radar tracking station. Let $s$ represent the distance (measured in miles) between the radar station and the plane. If $s$ is decreasing at a rate of 400 mi/hr when $s$ is 10 miles, what is the velocity of the plane?

**SOLUTION**

1. We label the distances $x$ and $s$ in Figure 2.43.

$$\text{Given:} \quad \frac{ds}{dt} = -400 \text{ when } s = 10$$

$$\text{Find:} \quad \frac{dx}{dt} \text{ when } s = 10$$

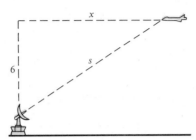

FIGURE 2.43

2. We can use the Pythagorean Theorem to form an equation relating $s$ and $x$.

$$x^2 + 6^2 = s^2$$

3. Differentiating implicitly with respect to $t$, we have

$$2x\frac{dx}{dt} = 2s\frac{ds}{dt}$$

$$\frac{dx}{dt} = \frac{s}{x}\left(\frac{ds}{dt}\right)$$

4. To find $dx/dt$ we first must find $x$ when $s = 10$.

$$x = \sqrt{s^2 - 36} = \sqrt{100 - 36} = \sqrt{64} = 8$$

Finally, when $s = 10$ we have

$$\frac{dx}{dt} = \frac{10}{8}(-400) = -500 \text{ mi/hr}$$

**EXAMPLE 5**

**Increasing Production**

A company is increasing its production of a certain product at the rate of 200 units per week. The weekly demand function is given by

$$p = 100 - \frac{x}{1000}$$

where $p$ is the price per unit and $x$ is the number of units produced in a week. Find the rate of change of the revenue with respect to time (in weeks) when the weekly production is 2000.

**SOLUTION**

1. Using $R$ = weekly revenue, $x$ = weekly production, and $t$ = time in weeks, we have the following:

$$\text{Given: } \frac{dx}{dt} = 200 \qquad \text{Constant rate}$$

$$\text{Find: } \frac{dR}{dt} \text{ when } x = 2000$$

2. The revenue is related to the price by the equation $R = xp$. Therefore, we have

$$R = xp = x\left(100 - \frac{x}{1000}\right)$$

$$= 100x - \frac{x^2}{1000}$$

3. Differentiating implicitly with respect to $t$ we have

$$\frac{dR}{dt} = 100\left(\frac{dx}{dt}\right) - \frac{x}{500}\left(\frac{dx}{dt}\right)$$

$$= \left(100 - \frac{x}{500}\right)\left(\frac{dx}{dt}\right)$$

4. When $x = 2000$, we find the rate of change in revenue to be

$$\frac{dR}{dt} = \left(100 - \frac{2000}{500}\right)(200)$$

$$= 96(200)$$

$$= \$19{,}200 \text{ per week}$$

## SECTION EXERCISES 2.8

In Exercises 1–4, assume that $x$ and $y$ are both differentiable functions of $t$ and find the indicated values of $dy/dt$ and $dx/dt$.

| Equation | Find | Given |
|---|---|---|
| **1.** $y = \sqrt{x}$ | (a) $\dfrac{dy}{dt}$ | $x = 4, \dfrac{dx}{dt} = 3$ |
| | (b) $\dfrac{dx}{dt}$ | $x = 25, \dfrac{dy}{dt} = 2$ |
| **2.** $y = x^2 - 3x$ | (a) $\dfrac{dy}{dt}$ | $x = 3, \dfrac{dx}{dt} = 2$ |
| | (b) $\dfrac{dx}{dt}$ | $x = 1, \dfrac{dy}{dt} = 5$ |
| **3.** $xy = 4$ | (a) $\dfrac{dy}{dt}$ | $x = 8, \dfrac{dx}{dt} = 10$ |

| Equation | Find | Given |
|---|---|---|
| | (b) $\dfrac{dx}{dt}$ | $x = 1, \dfrac{dy}{dt} = -6$ |
| **4.** $x^2 + y^2 = 25$ | (a) $\dfrac{dy}{dt}$ | $x = 3, y = 4, \dfrac{dx}{dt} = 8$ |
| | (b) $\dfrac{dx}{dt}$ | $x = 4, y = 3, \dfrac{dy}{dt} = -2$ |

**5.** The radius $r$ of a circle is increasing at a rate of 2 in./min. Find the rate of change of the area when
   (a) $r = 6$ inches   (b) $r = 24$ inches

**6.** The radius $r$ of a sphere is increasing at a rate of 2 in./min. Find the rate of change of the volume when
   (a) $r = 6$ inches   (b) $r = 24$ inches

7. Let $A$ be the area of a circle of radius $r$ that is changing with respect to time. If $dr/dt$ is constant, is $dA/dt$ constant? Explain why or why not.

8. Let $V$ be the volume of a sphere of radius $r$ that is changing with respect to time. If $dr/dt$ is constant, is $dV/dt$ constant? Explain why or why not.

9. A spherical balloon is inflated with gas at the rate of 20 ft$^3$/min. How fast is the radius of the balloon changing at the instant the radius is
   (a) 1 foot       (b) 2 feet

10. The formula for the volume of a cone is $V = \frac{1}{3}\pi r^2 h$. Find the rate of change of the volume if $dr/dt$ is 2 in./min and $h = 3r$ when
    (a) $r = 6$ inches       (b) $r = 24$ inches

11. At a sand and gravel plant, sand is falling off a conveyer and onto a conical pile at the rate of 10 ft$^3$/min. The diameter of the base of the cone is approximately three times the altitude. At what rate is the height of the pile changing when it is 15 feet high?

12. A conical tank (with vertex down) is 10 feet across the top and 12 feet deep. If water is flowing into the tank at the rate of 10 ft$^3$/min, find the rate of change of the depth of the water the instant it is 8 feet deep.

13. All edges of a cube are expanding at the rate of 3 cm/sec. How fast is the volume changing when each edge is
    (a) 1 centimeter       (b) 10 centimeters

14. The conditions are the same as in Exercise 13. Now measure how fast the *surface area* is changing when each edge is
    (a) 1 centimeter       (b) 10 centimeters

15. A point is moving along the graph of $y = x^2$ so that $dx/dt$ is 2 cm/min. Find $dy/dt$ when
    (a) $x = -3$       (b) $x = 0$
    (c) $x = 1$       (d) $x = 3$

16. A point is moving along the graph of $y = 1/(1 + x^2)$ so that $dx/dt = 2$ cm/min. Find $dy/dt$ when
    (a) $x = -2$       (b) $x = 2$
    (c) $x = 0$       (d) $x = 10$

17. A ladder 25 feet long is leaning against a house, as shown in Figure 2.44. The base of the ladder is pulled away from the house wall at a rate of 2 ft/sec. How fast is the top of the ladder moving down the wall when the base of the ladder is
    (a) 7 feet       (b) 15 feet
    (c) 24 feet
        from the wall.

18. A boat is pulled in by means of a winch on the dock 12 feet above the deck of the boat, as shown in Figure 2.45. The winch pulls in rope at the rate of 4 ft/sec. Determine the speed of the boat when there

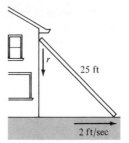

FIGURE 2.44

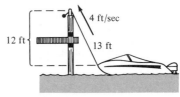

FIGURE 2.45

are 13 feet of rope out. What happens to the speed of the boat as it gets closer to the dock?

19. An air traffic controller spots two planes at the same altitude converging on a point as they fly at right angles to one another, as shown in Figure 2.46. One plane is 150 miles from the point and is moving at 450 mi/hr. The other plane is 200 miles from the point and has a speed of 600 mi/hr.
    (a) At what rate is the distance between the planes changing?
    (b) How much time does the traffic controller have to get one of the planes on a different flight path?

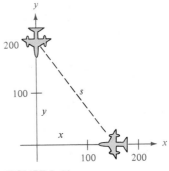

FIGURE 2.46

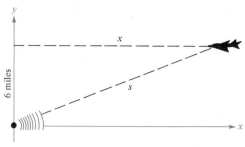

FIGURE 2.47

**20.** An airplane flying at an altitude of 6 miles passes directly over a radar antenna, as shown in Figure 2.47. When the plane is 10 miles away ($s = 10$), the radar detects that the distance $s$ is changing at a rate of 240 mi/hr. What is the speed of the plane?

**21.** A baseball diamond has the shape of a square with sides 90 feet long, as shown in Figure 2.48. A player 30 feet from third base $t$ is running at a speed of 28 ft/sec. At what rate is the player's distance from home plate changing?

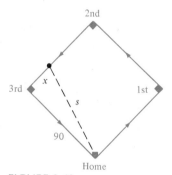

FIGURE 2.48

**22.** For the baseball diamond in Exercise 21, suppose the player is running from first to second at a speed of 28 ft/sec. Find the rate at which the distance from home plate is changing when the player is 30 feet from second.

**23.** A company is increasing its production of a certain product at the rate of 25 units per week. The demand and cost functions for this product are given, respectively, by

$$p = 50 - \frac{x}{100} \quad \text{and}$$

$$C = 4000 + 40x - 0.02x^2$$

Find the rate of change of the profit with respect to time when weekly sales are $x = 800$ units.

**24.** The profit on a certain product is growing at the rate of $6,384 per week. Find the rate of change in sales when sales are $x = 44$ per week if the demand and cost functions are, respectively,

$$p = 6000 - 0.4x^2 \quad \text{and}$$

$$C = 2400x + 5200$$

**25.** An accident at an oil drilling platform in coastal waters is causing a circular oil slick to form. Engineers determine that the slick is 0.08 feet thick, and when the radius is 750 feet it is increasing at the rate of $\frac{1}{2}$ ft/min. Use this information to estimate (in cubic feet) the rate at which oil is flowing from the site of the accident.

# CHAPTER 2 SUMMARY

## Important Terms

Tangent line
Slope of a curve at a point
Secant line
Difference quotient
Change in $x$, $\Delta x$
Change in $y$, $\Delta y$
Derivative
Differentiable
Vertical tangent line
Constant Rule
Binomial expansion
Simple Power Rule
Constant Multiple Rule
Sum Rule

Difference Rule
Average rate of change
Instantaneous rate of change
Marginal profit
Marginal revenue
Marginal cost
Discrete variable
Continuous variable
Demand function
Product Rule
Quotient Rule
Chain Rule
General Power Rule

Higher-order derivative
First derivative
Second derivative
Third derivative
Position function, $s$
Velocity, velocity function, $v$
Acceleration, acceleration function, $a$
Gravity, acceleration due to gravity
Speed
Function given in explicit form
Function defined implicitly
Implicit differentiation
Related rates

## Important Techniques

Finding derivatives by the four-step process
Using differentiation rules (rewriting before
  differentiating)
Finding an average rate of change over an interval
Finding an instantaneous rate of change at a point
Simplifying derivatives

Finding higher-order derivatives
Finding derivatives implicitly
Setting up a mathematical model for a related-rate
  problem
Solving a related-rate problem

## Important Formulas

Derivative of $f(x)$:

$$f'(x) = \lim_{\Delta x \to 0} \frac{f(x + \Delta x) - f(x)}{\Delta x}$$

Sum and Difference Rule:

$$\frac{d}{dx}[u \pm v] = \frac{du}{dx} \pm \frac{dv}{dx}$$

Quotient Rule:

$$\frac{d}{dx}\left[\frac{u}{v}\right] = \frac{v\dfrac{du}{dx} - u\dfrac{dv}{dx}}{v^2}$$

Constant Rule:

$$\frac{d}{dx}[c] = 0$$

Simple Power Rule:

$$\frac{d}{dx}[x^n] = nx^{n-1}$$

Chain Rule:

$$\frac{dy}{dx} = \frac{dy}{du}\frac{du}{dx}$$

General Power Rule:

$$\frac{d}{dx}[u^n] = nu^{n-1}\frac{du}{dx}$$

Constant Multiple Rule:

$$\frac{d}{dx}[cu] = c\frac{du}{dx}$$

Product Rule:

$$\frac{d}{dx}[uv] = u\frac{dv}{dx} + v\frac{du}{dx}$$

Position function $= s(t)$
Velocity $= v(t) = s'(t)$
Acceleration $= a(t) = s''(t)$

In Exercises 1–4, find the derivative of the function by the four-step process.

**1.** $f(x) = 7x + 3$

**2.** $f(x) = 10 - x^2$

**3.** $h(t) = \sqrt{t + 9}$

**4.** $g(x) = \dfrac{x + 1}{x}$

In Exercises 5–30, find the derivative of the given function.

**5.** $f(x) = x^3 - 3x^2$

**6.** $f(x) = 11x^4 - 5x^2 + 1$

**7.** $f(x) = x^3 - 5 + 3x^{-3}$

**8.** $f(x) = \dfrac{2x^3 - 1}{x^2}$

**9.** $f(x) = x^{1/2} - x^{-1/2}$

**10.** $f(x) = \dfrac{x + 1}{x - 1}$

**11.** $f(x) = (3x^2 + 7)(x^2 - 2x + 3)$

**12.** $g(x) = (5x + 3)(x^3 - 3x^2 + 7x + 4)$

**13.** $g(t) = \dfrac{2}{3t^2}$

**14.** $h(x) = \dfrac{2}{(3x)^2}$

**15.** $f(x) = \dfrac{x^2 + x - 1}{x^2 - 1}$

**16.** $f(x) = \dfrac{6x - 5}{x^2 + 1}$

**17.** $f(x) = \dfrac{1}{4 - 3x^2}$

**18.** $f(x) = \dfrac{9}{3x^2 - 2x}$

**19.** $g(x) = \dfrac{2}{\sqrt{x + 1}}$

**20.** $f(x) = -\dfrac{2x^2}{x - 1}$

**21.** $f(x) = \sqrt{x^3 + 1}$

**22.** $f(x) = \sqrt[3]{x^2 - 1}$

**23.** $g(x) = x\sqrt{x^2 + 1}$

**24.** $h(\theta) = \dfrac{\theta}{(1 - \theta)^3}$

**25.** $f(t) = (t + 1)\sqrt[3]{t + 1}$

**26.** $y = \sqrt{3x}(x + 2)^3$

**27.** $f(x) = -2(1 - 4x^2)^2$

**28.** $f(x) = \left(x^2 + \dfrac{1}{x}\right)^5$

**29.** $h(x) = [x^2(2x + 3)]^3$

**30.** $f(x) = [(x - 2)(x + 4)]^2$

In Exercises 31–40, find the second derivative of the given function.

**31.** $f(x) = x^2 + 9$

**32.** $h(x) = x(x^2 - 1)$

**33.** $f(t) = \dfrac{5}{(1 - t)^2}$

**34.** $h(x) = x^2 + \dfrac{3}{x}$

**35.** $f(x) = (3x^2 + 7)(x^2 - 2x + 3)$

**36.** $f(x) = \dfrac{1}{\sqrt{x}}$

**37.** $f(x) = (1 - x^2)^4$

**38.** $f(x) = 8x^{5/2}$

**39.** $f(x) = 18\sqrt[3]{x}$

**40.** $f(x) = \sqrt{1 - x^2}$

In Exercises 41–44, use implicit differentiation to find $dy/dx$.

**41.** $x^2 + 3xy + y^3 = 10$

**42.** $x^2 + 9y^2 - 4x + 3y - 7 = 0$

**43.** $y^2 - x^2 = 25$

**44.** $y^2 + x^2 - 6y - 2x - 5 = 0$

In Exercises 45–50, find an equation of the tangent line to the graph of the given equation at the indicated point.

| Equation | Point |
| --- | --- |
| **45.** $y = (x + 3)^3$ | $(-2, 1)$ |
| **46.** $y = (x - 2)^2$ | $(2, 0)$ |
| **47.** $x^2 + y^2 = 20$ | $(2, 4)$ |
| **48.** $x^2 - y^2 = 16$ | $(5, 3)$ |
| **49.** $y = \sqrt[3]{(x - 2)^2}$ | $(3, 1)$ |
| **50.** $y = \dfrac{2x}{1 - x^2}$ | $(0, 0)$ |

In Exercises 51 and 52, use the cost function to find the marginal cost.

**51.** $C = 5000 + 650x$

**52.** $C = 475 + 5.25x^{2/3}$

In Exercises 53 and 54, use the revenue function to find the marginal revenue.

**53.** $R = \dfrac{50x}{\sqrt{x - 2}}, \ x \geq 6$

**54.** $R = x\left(5 + \dfrac{10}{\sqrt{x}}\right)$

In Exercises 55 and 56, use the profit function to find the marginal profit.

**55.** $P = -0.0005x^3 + 5x^2 - x - 2500$

**56.** $P = -\dfrac{1}{15}x^3 + 4000x^2 - 120x - 144,000$

**57.** Find the points on the graph of

$$f(x) = \frac{1}{3}x^3 + x^2 - x - 1$$

at which the slope is

(a) $-1$    (b) $2$    (c) $0$

**58.** Find the points on the graph of

$$f(x) = x^2 + 1$$

at which the slope is

(a) $-1$    (b) $0$    (c) $1$

**59.** Derive the equations for the velocity and acceleration of a particle whose position function is

$$s(t) = t + \frac{2}{t + 1}$$

**60.** Derive the equations for the velocity and acceleration of a particle whose position function is

$$s(t) = \frac{1}{t^2 + 2t + 1}$$

**61.** Suppose that the temperature $T$ of food placed in a freezer is given by the equation

$$T = \frac{700}{t^2 + 4t + 10}$$

where $t$ is the time in hours. Find the rate of change of $T$ with respect to $t$ when

(a) $t = 1$   (b) $t = 3$   (c) $t = 5$   (d) $t = 10$

**62.** Based on present performance, a company predicts that its profit per unit during the next five years will be given by

$$P = \sqrt{0.05t^3 + 40.75t}$$

where $t$ is time in years. Find the rate of change of the profit for each of the next five years.

**63.** The **Doyle Log Rule** is a mathematical model for estimating the volume (in board feet) of a log of length $L$ feet and diameter $D$ inches at the small end. According to this model the volume is

$$V = \left(\frac{D - 4}{4}\right)^2 L$$

Find the rate at which the volume is changing for a 12-foot log whose smallest diameter is

(a) 8 inches   (b) 16 inches   (c) 24 inches
(d) 36 inches

**64.** A forester has studied the growth curve for a certain species of tree and has found that the expected height after $t$ years is

$$h = 62.9 + 27.6\left(\frac{t - 120}{40}\right)$$

$$- 0.4\left(\frac{t - 120}{40}\right)^2 - 0.8\left(\frac{t - 120}{40}\right)^3$$

where $0 \le t \le 240$. Complete the following table showing the height and the rate of change of the height.

| $t$ | 40 | 120 | 200 |
|---|---|---|---|
| $h$ | | | |
| $\dfrac{dh}{dt}$ | | | |

# Applications of
# the Derivative

# CHAPTER 3

# Increasing and Decreasing Functions

## INTRODUCTORY EXAMPLE
### Egg Yolk Growth Curve

The normal time period between the fertilization and the laying of a chicken egg is about 19 days. Although the egg yolk grows continuously during this time, the rate of growth varies significantly. Figure 3.1 shows that almost all of the yolk's growth takes place in the final 7 days, with the maximum growth rate occurring during the third and fourth days before laying.

We say that this growth function is **increasing** up until the egg is laid, since for each increase in time there is a corresponding *increase* in weight. Letting $f(t)$ represent the weight of the yolk at time $t$, we can observe that for this function the following three conditions are equivalent:

1. The function $f$ is increasing.
2. The graph of $f$ has a positive slope.
3. The derivative $f'$ is positive.

After the egg is laid and the chick begins to develop, the amount of egg yolk decreases. During this period of time the function $f$ is **decreasing,** since for each increase in time there is a corresponding *decrease* in weight. Paralleling our observation for increasing functions, we note that for this particular function the following three conditions are equivalent:

1. The function $f$ is decreasing.
2. The graph of $f$ has a negative slope.
3. The derivative $f'$ is negative.

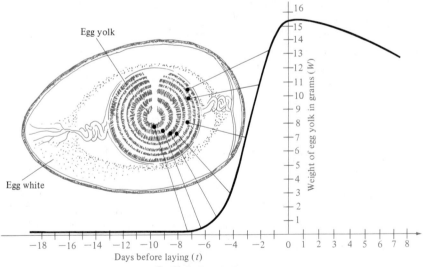

Egg Yolk Growth Curve

FIGURE 3.1

■ **Increasing and Decreasing Functions**
■ **Test for Increasing or Decreasing Functions**
■ **Critical Numbers**

We say that a function $y = f(x)$ is **increasing** if its graph moves up as $x$ moves to the right or **decreasing** if its graph moves down as $x$ moves to the right. The following definition states this more formally.

| | |
|---|---|
| **Definition of increasing and decreasing functions** | A function $f$ is **increasing** on an interval if for any $x_1$ and $x_2$ in the interval $$x_2 > x_1 \quad \text{implies} \quad f(x_2) > f(x_1)$$ A function $f$ is **decreasing** on an interval if for any $x_1$ and $x_2$ in the interval $$x_2 > x_1 \quad \text{implies} \quad f(x_2) < f(x_1)$$ |

For instance, the function shown in Figure 3.2 is decreasing on the interval $(-\infty, a)$, is constant on the interval $(a, b)$, and is increasing on the interval $(b, \infty)$. Actually, from the definition of increasing and decreasing, the function shown in Figure 3.2 is decreasing on the interval $(-\infty, a]$ and increasing on the interval $[b, \infty)$. However, for the purpose of this course open intervals are sufficient, and in the examples and exercises we will restrict our discussion to finding *open* intervals on which a function is increasing or decreasing.

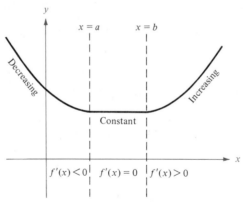

FIGURE 3.2

The derivative is useful in determining whether a function is increasing or decreasing on an interval. Specifically, as Figure 3.2 shows, a positive derivative implies that the graph slopes upward and the function is increasing. Similarly, a negative derivative implies that the function is decreasing. Finally, a zero derivative on an entire interval implies that the function is constant on the interval. This use of the derivative is summarized in the following test for increasing or decreasing functions.

| **Test for increasing or decreasing functions** | Let $f$ be differentiable on the interval $(a, b)$. |
|---|---|

1. If $f'(x) > 0$ for all $x$ in $(a, b)$, then $f$ is increasing on $(a, b)$.
2. If $f'(x) < 0$ for all $x$ in $(a, b)$, then $f$ is decreasing on $(a, b)$.
3. If $f'(x) = 0$ for all $x$ in $(a, b)$, then $f$ is constant on $(a, b)$.

To apply this test, we note in Figure 3.3 that for a continuous function, $f'(x)$ can change sign only at $x$-values where $f'(x) = 0$ or at $x$-values where $f'(x)$ is undefined. We call these two types of $x$-values **critical numbers** of $f$.

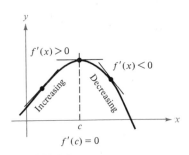

FIGURE 3.3

| **Definition of a critical number** | If $f$ is defined at $c$, then $c$ is a **critical number** of $f$ if $f'(c) = 0$ or if $f'$ is undefined at $c$. |
|---|---|

To determine the intervals on which a continuous function $f$ is increasing or decreasing, we suggest the following guidelines.

1. Find the derivative of $f$.
2. Locate the critical numbers of $f$ and use these numbers to determine test intervals. [Find all $x$ such that $f'(x) = 0$ or $f'(x)$ is undefined.]
3. Test the sign of $f'(x)$ at an arbitrary number in each of the intervals determined in the second step.
4. Use the test for increasing or decreasing functions to decide whether $f$ is increasing or decreasing on each interval.

**EXAMPLE 1**

**Determining Intervals on Which $f$ Is Increasing or Decreasing**

Find the intervals on which the following function is increasing or decreasing:

$$f(x) = x^3 - \frac{3}{2}x^2$$

**SOLUTION**

We begin by finding $f'(x)$ and then setting $f'(x)$ equal to zero.

$$f'(x) = 3x^2 - 3x = 0 \qquad \text{Let } f'(x) = 0$$

$$3(x)(x - 1) = 0 \qquad \text{Factor}$$

$$x = 0, 1 \qquad \text{Critical numbers}$$

Since there are no $x$-values for which $f'$ is undefined, we conclude that $x = 0$ and $x = 1$ are the only critical numbers. Thus, the intervals to be tested are $(-\infty, 0)$, $(0, 1)$, and $(1, \infty)$. Table 3.1 summarizes the testing of these three intervals.

TABLE 3.1

| Interval | $-\infty < x < 0$ | $0 < x < 1$ | $1 < x < \infty$ |
|---|---|---|---|
| Test value | $x = -1$ | $x = \dfrac{1}{2}$ | $x = 2$ |
| Sign of $f'(x)$ | $f'(-1) = 6 > 0$ | $f'\left(\dfrac{1}{2}\right) = -\dfrac{3}{4} < 0$ | $f'(2) = 6 > 0$ |
| Conclusion | increasing | decreasing | increasing |

The graph of $f$ is shown in Figure 3.4. Note that the test values in the intervals were chosen for convenience; other $x$-values could have been used.

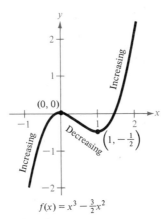

$$f(x) = x^3 - \tfrac{3}{2}x^2$$

FIGURE 3.4

Not only is the function in Example 1 continuous on the entire real line, it is differentiable there. For such functions, the only critical numbers are those for which $f'(x) = 0$. In the next example, we look at a continuous function that has both types of critical numbers—those for which $f'(x) = 0$ and those for which $f'$ is undefined.

**EXAMPLE 2**

**Determining Intervals on Which $f$ Is Increasing or Decreasing**

Find the intervals on which the following function is increasing or decreasing:

$$f(x) = (x^2 - 4)^{2/3}$$

**SOLUTION**      Since

$$f'(x) = \frac{2}{3}(x^2 - 4)^{-1/3}(2x)$$

$$= \frac{4x}{3(x^2 - 4)^{1/3}}$$

we see that $f'(x)$ is zero at $x = 0$ and, furthermore, $f'$ is undefined at $x = \pm 2$. Thus, the critical numbers are

$$x = -2, \qquad x = 0, \qquad x = 2$$

and the resulting test intervals are

$$(-\infty, -2), \qquad (-2, 0), \qquad (0, 2), \qquad (2, \infty)$$

Table 3.2 summarizes the testing of these four intervals, and the graph of the function is shown in Figure 3.5.

**TABLE 3.2**

| Interval | $-\infty < x < -2$ | $-2 < x < 0$ | $0 < x < 2$ | $2 < x < \infty$ |
|----------|-----|-----|-----|-----|
| Test value | $x = -3$ | $x = -1$ | $x = 1$ | $x = 3$ |
| Sign of $f'(x)$ | $f'(-3) < 0$ | $f'(-1) > 0$ | $f'(1) < 0$ | $f'(3) > 0$ |
| Conclusion | decreasing | increasing | decreasing | increasing |

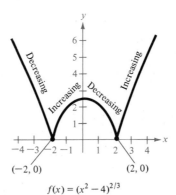

$$f(x) = (x^2 - 4)^{2/3}$$

**FIGURE 3.5**

Note in Table 3.2 that we do not need to *evaluate* $f'(x)$ at the test values—we only need to determine its sign. For example, we can determine that $f'(-3)$ is negative as follows:

$$f'(-3) = \frac{4(-3)}{3[(-3)^2 - 4]^{1/3}} = \frac{4(-3)}{3(9 - 4)^{1/3}} = \frac{\text{negative}}{\text{positive}} = \text{negative}$$

**Remark:** Note in Figure 3.5 that the two types of critical numbers correspond to horizontal and vertical tangent lines of the graph. Specifically, for the critical number $x = 0$, we have $f'(0) = 0$ and the graph has a *horizontal* tangent line. But for the critical numbers $x = \pm 2$, the derivative is undefined and the graph has *vertical* tangent lines.

The functions in Examples 1 and 2 are continuous on the entire real line. If the domain of a function has $x$-values at which the function is discontinuous, or if there are $x$-values at which the function is not defined, then these numbers should be used along with the critical numbers to determine the test intervals. This is demonstrated in the next example.

**EXAMPLE 3**

**Finding Test Intervals for a Discontinuous Function**

Find the intervals on which the following function is increasing or decreasing:

$$f(x) = \frac{x^4 + 1}{x^2}$$

**SOLUTION**

To begin, we note that $x = 0$ is not in the domain of $f$. Also, before differentiating it is helpful to rewrite $f(x)$ as

$$f(x) = \frac{x^4 + 1}{x^2} = x^2 + \frac{1}{x^2} = x^2 + x^{-2}$$

The derivative of $f$ is then

$$f'(x) = 2x - 2x^{-3} = 2x - \frac{2}{x^3}$$

$$= \frac{2(x^4 - 1)}{x^3}$$

$$= \frac{2(x^2 + 1)(x - 1)(x + 1)}{x^3}$$

Since $f'(x)$ is zero at $x = 1$ and $x = -1$, and $f$ is undefined at $x = 0$, we use these three $x$-values to determine the test intervals:

$$x = -1, 1 \qquad \text{Critical numbers}$$

$$x = 0 \qquad \text{Discontinuity}$$

Table 3.3 summarizes the testing of the four intervals determined by these three numbers, and the graph of $f$ is shown in Figure 3.6.

TABLE 3.3

| Interval | $-\infty < x < -1$ | $-1 < x < 0$ | $0 < x < 1$ | $1 < x < \infty$ |
|---|---|---|---|---|
| Test value | $x = -2$ | $x = -\dfrac{1}{2}$ | $x = \dfrac{1}{2}$ | $x = 2$ |
| Sign of $f'(x)$ | $f'(-2) < 0$ | $f'\left(-\dfrac{1}{2}\right) > 0$ | $f'\left(\dfrac{1}{2}\right) < 0$ | $f'(2) > 0$ |
| Conclusion | decreasing | increasing | decreasing | increasing |

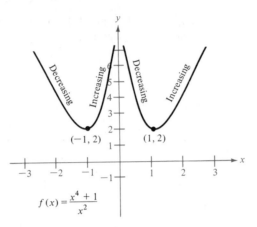

$$f(x) = \frac{x^4 + 1}{x^2}$$

FIGURE 3.6

The converse of the test for increasing and decreasing functions is not true. This means that it is possible for a function to be increasing on an interval even though the derivative is not *positive* at every point in the interval. The next example illustrates such a case.

**EXAMPLE 4**

**An Increasing Function Whose Derivative Is Zero at a Point**

Show that the following function is increasing on the entire real line:

$$f(x) = x^3 - 3x^2 + 3x$$

**SOLUTION**

Since

$$f'(x) = 3x^2 - 6x + 3 = 0$$

$$3(x^2 - 2x + 1) = 0$$

$$3(x - 1)^2 = 0$$

$f'$ is defined everywhere and has only one critical number, $x = 1$. Thus the intervals to be tested are $(-\infty, 1)$ and $(1, \infty)$. Table 3.4 summarizes the testing of these two intervals.

FIGURE 3.7

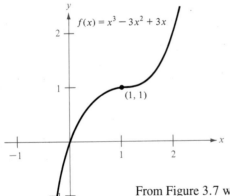

TABLE 3.4

| Interval | $-\infty < x < 1$ | $1 < x < \infty$ |
|---|---|---|
| Test value | $x = 0$ | $x = 2$ |
| Test | $f'(0) > 0$ | $f'(2) > 0$ |
| Conclusion | increasing | increasing |

From Figure 3.7 we can see that $f$ is increasing on the entire real line—even though $f'(1) = 0$. To see this, look back at the definition of an increasing function.

---

**EXAMPLE 5**

**A Business Application**

A national toy distributor sells a certain game with the following cost and revenue functions:

$$C(x) = 2.4x - 0.0002x^2, \quad 0 \le x \le 6000$$

$$R(x) = 7.2x - 0.001x^2, \quad 0 \le x \le 6000$$

Determine the interval for which the profit function is increasing.

**SOLUTION**

The profit for producing $x$ units is given by

$$P(x) = R(x) - C(x)$$

$$= (7.2x - 0.001x^2) - (2.4x - 0.0002x^2)$$

$$= 4.8x - 0.0008x^2$$

To find the interval for which this function is increasing, we set the marginal profit equal to zero. (Recall that the marginal profit $P'$ is the derivative of the profit function.)

$$P'(x) = 4.8 - 0.0016x$$

$P'(x) = 0$ when

$$x = \frac{4.8}{0.0016} = 3000 \text{ units}$$

In the interval $(0, 3000)$, $P'$ is positive and the profit is *increasing,* and in the interval $(3000, 6000)$, $P'$ is negative and the profit is *decreasing,* as shown in Figure 3.8.

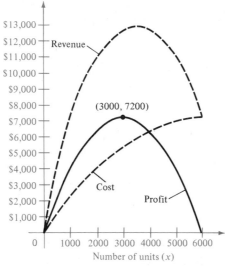

FIGURE 3.8

Note in Figure 3.8 that the profit is increasing when the marginal revenue (the slope of the revenue curve) exceeds the marginal cost (the slope of the cost curve). In other words, as long as the revenue function is increasing at a more rapid rate than the cost function, increased production will yield a greater profit.

## SECTION EXERCISES 3.1

In Exercises 1–4, evaluate the derivative of the function at the indicated points on its graph. Observe the relationship between the sign of the derivative and the increasing or decreasing behavior of the graph.

**1.** $f(x) = \dfrac{x^2}{x^2 + 4}$

**2.** $f(x) = x + \dfrac{32}{x^2}$

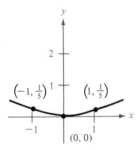

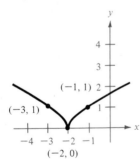

**3.** $f(x) = (x + 2)^{2/3}$

**4.** $f(x) = -3x\sqrt{x + 1}$

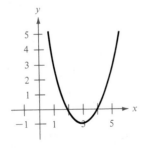

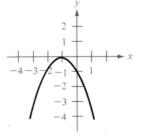

In Exercises 5–10, identify the open intervals on which the function is increasing or decreasing.

**5.** $f(x) = x^2 - 6x + 8$

**6.** $y = -(x + 1)^2$

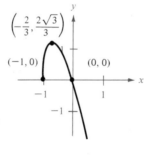

**7.** $y = \dfrac{x^3}{4} - 3x$

**8.** $f(x) = x^4 - 2x^2$

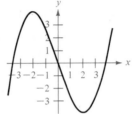

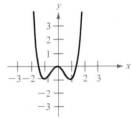

**9.** $f(x) = \dfrac{1}{x^2}$

**10.** $y = \dfrac{x^2}{x + 1}$

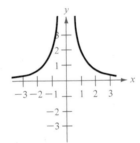

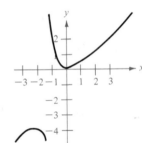

In Exercises 11–20, find the critical numbers (if any) and the open intervals on which the function is increasing or decreasing. Sketch the graph of the function.

**11.** $f(x) = 2x - 3$
**12.** $f(x) = 5 - 3x$
**13.** $g(x) = -(x - 1)^2$
**14.** $g(x) = (x + 2)^2$
**15.** $y = x^2 - 2x$
**16.** $y = -(x^2 - 2x)$
**17.** $y = x^3 - 6x^2$
**18.** $y = (x - 2)^3$
**19.** $f(x) = -(x + 1)^3$
**20.** $f(x) = \sqrt{4 - x^2}$

In Exercises 21–32, find the critical numbers (if any) and the open intervals on which the function is increasing or decreasing.

**21.** $f(x) = -2x^2 + 4x + 3$
**22.** $f(x) = x^2 + 8x + 10$
**23.** $f(x) = 2x^3 + 3x^2 - 12x$
**24.** $f(x) = (x - 1)^2(x + 2)$
**25.** $h(x) = x^{2/3}$
**26.** $h(x) = \sqrt[3]{x - 1}$
**27.** $f(x) = x^4 - 2x^3$
**28.** $f(x) = x\sqrt{x + 1}$
**29.** $f(x) = 2x\sqrt{3 - x}$
**30.** $f(x) = 2(x + 2)\sqrt{1 - x}$
**31.** $y = \dfrac{x}{x^2 + 4}$
**32.** $y = \dfrac{x^2}{x^2 + 4}$

In Exercises 33–38, find the critical numbers (if any) and the open intervals on which the *discontinuous* function is increasing or decreasing.

**33.** $f(x) = x + \dfrac{1}{x}$

**34.** $f(x) = \dfrac{x}{x + 1}$

**35.** $f(x) = \dfrac{x^2}{x^2 - 9}$

**36.** $f(x) = \begin{cases} 2x + 1, & x \le -1 \\ x^2 - 2, & x > -1 \end{cases}$

**37.** $f(x) = \begin{cases} 4 - x^2, & x \le 0 \\ -2x - 2, & x > 0 \end{cases}$

**38.** $f(x) = \begin{cases} -x^3 + 1, & x \le 0 \\ -x^2 + 2x + 1, & x > 0 \end{cases}$

In Exercises 39 and 40, the position function gives the height (in feet) of a ball, where the time $t$ is measured in seconds. Find the time interval in which the ball is moving up and the interval in which it is moving down.

**39.** $s(t) = 96t - 16t^2, \quad 0 \le t \le 6$

**40.** $s(t) = -16t^2 + 64t, \quad 0 \le t \le 4$

**41.** A drug is administered to a patient. A model giving the drug concentration in the patient's bloodstream over a two-hour period is

$$C = 0.29483t + 0.04253t^2 - 0.00035t^3$$

where $0 \le t \le 120$. $C$ is measured in milligrams and $t$ is the time in minutes. Find the intervals on which $C$ is increasing or decreasing.

**42.** A fast-food restaurant sells $x$ hamburgers to make a profit $P$ given by

$$P = 2.44x - \frac{x^2}{20{,}000} - 5000$$

where $0 \le x \le 35{,}000$. Find the intervals on which $P$ is increasing or decreasing.

**43.** After birth, an infant will normally lose weight for a few days and then start gaining. A model for the average weight $W$ of infants over the first two weeks following birth is

$$W = 0.033t^2 - 0.3974t + 7.3032$$

where $0 \le t \le 14$ and $t$ is measured in days. Find the intervals on which $W$ is increasing or decreasing.

**44.** The ordering and transportation cost $C$ of components used in a manufacturing firm is given by

$$C = 10\left(\frac{1}{x} + \frac{x}{x + 3}\right), \quad 1 \le x$$

where $C$ is measured in thousands of dollars and $x$ is the order size in hundreds. Find the intervals on which $C$ is increasing or decreasing.

In Exercises 45–50, assume that $f$ is differentiable for all $x$. The sign of $f'$ is as follows:

$f'(x) > 0$ on $(-\infty, -4)$

$f'(x) < 0$ on $(-4, 6)$

$f'(x) > 0$ on $(6, \infty)$

In each exercise, supply the appropriate inequality for the indicated value of $c$.

| Function | Sign of g'(c) |
|---|---|
| **45.** $g(x) = f(x) + 5$ | $g'(0)$ ☐ $0$ |
| **46.** $g(x) = 3f(x) - 3$ | $g'(-5)$ ☐ $0$ |
| **47.** $g(x) = -f(x)$ | $g'(-6)$ ☐ $0$ |
| **48.** $g(x) = -f(x)$ | $g'(0)$ ☐ $0$ |
| **49.** $g(x) = f(x - 10)$ | $g'(0)$ ☐ $0$ |
| **50.** $g(x) = f(x - 10)$ | $g'(8)$ ☐ $0$ |

# Extrema and the First-Derivative Test

The official birth rate in the United States is prepared by the National Center for Health Studies and is based on the vital records received from all states and the District of Columbia. The birth rate is listed in terms of the number of live births per thousand people. In 1800, the U.S. birth rate was 55.0 per thousand. During the nineteenth century this figure gradually dropped, reaching 32.3 in 1900. The rate since 1900 is shown graphically in Figure 3.9.

Note that although the birth rate has decreased overall since 1900, it has done so through several rises and falls rather than through a steady decrease. For instance, shortly after U.S. involvement in World War I (1917–1918), the U.S. birth rate increased for a couple of years. When the rate changes from decreasing to increasing, we say it has a **relative minimum.** Similarly, when the rate changes from increasing to decreasing, we say it has a **relative maximum.**

The lowest birth rate between 1900 and 1983 occurred during 1975, when the rate fell to 14.6 per thousand. We call this value the **absolute minimum** for the interval [1900, 1983].

FIGURE 3.9

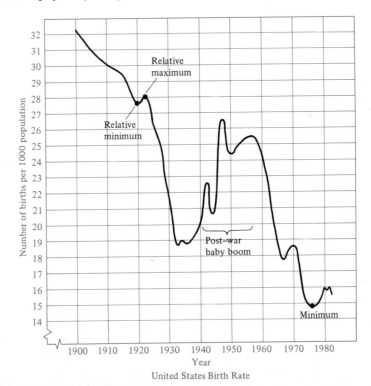

United States Birth Rate

■ **Relative Minimum and Relative Maximum**
■ **The First-Derivative Test**
■ **Absolute Minimum and Absolute Maximum**

In the preceding section, we used the derivative of a function to determine the intervals in which the function is increasing or decreasing. In this section, we examine the points at which a function changes from increasing to decreasing, or vice versa, as shown in Figure 3.10. At such points we say the function achieves a **relative maximum** or **relative minimum.** The term *extremum* (the plural is *extrema*) is used to describe a value that is a minimum or a maximum. Thus, the relative extrema of a function include both the relative minima and the relative maxima.

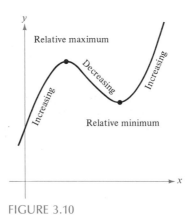

FIGURE 3.10

| **Definition of relative extrema** | Let $f$ be a function defined at $c$. |
|---|---|

1. $f(c)$ is called a **relative maximum** of $f$ if there exists an interval $(a, b)$ containing $c$ such that for all $x$ in $(a, b)$

$$f(x) \le f(c)$$

2. $f(c)$ is called a **relative minimum** of $f$ if there exists an interval $(a, b)$ containing $c$ such that for all $x$ in $(a, b)$

$$f(x) \ge f(c)$$

(The plurals of maximum and minimum are *maxima* and *minima*.)

For a continuous function, the relative maxima and minima must occur at a critical number of the function, as shown in Figure 3.11.

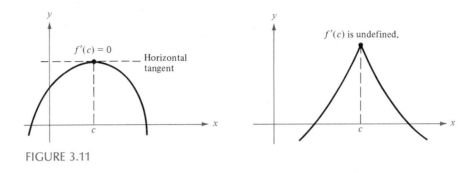

FIGURE 3.11

A quick reading of this result sometimes leads students to think that every critical number must produce a relative minimum or maximum. This is not what the statement says. What it does say is that in our search for relative extrema of a continuous function, we need only test the critical numbers of $f$. Once we have determined that $c$ is a critical number of $f$, the following test enables us to classify $f(c)$ as a relative minimum, relative maximum, or neither. We call this test the **First-Derivative Test for Relative Extrema.**

| $f(c)$ | Sign of $f'$ in $(a, c)$ | Sign of $f'$ in $(c, b)$ | Graph |
|---|---|---|---|
| 1. Relative minimum | $-$ | $+$ | $(-) \quad (+)$ |
| 2. Relative maximum | $+$ | $-$ | $(+) \quad (-)$ |
| 3. Neither | $+$ | $+$ | $(+) \quad (+)$ |
| 4. Neither | $-$ | $-$ | $(-) \quad (-)$ |

**EXAMPLE 1**

**Finding Relative Extrema**

Find all relative extrema of the function

$$f(x) = 2x^3 - 3x^2 - 36x + 14$$

**SOLUTION**

Setting the derivative of $f$ equal to zero, we have

$$f'(x) = 6x^2 - 6x - 36 = 0$$
$$6(x^2 - x - 6) = 0$$
$$6(x - 3)(x + 2) = 0$$

Since $f'$ is defined for all real numbers, the only critical numbers of $f$ are $x = -2$ and $x = 3$. Using the pattern we used to test for increasing and decreasing functions, we test the sign of $f'$ in each of the intervals $(-\infty, -2)$, $(-2, 3)$, and $(3, \infty)$. The result of this testing is shown in Table 3.5.

TABLE 3.5

| Interval | $-\infty < x < -2$ | $-2 < x < 3$ | $3 < x < \infty$ |
|---|---|---|---|
| Test value | $x = -3$ | $x = 0$ | $x = 4$ |
| Sign of $f'(x)$ | $f'(-3) > 0$ ($f$ is increasing) | $f'(0) < 0$ ($f$ is decreasing) | $f'(4) > 0$ ($f$ is increasing) |

Using the First-Derivative Test, we conclude that the critical number $-2$ yields a relative maximum ($f'$ changes sign from $+$ to $-$) and 3 yields a relative minimum ($f'$ changes sign from $-$ to $+$). The graph of $f$ is shown in Figure 3.12.

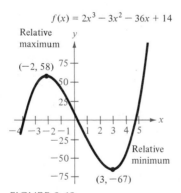

FIGURE 3.12

**Remark:** In Section 2.2, Example 5, we looked at the graph of $f(x) = x^3 - 4x + 2$ and saw that it does *not* have a relative minimum at $(1, -1)$. Try looking at that example now to find the point at which the graph *does* have a relative minimum.

In Example 1 both critical numbers yielded relative extrema of the given function. In the next example, we look at a function that also has two critical numbers. However, only one of the critical numbers yields a relative extremum.

**EXAMPLE 2**

**Finding Relative Extrema**

Find all relative extrema of the function

$$f(x) = x^4 - x^3$$

**SOLUTION**

Setting the derivative of $f$ equal to zero, we have

$$f'(x) = 4x^3 - 3x^2 = 0$$

$$x^2(4x - 3) = 0$$

Since $f'$ is defined everywhere, the only critical numbers of $f$ are 0 and $\frac{3}{4}$. Thus, we test the sign of the derivative in the intervals $(-\infty, 0)$, $(0, \frac{3}{4})$, and $(\frac{3}{4}, \infty)$, as shown in Table 3.6.

TABLE 3.6

| Interval | $-\infty < x < 0$ | $0 < x < \dfrac{3}{4}$ | $\dfrac{3}{4} < x < \infty$ |
|---|---|---|---|
| Test value | $x = -1$ | $x = \dfrac{1}{2}$ | $x = 1$ |
| Sign of $f'(x)$ | $f'(-1) < 0$ ($f$ is decreasing) | $f'\left(\dfrac{1}{2}\right) < 0$ ($f$ is decreasing) | $f'(1) > 0$ ($f$ is increasing) |

Using the First-Derivative Test, we conclude that the critical number $\frac{3}{4}$ yields a relative minimum, whereas 0 yields neither a relative minimum nor a relative maximum. The graph of $f$ is shown in Figure 3.13.

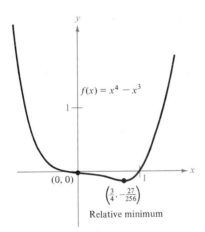

FIGURE 3.13

**EXAMPLE 3**

**Finding Relative Extrema**

Find all relative extrema of the function

$$f(x) = 2x - 3x^{2/3}$$

**SOLUTION**    Differentiating, we have

$$f'(x) = 2 - \frac{2}{x^{1/3}} = \frac{2(x^{1/3} - 1)}{x^{1/3}}$$

Since $f'(x) = 0$ when $x = 1$ and $f'$ is undefined when $x = 0$, the critical numbers of $f$ are 0 and 1. Therefore, we test the sign of the derivative of $f$ in the intervals $(-\infty, 0)$, $(0, 1)$, and $(1, \infty)$, as shown in Table 3.7.

TABLE 3.7

| Interval | $-\infty < x < 0$ | $0 < x < 1$ | $1 < x < \infty$ |
|---|---|---|---|
| Test value | $x = -1$ | $x = \dfrac{1}{2}$ | $x = 2$ |
| Sign of $f'(x)$ | $f'(-1) > 0$ | $f'\left(\dfrac{1}{2}\right) < 0$ | $f'(2) > 0$ |
| | ($f$ is increasing) | ($f$ is decreasing) | ($f$ is increasing) |

Using the First-Derivative Test, we conclude that 0 yields a relative maximum and 1 yields a relative minimum, as shown in Figure 3.14.

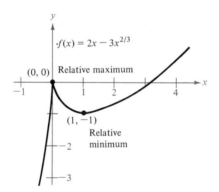

FIGURE 3.14

## Absolute Extrema

Just as we use the terms relative maximum and relative minimum to describe the local behavior of a function, we use the terms *maximum* and *minimum,* or *absolute maximum* and *absolute minimum,* to describe a function's global or overall behavior on an *entire* interval.

| **Definition of (absolute) extrema** | Let $f$ be defined on an interval $I$ containing $c$.<br><br>1. $f(c)$ is a **minimum of $f$ on $I$** if $f(c) \le f(x)$ for every $x$ in $I$.<br>2. $f(c)$ is a **maximum of $f$ on $I$** if $f(c) \ge f(x)$ for every $x$ in $I$.<br><br>The minimum and maximum values of a function on an interval are sometimes referred to as the *absolute minimum* and *absolute maximum* of $f$ on $I$. |
|---|---|

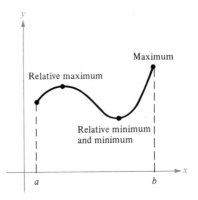

FIGURE 3.15

Be sure that you see the distinction between relative extrema and absolute extrema. For instance, in Figure 3.15, the function has a relative minimum that also happens to be an absolute minimum on the interval $[a, b]$. However, the relative maximum of $f$ is not the absolute maximum on the interval $[a, b]$.

Reviewing Examples 1 through 3, we can see that the only instance of an *absolute* extremum is the point $(\frac{3}{4}, -\frac{27}{256})$ encountered in Example 2. In practical applications absolute extrema occur more often, because practical applications often involve restricted domains that consist of only a portion of the real numbers. The next theorem points out that if a continuous function has a closed interval as its domain then it *must* have both a minimum and a maximum on the interval.

| **Extreme Value Theorem** | If $f$ is continuous on $[a, b]$, then $f$ takes on both a minimum value and a maximum value on $[a, b]$. |

**Remark:** Note that although a continuous function has just one minimum and one maximum on a closed interval $[a, b]$, this value may occur for more than one value of $x$. For example, on the interval $[-3, 3]$, the function $f(x) = 9 - x^2$ has a minimum value of zero when $x = -3$ *and* when $x = 3$.

In Figure 3.15, the maximum value of the function occurs at the right endpoint of the interval. When looking for the extreme values of a function on a *closed* interval, you need to consider the values of $f$ at the endpoints as well as at the critical numbers of the function. We suggest the following guidelines.

| **Guidelines for finding extrema on a closed interval** | To find the extrema of a continuous function $f$ on a closed interval $[a, b]$, use the following steps. |

1. Evaluate $f$ at each of its critical numbers in $(a, b)$.
2. Evaluate $f$ at each endpoint $a$ and $b$.
3. The least of these values is the minimum, and the greatest is the maximum.

EXAMPLE 4

**Finding Extrema on a Closed Interval**

Find the minimum and maximum values of

$$f(x) = x^2 - 6x + 2$$

on the interval $[0, 5]$.

**SOLUTION**

Setting the derivative of $f$ equal to zero, we have

$$f'(x) = 2x - 6 = 0$$

$$2(x - 3) = 0$$

Thus, the only critical number of $f$ is $x = 3$. To find the minimum and maximum values of $f$ on the interval $[0, 5]$, we simply compute the values of $f$ at this critical number *and* at the endpoints and then compare them, as shown in Table 3.8.

TABLE 3.8

| $x$-value | endpoint $x = 0$ | critical number $x = 3$ | endpoint $x = 5$ |
|---|---|---|---|
| $f(x)$ | $f(0) = 2$ | $f(3) = -7$ | $f(5) = -3$ |
| Conclusion | maximum | minimum | |

Thus, the minimum of $f$ on the interval $[0, 5]$ occurs at the critical number $x = 3$, and the maximum occurs at the left endpoint $x = 0$, as shown in Figure 3.16.

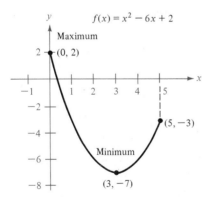

FIGURE 3.16

## Applications of Extrema

Finding the minimum and maximum values of a function is one of the most common applications of calculus. We conclude this section with two straightforward applications.

**EXAMPLE 5**

**Finding the Maximum Profit**

In Example 6 of Section 2.3, we looked at a fast-food restaurant whose profit function for hamburgers is

$$P(x) = 2.44x - \frac{x^2}{20,000} - 5000$$

Find the production level that produces a maximum profit.

**SOLUTION**

Setting the marginal profit equal to zero, we have

$$P'(x) = 2.44 - \frac{x}{10,000} = 0$$

$$-\frac{x}{10,000} = -2.44$$

$$x = 24,400 \text{ units}$$

From Figure 3.17 we can see that the critical number $x = 24,400$ corresponds to the production level that produces a maximum profit.

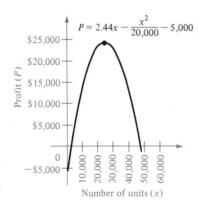

FIGURE 3.17

In Example 5, we concluded that the maximum profit occurred at the critical number $x = 24,400$ by making a sketch of the function. It is often the case that the nature of an application clearly indicates whether a critical number yields a maximum or minimum without a formal test. This is further demonstrated in the next example.

**EXAMPLE 6**

**Finding the Minimum and Maximum Oxygen Levels**

Suppose that $f(t)$ measures the level of oxygen in a pond, where $f(t) = 1$ is the normal level and the time $t$ is measured in weeks. When $t = 0$, some organic waste is dumped into the pond, and as the waste material oxidizes, the amount of oxygen in the pond is given by

$$f(t) = \frac{t^2 - t + 1}{t^2 + 1}, \quad 0 \le t < \infty$$

(a) When is the oxygen level lowest?

(b) When is the oxygen level highest?

**SOLUTION**    (a) Since

$$f'(t) = \frac{(t^2 + 1)(2t - 1) - (t^2 - t + 1)(2t)}{(t^2 + 1)^2}$$

$$= \frac{2t^3 - t^2 + 2t - 1 - 2t^3 + 2t^2 - 2t}{(t^2 + 1)^2}$$

$$= \frac{t^2 - 1}{(t^2 + 1)^2}$$

is zero when $t^2 - 1 = 0$, and since $t = -1$ is not in the domain, the only critical number is $t = 1$. The nature of the problem indicates that this value ($t = 1$ week) yields the minimum oxygen level, as shown in Figure 3.18.

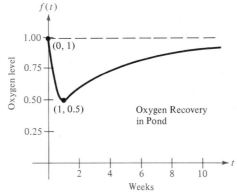

FIGURE 3.18

(b) Since $y = 1$ is an upper bound for $f(t)$, it is clear that the maximum level of oxygen occurs at the endpoint $t = 0$.

## SECTION EXERCISES 3.2

In Exercises 1–20, find all relative extrema of the given function.

**1.** $f(x) = -2x^2 + 4x + 3$
**2.** $f(x) = x^2 + 8x + 10$
**3.** $f(x) = x^2 - 6x$
**4.** $f(x) = (x - 1)^2(x + 2)$
**5.** $g(x) = 2x^3 + 3x^2 - 12x$
**6.** $g(x) = \dfrac{x^5 - 5x}{5}$
**7.** $h(x) = -(x + 4)^3$
**8.** $h(x) = (x - 3)^3$
**9.** $f(x) = x^3 - 6x^2 + 15$
**10.** $f(x) = x^4 - 1$
**11.** $f(x) = x^4 - 2x^3$
**12.** $f(x) = x^4 - 32x + 4$

**13.** $f(t) = t^{1/3} + 1$
**14.** $f(t) = (t - 1)^{1/3}$
**15.** $g(t) = t^{2/3}$
**16.** $h(t) = (t - 1)^{2/3}$
**17.** $f(x) = x + \dfrac{1}{x}$
**18.** $f(x) = \dfrac{x}{x + 1}$
**19.** $h(x) = \dfrac{4}{x^2 + 1}$
**20.** $g(x) = 4\left(1 + \dfrac{1}{x} + \dfrac{1}{x^2}\right)$

In Exercises 21–26, determine from the graph of $f$ if $f$ possesses a relative minimum in the interval $(a, b)$.

**21.**

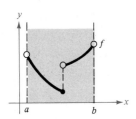

**22.**

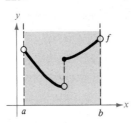

**23.**

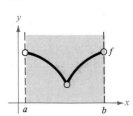

**24.**

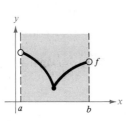

**25.**

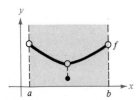

**26.**

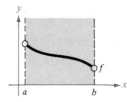

In Exercises 27–30, locate the extrema of the function (if any exist) over the indicated interval.

**27.** $f(x) = 5 - x$
- (a) $[1, 4]$
- (b) $[1, 4)$
- (c) $(1, 4]$
- (d) $(1, 4)$

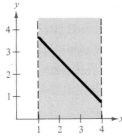

**28.** $f(x) = x^2 - 2x$
- (a) $[-1, 2]$
- (b) $(1, 3]$
- (c) $(0, 2)$
- (d) $[1, 3]$

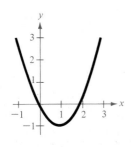

**29.** $f(x) = \sqrt{4 - x^2}$
- (a) $[-2, 2]$
- (b) $[-2, 0)$
- (c) $(-2, 2)$
- (d) $[1, 2)$

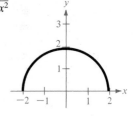

**30.** $f(x) = \frac{1}{2}(x - 1)^2(4 - x)$
- (a) $(-1, 5)$
- (b) $[-1, 4)$
- (c) $(2, 4)$
- (d) $(-1, 0]$

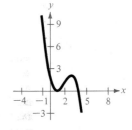

In Exercises 31–40, locate the extrema of the given function on the indicated interval.

| | Function | Interval |
|---|---|---|
| **31.** | $f(x) = 2(3 - x)$ | $[-1, 2]$ |
| **32.** | $f(x) = \dfrac{2x + 5}{3}$ | $[0, 5]$ |
| **33.** | $f(x) = -x^2 + 4x$ | $[0, 3]$ |
| **34.** | $f(x) = x^2 + 2x - 4$ | $[-1, 1]$ |
| **35.** | $f(x) = x^3 - 3x^2$ | $[-1, 3]$ |
| **36.** | $f(x) = x^3 - 12x$ | $[0, 4]$ |
| **37.** | $f(x) = 3x^{2/3} - 2x$ | $[-1, 1]$ |
| **38.** | $g(t) = \dfrac{t^2}{t^2 + 3}$ | $[-1, 1]$ |
| **39.** | $h(s) = \dfrac{1}{s - 2}$ | $[0, 1]$ |
| **40.** | $h(t) = \dfrac{t}{t - 2}$ | $[3, 5]$ |

The error estimate for the Trapezoidal Rule (given in Section 6.5) involves the maximum of the absolute value of the second derivative in an interval. In Exercises 41 and 42, find the maximum value of $|f''(x)|$ in the indicated interval.

| | Function | Interval |
|---|---|---|
| **41.** | $f(x) = x^3(3x^2 - 10)$ | $[0, 1]$ |
| **42.** | $f(x) = \dfrac{1}{x^2 + 1}$ | $[0, 3]$ |

The error estimate for Simpson's Rule (given in Section 6.5) involves the maximum of the absolute value of the fourth derivative in an interval. In Exercises 43 and 44, find the maximum value of $|f^{(4)}(x)|$ in the indicated interval.

| Function | Interval |
| --- | --- |
| **43.** $f(x) = 15x^4 - \left(\dfrac{2x-1}{2}\right)^6$ | $[0, 1]$ |
| **44.** $f(x) = \dfrac{1}{x^2}$ | $[1, 2]$ |

**45.** A retailer has determined that the cost $C$ for ordering and storing $x$ units of a certain product is

$$C = 2x + \frac{300{,}000}{x}, \quad 0 < x \le 300$$

Find the order size that will minimize cost given that the delivery truck can bring a maximum of 300 units per order.

**46.** The concentration $C$ of a certain chemical in the bloodstream $t$ hours after injection into muscle tissue is given by

$$C = \frac{3t}{27 + t^3}, \quad 0 \le t$$

When is the concentration greatest?

**47.** Coughing forces the trachea (windpipe) to contract, which in turn affects the velocity $v$ of the air through the trachea. Suppose the velocity of the air during coughing is

$$v = k(R - r)r^2, \quad 0 \le r < R$$

where $k$ is a constant, $R$ is the normal radius of the trachea, and $r$ is the radius during coughing. What radius will produce the maximum air velocity?

**48.** Find $a$, $b$, and $c$ so that the function given by $f(x) = ax^2 + bx + c$ has a relative maximum at $(5, 20)$ and passes through the point $(2, 10)$.

# Concavity and the Second-Derivative Test

The divorce rate in the United States is based on the annual number of divorces per 1000 marriages. Figure 3.19 graphically depicts the changes in this rate between 1920 and 1981. Note that the divorce rate increased from the early part of the Great Depression until the end of World War II. Between 1932 and 1946 there are no relative minimum or relative maximum points indicated on the graph, and yet the graph does have definite "turning points." There are some intervals in which the *slope* of the graph is increasing and other intervals in which the *slope* is decreasing.

During a period in which the slope is increasing (such as 1937–1942), the graph is said to be **concave upward**. During a period in which the slope is decreasing (such as 1933–1937), the graph is said to be **concave downward**. The turning point between these two types of concavity is called a **point of inflection.**

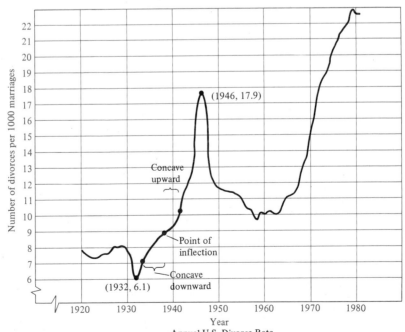

FIGURE 3.19

Annual U.S. Divorce Rate

■ **Concavity**
■ **Points of Inflection**
■ **Second-Derivative Test**

We have seen that locating the intervals in which a function $f$ increases or decreases is helpful in determining its graph. In this section, we extend this idea. We show that, by locating the intervals in which $f'$ increases or decreases, we can determine where the graph of $f$ is curving upward or curving downward. We define this notion of curving upward or downward as **concavity.**

| | |
|---|---|
| **Definition of concavity** | Let $f$ be differentiable on an open interval $I$. The graph of $f$ is<br><br>1. **concave upward** on $I$ if $f'$ is increasing on the interval<br>2. **concave downward** on $I$ if $f'$ is decreasing on the interval |

From Figure 3.20 we can obtain the following graphical interpretation of concavity.

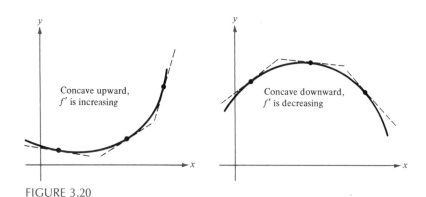

Concave upward, $f'$ is increasing

Concave downward, $f'$ is decreasing

FIGURE 3.20

1. A curve that lies *above* its tangent lines is concave upward.
2. A curve that lies *below* its tangent lines is concave downward.

This visual test for concavity is useful when the graph of a function is given. To determine concavity without seeing a graph, we need an analytic test for finding the intervals on which the derivative is increasing or decreasing. It turns out that we can use the second derivative to determine these intervals in much the same way we used the first derivative to determine the intervals in which $f$ is increasing or decreasing.

Let $f$ be a function whose second derivative exists on an open interval $I$.

1. If $f''(x) > 0$ for all $x$ in $I$, then the graph of $f$ is concave upward on $I$.
2. If $f''(x) < 0$ for all $x$ in $I$, then the graph of $f$ is concave downward on $I$.

For a *continuous* function $f$ we can find the intervals on which the graph of $f$ is concave upward and concave downward as follows.

1. Find the second derivative.
2. Locate the $x$-values at which $f''(x) = 0$ or $f''(x)$ is undefined.
3. Use these $x$-values to determine the test intervals.
4. Test the sign of $f''(x)$ in each of the test intervals.

For *discontinuous* functions the test intervals should be formed using the points of discontinuity (along with the points at which $f''$ is zero or undefined).

**EXAMPLE 1**

**Determining Concavity**

Determine the intervals in which the graph of the following function is concave upward or downward:

$$f(x) = \frac{6}{x^2 + 3}$$

**SOLUTION**

We begin by finding the second derivative of $f$.

$$f(x) = 6(x^2 + 3)^{-1}$$

$$f'(x) = (-6)(2x)(x^2 + 3)^{-2} = \frac{-12x}{(x^2 + 3)^2}$$

$$f''(x) = \frac{(x^2 + 3)^2(-12) - (-12x)(2)(2x)(x^2 + 3)}{(x^2 + 3)^4}$$

$$= \frac{-12(x^2 + 3) + (48x^2)}{(x^2 + 3)^3}$$

$$= \frac{36(x^2 - 1)}{(x^2 + 3)^3}$$

Since $f''(x) = 0$ when $x = \pm 1$ and since $f''(x)$ is defined for all $x$, we test $f''$ in the intervals $(-\infty, -1)$, $(-1, 1)$, and $(1, \infty)$, as shown in Table 3.9.

TABLE 3.9

| Interval | $(-\infty, -1)$ | $(-1, 1)$ | $(1, \infty)$ |
|---|---|---|---|
| Sign of $f''$ | $+$ | $-$ | $+$ |
| $f'$ | increasing | decreasing | increasing |
| Graph of $f$ | concave upward | concave downward | concave upward |

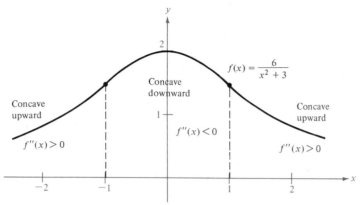

FIGURE 3.21

The graph of $f$ is shown in Figure 3.21.

**Remark:** In Example 1, $f'$ is increasing on the interval $(1, \infty)$ even though $f$ is decreasing there. Be sure you see that the increasing or decreasing of $f'$ does not necessarily correspond to the increasing or decreasing of $f$.

### Points of Inflection

The graph in Figure 3.21 has two points at which the concavity changes. If at such a point the tangent line to the graph exists, we call the point a **point of inflection.** Three types of inflection points are shown in Figure 3.22. (Note that the third graph has a vertical tangent line at its point of inflection.)

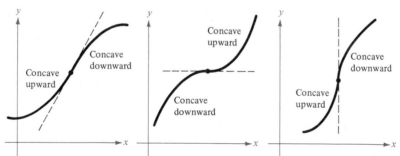

The graph *crosses* its tangent line at a point of inflection.

FIGURE 3.22

| **Definition of a point of inflection** | If the graph of a continuous function possesses a tangent line at a point where its concavity changes from upward to downward (or vice versa), then the point is called a **point of inflection.** |
|---|---|

**Remark:** Note in Figure 3.22 that at a point of inflection the graph crosses its tangent line.

Since a point of inflection occurs where the concavity of a graph changes, it must be true that at such points the sign of $f''$ changes. Thus, to locate possible points of inflection, we need only determine the values of $x$ for which $f''(x) = 0$ or for which $f''(x)$ does not exist. This parallels the procedure for locating relative extrema of $f$ by determining the critical numbers of $f$.

| **Property of points of inflection** | If $(c, f(c))$ is a point of inflection of the graph of $f$, then either $$f''(c) = 0 \quad \text{or} \quad f''(c) \text{ does not exist}$$ |
| --- | --- |

**EXAMPLE 2**

**Finding Points of Inflection**

Determine the points of inflection and discuss the concavity of the graph of

$$f(x) = x^4 + x^3 - 3x^2 + 1$$

**SOLUTION**

Differentiating twice, we have

$$f'(x) = 4x^3 + 3x^2 - 6x$$

$$f''(x) = 12x^2 + 6x - 6$$

$$= 6(2x^2 + x - 1)$$

$$= 6(2x - 1)(x + 1)$$

By setting $f''(x)$ equal to zero, we see that the possible points of inflection occur at $x = -1$ and $x = \frac{1}{2}$. In Table 3.10, we test the intervals

$$(-\infty, -1), \qquad \left(-1, \frac{1}{2}\right), \qquad \left(\frac{1}{2}, \infty\right)$$

TABLE 3.10

| Interval | $(-\infty, -1)$ | $\left(-1, \frac{1}{2}\right)$ | $\left(\frac{1}{2}, \infty\right)$ |
| --- | --- | --- | --- |
| Sign of $f''$ | $+$ | $-$ | $+$ |
| $f'$ | increasing | decreasing | increasing |
| Graph of $f$ | concave upward | concave downward | concave upward |

From this table we see that $f$ is concave upward in $(-\infty, -1)$, concave downward in $(-1, \frac{1}{2})$, and concave upward again in $(\frac{1}{2}, \infty)$. Thus, the numbers $-1$ and $\frac{1}{2}$ both yield points of inflection, as shown in Figure 3.23.

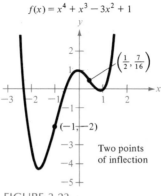

$$f(x) = x^4 + x^3 - 3x^2 + 1$$

$\left(\frac{1}{2}, \frac{7}{16}\right)$

$(-1, -2)$

Two points
of inflection

FIGURE 3.23

We should point out that it is possible for the second derivative to be zero at a point that is *not* a point of inflection. For example, compare the graphs of $f(x) = x^3$ and $g(x) = x^4$, as shown in Figure 3.24. Both second derivatives are zero when $x = 0$, but only the graph of $f$ has a point of inflection at $x = 0$. This shows that, before concluding that a point of inflection exists at a value of $x$ for which $f''(x) = 0$, we should test to be certain that the concavity actually changes there.

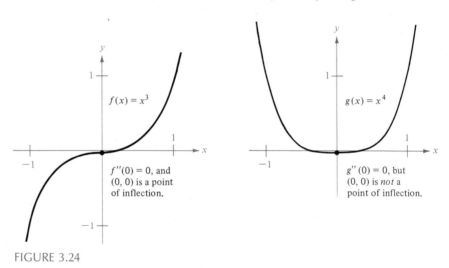

$f(x) = x^3$

$f''(0) = 0$, and
$(0, 0)$ is a point
of inflection.

$g(x) = x^4$

$g''(0) = 0$, but
$(0, 0)$ is *not* a
point of inflection.

FIGURE 3.24

### Second-Derivative Test

If the second derivative exists, we can often use it as a simple test for relative extrema. The test is based on the fact that if $f(c)$ is a relative maximum of a *differentiable* function $f$, then its graph is concave downward in some interval containing $c$. Similarly, if $f(c)$ is a relative minimum, then the graph of $f$ is concave upward in some interval containing $c$, as shown in Figure 3.25. This test is referred to as the **Second-Derivative Test for Relative Extrema.**

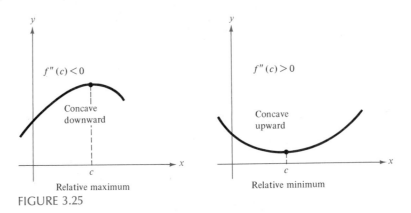

FIGURE 3.25

| Second-Derivative Test for Relative Extrema | Let $f''$ exist on some open interval containing $c$ and $f'(c) = 0$. |
|---|---|

1. If $f''(c) > 0$, then $f(c)$ is a relative minimum.
2. If $f''(c) < 0$, then $f(c)$ is a relative maximum.
3. If $f''(c) = 0$, then the test fails.

■■■ **Remark:** Be sure you see that if $f''(c) = 0$, the Second-Derivative Test does not apply. In such cases, we can use the First-Derivative Test.

**EXAMPLE 3**

**Using the Second-Derivative Test**

Find the relative extrema for the function

$$f(x) = -3x^5 + 5x^3$$

**SOLUTION**

We begin by finding the critical numbers of $f$.

$$f'(x) = -15x^4 + 15x^2 = 15(-x^4 + x^2) = 15x^2(1 - x^2)$$

Since $f'$ is defined for all $x$ we conclude that the only critical numbers are $x = 0$, $\pm 1$. Now, to apply the Second-Derivative Test, we must find $f''(x)$:

$$f''(x) = 15(-4x^3 + 2x)$$

Then we evaluate $f''$ at each of the three critical numbers as follows.

| Point | Sign of $f''$ | Conclusion |
|---|---|---|
| $(-1, -2)$ | $f''(-1) = 30 > 0$ | Relative minimum |
| $(1, 2)$ | $f''(1) = -30 < 0$ | Relative maximum |
| $(0, 0)$ | $f''(0) = 0$ | Test fails |

Since the test fails at $(0, 0)$, we use the First-Derivative Test and observe that $f$ increases to the right and left of $x = 0$. Thus, $(0, 0)$ is neither a relative minimum nor a relative maximum. [A test of concavity would show that $(0, 0)$ is a point of inflection.] The graph of $f$ is shown in Figure 3.26.

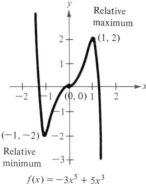

FIGURE 3.26  $f(x) = -3x^5 + 5x^3$

## Applications

In economics, the notion of concavity is related to the **Law of Diminishing Returns** (also called the Law of Diminishing Marginal Returns). Suppose the function

$$y = f(x)$$

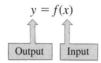

measures the output $y$ obtained from a given input $x$. (For instance, the input could represent advertising costs for a product and the output could represent revenue from sales.) In Figure 3.27, the input-output function is concave upward on the interval $[a, c]$ and concave downward on the interval $[c, b]$. In the interval $[a, c]$ each additional dollar of input returns more than the previous input dollar. This is true because the slope of the curve is increasing. By contrast, in the interval $[c, b]$ the slope of the curve is decreasing and an additional dollar of input returns less than the previous input dollar. The point of inflection separating the two parts of the graph is called the **point of diminishing returns.** For a curve like that pictured in Figure 3.27, an increased investment beyond the point of diminishing returns would usually not be considered a good use of capital.

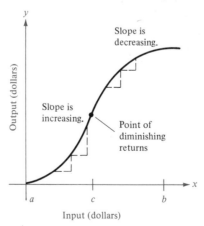

FIGURE 3.27

EXAMPLE 4

**Finding the Point of Diminishing Returns**

By increasing its advertising cost $x$ for a certain product, a company discovers that it can increase the sales $y$ of the product according to the model

$$y = \frac{1}{10,000}(300x^2 - x^3), \quad 0 \le x \le 200$$

where $x$ and $y$ are measured in 1000s of dollars. Find the intervals on which the graph of this function are concave upward and concave downward. What is the point of diminishing returns for this function?

**SOLUTION**

We begin by finding the first and second derivatives.

$$y' = \frac{1}{10,000}(600x - 3x^2)$$

$$y'' = \frac{1}{10,000}(600 - 6x)$$

Since the second derivative is zero only when $x = 100$, we test the concavity in the intervals $(0, 100)$ and $(100, 200)$. Since the second derivative is positive in the interval $(0, 100)$, we conclude that the graph is concave upward there. Similarly, since the second derivative is negative in the interval $(100, 200)$, the graph is concave downward there. Finally, the point of diminishing returns occurs at the point of inflection

$$(100, 200) \qquad \text{Point of diminishing returns}$$

The graph of this function is shown in Figure 3.28.

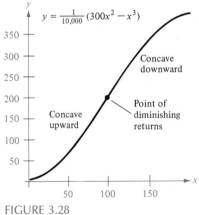

FIGURE 3.28

# SECTION EXERCISES 3.3

In Exercises 1–6, find the intervals on which the given function is concave upward and those on which it is concave downward.

**1.** $y = x^2 - x - 2$

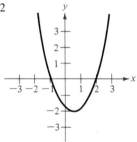

**2.** $y = -x^3 + 3x^2 - 2$

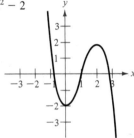

**3.** $f(x) = \dfrac{24}{x^2 + 12}$

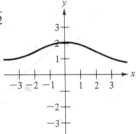

**4.** $f(x) = \dfrac{x^2 - 1}{2x + 1}$

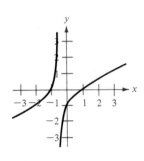

**5.** $f(x) = \dfrac{x^2 + 1}{x^2 - 1}$

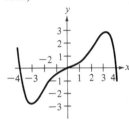

**6.** $y = \dfrac{1}{270}(-3x^5 + 40x^3 + 135x)$

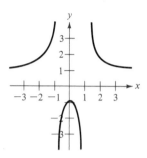

In Exercises 7–18, identify all relative extrema. Use the Second-Derivative Test when applicable.

**7.** $f(x) = 6x - x^2$ 　　　　　**8.** $f(x) = x^2 + 3x - 8$
**9.** $f(x) = (x - 5)^2$ 　　　　**10.** $f(x) = -(x - 5)^2$
**11.** $f(x) = x^3 - 3x^2 + 3$ 　**12.** $f(x) = 5 + 3x^2 - x^3$
**13.** $f(x) = x^4 - 4x^3 + 2$
**14.** $f(x) = x^3 - 9x^2 + 27x - 26$
**15.** $f(x) = x^{2/3} - 3$ 　　　　**16.** $f(x) = \sqrt{x^2 + 1}$

**17.** $f(x) = x + \dfrac{4}{x}$ 　　　**18.** $f(x) = \dfrac{x}{x - 1}$

In Exercises 19–30, sketch the graph of the given function and identify all relative extrema and points of inflection.

**19.** $f(x) = x^3 - 12x$ 　　　**20.** $f(x) = x^3 + 1$
**21.** $f(x) = x^3 - 6x^2 + 12x - 8$
**22.** $f(x) = 2x^3 - 3x^2 - 12x + 8$

**23.** $f(x) = \dfrac{1}{4}x^4 - 2x^2$ 　　　**24.** $f(x) = 2x^4 - 8x + 3$

**25.** $g(x) = (x - 1)(x + 2)^2$
**26.** $g(x) = (x - 6)(x + 2)^3$

**27.** $g(x) = x\sqrt{x + 3}$ 　　　**28.** $g(x) = \dfrac{x^3}{16}(x + 4)$

**29.** $f(x) = \dfrac{4}{1 + x^2}$ 　　　**30.** $f(x) = x - 4\sqrt{x + 1}$

In Exercises 31 and 32, identify the point of diminishing returns for the given function, where $R$ is revenue and $x$ is the amount spent on advertising. Assume that $R$ and $x$ are measured in 1000s of dollars.

**31.** $R = -\dfrac{4}{9}(x^3 - 9x^2 - 27), \quad 0 \le x < 5$

**32.** $R = -\dfrac{4}{27}(x^3 - 33x^2 + 120x - 845), \quad 0 \le x < 20$

In Exercises 33 and 34, given the total cost function for producing $x$ units, determine the production level that minimizes the average cost per unit. (The average cost per unit is given by $\overline{C} = C/x$.)

**33.** $C = 0.5x^2 + 15x + 5000$

**34.** $C = 0.001x^3 + 0.10x + 750$

In Exercises 35 and 36, sketch a graph of a function $f$ having the given characteristics.

| Function | First Derivative | Second Derivative |
|---|---|---|
| **35.** $f(2) = 0$ | $f'(x) < 0, \ x < 3$ | $f''(x) > 0$ |
| $f(4) = 0$ | $f'(3) = 0$ | |
| | $f'(x) > 0, \ x > 3$ | |
| **36.** $f(2) = 0$ | $f'(x) > 0, \ x < 3$ | $f''(x) < 0, \ x \ne 3$ |
| $f(4) = 0$ | $f'(3)$ is undefined | |
| | $f'(x) < 0, \ x > 3$ | |

In Exercises 37 and 38, use the given graph to sketch the graph of $f'$. Find the intervals (if any) on which (a) $f'(x)$ is positive, (b) $f'(x)$ is negative, (c) $f'$ is increasing, and (d) $f'$ is decreasing. For each of these intervals describe the corresponding behavior of $f$.

**37.** $f(x) = 4 - x^2$

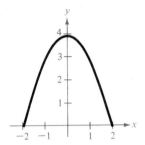

**38.** $f(x) = -\dfrac{1}{3}x^3 + x^2$

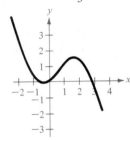

**39.** Show that the point of inflection of the graph of $f(x) = x(x - 6)^2$ lies midway between the relative extrema of $f$.

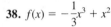

**40.** The deflection $D$ of a particular beam of length $L$ is given by

$$D = 2x^4 - 5Lx^3 + 3L^2x^2$$

where $x$ is the distance from one end of the beam. Find the value of $x$ that yields the maximum deflection.

# Optimization Problems

The federal government obtains its income primarily by taxing individuals and businesses within the country. For example, in 1980 these two sources of income totaled 527 billion dollars (421 billion from individuals and 106 billion from businesses)—roughly 20% of the gross national product for 1980. It seems clear that in order to increase its revenue the government could simply increase its percentage of the gross national product. The fallacy in this reasoning is that increases in tax rates tend to discourage productivity. In other words, if it greatly increased the tax rate, the government might actually end up getting less in tax dollars.

What tax rate will produce the maximum income for the government? We call this problem one of **optimization.** Optimization problems are typically characterized by two conflicting phenomena. For instance, the two conflicting phenomena in the government's tax problem are (1) increased rates produce more revenue *but* (2) increased rates discourage production, which in turn decreases revenue. In an effort to encourage production and at the same time increase revenues, the federal government has been gradually shifting the tax burden from business and corporate taxes to personal income and social insurance taxes, as shown in Table 3.11 and Figure 3.29.

FIGURE 3.29

TABLE 3.11

| Year | Total receipts (billions of dollars) | Individual | | Business | |
| | | Amount (billions of dollars) | Percentage of total | Amount (billions of dollars) | Percentage of total |
|---|---|---|---|---|---|
| 1965 | 120.0 | 76.9 | 64.1 | 43.1 | 35.9 |
| 1970 | 194.8 | 142.8 | 73.3 | 52.0 | 26.7 |
| 1975 | 283.4 | 219.4 | 77.4 | 64.0 | 22.6 |
| 1980 | 527.3 | 421.0 | 79.8 | 106.3 | 20.2 |
| 1984 | 691.3 | 559.9 | 81.0 | 131.4 | 19.0 |

## SECTION TOPIC

### ■ Optimization Problems

Up to this point in our study of calculus, most of the problems have been given in the form of mathematical equations. Our primary task has been to choose and then apply an appropriate calculus procedure to arrive at a solution. In this section, we focus on application problems that are not originally stated in terms of mathematical equations; hence an additional *problem formulation stage* is required in the solution process. In this and the next section, we provide some guidelines to assist you in constructing mathematical formulations for several common applications of calculus.

One of the most common applications of calculus is the determination of minimum or maximum values. Consider how frequently you hear or read terms like greatest profit, least cost, cheapest product, least time, greatest voltage, optimum size, least area, greatest strength, or greatest distance. Before outlining a general method of solution for such problems, we present an example.

**EXAMPLE 1**

**Finding the Maximum Volume**

An open box having a square base is to be constructed from 108 square inches of material. What dimensions should be used for the box to obtain a maximum volume?

**SOLUTION**

Since the box has a square base (as shown in Figure 3.30), its volume is given by

$$V = x^2h \qquad \text{Primary equation}$$

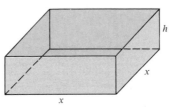

FIGURE 3.30

Open Box with Square Base:
$S = x^2 + 4xh = 108$

Furthermore, since the box is open at the top, its surface area is given by

$$S = \text{(area of base)} + \text{(area of four sides)}$$

$$S = x^2 + 4xh = 108 \qquad \text{Secondary equation}$$

Now since $V$ is to be maximized, it is helpful to express $V$ as a function of just one variable. To do this, we solve the equation $x^2 + 4xh = 108$ for $h$ in terms of $x$ to obtain

$$4xh = 108 - x^2$$

$$h = \frac{108 - x^2}{4x}$$

Substituting for $h$ in the equation for volume, we get

$$V = x^2 h$$

$$= x^2 \left( \frac{108 - x^2}{4x} \right)$$

$$= 27x - \frac{x^3}{4} \qquad \text{Function of one variable}$$

Before we try to find which $x$-value will yield a maximum value of $V$, we should determine the *feasible domain*. That is, what values of $x$ make sense in this problem? We know that $x$ should be nonnegative and that the area of the base ($A = x^2$) is at most 108. Thus, we have

$$0 \le x \le \sqrt{108} \qquad \text{Feasible domain}$$

On the interval $0 < x < \sqrt{108}$, the critical numbers for $V$ are the solutions to

$$\frac{dV}{dx} = 27 - \frac{3x^2}{4} = 0$$

$$3x^2 = 108$$

$$x = 6 \qquad \text{Critical number}$$

Evaluating $V$ at this critical number and at the endpoints of the domain, we have

$$V(0) = 0, \qquad V(6) = 108, \qquad V(\sqrt{108}) = 0$$

Thus, we conclude that $V$ is maximum when $x = 6$ and the height is

$$h = \frac{108 - (6)^2}{4(6)} = \frac{72}{24} = 3$$

Therefore, the maximum volume of the box occurs when

$$x = 6 \text{ inches} \qquad \text{and} \qquad h = 3 \text{ inches}$$

Before taking a closer look at the actual steps involved in Example 1, be sure that you understand the basic question it asks. Many students have trouble with word problems because they are too eager to start solving the problem by using a pat formula. For instance, in Example 1 you should realize that there are infinitely many open boxes having 108 square inches of surface area. You might begin this problem by asking yourself which basic shape would seem to yield a maximum volume. Should the box be tall, squat, or cubical? You might even try calculating a few volumes as shown in Figure 3.31, to see if you can get a better feeling for what the optimum dimensions should be.

Remember that you are not ready to begin solving a problem until you have clearly identified what the problem is. Once you are sure you understand what is being asked in an optimization problem, you are ready to begin considering a method for solving the problem.

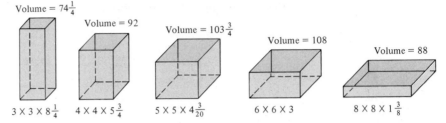

Volume = $74\frac{1}{4}$

Volume = 92

Volume = $103\frac{3}{4}$

Volume = 108

Volume = 88

$3 \times 3 \times 8\frac{1}{4}$    $4 \times 4 \times 5\frac{3}{4}$    $5 \times 5 \times 4\frac{3}{20}$    $6 \times 6 \times 3$    $8 \times 8 \times 1\frac{3}{8}$

Which size box has the maximum volume?

FIGURE 3.31

There are several stages in the solution of Example 1. First, we made a sketch and assigned symbols to all *known* quantities and quantities *to be determined*. Second, we identified an equation for the quantity to be maximized. Then we reduced this equation to obtain a function of one independent variable, and found the feasible domain for the function. Finally, we applied the techniques of calculus to find the value of $x$ that yielded the desired maximum.

**Guidelines for solving optimization problems**

1. Assign symbols to all given quantities and quantities to be determined. When feasible, make a sketch.
2. Write a **primary equation** for the quantity to be maximized or minimized.
3. Reduce the primary equation to one having a single independent variable—this may involve the use of **secondary equations** relating the independent variables of the primary equation.
4. Determine the domain of the primary equation. That is, determine the values for which the stated problem makes sense.
5. Determine the desired maximum or minimum value by the techniques of calculus.

▇▇▇ **Remark:** When performing step 5, recall that to determine the maximum or minimum value of a continuous function $f$ on a closed interval, we compare the values of $f$ at its critical numbers to the values of $f$ at the endpoints of the interval. The largest (smallest) of these values is the desired maximum (minimum).

**EXAMPLE 2**

**Finding the Minimum Sum**

The product of two positive numbers is 288. Minimize the sum of twice the first number plus the second.

**SOLUTION**

1. Let $x$ be the first number, $y$ the second, and $S$ the sum to be minimized.
2. Since we wish to minimize $S$, the *primary* equation is

$$S = 2x + y \qquad \text{Primary equation}$$

3. Since the product of the two numbers is 288, we have the *secondary* equation

$$xy = 288 \qquad \text{Secondary equation}$$

$$y = \frac{288}{x}$$

Thus, we can rewrite the primary equation in terms of $x$ alone as

$$S = 2x + \frac{288}{x} \qquad \text{Function of one variable}$$

4. Since $x$ must be positive, the feasible domain is

$$0 < x \qquad \text{Feasible domain}$$

5. To find the critical numbers, we set the derivative equal to 0.

$$\frac{dS}{dx} = 2 - \frac{288}{x^2} = 0$$

$$x^2 = 144$$

$$x = \pm 12$$

Choosing the positive $x$-value, we use the First-Derivative Test to conclude that $S$ is decreasing in the interval $(0, 12)$ and increasing in the interval $(12, \infty)$, as shown in Table 3.12. Therefore, $x = 12$ yields a minimum, and the two numbers are

$$x = 12$$

and

$$y = \frac{288}{12} = 24$$

TABLE 3.12

| Interval | $(0, 12)$ | $(12, \infty)$ |
|---|---|---|
| Test value | $x = 1$ | $x = 13$ |
| Sign of $\dfrac{dS}{dx}$ | $\dfrac{dS}{dx} < 0$ (S is decreasing) | $\dfrac{dS}{dx} > 0$ (S is increasing) |

**EXAMPLE 3**

**Finding the Minimum Distance**

Find the points on the graph of $y = 4 - x^2$ that are closest to the point $(0, 2)$.

**SOLUTION**

1. Figure 3.32 indicates that there are two points at a minimum distance from $(0, 2)$.

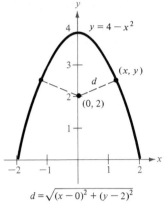

$$d = \sqrt{(x-0)^2 + (y-2)^2}$$

FIGURE 3.32

2. We are asked to minimize the distance $d$; hence we use the Distance Formula to obtain the primary equation

$$d = \sqrt{(x-0)^2 + (y-2)^2} \qquad \text{Primary equation}$$

3. Using the secondary equation $y = 4 - x^2$, we can rewrite the primary equation as

$$d = \sqrt{(x)^2 + (4 - x^2 - 2)^2}$$
$$= \sqrt{x^2 + (2 - x^2)^2}$$
$$= \sqrt{x^4 - 3x^2 + 4}$$

Since $d$ is smallest when the expression under the radical is smallest, we need only find the critical numbers of

$$f(x) = x^4 - 3x^2 + 4$$

4. The domain of $f$ is the entire real line.
5. Differentiation yields the following critical numbers:

$$f'(x) = 4x^3 - 6x = 0$$
$$2x(2x^2 - 3) = 0$$
$$x = 0, \ \sqrt{\frac{3}{2}}, \ -\sqrt{\frac{3}{2}}$$

The First-Derivative Test verifies that $x = 0$ yields a relative maximum, and both $x = \sqrt{\frac{3}{2}}$ and $x = -\sqrt{\frac{3}{2}}$ yield a minimum distance. Hence, on the graph of $y = 4 - x^2$, the points closest to $(0, 2)$ are

$$\left(\sqrt{\frac{3}{2}}, \frac{5}{2}\right) \qquad \text{and} \qquad \left(-\sqrt{\frac{3}{2}}, \frac{5}{2}\right)$$

**EXAMPLE 4**

**Finding the Minimum Area**

A rectangular page is to contain 24 square inches of print. The margins at the top and bottom of the page are each $1\frac{1}{2}$ inches wide. The margins on each side are 1 inch. What should the dimensions of the page be so that the least amount of paper is used?

**SOLUTION**

1. See Figure 3.33.

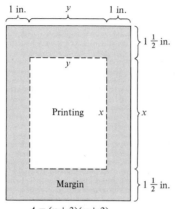

$$A = (x + 3)(y + 2)$$

**FIGURE 3.33**

2. Letting $A$ be the area to be minimized, we have for our primary equation

$$A = (x + 3)(y + 2) \qquad \text{Primary equation}$$

3. The printed area inside the margins is given by

$$24 = xy \qquad \text{Secondary equation}$$

Solving this equation for $y$, we have

$$y = \frac{24}{x}$$

Thus, the primary equation becomes

$$A = (x + 3)\left(\frac{24}{x} + 2\right) \qquad \text{Function of one variable}$$

$$= 30 + 2x + \frac{72}{x}$$

4. Since $x$ must be positive, the feasible domain is $x > 0$.

5. Now, to find the minimum area, we differentiate with respect to $x$ and obtain

$$\frac{dA}{dx} = 2 - \frac{72}{x^2} = 0$$

$$x^2 = 36$$

Thus, the critical numbers are $x = \pm 6$, and since $-6$ is not in the domain, we choose $x = 6$. The First-Derivative Test will confirm that $A$ is a minimum at this point. Therefore, the dimensions of the page should be

$$x + 3 = 9 \text{ inches}$$

by

$$y + 2 = 6 \text{ inches}$$

**EXAMPLE 5**

**Finding the Minimum Length**

Two posts, one 12 feet high and the other 28 feet high, stand 30 feet apart. They are to be stayed by wires attached to a single stake, running from ground level to the tops of the posts. Where should the stake be placed to use the least wire?

**SOLUTION**

1. See Figure 3.34.

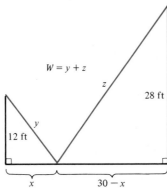

FIGURE 3.34

2. Let $W$ be the length of the wire to be minimized. Then we have

$$W = y + z \qquad \text{Primary equation}$$

3. In this problem, rather than solving for $y$ in terms of $z$ (or vice versa), it is more convenient to solve for both $y$ and $z$ in terms of a third variable $x$, as shown in Figure 3.34.

$$x^2 + 12^2 = y^2 \quad \Longrightarrow \quad y = \sqrt{x^2 + 144}$$

$$(30 - x)^2 + 28^2 = z^2 \quad \Longrightarrow \quad z = \sqrt{x^2 - 60x + 1684}$$

Thus, we have

$$W = y + z$$

$$= \sqrt{x^2 + 144} + \sqrt{x^2 - 60x + 1684}$$

4. The feasible domain for $W$ is $0 \le x \le 30$.

5. Differentiating $W$ with respect to $x$ yields

$$\frac{dW}{dx} = \frac{x}{\sqrt{x^2 + 144}} + \frac{x - 30}{\sqrt{x^2 - 60x + 1684}}.$$

Setting $dW/dx = 0$, we obtain

$$x\sqrt{x^2 - 60x + 1684} = (30 - x)\sqrt{x^2 + 144}$$

Squaring both sides, we have

$$x^4 - 60x^3 + 1684x^2 = x^4 - 60x^3 + 1044x^2 - 8640x + 129{,}600$$

Finally, collecting terms, we have

$$640x^2 + 8640x - 129{,}600 = 0$$

$$320(2x^2 + 27x - 405) = 0$$

$$320(x - 9)(2x + 45) = 0$$

$$x = -22.5, \ 9$$

Since $x = -22.5$ is not in the interval $[0, 30]$ and since

$$W(0) \approx 53.04, \qquad W(9) = 50, \qquad W(30) \approx 60.31$$

we conclude that the stake should be placed at $x = 9$ feet from the 12-foot pole.

Let's review the primary equations developed in the five examples in this section.

| Example | Primary equation | Domain |
|---------|------------------|--------|
| 1 | $V = 27x - \dfrac{x^3}{4}$ | $0 \le x \le \sqrt{108}$ |
| 2 | $S = 2x + \dfrac{288}{x}$ | $0 < x$ |
| 3 | $d = \sqrt{x^4 - 3x^2 + 4}$ | all real numbers |
| 4 | $A = 30 + 2x + \dfrac{72}{x}$ | $0 < x$ |
| 5 | $W = \sqrt{x^2 + 144} + \sqrt{x^2 - 60x + 1684}$ | $0 \le x \le 30$ |

As applications go, these five examples are fairly simple, and yet the resulting primary equations are quite complicated. The point is that real applications often involve equations that are at least as messy as these five, and you should expect that. Remember that one of the main goals of this course is to enable you to use the power of calculus to analyze equations that at first glance seem formidable. (See Figure 3.35 for the graphs of these five equations.)

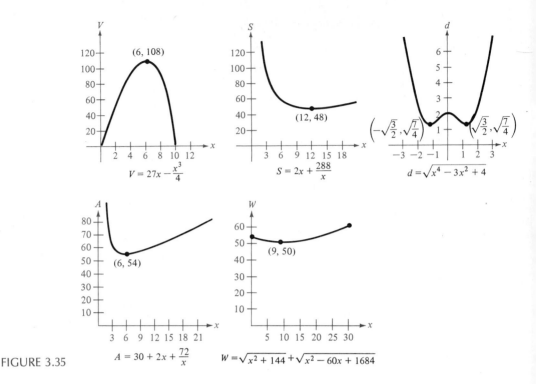

$$V = 27x - \frac{x^3}{4}$$

$$S = 2x + \frac{288}{x}$$

$$d = \sqrt{x^4 - 3x^2 + 4}$$

$$A = 30 + 2x + \frac{72}{x}$$

$$W = \sqrt{x^2 + 144} + \sqrt{x^2 - 60x + 1684}$$

FIGURE 3.35

## SECTION EXERCISES 3.4

In Exercises 1–6, find two positive numbers satisfying the given requirements.

**1.** The sum is 110 and the product is maximum.

**2.** The sum is $S$ and the product is maximum.

**3.** The sum of the first and twice the second is 24 and the product is maximum.

**4.** The sum of the first and twice the second is 100 and the product is maximum.

**5.** The product is 192 and the sum is minimum.

**6.** The product is 192 and the sum of the first plus three times the second is minimum.

**7.** What positive number $x$ minimizes the sum of $x$ and its reciprocal?

**8.** The difference of two numbers is 50. Find the two numbers so that their product is as small as possible.

In Exercises 9 and 10, find the length and width of a rectangle of maximum area for the given perimeter.

**9.** Perimeter: 100 feet      **10.** Perimeter: $P$ units

In Exercises 11 and 12, find the length and width of a rectangle of minimum perimeter for the given area.

**11.** Area: 64 square feet      **12.** Area: $A$ square units

**13.** A rancher has 200 feet of fencing to enclose two adjacent rectangular corrals, as shown in Figure 3.36. What dimensions should be used so that the enclosed area will be a maximum?

**14.** A dairy farmer plans to fence in a rectangular pasture adjacent to a river. The pasture must contain 180,000 square meters in order to provide enough grass for the herd. What dimensions require the least amount of fencing if no fencing is needed along the river?

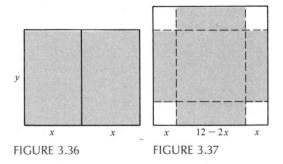

FIGURE 3.36        FIGURE 3.37

**15.** An open box is to be made from a square piece of material, 12 inches on a side, by cutting equal squares from each corner and turning up the sides, as shown in Figure 3.37. Find the volume of the largest box that can be made in this manner.

**16.** (a) Solve Exercise 15 if the square of material is $s$ inches on a side.

(b) If the dimensions of the square piece of material are doubled, how does the volume change?

**17.** An open box is to be made from a rectangular piece of material by cutting equal squares from each corner and turning up the sides. Find the dimensions of the box of maximum volume if the material has dimensions 2 feet by 3 feet.

**18.** A net enclosure for golf practice is open at one end, as shown in Figure 3.38. Find the dimensions that require the least amount of netting if the volume of the enclosure is to be $83\frac{1}{3}$ cubic meters.

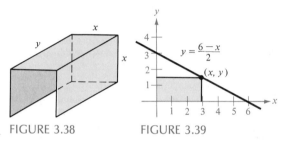

FIGURE 3.38          FIGURE 3.39

**19.** An indoor physical fitness room consists of a rectangular region with a semicircle on each end. If the perimeter of the room is to be a 200-meter running track, find the dimensions that will make the area of the rectangular region as large as possible.

**20.** A page is to contain 30 square inches of print. The margins at the top and the bottom of the page are each 2 inches wide. The margins on each side are only 1 inch wide. Find the dimensions of the page so that the least paper is used.

**21.** A rectangle is bounded by the $x$- and $y$-axes and the graph of $y = (6 - x)/2$, as shown in Figure 3.39. What length and width should the rectangle have so that its area is a maximum?

**22.** A right triangle is formed in the first quadrant by the $x$- and $y$-axes and a line through the point $(2, 3)$. Find the vertices of the triangle so that its area is minimum.

**23.** A right triangle is formed in the first quadrant by the $x$- and $y$-axes and a line through the point $(1, 2)$, as shown in Figure 3.40. Find the vertices of the triangle so that the length of the hypotenuse is minimum.

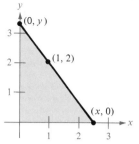

FIGURE 3.40

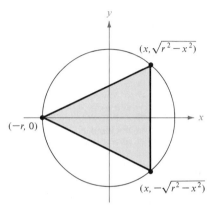

FIGURE 3.41

**24.** Find the dimensions of the largest isosceles triangle that can be inscribed in a circle of radius $r$. (See Figure 3.41.)

**25.** A rectangle is bounded by the $x$-axis and the semicircle $y = \sqrt{25 - x^2}$, as shown in Figure 3.42. What length and width should the rectangle have so that its area is a maximum?

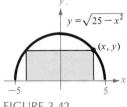

FIGURE 3.42

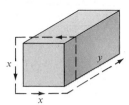

FIGURE 3.43

**26.** Find the dimensions of the largest rectangle that can be inscribed in a semicircle of radius $r$. (See Exercise 25.)

**27.** Find the point on the graph of $y = \sqrt{x}$ that is closest to the point $(4, 0)$.

**28.** Find the point on the graph of $y = x^2$ that is closest to the point $(2, \frac{1}{2})$.

**29.** A right circular cylinder is to be designed to hold 12 fluid ounces of a soft drink and to use a minimal amount of material in its construction. Find the required dimensions for the container (1 fl oz $\approx$ 1.80469 in$^3$).

**30.** A figure is formed by adjoining two hemispheres to each end of a right circular cylinder. The total volume of the figure is 12 cubic inches. Find the radius of the cylinder that produces the minimum surface area.

**31.** A rectangular package to be sent by a postal service can have a maximum combined length and girth (perimeter of a cross section) of 108 inches, as shown in Figure 3.43. Find the dimensions of the package of maximum volume that can be sent. (Assume the cross section is square.)

**32.** Find the volume of the largest right circular cylinder that can be inscribed in a sphere of radius $r$. (See Figure 3.44.)

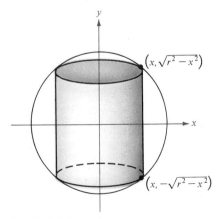

FIGURE 3.44

**33.** The combined perimeter of a circle and a square is 16. Find the dimensions of the circle and square that produce a minimum total area.

**34.** The combined perimeter of an equilateral triangle and a square is 10. Find the dimensions of the triangle and square that produce a minimum total area.

**35.** A man is in a boat 2 miles from the nearest point on the coast. He is to go to a point $Q$, 3 miles down the coast and 1 mile inland, as shown in Figure 3.45. If he can row at 2 mi/hr and walk at 4 mi/hr, toward what point on the coast should he row in order to reach point $Q$ in the least time?

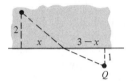

FIGURE 3.45

**36.** The conditions are the same as in Exercise 35 except that the man can row at 4 mi/hr. How does this change the solution?

**37.** A wooden beam has a rectangular cross section of height $h$ and width $w$, as shown in Figure 3.46. The strength $S$ of the beam is directly proportional to the width and the square of the height. What are the dimensions of the strongest beam that can be cut from a round log of diameter 24 inches? (Hint: $S = kh^2w$, where $k$ is the proportionality constant.)

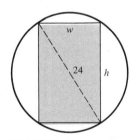

FIGURE 3.46

**38.** Show that among all positive numbers $x$ and $y$ with $x^2 + y^2 = r^2$, the sum $x + y$ is largest when $x = y$.

# Business and Economics Applications

An airline company offers flights daily between Chicago and Denver. The total monthly revenue and cost for these flights are given by

$$R = \frac{240x}{400 + x^2} \quad \text{and} \quad C = \sqrt{0.2x + 1}$$

where $R$ and $C$ are measured in millions of dollars and $x$ is measured in thousands of passengers. The price per ticket is given by

$$p = \frac{1,000,000R}{1000x} = \frac{1000R}{x}$$

What price will yield a maximum profit for the airline?

One way to find the maximum profit is to examine the marginal revenue and marginal cost functions:

$$\frac{dR}{dx} = \frac{240(400 - x^2)}{(400 + x^2)^2}$$

and

$$\frac{dC}{dx} = \frac{0.1}{\sqrt{0.2x + 1}}$$

Recall that these marginals measure the *rate of change* of $R$ and $C$ with respect to $x$. As long as the rate of change of $R$ exceeds that of $C$, greater sales will yield greater profits. However, when the rate of change of $R$ falls below that of $C$, greater sales will lessen profits. Thus, the maximum profit will occur at the point where these two marginals are equal, as shown in Figure 3.47. By solving the equation $dR/dx = dC/dx$, we can determine that the maximum profit occurs at $x \approx 17.4$, or roughly 17,400 passengers. This corresponds to a ticket price of about $341.50. Table 3.13 lists the monthly profit for ticket prices ranging from $100.00 to $450.00.

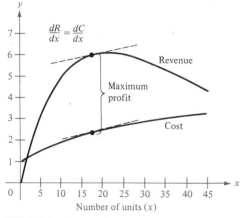

FIGURE 3.47

TABLE 3.13

| Price (dollars) | Number of passengers | Monthly profit (dollars) |
|---|---|---|
| 100.00 | 44,721 | 1,318,682 |
| 150.00 | 34,641 | 2,380,446 |
| 200.00 | 28,284 | 3,076,766 |
| 250.00 | 23,664 | 3,521,740 |
| 300.00 | 20,000 | 3,763,932 |
| 341.50 | 17,401 | 3,825,684 |
| 350.00 | 16,903 | 3,823,087 |
| 400.00 | 14,142 | 3,700,218 |
| 450.00 | 11,547 | 3,376,976 |

■ **Optimization Problems in Business and Economics**
■ **Price Elasticity of Demand**

Recall from Section 2.3 that many business applications involve the equation

$$P = R - C$$

where $P$ is total profit, $R$ is total revenue, and $C$ is total cost. The problems in this section are primarily optimization problems. Hence, the five-step procedure used in Section 3.4 is an appropriate model to follow.

**EXAMPLE 1**

**Finding the Maximum Revenue**

Suppose a company has determined that its total revenue $R$ for a particular product is given by

$$R = -x^3 + 450x^2 + 52{,}500x$$

where $R$ is measured in dollars and $x$ is the number of units produced. What production level will yield a maximum revenue?

**SOLUTION**

1. A sketch is given in Figure 3.48.

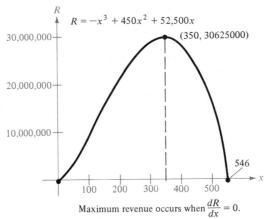

FIGURE 3.48

2. The primary equation is the given revenue equation:

$$R = -x^3 + 450x^2 + 52{,}500x$$

3. Since $R$ is already given as a function of one variable, we do not need a secondary equation.
4. The feasible domain of this function is $0 \le x \le 546$. (This is determined by finding the $x$-intercepts of the revenue function, as shown in Figure 3.48.)
5. Differentiating and setting the derivative equal to zero yield

$$\frac{dR}{dx} = -3x^2 + 900x + 52{,}500 = 0$$

$$-x^2 + 300x + 17{,}500 = 0$$

$$(-x + 350)(x + 50) = 0$$

$$x = 350, \ -50$$

The critical numbers are $x = 350$ and $x = -50$. Choosing the positive value of $x$, we conclude that the maximum revenue is obtained when 350 units are produced.

To study the effect of production levels on cost, economists use the **average cost function** $\overline{C}$, defined as

$$\overline{C} = \frac{C}{x}$$

where $C = f(x)$ is the total cost function.

**EXAMPLE 2**

**Finding the Minimum Average Cost**

A company estimates that the cost (in dollars) of producing $x$ units of a certain product is given by

$$C = 800 + 0.04x + 0.0002x^2$$

Find the production level that minimizes the average cost per unit. Compare this minimal average cost to the average cost when 400 units are produced.

**SOLUTION**

1. We let $\overline{C}$ be the average cost and $x$ be the number of units produced, as shown in Figure 3.49.

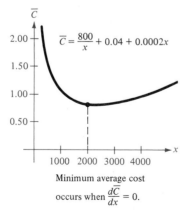

FIGURE 3.49

2. The primary equation representing the quantity to be minimized is

$$\overline{C} = \frac{C}{x}$$

3. Substituting from the given equation for $C$, we have

$$\overline{C} = \frac{800 + 0.04x + 0.0002x^2}{x}$$

$$= \frac{800}{x} + 0.04 + 0.0002x$$

4. The feasible domain for this function is $0 < x$.
5. Setting the derivative equal to zero yields

$$\frac{d\overline{C}}{dx} = -\frac{800}{x^2} + 0.0002 = 0$$

which implies that

$$x^2 = \frac{800}{0.0002} = 4{,}000{,}000$$

$$x = \pm 2000$$

Choosing the positive $x$-value, we determine that the minimum average cost occurs when 2000 units are produced.

For 2000 units the average cost (per unit) is

$$\overline{C} = \frac{800}{2000} + 0.04 + (0.0002)(2000) = \$0.84$$

For 400 units the average cost (per unit) is

$$\overline{C} = \frac{800}{400} + 0.04 + (0.0002)(400) = \$2.12$$

**EXAMPLE 3**

**Finding the Maximum Revenue**

A certain business sells 2000 items per month at a price of $10 each. It can sell 250 more items per month for each $0.25 reduction in price. What price per item will maximize its monthly revenue?

**SOLUTION**

1. Let

$$x = \text{number of items sold per month}$$

$$p = \text{price of each item (in dollars)}$$

$$R = \text{monthly revenue (in dollars)}$$

2. Since the monthly revenue is to be maximized, the primary equation is

$$\text{Revenue} = (\text{number of units sold})(\text{price per unit})$$

$$R = xp$$

3. The number of items $x$ is related to $p$ in that $x$ increases 250 units every time $p$ drops \$0.25 from the original price of \$10. This is described by the equation

$$x = 2000 + 250\left(\frac{10 - p}{0.25}\right)$$

$$= 12{,}000 - 1000p$$

or

$$p = 12 - \frac{x}{1000}$$

Substituting this result into the revenue equation, we have

$$R = x\left(12 - \frac{x}{1000}\right)$$

$$= 12x - \frac{x^2}{1000}$$

(See Figure 3.50.)

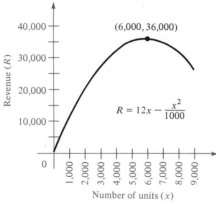

FIGURE 3.50

4. The feasible domain for this function is $0 \le x \le 12{,}000$.
5. Setting the derivative equal to zero, we obtain

$$\frac{dR}{dx} = 12 - \frac{x}{500} = 0$$

Therefore, $dR/dx = 0$ if $x = 6000$, and we conclude that the price that maximizes the revenue is

$$p = 12 - \frac{6000}{1000} = \$6$$

From the equation

$$p = 12 - \frac{x}{1000}$$

in Example 3 we can determine the number of units that can be sold—that is, the *quantity demanded*—at a given price. We call the function given by $p = f(x)$ the **demand function.** Note that as the price increases, the quantity demanded decreases, and vice versa.

**EXAMPLE 4**

**Finding the Maximum Profit**

The marketing department for a business has determined that the demand for a certain product is given by the model

$$p = \frac{50}{\sqrt{x}}$$

The cost of producing $x$ items is given by $C = 0.5x + 500$. What price $p$ will yield a maximum profit?

**SOLUTION**

1. Let $P$ represent the profit.
2. Since we are seeking a maximum profit $P$, the primary equation is

$$P = R - C$$

3. Since the total revenue for this product is given by $R = xp$, we can write the profit function as

$$P = R - C = xp - (0.5x + 500)$$

$$= x\left(\frac{50}{\sqrt{x}}\right) - (0.5x + 500)$$

$$= 50\sqrt{x} - 0.5x - 500$$

4. The feasible domain is $0 \leq x \leq 7{,}872$. (When $x$ is greater than 7,872, the profit is negative.)
5. Setting the derivative equal to zero, we obtain

$$\frac{dP}{dx} = \frac{25}{\sqrt{x}} - 0.5 = 0$$

which implies that

$$\sqrt{x} = \frac{25}{0.5} = 50$$

$$x = 2500$$

Finally, we conclude that the maximum profit occurs when the price is

$$p = \frac{50}{\sqrt{2500}} = \frac{50}{50} = \$1.00$$

**Remark:** To find the maximum profit in Example 4, we differentiated the equation $P = R - C$ and set $dP/dx$ equal to zero. From the equation

$$\frac{dP}{dx} = \frac{dR}{dx} - \frac{dC}{dx} = 0$$

it follows that the maximum profit occurs when the marginal revenue is equal to the marginal cost, as shown in Figure 3.51.

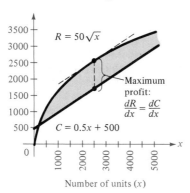

FIGURE 3.51

Number of units $(x)$

## Price Elasticity of Demand

One way economists describe the behavior of a demand function is by a term called the **price elasticity of demand.** It describes the relative responsiveness of consumers to a change in the price of an item. If $p = f(x)$ is a differentiable demand function, then the price elasticity of demand is given by

$$\eta = \frac{p/x}{dp/dx}$$

(where $\eta$ is the lowercase Greek letter eta). For a given price, if $|\eta| < 1$ the demand is said to be **inelastic;** if $|\eta| > 1$ the demand is said to be **elastic.**

**EXAMPLE 5**

**Test for Elasticity**

Show that the demand function in Example 4, $p = 50x^{-1/2}$, is elastic.

**SOLUTION**

The price elasticity of demand is given by

$$\eta = \frac{p/x}{dp/dx}$$

$$= \frac{(50x^{-1/2})/x}{-25x^{-3/2}}$$

$$= \frac{50x^{-3/2}}{-25x^{-3/2}}$$

$$= -2$$

Since $|\eta| = 2$, we conclude that the demand is elastic for any price.

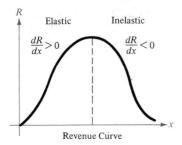

FIGURE 3.52                           Revenue Curve

Elasticity has an interesting relationship to the total revenue, as shown in Figure 3.52.

1. If the demand is *elastic,* then a decrease in the price per unit is accompanied by a sufficient increase in unit sales to cause the total revenue to increase.
2. If the price is *inelastic,* then the increase in unit sales accompanying a decrease in price per unit is *not* sufficient to increase revenue, and the total revenue decreases.

The next example verifies this relationship between elasticity and total revenue.

**EXAMPLE 6**

**Elasticity and Marginal Revenue**

Show that for a differentiable demand function, the marginal revenue is positive when the demand is elastic and negative when the demand is inelastic. (Assume that the quantity demanded increases as the price decreases, and thus $dp/dx$ is negative.)

**SOLUTION**

Since the revenue is given by $R = xp$, we calculate the marginal revenue to be

$$\frac{dR}{dx} = x\left(\frac{dp}{dx}\right) + p$$

$$= p\left[\left(\frac{x}{p}\right)\left(\frac{dp}{dx}\right) + 1\right]$$

$$= p\left[\frac{dp/dx}{p/x} + 1\right]$$

$$= p\left(\frac{1}{\eta} + 1\right)$$

If the demand is inelastic, then $|\eta| < 1$ implies that $1/\eta < -1$, since $dp/dx$ is negative. (We assume that $x$ and $p$ are positive.) Therefore,

$$\frac{dR}{dx} = p\left(\frac{1}{\eta} + 1\right)$$

is negative. Similarly, if the demand is elastic, then $|\eta| > 1$ and $-1 < 1/\eta$, which implies that $dR/dx$ is positive.

We conclude this section with a summary of the basic business terms and formulas used in this section.

## Summary of business terms and formulas

**Basic Terms**

$x$ = number of units produced (or sold)
$p$ = price per unit
$R$ = total revenue from selling $x$ units

$$R = xp$$

$C$ = total cost of producing $x$ units
$\overline{C}$ = average cost per unit

$$\overline{C} = \frac{C}{x}$$

$P$ = total profit from selling $x$ units

$$P = R - C$$

**Demand Function**

$p = f(x)$ = price required to sell $x$ units
$\eta$ = price elasticity of demand

$$\eta = \frac{p/x}{dp/dx}$$

If $|\eta| < 1$, the demand is inelastic.
If $|\eta| > 1$, the demand is elastic.

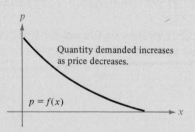

Quantity demanded increases as price decreases.

$p = f(x)$

**Typical Graphs of Revenue, Cost, and Profit Functions**

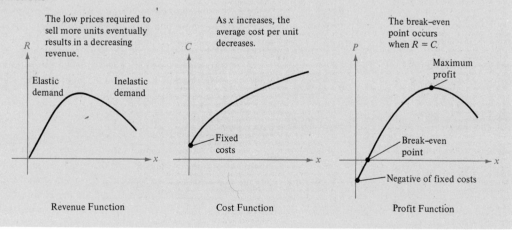

The low prices required to sell more units eventually results in a decreasing revenue.

Elastic demand    Inelastic demand

Revenue Function

As $x$ increases, the average cost per unit decreases.

Fixed costs

Cost Function

The break–even point occurs when $R = C$.

Maximum profit

Break–even point

Negative of fixed costs

Profit Function

**Marginals**

$$\frac{dR}{dx} = \text{marginal revenue} = \text{the rate of change of the revenue}$$

$$\approx \text{the } \textit{extra} \text{ revenue for selling one additional unit}$$

$$\frac{dC}{dx} = \text{marginal cost} \quad = \text{the rate of change of the cost}$$

$$\approx \text{the } \textit{extra} \text{ cost of producing one additional unit}$$

$$\frac{dP}{dx} = \text{marginal profit} \quad = \text{the rate of change of the profit}$$

$$\approx \text{the } \textit{extra} \text{ profit for selling one additional unit}$$

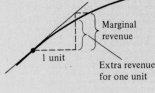

Revenue Function

## SECTION EXERCISES 3.5

In Exercises 1–4, find the number of units $x$ that produces a maximum revenue $R$.

**1.** $R = 900x - 0.1x^2$

**2.** $R = 600x^2 - 0.02x^3$

**3.** $R = \dfrac{1{,}000{,}000x}{0.02x^2 + 1800}$

**4.** $R = 30x^{2/3} - 2x$

In Exercises 5–8, find the number of units $x$ that produces the minimum average cost per unit $\overline{C}$.

**5.** $C = 0.125x^2 + 20x + 5000$

**6.** $C = 0.001x^3 - 5x + 250$

**7.** $C = 3000x - x^2\sqrt{300 - x}$

**8.** $C = (2x^3 - x^2 + 5000x)/(x^2 + 2500)$

In Exercises 9–12, find the price per unit $p$ that produces the maximum profit $P$.

|  | Cost function | Demand function |
|---|---|---|
| **9.** | $C = 100 + 30x$ | $p = 90 - x$ |
| **10.** | $C = 2400x + 5200$ | $p = 6000 - 0.4x^2$ |
| **11.** | $C = 4000 - 40x + 0.02x^2$ | $p = 50 - \dfrac{x}{100}$ |
| **12.** | $C = 35x + 2\sqrt{x - 1}$ | $p = 40 - \sqrt{x - 1}$ |

In Exercises 13 and 14, use the given cost function to find the value of $x$ for which the average cost is minimum. For this value of $x$, show that the marginal cost and average cost are equal.

**13.** $C = 2x^2 + 5x + 18$

**14.** $C = x^3 - 6x^2 + 13x$

**15.** A given commodity has a demand function given by $p = 100 - \frac{1}{2}x^2$ and a total cost function given by $C = 40x + 37.5$.

(a) What price gives the maximum profit?

(b) What is the average cost per unit if production is set to give maximum profit?

**16.** Rework Exercise 15 using the cost function $C = 50x + 37.5$. (In other words, how would the answer to Exercise 15 change if the marginal cost of producing a unit rose from \$40 to \$50?)

In Exercises 17 and 18, find the amount of advertising $s$ (in 1000s of dollars) that yields the maximum profit $P$ (in 1000s of dollars) and find the point of diminishing returns.

**17.** $P = -2s^3 + 35s^2 - 100s + 200$

**18.** $P = -\dfrac{1}{10}s^3 + 6s^2 + 400$

**19.** A manufacturer of radios charges \$90 per unit when the average production cost per unit is \$60. However, to encourage large orders from distributors, the manufacturer will reduce the charge by \$0.10 per unit for each unit ordered in excess of 100 (for example, there would be a charge of \$88 per radio for an order size of 120). Find the largest order the manufacturer should allow so as to realize maximum profit.

20. A real estate office handles 50 apartment units. When the rent is $270 per month, all units are occupied. However, on the average, for each $15 increase in rent one unit becomes vacant. Each occupied unit requires an average of $18 per month for service and repairs. What rent should be charged to realize the most profit?

21. When a wholesaler sold a certain product at $25 per unit, sales were 800 units each week. However, after a price increase of $5, the average number of units sold dropped to 775 per week. Assuming that the demand function is linear, find the price that will maximize the total revenue.

22. Assume that the amount of money deposited in a bank is proportional to the square of the interest rate the bank pays on this money. Furthermore, the bank can reinvest this money at 12%. Find the interest rate the bank should pay to maximize profit. (Use the simple interest formula.)

23. A power station is on one side of a river that is $\frac{1}{2}$ mile wide, and a factory is 6 miles downstream on the other side. It costs $6 per foot to run power lines overland and $8 per foot to run them underwater. Find the most economical path for the transmission line from the power station to the factory.

24. An offshore oil well is 1 mile off the coast. The refinery is 2 miles down the coast. If laying pipe in the ocean is twice as expensive as laying it on land, in what path should the pipe be constructed in order to minimize the cost?

25. A small business has a minivan that it uses for making deliveries. The cost per hour for fuel for the van is

$$C = \frac{v^2}{600}$$

where $v$ is the speed in miles per hour. If the driver is paid $5 per hour, find the speed that minimizes cost on a 110-mile trip. (Assume there are no costs other than wages and fuel.)

26. Repeat Exercise 25 if the fuel cost per hour is

$$C = \frac{v^2 + 360}{720}$$

and the driver is paid $3.75 per hour.

In Exercises 27–32, find $\eta$ (the price elasticity of demand) for the given demand function at the indicated $x$-value. Is the demand elastic or inelastic (or neither) at the indicated $x$-value?

| Demand Function | Quantity Demanded |
|---|---|
| 27. $p = 400 - 3x$ | $x = 20$ |
| 28. $p = 5 - 0.03x$ | $x = 100$ |
| 29. $p = 400 - 0.5x^2$ | $x = 20$ |
| 30. $p = \dfrac{500}{x + 2}$ | $x = 23$ |
| 31. $p = \dfrac{100}{x^2} + 2$ | $x = 10$ |
| 32. $p = 100 - \sqrt{0.2x}$ | $x = 125$ |

33. The demand function for a certain product is given by

$$x = 20 - 2p^2$$

(a) Consider a price of $2. If the price increases by 5%, determine the corresponding percentage change in quantity demanded.

(b) Average elasticity of demand is defined to be the percentage change in quantity divided by the percentage change in price. Use the percentage of part (a) to find the average elasticity over the interval [2, 2.1].

(c) Find the elasticity for a price of $2 and compare the result with that of part (b).

(d) Find an expression for total revenue and find the values of $x$ and $p$ that maximize $R$.

(e) For the value of $x$ found in part (d), show that $|\eta| = 1$.

34. The demand function for a particular commodity is given by $p^3 + x^3 = 9$.

(a) Find $\eta$ when $x = 2$.

(b) Find the values of $x$ and $p$ that maximize the total revenue.

(c) Show that $|\eta| = 1$ for the value of $x$ in part (b).

35. The demand function for a particular commodity is

$$p = (16 - x)^{1/2}, \quad 0 < x < 16$$

(a) Find $\eta$ when $x = 9$.

(b) Find the values of $x$ and $p$ that maximize the total revenue.

(c) Show that $|\eta| = 1$ for the value of $x$ in part (b).

36. A demand function is given by the equation

$$x = \frac{a}{p^m}, \quad m > 1$$

Show that $\eta = -m$. (In other words, in terms of approximate price changes, a 1% increase in price results in an $m\%$ decrease in quantity demanded.)

# Asymptotes

During the 1960s and 1970s, the federal government became increasingly sensitive to environmental concerns. One of the major forms of pollution that came under legislative attack was pollution of the air. Regulations governing auto exhaust, smokestack emissions, aerial pesticides, and aerosol sprays were the topic of heated controversy. The industries affected by the regulations tended to argue for moderate regulations building up over a period of years, whereas environmentalists tended to push for strict regulatory standards realized over a short period of time. The basic objection from industry centered around the cost of meeting the pollution regulations.

As a case in point, consider the air pollutants in the stack emission of a utility company that burns coal to generate electricity. The cost of removing a certain percentage of the pollutants is typically *not* a linear function. That is, if it would cost $C$ dollars to remove 25% of the pollutants, it would cost more than $2C$ dollars to remove 50% of the pollutants. Ultimately, as the percentage approaches 100%, the cost tends to become prohibitive. Suppose that the cost $C$ of removing $p\%$ of the smokestack pollutants is given by

$$C = \frac{80,000p}{100 - p}$$

Observe in Figure 3.53 the significant difference in cost between 85% and 90% removal, as contrasted to the minimal difference between 10% and 15% removal. Also, note that the cost increases without bound as $p$ approaches 100%. We call the line $p = 100$ a **vertical asymptote** of this graph.

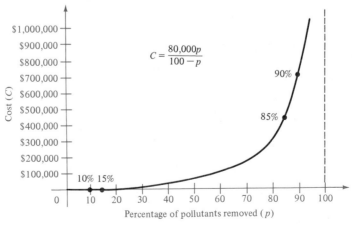

FIGURE 3.53

■ **Vertical Asymptotes**
■ **Infinite Limits**
■ **Horizontal Asymptotes**
■ **Limits at Infinity**

In the first three sections of this chapter, we looked at different ways we can use calculus as an aid to sketching the graph of a function. In this section, we look at another valuable aid to curve sketching—the determination of vertical and horizontal asymptotes.

Recall from Section 1.5, Example 10 that the function

$$f(x) = \frac{3}{x - 2}$$

is unbounded as $x$ approaches 2. (See Figure 3.54.) We call the line $x = 2$ a **vertical asymptote** of the graph of $f$. The graph of $f$ becomes arbitrarily close to the asymptote as $x$ approaches 2. The type of limit where $f(x)$ approaches infinity (or negative infinity) as $x$ approaches $c$ from the left or right is called an **infinite limit.**

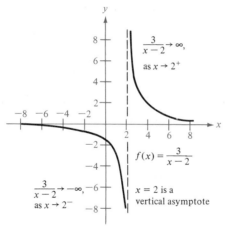

FIGURE 3.54

---

**Definition of
vertical asymptote**

If $f(x)$ approaches infinity (or negative infinity) as $x$ approaches $c$ from the right or left, then the line $x = c$ is called a **vertical asymptote** of $f$.

One of the most common instances of a vertical asymptote is with the graph of a *rational function*—that is, a function of the form

$$f(x) = \frac{p(x)}{q(x)}$$

where $p(x)$ and $q(x)$ are polynomials. If $c$ is a real number such that $q(c) = 0$ and $p(c) \neq 0$, then the graph of $f$ has a vertical asymptote at $x = c$. Figure 3.55 shows four typical cases. In each case the line $x = 1$ is a vertical asymptote of the graph.

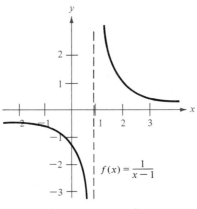

$$\lim_{x \to 1^-} \frac{1}{x-1} = -\infty \qquad \lim_{x \to 1^+} \frac{1}{x-1} = \infty$$

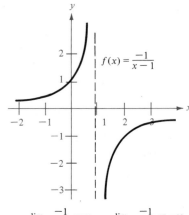

$$\lim_{x \to 1^-} \frac{-1}{x-1} = \infty \qquad \lim_{x \to 1^+} \frac{-1}{x-1} = -\infty$$

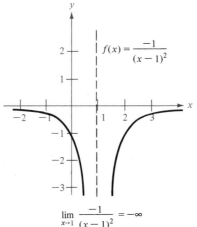

$$\lim_{x \to 1} \frac{-1}{(x-1)^2} = -\infty$$

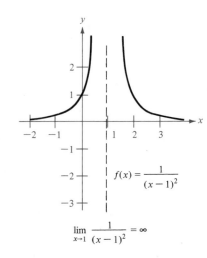

$$\lim_{x \to 1} \frac{1}{(x-1)^2} = \infty$$

FIGURE 3.55

Each of the graphs shown in Figure 3.55 has only one vertical asymptote. It is possible for the graph of a rational function to have more than one vertical asymptote, as illustrated in Example 1.

**EXAMPLE 1**

**Finding Vertical Asymptotes**

Determine the vertical asymptotes of the graph of

$$f(x) = \frac{x+2}{x^2 - 2x}$$

**SOLUTION**

The possible vertical asymptotes occur where the denominator of $f(x)$ is zero. By setting the denominator equal to zero, we have

$$x^2 - 2x = 0$$

$$x(x - 2) = 0$$

$$x = 0, 2$$

Since the numerator of this rational function is not zero at either of these $x$-values, we conclude that the graph of $f$ has two vertical asymptotes—one at $x = 0$ and the other at $x = 2$, as shown in Figure 3.56.

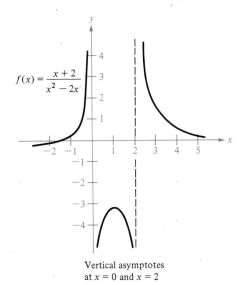

$$f(x) = \frac{x + 2}{x^2 - 2x}$$

Vertical asymptotes
at $x = 0$ and $x = 2$

FIGURE 3.56

If both the numerator and denominator of a rational function are zero at $x = c$, then we cannot determine the limit at $x = c$ without further investigation, as shown in the next example.

**EXAMPLE 2**

**A Rational Function with Common Factors**

Determine all vertical asymptotes of the graph of

$$f(x) = \frac{x^2 + 2x - 8}{x^2 - 4}$$

**SOLUTION**

Factoring both numerator and denominator, we have

$$f(x) = \frac{x^2 + 2x - 8}{x^2 - 4}$$

$$= \frac{(x + 4)(x - 2)}{(x + 2)(x - 2)}$$

Now we see that the denominator is zero when $x = \pm 2$. However, since the numerator is also zero when $x = 2$, we cancel the common factor $(x - 2)$ from the numerator and denominator. Then, for all $x$ other than 2, the function is given by

$$f(x) = \frac{(x + 4)(x - 2)}{(x + 2)(x - 2)}$$

$$= \frac{x + 4}{x + 2}, \quad x \neq 2$$

Thus, we determine that the graph of $f$ has one vertical asymptote at $x = -2$, as shown in Figure 3.57.

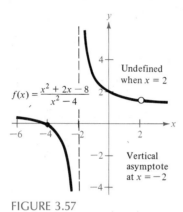

FIGURE 3.57

In Example 2, we know that the limit as $x \to -2$ from the right and left is plus *or* minus infinity. But without looking at the graph, how can we determine the following limits?

$$\lim_{x \to -2^-} \frac{x^2 + 2x - 8}{x^2 - 4} = -\infty \qquad \text{Limit from the left}$$

$$\lim_{x \to -2^+} \frac{x^2 + 2x - 8}{x^2 - 4} = \infty \qquad \text{Limit from the right}$$

It is cumbersome to answer this question analytically, and we have found the graphical approach outlined in the following example to be more efficient.

**EXAMPLE 3**

**Determining Infinite Limits**

Find the following limits:

$$\lim_{x \to 1^-} \frac{x^2 - 3x}{x - 1} \qquad \text{and} \qquad \lim_{x \to 1^+} \frac{x^2 - 3x}{x - 1}$$

**SOLUTION**

1. Factoring both numerator and denominator, we have

$$f(x) = \frac{x^2 - 3x}{x - 1} = \frac{x(x - 3)}{x - 1}$$

2. The (remaining) factors of the numerator yield the *x*-intercepts of the function. In this case, the *x*-intercepts occur at $x = 0$ and $x = 3$.
3. The (remaining) factors of the denominator yield the vertical asymptotes of the function. In this case, a vertical asymptote occurs at $x = 1$.
4. We determine a few additional points on the graph, making sure to choose at least one point between each intercept and asymptote, as shown in the accompanying table.

| $x$ | $-2$ | $\dfrac{1}{2}$ | $2$ | $4$ |
|---|---|---|---|---|
| $f(x)$ | $-\dfrac{10}{3}$ | $\dfrac{5}{2}$ | $-2$ | $\dfrac{4}{3}$ |

Then we plot the information obtained in the first four steps, as shown in Figure 3.58.
5. Finally, we complete the sketch of the graph, as shown in Figure 3.59, and conclude that

$$\lim_{x \to 1^-} \frac{x^2 - 3x}{x - 1} = \infty \qquad \text{Limit from the left}$$

and

$$\lim_{x \to 1^+} \frac{x^2 - 3x}{x - 1} = -\infty \qquad \text{Limit from the right}$$

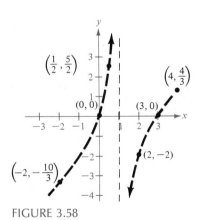

FIGURE 3.58

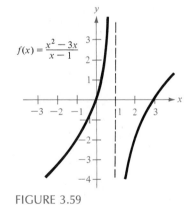

FIGURE 3.59

## Horizontal Asymptotes

Another type of limit, called a **limit at infinity,** specifies a finite value approached by a function as *x* increases (or decreases) without bound.

| **Definition of horizontal asymptote** | If $f$ is a function and $L$ is some number, then the statements |
|---|---|

$$1.\ \lim_{x \to \infty} f(x) = L \qquad \text{and} \qquad 2.\ \lim_{x \to -\infty} f(x) = L$$

denote **limits at infinity.** In either case the line $y = L$ is called a **horizontal asymptote** of the graph of $f$.

Figure 3.60 suggests some ways in which the graph of a function $f$ may approach one or more horizontal asymptotes. Observe in Figure 3.60 that it is possible for a function to cross its horizontal asymptote.

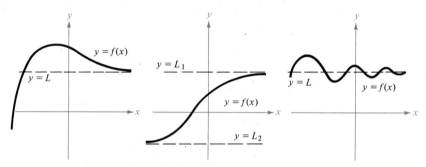

The graph of a function may approach
a horizontal asymptote in various ways.

FIGURE 3.60

The following theorem is useful for evaluating limits at infinity.

| **Limits at infinity** | If $r$ is a positive real number, then |
|---|---|

$$\lim_{x \to \infty} \frac{1}{x^r} = 0 \qquad \text{Limit toward the right}$$

Furthermore, if $x^r$ is defined when $x < 0$, then

$$\lim_{x \to -\infty} \frac{1}{x^r} = 0 \qquad \text{Limit toward the left}$$

Limits at infinity share many of the properties of limits given in Section 1.5. Some of these are demonstrated in the next example.

**EXAMPLE 4**

**Evaluating Limits at Infinity**

Evaluate the following limit:

$$\lim_{x \to \infty} \left( 5 - \frac{2}{x^2} \right)$$

**SOLUTION**  Try to identify the properties of limits used in the following steps.

$$\lim_{x\to\infty}\left(5-\frac{2}{x^2}\right) = \lim_{x\to\infty}5 - \lim_{x\to\infty}\frac{2}{x^2}$$

$$= \lim_{x\to\infty}5 - 2\left[\lim_{x\to\infty}\frac{1}{x^2}\right]$$

$$= 5 - 2(0)$$

$$= 5$$

Therefore, the graph of the function

$$f(x) = 5 - \frac{2}{x^2}$$

has $y = 5$ as a horizontal asymptote to the right, as shown in Figure 3.61.

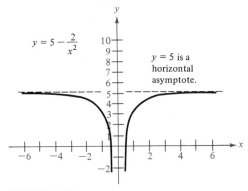

FIGURE 3.61

In Example 4 we showed that the graph of $f$ has $y = 5$ as a horizontal asymptote *to the right*. From Figure 3.61 it appears that the line $y = 5$ is also a horizontal asymptote *to the left*. You can verify this by showing that

$$\lim_{x\to-\infty}\left(5-\frac{2}{x^2}\right) = 5$$

The graph of a rational function need not have a horizontal asymptote. However, if it does, then it must approach the same asymptote to the left and to the right. This is not true of other types of functions. For instance, the graph of

$$f(x) = \frac{x}{\sqrt{x^2 + 1}}$$

has $y = 1$ as a horizontal asymptote to the right and $y = -1$ as a horizontal asymptote to the left, as shown in Figure 3.62.

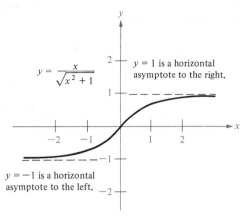

$y = \dfrac{x}{\sqrt{x^2 + 1}}$

$y = 1$ is a horizontal asymptote to the right.

$y = -1$ is a horizontal asymptote to the left.

FIGURE 3.62

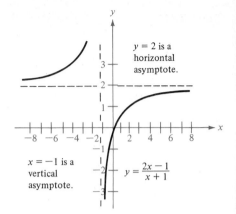

$y = 2$ is a horizontal asymptote.

$x = -1$ is a vertical asymptote.

$y = \dfrac{2x - 1}{x + 1}$

FIGURE 3.63

**EXAMPLE 5**

**Evaluating Limits at Infinity**

Evaluate the following limit:

$$\lim_{x \to \infty} \frac{2x - 1}{x + 1}$$

**SOLUTION**

Since both the numerator and denominator approach infinity as $x$ approaches infinity, it is not immediately clear what the limit of the quotient will be (or if it will even exist). Thus, we begin by rewriting the quotient in a different form. To do this, we divide both numerator and denominator by the highest power of $x$ in the denominator—in this case, $x$.

$$\lim_{x \to \infty} \frac{2x - 1}{x + 1} = \lim_{x \to \infty} \frac{2 - (1/x)}{1 + (1/x)}$$

$$= \frac{2 - 0}{1 + 0}$$

$$= 2$$

Thus, the line $y = 2$ is a horizontal asymptote to the right. (You may wish to test the reasonableness of this result by evaluating $f(x)$ at large positive values for $x$.) By taking the limit as $x \to -\infty$, you can see that $y = 2$ is also a horizontal asymptote to the left. Moreover, $x = -1$ is a vertical asymptote; the graph of this function is shown in Figure 3.63.

There is an easy way to determine whether the graph of a rational function has a horizontal asymptote. This shortcut is based on a comparison of the degrees of the numerator and the denominator of the rational function. Before describing this result, we give an example that compares the limits at infinity of three similar rational functions.

**EXAMPLE 6**

**Evaluating Limits at Infinity**

Find the limit as $x$ approaches infinity for the following three functions:

(a) $y = \dfrac{-2x + 3}{3x^2 + 1}$

(b) $y = \dfrac{-2x^2 + 3}{3x^2 + 1}$

(c) $y = \dfrac{-2x^3 + 3}{3x^2 + 1}$

**SOLUTION**

(a) We begin by dividing both numerator and denominator by $x^2$. Thus, it follows that

$$\lim_{x \to \infty} \frac{-2x + 3}{3x^2 + 1} = \lim_{x \to \infty} \frac{(-2/x) + (3/x^2)}{3 + (1/x^2)}$$

$$= \frac{-0 + 0}{3 + 0} = \frac{0}{3} = 0$$

(b) In this case, we also divide by $x^2$ (the highest power of $x$ in the denominator).

$$\lim_{x \to \infty} \frac{-2x^2 + 3}{3x^2 + 1} = \lim_{x \to \infty} \frac{-2 + (3/x^2)}{3 + (1/x^2)}$$

$$= \frac{-2 + 0}{3 + 0} = -\frac{2}{3}$$

(c) Again, we divide by $x^2$ to obtain

$$\lim_{x \to \infty} \frac{-2x^3 + 3}{3x^2 + 1} = \lim_{x \to \infty} \frac{-2x + (3/x^2)}{3 + (1/x^2)}$$

In this case, we conclude that the limit does not exist because the numerator increases without bound as the denominator approaches 3.

Observe that in part (a) of Example 6 the degree of the numerator was *less* than the degree of the denominator and the limit of the ratio was zero. In part (b), the degrees of the numerator and denominator were *equal* and the limit was given by the ratio of the coefficients of the highest-powered terms. Finally, in part (c), the degree of the numerator was *greater* than that of the denominator and the limit did not exist (it was infinite). These results suggest the following alternative way of determining the limits at infinity for rational functions.

**Limits at infinity for rational functions**

For the rational function given by $y = f(x)/g(x)$, where

$$f(x) = a_n x^n + a_{n-1} x^{n-1} + \cdots + a_0$$

and

$$g(x) = b_m x^m + b_{m-1} x^{m-1} + \cdots + b_0$$

we have

$$\lim_{x \to \pm\infty} \frac{f(x)}{g(x)} = \begin{cases} 0, & \text{if } n < m \\ \dfrac{a_n}{b_m}, & \text{if } n = m \\ \pm\infty, & \text{if } n > m \end{cases}$$

This result seems reasonable if we realize that for large values of $x$ the highest-powered term of a polynomial is the most "influential" term. That is, a polynomial tends to behave like its highest-powered term as $x$ approaches positive or negative infinity.

## Applications

There are many examples of asymptotic behavior in business and biology. For instance, the following example describes the asymptotic behavior of an average cost function.

**EXAMPLE 7**

**An Application Involving a Horizontal Asymptote**

A business has a cost function of $C = 0.5x + 5000$, where $C$ is measured in dollars and $x$ is the number of units produced. The *average cost per unit* is given by

$$\overline{C} = \frac{C}{x}$$

Find the average cost per unit when $x = 1000$, $10{,}000$, and $100{,}000$. What is the limit of $\overline{C}$ when $x$ approaches infinity?

**SOLUTION**

When $x = 1000$, the average cost per unit is

$$\overline{C} = \frac{0.5(1000) + 5000}{1000} = \$5.50$$

When $x = 10{,}000$, the average cost per unit is

$$\overline{C} = \frac{0.5(10{,}000) + 5000}{10{,}000} = \$1.00$$

When $x = 100{,}000$, the average cost per unit is

$$\overline{C} = \frac{0.5(100{,}000) + 5000}{100{,}000} = \$0.55$$

As $x$ approaches infinity, the limiting average cost per unit is

$$\lim_{x \to \infty} \frac{0.5x + 5000}{x} = \$0.50$$

As shown in Figure 3.64, this example points out one of the major problems of a small business. That is, it is difficult to have competitively low prices when the production level is low.

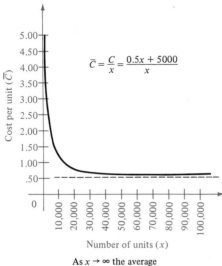

FIGURE 3.64

## SECTION EXERCISES 3.6

In Exercises 1–6, find the vertical and horizontal asymptotes.

**1.** $f(x) = \dfrac{1}{x^2}$

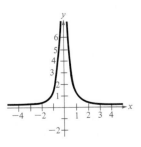

**2.** $f(x) = \dfrac{4}{(x - 2)^3}$

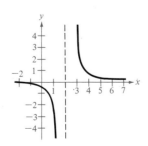

**3.** $f(x) = \dfrac{x^2 - 2}{x^2 - x - 2}$

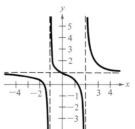

**4.** $f(x) = \dfrac{2 + x}{1 - x}$

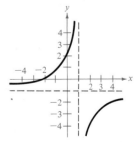

**5.** $f(x) = \dfrac{x^3}{x^2 - 1}$

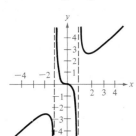

**6.** $f(x) = \dfrac{-4x}{x^2 + 4}$

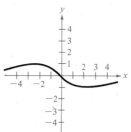

**(e)**

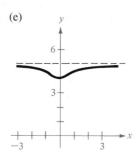

**(f)**

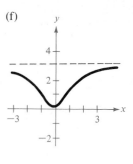

In Exercises 7–12, match the given function with the correct graph, using horizontal asymptotes as an aid. [Graphs are labeled (a), (b), (c), (d), (e), and (f).]

**7.** $f(x) = \dfrac{3x^2}{x^2 + 2}$

**8.** $f(x) = \dfrac{2x}{\sqrt{x^2 + 2}}$

**9.** $f(x) = \dfrac{x}{x^2 + 2}$

**10.** $f(x) = 2 + \dfrac{x^2}{x^4 + 1}$

**11.** $f(x) = 5 - \dfrac{1}{x^2 + 1}$

**12.** $f(x) = \dfrac{2x^2 - 3x + 5}{x^2 + 1}$

**(a)**

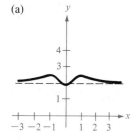

**(b)**

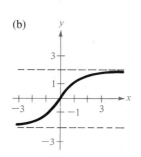

**(c)**

**(d)**

In Exercises 13–20, find the indicated limit.

**13.** $\displaystyle\lim_{x \to -2^-} \dfrac{1}{(x + 2)^2}$

**14.** $\displaystyle\lim_{x \to -2^-} \dfrac{1}{x + 2}$

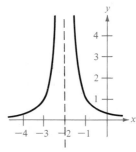

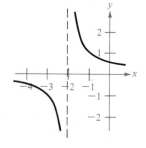

**15.** $\displaystyle\lim_{x \to 2^+} \dfrac{x - 3}{x - 2}$

**16.** $\displaystyle\lim_{x \to 1^+} \dfrac{2 + x}{1 - x}$

**17.** $\displaystyle\lim_{x \to 4^-} \dfrac{x^2}{x^2 - 16}$

**18.** $\displaystyle\lim_{x \to 4} \dfrac{x^2}{x^2 + 16}$

**19.** $\displaystyle\lim_{x \to 0^-} \left(1 + \dfrac{1}{x}\right)$

**20.** $\displaystyle\lim_{x \to 0^-} \left(x^2 - \dfrac{1}{x}\right)$

In Exercises 21–30, find the indicated limit.

**21.** $\displaystyle\lim_{x \to \infty} \dfrac{2x - 1}{3x + 2}$

**22.** $\displaystyle\lim_{x \to \infty} \dfrac{5x^3 + 1}{10x^3 - 3x^2 + 7}$

**23.** $\displaystyle\lim_{x \to \infty} \dfrac{x}{x^2 - 1}$

**24.** $\displaystyle\lim_{x \to \infty} \dfrac{2x^{10} - 1}{10x^{11} - 3}$

**25.** $\displaystyle\lim_{x \to -\infty} \dfrac{5x^2}{x + 3}$

**26.** $\displaystyle\lim_{x \to \infty} \dfrac{x^3 - 2x^2 + 3x + 1}{x^2 - 3x + 2}$

**27.** $\displaystyle\lim_{x \to \infty} \left(2x - \dfrac{1}{x^2}\right)$

**28.** $\displaystyle\lim_{x \to \infty} (x + 3)^{-2}$

**29.** $\displaystyle\lim_{x \to -\infty} \left(\dfrac{2x}{x - 1} + \dfrac{3x}{x + 1}\right)$

**30.** $\displaystyle\lim_{x \to \infty} \left(\dfrac{2x^2}{x - 1} + \dfrac{3x}{x + 1}\right)$

In Exercises 31–34, complete the given table and estimate the limit of $f(x)$ as $x$ approaches infinity.

**31.** $f(x) = \dfrac{x + 1}{x\sqrt{x}}$

| $x$ | $10^0$ | $10^1$ | $10^2$ | $10^3$ | $10^4$ | $10^5$ | $10^6$ |
|-----|--------|--------|--------|--------|--------|--------|--------|
| $f(x)$ | | | | | | | |

**32.** $f(x) = x - \sqrt{x(x - 1)}$

| $x$ | $10^0$ | $10^1$ | $10^2$ | $10^3$ | $10^4$ | $10^5$ | $10^6$ |
|-----|--------|--------|--------|--------|--------|--------|--------|
| $f(x)$ | | | | | | | |

**33.** $f(x) = x^2 - x\sqrt{x(x - 1)}$

| $x$ | $10^0$ | $10^1$ | $10^2$ | $10^3$ | $10^4$ | $10^5$ | $10^6$ |
|-----|--------|--------|--------|--------|--------|--------|--------|
| $f(x)$ | | | | | | | |

**34.** $f(x) = \dfrac{x}{\sqrt{x^2 + 1}}$

| $x$ | $10^0$ | $10^1$ | $10^2$ | $10^3$ | $10^4$ | $10^5$ | $10^6$ |
|-----|--------|--------|--------|--------|--------|--------|--------|
| $f(x)$ | | | | | | | |

In Exercises 35–50, sketch the graph of each equation, using intercepts, relative extrema, and asymptotes as sketching aids.

**35.** $y = \dfrac{2 + x}{1 - x}$

**36.** $y = \dfrac{x - 3}{x - 2}$

**37.** $f(x) = \dfrac{x^2}{x^2 + 9}$

**38.** $f(x) = \dfrac{x^2}{x^2 - 9}$

**39.** $g(x) = \dfrac{x^2}{x^2 - 4}$

**40.** $g(x) = \dfrac{x}{x^2 - 4}$

**41.** $xy^2 = 4$

**42.** $x^2y = 4$

**43.** $y = \dfrac{2x}{1 - x}$

**44.** $y = \dfrac{2x}{1 - x^2}$

**45.** $y = 3\left(1 - \dfrac{1}{x^2}\right)$

**46.** $y = 1 + \dfrac{1}{x}$

**47.** $f(x) = \dfrac{1}{x^2 - x - 2}$

**48.** $f(x) = \dfrac{x - 2}{x^2 - 4x + 3}$

**49.** $g(x) = \dfrac{x^2 - x - 2}{x - 2}$

**50.** $g(x) = \dfrac{x^2 - 9}{x + 3}$

**51.** The cost function for a certain product is given by

$$C = 1.35x + 4570$$

where $C$ is measured in dollars and $x$ is the number of units produced.

(a) Find the average cost per unit when $x = 100$ and when $x = 1000$.

(b) What is the limit of the average cost function as $x$ approaches infinity?

**52.** A business has a cost of $C = 0.5x + 500$ for producing $x$ units. The average cost per unit is given by $\overline{C} = C/x$. Find the limit of $\overline{C}$ as $x$ approaches infinity.

**53.** The cost in millions of dollars for the federal government to seize $p\%$ of a certain illegal drug as it enters the country is given by

$$C = \dfrac{528p}{100 - p}, \quad 0 \le p < 100$$

(a) Find the cost of seizing 25%.

(b) Find the cost of seizing 50%.

(c) Find the cost of seizing 75%.

(d) Find the limit of $C$ as $p \to 100^-$.

**54.** The cost in dollars of removing $p\%$ of the air pollutants in the stack emission of a utility company that burns coal to generate electricity is

$$C = \dfrac{80{,}000p}{100 - p}, \quad 0 \le p < 100$$

(a) Find the cost of removing 15%.

(b) Find the cost of removing 50%.

(c) Find the cost of removing 90%.

(d) Find the limit of $C$ as $x \to 100^-$.

**55.** Psychologists have developed mathematical models to predict performance as a function of the number of trials $n$ for a certain task. One such model is

$$P = \dfrac{b + \theta a(n - 1)}{1 + \theta(n - 1)}$$

where $P$ is the percentage of correct responses after $n$ trials and $a$, $b$, and $\theta$ are constants depending on the actual learning situation. Find the limit of $P$ as $n$ approaches infinity.

**56.** Consider the learning curve given by

$$P = \dfrac{0.5 + 0.9(n - 1)}{1 + 0.9(n - 1)}, \quad 0 < n$$

(a) Complete the following table for this model.

| $n$ | 1 | 2 | 3 | 4 | 5 | 6 | 7 | 8 | 9 | 10 |
|-----|---|---|---|---|---|---|---|---|---|----|
| $P$ | | | | | | | | | | |

(b) Find the limit as $n$ approaches infinity.

(c) Sketch a graph of this learning curve.

**57.** The game commission in a certain state introduces 50 deer into newly acquired state game lands. The population $N$ of the herd is given by the model

$$N = \frac{10(5 + 3t)}{1 + 0.04t}$$

where $t$ is time in years.

(a) Find the number in the herd when $t$ is 5, 10, and 25.

(b) According to this model, what is the limiting size of the herd as time increases?

**58.** The cost and revenue functions for a particular item are given by

$$C = 34.5x + 15,000$$

and $R = 69.9x$

(a) Find the average profit function

$$\overline{P} = \frac{R - C}{x}$$

(b) Find the average profit per unit when $x$ is 1000, 10,000, and 100,000.

(c) What is the limit of the average profit function as $x$ approaches infinity?

# SECTION 3.7

# Curve Sketching: A Summary

**INTRODUCTORY EXAMPLE**
**United States Population Predictions**

The U.S. Census Bureau conducts a nationwide census every 10 years, as shown in Table 3.14. In addition to tallying the population of the United States, the Census Bureau compiles numerous population characteristics regarding age, sex, ethnic background, marital status, immigration patterns, and so on. With these data, the Census Bureau is able to predict various population growth patterns.

Four of the current predictions are shown in Figure 3.65. Note that since 1950 the population curve seems to have changed its rate of growth slightly, and the three lower predictions reflect this change. All four of the "series" predictions assume a slightly increased life expectancy and (except for Series III) an annual immigration of 400,000. The difference in the predictions stems from different assumptions regarding the birth rate per "female lifetime."

| Series | Birth rate | Immigration |
|---|---|---|
| I | 2.7 children | 400,000 |
| II | 2.1 children | 400,000 |
| III | 2.1 children | none |
| IV | 1.7 children | 400,000 |

TABLE 3.14

| U.S. population | 75,994,575 | 91,972,266 | 105,710,620 |
|---|---|---|---|
| Year | 1900 | 1910 | 1920 |

| U.S. population | 122,775,048 | 131,669,275 | 150,697,361 |
|---|---|---|---|
| Year | 1930 | 1940 | 1950 |

| U.S. population | 179,323,175 | 203,235,298 | 223,889,000 |
|---|---|---|---|
| Year | 1960 | 1970 | 1980 |

FIGURE 3.65

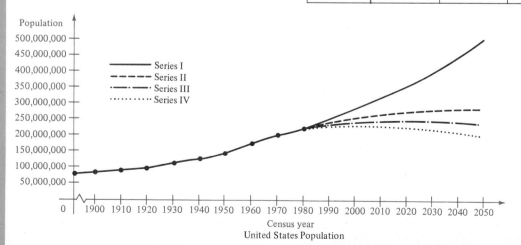

Census year
United States Population

### ■ Curve Sketching

The saying "a picture is worth a thousand words" properly identifies the importance of curve sketching in mathematics. Descartes's introduction of this fruitful concept not only preceded but to a great extent was responsible for the rapid advances in mathematics during the last half of the seventeenth century. Descartes's contribution was described this way by the French mathematician Joseph Lagrange (1736–1813): "As long as algebra and geometry traveled separate paths their advance was slow and their applications limited. But when these two sciences joined company, they drew from each other fresh vitality and thenceforth marched on at a rapid pace towards perfection."

Today government, science, industry, business, education, and the social and health sciences all make use of graphs to describe and predict relationships between variables. However, we have seen that sometimes sketching a graph can require considerable ingenuity.

Up to this point in the text we have discussed several concepts that are useful in sketching the graph of a function. For instance, the following properties of a function $f$ and its graph have been discussed at some length: domain and range of $f$, continuity, $x$- and $y$-intercepts, horizontal and vertical asymptotes, relative minima and maxima, concavity, and points of inflection. In this section, we incorporate these concepts into an effective procedure for sketching curves.

**Suggestions for sketching the graph of a function**

1. Make a rough preliminary sketch, using any easily determined intercepts, vertical or horizontal asymptotes, and points of discontinuity.
2. Locate the $x$-values where $f'(x)$ and $f''(x)$ are either zero or undefined.
3. Test the behavior of $f$ at each of these $x$-values as well as *within* each interval determined by them. (See Table 3.15 in Example 1.)
4. Sharpen the accuracy of the final sketch by plotting the relative extrema points, the points of inflection, and a few points in between.

All of the functions we have studied up to this point in the text are *algebraic*—that is, they are

Polynomial functions, as in Examples 1 and 2

Rational functions, as in Examples 3 and 4

Functions involving radicals, as in Example 5

Later in the text we will study nonalgebraic functions (called **transcendental** functions) such as exponential and logarithmic functions.

**EXAMPLE 1**

**Analyzing a Graph**

Sketch the graph of

$$f(x) = x^3 + 3x^2 - 9x + 5$$

---

**SOLUTION**

We begin by finding the intercepts of the graph of $f$. Factoring,* we have

$$f(x) = (x - 1)^2(x + 5)$$

Thus, the $x$-intercepts occur when $x = 1$ and $x = -5$. The derivative of $f$ is given by

$$f'(x) = 3x^2 + 6x - 9$$
$$= 3(x - 1)(x + 3)$$

Therefore, the critical numbers of $f$ are $x = 1$ and $x = -3$. The second derivative of $f$ is given by

$$f''(x) = 6x + 6$$

which implies that the second derivative is zero when $x = -1$. Testing the values of $f'$ and $f''$, as shown in Table 3.15, we see that $f$ has one relative minimum, one relative maximum, and one point of inflection. The graph of $f$ is shown in Figure 3.66.

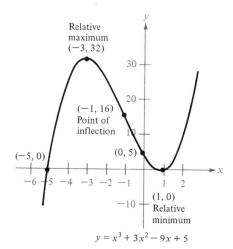

**FIGURE 3.66**

$$y = x^3 + 3x^2 - 9x + 5$$

**TABLE 3.15**

| | $f(x)$ | $f'(x)$ | $f''(x)$ | Shape of graph |
|---|---|---|---|---|
| $x$ in $(-\infty, -3)$ | | + | − | increasing, concave down |
| $x = -3$ | 32 | 0 | − | relative maximum |
| $x$ in $(-3, -1)$ | | − | − | decreasing, concave down |
| $x = -1$ | 16 | − | 0 | point of inflection |
| $x$ in $(-1, 1)$ | | − | + | decreasing, concave up |
| $x = 1$ | 0 | 0 | + | relative minimum |
| $x$ in $(1, \infty)$ | | + | + | increasing, concave up |

*This polynomial can be factored by using the Rational Zero Theorem, as described in Section 0.4.

**Remark:** In the summary at the end of this section, you can see that this graph is representative of one of the four basic shapes for cubic polynomials.

**EXAMPLE 2**

**Analyzing a Graph**

Sketch the graph of

$$f(x) = x^4 - 12x^3 + 48x^2 - 64x$$

**SOLUTION**

Factoring, we see that

$$f(x) = x(x^3 - 12x^2 + 48x - 64)$$
$$= x(x - 4)^3$$

Thus, the $x$-intercepts occur when $x = 0$ and $x = 4$. The first derivative is given by

$$f'(x) = 4x^3 - 36x^2 + 96x - 64$$
$$= 4(x - 1)(x - 4)^2$$

which implies that the critical numbers are $x = 1$ and $x = 4$. The second derivative is

$$f''(x) = 12x^2 - 72x + 96$$
$$= 12(x - 4)(x - 2)$$

which implies that there are possible points of inflection at $x = 2$ and $x = 4$. Testing these values, we have the results shown in Table 3.16. The graph of this function is shown in Figure 3.67.

TABLE 3.16

|  | $f(x)$ | $f'(x)$ | $f''(x)$ | Shape of graph |
|---|---|---|---|---|
| $x$ in $(-\infty, 1)$ |  | $-$ | $+$ | decreasing, concave up |
| $x = 1$ | $-27$ | $0$ | $+$ | minimum |
| $x$ in $(1, 2)$ |  | $+$ | $+$ | increasing, concave up |
| $x = 2$ | $-16$ | $+$ | $0$ | point of inflection |
| $x$ in $(2, 4)$ |  | $+$ | $-$ | increasing, concave down |
| $x = 4$ | $0$ | $0$ | $0$ | point of inflection |
| $x$ in $(4, \infty)$ |  | $+$ | $+$ | increasing, concave up |

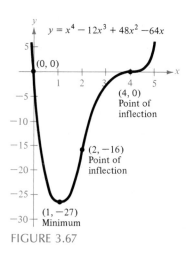

$y = x^4 - 12x^3 + 48x^2 - 64x$

$(0, 0)$

$(4, 0)$
Point of
inflection

$(2, -16)$
Point of
inflection

$(1, -27)$
Minimum

FIGURE 3.67

**EXAMPLE 3**

**Analyzing a Graph**

Sketch the graph of

$$f(x) = \frac{x^2 - 2x + 4}{x - 2}$$

**SOLUTION**

The $y$-intercept occurs at $(0, -2)$. Using the Quadratic Formula on the numerator, we see that there are no $x$-intercepts. Moreover, since the denominator is zero when $x = 2$, we determine that $x = 2$ is a vertical asymptote. (There are no horizontal asymptotes since the degree of the numerator is greater than the degree of the denominator.) The first derivative is given by

$$f'(x) = \frac{(x - 2)(2x - 2) - (x^2 - 2x + 4)}{(x - 2)^2}$$

$$= \frac{x(x - 4)}{(x - 2)^2}$$

and the second derivative is

$$f''(x) = \frac{(x - 2)^2(2x - 4) - (x^2 - 4x)(2)(x - 2)}{(x - 2)^4}$$

$$= \frac{8}{(x - 2)^3}$$

The critical numbers are $x = 0$ and $x = 4$. There are no possible points of inflection since $f''(x)$ is never zero and $f$ is not defined at $x = 2$. Testing, we have the results shown in Table 3.17. The graph is shown in Figure 3.68.

TABLE 3.17

| | $f(x)$ | $f'(x)$ | $f''(x)$ | Shape of graph |
|---|---|---|---|---|
| $x$ in $(-\infty, 0)$ | | $+$ | $-$ | increasing, concave down |
| $x = 0$ | $-2$ | $0$ | $-$ | relative maximum |
| $x$ in $(0, 2)$ | | $-$ | $-$ | decreasing, concave down |
| $x = 2$ | does not exist | does not exist | does not exist | vertical asymptote |
| $x$ in $(2, 4)$ | | $-$ | $+$ | decreasing, concave up |
| $x = 4$ | $6$ | $0$ | $+$ | relative minimum |
| $x$ in $(4, \infty)$ | | $+$ | $+$ | increasing, concave up |

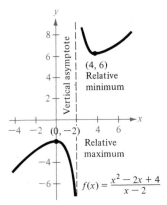

FIGURE 3.68

## EXAMPLE 4

**Analyzing a Graph**

Sketch the graph of

$$f(x) = \frac{2(x^2 - 9)}{x^2 - 4}$$

**SOLUTION**

The $y$-intercept is $(0, \frac{9}{2})$ and the $x$-intercepts occur at $(-3, 0)$ and $(3, 0)$. There are vertical asymptotes at $x = \pm 2$, and a horizontal asymptote occurs at $y = 2$. Since the first derivative is

$$f'(x) = 2\frac{(x^2 - 4)(2x) - (x^2 - 9)(2x)}{(x^2 - 4)^2}$$

$$= \frac{20x}{(x^2 - 4)^2}$$

we see that $x = 0$ is a critical number of the function. Furthermore, the second derivative

$$f''(x) = \frac{(x^2 - 4)^2(20) - (20x)(2)(2x)(x^2 - 4)}{(x^2 - 4)^4}$$

$$= -\frac{20(3x^2 + 4)}{(x^2 - 4)^3}$$

is never zero, which implies that there are no points of inflection. Testing, we have the results shown in Table 3.18. The graph is shown in Figure 3.69.

TABLE 3.18

| | $f(x)$ | $f'(x)$ | $f''(x)$ | Shape of graph |
|---|---|---|---|---|
| $x$ in $(-\infty, -2)$ | | $-$ | $-$ | decreasing, concave down |
| $x = -2$ | does not exist | does not exist | does not exist | vertical asymptote |
| $x$ in $(-2, 0)$ | | $-$ | $+$ | decreasing, concave up |
| $x = 0$ | $\dfrac{9}{2}$ | $0$ | $+$ | relative minimum |
| $x$ in $(0, 2)$ | | $+$ | $+$ | increasing, concave up |
| $x = 2$ | does not exist | does not exist | does not exist | vertical asymptote |
| $x$ in $(2, \infty)$ | | $+$ | $-$ | increasing, concave down |

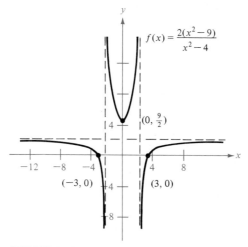

$$f(x) = \frac{2(x^2 - 9)}{x^2 - 4}$$

$(0, \frac{9}{2})$

$(-3, 0)$   $(3, 0)$

FIGURE 3.69

EXAMPLE 5

**Analyzing a Graph**

Sketch the graph of

$$f(x) = 2x^{5/3} - 5x^{4/3}$$

**SOLUTION**

Since

$$f(x) = x^{4/3}(2x^{1/3} - 5)$$

we see that one intercept occurs at $(0, 0)$. Moreover, a second $x$-intercept occurs when $2x^{1/3} = 5$, that is, when

$$x = \left(\frac{5}{2}\right)^3 = \frac{125}{8}$$

The first and second derivatives are

$$f'(x) = \frac{10}{3}x^{2/3} - \frac{20}{3}x^{1/3}$$

$$= \frac{10}{3}x^{1/3}(x^{1/3} - 2)$$

and

$$f''(x) = \frac{20}{9}x^{-1/3} - \frac{20}{9}x^{-2/3}$$

$$= \frac{20(x^{1/3} - 1)}{9x^{2/3}}$$

Thus, the critical numbers of $f$ are $x = 0$ and $x = 8$. Moreover, a possible point of inflection occurs at $x = 1$. Testing, we have the results given in Table 3.19. The graph is shown in Figure 3.70.

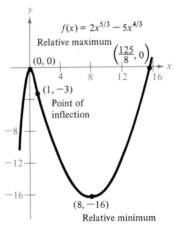

FIGURE 3.70

**TABLE 3.19**

| | $f(x)$ | $f'(x)$ | $f''(x)$ | Shape of graph |
|---|---|---|---|---|
| $x$ in $(-\infty, 0)$ | | $+$ | $-$ | increasing, concave down |
| $x = 0$ | $0$ | $0$ | does not exist | relative maximum |
| $x$ in $(0, 1)$ | | $-$ | $-$ | decreasing, concave down |
| $x = 1$ | $-3$ | $-$ | $0$ | point of inflection |
| $x$ in $(1, 8)$ | | $-$ | $+$ | decreasing, concave up |
| $x = 8$ | $-16$ | $0$ | $+$ | relative minimum |
| $x$ in $(8, \infty)$ | | $+$ | $+$ | increasing, concave up |

We conclude this section with a summary of the graphs of polynomial functions of degrees 0, 1, 2, and 3. Because of their simplicity, lower-degree polynomial functions are commonly used as mathematical models.

**Graphs of polynomial functions**

Constant function (degree 0):   Horizontal line:  ⎯⎯⎯⎯⎯⎯

$$y = a$$

Linear function (degree 1):    Line (of slope $a$):

$$y = ax + b$$

$a < 0$        $a > 0$

Quadratic function (degree 2):  Parabola:

$$y = ax^2 + bx + c$$

$a < 0$        $a > 0$

Cubic function (degree 3):    Cubic curve:

$$y = ax^3 + bx^2 + cx + d$$

$a < 0$        $a > 0$

In Exercises 1–4, determine the sign of $a$ for which the graph of $f(x) = ax^3 + bx^2 + cx + d$ will resemble the given graph.

**1.**

**2.**

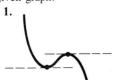

**3.**

**4.**

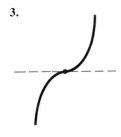

In Exercises 5–30, sketch the graph of the given function. Choose a scale that allows all relative extrema and points of inflection to be identified on the sketch.

**5.** $y = -x^2 - 2x + 3$     **6.** $y = 2x^2 - 4x + 1$

**7.** $y = x^3 - 3x^2 + 3$     **8.** $y = -\frac{1}{3}(x^3 - 3x + 2)$

**9.** $y = 2 - x - x^3$     **10.** $y = x^3 + 3x^2 + 3x + 2$

**11.** $f(x) = 3x^3 - 9x + 1$

**12.** $f(x) = (x + 1)(x - 2)(x - 5)$

**13.** $f(x) = -x^3 + 3x^2 + 9x - 2$

**14.** $f(x) = \frac{1}{3}(x - 1)^2 + 2$

**15.** $y = 3x^4 + 4x^3$     **16.** $y = 3x^4 - 6x^2$

**17.** $f(x) = x^4 - 4x^3 + 16x$

**18.** $f(x) = x^4 - 8x^3 + 18x^2 - 16x + 5$

**19.** $f(x) = x^4 - 4x^3 + 16x - 16$

**20.** $f(x) = x^5 + 1$

**21.** $y = x^5 - 5x$     **22.** $y = (x - 1)^5$

**23.** $y = |2x - 3|$     **24.** $y = |x^2 - 6x + 5|$

**25.** $y = \frac{x^2}{x^2 + 3}$     **26.** $y = \frac{x}{x^2 + 1}$

**27.** $y = 3x^{2/3} - 2x$     **28.** $y = 3x^{2/3} - x^2$

**29.** $y = 1 - x^{2/3}$     **30.** $y = (1 - x)^{2/3}$

In Exercises 31–38, sketch the graph of the given function. In each case label the intercepts, relative extrema, points of inflection, and asymptotes, and give the domain of the function.

**31.** $y = \dfrac{1}{x - 2} - 3$     **32.** $y = \dfrac{x^2 + 1}{x^2 - 2}$

**33.** $y = \dfrac{2x}{x^2 - 1}$     **34.** $y = \dfrac{x^2 - 6x + 12}{x - 4}$

**35.** $y = x\sqrt{4 - x}$     **36.** $y = x\sqrt{4 - x^2}$

**37.** $y = \dfrac{x + 2}{x}$     **38.** $y = x + \dfrac{32}{x^2}$

In Exercises 39–42, use the given graph (of $f'$ or $f''$) to sketch a graph of the function $f$. (The solutions are not unique.)

**39.**

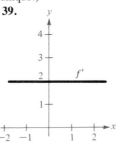

**40.**

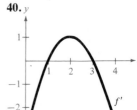

**41.**

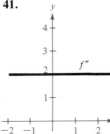

**42.**

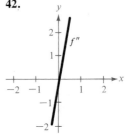

# Differentials

**INTRODUCTORY EXAMPLE**
**Estimating Profit Using Differentials**

In Example 4 of Section 2.3, we looked at a profit function given by the model

$$P = 0.0002x^3 + 10x$$

In that example we compared the marginal profit for a production level of 50 units to the actual gain in profit obtained by increasing the production level from 50 to 51 units. We found that the marginal profit is $dP/dx = 0.0006x^2 + 10$, which means that for a production level of 50 units the marginal profit is

$$\frac{dP}{dx} = (0.0006)(50^2) + 10 = \$11.50 \text{ per unit}$$

Moreover, by comparing the actual profit for production levels of 50 and 51 units, we saw that the actual gain in profit from increasing the production level 1 unit is

$$P(51) - P(50) = 536.53 - 525.00 = \$11.53$$

We call this difference between two levels of profit the *change in profit* corresponding to a *change in production level* of 1 unit. These two changes are denoted by

$$\Delta P = 11.53 \qquad \text{Change in profit}$$
and $\quad \Delta x = 1 \qquad$ Change in production

The ratio of these two changes is approximately equal to the marginal profit, and we can write

$$\frac{\Delta P}{\Delta x} \approx \frac{dP}{dx}$$

which means that the change in profit can be approximated by the formula

$$\Delta P \approx \frac{dP}{dx}\Delta x$$

In approximation problems, it is customary to define the **differential of $x$** to be $dx = \Delta x$ and the **differential of $P$** to be

$$dP = \frac{dP}{dx}dx$$

This means that the actual change in profit can be approximated by the differential $dP$. That is,

$$\Delta P \approx dP$$

as shown in Figure 3.71.

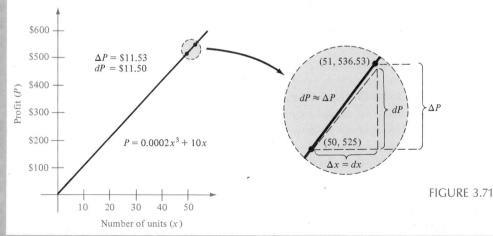

FIGURE 3.71

## SECTION TOPICS

■ **Differentials**
■ **Error Propagation**
■ **Differential Formulas**

We previously defined the derivative as the limit (as $\Delta x \to 0$) of the ratio $\Delta y/\Delta x$, and it seemed natural to retain the quotient symbolism for the limit itself. Thus, we denoted the derivative by

$$\frac{dy}{dx} = \lim_{\Delta x \to 0} \frac{\Delta y}{\Delta x}$$

*even though* we did not think of $dy/dx$ as the quotient of the two separate quantities $dy$ and $dx$. In this section, we give separate meanings to $dy$ and $dx$ in such a way that their quotient, when $dx \neq 0$, is equal to the derivative of $y$ with respect to $x$. We do this in the following manner.

**Definition of differentials**

Let $y = f(x)$ represent a differentiable function. The **differential of $x$** (denoted by $dx$) is any nonzero real number. The **differential of $y$** (denoted by $dy$) is given by

$$dy = f'(x)\ dx$$

■ **Remark:** Note that in this definition $dx$ can have any nonzero value. However, in most applications of differentials we choose $dx$ to be small and we denote this choice by $dx = \Delta x$.

One use of differentials is in approximating the change in $f(x)$ that corresponds to a change in $x$, as shown in Figure 3.72. We denote this change by

$$\Delta y = f(x + \Delta x) - f(x)$$

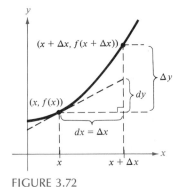

FIGURE 3.72

The following example compares the values of $dy$ and $\Delta y$.

**EXAMPLE 1**

**Comparing $\Delta y$ and $dy$**

Let $y = x^2$. Find $dy$ when $x = 1$ and $dx = 0.01$. Compare this value to $\Delta y$ when $x = 1$ and $\Delta x = 0.01$.

**SOLUTION**

Since

$$y = f(x) = x^2$$

we have

$$f'(x) = 2x$$

and the differential $dy$ is given by

$$dy = f'(x)\, dx \qquad \text{Differential of } y: dy$$
$$= f'(1)(0.01)$$
$$= 2(0.01)$$
$$= 0.02$$

Now, since $\Delta x = 0.01$, the change in $y$ is

$$\Delta y = f(x + \Delta x) - f(x) \qquad \text{Change in } y: \Delta y$$
$$= f(1.01) - f(1)$$
$$= (1.01)^2 - 1^2$$
$$= 0.0201$$

Figure 3.73 shows the graphical relation between $dy$ and $\Delta y$.

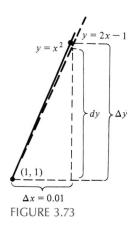

$y = 2x - 1$

$y = x^2$

$dy$  $\Delta y$

$(1, 1)$

$\Delta x = 0.01$

FIGURE 3.73

For the function in Example 1 we can see in Table 3.20 that $dy$ approximates $\Delta y$ more and more closely as $dx$ approaches zero.

TABLE 3.20

| $dx = \Delta x$ | $dy$ | $\Delta y$ | $\Delta y - dy$ |
|---|---|---|---|
| 0.1000 | 0.2000 | 0.21000000 | 0.01000000 |
| 0.0100 | 0.0200 | 0.02010000 | 0.00010000 |
| 0.0010 | 0.0020 | 0.00200100 | 0.00000100 |
| 0.0001 | 0.0002 | 0.00020001 | 0.00000001 |

Actually, this is not the first time we have encountered an approximation of this type. In Section 2.3 (Example 4), we used the *marginal profit* to approximate the extra profit gained by selling one more unit. We call this type of approximation a **linear approximation.** More specifically, it is called a **tangent line approximation** since we are using the tangent line at a point to approximate the graph of the function at the point.

The validity of using a tangent line to approximate a curve stems from the limit definition of the tangent line. That is, the existence of the limit

$$f'(x) = \lim_{\Delta x \to 0} \frac{f(x + \Delta x) - f(x)}{\Delta x}.$$

implies that when $\Delta x$ is close to zero then $f'(x)$ is close to the difference quotient, and we have

$$\frac{f(x + \Delta x) - f(x)}{\Delta x} \approx f'(x), \quad \Delta x \neq 0$$

$$f(x + \Delta x) - f(x) \approx f'(x) \, \Delta x$$

$$\Delta y \approx f'(x) \, \Delta x$$

Substituting $dx$ for $\Delta x$, we have

$$\Delta y \approx f'(x) \, dx = dy, \quad dx \neq 0$$

## Error Propagation

One practical use of an approximation of $\Delta y$ by $dy$ is in the estimation of errors propagated by physical measuring devices. For example, if we let $x$ represent the measured value of a variable and let $x + \Delta x$ represent the exact value, then $\Delta x$ is the *error in measurement*. Finally, if the measured value $x$ is used to compute the value $f(x)$, then the difference between $f(x + \Delta x)$ and $f(x)$ is called the **propagated error.**

$$\underbrace{f(x + \underbrace{\Delta x}_{\substack{\text{Measurement} \\ \text{error}}}) - f(x)}_{} = \underbrace{\Delta y}_{\substack{\text{Propagated} \\ \text{error}}}$$

Exact    Measured
value    value

**EXAMPLE 2**

**Estimation of Error**

The radius of a ball bearing is measured to be 0.7 inch, as shown in Figure 3.74. If the measurement is correct to within 0.01 inch, estimate the propagated error in the volume $V$ of the ball bearing.

**SOLUTION**

FIGURE 3.74

The formula for the volume of a sphere is

$$V = \frac{4}{3}\pi r^3$$

where $r$ is the radius of the sphere. Thus, we have

$$r = 0.7 \qquad \text{Measured radius}$$

and

$$-0.01 \le \Delta r \le 0.01 \qquad \text{Possible error}$$

To approximate the propagated error in the volume we differentiate $V$ to obtain

$$\frac{dV}{dr} = 4\pi r^2$$

and write

$$\Delta V \approx dV = 4\pi r^2 \, dr$$

$$= 4\pi (0.7)^2 (\pm 0.01)$$

$$\approx \pm 0.06158 \text{ cubic inch}$$

Would you say that the propagated error in Example 2 is large or small? The answer is best given in *relative* terms by comparing the ratio of $dV$ to $V$. This ratio is

$$\frac{dV}{V} = \frac{4\pi r^2 \, dr}{\frac{4}{3}\pi r^3}$$

$$= \frac{3 \, dr}{r}$$

$$\approx \frac{3}{0.7}(\pm 0.01)$$

$$\approx \pm 0.0429$$

which is called the **relative error**. The corresponding **percentage error** is

$$\frac{dV}{V}(100) \approx 4.29\%$$

We can use the definition of differentials to rewrite each of the derivative rules in **differential form.** For example, suppose $u$ and $v$ are differentiable functions of $x$. By the definition of differentials, we have $du = u' \, dx$ and $dv = v' \, dx$. Therefore, we can write the differential form of the Product Rule as follows:

$$d[uv] = \frac{d}{dx}[uv] \, dx \qquad \text{Differential of } uv$$

$$= [uv' + vu'] \, dx \qquad \text{Product Rule}$$

$$= uv' \, dx + vu' \, dx$$

$$= u \, dv + v \, du$$

The following summary lists the differential forms corresponding to the derivative rules we have studied so far in the text.

---

**General differential formulas**

($u$ and $v$ are differentiable functions of $x$.)

1. Constant Multiple Rule: $\quad d[cu] = c \, du$
2. Sum or Difference Rule: $\quad d[u \pm v] = du \pm dv$
3. Product Rule: $\quad d[uv] = u \, dv + v \, du$
4. Quotient Rule: $\quad d\left[\dfrac{u}{v}\right] = \dfrac{v \, du - u \, dv}{v^2}$
5. Constant Rule: $\quad d[c] = 0$
6. Power Rule: $\quad d[x^n] = nx^{n-1} \, dx$
$\qquad\qquad\qquad\qquad d[x] = dx$

---

In the next example, we compare the derivatives and differentials of several functions.

**EXAMPLE 3**

**Finding Differentials**

| *Function* | *Derivative* | *Differential* |
|---|---|---|
| (a) $y = x^2$ | $\dfrac{dy}{dx} = 2x$ | $dy = 2x \, dx$ |
| (b) $y = \dfrac{3x + 2}{5}$ | $\dfrac{dy}{dx} = \dfrac{3}{5}$ | $dy = \dfrac{3}{5} \, dx$ |
| (c) $y = 2x^2 - 3x$ | $\dfrac{dy}{dx} = 4x - 3$ | $dy = (4x - 3) \, dx$ |
| (d) $y = \dfrac{1}{x}$ | $\dfrac{dy}{dx} = -\dfrac{1}{x^2}$ | $dy = -\dfrac{dx}{x^2}$ |

The notation in Example 3 is called the **Leibniz notation** for derivatives and differentials and is named after the German mathematician Gottfried Wilhelm Leibniz (1646–1716). The beauty of this notation is that it provides us with an easy way to remember several important calculus formulas by making it seem as though the formulas were derived from algebraic manipulations of differentials. For example, in Leibniz notation, the Chain Rule is

$$\frac{dy}{dx} = \frac{dy}{du}\frac{du}{dx}$$

This formula would probably appear to be true to a student in elementary algebra. Of course, it appears to be true for the *wrong reason*—that we have simply canceled the *du*'s. Even so, the notation's many advantages overshadow the potential problem of drawing valid conclusions from invalid arguments.

**EXAMPLE 4**

**Differential of a Composite Function**

Use the Chain Rule to find $dy$ for

$$y = \sqrt{x^2 + 1}$$

**SOLUTION**

$$y = (x^2 + 1)^{1/2}$$

$$\frac{dy}{dx} = \left(\frac{1}{2}\right)(2x)(x^2 + 1)^{-1/2}$$

$$= \frac{x}{\sqrt{x^2 + 1}}$$

$$dy = \frac{x\,dx}{\sqrt{x^2 + 1}}$$

Differentials can be used to approximate function values. To do this for the function given by $y = f(x)$, we make use of the approximation

$$\Delta y = f(x + \Delta x) - f(x) \approx dy$$

to obtain

$$f(x + \Delta x) \approx f(x) + dy$$

$$= f(x) + f'(x)\,dx$$

The key to the use of this approximation technique lies in choosing a value for $x$ that makes the calculations easier. For instance, to approximate $\sqrt{101}$ we would let $x = 100$ and $\Delta x = 1$. This type of approximation is illustrated in the next example.

**EXAMPLE 5**

**Approximating Function Values**

Use differentials to approximate $\sqrt{16.5}$.

**SOLUTION**

We use the function

$$f(x) = \sqrt{x}$$

Then we write the approximation

$$f(x + \Delta x) \approx f(x) + f'(x)\, dx$$

$$= \sqrt{x} + \frac{1}{2\sqrt{x}}\, dx$$

Now, choosing $x = 16$ and $dx = 0.5$, we obtain

$$\sqrt{16.5} \approx \sqrt{16} + \frac{1}{2\sqrt{16}}(0.5)$$

$$\approx 4 + \left(\frac{1}{8}\right)\left(\frac{1}{2}\right)$$

$$= 4 + \frac{1}{16}$$

$$= 4.0625$$

**Remark:** Using a calculator, we obtain the value $\sqrt{16.5} \approx 4.0620$ which indicates the accuracy of the differential method in Example 5.

## SECTION EXERCISES 3.8

In Exercises 1–10, find the differential $dy$ of each function.

**1.** $y = 3x^2 - 4$  
**3.** $y = (2x + 5)^3$  
**5.** $y = \sqrt{x^2 + 1}$  
**7.** $y = x\sqrt{1 - x^2}$  
**9.** $y = \dfrac{x + 1}{2x - 1}$

**2.** $y = 2x^{3/2}$  
**4.** $y = (1 - 2x^2)^{-2}$  
**6.** $y = \sqrt[3]{6x^2}$  
**8.** $y = -4x^{-1}$  
**10.** $y = \dfrac{x}{x^2 + 1}$

In Exercises 11–14, let $x = 2$ and use the given function and value of $\Delta x = dx$ to complete the table.

| $dx = \Delta x$ | $dy$ | $\Delta y$ | $\Delta y - dy$ | $\dfrac{dy}{\Delta y}$ |
|---|---|---|---|---|
| 1.000 | | | | |
| 0.500 | | | | |
| 0.100 | | | | |
| 0.010 | | | | |
| 0.001 | | | | |

**11.** $y = x^2$  
**12.** $y = \dfrac{1}{x^2}$  
**13.** $y = x^5$  
**14.** $y = \sqrt{x}$

In Exercises 15–18, approximate the root of the given quantity using differentials. Compare the result with the value obtained using a calculator.

**15.** $\sqrt[4]{83}$  
**16.** $\sqrt[3]{63}$  
**17.** $\dfrac{1}{\sqrt{25}}$  
**18.** $\dfrac{1}{\sqrt{101}}$

**19.** The area of a square of side $x$ is given by $A = x^2$.
(a) Compute $dA$ and $\Delta A$ in terms of $x$ and $\Delta x$.
(b) In Figure 3.75, identify the region whose area is $dA$.
(c) In Figure 3.75, identify the region whose area is $\Delta A - dA$.

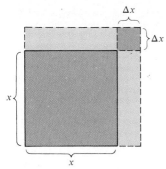

FIGURE 3.75

**20.** The measurement of the side of a square is found to be 12 inches, with a possible error of $\frac{1}{64}$ inch. Use differentials to approximate the possible error and the percentage error in computing the area of the square.

**21.** The measurement of the radius of a circle is found to be 14 inches, with a possible error of $\frac{1}{4}$ inch. Use differentials to approximate the possible error and the percentage error in computing the area of the circle.

**22.** The measurement of the edge of a cube is found to be 12 inches, with a possible error of 0.03 inch. Use differentials to approximate the possible error and the relative error in computing the following:
(a) the volume of the cube
(b) the surface area of the cube

**23.** The radius of a sphere is measured to be 6 inches, with a possible error of 0.02 inch. Use differentials to approximate the possible error and the relative error in calculating the volume of the sphere.

**24.** The revenue $R$ for a company selling $x$ units is given by

$$R = 900x - 0.1x^2$$

Use differentials to approximate the change and the percentage change in revenue if sales increase from $x = 3000$ to $x = 3100$ units.

**25.** The profit $P$ for a company selling $x$ units is given by

$$P = (500x - x^2) - \left(\frac{1}{2}x^2 - 77x + 3000\right)$$

Approximate the change and the percentage change in profit as production changes from $x = 115$ to $x = 120$ units.

**26.** The cost in dollars of removing $p\%$ of the air pollutants in the stack emission of a utility company that burns coal to generate electricity is

$$C = \frac{80,000p}{100 - p}, \quad 0 \le p < 100$$

Use differentials to approximate the increase in cost if the government requires the utility company to remove 2% more of the pollutants and $p$ is currently
(a) 40%     (b) 75%

**27.** The game commission in a certain state introduces 50 deer into newly acquired state game lands. The population $N$ of the herd is given by the model

$$N = \frac{10(5 + 3t)}{1 + 0.04t}$$

where $t$ is time in years. Use differentials to approximate the change in the herd size from year $t = 5$ to year $t = 6$.

**28.** The concentration $C$ of a certain chemical in the bloodstream $t$ hours after injection into muscle tissue is given by

$$C = \frac{3t}{27 + t^3}$$

Use differentials to approximate the change in the concentration when $t$ changes from $t = 1$ to $t = \frac{3}{2}$.

## Important Terms

Increasing function
Decreasing function
Critical number
Relative maximum
Relative minimum
Relative extremum
Absolute extremum
Concave upward
Concave downward
Point of inflection
Optimization problem
Primary equation
Secondary equation
Total profit, $P$
Total revenue, $R$
Total cost, $C$
Average cost, $\overline{C}$
Price per unit, $p$

Number of units, $x$
Demand function, $p = f(x)$
Price elasticity of demand, $\eta$
Inelastic demand, $|\eta| < 1$
Elastic demand, $|\eta| > 1$
Vertical asymptote
Infinite limit
Horizontal asymptote
Limit at infinity
Differential of $x$, $dx$
Differential of $y$, $dy$
Change in $x$, $\Delta x$
Change in $y$, $\Delta y$
Linear approximation
Propagated error
Relative error
Percentage error

## Important Techniques

Finding intervals on which a function is increasing or
   decreasing
Finding critical numbers of a function
Finding relative extrema of a function
Finding extrema on a closed interval
Determining concavity and finding points of inflection
Solving optimization problems
Finding vertical and horizontal asymptotes
Finding infinite limits and limits at infinity
Sketching the graph of a function
Finding differentials
Using the differential $dy$ to approximate $\Delta y$

## Important Formulas

Test for an increasing or decreasing function, differentiable on $(a, b)$:

1. If $f'(x) > 0$ for all $x$ in $(a, b)$, then $f$ is increasing on $(a, b)$.
2. If $f'(x) < 0$ for all $x$ in $(a, b)$, then $f$ is decreasing on $(a, b)$.
3. If $f'(x) = 0$ for all $x$ in $(a, b)$, then $f$ is constant on $(a, b)$.

Relative extrema of $f$ can occur only at critical numbers of $f$

First-Derivative Test for Relative Extrema

Guidelines for finding (absolute) extrema on a closed interval

Test for concavity

Points of inflection of $f$ can occur only at critical numbers of $f'$

Second-Derivative Test for Relative Extrema ($c$ is a critical number of $f$):

1. If $f''(c) > 0$, then $f(c)$ is a relative minimum.
2. If $f''(c) < 0$, then $f(c)$ is a relative maximum.
3. If $f''(c) = 0$, then the test fails.

Guidelines for solving optimization problems
Suggestions for sketching the graph of a function:

1. Make a rough preliminary sketch, using any easily determined intercepts, vertical or horizontal asymptotes, and points of discontinuity.
2. Locate the $x$-values where $f'(x)$ and $f''(x)$ are either zero or undefined.
3. Test the behavior of $f$ at each of these $x$-values as well as *within* each interval determined by them.
4. Sharpen the accuracy of the final sketch by plotting the relative extrema points, the points of inflection, and a few points between.

$$R = xp$$

$$\overline{C} = \frac{C}{x}$$

$$P = R - C$$

$$\eta = \frac{p/x}{dp/dx}$$

$$\frac{dR}{dx} = \text{marginal revenue}$$

$$\frac{dC}{dx} = \text{marginal cost}$$

$$\frac{dP}{dx} = \text{marginal profit}$$

Limits at infinity:

If $r$ is a positive real number, then

$$\lim_{x \to \infty} \frac{1}{x^r} = 0 \quad \text{and} \quad \lim_{x \to -\infty} \frac{1}{x^r} = 0$$

Limits at infinity for rational functions:

$$\lim_{x \to \pm\infty} \frac{a_n x^n + a_{n-1}x^{n-1} + \cdots + a_0}{b_m x^m + b_{m-1}x^{m-1} + \cdots + b_0} = \begin{cases} 0, & \text{if } n < m \\ \dfrac{a_n}{b_m}, & \text{if } n = m \\ \pm\infty, & \text{if } n > m \end{cases}$$

Differential of $y$:

$$dy = f'(x)\, dx$$

Differential formulas:

Constant Multiple Rule: $d[cu] = c\, du$

Sum or Difference Rule: $d[u \pm v] = du \pm dv$

Product Rule: $d[uv] = u\, dv + v\, du$

Quotient Rule: $d\left[\dfrac{u}{v}\right] = \dfrac{v\, du - u\, dv}{v^2}$

Constant Rule: $d[c] = 0$

Power Rule: $d[x^n] = nx^{n-1}\, dx,\ d[x] = dx$

In Exercises 1–8, make use of domain, range, asymptotes, and intercepts to match the given function with the correct graph. [Graphs are labeled (a), (b), (c), (d), (e), (f), (g) and (h).]

**1.** $f(x) = \dfrac{4}{x^2 + 4}$

**2.** $f(x) = \dfrac{x}{x^2 + 4}$

**3.** $g(x) = \dfrac{x}{x^2 - 4}$

**4.** $g(x) = \dfrac{4}{x^2 - 4}$

**5.** $f(x) = \dfrac{3x^2}{x^2 + 4}$

**6.** $f(x) = \dfrac{x^2 + 4}{x^2}$

**7.** $h(x) = \sqrt{x^2 - 4}$

**8.** $h(x) = \sqrt{4 - x^2}$

(a)

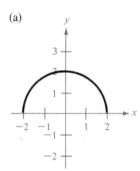

(b)

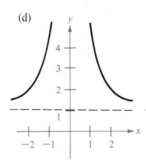

(c)

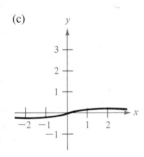

(d)

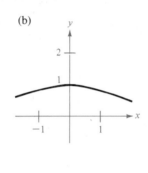

(e)

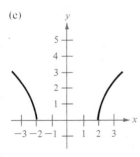

(f)

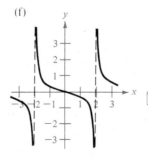

(g)

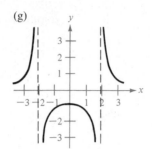

(h)

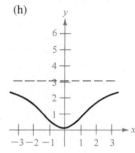

In Exercises 9–14, find the limit (if it exists).

**9.** $\displaystyle\lim_{x \to 0^+} \left( x - \dfrac{1}{x^3} \right)$

**10.** $\displaystyle\lim_{x \to -1^+} \dfrac{x^2 - 2x + 1}{x + 1}$

**11.** $\displaystyle\lim_{x \to 1/2} \dfrac{2x - 1}{6x - 3}$

**12.** $\displaystyle\lim_{x \to \infty} \dfrac{2x - 1}{6x - 3}$

**13.** $\displaystyle\lim_{x \to \infty} \dfrac{5x^2 + 3}{2x^2 - x + 1}$

**14.** $\displaystyle\lim_{x \to \infty} \dfrac{3x^2 - 2x + 3}{x + 1}$

In Exercises 15–30, make use of domain, range, asymptotes, intercepts, relative extrema, and points of inflection (when appropriate) to sketch the graph of the given function.

**15.** $f(x) = 4x - x^2$

**16.** $f(x) = 4x^3 - x^4$

**17.** $f(x) = x\sqrt{16 - x^2}$

**18.** $f(x) = x + \dfrac{4}{x^2}$

**19.** $f(x) = \dfrac{x + 1}{x - 1}$

**20.** $f(x) = x^2 + \dfrac{2}{x}$

**21.** $f(x) = x^3 + x + \dfrac{4}{x}$

**22.** $f(x) = x^3(x + 1)$

**23.** $f(x) = (x - 1)(x - 4)^2$

**24.** $f(x) = (x - 3)(x + 3)^2$

**25.** $f(x) = (5 - x)^3$

**26.** $f(x) = (x^2 - 4)^2$

**27.** $f(x) = x^3 + \dfrac{243}{x}$

**28.** $f(x) = \dfrac{2x}{1 + x^2}$

**29.** $f(x) = x^{4/5}$

**30.** $f(x) = |9 - x^2|$

**31.** The sides of a right triangle (in the first quadrant) lie on the coordinate axes, and its hypotenuse passes through the point (1, 8). Find the vertices of the triangle so that the length of the hypotenuse is minimum.

**32.** The wall of a building is to be braced by a beam that must pass over a parallel fence 5 feet high and 4 feet from the building. Find the length of the shortest beam that can be used.

**33.** Find the dimensions of the rectangle of maximum area, with sides parallel to the coordinate axes, that can be inscribed in the ellipse given by

$$\frac{x^2}{144} + \frac{y^2}{16} = 1$$

**34.** Show that the area of a rectangle inscribed in a triangle is at most one-half that of the triangle.

**35.** Find the maximum profit corresponding to a demand function of $p = 36 - 4x$ and a total cost function of $C = 2x^2 + 6$.

**36.** The cost $C$ of producing $x$ units per day is $C = \frac{1}{3}x^2 + 62x + 125$ and the price $p$ per unit is $p = 75 - x$.
  (a) What should be the daily output to obtain maximum profit?
  (b) What should be the daily output to obtain minimum average cost?
  (c) Find the elasticity of demand.

**37.** For groups of 80 or more, a charter bus company determines the rate per person according to the following formula:

$$\text{Rate} = \$8.00 - \$0.05(n - 80), \quad n \geq 80$$

What number of passengers will give the bus company maximum revenue?

**38.** The cost of fuel to run a locomotive is proportional to the $\frac{3}{2}$ power of the speed and is $50 per hour for a speed of 25 mi/hr. Other fixed costs amount to an average of $100 per hour. Find the speed that will minimize the cost per mile.

**39.** The cost of inventory depends on ordering cost and storage cost. In the following inventory model, assume that sales occur at a constant rate, $Q$ is the number of units sold per year, $r$ is the cost of storing 1 unit for 1 year, $s$ is the cost of placing an order, and $x$ is the number of units per order.

$$C = \left(\frac{Q}{x}\right)s + \left(\frac{x}{2}\right)r$$

Determine the order size that will minimize the cost.

**40.** The demand and cost equations for a certain product are $p = 600 - 3x$ and $C = 0.3x^2 + 6x + 600$, where $p$ is the price per unit, $x$ is the number of units, and $C$ is the cost of producing $x$ units. If $t$ is the excise tax per unit, the profit for producing $x$ units is $P = xp - C - xt$. Find the maximum profit for
  (a) $t = 5$     (b) $t = 10$     (c) $t = 20$

**41.** If a 1% error is made in measuring the edge of a cube, approximately what percentage error will be made in calculating the surface area and the volume of the cube?

**42.** The diameter of a sphere is measured to be 18 inches with a maximum possible error of 0.05 inch. Use differentials to approximate the possible error in the surface area and the volume of the sphere.

**43.** A company finds that the demand for its commodity is given by $p = 75 - \frac{1}{4}x$. If $x$ changes from 7 to 8, find the corresponding change in $p$. Compare the values of $\Delta p$ and $dp$.

# Integration

# Antiderivatives and the Indefinite Integral

## INTRODUCTORY EXAMPLE
### Average Weekly Salary for U.S. Workers

The average weekly earnings for full-time (nonsupervisory) workers in the United States increased from $85.91 in 1962 to $235.10 in 1980. If we represent the weekly earnings by $S$, then the rate of change of $S$ during this period is approximated by the mathematical model

$$\frac{dS}{dt} = 0.0273t^2 + 0.2708t + 2.1336$$

where $S$ is measured in dollars and $t$ in years, with $t = 1$ corresponding to 1962. Thus, at the beginning of 1962, weekly salaries were increasing at a rate of $2.43 per year. To estimate the value of $S$ for 1963, we could add the mid-year ($t = 1.5$) estimated increase to the weekly earnings for 1962 to obtain an average weekly salary in 1963:

1963 salary
$$= \text{(1962 salary)} + \text{(mid-year raise estimate)}$$
$$= 85.91 + 0.0273(1.5)^2 + 0.2708(1.5) + 2.1336$$
$$= 85.91 + 2.60$$
$$= \$88.51$$

Suppose that we wanted to use this model to estimate the average weekly salary in the year 2000. We could continue the process shown for estimating the 1963 salary, but that would be tedious. A simpler way would be to find the weekly salary function $S$ from its derivative (rate of change) $dS/dt$. We call the function $S$ an **antiderivative** of $dS/dt$. Using techniques presented in this section, we find the weekly salary function to be

$$S = 0.0091t^3 + 0.1354t^2 + 2.1336t + 83.2653$$

Finally, letting $t = 39$ (for the year 2000), we approximate the weekly salary in the year 2000 to be $912.22, as shown in Figure 4.1.

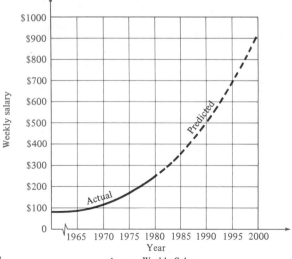

FIGURE 4.1                    Average Weekly Salary

- **Antiderivatives**
- **Notation for Antidifferentiation**
- **Basic Integration Rules**
- **Initial Conditions and Particular Solutions**

Up to this point in our study of calculus, we have been concerned primarily with this problem: Given a function, find its derivative. Many important applications of calculus involve the inverse problem: Given the derivative of a function, find the original function. For example, suppose we are given the following derivatives:

$$f'(x) = 2, \qquad g'(x) = 3x^2, \qquad s'(t) = 4t$$

Our problem is to determine the functions $f$, $g$, and $s$ that have these respective derivatives. If we make some educated guesses, we might come up with the following functions:

$$f(x) = 2x \qquad \text{because} \qquad \frac{d}{dx}[2x] = 2$$

$$g(x) = x^3 \qquad \text{because} \qquad \frac{d}{dx}[x^3] = 3x^2$$

$$s(t) = 2t^2 \qquad \text{because} \qquad \frac{d}{dt}[2t^2] = 4t$$

This operation of determining the original function from its derivative is the inverse operation of differentiation, and we call it **antidifferentiation.**

| **Definition of an antiderivative** | A function $F$ is called an **antiderivative** of a function $f$ if for every $x$ in the domain of $f$ $$F'(x) = f(x)$$ |
|---|---|

**Remark:** We will use the phrase "$F(x)$ is an antiderivative of $f(x)$" synonymously with "$F$ is an antiderivative of $f$."

If $F(x)$ is an antiderivative of $f(x)$, then $F(x) + C$, where $C$ is any constant, is also an antiderivative of $f(x)$. For instance,

$$F(x) = x^3, \qquad G(x) = x^3 - 5, \qquad H(x) = x^3 + 0.3$$

are all antiderivatives of $3x^2$, because

$$\frac{d}{dx}[x^3] = \frac{d}{dx}[x^3 - 5] = \frac{d}{dx}[x^3 + 0.3] = 3x^2$$

As it turns out, *all* the antiderivatives of $3x^2$ are of the form $x^3 + C$. The point is that the process of antidifferentiation does not determine a unique function but rather it determines a *family* of functions, each differing from the others by a constant.

## Notation for Antiderivatives

The antidifferentiation process is also called **integration** and is denoted by the symbol

$$\int$$

called an **integral sign.** The symbol

$$\int f(x)\, dx$$

is called the **indefinite integral** of $f(x)$, and it denotes the family of antiderivatives of $f(x)$. That is, if $F'(x) = f(x)$ for all $x$, then

$$\int f(x)\, dx = F(x) + C$$

where $f(x)$ is called the **integrand** and $C$ the **constant of integration.**

The differential $dx$ in the indefinite integral identifies the variable of integration. That is, the symbol $\int f(x)\, dx$ denotes the "antiderivative of $f$ *with respect to x*" just as the symbol $dy/dx$ denotes the "derivative of $y$ *with respect to x.*"

**Definition of integral notation for antiderivatives**

The notation

$$\int f(x)\, dx = F(x) + C$$

where $C$ is an arbitrary constant, means that $F$ is an antiderivative of $f$. That is, $F'(x) = f(x)$ for all $x$ in the domain of $f$.

The inverse nature of the operations of integration and differentiation can be shown symbolically as follows.

$$\frac{d}{dx}\left[\int f(x)\, dx\right] = f(x)$$ 
   Differentiation is the inverse of integration

$$\int f'(x)\, dx = f(x) + C$$ 
   Integration is the inverse of differentiation

This inverse relationship between integration and differentiation allows us to obtain integration formulas directly from differentiation formulas. In the following summary we list the integration formula that corresponds to each of the differentiation formulas we have studied up to this point.

| **Basic integration rules** | 1. Constant Rule: | $\int k\,dx = kx + C$ |
| | 2. Constant Multiple Rule: | $\int kf(x)\,dx = k\int f(x)\,dx$ |
| | 3. Sum Rule: | $\int [f(x) \pm g(x)]\,dx = \int f(x)\,dx \pm \int g(x)\,dx$ |
| | 4. Simple Power Rule: | $\int x^n\,dx = \dfrac{x^{n+1}}{n+1} + C,\ n \neq -1$ |

**Remark:** Be sure you see that the Simple Power Rule has the restriction that $n$ cannot be $-1$. This means that we *cannot* use the Simple Power Rule to evaluate the integral

$$\int \frac{1}{x}\,dx$$

To evaluate this integral, we must wait until the natural logarithmic function is introduced in Section 5.4.

Applications of these rules are demonstrated in the following examples.

**EXAMPLE 1**

**Evaluating Indefinite Integrals**

Evaluate the indefinite integral

$$\int 3x\,dx$$

**SOLUTION**

$$\int 3x\,dx = 3\int x\,dx \qquad \text{Constant Multiple Rule}$$

$$= 3\int x^1\,dx \qquad \text{Rewrite } x = x^1$$

$$= 3\left(\frac{x^2}{2}\right) + C \qquad \text{Power Rule } (n = 1)$$

$$= \frac{3}{2}x^2 + C \qquad \text{Simplify}$$

**Remark:** In evaluating indefinite integrals a strict application of the basic integration rules tends to produce messy constants of integration. For instance, in Example 1, we could have written

$$\int 3x\,dx = 3\int x\,dx = 3\left(\frac{x^2}{2} + C\right) = \frac{3}{2}x^2 + 3C$$

However, since $C$ represents *any* constant, it is both cumbersome and unnecessary to write $3C$ as the constant of integration, and we opt for the simpler $\frac{3}{2}x^2 + C$.

In Example 1, note that the general pattern of integration is similar to that of differentiation.

$$\boxed{\text{Given integral}} \rightarrow \boxed{\text{Rewrite}} \rightarrow \boxed{\text{Integrate}} \rightarrow \boxed{\text{Simplify}}$$

This pattern is followed in the next example.

**EXAMPLE 2**

**Applying the Basic Integration Rules**

| Given integral | Rewrite | Integrate | Simplify |
|---|---|---|---|
| (a) $\displaystyle\int \frac{1}{x^3}\, dx$ | $\displaystyle\int x^{-3}\, dx$ | $\dfrac{x^{-2}}{-2} + C$ | $-\dfrac{1}{2x^2} + C$ |
| (b) $\displaystyle\int \sqrt{x}\, dx$ | $\displaystyle\int x^{1/2}\, dx$ | $\dfrac{x^{3/2}}{3/2} + C$ | $\dfrac{2}{3}x^{3/2} + C$ |

With the four basic integration rules, we can integrate *any* polynomial function. This is demonstrated in the next example.

**EXAMPLE 3**

**Integrating Polynomial Functions**

Evaluate the following integrals:

(a) $\displaystyle\int 1\, dx$   (b) $\displaystyle\int (x + 2)\, dx$   (c) $\displaystyle\int (3x^4 - 5x^2 + x)\, dx$

**SOLUTION**

(a) By the Constant Rule we have

$$\int 1\, dx = x + C$$

(b) Using the Sum Rule, we integrate each part separately.

$$\int (x + 2)\, dx = \int x\, dx + \int 2\, dx$$

$$= \frac{x^2}{2} + 2x + C$$

(c) Try to identify each basic integration rule used to evaluate this integral.

$$\int (3x^4 - 5x^2 + x)\, dx = 3\left(\frac{x^5}{5}\right) - 5\left(\frac{x^3}{3}\right) + \frac{x^2}{2} + C$$

$$= \frac{3}{5}x^5 - \frac{5}{3}x^3 + \frac{1}{2}x^2 + C$$

**Remark:** The integral in Example 3(a) is usually simplified to the form

$$\int 1 \, dx = \int dx$$

## EXAMPLE 4

### Rewriting Before Integrating

Evaluate the indefinite integral

$$\int \frac{x + 1}{\sqrt{x}} \, dx$$

**SOLUTION**

We begin by rewriting the quotient in the integrand as a sum.

$$\int \frac{x + 1}{\sqrt{x}} \, dx = \int \left( \frac{x}{\sqrt{x}} + \frac{1}{\sqrt{x}} \right) dx$$

Then, rewriting each term using fractional exponents, we have

$$\int \left( \frac{x}{\sqrt{x}} + \frac{1}{\sqrt{x}} \right) dx = \int (x^{1/2} + x^{-1/2}) \, dx$$

$$= \frac{x^{3/2}}{3/2} + \frac{x^{1/2}}{1/2} + C$$

$$= \frac{2}{3} x^{3/2} + 2x^{1/2} + C$$

**Remark:** When integrating quotients, don't make the mistake of integrating the numerator and the denominator separately. This is no more valid in integration than it is in differentiation. For instance, in Example 4 be sure you see that

$$\int \frac{x + 1}{\sqrt{x}} \, dx \neq \frac{\int (x + 1) \, dx}{\int \sqrt{x} \, dx}$$

### Particular Solutions

We have already seen that the equation $y = \int f(x) \, dx$ has many solutions, each differing by a constant. This means that the graphs of any two antiderivatives of $f$ are vertical translations of each other. For example, Figure 4.2 shows the graphs of several antiderivatives of the form

$$y = \int (3x^2 - 1) \, dx = x^3 - x + C$$

for various integer values of $C$. Each of these antiderivatives is a solution of

$$\frac{dy}{dx} = 3x^2 - 1$$

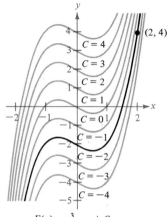

FIGURE 4.2     $F(x) = x^3 - x + C$

In many applications of integration, we are given enough information to determine a **particular solution.** To do so, we need only know the value of $F(x)$ for one value of $x$. (This information is called an **initial condition.**) For example, in Figure 4.2, there is only one curve that passes through the point $(2, 4)$. To find this particular curve, we use the information

$$F(x) = x^3 - x + C \qquad \text{General solution}$$

$$F(2) = 4 \qquad \text{Initial condition}$$

By using the initial condition in the general solution, we determine that $F(2) = 8 - 2 + C = 4$, which implies that $C = -2$. Thus, we obtain

$$F(x) = x^3 - x - 2 \qquad \text{Particular solution}$$

**EXAMPLE 5**

**Finding a Particular Solution**

Find the general solution of the equation

$$F'(x) = \frac{1}{x^2}$$

and find the particular solution that satisfies the initial condition $F(1) = 2$.

**SOLUTION**

To find the general solution, we integrate as follows.

$$F(x) = \int \frac{1}{x^2}\, dx$$

$$= \int x^{-2}\, dx$$

$$= \frac{x^{-1}}{-1} + C$$

$$= -\frac{1}{x} + C$$

Now, since $F(1) = 2$, we write

$$F(1) = -\frac{1}{1} + C = 2$$

which implies that $C = 3$. Thus, the particular solution, as shown in Figure 4.3, is

$$F(x) = -\frac{1}{x} + 3$$

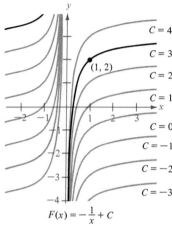

$$F(x) = -\frac{1}{x} + C$$

FIGURE 4.3

**EXAMPLE 6**

**A Business Application**

Suppose the marginal cost for producing $x$ units of a commodity is given by

$$\frac{dC}{dx} = 32 - 0.04x$$

If it costs $50 to make 1 unit, find the total cost of making 200 units.

**SOLUTION**

Since $dC/dx$ is the derivative of the total cost function, we have

$$C = \int (32 - 0.04x)\, dx$$

$$= 32x - 0.04\left(\frac{x^2}{2}\right) + K$$

$$= \underbrace{32x - 0.02x^2}_{\text{Total variable cost}} + \underbrace{K}_{\text{Total fixed cost}}$$

When $x = 1$, $C = 50$. Thus

$$50 = 32(1) - 0.02(1)^2 + K$$

$$18.02 = K$$

Therefore, the total cost function is

$$C = 32x - 0.02x^2 + 18.02$$

The cost of making 200 units is therefore

$$C = 32(200) - 0.02(40,000) + 18.02$$

$$= \$5,618.02$$

---

**EXAMPLE 7**

**An Application Involving Gravity**

A ball is thrown upward with an initial velocity of 64 ft/sec from a height of 80 feet, as shown in Figure 4.4. Find the position function for this motion.

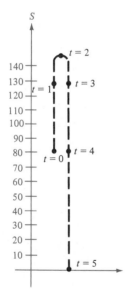

FIGURE 4.4

**SOLUTION**

To begin we let $t = 0$ represent the initial time. Then the two stated conditions in the problem can be written as follows:

Initial height is 80 feet  ⟹  $s(0) = 80$

Initial velocity is 64 ft/sec  ⟹  $s'(0) = 64$

Furthermore, we know that the acceleration due to gravity is $-32$ ft/sec$^2$, so we can write the following equation for the second derivative of $s$.

Acceleration due to gravity is $-32$ ft/s$^2$  ⟹  $s''(t) = -32$

---

Now, integrating $s''(t)$, we have

$$s'(t) = \int s''(t)\, dt$$

$$= \int -32\, dt$$

$$= -32t + C_1$$

where $C_1 = s'(0) = 64$. Similarly, integrating $s'(t)$, we have

$$s(t) = \int s'(t)\, dt$$

$$= \int (-32t + 64)\, dt$$

$$= -16t^2 + 64t + C_2$$

where $C_2 = s(0) = 80$. Therefore, the position function is

$$s(t) = -16t^2 + 64t + 80$$

Before beginning the exercise set for this section, be sure you realize that one of the most important steps in finding antiderivatives is *rewriting the integrand* in a form that fits the basic integration rules. To further illustrate this point, we list several additional examples in Table 4.1.

TABLE 4.1

| Given | Rewrite | Integrate | Simplify |
|---|---|---|---|
| $\int \dfrac{2}{\sqrt{x}}\, dx$ | $2\int x^{-1/2}\, dx$ | $2\left(\dfrac{x^{1/2}}{1/2}\right) + C$ | $4x^{1/2} + C$ |
| $\int (x^2 + 1)^2\, dx$ | $\int (x^4 + 2x^2 + 1)\, dx$ | $\dfrac{x^5}{5} + 2\left(\dfrac{x^3}{3}\right) + x + C$ | $\dfrac{x^5}{5} + \dfrac{2x^3}{3} + x + C$ |
| $\int \dfrac{x^3 + 3}{x^2}\, dx$ | $\int (x + 3x^{-2})\, dx$ | $\dfrac{x^2}{2} + 3\left(\dfrac{x^{-1}}{-1}\right) + C$ | $\dfrac{x^2}{2} - \dfrac{3}{x} + C$ |
| $\int \sqrt[3]{x}(x - 4)\, dx$ | $\int (x^{4/3} - 4x^{1/3})\, dx$ | $\dfrac{x^{7/3}}{7/3} - 4\left(\dfrac{x^{4/3}}{4/3}\right) + C$ | $\dfrac{3x^{4/3}}{7}(x - 7) + C$ |

# SECTION EXERCISES 4.1

In Exercises 1–10, evaluate the indefinite integral and check your result by differentiation.

**1.** $\int 6\, dx$

**2.** $\int -4\, dx$

**3.** $\int 3t^2\, dt$

**4.** $\int t^4\, dt$

**5.** $\int 3x^{-4} \, dx$

**6.** $\int 4y^{-3} \, dy$

**7.** $\int du$

**8.** $\int \pi \, dt$

**9.** $\int x^{3/2} \, dx$

**10.** $\int v^{-1/2} \, dv$

In Exercises 11–16, complete the table using Example 2 as a model.

| Given | Rewrite | Integrate | Simplify |
|---|---|---|---|

**11.** $\int \sqrt[3]{x} \, dx$

**12.** $\int \frac{1}{x^2} \, dx$

**13.** $\int \frac{1}{x\sqrt{x}} \, dx$

**14.** $\int x(x^2 + 3) \, dx$

**15.** $\int \frac{1}{2x^3} \, dx$

**16.** $\int \frac{1}{(2x)^3} \, dx$

In Exercises 17–34, evaluate the indefinite integral and check your result by differentiation.

**17.** $\int (x^3 + 2) \, dx$

**18.** $\int (x^2 - 2x + 3) \, dx$

**19.** $\int (x^{3/2} + 2x + 1) \, dx$

**20.** $\int \left( \sqrt{x} + \frac{1}{2\sqrt{x}} \right) dx$

**21.** $\int \sqrt[3]{x^2} \, dx$

**22.** $\int (\sqrt[4]{x^3} + 1) \, dx$

**23.** $\int \frac{1}{x^3} \, dx$

**24.** $\int \frac{1}{x^4} \, dx$

**25.** $\int \frac{1}{4x^2} \, dx$

**26.** $\int (2x + x^{-1/2}) \, dx$

**27.** $\int \frac{t^2 + 2}{t^2} \, dt$

**28.** $\int \frac{x^2 + 1}{x^2} \, dx$

**29.** $\int u(3u^2 + 1) \, du$

**30.** $\int \sqrt{x}(x + 1) \, dx$

**31.** $\int (x + 1)(3x - 2) \, dx$

**32.** $\int (2t^2 - 1)^2 \, dt$

**33.** $\int y^2 \sqrt{y} \, dy$

**34.** $\int (1 + 3t)t^2 \, dt$

In Exercises 35 and 36, find the equation of the curve, given the derivative and the indicated point on the curve.

**35.** $\frac{dy}{dx} = 2x - 1$

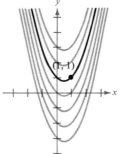
(1, 1)

**36.** $\frac{dy}{dx} = 2(x - 1)$

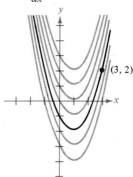
(3, 2)

In Exercises 37–40, find $y = f(x)$ satisfying the given conditions.

**37.** $f''(x) = 2, \, f'(2) = 5, \, f(2) = 10$

**38.** $f''(x) = x^2, \, f'(0) = 6, \, f(0) = 3$

**39.** $f''(x) = x^{-3/2}, \, f'(4) = 2, \, f(0) = 0$

**40.** $f''(x) = x^{-3/2}, \, f'(1) = 2, \, f(9) = -4$

In Exercises 41 and 42, find the revenue and demand functions for the given marginal revenue. (Use the fact that $R = 0$ when $x = 0$.)

**41.** $\frac{dR}{dx} = 500 - 5x$

**42.** $\frac{dR}{dx} = 100 - 6x - 2x^2$

**43.** A company produces a product for which the marginal cost of producing $x$ units is

$$\frac{dC}{dx} = 2x - 12$$

and fixed costs are $125.
   (a) Find the total cost function and the average cost function.
   (b) Find the total cost of producing 50 units.

**44.** A company produces a product for which the marginal cost of producing $x$ units is

$$\frac{dC}{dx} = k, \quad k \text{ is a constant}$$

(That is, for each unit increase in output the increase in cost is always $k$.) Describe the cost function for this product.

**45.** An evergreen nursery usually sells a certain type of shrub after 6 years of growth and shaping. The growth rate after $t$ years is given by

$$\frac{dh}{dt} = 0.5t + 2$$

where $t = 0$ represents the time when the shrubs are 5-inch seedlings ($h = 5$ when $t = 0$).
(a) Find the height $h$ after $t$ years.
(b) How tall are the shrubs when they are sold?

**46.** It is predicted that the growth rate of the population $P$ of a city $t$ years from now will be

$$\frac{dP}{dt} = 500t^{1.06}$$

Estimate the population of the city 10 years from now if the present population is 50,000.

In Exercises 47–50, use $a(t) = -32$ ft/s$^2$ as the acceleration due to gravity. (Neglect air resistance.)

**47.** A ball is thrown vertically upward with an initial velocity of 60 ft/sec. How high will the ball go?

**48.** An object is dropped from a balloon, which is stationary at a height of 1600 feet. Express the height of the object as a function of $t$. How long will it take the object to reach the ground?

**49.** With what initial velocity must an object be thrown upward from the ground to reach the height of the Washington Monument (550 ft)?

**50.** A balloon, rising vertically with a velocity of 16 ft/sec, releases a sandbag at an instant when the balloon is 64 feet above the ground.
(a) How many seconds after its release will the bag strike the ground?
(b) With what velocity will it strike the ground?

# The General Power Rule

In 1980, the U.S. poverty level for a family of four was around $10,000 (after taxes). Families at or below this income level tend to consume 100% of their income—that is, they use all their income to purchase necessities such as food, clothing, and shelter. As income level increases, the average consumption begins to drop below 100%. For instance, a family earning $11,000 (after taxes) may be able to save $185 and thus consume only $10,815 (98.3%) of their income. Although both the consumption $Q$ and savings tend to increase as the income $x$ increases, the ratio of consumption to income tends to decrease, and we call the rate of change of consumption with respect to income the *marginal propensity to consume*.

Suppose that the marginal propensity to consume an *after-tax* income $x$ is given by the model

$$\frac{dQ}{dx} = \frac{0.97}{(x - 9999)^{0.03}}, \quad 10,000 \le x$$

where $Q$ represents the income consumed. Note that as the income $x$ increases, $dQ/dx$ decreases. How can we use this model to predict the amount a family with an after-tax income of $20,000 will consume? To do this, we observe that $Q$, the income consumed, is an antiderivative of $dQ/dx$. Furthermore, since $Q = \$10,000$ when $x = \$10,000$, we can apply the **General Power Rule** introduced in this section to obtain

$$Q = 9999 + (x - 9999)^{0.97}, \quad 10,000 \le x$$

Thus, from this model we predict that a family with an after-tax income of $20,000 will consume roughly $17,586 and save the remaining $2,414. Figure 4.5 shows this consumption function for after-tax incomes between $10,000 and $150,000.

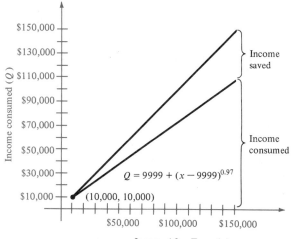

FIGURE 4.5

■ **The General Power Rule**
■ **Formal Substitution**

In Section 4.1, we used the Simple Power Rule

$$\int x^n \, dx = \frac{x^{n+1}}{n+1} + C, \quad n \neq -1$$

to find antiderivatives of functions expressed as powers of $x$ alone. In this section, we look at a technique for finding antiderivatives for more general functions.

### Development of the General Power Rule

To begin, consider how you might evaluate the following integral:

$$\int 2x(x^2 + 1)^3 \, dx$$

You could expand the integrand into polynomial form and then use the Simple Power Rule on the individual powers of $x$. *But* there is a simpler way to evaluate this integral! To see it, you need to remember that what we are hunting for is a function $f$ such that

$$f'(x) = 2x(x^2 + 1)^3$$

After some experimentation you might observe that

$$\frac{d}{dx}[(x^2 + 1)^4] = 4(x^2 + 1)^3(2x)$$

and that by dividing by 4 you could obtain

$$\frac{d}{dx}\left[\frac{(x^2 + 1)^4}{4}\right] = (x^2 + 1)^3(2x)$$

Now, having found a function with the desired derivative, you could conclude that

$$\int 2x(x^2 + 1)^3 \, dx = \frac{(x^2 + 1)^4}{4} + C$$

Be sure you see that the key to this solution is the presence of the factor $2x$ in the integrand. The importance of $2x$ as a factor of the integrand lies in the fact that it is precisely the derivative of $x^2 + 1$. That is, if

$$u = x^2 + 1$$

then

$$\frac{du}{dx} = 2x$$

Thus, written in terms of $u$, the original integral becomes

$$\int \underbrace{2x(x^2 + 1)^3} \, dx = \int u^3 \frac{du}{dx} \, dx = \frac{u^4}{4} + C$$

This is just one example of an important integration rule called the **General Power Rule.**

---

**General Power Rule for integration**

If $u$ is a differentiable function of $x$, then

$$\int u^n \frac{du}{dx} \, dx = \frac{u^{n+1}}{n + 1} + C, \quad n \neq -1$$

---

An important consideration that is often overlooked in using the General Power Rule is the existence of $du/dx$ as a factor of the integrand. We must first determine $u$ by identifying within the integrand a function $u$ that is raised to a power. Then we must show that its derivative $du/dx$ is also a factor of the integrand. Simply stated, the clues for using the General Power Rule for integration are as follows:

1. Identify the function $u$ that is raised to a power.
2. Check to see if $du/dx$ is also a factor of the integrand.

We demonstrate these steps in the next example.

**EXAMPLE 1**

**Applying the General Power Rule**

Use the General Power Rule to evaluate the following integrals:

(a) $\displaystyle\int 3(3x - 1)^4 \, dx$

(b) $\displaystyle\int (2x + 1)(x^2 + x) \, dx$

(c) $\displaystyle\int 3x^2\sqrt{x^3 - 2} \, dx$

(d) $\displaystyle\int \frac{-4x}{(1 - 2x^2)^2} \, dx$

**SOLUTION**

(a) Letting $u = 3x - 1$, we have $du/dx = 3$.

$$\int 3(3x - 1)^4 \, dx = \int \overbrace{(3x - 1)^4}^{u^n} \overbrace{(3)}^{\frac{du}{dx}} \, dx$$

$$= \frac{(3x - 1)^5}{5} + C$$

(b) Letting $u = x^2 + x$, we have $du/dx = 2x + 1$.

$$\int (2x + 1)(x^2 + x) \, dx = \int \overbrace{(x^2 + x)^1}^{u^n}\overbrace{(2x + 1)}^{\frac{du}{dx}} \, dx$$

$$= \frac{(x^2 + x)^2}{2} + C$$

(c) Letting $u = x^3 - 2$, we have $du/dx = 3x^2$.

$$\int 3x^2\sqrt{x^3 - 2} \, dx = \int \overbrace{(x^3 - 2)^{1/2}}^{u^n}\overbrace{(3x^2)}^{\frac{du}{dx}} \, dx$$

$$= \frac{(x^3 - 2)^{3/2}}{3/2} + C$$

$$= \frac{2}{3}(x^3 - 2)^{3/2} + C$$

(d) Letting $u = 1 - 2x^2$, we have $du/dx = -4x$.

$$\int \frac{-4x}{(1 - 2x^2)^2} \, dx = \int \overbrace{(1 - 2x^2)^{-2}}^{u^n}\overbrace{(-4x)}^{\frac{du}{dx}} \, dx$$

$$= \frac{(1 - 2x^2)^{-1}}{-1} + C$$

$$= -\frac{1}{1 - 2x^2} + C$$

■■■ **Remark:** Part (b) of Example 1 illustrates a case of the General Power Rule that is sometimes overlooked—when the power is $n = 1$. In this case the rule takes the form

$$\int u\frac{du}{dx} \, dx = \frac{u^2}{2} + C$$

Many times part of the derivative $du/dx$ is missing from the integrand, and in *some* such cases we can make the necessary adjustments in order to apply the General Power Rule. In other instances we cannot adjust appropriately and therefore cannot apply the General Power Rule.

**EXAMPLE 2**

**Multiplying and Dividing by a Constant**

Evaluate the integral

$$\int x(3 - 4x^2)^2 \, dx$$

**SOLUTION**

If we let $u = 3 - 4x^2$, it follows that $du/dx = -8x$. Now we see that the factor $-8$ is not a part of the integrand. However, we can adjust the integrand by multiplying and dividing by $-8$ as follows:

$$\int x(3 - 4x^2)^2 \, dx = \int \frac{1}{-8}(3 - 4x^2)^2(-8x) \, dx$$

Now, because $-\frac{1}{8}$ is a constant, we can factor it out of the right-hand integral to obtain

$$-\frac{1}{8} \int \overbrace{(3 - 4x^2)^2}^{u^2}\overbrace{(-8x)}^{\frac{du}{dx}} \, dx$$

to which we apply the General Power Rule to get

$$\left(-\frac{1}{8}\right)\frac{(3 - 4x^2)^3}{3} + C = -\frac{(3 - 4x^2)^3}{24} + C$$

**EXAMPLE 3**

**An Example for Which the General Power Rule Fails**

Evaluate

$$\int -8(3 - 4x^2)^2 \, dx$$

**SOLUTION**

If we let $u = 3 - 4x^2$, then $u' = -8x$, and again part of $u'$ is missing from the integrand. Since the missing part of $u'$ is a *variable* rather than a constant, the adjustment would require that we move the variable quantity $1/x$ outside the integral sign. However, this is not possible. That is,

$$\int -8(3 - 4x^2)^2 \, dx = \int \left(\frac{1}{x}\right)(3 - 4x^2)^2(-8x) \, dx$$

$$\neq \frac{1}{x} \int (3 - 4x^2)^2(-8x) \, dx$$

(Note: *If* we were permitted to move variable quantities outside the integral sign, we would be able to move the entire integrand out and eliminate the problem entirely.) In this example, we cannot apply the General Power Rule since we cannot make the necessary adjustments for $u'$. However, we can (in this particular case) expand the integrand and write

$$\int -8(3 - 4x^2)^2 \, dx = \int -8(9 - 24x^2 + 16x^4) \, dx$$

$$= \int (-72 + 192x^2 - 128x^4) \, dx$$

$$= -72x + 64x^3 - \frac{128}{5}x^5 + C$$

Sometimes an integrand contains an extra factor that is *not* needed as part of $u'$. In such cases we simply move this factor outside the integral sign, insert the necessary factor, and adjust accordingly. The next example illustrates this situation.

**EXAMPLE 4**

**Applying the General Power Rule to Fractional Powers**

Evaluate

$$\int \frac{7x^2 \, dx}{\sqrt{4x^3 - 5}}$$

**SOLUTION**

We begin by writing the integral as

$$\int \frac{7x^2 \, dx}{\sqrt{4x^3 - 5}} = \int 7x^2(4x^3 - 5)^{-1/2} \, dx$$

and letting $u = 4x^3 - 5$. Then $u' = 12x^2$. We need the factor 12, rather than 7, so we write

$$\int 7x^2(4x^3 - 5)^{-1/2} \, dx = 7 \int \frac{1}{12}(4x^3 - 5)^{-1/2}(12x^2) \, dx$$

$$= \frac{7}{12} \int (4x^3 - 5)^{-1/2}(12x^2) \, dx$$

$$= \left(\frac{7}{12}\right)\frac{(4x^3 - 5)^{1/2}}{1/2} + C$$

$$= \frac{7}{6}\sqrt{4x^3 - 5} + C$$

**Remark:** Be sure you see the distinction between the two types of constant factors moved outside the integral sign in Example 4. In the equation

$$\int 7x^2(4x^3 - 5)^{-1/2} \, dx = (7)\left(\frac{1}{12}\right) \int (4x^3 - 5)^{-1/2}(12x^2) \, dx$$

the 7 is an unnecessary factor that is moved out *as is,* whereas the $\frac{1}{12}$ is the *reciprocal* adjustment for the factor 12 used to create $du/dx = 12x^2$.

### Substitution

The integration technique used in Examples 1, 2, and 4 depends on the ability to recognize (or create) integrands of the form $u^n(du/dx)$. With more complicated integrands it can be difficult to recognize the steps needed to rewrite the integrand in a form that fits a basic integration formula. When this occurs, an alternative procedure called **substitution** or **change of variables** can be helpful. With a formal change of variables, we completely rewrite the integral in terms of $u$ and $du$. This technique uses the Leibniz notation for the differential. That is, if $u = f(x)$, we write $du = f'(x) \, du$, and the General Power Rule takes the form

$$\int u^n \frac{du}{dx} \, dx = \int u^n \, du$$

**EXAMPLE 5**

**Change of Variables**

Evaluate the integral

$$\int \sqrt{1 - 3x} \, dx$$

**SOLUTION**

By writing the integral as

$$\int (1 - 3x)^{1/2} \, dx$$

we choose to let $u = 1 - 3x$. Then the differential $du$ is given by

$$du = -3 \, dx$$

which implies that $dx$ is given by

$$dx = -\frac{1}{3} \, du$$

Now, for each part of the integral containing $x$, we *substitute* the corresponding values for $u$ to obtain

$$\int (1 - 3x)^{1/2} \, dx = \int u^{1/2} \left( -\frac{1}{3} \, du \right)$$

$$= -\frac{1}{3} \int u^{1/2} \, du$$

$$= -\frac{1}{3} \frac{u^{3/2}}{3/2} + C$$

$$= -\frac{2}{9} u^{3/2} + C$$

Finally, using the substitution $u = 1 - 3x$, we write the antiderivative in terms of $x$ to obtain

$$\int (1 - 3x)^{1/2} \, dx = -\frac{2}{9}(1 - 3x)^{3/2} + C$$

To become efficient at integration it is necessary for you to learn to use *both* of the techniques discussed in this section. For simpler integrals we use pattern recognition and create the needed *du/dx* factor by multiplying and dividing by an appropriate constant. For more complicated integrals we use a formal change of variables (this technique will be discussed in more detail in later sections of the text). For this exercise set we suggest that you try working several of the problems twice—once using pattern recognition and once using a formal change of variables.

## SECTION EXERCISES 4.2

In Exercises 1–6, complete the table by identifying $u$ and $du/dx$ for the given integral.

$$\underline{\int u^n \frac{du}{dx} \, dx} \qquad\qquad \underline{u} \qquad\qquad \underline{\frac{du}{dx}}$$

**1.** $\displaystyle\int (5x^2 + 1)^2(10x) \, dx$

**2.** $\displaystyle\int (3 - 4x^2)^3(-8x) \, dx$

**3.** $\displaystyle\int \sqrt{1 - x^2}(-2x) \, dx$

**4.** $\displaystyle\int 3x^2\sqrt{x^3 + 1} \, dx$

**5.** $\displaystyle\int \left(4 + \frac{1}{x^2}\right)\left(\frac{-2}{x^3}\right) dx$

**6.** $\displaystyle\int \frac{1}{(1 + 2x)^2}(2) \, dx$

In Exercises 7–36, find the indefinite integral and check the result by differentiation.

**7.** $\displaystyle\int (1 + 2x)^4(2) \, dx$

**8.** $\displaystyle\int (x^2 - 1)^3(2x) \, dx$

**9.** $\displaystyle\int \sqrt{3x^2 + 4}(6x) \, dx$

**10.** $\displaystyle\int \sqrt{3 - x^3}(3x^2) \, dx$

**– 11.** $\displaystyle\int x^2(x^3 - 1)^4 \, dx$

**12.** $\displaystyle\int 3(x - 3)^{5/2} \, dx$

**– 13.** $\displaystyle\int x(x^2 - 1)^7 \, dx$

**14.** $\displaystyle\int x(1 - 2x^2)^3 \, dx$

**15.** $\displaystyle\int \frac{x^2}{(1 + x^3)^2} \, dx$

**16.** $\displaystyle\int \frac{x^2}{(x^3 - 1)^2} \, dx$

**17.** $\displaystyle\int \frac{x + 1}{(x^2 + 2x - 3)^2} \, dx$

**18.** $\displaystyle\int \frac{6x}{(1 + x^2)^3} \, dx$

**19.** $\displaystyle\int \frac{x - 4}{\sqrt{x^2 - 8x + 1}} \, dx$

**20.** $\displaystyle\int \frac{4x + 6}{(x^2 + 3x + 7)^3} \, dx$

**21.** $\displaystyle\int 5x\sqrt{1 + x^2} \, dx$

**22.** $\displaystyle\int u^3\sqrt{u^4 + 2} \, du$

**23.** $\displaystyle\int \frac{4x}{\sqrt{1 + x^2}} \, dx$

**24.** $\displaystyle\int \frac{x^2}{\sqrt{1 + x^3}} \, dx$

**25.** $\displaystyle\int \frac{-3}{\sqrt{2x + 3}} \, dx$

**26.** $\displaystyle\int \frac{t + 2t^2}{\sqrt{t}} \, dt$

**27.** $\displaystyle\int \frac{x^3}{\sqrt{1 + x^4}} \, dx$

**28.** $\displaystyle\int \left(1 + \frac{1}{t}\right)^3\left(\frac{1}{t^2}\right) dt$

**29.** $\displaystyle\int \frac{1}{2\sqrt{x}} \, dx$

**30.** $\displaystyle\int \frac{1}{(3x)^2} \, dx$

**31.** $\displaystyle\int \frac{1}{\sqrt{2x}} \, dx$

**32.** $\displaystyle\int \frac{1}{3x^2} \, dx$

**33.** $\displaystyle\int t^2\left(t - \frac{2}{t}\right) dt$

**34.** $\displaystyle\int \left(\frac{t^3}{3} + \frac{1}{4t^2}\right) dt$

**35.** $\displaystyle\int (9 - y)\sqrt{y} \, dy$

**36.** $\displaystyle\int 2\pi y(8 - y^{3/2}) \, dy$

In Exercises 37 and 38, perform the integration in two ways and explain the difference in appearance of the answers.

**37.** $\displaystyle\int (2x - 1)^2 \, dx$

**38.** $\displaystyle\int x(x^2 - 1)^2 \, dx$

In Exercises 39–46, use formal substitution (as illustrated in Example 5) to find the indefinite integral.

**39.** $\displaystyle\int x(3x^2 - 5)^3 \, dx$

**40.** $\displaystyle\int x^2(1 - x^3)^2 \, dx$

**41.** $\displaystyle\int x^2(2 - 3x^3)^{3/2} \, dx$

**42.** $\displaystyle\int t\sqrt{t^2 + 1} \, dt$

**43.** $\displaystyle\int \frac{x}{\sqrt{x^2 + 25}} \, dx$

**44.** $\displaystyle\int \frac{3}{\sqrt{2x + 1}} \, dx$

**45.** $\displaystyle\int \frac{x^2 + 1}{\sqrt{x^3 + 3x + 4}} \, dx$

**46.** $\displaystyle\int \sqrt{x}(4 - x^{3/2})^2 \, dx$

**47.** Find the equation of the function $f$ whose graph passes through the point $(0, \frac{4}{3})$ and whose derivative is $f'(x) = x\sqrt{1 - x^2}$.

**48.** Find the equation of the function $f$ whose graph passes through the point $(0, \frac{7}{3})$ and whose derivative is $f'(x) = x\sqrt{1 - x^2}$.

**49.** A company has determined the marginal cost for a particular product to be

$$\frac{dC}{dx} = \frac{4}{\sqrt{x+1}}$$

(a) Find the cost function if $C = 50$ when $x = 15$.

(b) Graph the marginal cost function and the cost function on the same set of axes.

**50.** The marginal cost for a certain commodity has been determined to be

$$\frac{dC}{dx} = \frac{12}{\sqrt[3]{12x+1}}$$

(a) Find the cost function if $C = 100$ when $x = 13$.

(b) Graph the marginal cost function and the cost function on the same set of axes.

In Exercises 51 and 52, find the supply function $x = f(p)$ that satisfies the given conditions.

**51.** $\dfrac{dx}{dp} = p\sqrt{p^2 - 16}$, $\quad x = 50$ when $p = \$5$

**52.** $\dfrac{dx}{dp} = \dfrac{10}{\sqrt{p-3}}$, $\quad x = 100$ when $p = \$3$

In Exercises 53 and 54, find the demand function $x = f(p)$ that satisfies the given conditions.

**53.** $\dfrac{dx}{dp} = -\dfrac{6000p}{(p^2 - 16)^{3/2}}$, $\quad x = 5000$ when $p = \$5$

**54.** $\dfrac{dx}{dp} = -\dfrac{400}{(0.02p - 1)^3}$, $\quad x = 10,000$ when $p = \$100$

**55.** A lumber company is seeking a model that yields the average weight loss $W$ per ponderosa pine log as a function of the number of days of drying time $t$. The model is to be reliable up to 100 days after the log is cut. Based on the weight loss during the first 30 days, it was determined that

$$\frac{dW}{dt} = \frac{12}{\sqrt{16t + 9}}, \quad 0 \le t \le 100$$

(a) Find $W$ as a function of $t$. Note that no weight loss occurs until the tree is cut.

(b) Find the total weight loss after 100 days.

# Area and the Fundamental Theorem of Calculus

## INTRODUCTORY EXAMPLE
### The Average Cost of Fuel

A large trucking firm estimates that the price of a gallon of diesel fuel during the next 5 years will rise according to the model

$$p = 1.21 + 0.11t + 0.02t^2$$

where $p$ is measured in dollars per gallon and $t$ is measured in years. To construct its 5-year budget plan, the firm wants to estimate the average cost of fuel over this 5-year period. The problem in estimating this average is that the cost of the fuel is not increasing linearly. (If the increase were linear we could use the midpoint $t = 2.5$ years and $p = \$1.61$ to estimate the average.)

The solution to this problem is given by the **definite integral**

$$\text{Average price} = \frac{1}{5} \int_0^5 [1.21 + 0.11t + 0.02t^2] \, dt$$
$$\approx \$1.65$$

Graphically this average corresponds to one-fifth of the **area** of the region under the price curve shown in Figure 4.6.

Now since the trucking firm purchases 100,000 gallons of fuel per year, it should plan to spend

$$(1.65)(100,000)(5) = \$825,000$$

on fuel during the next 5 years.

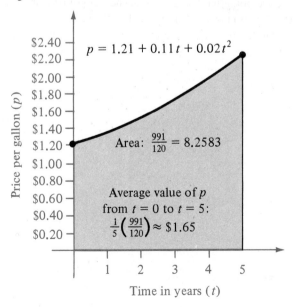

FIGURE 4.6

■ **The Fundamental Theorem of Calculus**
■ **Area**
■ **Definite Integrals**
■ **Average Value of a Function on an Interval**

Area is a concept familiar to all of us from our study of various geometric figures such as rectangles, squares, triangles, and circles. We generally think of area as a number that in some way suggests the size of a bounded region. Of course, for simple geometric figures we have specific formulas for calculating their areas.

Our problem here is to develop a way to calculate the area of a plane region $R$, bounded by the $x$-axis, the lines $x = a$ and $x = b$, and the graph of a nonnegative continuous function $f$, as shown in Figure 4.7(a).

The solution to this problem is given by the Fundamental Theorem of Calculus and represents one of the most famous discoveries in mathematics. From this theorem we can see that, just as the derivative can be used to find slope, the antiderivative can be used to find area. In anticipation of the connection between antiderivatives and area, we denote the area of the region shown in Figure 4.7(a) by

$$\text{Area} = \int_a^b f(x)\ dx$$

The symbol $\int_a^b f(x)\ dx$ is called the **definite integral from $a$ to $b$,** where $a$ is the **lower limit of integration** and $b$ is the **upper limit of integration.**

### Development of the Fundamental Theorem of Calculus

In developing the Fundamental Theorem of Calculus, we temporarily introduce the area function shown in Figure 4.7(b). Specifically, if $f$ is continuous and nonnegative on $[a, b]$, we denote the area of the region under the graph of $f$ from $a$ to $x$ by $A(x)$.

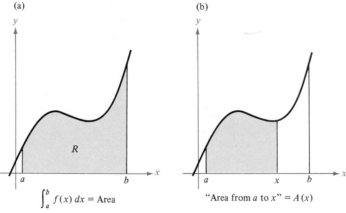

(a)

$$\int_a^b f(x)\ dx = \text{Area}$$

(b)

"Area from $a$ to $x$" $= A(x)$

FIGURE 4.7

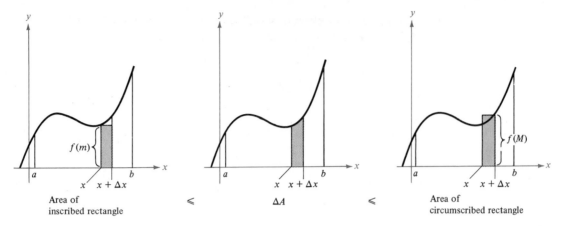

FIGURE 4.8

Now, if we let $x$ increase by an amount $\Delta x$, then the area of the region under the graph of $f$ increases by $\Delta A$. Furthermore, if $f(m)$ and $f(M)$ denote the minimum and maximum values of $f$ on the interval $[x, x + \Delta x]$, then we have the relationship

$$\overbrace{f(m)\ \Delta x}^{\substack{\text{Area of} \\ \text{inscribed} \\ \text{rectangle}}} \le \Delta A \le \overbrace{f(M)\ \Delta x}^{\substack{\text{Area of} \\ \text{circumscribed} \\ \text{rectangle}}}$$

as shown in Figure 4.8.

Dividing each term in the inequality

$$f(m)\ \Delta x \le \Delta A \le f(M)\ \Delta x$$

by the positive number $\Delta x$, we have

$$f(m) \le \frac{\Delta A}{\Delta x} \le f(M)$$

Since both $f(m)$ and $f(M)$ approach $f(x)$ as $\Delta x$ approaches zero and since

$$\lim_{\Delta x \to 0} \frac{\Delta A}{\Delta x} = A'(x)$$

it follows that

$$f(x) \le A'(x) \le f(x)$$

which means that

$$f(x) = A'(x)$$

Thus, we have established that the area function $A(x)$ is an antiderivative of $f$ and consequently must be of the form

$$A(x) = F(x) + C$$

where $F(x)$ is any antiderivative of $f$. To solve for $C$, we note that $A(a) = 0$ and thus $C = -F(a)$. Furthermore, evaluating $A(b)$, we have

$$A(b) = F(b) + C = F(b) - F(a)$$

Finally, replacing $A(b)$ by its integral form, we have

$$\int_a^b f(x) \, dx = F(b) - F(a)$$

This equation tells us that *if we can find an antiderivative for f,* then we can use the antiderivative to evaluate the definite integral $\int_a^b f(x) \, dx$. We summarize this result in the following theorem.

| | |
|---|---|
| **Fundamental Theorem of Calculus** | If a function $f$ is continuous on the interval $[a, b]$, then $$\int_a^b f(x) \, dx = F(b) - F(a)$$ where $F$ is any function such that $F'(x) = f(x)$ for all $x$ in $[a, b]$. |

Several comments regarding the Fundamental Theorem of Calculus are in order. First, this theorem describes a means for *evaluating* a definite integral, not a procedure for finding antiderivatives. Second, when applying this theorem we find it helpful to use the formulation

$$\int_a^b f(x) \, dx = F(x) \Big]_a^b = F(b) - F(a)$$

For instance, we write

$$\int_1^3 x^3 \, dx = \frac{x^4}{4} \Big]_1^3 = \frac{(3)^4}{4} - \frac{(1)^4}{4} = \frac{81}{4} - \frac{1}{4} = 20$$

Third, we observe that the constant of integration $C$ can be dropped from the antiderivative, because

$$\int_a^b f(x) \, dx = \left[ F(x) + C \right]_a^b$$
$$= [F(b) + C] - [F(a) + C]$$
$$= F(b) - F(a) + C - C$$
$$= F(b) - F(a)$$

Although it was convenient to assume $f$ to be nonnegative in our development of the Fundamental Theorem, this is not necessary, and we can apply the Fundamental Theorem in cases where $f$ is negative. We will say more about this possibility later.

We now list some useful properties of the definite integral. In each case we assume that $f$ and $g$ are integrable on $[a, b]$.

---

**Properties of definite integrals**

1. $\displaystyle\int_a^b kf(x)\,dx = k\int_a^b f(x)\,dx, \quad k \text{ is a constant}$

2. $\displaystyle\int_a^b [f(x) \pm g(x)]\,dx = \int_a^b f(x)\,dx \pm \int_a^b g(x)\,dx$

3. $\displaystyle\int_a^b f(x)\,dx = \int_a^c f(x)\,dx + \int_c^b f(x)\,dx, \quad a < c < b$

4. $\displaystyle\int_a^a f(x)\,dx = 0$

5. $\displaystyle\int_b^a f(x)\,dx = -\int_a^b f(x)\,dx$

## EXAMPLE 1

**Finding Area by the Fundamental Theorem**

Use the Fundamental Theorem to find the area of the region bounded by the $x$-axis and the graph of

$$f(x) = x^2 - 1, \quad 1 \le x \le 2$$

as shown in Figure 4.9.

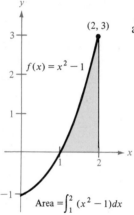

Area $= \displaystyle\int_1^2 (x^2 - 1)\,dx$    FIGURE 4.9

**SOLUTION**

Note that $f(x) \ge 0$ in the interval $1 \le x \le 2$, and therefore we can represent the area of the region by the definite integral

$$\text{Area} = \int_1^2 (x^2 - 1)\,dx$$

Now, applying the Fundamental Theorem, we have

$$\int_1^2 (x^2 - 1)\,dx = \left[\frac{x^3}{3} - x\right]_1^2$$

$$= \left(\frac{8}{3} - 2\right) - \left(\frac{1}{3} - 1\right)$$

$$= \frac{4}{3}$$

**Remark:** Note the use of parentheses in Example 1. We have found that it is easy to make errors in signs when evaluating definite integrals. To avoid such errors it helps to enclose the values of the antiderivative at the upper and lower limits in separate sets of parentheses.

**EXAMPLE 2**

**Evaluating Definite Integrals**

Evaluate the definite integral

$$\int_0^1 (4t + 1)^2 \, dt$$

and sketch the region whose area is represented by this integral.

**SOLUTION**

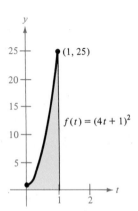

$f(t) = (4t + 1)^2$

(1, 25)

FIGURE 4.10

$$\int_0^1 (4t + 1)^2 \, dt = \frac{1}{4} \int_0^1 (4t + 1)^2 (4) \, dt$$

$$= \frac{1}{4} \left[ \frac{(4t + 1)^3}{3} \right]_0^1$$

$$= \frac{1}{4} \left[ \left( \frac{5^3}{3} \right) - \left( \frac{1}{3} \right) \right]$$

$$= \frac{1}{4} \left[ \frac{125}{3} - \frac{1}{3} \right]$$

$$= \frac{31}{3}$$

(Note the use of Property 1 in the first line of the solution.) The region whose area is represented by this integral is shown in Figure 4.10.

**EXAMPLE 3**

**Evaluating a Definite Integral**

Evaluate the definite integral

$$\int_1^4 3\sqrt{x} \, dx$$

**SOLUTION**

$$\int_1^4 3\sqrt{x} \, dx = 3 \int_1^4 x^{1/2} \, dx$$

$$= 3 \left[ \frac{x^{3/2}}{3/2} \right]_1^4$$

$$= 2(4)^{3/2} - 2(1)^{3/2}$$

$$= 16 - 2$$

$$= 14$$

## EXAMPLE 4

**Integrals Involving Absolute Values**

Evaluate the definite integral

$$\int_0^2 |\, 2x - 1 \,|\ dx$$

## SOLUTION

Since we have no rule for finding the antiderivative of a function involving an absolute value, we use Figure 4.11 and the definition of absolute value to write

$$|\, 2x - 1 \,| = \begin{cases} -(2x - 1), & x < \dfrac{1}{2} \\[2mm] (2x - 1), & x \geq \dfrac{1}{2} \end{cases}$$

Now we can rewrite the integral in two parts as follows:

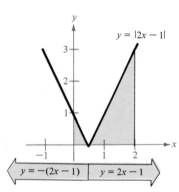

$$\int_0^2 |\, 2x - 1 \,|\ dx = \int_0^{1/2} -(2x - 1)\ dx + \int_{1/2}^2 (2x - 1)\ dx$$

$$= \Big[ -x^2 + x \Big]_0^{1/2} + \Big[ x^2 - x \Big]_{1/2}^2$$

$$= \left( -\frac{1}{4} + \frac{1}{2} \right) - (0 + 0) + (4 - 2) - \left( \frac{1}{4} - \frac{1}{2} \right)$$

$$= \frac{5}{2}$$

**FIGURE 4.11**

## Average Value of a Function

In addition to being used to find areas, definite integrals can be used to find the *average value* of a function over an interval.

---

**Definition of the average value of a function**

If $f$ is continuous on $[a, b]$, then the **average value** of $f$ on this interval is given by

$$\text{Average of } f \text{ on } [a, b] = \frac{1}{b - a} \int_a^b f(x)\ dx$$

---

## EXAMPLE 5

**Finding the Average Value of a Function**

What is the average value of

$$f(x) = 3x^2 - 2x$$

on the interval $[1, 4]$?

The average value of $f(x) = 3x^2 - 2x$ on $[1, 4]$ is given by

$$\frac{1}{4-1} \int_1^4 (3x^2 - 2x) \, dx = \frac{1}{3}\left[ \frac{3x^3}{3} - \frac{2x^2}{2} \right]_1^4$$

$$= \frac{1}{3}[(64 - 16) - (1 - 1)]$$

$$= \frac{48}{3}$$

$$= 16$$

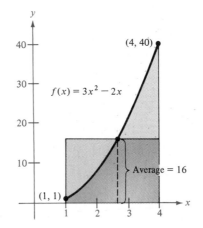

$f(x) = 3x^2 - 2x$

(4, 40)

(1, 1)

Average = 16

FIGURE 4.12

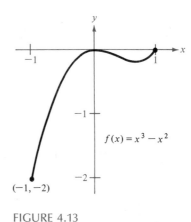

$f(x) = x^3 - x^2$

$(-1, -2)$

FIGURE 4.13

Note in Figure 4.12 that the area under the curve is equal to the area of the rectangle whose height is the average value. In fact, we could have defined the average value of $f$ on the interval $[a, b]$ to be that value $K$ such that

$$\int_a^b f(x) \, dx = \int_a^b K \, dx = K(b - a)$$

**EXAMPLE 6**

**Finding the Average Value of a Function**

Find the average value of

$$f(x) = x^3 - x^2$$

on the interval $[-1, 1]$.

**SOLUTION**

Note from Figure 4.13 that $f(x)$ is nonpositive over the entire interval $[-1, 1]$. Therefore, it is not surprising that the average value of $f$ over this interval turns out to be negative. The average of $f$ is

$$\frac{1}{1 - (-1)} \int_{-1}^1 (x^3 - x^2) \, dx = \frac{1}{2}\left[ \frac{x^4}{4} - \frac{x^3}{3} \right]_{-1}^1 = -\frac{1}{3}$$

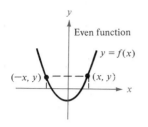

Even function

$y = f(x)$

$(-x, y)$   $(x, y)$

y-Axis Symmetry

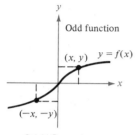

Odd function

$(x, y)$   $y = f(x)$

$(-x, -y)$

Origin Symmetry

FIGURE 4.14

## Integration of Even and Odd Functions

Several common types of functions have graphs that are symmetrical with respect to the y-axis or the origin, as shown in Figure 4.14. If the graph of $y = f(x)$ is symmetrical with respect to the y-axis, then we call $f$ an **even** function. Similarly, if the graph of $y = f(x)$ is symmetrical with respect to the origin, then we call $f$ an **odd** function.

We can test for even and odd functions as follows:

1. The function $f$ is **even** if $f(-x) = f(x)$.
2. The function $f$ is **odd** if $f(-x) = -f(x)$.

For example, the function given by $f(x) = x^2 + 1$ is even since

$$f(-x) = (-x)^2 + 1 = x^2 + 1 = f(x)$$

and the function given by $f(x) = x^3 + x$ is odd since

$$f(-x) = (-x)^3 + (-x) = -(x^3 + x) = -f(x)$$

---

**Integration of even and odd functions**

Let $f$ be integrable on the interval $[-a, a]$.

1. If $f$ is an *even* function, then

$$\int_{-a}^{a} f(x)\, dx = 2 \int_{0}^{a} f(x)\, dx$$

2. If $f$ is an *odd* function, then

$$\int_{-a}^{a} f(x)\, dx = 0$$

---

**EXAMPLE 7**

**Integrating Even and Odd Functions**

Evaluate the following definite integrals:

(a) $\displaystyle\int_{-2}^{2} x^2\, dx$   (b) $\displaystyle\int_{-2}^{2} x^3\, dx$

**SOLUTION**

(a) Since $f(x) = x^2$ is an even function, we can write

$$\int_{-2}^{2} x^2\, dx = 2 \int_{0}^{2} x^2\, dx = 2 \left[ \frac{x^3}{3} \right]_{0}^{2} = 2 \left( \frac{8}{3} - 0 \right) = \frac{16}{3}$$

(b) Since $f(x) = x^3$ is an odd function, we can write

$$\int_{-2}^{2} x^3\, dx = 0$$

Before concluding this section, we make two important observations regarding integrals.

1. Be sure you realize from Example 6 that a definite integral can have a negative value. Even though one of the main applications of definite integrals is finding area (which must be positive), this is not the only application, and it is common to encounter a definite integral whose value is negative.

2. Be sure you recognize the distinction between indefinite and definite integrals. The *indefinite integral*

$$\int f(x)\, dx$$

denotes a family of *functions* (the antiderivatives of $f(x)$), whereas the *definite integral*

$$\int_a^b f(x)\, dx$$

is a *number*.

## SECTION EXERCISES 4.3

In Exercises 1–34, evaluate the definite integral.

**1.** $\displaystyle\int_0^1 2x\, dx$

**2.** $\displaystyle\int_2^7 3\, dv$

**3.** $\displaystyle\int_{-1}^0 (x - 2)\, dx$

**4.** $\displaystyle\int_2^5 (-3v + 4)\, dv$

**5.** $\displaystyle\int_{-1}^1 (t^2 - 2)\, dt$

**6.** $\displaystyle\int_0^3 (3x^2 + x - 2)\, dx$

**7.** $\displaystyle\int_0^1 (2t - 1)^2\, dt$

**8.** $\displaystyle\int_{-1}^1 (t^3 - 9t)\, dt$

**9.** $\displaystyle\int_1^2 \left(\frac{3}{x^2} - 1\right) dx$

**10.** $\displaystyle\int_0^1 (3x^3 - 9x + 7)\, dx$

**11.** $\displaystyle\int_1^2 (5x^4 + 5)\, dx$

**12.** $\displaystyle\int_{-3}^3 v^{1/3}\, dv$

**13.** $\displaystyle\int_{-1}^1 (\sqrt[3]{t} - 2)\, dt$

**14.** $\displaystyle\int_1^4 \sqrt{\frac{2}{x}}\, dx$

**15.** $\displaystyle\int_1^4 \frac{u - 2}{\sqrt{u}}\, du$

**16.** $\displaystyle\int_{-2}^{-1} \left(\frac{-1}{u^2} + u\right) du$

**17.** $\displaystyle\int_0^1 \frac{x - \sqrt{x}}{3}\, dx$

**18.** $\displaystyle\int_0^2 (2 - t)\sqrt{t}\, dt$

**19.** $\displaystyle\int_{-1}^0 (t^{1/3} - t^{2/3})\, dt$

**20.** $\displaystyle\int_0^4 (x^{1/2} + x^{1/4})\, dx$

**21.** $\displaystyle\int_0^4 \frac{1}{\sqrt{2x + 1}}\, dx$

**22.** $\displaystyle\int_0^1 x\sqrt{1 - x^2}\, dx$

**23.** $\displaystyle\int_{-1}^1 x(x^2 + 1)^3\, dx$

**24.** $\displaystyle\int_0^2 \frac{x}{\sqrt{1 + 2x^2}}\, dx$

**25.** $\displaystyle\int_{-2}^2 x\sqrt[3]{4 + x^2}\, dx$

**26.** $\displaystyle\int_1^9 \frac{1}{\sqrt{x}(1 + \sqrt{x})^2}\, dx$

**27.** $\displaystyle\int_{-1}^1 |x|\, dx$

**28.** $\displaystyle\int_0^3 |2x - 3|\, dx$

**29.** $\displaystyle\int_1^2 (x - 1)(2 - x)\, dx$

**30.** $\displaystyle\int_0^4 \frac{1}{\sqrt{2x + 1}}\, dx$

**31.** $\displaystyle\int_0^3 x(x - 3)^2\, dx$

**32.** $\displaystyle\int_1^5 x^2\sqrt{x^3 - 1}\, dx$

**33.** $\displaystyle\int_0^7 x\sqrt[3]{x^2 + 1}\, dx$

**34.** $\displaystyle\int_{-2}^6 \sqrt[3]{x + 2}\, dx$

In Exercises 35–40, determine the area of the region having the given boundaries.

**35.** $y = x - x^2$, $y = 0$  **36.** $y = -x^2 + 2x + 3$, $y = 0$

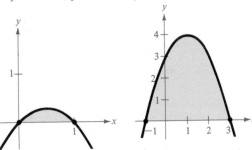

**37.** $y = 1 - x^4,\ y = 0$

**38.** $y = \dfrac{1}{x^2},\ x = 1,$
$x = 2,\ y = 0$

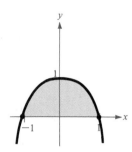

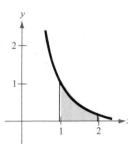

**39.** $y = \sqrt[3]{2x},\ x = 0,$

**40.** $y = (3 - x)\sqrt{x},$
$y = 0$

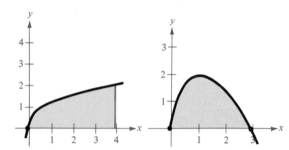

In Exercises 41–46, evaluate the definite integral and make a sketch of the region whose area is given by the integral.

**41.** $\displaystyle\int_1^3 (2x - 1)\ dx$

**42.** $\displaystyle\int_0^2 (x + 4)\ dx$

**43.** $\displaystyle\int_3^4 (x^2 - 9)\ dx$

**44.** $\displaystyle\int_{-1}^2 (-x^2 + x + 2)\ dx$

**45.** $\displaystyle\int_0^1 (x - x^3)\ dx$

**46.** $\displaystyle\int_0^1 \sqrt{x}(1 - x)\ dx$

In Exercises 47–50, determine the area of the region having the given boundaries.

**47.** $y = 3x^2 + 1,\ x = 0,\ x = 2,\ y = 0$

**48.** $y = 1 + \sqrt{x},\ x = 0,\ x = 4,\ y = 0$

**49.** $y = x^3 + x,\ x = 2,\ y = 0$

**50.** $y = -x^2 + 3x,\ y = 0$

In Exercises 51–56, sketch the graph of the function over the given interval. Find the average value of the function over the given interval and all values of $x$ where the function equals its average.

|  | Function | Interval |
|---|---|---|
| **51.** | $f(x) = 4 - x^2$ | $[-2, 2]$ |
| **52.** | $f(x) = x^2 - 2x + 1$ | $[0, 1]$ |
| **53.** | $f(x) = x\sqrt{4 - x^2}$ | $[0, 2]$ |
| **54.** | $f(x) = \dfrac{x^2 + 1}{x^2}$ | $\left[\dfrac{1}{2}, 2\right]$ |
| **55.** | $f(x) = x - 2\sqrt{x}$ | $[0, 4]$ |
| **56.** | $f(x) = \dfrac{1}{(x - 3)^2}$ | $[0, 2]$ |

**57.** Use the fact that $\int_0^2 x^2\ dx = \frac{8}{3}$ to evaluate the following definite integrals without using the Fundamental Theorem of Calculus:

(a) $\displaystyle\int_{-2}^0 x^2\ dx$  (b) $\displaystyle\int_{-2}^2 x^2\ dx$

(c) $\displaystyle\int_0^2 -x^2\ dx$  (d) $\displaystyle\int_{-2}^0 3x^2\ dx$

**58.** Use the fact that $\int_0^2 x^3\ dx = 4$ to evaluate the following definite integrals without using the Fundamental Theorem of Calculus:

(a) $\displaystyle\int_{-2}^0 x^3\ dx$  (b) $\displaystyle\int_{-2}^2 x^3\ dx$

(c) $\displaystyle\int_0^2 -x^3\ dx$  (d) $\displaystyle\int_0^2 3x^3\ dx$

**59.** The total cost of purchasing and maintaining a piece of equipment for $x$ years is given by

$$C = 5000\left(25 + 3\int_0^x t^{1/4}\ dt\right)$$

Find

(a) $C(1)$  (b) $C(5)$  (c) $C(10)$

**60.** The rate of flow of revenue for a small home business is given by

$$\frac{dR}{dt} = \frac{12{,}000t}{\sqrt{t^2 + 2}}$$

where $t$ is time in years. Find the total revenue for 5 years.

[Total revenue $= \int_0^5$ (rate of flow) $dt$.]

**61.** A company purchases a new machine for which the rate of depreciation is given by

$$\frac{dV}{dt} = 10{,}000(t - 6),\quad 0 \le t \le 5$$

where $V$ is the value of the machine after $t$ years. Set up and evaluate the definite integral that yields the total loss of value of the machine over the first 3 years.

**62.** The velocity $v$ of the flow of blood at a distance $r$ from the central axis of an artery of radius $R$ is given by

$$v = k(R^2 - r^2)$$

where $k$ is a constant of proportionality. Find the average rate of flow of blood along a radius of the artery. (Use zero and $R$ as the limits of integration.)

**63.** The air temperature during a period of 12 hours is given by the model

$$T = 53 + 5t - 0.3t^2, \quad 0 \leq t \leq 12$$

where $t$ is measured in hours and $T$ in degrees Fahrenheit. (See Figure 4.15.) Find the average temperature during

(a) the first 6-hour period
(b) the entire period

**64.** The annual death rate (per 1000 people $x$ years of age) can be approximated by the model

$$R = 0.036x^2 - 2.8x + 58.14, \quad 40 \leq x \leq 60$$

(See Figure 4.16.) Find the average death rate for people between

(a) 40 and 50 years of age
(b) 50 and 60 years of age

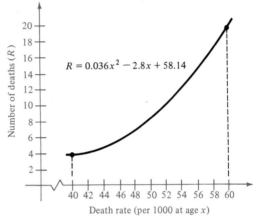

FIGURE 4.16

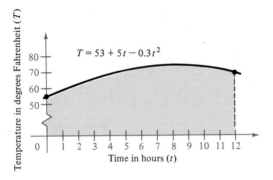

FIGURE 4.15

# The Area of a Region between Two Curves

## INTRODUCTORY EXAMPLE
### Consumer Surplus

During the 1970s the price of hand-held calculators dropped dramatically, and with the decline in price the sale of these electronic wonders followed the classic demand pattern. That is, as the price dropped more and more people were willing to purchase calculators. Suppose that the annual demand function for a particular model of calculator is

$$p = 0.3625x^3 - 4.35x^2 - 21.75x + 300, \quad 0 \le x \le 10$$

where the price $p$ is measured in dollars and the number sold $x$ is measured in units of 10,000. From Figure 4.17 we can see that no one would be willing to pay $300 for this model of calculator, but 100,000 people would be willing to pay $10.

Suppose that the actual selling price is $91.20. (This corresponds to sales of 60,000 units.) Then the total revenue $R = xp$ is represented by the area of the rectangle shown in Figure 4.17. The area of the region lying between this rectangle and the demand curve represents a quantity that economists call the *consumer surplus*. The consumer surplus is the extra amount of money consumers would have been willing to spend but in fact did not have to spend because the price was low. The formula for the **area of the region** between the demand curve and the line representing the actual price is

$$\text{Area} = \int_0^6 [(0.3625x^3 - 4.35x^2 - 21.75x + 300) \\ - (91.2)] \, dx$$

$$= 665.55$$

Since $x$ is measured in units of 10,000, the consumer surplus for this particular price and demand function is

$$\text{Consumer surplus} = (10,000)(665.55) = \$6,655,500.00$$

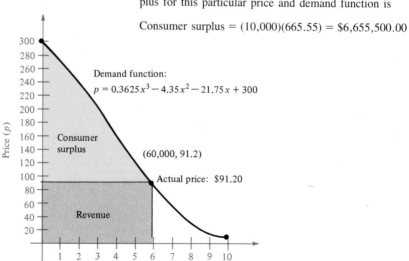

FIGURE 4.17

Demand function:
$$p = 0.3625x^3 - 4.35x^2 - 21.75x + 300$$

Consumer surplus

(60,000, 91.2)

Actual price: $91.20

Revenue

Price ($p$)

Number of units (in units of 10,000, $x$)

■ **Area of a Region between Two Curves**
■ **Points of Intersection of Two Curves**

With a few modifications we can extend the application of definite integrals from the area of a region *under* a curve to the area of a region *between* two curves. Let us consider the region bounded by the graphs of $y = f(x)$, $y = g(x)$, $x = a$, and $x = b$. If, as in Figure 4.18, the graphs of both $f$ and $g$ lie above the $x$-axis, we can geometrically interpret the area of the region between the graphs as the area of the region under the graph of $g$ subtracted from the area of the region under the graph of $f$, as shown in Figure 4.19.

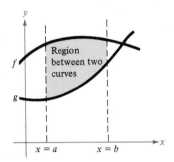

FIGURE 4.18

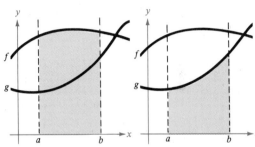

(Area of region between $f$ and $g$) = (Area of region under $f$) − (Area of region under $g$)

$$\int_a^b [f(x) - g(x)] \, dx \quad = \quad \int_a^b f(x) \, dx \quad - \quad \int_a^b g(x) \, dx$$

FIGURE 4.19

Although Figure 4.18 depicts the graphs of $f$ and $g$ lying above the $x$-axis, this is not necessary, and the same integrand $[f(x) - g(x)]$ can be used as long as $f$ and $g$ are continuous and

$$g(x) \le f(x)$$

on the interval $[a, b]$. This result is summarized as follows.

**Area between two curves**

If $f$ and $g$ are continuous on $[a, b]$ and $g(x) \le f(x)$ for all $x$ in $[a, b]$, then the area of the region bounded by $y = f(x)$, $y = g(x)$, $x = a$, and $x = b$ is given by

$$A = \int_a^b [f(x) - g(x)] \, dx$$

**EXAMPLE 1**

**Finding the Area between Two Graphs**

Find the area of the region bounded by the graphs of

$$y = x^2 + 2 \quad \text{and} \quad y = x, \quad 0 \le x \le 1$$

**SOLUTION**

We begin by sketching the graphs of both functions, as shown in Figure 4.20. Then we can see that

$$x \leq x^2 + 2$$

for all $x$ in [0, 1]. Therefore, we let $f(x) = x^2 + 2$ and $g(x) = x$, and the area is given by

$$\text{Area} = \int_a^b [f(x) - g(x)] \, dx = \int_0^1 [(x^2 + 2) - (x)] \, dx$$

$$= \int_0^1 (x^2 - x + 2) \, dx$$

$$= \left[ \frac{x^3}{3} - \frac{x^2}{2} + 2x \right]_0^1$$

$$= \left( \frac{1}{3} - \frac{1}{2} + 2 \right) - (0)$$

$$= \frac{11}{6}$$

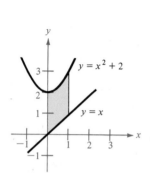

FIGURE 4.20

In Example 1, the graphs of $y = x$ and $y = x^2 + 2$ do not intersect, and the values of $a$ and $b$ are explicitly given. A more common type of problem involves the area of the region bounded by two *intersecting* graphs. In this type of problem, the values of $a$ and $b$ must be calculated.

**EXAMPLE 2**

**Finding the Area between Two Intersecting Graphs**

Find the area of the region bounded by the graphs of

$$y = 2 - x^2 \qquad \text{and} \qquad y = x$$

**SOLUTION**

In this case $a$ and $b$ are determined by the points of intersection of $y = x$ and $y = 2 - x^2$. In order to find these points, we set these two functions equal to each other and solve for $x$.

$$x = 2 - x^2 \qquad \text{Equate } y\text{-values}$$

$$x^2 + x - 2 = 0$$

$$(x + 2)(x - 1) = 0 \qquad \text{Factor}$$

$$x = -2, 1 \qquad \text{Solve for } x$$

After sketching the graphs of both functions, as shown in Figure 4.21, we can see that $x \leq 2 - x^2$ on the interval $[-2, 1]$. Therefore, we let $f(x) = 2 - x^2$ and $g(x) = x$, and the area of the region is given by

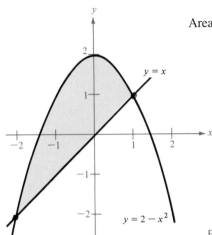

$$\text{Area} = \int_a^b [f(x) - g(x)] \, dx = \int_{-2}^1 [(2 - x^2) - (x)] \, dx$$

$$= \int_{-2}^1 (2 - x^2 - x) \, dx$$

$$= \left[ 2x - \frac{x^3}{3} - \frac{x^2}{2} \right]_{-2}^1$$

$$= \left( 2 - \frac{1}{3} - \frac{1}{2} \right) - \left( -4 + \frac{8}{3} - 2 \right)$$

$$= \frac{9}{2}$$

FIGURE 4.21

## EXAMPLE 3

**Finding the Area of a Region Below the x-Axis**

Find the area of the region bounded by the graph of

$$y = x^2 - 3x - 4$$

and the x-axis.

**SOLUTION**

To find the points at which the graph of $y = x^2 - 3x - 4$ intersects the x-axis, we set $x^2 - 3x - 4$ equal to zero and solve for x.

$$x^2 - 3x - 4 = 0$$

$$(x - 4)(x + 1) = 0$$

$$x = 4, \; -1$$

Moreover, from Figure 4.22 we see that $x^2 - 3x - 4 \le 0$ for all x in the interval $[-1, 4]$. Therefore, we let $f(x) = 0$ and $g(x) = x^2 - 3x - 4$, and the area of the region is given by

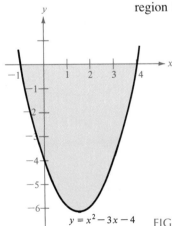

$$\text{Area} = \int_a^b [f(x) - g(x)] \, dx = \int_{-1}^4 [(0) - (x^2 - 3x - 4)] \, dx$$

$$= \int_{-1}^4 (-x^2 + 3x + 4) \, dx$$

$$= \left[ -\frac{x^3}{3} + \frac{3x^2}{2} + 4x \right]_{-1}^4$$

$$= \left( -\frac{64}{3} + 24 + 16 \right) - \left( \frac{1}{3} + \frac{3}{2} - 4 \right)$$

$$= \frac{125}{6}$$

$y = x^2 - 3x - 4$    FIGURE 4.22

Occasionally two graphs intersect in more than two points. In determining the area of the region between two such graphs, we must find *all* points of intersection and check to see which graph is above the other in each interval determined by these points.

**EXAMPLE 4**

**Graphs Having Multiple Points of Intersection**

Find the area of the region between the graphs of

$$f(x) = 3x^3 - x^2 - 10x \quad \text{and} \quad g(x) = -x^2 + 2x$$

**SOLUTION**

To find the points of intersection of these two graphs, we set the functions equal to each other and solve for $x$.

$$3x^3 - x^2 - 10x = -x^2 + 2x \qquad \text{Let } f(x) = g(x)$$

$$3x^3 - 12x = 0$$

$$3x(x^2 - 4) = 0$$

$$3x(x - 2)(x + 2) = 0 \qquad \text{Factor}$$

$$x = 0, 2, -2 \qquad \text{Solve for } x$$

In Figure 4.23, we see that $g(x) \le f(x)$ in the interval $[-2, 0]$, but the two curves switch at the point $(0, 0)$ and $f(x) \le g(x)$ in the interval $[0, 2]$. Therefore, we need to use two integrals to determine the area of the region between the graphs of $f$ and $g$, one for the interval $[-2, 0]$ and one for the interval $[0, 2]$.

$$\text{Area} = \int_{-2}^{0} [f(x) - g(x)] \, dx + \int_{0}^{2} [g(x) - f(x)] \, dx$$

$$= \int_{-2}^{0} (3x^3 - 12x) \, dx + \int_{0}^{2} (-3x^3 + 12x) \, dx$$

$$= \left[ \frac{3x^4}{4} - 6x^2 \right]_{-2}^{0} + \left[ \frac{-3x^4}{4} + 6x^2 \right]_{0}^{2}$$

$$= -(12 - 24) + (-12 + 24)$$

$$= 24$$

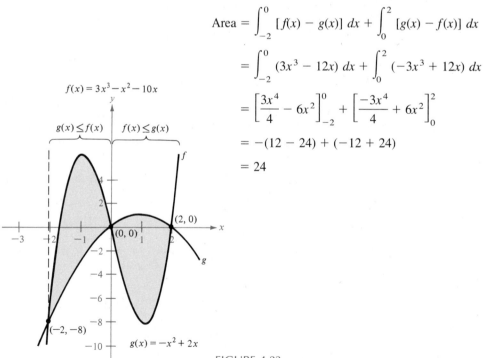

$f(x) = 3x^3 - x^2 - 10x$

$g(x) \le f(x) \qquad f(x) \le g(x)$

$(2, 0)$

$(0, 0)$

$(-2, -8)$

$g(x) = -x^2 + 2x$

FIGURE 4.23

**EXAMPLE 5**

**A Business Application**

According to U.S. Department of Energy statistics, the total consumption of petroleum fuel for transportation in the United States from 1950 to 1979 followed a growth pattern given by the model

$$f(t) = 0.000433t^2 + 0.0962t + 2.76, \quad -20 \leq t \leq 9$$

where $f(t)$ is measured in billions of barrels and $t$ in years, with $t = 0$ corresponding to January 1, 1970. With the onset of dramatic increases in crude oil prices in the late 1970s, the growth pattern for fuel consumption changed and began following the pattern described by the model

$$g(t) = -0.00831t^2 + 0.152t + 2.81, \quad 9 \leq t \leq 16$$

as shown in Figure 4.24. Find the total amount of fuel saved from 1974 to 1985 as a result of consuming fuel at the post-1979 rate rather than the pre-1979 rate.

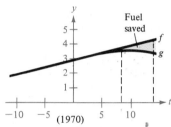

f: Pre–1979 consumption rate
g: Post–1979 consumption rate

FIGURE 4.24

**SOLUTION**

Since the graph of the pre-1979 model $f$ lies above the graph of the post-1979 model $g$ on the interval [9, 16], the amount of fuel saved is given by the following integral:

$$\int_9^{16} [\overbrace{(0.000433t^2 + 0.0962t + 2.76)}^{f(t)} - \overbrace{(-0.00831t^2 + 0.152t + 2.81)}^{g(t)}] \, dt$$

$$= \int_9^{16} (0.008743t^2 - 0.0558t - 0.05) \, dt$$

$$= \left[ \frac{0.008743t^3}{3} - \frac{0.0558t^2}{2} - 0.05t \right]_9^{16}$$

$$\approx 4.58 \text{ billion barrels}$$

Therefore, approximately 4.58 billion barrels of fuel were saved. (At 42 gallons per barrel, that is a savings of about 200 billion gallons!)

# SECTION EXERCISES 4.4

In Exercises 1–6, sketch the region whose area is given by the definite integral. When the integrand is given as a difference of two functions, sketch the graph of each function individually and show the required region between them.

**1.** $\displaystyle\int_0^4 x \, dx$

**2.** $\displaystyle\int_0^4 \frac{x}{2} \, dx$

**3.** $\displaystyle\int_{-3}^3 \sqrt{9 - x^2} \, dx$

**4.** $\displaystyle\int_0^4 \sqrt{4 - x} \, dx$

**5.** $\displaystyle\int_0^4 \left[ (x + 1) - \frac{x}{2} \right] dx$

**6.** $\displaystyle\int_{-1}^1 [(1 - x^2) - (x^2 - 1)] \, dx$

In Exercises 7–12, find the area of the given region.

**7.** $f(x) = x^2 - 6x,$
$g(x) = 0$

**8.** $f(x) = x^2 + 2x + 1,$
$g(x) = 2x + 5$

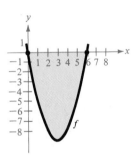

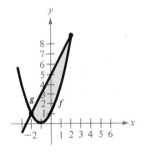

**9.** $f(x) = x^2 - 4x + 3,$
$g(x) = -x^2 + 2x + 3$

**10.** $f(x) = x^2,$
$g(x) = x^3$

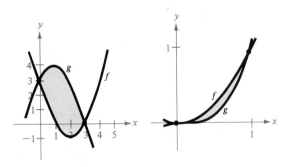

**11.** $f(x) = 3(x^3 - x),$
$g(x) = 0$

**12.** $f(x) = (x - 1)^3,$
$g(x) = x - 1$

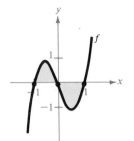

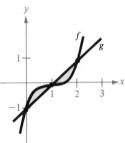

In Exercises 13–30, sketch the region bounded by the graphs of the given functions and find the area of the region.

**13.** $f(x) = x^2 - 4x, \ g(x) = 0$
**14.** $f(x) = 3 - 2x - x^2, \ g(x) = 0$
**15.** $f(x) = x^2 + 2x + 1, \ g(x) = 3x + 3$
**16.** $f(x) = -x^2 + 4x + 2, \ g(x) = x + 2$
**17.** $y = x, \ y = 2 - x, \ y = 0$
**18.** $y = \dfrac{1}{x^2}, \ y = 0, \ x = 1, \ x = 5$
**19.** $f(x) = x^2 - x, \ g(x) = 2(x + 2)$
**20.** $f(x) = x(x^2 - 3x + 3), \ g(x) = x^2$
**21.** $y = x^3 - 2x + 1, \ y = -2x, \ x = 1$
**22.** $f(x) = \sqrt{x}, \ g(x) = x$
**23.** $f(x) = \sqrt{3x + 1}, \ g(x) = x + 1$
**24.** $f(x) = x^2 + 5x - 6, \ g(x) = 6x - 6$
**25.** $y = x^2 - 4x + 3, \ y = 3 + 4x - x^2$
**26.** $y = x^4 - 2x^2, \ y = 2x^2$
**27.** $f(y) = y^2, \ g(y) = y + 2$
**28.** $f(y) = y(2 - y), \ g(y) = -y$
**29.** $x = y^2 + 1, \ x = 0, \ y = -1, \ y = 2$
**30.** $f(y) = \sqrt{y}, \ y = 9, \ x = 0$

In Exercises 31 and 32, use integration to find the area of the triangle having the given vertices.

**31.** $(0, 0), (4, 0), (4, 4)$    **32.** $(0, 0), (4, 0), (6, 4)$

In Exercises 33 and 34, two models $R_1$ and $R_2$ are given for revenue (in billions of dollars) for a large corporation. Model $R_1$ gives projected annual revenues from 1985 to 1990, with $t = 0$ corresponding to 1985, and model $R_2$ gives projected revenues if there is a decrease in growth of corporate sales over the period. Approximate the total reduction in revenue if corporate sales are actually closer to model $R_2$.

**33.** $R_1 = 7.21 + 0.58t$
$R_2 = 7.21 + 0.45t$

**34.** $R_1 = 7.21 + 0.26t + 0.02t^2$
$R_2 = 7.21 + 0.1t + 0.01t^2$

**35.** According to U.S. Department of Agriculture statistics, the total consumption of beef in the United States from 1950 to 1970 followed a growth pattern approximated by

$$f(t) = 23.703 + 1.002t + 0.015t^2$$

where $f(t)$ is measured in billions of pounds and $t$ is measured in years, with $t = 0$ representing 1970. From 1970 to 1980, the growth pattern was more closely approximated by

$$g(t) = 22.93 + 0.678t - 0.037t^2$$

Estimate the total reduction in consumption of beef from 1970 to 1980 due to the change in consumption rate.

**36.** For the years from 1980 to 1990, the projected fuel cost $C$ (in millions of dollars) for a large corporation was given by

$$C_1 = 568.50 + 7.15t$$

where $t$ is time in years, with $t = 0$ corresponding to 1980. Because of the installation of fuel-saving equipment, a more accurate model for the fuel cost for the period is

$$C_2 = 525.60 + 6.43t$$

Approximate the total savings for the 10-year period due to the installation of the new equipment.

In Exercises 37–46, find the consumer surplus and producer surplus for the given supply and demand curves. The consumer surplus and producer surplus are represented by the areas shown in Figure 4.25.

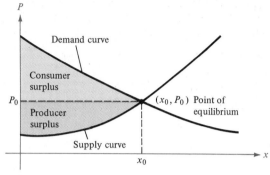

FIGURE 4.25

| Demand function | Supply function |
|---|---|
| **37.** $p_1(x) = 50 - 0.5x$ | $p_2(x) = 0.125x$ |
| **38.** $p_1(x) = 60 - x$ | $p_2(x) = 10 + \frac{7}{3}x$ |
| **39.** $p_1(x) = 300 - x$ | $p_2(x) = 100 + x$ |
| **40.** $p_1(x) = \frac{1}{20}(-x + 2000)$ | $p_2(x) = \frac{1}{10}(x + 250)$ |
| **41.** $p_1(x) = 300 - \frac{1}{100}x^2$ | $p_2(x) = 100 + x$ |
| **42.** $p_1(x) = 1000 - 0.4x^2$ | $p_2(x) = 42x$ |
| **43.** $p_1(x) = \frac{1}{8}(-x + 400)$ | $p_2(x) = 10 + \frac{1}{40}x + \frac{1}{4000}x^2$ |
| **44.** $p_1(x) = 10 - 0.00001x^2$ | $p_2(x) = 5 + 0.005x$ |
| **45.** $p_1(x) = \dfrac{10,000}{\sqrt{x + 100}}$ | $p_2(x) = 100\sqrt{0.05x + 10}$ |
| **46.** $p_1(x) = \sqrt{25 - 0.1x}$ | $p_2(x) = \sqrt{9 + 0.1x} - 2$ |

# The Definite Integral as the Limit of a Sum

## INTRODUCTORY EXAMPLE
### The Method of Exhaustion

Have you ever wondered how the formulas for the area and volume of some common geometrical figures were discovered? For example, we know from geometry that the area of a circular region is given by

$$\text{Area} = \pi r^2$$

where $\pi \approx 3.1416$ is defined to be the ratio of the circumference of a circle to its diameter. One method that ancient Greek mathematicians used to discover such formulas is called the *method of exhaustion*. As applied to finding the area of a circular region, the method consists of approximating the circular region by a regular $n$-sided polygon, as shown in Figure 4.26. By dividing the region bounded by the polygon into $n$ triangular regions, we see that the height of each triangular region is approximately equal to the radius of the circle. That is,

$$h \approx r$$

Also, the base of each triangular region is approximately equal to one-$n$th of the circumference of the circle. That is,

$$b \approx \frac{2\pi r}{n}$$

which means that the approximate area of each triangular region is given by

$$\text{Area of triangular region} = \frac{1}{2}bh \approx \frac{1}{2}\left(\frac{2\pi r}{n}\right)(r)$$

$$= \frac{\pi r^2}{n}$$

An important aspect of this approximation is that as $n$ increases, the approximation improves. Now, summing the areas of the $n$ triangular regions, we have

$$\text{Area of circle} \approx \text{area of } n \text{ triangular regions}$$

$$= n\left(\frac{\pi r^2}{n}\right)$$

$$= \pi r^2$$

Circumference $= 2\pi r$

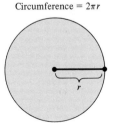
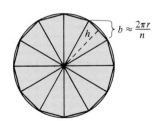

$b \approx \dfrac{2\pi r}{n}$

FIGURE 4.26

■ **Approximating a Definite Integral by the Midpoint Rule**
■ **The Definite Integral as the Limit of a Sum**

For some types of applications involving definite integrals, it is helpful to look at definite integration as a *summation process*. In this section, we take such an approach.

**EXAMPLE 1**

**Approximating the Area of a Plane Region**

Use the five rectangles in Figure 4.27 to approximate the area of the region lying between the graph of

$$f(x) = -x^2 + 5$$

and the *x*-axis between $x = 0$ and $x = 2$.

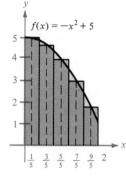

$f(x) = -x^2 + 5$

FIGURE 4.27

**SOLUTION**

We can find the height of the five rectangles shown in Figure 4.27 by evaluating the function $f$ at the midpoint of each of the following intervals:

$$\left[0, \frac{2}{5}\right], \quad \left[\frac{2}{5}, \frac{4}{5}\right], \quad \left[\frac{4}{5}, \frac{6}{5}\right], \quad \left[\frac{6}{5}, \frac{8}{5}\right], \quad \left[\frac{8}{5}, \frac{10}{5}\right]$$

Evaluate $f(x)$ at the midpoints of these intervals.

Since the width of each rectangle is $\frac{2}{5}$, the sum of the areas of the five rectangles is

$$\text{Area} \approx f\left(\frac{1}{5}\right)\left(\frac{2}{5}\right) + f\left(\frac{3}{5}\right)\left(\frac{2}{5}\right) + f\left(\frac{5}{5}\right)\left(\frac{2}{5}\right) + f\left(\frac{7}{5}\right)\left(\frac{2}{5}\right) + f\left(\frac{9}{5}\right)\left(\frac{2}{5}\right)$$

$$= \frac{2}{5}\left[f\left(\frac{1}{5}\right) + f\left(\frac{3}{5}\right) + f\left(\frac{5}{5}\right) + f\left(\frac{7}{5}\right) + f\left(\frac{9}{5}\right)\right]$$

$$= \frac{2}{5}\left[\frac{124}{25} + \frac{116}{25} + \frac{100}{25} + \frac{76}{25} + \frac{44}{25}\right]$$

$$= \frac{2}{125}[124 + 116 + 100 + 76 + 44]$$

$$= \frac{920}{125} = 7.36$$

For the region in Example 1 we can find the exact area with a definite integral. That is,

$$\text{Area} = \int_0^2 (-x^2 + 5) \, dx = \left[ -\frac{x^3}{3} + 5x \right]_0^2$$

$$= \left( -\frac{8}{3} + 10 \right) - (0)$$

$$= \frac{22}{3} \approx 7.3333$$

We call the approximation procedure used in Example 1 the **Midpoint Rule.** We can apply the Midpoint Rule to approximate any definite integral—not just those representing area. The basic steps are summarized as follows.

---

**The midpoint rule for approximating a definite integral**

To approximate the definite integral

$$\int_a^b f(x) \, dx$$

by the Midpoint Rule, use the following steps.

1. Divide the interval $[a, b]$ into $n$ subintervals, each of width

$$\Delta x = \frac{b - a}{n}$$

2. Find the midpoint of each subinterval.

$$\text{Midpoints} = \{x_1, x_2, x_3, \ldots, x_n\}$$

3. Evaluate $f$ at each of these midpoints and form the following sum:

$$\int_a^b f(x) \, dx \approx \frac{b - a}{n} [f(x_1) + f(x_2) + f(x_3) + \cdots + f(x_n)]$$

---

An important characteristic of the Midpoint Rule is that the approximation tends to improve as $n$ increases. Table 4.2 shows the approximations for the area of the region described in Example 1 for various values of $n$. For example, for $n = 10$, the Midpoint Rule yields the following approximation:

$$\int_0^2 (-x^2 + 5) \, dx \approx \frac{2}{10} \left[ f\!\left(\frac{1}{10}\right) + f\!\left(\frac{3}{10}\right) + f\!\left(\frac{5}{10}\right) + \cdots + f\!\left(\frac{19}{10}\right) \right]$$

$$\approx 7.3400$$

TABLE 4.2

| $n$ | 5 | 10 | 15 | 20 | 25 | 30 |
|---|---|---|---|---|---|---|
| Approximation | 7.3600 | 7.3400 | 7.3363 | 7.3350 | 7.3344 | 7.3341 |

Note from this table that as $n$ increases the approximation appears to be getting closer and closer to the exact value of the integral, which we found to be

$$\frac{22}{3} \approx 7.3333$$

In fact, taking the limit as $n$ approaches infinity makes the approximation exact. That is, if $\Delta x = (b - a)/n$, then

$$\int_a^b f(x)\, dx = \lim_{n \to \infty} [f(x_1)\, \Delta x + f(x_2)\, \Delta x + \cdots + f(x_n)\, \Delta x]$$

For the Midpoint Rule we choose $x_i$ to be the midpoint of the $i$th subinterval. However, in the limit form, $x_i$ can be any number in the $i$th interval. We summarize this result as follows.

**Definite integral as the limit of a sum**

If $f$ is continuous on the interval $[a, b]$, then

$$\int_a^b f(x)\, dx = \lim_{n \to \infty} [f(x_1)\, \Delta x + f(x_2)\, \Delta x + \cdots + f(x_n)\, \Delta x]$$

where $\Delta x = (b - a)/n$ and $x_i$ is any number in the $i$th interval.

In Example 1, we used the Midpoint Rule to approximate the area of the region bounded by the graph of $f(x) = -x^2 + 5$ and the $x$-axis between $x = 0$ and $x = 2$. Of course, for this particular region we wouldn't normally go to the trouble of approximating the area, since we can find the exact area by evaluating the definite integral

$$\int_0^2 (-x^2 + 5)\, dx$$

The real usefulness of the Midpoint Rule lies in its application in problems where we encounter definite integrals that we don't know how to evaluate. This is demonstrated in the next two examples.

**EXAMPLE 2**

**Using the Midpoint Rule to Approximate a Definite Integral**

Use the Midpoint Rule with $n = 4$ to approximate the definite integral

$$\int_1^2 \frac{1}{x}\, dx$$

**SOLUTION**

Using four subintervals, each of width

$$\Delta x = \frac{2 - 1}{4} = \frac{1}{4}$$

we divide the interval $[1, 2]$ into the following subintervals:

$$\left[1, \frac{5}{4}\right], \quad \left[\frac{5}{4}, \frac{3}{2}\right], \quad \left[\frac{3}{2}, \frac{7}{4}\right], \quad \left[\frac{7}{4}, 2\right]$$

Since the midpoints of these subintervals are given by

$$\overset{x_1 \quad x_2 \quad x_3 \quad x_4}{\text{Midpoints} = \left\{\frac{9}{8}, \frac{11}{8}, \frac{13}{8}, \frac{15}{8}\right\}}$$

we approximate the definite integral as follows:

$$\int_1^2 \frac{1}{x}\, dx \approx \frac{b - a}{n}[f(x_1) + f(x_2) + f(x_3) + f(x_4)]$$

$$= \frac{1}{4}\left[\frac{8}{9} + \frac{8}{11} + \frac{8}{13} + \frac{8}{15}\right]$$

$$\approx 0.691$$

The area represented by this integral is shown in Figure 4.28. (In Section 5.4, we will see that the actual area is given by $\ln 2 \approx 0.693$. Thus, the approximation is off by only 0.002.)

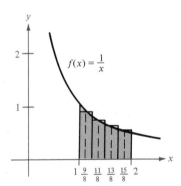

$f(x) = \dfrac{1}{x}$

FIGURE 4.28

**EXAMPLE 3**

**Using the Midpoint Rule to Approximate a Definite Integral**

Use the Midpoint Rule with $n = 5$ to approximate the definite integral

$$\int_0^1 \frac{1}{x^2 + 1}\, dx$$

**SOLUTION**

Using five subintervals, each of width

$$\Delta x = \frac{1 - 0}{5} = \frac{1}{5}$$

we divide the interval $[0, 1]$ into the following subintervals:

$$\left[0, \frac{1}{5}\right], \quad \left[\frac{1}{5}, \frac{2}{5}\right], \quad \left[\frac{2}{5}, \frac{3}{5}\right], \quad \left[\frac{3}{5}, \frac{4}{5}\right], \quad \left[\frac{4}{5}, 1\right]$$

Since the midpoints of these subintervals are given by

$$\text{Midpoints} = \left\{ \overbrace{\frac{1}{10}}^{x_1}, \overbrace{\frac{3}{10}}^{x_2}, \overbrace{\frac{5}{10}}^{x_3}, \overbrace{\frac{7}{10}}^{x_4}, \overbrace{\frac{9}{10}}^{x_5} \right\}$$

we approximate the definite integral as follows:

$$\int_0^1 \frac{1}{x^2+1} \, dx \approx \frac{b-a}{n}[f(x_1) + f(x_2) + f(x_3) + f(x_4) + f(x_5)]$$

$$= \frac{1}{5}\left[\frac{1}{1.01} + \frac{1}{1.09} + \frac{1}{1.25} + \frac{1}{1.49} + \frac{1}{1.81}\right]$$

$$\approx 0.786$$

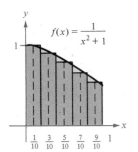

$f(x) = \dfrac{1}{x^2+1}$

The area represented by this integral is shown in Figure 4.29. (The actual area of this region is given by $\pi/4 \approx 0.785$. Thus, the approximation is off by only 0.001.)

FIGURE 4.29

---

## SECTION EXERCISES 4.5

In Exercises 1–4, use the Midpoint Rule with $n = 4$ to approximate the area of the region bounded by the graph of the given function and the $x$-axis over the indicated interval. Compare this result with the exact area obtained by using the definite integral.

**1.** $f(x) = -2x + 3$, $[0, 1]$

**2.** $f(x) = 1 - x^2$, $[-1, 1]$

**3.** $y = \sqrt{x}$, $[0, 1]$

**4.** $y = \sqrt{x} + 1$, $[0, 2]$

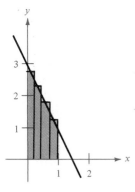

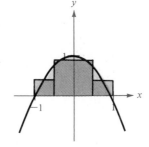

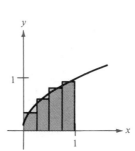

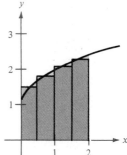

In Exercises 5–14, use the Midpoint Rule with $n = 4$ to approximate the area of the region bounded by the graph of the given function and the $x$-axis over the indicated interval. Compare this result with the exact area obtained by using the definite integral. Sketch the region.

|  Function  | Interval |
|------------|----------|
| **5.** $y = x^2 + 2$ | $[-1, 1]$ |
| **6.** $y = 3x - 4$ | $[2, 5]$ |
| **7.** $g(x) = 2x^2$ | $[1, 3]$ |
| **8.** $g(x) = 2x - x^3$ | $[0, 1]$ |
| **9.** $f(x) = 1 - x^3$ | $[0, 1]$ |
| **10.** $f(x) = x^2 - x^3$ | $[0, 1]$ |
| **11.** $y = x^2 - x^3$ | $[-1, 0]$ |
| **12.** $y = 2x^2 - x + 1$ | $[0, 2]$ |
| **13.** $y = x(1 - x)^2$ | $[0, 1]$ |
| **14.** $y = x^2(3 - x)$ | $[0, 3]$ |

In Exercises 15 and 16, use the Midpoint Rule with $n = 4$ to approximate the area of the region between the graph of the function and the $y$-axis over the indicated interval. Compare this result with the exact area obtained by using the definite integral.

**15.** $f(y) = 3y$, $[0, 2]$        **16.** $f(y) = 4y - y^2$, $[0, 4]$

In Exercises 17–20, complete the following table using the Midpoint Rule to approximate the given definite integral.

| $n$ | 2 | 4 | 8 |
|-----|---|---|---|
| Approximation | | | |

**17.** $\displaystyle\int_0^4 \sqrt{16 - x^2}\, dx$   (Exact value is $4\pi$)

**18.** $\displaystyle\int_{-1}^1 (3 - x)\sqrt{1 - x^2}\, dx$  $\left(\text{Exact value is } \dfrac{3\pi}{2}\right)$

**19.** $\displaystyle\int_0^2 \sqrt{1 + x^3}\, dx$

**20.** $\displaystyle\int_0^2 x\sqrt{1 + x^3}\, dx$

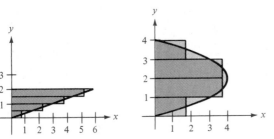

# Volumes of Solids of Revolution

A manufacturer plans to make gold wedding rings out of a gold alloy that costs $1,200 per cubic inch. To estimate the cost of material for one ring, we find the volume of the ring using the dimensions shown in Figure 4.30. Mathematically, we classify the ring shown in this figure as a **solid of revolution,** and we imagine that it is formed by revolving a plane region about a line. For this particular solid, the plane region is bounded above by the graph of

$$f(x) = \frac{7}{16} - 16x^2$$

and below by the straight line given by $g(x) = \frac{3}{8}$. As this region is revolved about the x-axis, it generates the ring shown in the figure.

In this section, we will see that the **volume** of this particular solid of revolution is given by

$$\text{Volume} = \pi \int_{-1/16}^{1/16} ([f(x)]^2 - [g(x)]^2) \, dx$$

$$= \pi \int_{-1/16}^{1/16} \left[ \left( \frac{7}{16} - 16x^2 \right)^2 - \left( \frac{3}{8} \right)^2 \right] dx$$

$$= \pi \int_{-1/16}^{1/16} \left( \frac{13}{256} - 14x^2 + 256x^4 \right) dx$$

$$= \pi \left[ \frac{13x}{256} - \frac{14x^3}{3} + \frac{256x^5}{5} \right]_{-1/16}^{1/16}$$

$$\approx 0.01309 \text{ cubic inches}$$

Now, since the alloy for the ring costs $1,200 per cubic inch, the total cost of material for the ring is given by

$$\text{Cost} = 0.01309(1200) \approx \$15.71$$

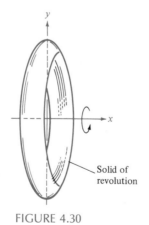

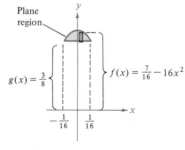

FIGURE 4.30

### Disc Method

In Sections 4.3 and 4.4, we saw that a definite integral can be used to find the *area* of a region and the *average value* of a function. These are only two of a wide variety of applications of definite integrals. For example, definite integrals can be used to find the *work* done by a force, the *arc length* of a curve, the *surface area* of a solid, the *center of mass* of a solid, and many other quantities that are useful in the physical sciences.

In this section, we will see how integration can be used to find the volume of a solid. Specifically, we will look at solids that are formed by revolving plane regions about an axis.

---

**Definition of a solid of revolution**

If a region in the plane is revolved about a line, the resulting solid is called a **solid of revolution** and the line is called the **axis of revolution,** as shown in Figure 4.31.

---

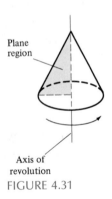

Plane region

Axis of revolution

FIGURE 4.31

To develop a formula for finding the volume of a solid of revolution, we consider a continuous function $y = f(x)$ that is nonnegative on the interval $[a, b]$. Figure 4.32 shows that the area of the region bounded by the graph of $f$ and the $x$-axis ($a \leq x \leq b$) can be approximated by $n$ rectangles, each of width

$$\Delta x = \frac{b - a}{n}$$

By revolving each of the rectangles in Figure 4.32 about the $x$-axis, we form a circular disc (see Figure 4.33) whose volume is given by

$$\text{Volume of } i\text{th disc} = \pi(\text{radius})^2(\text{width})$$
$$= \pi[f(x_i)]^2 \, \Delta x$$

and by summing the volumes of all $n$ discs, we have

$$\begin{array}{l} \text{Volume of solid} \\ \text{of revolution} \end{array} \approx \pi[f(x_1)]^2 \, \Delta x + \pi[f(x_2)]^2 \, \Delta x + \cdots + \pi[f(x_n)]^2 \, \Delta x$$

as shown in Figure 4.33. Finally, by letting $n$ approach infinity, we can conclude that

$$\text{Volume of solid of revolution} = \pi \int_a^b [f(x)]^2 \, dx$$

---

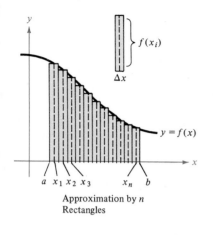

$f(x_i)$

$\Delta x$

$y = f(x)$

$a$ $x_1$ $x_2$ $x_3$     $x_n$ $b$

Approximation by $n$
Rectangles

FIGURE 4.32

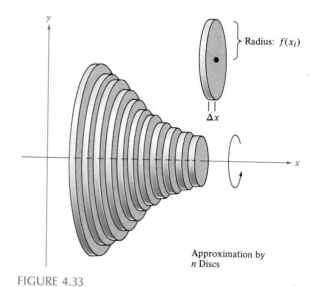

Radius: $f(x_i)$

$\Delta x$

Approximation by
$n$ Discs

FIGURE 4.33

| **Disc method for finding the volume of a solid of revolution** | The volume of the solid formed by revolving the region bounded by $y = f(x)$ and the $x$-axis ($a \le x \le b$) about the $x$-axis is given by <br><br> $$\text{Volume} = \pi \int_a^b [f(x)]^2 \, dx$$ |
|---|---|

In Figure 4.34, note that the *representative rectangle* shown in the plane region is useful in visualizing the radius $f(x)$.

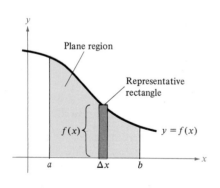

Plane region

Representative
rectangle

$f(x)$

$a$    $\Delta x$   $b$

$y = f(x)$

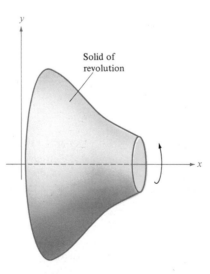

Solid of
revolution

FIGURE 4.34

EXAMPLE 1

**Finding the Volume of a Solid of Revolution**

Find the volume of the solid formed by revolving the region bounded by the graph of

$$y = -x^2 + x$$

and the line $y = 0$ about the $x$-axis.

SOLUTION

From the representative rectangle in Figure 4.35 we see that the radius of this solid is given by

$$\text{Radius} = f(x)$$
$$= -x^2 + x$$

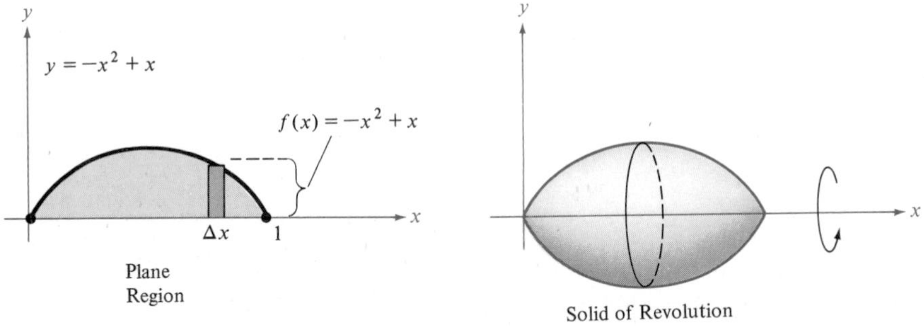

FIGURE 4.35

Plane Region

Solid of Revolution

and it follows that its volume is

$$\text{Volume} = \pi \int_0^1 [f(x)]^2 \, dx = \pi \int_0^1 (-x^2 + x)^2 \, dx$$

$$= \pi \int_0^1 (x^4 - 2x^3 + x^2) \, dx$$

$$= \pi \left[ \frac{x^5}{5} - \frac{x^4}{2} + \frac{x^3}{3} \right]_0^1$$

$$= \frac{\pi}{30}$$

$$\approx 0.105$$

■ **Remark:** Note in Example 1 that the entire problem was worked *without* any reference to the three-dimensional sketch given in Figure 4.35. In general, to set up the integral for calculating the volume of a solid of revolution, a sketch of the plane region is more useful than a sketch of the solid, since the radius is more readily visualized in the plane region.

We can extend the Disc Method to find the volume of a solid of revolution with a *hole*, as follows. Suppose that a region is bounded by $y = f(x)$ and $y = g(x)$, as shown in Figure 4.36. If this region is revolved about the $x$-axis, then the volume of the resulting solid is given by

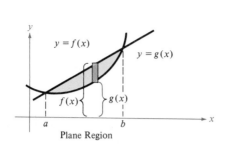

FIGURE 4.36

$$\text{Volume} = \pi \int_a^b [f(x)]^2 \, dx - \pi \int_a^b [g(x)]^2 \, dx$$

$$= \pi \int_a^b [[f(x)]^2 - [g(x)]^2] \, dx$$

We call $f(x)$ the *outer radius* of the solid and $g(x)$ the *inner radius*. Note that the integral involving the inner radius represents the volume of the hole and is *subtracted* from the integral involving the outer radius.

**EXAMPLE 2**

**Finding the Volume of a Solid of Revolution with a Hole**

Find the volume of the solid formed by revolving the region bounded by the graphs of

$$f(x) = \sqrt{25 - x^2} \qquad \text{and} \qquad g(x) = 3$$

about the *x*-axis, as shown in Figure 4.37.

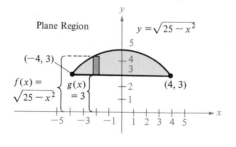

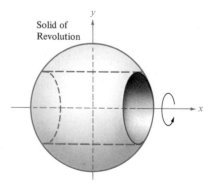

FIGURE 4.37

**SOLUTION**

We begin by finding the points of intersection of $f$ and $g$ by setting $f(x)$ equal to $g(x)$ and solving for $x$ as follows.

$$f(x) = g(x)$$

$$\sqrt{25 - x^2} = 3$$

$$25 - x^2 = 9$$

$$x^2 = 16$$

$$x = \pm 4$$

Now, using $f(x)$ as the outer radius and $g(x)$ as the inner radius, we integrate from $-4$ to $4$ to obtain the volume.

$$\text{Volume} = \pi \int_{-4}^{4} [[f(x)]^2 - [g(x)]^2] \, dx$$

$$= \pi \int_{-4}^{4} ([\sqrt{25 - x^2}]^2 - [3]^2) \, dx$$

$$= \pi \int_{-4}^{4} (16 - x^2) \, dx$$

$$= \pi \left[ 16x - \frac{x^3}{3} \right]_{-4}^{4}$$

$$= \frac{256\pi}{3} \approx 268.08$$

Occasionally an application of the Disc Method involves a vertical axis of revolution. In such cases we must find the radius of the solid as a function of $y$ and integrate along the $y$-axis, as demonstrated in the next example.

**EXAMPLE 3**

**An Application to Biology**

A pond is to be stocked with a certain species of fish. It has been determined that the food supply in 500 cubic feet of this pond water can adequately support one fish. The pond is nearly circular in circumference, is 20 feet deep at its center, and has a radius of 200 feet. Assuming that the bottom of the pond can be approximated by the model

$$y = 20\left[ \left( \frac{x}{200} \right)^2 - 1 \right]$$

find the volume of water in the pond and then estimate the maximum number of fish that the pond can support.

**SOLUTION**

We see from Figure 4.38 that the radius of the pond is measured *horizontally* rather than vertically. Hence, we must adapt the Disc Method by integrating along the $y$-axis from $-20$ to $0$. To find the radius, we solve for $x$ in the given equation as follows:

$$20\left[\left(\frac{x}{200}\right)^2 - 1\right] = y$$

$$\left(\frac{x}{200}\right)^2 - 1 = \frac{y}{20}$$

$$\left(\frac{x}{200}\right)^2 = \frac{y}{20} + 1$$

$$\frac{x}{200} = \sqrt{\frac{y}{20} + 1}$$

$$x = 200\sqrt{\frac{y}{20} + 1}$$

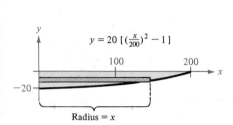

$y = 20\left[\left(\frac{x}{200}\right)^2 - 1\right]$

Radius = $x$

FIGURE 4.38

Now, using $x$ as the radius, we find the volume of the pond:

$$\text{Volume} = \pi \int_{-20}^{0} \left(200\sqrt{\frac{y}{20} + 1}\right)^2 dy = 40{,}000\pi \int_{-20}^{0} \left(\frac{y}{20} + 1\right) dy$$

$$= 40{,}000\pi \left[\frac{y^2}{40} + y\right]_{-20}^{0} = 40{,}000\pi \left[0 - \left(\frac{400}{40} - 20\right)\right]$$

$$\approx 1{,}256{,}637 \text{ cubic feet}$$

Finally, since each fish requires 500 cubic feet of water, the maximum number of fish that this pond can support is

$$\frac{1{,}256{,}637}{500} \approx 2513 \text{ fish}$$

## SECTION EXERCISES 4.6

In Exercises 1–16, find the volume of the solid formed by revolving the region bounded by the graphs of the given equations about the $x$-axis.

**1.** $y = -x + 1$

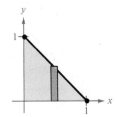

**2.** $y = 4 - x^2$

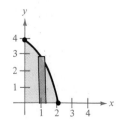

**3.** $y = \sqrt{4 - x^2}$

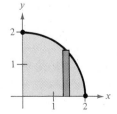

**4.** $y = x^2$

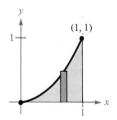

(1, 1)

**5.** $y = \sqrt{x}$

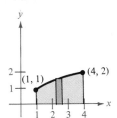

**6.** $y = \sqrt{4 - x^2}$

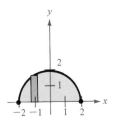

**7.** $y = x^2$, $y = x^3$

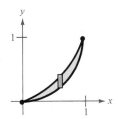

**8.** $y = 4 - \dfrac{x^2}{2}$, $y = 2$

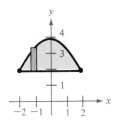

**9.** $y = x$, $y = 0$, $x = 4$   **10.** $y = \sqrt{x}$, $y = 0$, $x = 4$

**11.** $y = 2x^2$, $y = 0$, $x = 2$   **12.** $y = 1 - \dfrac{x^2}{4}$, $y = 0$

**13.** $y = \dfrac{1}{x}$, $y = 0$, $x = 1$, $x = 3$

**14.** $y = x^2 + 1$, $y = 5$

**15.** $y = 6 - 2x - x^2$, $y = x + 6$

**16.** $y = x^2$, $y = 4x - x^2$

In Exercises 17–24, find the volume of the solid formed by revolving the region bounded by the graphs of the given equations about the y-axis.

**17.** $y = x^2$

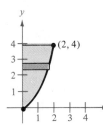

**18.** $y = \sqrt{16 - x^2}$

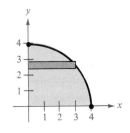

**19.** $y = x^{2/3}$

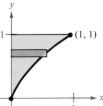

**20.** $x = -y^2 + 4y$

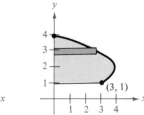

**21.** $x = y - 1$, $x = 0$, $y = 0$   **22.** $x = y(y - 1)$, $x = 0$
**23.** $y = \sqrt{4 - x}$, $y = 0$, $x = 0$
**24.** $y = 4$, $y = 0$, $x = 2$, $x = 0$

**25.** If the portion of the line $y = \frac{1}{2}x$ lying in the first quadrant is revolved about the x-axis, a cone is generated. Find the volume of the cone formed by the line extending from $x = 0$ to $x = 6$.

**26.** Use the Disc Method to verify that the volume of a right circular cone is $\frac{1}{3}\pi r^2 h$, where $r$ is the radius of the base and $h$ is the height.

**27.** Use the Disc Method to verify that the volume of a sphere of radius $r$ is $\frac{4}{3}\pi r^3$.

**28.** The right half of the ellipse $9x^2 + 25y^2 = 225$ is revolved about the y-axis to form an oblate spheroid (shaped like an M&M candy). Find the volume of the spheroid.

**29.** The upper half of the ellipse $9x^2 + 25y^2 = 225$ is revolved about the x-axis to form a prolate spheroid (shaped like a football). Find the volume of the spheroid.

**30.** A tank on the wing of a jet is formed by revolving the region bounded by the graph of $y = \frac{1}{8}x^2\sqrt{2 - x}$ and the x-axis about the x-axis, as shown in Figure 4.39, where $x$ and $y$ are measured in meters. Find the volume of the tank.

In Exercises 31 and 32, solve the problem given in Example 3 of this section using the given dimensions of the pond.

**31.** 15 feet deep at its center, radius of 150 feet
**32.** 30 feet deep at its center, radius of 300 feet

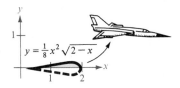

FIGURE 4.39

# CHAPTER 4 SUMMARY

## Important Terms

Antiderivative
Integration
Integral sign
Indefinite integral
Integrand
Constant of integration
Particular solution
Initial condition
Substitution
Change of variables
Area of a plane region
Definite integral from $a$ to $b$
Lower limit of integration
Upper limit of integration
Average value of a function
Area of a region between two curves
Summation process
Solid of revolution
Axis of revolution
Representative rectangle
Outer radius
Inner radius

## Important Techniques

Finding an antiderivative of a function
Using an initial condition to find a particular solution
Using pattern recognition or substitution to find antiderivatives
Finding the area of a region in the plane
Evaluating a definite integral
Finding the average value of a function over an interval
Approximating a definite integral by the Midpoint Rule
Finding the volume of a solid of revolution

## Important Formulas

General pattern for integration:

Given integral → Rewrite → Integrate → Simplify

Basic integration rules:

Constant Rule: $\int k\,dx = kx + C$

Constant Multiple Rule: $\int kf(x)\,dx = k\int f(x)\,dx$

Sum Rule:

$$\int [f(x) \pm g(x)]\,dx = \int f(x)\,dx \pm \int g(x)\,dx$$

Simple Power Rule: $\int x^n\,dx = \dfrac{x^{n+1}}{n+1} + C,\ n \neq -1$

General Power Rule: $\int u^n\dfrac{du}{dx}\,dx = \dfrac{u^{n+1}}{n+1} + C,\ n \neq -1$

Fundamental Theorem of Calculus:
  If a function $f$ is continuous on the interval $[a, b]$, then

$$\int_a^b f(x)\,dx = F(b) - F(a)$$

  where $F'(x) = f(x)$ for all $x$ in $[a, b]$.

Properties of definite integrals:

1. $\int_a^b kf(x)\,dx = k\int_a^b f(x)\,dx,\quad k$ is a constant

2. $\int_a^b [f(x) \pm g(x)]\,dx = \int_a^b f(x)\,dx \pm \int_a^b g(x)\,dx$

3. $\int_a^b f(x)\,dx = \int_a^c f(x)\,dx + \int_c^b f(x)\,dx,\ a < c < b$

4. $\int_a^a f(x)\,dx = 0$

5. $\int_b^a f(x)\,dx = -\int_a^b f(x)\,dx$

Average of $f$ on $[a, b] = \dfrac{1}{b-a}\int_a^b f(x)\,dx$

If $g(x) \leq f(x)$ for all $x$ in $[a, b]$, then the area of the region bounded by $y = f(x)$, $y = g(x)$, $x = a$, and $x = b$ is given by

$$A = \int_a^b [f(x) - g(x)]\,dx$$

Midpoint Rule:

$$\int_a^b f(x)\,dx \approx \frac{b-a}{n}[f(x_1) + f(x_2) + f(x_3)$$
$$+ \cdots + f(x_n)]$$

Definite integral as the limit of a sum:
  If $f$ is continuous on the interval $[a, b]$, then

$$\int_a^b f(x)\,dx = \lim_{n\to\infty} [f(x_1)\,\Delta x + f(x_2)\,\Delta x$$
$$+ \cdots + f(x_n)\,\Delta x]$$

  where $\Delta x = (b - a)/n$ and $x_i$ is any number in the $i$th interval.

Disc Method for the volume of a solid of revolution:
  The volume of the solid formed by revolving the region bounded by $y = f(x)$ and the $x$-axis ($a \leq x \leq b$) about the $x$-axis is

$$\text{Volume} = \pi\int_a^b [f(x)]^2\,dx$$

Disc Method for the volume of a solid of revolution with a hole:

$$\text{Volume} = \pi\int_a^b [[f(x)]^2 - [g(x)]^2]\,dx$$

In Exercises 1–20, find the indefinite integral.

**1.** $\int (2x^2 + x - 1)\, dx$

**2.** $\int (5 - 6x^2 - x^3)\, dx$

**3.** $\int \dfrac{x^2 + 3}{x^2}\, dx$

**4.** $\int \dfrac{x^3 + 1}{x^2}\, dx$

**5.** $\int \dfrac{x^2}{(x^3 - 1)^2}\, dx$

**6.** $\int \left(x + \dfrac{1}{x}\right)^2 dx$

**7.** $\int \dfrac{x^3 - 2x^2 + 1}{x^2}\, dx$

**8.** $\int \dfrac{x + 3}{(x^2 + 6x - 5)^2}\, dx$

**9.** $\int \dfrac{2}{3\sqrt[3]{x}}\, dx$

**10.** $\int \dfrac{2}{\sqrt[3]{3x}}\, dx$

**11.** $\int \dfrac{(1 + x)^2}{\sqrt{x}}\, dx$

**12.** $\int x^2\sqrt{x^3 + 3}\, dx$

**13.** $\int \dfrac{x^2}{\sqrt{x^3 + 3}}\, dx$

**14.** $\int \dfrac{1}{\sqrt{5x - 1}}\, dx$

**15.** $\int \dfrac{3x}{\sqrt{1 - 2x^2}}\, dx$

**16.** $\int \dfrac{3}{\sqrt{1 - 2x}}\, dx$

**17.** $\int (x^2 + 1)^3\, dx$

**18.** $\int \sqrt{2 - 5x}\, dx$

**19.** $\int \sqrt{x}(x + 3)\, dx$

**20.** $\int \sqrt{x}(x + 1)^2\, dx$

In Exercises 21–30, use the Fundamental Theorem of Calculus to evaluate the definite integral.

**21.** $\displaystyle\int_0^4 (2 + x)\, dx$

**22.** $\displaystyle\int_{-1}^1 (t^2 + 2)\, dt$

**23.** $\displaystyle\int_{-1}^1 (4t^3 - 2t)\, dt$

**24.** $\displaystyle\int_3^6 \dfrac{x}{3\sqrt{x^2 - 8}}\, dx$

**25.** $\displaystyle\int_0^3 \dfrac{1}{\sqrt{1 + x}}\, dx$

**26.** $\displaystyle\int_0^1 x^2(x^3 + 1)^3\, dx$

**27.** $\displaystyle\int_4^9 x\sqrt{x}\, dx$

**28.** $\displaystyle\int_1^2 \left(\dfrac{1}{x^2} - \dfrac{1}{x^3}\right) dx$

**29.** $2\pi \displaystyle\int_{-1}^0 x^2(x + 1)^2\, dx$

**30.** $2\pi \displaystyle\int_0^1 (y + 1)^2(1 - y)\, dy$

In Exercises 31–38, set up a definite integral that yields the area of the given region. (Do not evaluate the integral.)

**31.** $f(x) = 3$

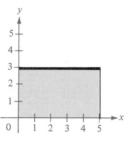

**32.** $f(x) = 4 - 2x$

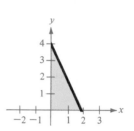

**33.** $f(x) = x^2$

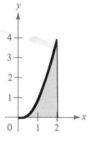

**34.** $f(x) = 4 - |x|$

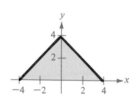

**35.** $f(x) = 4 - x^2$

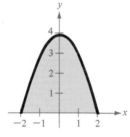

**36.** $f(y) = (y - 2)^2$

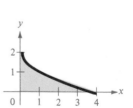

**37.** $f(y) = y^3$

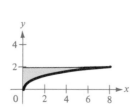

**38.** $f(x) = \sqrt{9 - x^2}$

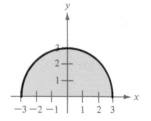

In Exercises 39–44, sketch the graph of the region whose area is given by the integral, and find the area.

**39.** $\int_0^2 (2 - x)\, dx$    **40.** $\int_0^2 2\, dx$

**41.** $\int_1^4 \frac{1}{\sqrt{x}}\, dx$    **42.** $\int_1^8 x^{2/3}\, dx$

**43.** $\int_0^4 \frac{y}{2}\, dy$    **44.** $\int_0^2 (2y - y^2)\, dy$

**45.** If $\int_2^6 f(x)\, dx = 10$ and $\int_2^6 g(x)\, dx = -2$, find

(a) $\int_2^6 [f(x) + g(x)]\, dx$    (b) $\int_2^6 [g(x) - f(x)]\, dx$

(c) $\int_2^6 [2f(x) - 3g(x)]\, dx$    (d) $\int_2^6 3f(x)\, dx$

**46.** If $\int_0^3 f(x)\, dx = 4$ and $\int_3^6 f(x)\, dx = -1$, find

(a) $\int_0^6 f(x)\, dx$    (b) $\int_6^3 f(x)\, dx$

(c) $\int_4^4 f(x)\, dx$    (d) $\int_3^6 -5f(x)\, dx$

In Exercises 47–50, approximate the definite integral by the Midpoint Rule, letting $n = 4$.

**47.** $\int_0^2 x^3\, dx$    **48.** $\int_0^2 4x^2\, dx$

**49.** $\int_0^1 \sqrt{1 - x^2}\, dx$    **50.** $\int_0^1 \frac{1}{x + 1}\, dx$

In Exercises 51–60, sketch the region bounded by the graphs of the given equations and determine the area of the region.

**51.** $y = \frac{1}{x^2}$, $y = 0$, $x = 1$, $x = 5$

**52.** $y = \frac{1}{x^2}$, $y = 4$, $x = 5$    **53.** $y = x$, $y = 2 - x^2$

**54.** $y = 1 - \frac{x}{2}$, $y = x - 2$, $y = 1$

**55.** $x = y^2 - 2y$, $x = 0$    **56.** $x = y^2 + 1$, $x = y + 3$
**57.** $\sqrt{x} + \sqrt{y} = 1$, $y = 0$, $x = 0$

**58.** $y = \frac{4}{\sqrt{x} + 1}$, $y = 0$, $x = 0$, $x = 8$

**59.** $y = \sqrt{x}(x - 1)$, $y = 0$    **60.** $y = x$, $y = x^5$

In Exercises 61 and 62, find the average value of the function over the given interval. Find the values of $x$ where the function assumes its mean value and sketch the graph of the function.

| Function | Interval |
|---|---|
| **61.** $f(x) = \dfrac{1}{\sqrt{x - 1}}$ | $[5, 10]$ |
| **62.** $f(x) = x^3$ | $[0, 2]$ |

In Exercises 63 and 64, find the volume of the solid generated by revolving the region bounded by the graphs of the given equations about (a) the $x$-axis and (b) the $y$-axis.
**63.** $y = \sqrt{16 - x}$, $y = 0$, $x = 0$
**64.** $y = \sqrt{x}$, $y = 2$, $x = 0$

**65.** Find the function $f$ whose derivative is $f'(x) = -2x$ and whose graph passes through the point $(-1, 1)$.
**66.** A function $f$ has a second derivative $f''(x) = 6(x - 1)$. Find the function if its graph passes through the point $(2, 1)$ and at that point is tangent to the line $3x - y - 5 = 0$.
**67.** An airplane taking off from a runway travels 3600 feet before lifting off. If it starts from rest, moves with constant acceleration, and makes the run in 30 seconds, with what velocity does it lift off?
**68.** A ball is thrown vertically upward from the ground with an initial velocity of 96 ft/sec. (The acceleration due to gravity is $a(t) = -32$ ft/sec$^2$.)
(a) How long will the ball take to rise to its maximum height?
(b) What is the maximum height?
(c) When is the velocity of the ball one-half the initial velocity?
(d) What is the height of the ball when its velocity is one-half the initial velocity?
**69.** Suppose that gasoline is increasing in price according to the equation

$$p = 1.00 + 0.1t + 0.02t^2$$

where $p$ is the dollar price per gallon and $t = 0$ represents the year 1985. If an automobile is driven 15,000 miles a year and gets $M$ miles per gallon, then the annual fuel cost is

$$C = \frac{15,000}{M} \int_t^{t+1} p\, dt$$

Find the annual fuel cost for the years
(a) 1987    (b) 1990

# Exponential and Logarithmic Functions

# CHAPTER 5

# Exponential Functions

## INTRODUCTORY EXAMPLE
## Radioactive Carbon Dating

In living organic material the ratio of radioactive carbon isotopes to the total number of carbon atoms is about 1 to $10^{12}$. When organic material dies, its radioactive carbon isotopes begin to decay, with a half-life of about 5700 years. This means that after 5700 years the ratio of isotopes to atoms will have decreased to one-half of the original ratio, after a second 5700 years the ratio will have decreased to one-fourth of the original, and so on. Figure 5.1 shows this declining ratio. The formula for the ratio $R$ of carbon isotopes to carbon atoms is given by the **exponential function**

$$R = 10^{-12}2^{-t/5700}$$

where $t$ is measured in years.

Suppose that we want to estimate the age of a certain fossil, and that by testing the carbon makeup of the fossil we determine that the ratio is 1 isotope to $10^{13}$ atoms. Thus, we have $R = 10^{-13}$, and we can estimate the age of the fossil by solving the equation

$$10^{-13} = 10^{-12}2^{-t/5700}$$

Since $0.1 \times 10^{-12} = (10^{-1})(10^{-12}) = 10^{-13}$, it follows that

$$\frac{1}{10} = \frac{1}{2^{t/5700}}$$

$$10 = 2^{t/5700}$$

In Section 5.3, we will see that this equation can be solved by using **logarithms,** and the solution turns out to be

$$t = 5700 \log_2 10 \approx 18{,}935 \text{ years}$$

as indicated in Figure 5.1.

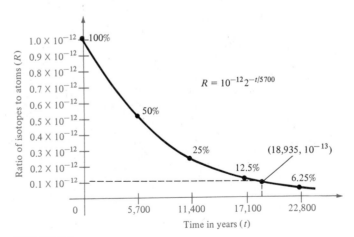

FIGURE 5.1

- **Properties of Exponents**
- **Exponential Functions**
- **The Number** *e*

You are already familiar with the behavior of functions such as

$$f(x) = x^2, \qquad g(x) = \sqrt{x} = x^{1/2}, \qquad h(x) = x^{-1}$$

that involve a variable raised to a constant power. By interchanging roles and raising a constant to a variable power, we obtain an important class of functions called **exponential functions.** Some simple examples are

$$f(x) = 2^x, \qquad g(x) = \left(\frac{1}{10}\right)^x = \frac{1}{10^x}, \qquad h(x) = 3^{2x} = 9^x$$

In general, we can use any positive base $a \neq 1$ for exponential functions.

| | |
|---|---|
| **Definition of an exponential function** | If $a > 0$ and $a \neq 1$, then the **exponential function** with base $a$ is given by $$y = a^x$$ |

Before discussing the behavior of exponential functions, we list some familiar properties of exponents.

| | |
|---|---|
| **Properties of exponents** $(a, b > 0)$ | 1. $a^0 = 1$           2. $a^x a^y = a^{x+y}$ <br><br> 3. $\dfrac{a^x}{a^y} = a^{x-y}$     4. $(a^x)^y = a^{xy}$ <br><br> 5. $(ab)^x = a^x b^x$    6. $\left(\dfrac{a}{b}\right)^x = \dfrac{a^x}{b^x}$ <br><br> 7. $a^{-x} = \dfrac{1}{a^x}$ |

**EXAMPLE 1**

**Applying the Properties of Exponents**

Use the properties of exponents to evaluate the following expressions.

(a) $(2^2)(2^3)$

(b) $2^2(2^{-3})$

(c) $(3^2)^3$

(d) $\left(\dfrac{1}{3}\right)^{-2}$

(e) $\dfrac{3^2}{3^3}$

(f) $(2^{1/2})(3^{1/2})$

**SOLUTION**

(a) By Property 2, we have
$$(2^2)(2^3) = 2^{2+3} = 2^5 = 32$$

(b) By Properties 2 and 7, we have
$$(2^2)(2^{-3}) = 2^{2-3} = 2^{-1} = \frac{1}{2}$$

(c) By Property 4, we have
$$(3^2)^3 = 3^{2(3)} = 3^6 = 729$$

(d) By Properties 6 and 7, we have
$$\left(\frac{1}{3}\right)^{-2} = \frac{1}{(1/3)^2} = \left(\frac{1}{1/3}\right)^2 = 3^2 = 9$$

(e) By Properties 3 and 7, we have
$$\frac{3^2}{3^3} = 3^{2-3} = 3^{-1} = \frac{1}{3}$$

(f) By Property 5, we have
$$(2^{1/2})(3^{1/2}) = [(2)(3)]^{1/2} = 6^{1/2} = \sqrt{6}$$

■■■ **Remark:** Although we demonstrated the seven properties of exponents with integer and rational values for $x$ and $y$, it is important to realize that the properties hold for *any* real values for $x$ and $y$. With a calculator we can obtain approximate values for $a^x$ when $x$ is any real number. For example,

$$\frac{1}{2^{0.6}} = 2^{-0.6} \approx 0.660$$

$$2^{3/4} = 2^{0.75} \approx 1.682$$

$$2^{\sqrt{2}} \approx 2^{1.4142} \approx 2.665$$

**EXAMPLE 2**

**Graphing Exponential Functions**

Sketch the graphs of the exponential functions

$$f(x) = 2^x, \qquad g(x) = \left(\frac{1}{2}\right)^x = 2^{-x}, \qquad h(x) = 3^x$$

**SOLUTION**

Table 5.1 lists some values for these functions, and Figure 5.2 shows their graphs.

TABLE 5.1

| $x$ | $-3$ | $-2$ | $-1$ | 0 | 1 | 2 | 3 | 4 |
|---|---|---|---|---|---|---|---|---|
| $2^x$ | $\dfrac{1}{8}$ | $\dfrac{1}{4}$ | $\dfrac{1}{2}$ | 1 | 2 | 4 | 8 | 16 |
| $2^{-x}$ | 8 | 4 | 2 | 1 | $\dfrac{1}{2}$ | $\dfrac{1}{4}$ | $\dfrac{1}{8}$ | $\dfrac{1}{16}$ |
| $3^x$ | $\dfrac{1}{27}$ | $\dfrac{1}{9}$ | $\dfrac{1}{3}$ | 1 | 3 | 9 | 27 | 81 |

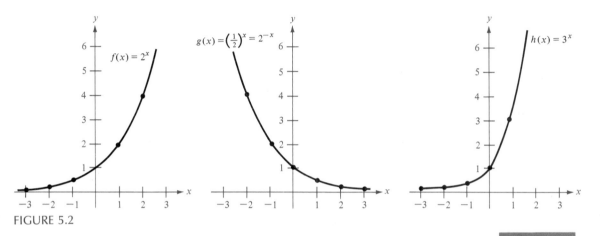

FIGURE 5.2

The forms of the graphs in Figure 5.2 are typical of those of the exponential functions $a^x$ and $a^{-x}$ $(a > 1)$. We summarize the characteristics of these graphs in Figure 5.3.

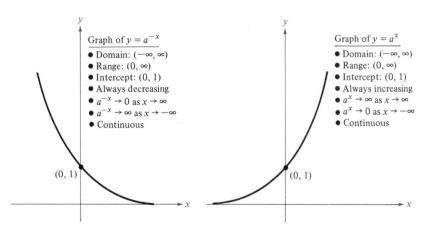

Characteristics of the Exponential Functions $a^{-x}$ and $a^x$ $(a > 1)$

FIGURE 5.3

**EXAMPLE 3**

**Graphing an Exponential Function**

Sketch the graph of

$$f(x) = 3^{-x} - 1$$

**SOLUTION**

By plotting the points listed in Table 5.2 and determining that $y = -1$ is a horizontal asymptote to the right, we obtain the graph shown in Figure 5.4.

$$\lim_{x \to \infty} (3^{-x} - 1) = \lim_{x \to \infty} 3^{-x} - \lim_{x \to \infty} 1$$

$$= \lim_{x \to \infty} \frac{1}{3^x} - \lim_{x \to \infty} 1$$

$$= 0 - 1 = -1$$

TABLE 5.2

| $x$ | $-2$ | $-1$ | 0 | 1 | 2 |
|------|------|------|---|-----|-----|
| $f(x)$ | 8 | 2 | 0 | $-\dfrac{2}{3}$ | $-\dfrac{8}{9}$ |

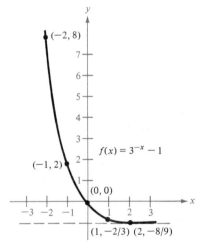

FIGURE 5.4

### The Natural Number e

We have introduced exponential functions using an unspecified base $a$. It turns out that in calculus the natural (or convenient) choice for a base is the irrational number $e$, whose decimal approximation is

$$e \approx 2.71828 \ \ldots$$

This choice may seem anything but natural. However, the convenience of this particular choice will become apparent as we develop the rules for differentiating and integrating exponential and logarithmic functions. In the development of these differentiation rules, we encounter the limit used in the following definition of $e$.

**Limit definition of e**

$$e = \lim_{x \to 0} (1 + x)^{1/x}$$

(To twelve significant digits, $e \approx 2.71828182846$.)

Table 5.3 lists some values of the function $f(x) = (1 + x)^{1/x}$ for $x$ near zero, and Figure 5.5 shows how the graph of $f$ approaches $e$ as $x$ approaches zero.

TABLE 5.3

| x approaches 0 from the left | | | | | x approaches 0 from the right | | | |
|---|---|---|---|---|---|---|---|---|

| $x$ | $-0.1$ | $-0.01$ | $-0.001$ | $-0.0001$ | $0$ | $0.0001$ | $0.001$ | $0.01$ | $0.1$ |
|---|---|---|---|---|---|---|---|---|---|
| $(1 + x)^{1/x}$ | 2.8680 | 2.7320 | 2.7196 | 2.7184 | $e \approx 2.71828$ | 2.7181 | 2.7169 | 2.7048 | 2.5937 |

| $(1 + x)^{1/x}$ approaches $e$ | | | | | $(1 + x)^{1/x}$ approaches $e$ | | | |
|---|---|---|---|---|---|---|---|---|

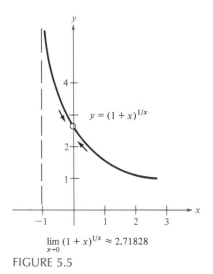

$$\lim_{x \to 0} (1 + x)^{1/x} \approx 2.71828$$

FIGURE 5.5

**EXAMPLE 4**

**Graphing an Exponential Function**

Sketch the graph of

$$f(x) = e^x$$

**SOLUTION**

Figure 5.6 shows the graph of $f(x) = e^x$ determined from the values in Table 5.4. (See the Appendix for a table of exponential values.)

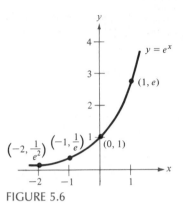

FIGURE 5.6

TABLE 5.4

| $x$ | $-2$ | $-1$ | 0 | 1 | 2 |
|-----|------|------|---|---|---|
| $e^x$ | $\dfrac{1}{e^2} \approx 0.135$ | $\dfrac{1}{e} \approx 0.368$ | 1 | $e \approx 2.718$ | $e^2 \approx 7.389$ |

## Compound Interest

To give you greater insight into the usefulness of the natural base $e$, we take another look at the compound interest problem discussed in the Introductory Examples of Sections 1.5 and 1.6.

Suppose $P$ dollars are deposited in a savings account at an annual interest rate $r$. If accumulated interest is deposited into the account, what is the balance in the account at the end of one year? The answer depends, of course, on the number of times the interest is compounded, according to the formula

$$A = P\left(1 + \frac{r}{n}\right)^n$$

where $n$ is the number of compoundings per year. For instance, the results for a deposit of \$1,000 at 8% interest compounded $n$ times a year are as shown in Table 5.5.

TABLE 5.5

| $n$ | Balance (dollars) |
|-----|-------------------|
| 1 (annually) | 1080.00 |
| 2 (semiannually) | 1081.60 |
| 4 (quarterly) | 1082.43 |
| 12 (monthly) | 1083.00 |
| 365 (daily) | 1083.28 |

It may surprise you to discover that as $n$ increases, the balance $A$ approaches a limit, as indicated in the following development. In this development, we let $x = r/n$. Then $x \to 0$ as $n \to \infty$, and we have

$$A = \lim_{n \to \infty} P\left(1 + \frac{r}{n}\right)^n = P \lim_{n \to \infty} \left[\left(1 + \frac{r}{n}\right)^{n/r}\right]^r$$

$$= P[\lim_{x \to 0} (1 + x)^{1/x}]^r$$

$$= Pe^r$$

We call this limit the balance after one year of **continuous compounding.** Thus, for a deposit of $1,000 at 8% interest compounded continuously, the balance at the end of one year would be

$$A = 1000e^{0.08} \approx \$1,083.29$$

We summarize these results as follows.

**Summary of compound interest formulas**

Let $P$ = amount of deposit, $t$ = number of years, $A$ = balance after $t$ years, and $r$ = annual interest rate (decimal form).

1. Compounded $n$ times per year: $A = P\left(1 + \dfrac{r}{n}\right)^{nt}$

2. Compounded continuously:    $A = Pe^{rt}$

**EXAMPLE 5**

**Finding Compound Interest**

A deposit of $2,500 is made in a savings account that pays an annual interest rate of 10%. Find the balance in the account at the end of 5 years if the interest is compounded

(a) quarterly                          (b) continuously

**SOLUTION**

(a) The balance after quarterly compounding is

$$A = P\left(1 + \frac{r}{n}\right)^{nt} = 2500\left(1 + \frac{0.1}{4}\right)^{(4)(5)}$$

$$= 2500(1.025)^{20}$$

$$= \$4,096.54$$

(b) The balance after continuous compounding is

$$A = Pe^{rt} = 2500[e^{(0.1)(5)}]$$

$$= 2500e^{0.5}$$

$$= \$4,121.80$$

**EXAMPLE 6**

**Evaluating Exponential Functions**

A certain bacterial culture is growing according to the *logistics growth function*

$$y = \frac{1.25}{1 + 0.25e^{-0.4t}}, \quad 0 \le t$$

where $y$ is the weight of the culture in grams and $t$ is the time in hours. Find the weight of the culture after

(a) 0 hours      (b) 1 hour      (c) 10 hours

What is the limit of this function as $t$ increases without bound?

**SOLUTION**

(a) When $t = 0$, we have

$$y = \frac{1.25}{1 + 0.25e^0} = \frac{1.25}{1.25} = 1 \text{ gram}$$

(b) When $t = 1$, we have

$$y = \frac{1.25}{1 + 0.25e^{-0.4}} \approx 1.071 \text{ grams}$$

(c) When $t = 10$, we have

$$y = \frac{1.25}{1 + 0.25e^{-4}} \approx 1.244 \text{ grams}$$

Finally, taking the limit as $t$ approaches $\infty$, we have

$$\lim_{t \to \infty} \frac{1.25}{1 + 0.25e^{-0.4t}} = \lim_{t \to \infty} \frac{1.25}{1 + (0.25/e^{0.4t})}$$

$$= \frac{1.25}{1 + 0}$$

$$= 1.25 \text{ grams}$$

The graph of this function is shown in Figure 5.7.

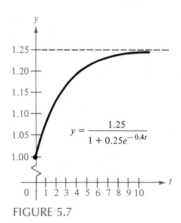

$$y = \frac{1.25}{1 + 0.25e^{-0.4t}}$$

FIGURE 5.7

In Exercises 1 and 2, evaluate each of the expressions.

**1.** (a) $5(5^3)$      (b) $27^{2/3}$
    (c) $64^{3/4}$      (d) $81^{1/2}$
    (e) $25^{3/2}$      (f) $32^{2/5}$

**2.** (a) $\left(\dfrac{1}{5}\right)^3$      (b) $\left(\dfrac{1}{8}\right)^{1/3}$

    (c) $64^{2/3}$      (d) $\left(\dfrac{5}{8}\right)^2$

    (e) $100^{3/2}$      (f) $4^{5/2}$

In Exercises 3–8, use the properties of exponents to simplify each expression.

**3.** (a) $(5^2)(5^3)$      (b) $(5^2)(5^{-3})$
    (c) $(5^2)^2$      (d) $5^{-3}$

**4.** (a) $\dfrac{5^3}{5^6}$      (b) $\left(\dfrac{1}{5}\right)^{-2}$

    (c) $(8^{1/2})(2^{1/2})$      (d) $(32^{3/2})\left(\dfrac{1}{2}\right)^{3/2}$

**5.** (a) $\dfrac{5^3}{25^2}$      (b) $(9^{2/3})(3)(3^{2/3})$

    (c) $[(25^{1/2})(25^2)]^{1/5}$      (d) $(8^2)(4^3)$

**6.** (a) $(4^3)(4^2)$      (b) $\left(\dfrac{1}{4}\right)^2(4^2)$

    (c) $(4^6)^{1/2}$      (d) $[(8^{-1})(8^{2/3})]^3$

**7.** (a) $e^2(e^4)$      (b) $(e^3)^4$

    (c) $(e^3)^{-2}$      (d) $\dfrac{e^5}{e^3}$

**8.** (a) $\left(\dfrac{1}{e}\right)^{-2}$      (b) $\left(\dfrac{e^5}{e^2}\right)^{-1}$

    (c) $e^0$      (d) $\dfrac{1}{e^{-3}}$

In Exercises 9–18, solve for $x$.

**9.** $3^x = 81$      **10.** $5^{x+1} = 125$

**11.** $\left(\dfrac{1}{3}\right)^{x-1} = 27$      **12.** $\left(\dfrac{1}{5}\right)^{2x} = 625$

**13.** $4^3 = (x + 2)^3$      **14.** $4^2 = (x + 2)^2$
**15.** $x^{3/4} = 8$      **16.** $(x + 3)^{4/3} = 16$
**17.** $e^{-2x} = e^5$      **18.** $e^x = 1$

In Exercises 19–24, match the given exponential function with the correct graph. [Graphs are labeled (a), (b), (c), (d), (e), and (f).]

**19.** $f(x) = 3^x$      **20.** $f(x) = 3^{-x/2}$
**21.** $f(x) = -3^x$      **22.** $f(x) = 3^{x-2}$
**23.** $f(x) = 3^{-x} - 1$      **24.** $f(x) = 3^x + 2$

(a)
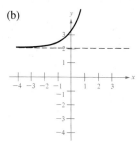

(b)
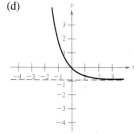

(c)
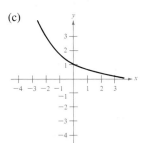

(d)
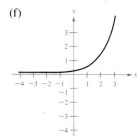

(e)
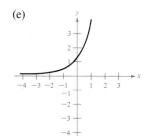

(f)

In Exercises 25–40, sketch the graph of the given exponential function.

**25.** $f(x) = 5^x$      **26.** $f(x) = 4^x$
**27.** $f(x) = \left(\dfrac{1}{5}\right)^x = 5^{-x}$      **28.** $f(x) = \left(\dfrac{1}{4}\right)^x = 4^{-x}$
**29.** $y = 3^{-x^2}$      **30.** $y = 2^{-x^2}$
**31.** $y = 3^{-|x|}$      **32.** $y = 3^{|x|}$
**33.** $s(t) = \dfrac{3^{-t}}{4}$      **34.** $s(t) = 2^{-t} + 3$
**35.** $h(x) = e^{x-2}$      **36.** $f(x) = e^{2x}$
**37.** $N(t) = 1000e^{-0.2t}$      **38.** $A(t) = 500e^{0.15t}$
**39.** $g(x) = \dfrac{2}{1 + e^{x^2}}$      **40.** $g(x) = \dfrac{10}{1 + e^{-x}}$

In Exercises 41 and 42, find the amount of an investment of $P$ dollars invested at $r$ percent for $t$ years if the interest is compounded (a) annually, (b) semiannually, (c) quarterly, (d) monthly, (e) daily, and (f) continuously.

**41.** $P = \$1,000$, $r = 10\%$, $t = 10$ years

**42.** $P = \$2,500$, $r = 12\%$, $t = 20$ years

In Exercises 43 and 44, find the investment that would be required at $r$ percent compounded continuously to yield an amount of $100,000 in (a) 1 year, (b) 10 years, (c) 20 years, and (d) 50 years.

**43.** $r = 12\%$      **44.** $r = 9\%$

**45.** The demand function for a certain product is given by

$$p = 5000\left(1 - \frac{4}{4 + e^{-0.002x}}\right)$$

(See Figure 5.8.) Find the price of the product if the quantity demanded is

(a) $x = 100$ units    (b) $x = 500$ units

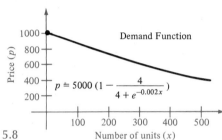

**FIGURE 5.8**

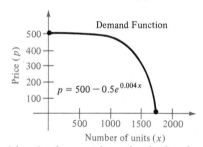

**FIGURE 5.9**

**46.** The demand function for a certain product is given by

$$p = 500 - 0.5e^{0.004x}$$

(See Figure 5.9.) Find the price of the product if the quantity demanded is

(a) $x = 1000$ units    (b) $x = 1500$ units

**47.** The average time between incoming calls at a switchboard is 3 minutes. If a call has just come in, the probability that the next call will come within the next $t$ minutes is given by

$$P(t) = 1 - e^{-t/3}$$

Find

(a) $P\left(\dfrac{1}{2}\right)$   (b) $P(2)$   (c) $P(5)$

**48.** A certain automobile gets 28 mi/gal at speeds of up to 50 mi/hr. At speeds of over 50 mi/hr, the number of miles per gallon drops at the rate of 12% for each 10 mi/hr. If $s$ is the speed (in miles per hour) and $y$ is the miles per gallon, then

$$y = 28e^{0.6-0.012s}, \quad s \geq 50$$

Use this function to complete the following table.

| Speed $(s)$ | 50 | 55 | 60 | 65 | 70 |
|---|---|---|---|---|---|
| Miles per gallon $(y)$ | | | | | |

**49.** The population of a bacterial culture is given by the logistics growth function

$$y = \frac{850}{1 + e^{-0.2t}}$$

where $y$ is the number of bacteria and $t$ is the time in days.

(a) Find the limit of this function as $t$ approaches infinity.

(b) Sketch the graph of this function.

**50.** The yield $V$ (in millions of cubic feet per acre) for a forest at age $t$ years is given by

$$V = 6.7e^{-48/t}$$

(a) Find the volume per acre after (i) 20 years and (ii) 50 years.

(b) Find the limiting volume of wood per acre as $t$ approaches infinity.

(c) Sketch the graph of this function.

**51.** In a group project in learning theory, a mathematical model for the proportion $P$ of correct responses after $n$ trials was found to be

$$P = \frac{0.83}{1 + e^{-0.2n}}$$

(a) Find the proportion of correct responses after 10 trials.

(b) Find the limiting proportion of correct responses as $n$ approaches infinity.

**52.** In a typing class, the average number $N$ of words per minute typed after $t$ weeks of lessons was found to be

$$N = \frac{157}{1 + 5.4e^{-0.12t}}$$

(a) Find the average number of words per minute after 10 weeks.

(b) Find the limiting number of words per minute as $t$ approaches infinity.

# Differentiation and Integration of Exponential Functions

In a given population how many people would you expect to be a certain height? One way to answer this question is to use the **normal probability density function** as a model. This model was developed by the German mathematician Karl Friedrich Gauss and is used to represent bell-shaped distributions that occur in a wide variety of fields. For example, suppose we wanted to use this model to represent the heights of American males between 18 and 24 years of age. If the mean height (average height) of this population is 70 inches, then the distribution of heights is given by

$$f(x) = 0.12e^{-0.045(x-70)^2}$$

Given this model we can use a definite integral to find the percentage of heights lying between any two given heights. For instance, the percentage of heights between 70 and 71 inches is given by

$$\text{Percentage} = 0.12 \int_{70}^{71} e^{-0.045(x-70)^2} \, dx$$

$$\approx 0.1182 \approx 11.82\%$$

Figure 5.10 shows the graph of $f$ together with the region whose area is represented by this integral.

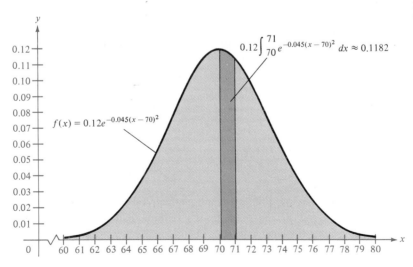

Normal Probability Density Function

FIGURE 5.10

■ **Differentiation of Exponential Functions**
■ **Integration of Exponential Functions**

In the preceding section we promised to defend the choice of $e$ as the most convenient base for exponential functions. Perhaps the most convincing argument is that by choosing $e$ as the exponential base we have the very nice result that the function $y = e^x$ *is its own derivative.* (This is not true of other exponential functions of the form $y = a^x$, $a \neq e$.) To see that this is true, let us return to the limit definition of the derivative as given in Section 2.1. Recall that

$$f'(x) = \lim_{\Delta x \to 0} \frac{f(x + \Delta x) - f(x)}{\Delta x}$$

Thus,

$$\frac{d}{dx}[e^x] = \lim_{\Delta x \to 0} \frac{e^{x + \Delta x} - e^x}{\Delta x}$$

$$= \lim_{\Delta x \to 0} \frac{e^x(e^{\Delta x} - 1)}{\Delta x}$$

Now the limit definition of $e$,

$$e = \lim_{\Delta x \to 0} (1 + \Delta x)^{1/\Delta x}$$

tells us that for small values of $\Delta x$ we have

$$e \approx (1 + \Delta x)^{1/\Delta x} \quad \Longrightarrow \quad e^{\Delta x} \approx 1 + \Delta x$$

Finally, replacing $e^{\Delta x}$ by this approximation, we have

$$\frac{d}{dx}[e^x] = \lim_{\Delta x \to 0} \frac{e^x[(1 + \Delta x) - 1]}{\Delta x}$$

$$= \lim_{\Delta x \to 0} \frac{e^x(\Delta x)}{\Delta x}$$

$$= e^x$$

Thus, we have the following result:

$$\frac{d}{dx}[e^x] = e^x$$

We can interpret this result geometrically by saying that the slope of the graph of $f(x) = e^x$ at any point $(x, e^x)$ is numerically equal to the $y$-coordinate of the point, as shown in Figure 5.11.

By applying the Chain Rule to the formula

$$\frac{d}{dx}[e^x] = e^x$$

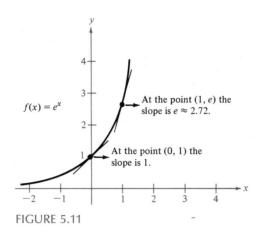

$f(x) = e^x$

At the point $(1, e)$ the slope is $e \approx 2.72$.

At the point $(0, 1)$ the slope is 1.

FIGURE 5.11

we can obtain the differentiation formula for $e^u$, where $u$ is a differentiable function of $x$.

**Derivatives of exponential functions**

$$\frac{d}{dx}[e^x] = e^x \quad \text{and} \quad \frac{d}{dx}[e^u] = e^u \frac{du}{dx}$$

**EXAMPLE 1**

**Differentiating Exponential Functions**

Differentiate the following functions:

(a) $y = e^{2x-1}$

(b) $y = e^{-3/x}$

**SOLUTION**

(a) Considering $u = 2x - 1$, we have $du/dx = 2$, and thus

$$\frac{dy}{dx} = e^u \frac{du}{dx}$$

$$= e^{2x-1}(2)$$

$$= 2e^{2x-1}$$

(b) Considering $u = -3/x$, we have $du/dx = 3/x^2$, and thus

$$\frac{dy}{dx} = e^u \frac{du}{dx}$$

$$= e^{-3/x}\left(\frac{3}{x^2}\right)$$

$$= \frac{3e^{-3/x}}{x^2}$$

**EXAMPLE 2**

**Combining Differentiation Rules**

Differentiate the following function:

$$f(x) = e^2 + \frac{e^{-2x}}{x^3}$$

**SOLUTION**

Note that $e^2$ is a constant. Thus, we have

$$f'(x) = 0 + \frac{d}{dx}\left[\frac{e^{-2x}}{x^3}\right]$$

Now, by the Quotient Rule we obtain

$$f'(x) = \frac{x^3[e^{-2x}(-2)] - e^{-2x}[3x^2]}{(x^3)^2}$$

$$= \frac{-2x^3e^{-2x} - 3x^2e^{-2x}}{x^6}$$

$$= \frac{-x^2e^{-2x}(2x + 3)}{x^6}$$

$$= \frac{-e^{-2x}(2x + 3)}{x^4}$$

**EXAMPLE 3**

**The Normal Probability Density Function**

Show that the *normal probability density function*

$$f(x) = \frac{1}{\sqrt{2\pi}}e^{-x^2/2}$$

has points of inflection when $x = \pm 1$.

**SOLUTION**

The first derivative of $f$ is

$$f'(x) = \frac{1}{\sqrt{2\pi}}(-x)e^{-x^2/2}$$

and by the Product Rule, the second derivative is

$$f''(x) = \frac{1}{\sqrt{2\pi}}[(-x)(-x)e^{-x^2/2} + (-1)e^{-x^2/2}]$$

$$= \frac{1}{\sqrt{2\pi}}(e^{-x^2/2})(x^2 - 1)$$

Therefore, $f''(x) = 0$ when $x = \pm 1$, and we can apply the techniques in Section 3.3 to conclude that these values of $x$ yield the two points of inflection shown in Figure 5.12.

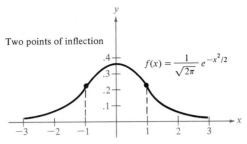

Two points of inflection

$$f(x) = \frac{1}{\sqrt{2\pi}} e^{-x^2/2}$$

Bell–Shaped Curve Given By Normal Probability
Density Function

FIGURE 5.12

■■■ **Remark:** The general form of a normal probability density function is given by

$$f(x) = \frac{1}{\sigma\sqrt{2\pi}} e^{-x^2/2\sigma^2}$$

where $\sigma$ is the standard deviation ($\sigma$ is the lowercase Greek letter sigma). By following the procedure of Example 3, we can show that the bell-shaped graph of $f$ has points of inflection when $x = \pm\sigma$.

### Integration of Exponential Functions

Each of the formulas for differentiating exponential functions has its corresponding integration formula, as shown next.

**Integrals of exponential functions**

$$\int e^x \, dx = e^x + C$$

Moreover, if $u$ is a differentiable function of $x$, then

$$\int e^u \frac{du}{dx} \, dx = \int e^u \, du = e^u + C$$

**EXAMPLE 4**
■■■

**Integrating Exponential Functions**

Evaluate the following integral:

$$\int e^{3x+1} \, dx$$

**SOLUTION**

Considering $u = 3x + 1$, we have $du/dx = 3$. Introducing the missing factor 3 in the integrand and then multiplying the integral by the reciprocal factor $\frac{1}{3}$, we write

$$\int e^{3x+1} \, dx = \frac{1}{3} \int e^{3x+1}(3) \, dx$$

$$= \frac{1}{3} \int e^u \frac{du}{dx} \, dx$$

$$= \frac{1}{3} e^u + C$$

$$= \frac{e^{3x+1}}{3} + C$$

**EXAMPLE 5**

**Integrating Exponential Functions**

Evaluate the following integral:

$$\int 5xe^{-x^2} \, dx$$

**SOLUTION**

If we let $u = -x^2$, then $du/dx = -2x$. First we adjust the integrand by moving the unneeded factor 5 outside the integral sign. Then we introduce the needed factor $-2$ and multiply by $-\frac{1}{2}$. Thus, we write

$$\int 5xe^{-x^2} \, dx = 5 \int e^{-x^2}(x) \, dx$$

$$= 5\left(\frac{-1}{2}\right) \int e^{-x^2}(-2x) \, dx$$

$$= -\frac{5}{2} \int e^u \frac{du}{dx} \, dx$$

$$= -\frac{5}{2} e^{-x^2} + C$$

**Remark:** Keep in mind that we cannot introduce a missing *variable* factor in the integrand. For instance, we cannot evaluate $\int 5e^{x^2} \, dx$ by introducing the factor $2x$ and then multiplying the integral by $1/(2x)$. That is,

$$\int 5e^{x^2} \, dx \neq \frac{5}{2x} \int e^{x^2}(2x) \, dx$$

**EXAMPLE 6**

**Integrating Exponential Functions**

Evaluate the following integral:

$$\int \frac{e^{1/x}}{x^2} \, dx$$

**SOLUTION**

By choosing $u = 1/x$, we have

$$\frac{du}{dx} = -\frac{1}{x^2}$$

Thus,

$$\int \frac{e^{1/x}}{x^2}\, dx = -\int e^{1/x}\left(-\frac{1}{x^2}\right) dx$$

$$= -e^{1/x} + C$$

## SECTION EXERCISES 5.2

In Exercises 1–6, find the slope of the tangent line to the given exponential function at the point $(0, 1)$.

**1.** $y = e^{3x}$       **2.** $y = e^{2x}$

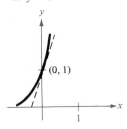

**3.** $y = e^x$       **4.** $y = e^{-3x}$

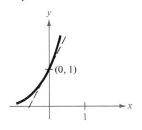

**5.** $y = e^{-2x}$       **6.** $y = e^{-x}$

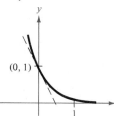

In Exercises 7–22, find the derivative of the exponential function.

**7.** $y = e^{2x}$       **8.** $y = e^{1-x}$
**9.** $y = e^{-2x+x^2}$       **10.** $y = e^{-x^2}$

**11.** $f(x) = e^{1/x}$       **12.** $f(x) = e^{-1/x^2}$
**13.** $g(x) = e^{\sqrt{x}}$       **14.** $g(x) = e^{x^3}$
**15.** $f(x) = (x + 1)e^{3x}$       **16.** $y = x^2 e^{-x}$
**17.** $y = (e^{-x} + e^x)^3$       **18.** $y = (1 - e^{-x})^2$

**19.** $f(x) = \dfrac{2}{e^x + e^{-x}}$       **20.** $f(x) = \dfrac{e^x - e^{-x}}{2}$

**21.** $y = xe^x - e^x$       **22.** $y = x^2 e^x - 2xe^x + 2e^x$

In Exercises 23–26, find the second derivative of the exponential function.

**23.** $f(x) = 2e^{3x} + 3e^{-2x}$       **24.** $f(x) = 5e^{-x} - 2e^{-5x}$
**25.** $g(x) = (1 + 2x)e^{4x}$       **26.** $g(x) = (3 + 2x)e^{-3x}$

In Exercises 27–30, find the extrema and the points of inflection (if any exist) and sketch the graph of the function.

**27.** $f(x) = \dfrac{2}{1 + e^{-x}}$       **28.** $f(x) = \dfrac{e^x - e^{-x}}{2}$

**29.** $f(x) = x^2 e^{-x}$       **30.** $f(x) = xe^{-x}$

**31.** Find an equation of the line tangent to the graph of $y = e^{-x}$ at $(0, 1)$.

**32.** Find the area of the largest rectangle that can be inscribed under the curve $y = e^{-x^2}$ in the first and second quadrants.

In Exercises 33 and 34, find the rate at which newly purchased equipment is depreciating (a) $t = 1$ year and (b) $t = 5$ years after it is purchased, if its value $V$ is given by the specified exponential function.

**33.** $V = 15,000e^{-0.6286t}$       **34.** $V = 500,000e^{-0.2231t}$

**35.** A lake is stocked with 500 fish, and their population is given by the **logistics curve**

$$p = \frac{10,000}{1 + 19e^{-t/5}}$$

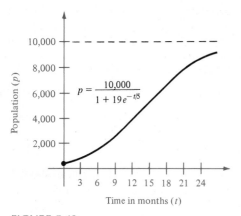

$$p = \frac{10{,}000}{1 + 19e^{-t/5}}$$

FIGURE 5.13

where $t$ is measured in months. At what rate is the fish population changing at the end of 1 month and at the end of 10 months? After how many months is the population increasing most rapidly? (See Figure 5.13.)

**36.** The **Ebbinghaus Model** for human memory is

$$p = (100 - a)e^{-bt} + a$$

where $p$ is the percentage retained after $t$ weeks. (The constants $a$ and $b$ vary from one person to another.) If $a = 20$ and $b = 0.5$, at what rate is information being retained after

(a) 1 week     (b) 3 weeks

**37.** The yield $V$ (in millions of cubic feet per acre) for a forest stand at age $t$ is given by

$$V = 6.7e^{-48.1/t}$$

where $t$ is measured in years. Find the rate at which the yield is changing when

(a) $t = 15$ years     (b) $t = 20$ years
(c) $t = 60$ years

**38.** The average typing speed $N$ (in words per minute) after $t$ weeks of lessons is given by

$$N = \frac{157}{1 + 5.4e^{-0.12t}}$$

Find the rate at which typing speed is changing when
(a) $t = 5$ weeks     (b) $t = 10$ weeks
(c) $t = 30$ weeks

**39.** The balance in a certain savings account is given by

$$A = (5000)e^{0.08t}$$

where $A$ is measured in dollars and $t$ is measured in years. Find the rate at which the balance is changing when

(a) $t = 1$ year     (b) $t = 10$ years     (c) $t = 50$ years

**40.** In a group project in learning theory, a mathematical model for the proportion $P$ of correct responses after $n$ trials was found to be

$$P = \frac{0.83}{1 + e^{-0.2n}}$$

Find the rate at which $P$ is changing after $n = 3$ and $n = 10$ trials.

In Exercises 41–56, evaluate the integral.

**41.** $\displaystyle\int e^{5x}(5)\,dx$     **42.** $\displaystyle\int e^{-3x}(-3)\,dx$

**43.** $\displaystyle\int e^{-x^4}(-4x^3)\,dx$     **44.** $\displaystyle\int e^{x/2}\left(\frac{1}{2}\right)dx$

**45.** $\displaystyle\int_0^1 e^{-2x}\,dx$     **46.** $\displaystyle\int_1^2 e^{1-x}\,dx$

**47.** $\displaystyle\int_0^2 (x^2 - 1)e^{x^3-3x+1}\,dx$     **48.** $\displaystyle\int x^2 e^{x^3}\,dx$

**49.** $\displaystyle\int xe^{ax^2}\,dx$     **50.** $\displaystyle\int_0^1 xe^{-x^2/2}\,dx$

**51.** $\displaystyle\int_1^3 \frac{e^{3/x}}{x^2}\,dx$     **52.** $\displaystyle\int (e^x - e^{-x})^2\,dx$

**53.** $\displaystyle\int \frac{e^{2x} + 2e^x + 1}{e^x}\,dx$     **54.** $\displaystyle\int (3^2 - 5^2)\,dx$

**55.** $\displaystyle\int e^x\sqrt{1 - e^x}\,dx$     **56.** $\displaystyle\int \frac{2(e^x - e^{-x})}{(e^x + e^{-x})^2}\,dx$

In Exercises 57–60, find the area of the region bounded by the graphs of the equations.

**57.** $y = e^x$, $y = 0$, $x = 0$, $x = 5$

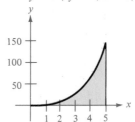

**58.** $y = e^{-x}$, $y = 0$, $x = a$, $x = b$

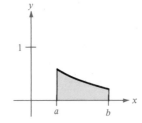

**59.** $y = xe^{-x^2}$, $y = 0$, $x = 0$, $x = 4$

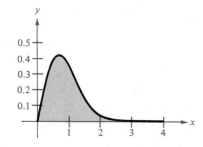

**60.** $y = 3e^{-x/2}$, $y = 0$, $x = 0$, $x = 4$

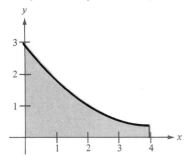

In Exercises 61 and 62, find the volume of the solid generated by revolving the region bounded by the graphs of the given equations about the *x*-axis.

**61.** $y = e^x$, $y = 0$, $x = 0$, $x = 1$

**62.** $y = e^{-x/2}$, $y = 0$, $x = 0$, $x = 4$

**63.** A deposit of $2,500 is made in a savings account at an annual percentage rate of 12% compounded continuously. Find the average balance in the account during the first 5 years.

**64.** In the Ebbinghaus Model for human memory (see Exercise 36), let $a = 20$ and $b = 0.5$. Then the percentage retained in memory after $t$ weeks is

$$p = 80e^{-0.5t} + 20$$

(See Figure 5.14.) Find the average percentage retained during

(a) the first 2 weeks     (b) the second 2 weeks

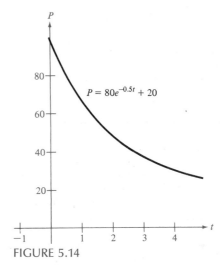

FIGURE 5.14

# The Natural Logarithmic Function

When money is deposited in a savings account with interest compounded continuously at $r$ percent, the balance after $t$ years is $A = Pe^{rt}$, where $P$ is the original deposit. How long would it take such a deposit to double? This depends, of course, on the value of $r$. If $r$ is large, the deposit will not take as long to double as it will if $r$ is small. To find the doubling time for a given value of $r$, we could try several different values of $t$ to find a time that doubles the original deposit. For instance, Table 5.6 shows the balances at the ends of successive years for an original deposit of $1,000 at 8%. From the table it appears that this balance doubles sometime between 8 and 9 years after being deposited.

The problem with this trial-and-error procedure is that it is time-consuming. Furthermore, it will usually not give the exact doubling time. In this section, we will see that there is a simple way to solve this problem by using the *inverse* of the exponential function. We call this inverse function the **natural logarithmic function,** and we can use it to solve our doubling time problem in the following way. Let $A = 2P$ to obtain

$$Pe^{rt} = 2P$$

$$e^{rt} = 2$$

Now, taking the natural logarithm of both sides of this equation, we have

$$\ln(e^{rt}) = \ln 2$$

$$rt = \ln 2$$

$$t = \frac{1}{r}\ln 2 \approx \frac{0.6931}{r} \text{ years}$$

Table 5.7 shows the time required to double a balance for several different interest rates.

TABLE 5.6

| $t$ | 1 | 2 | 3 | 4 | 5 | 6 | 7 | 8 | 9 |
|---|---|---|---|---|---|---|---|---|---|
| $A = 1000e^{0.08t}$ | $1,083.29 | $1,173.51 | $1,271.25 | $1,377.13 | $1,491.82 | $1,616.07 | $1,750.67 | $1,896.48 | $2,054.43 |

TABLE 5.7

| Rate (%) | 5 | 6 | 7 | 8 | 9 | 10 | 11 | 12 | 13 | 14 | 15 | 16 | 17 |
|---|---|---|---|---|---|---|---|---|---|---|---|---|---|
| Doubling time (years) | 13.86 | 11.55 | 9.90 | 8.66 | 7.70 | 6.93 | 6.30 | 5.78 | 5.33 | 4.95 | 4.62 | 4.33 | 4.08 |

■ **The Natural Logarithmic Function**
■ **Properties of the Natural Logarithmic Function**

A theorem from advanced calculus states that if a continuous function is always increasing (or decreasing) then it possesses an inverse. From Section 5.1 we know that the exponential function $f(x) = e^x$ is both increasing and continuous; hence it must possess an inverse. We call the inverse of the exponential function the **natural logarithmic function,** and we define it as follows.

| | |
|---|---|
| **Definition of the natural logarithmic function** | $\ln x = b$    if and only if    $e^b = x$ <br><br> ($\ln x$ is read "the natural log of $x$.") |

This definition suggests that logarithmic equations can be written in an equivalent exponential form and vice versa. For example,

| *Logarithmic form* | *Exponential form* |
|:---:|:---:|
| $\ln 1 = 0$ | $e^0 = 1$ |
| $\ln e = 1$ | $e^1 = e$ |
| $\ln e^{-1} = -1$ | $e^{-1} = \dfrac{1}{e}$ |
| $\ln 2 \approx 0.693$ | $e^{0.693} \approx 2$ |

Since the functions $f(x) = e^x$ and $f^{-1}(x) = \ln x$ are defined to be inverses of each other, their graphs should be reflections of each other in the line $y = x$. (See Section 1.4.) This reflective property is illustrated in Figure 5.15 along with a summary of some other properties of the natural logarithmic function.

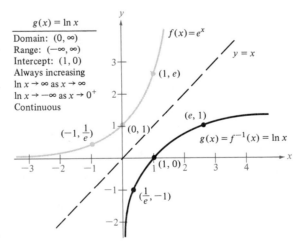

$g(x) = \ln x$

Domain: $(0, \infty)$
Range: $(-\infty, \infty)$
Intercept: $(1, 0)$
Always increasing
$\ln x \to \infty$ as $x \to \infty$
$\ln x \to -\infty$ as $x \to 0^+$
Continuous

FIGURE 5.15

Recall from Section 1.4 that inverse functions possess the property that

$$f(g(x)) = f(f^{-1}(x)) = x \qquad \text{and} \qquad g(f(x)) = f^{-1}(f(x)) = x$$

From this we conclude that for $f(x) = e^x$ and $f^{-1}(x) = \ln x$ we have

$$x = f(f^{-1}(x)) = f(\ln x) = e^{\ln x}$$

and

$$x = f^{-1}(f(x)) = f^{-1}(e^x) = \ln (e^x)$$

**Inverse properties of ln $x$ and $e^x$**

$$\ln e^x = x \qquad \text{and} \qquad e^{\ln x} = x$$

### EXAMPLE 1

**Applying the Inverse Properties of ln $x$ and $e^x$**

Simplify the following expressions:

(a) $\ln e^{\sqrt{2}}$ 　　　　　　　　　　(b) $e^{\ln 3x}$

**SOLUTION**

(a) Since the natural logarithmic function is the inverse of the exponential function, we have

$$\ln e^{\sqrt{2}} = \sqrt{2}$$

(b) By the same reasoning, we have

$$e^{\ln 3x} = 3x$$

In Section 5.1, we saw that we can multiply two exponential functions by adding exponents.

$$e^x e^y = e^{x+y}$$

The logarithmic version of this rule states that the log of the product of two numbers is equal to the sum of the logs of the numbers. That is,

$$\ln xy = \ln x + \ln y$$

This rule and the rules for evaluating the log of a quotient and the log of a power are summarized as follows.

**Properties of logarithms**

1. $\ln xy = \ln x + \ln y$

2. $\ln \dfrac{x}{y} = \ln x - \ln y$

3. $\ln x^y = y \ln x$

**EXAMPLE 2**

**Expanding Logarithmic Expressions**

Use the properties of logarithms to write each of the following expressions as a sum, difference, or multiple of logarithms.

(a) $\ln \dfrac{10}{9}$

(b) $\ln \sqrt{3x + 2}$

(c) $\ln \dfrac{xy}{5}$

(d) $\ln \dfrac{(x - 3)^2}{x\sqrt[3]{x - 1}}$

**SOLUTION**

(a) By Property 2, we have

$$\ln \frac{10}{9} = \ln 10 - \ln 9$$

(b) By Property 3, we have

$$\ln \sqrt{3x + 2} = \ln (3x + 2)^{1/2}$$

$$= \frac{1}{2} \ln (3x + 2)$$

(c) By Properties 1 and 2, we have

$$\ln \frac{xy}{5} = \ln (xy) - \ln 5$$

$$= \ln x + \ln y - \ln 5$$

(d) By Properties 1, 2, and 3, we have

$$\ln \frac{(x - 3)^2}{x\sqrt[3]{x - 1}} = \ln (x - 3)^2 - \ln x\sqrt[3]{x - 1}$$

$$= 2 \ln (x - 3) - [\ln x + \ln (x - 1)^{1/3}]$$

$$= 2 \ln (x - 3) - \ln x - \ln (x - 1)^{1/3}$$

$$= 2 \ln (x - 3) - \ln x - \frac{1}{3} \ln (x - 1)$$

**EXAMPLE 3**

**Condensing Logarithmic Expressions**

Use the properties of logarithms to rewrite each of the following expressions as the logarithm of a single quantity.

(a) $\ln x + 2 \ln y$

(b) $\ln (x + 1) - \dfrac{1}{2} \ln x - \ln (x^2 - 1)$

**SOLUTION**

(a) By Properties 1 and 3, we have

$$\ln x + 2 \ln y = \ln x + \ln y^2 = \ln xy^2$$

(b) By Properties 1, 2, and 3, we have

$$\ln (x + 1) - \frac{1}{2} \ln x - \ln (x^2 - 1) = \ln (x + 1) - \ln x^{1/2} - \ln (x^2 - 1)$$

$$= \ln (x + 1) - [\ln \sqrt{x} + \ln (x^2 - 1)]$$

$$= \ln (x + 1) - \ln [\sqrt{x}(x^2 - 1)]$$

$$= \ln \frac{x + 1}{\sqrt{x}(x^2 - 1)}$$

$$= \ln \frac{1}{\sqrt{x}(x - 1)}$$

## EXAMPLE 4

**Solving Equations**

Solve for $x$ in each of the following equations:

(a) $\ln x - \ln (x - 8) = 1$          (b) $y = e^{2x-5}$

**SOLUTION**

(a) We begin by writing the left side of the equation as the logarithm of a single quantity.

$$\ln x - \ln (x - 8) = 1$$

$$\ln \frac{x}{x - 8} = 1$$

The equivalent exponential form is

$$\frac{x}{x - 8} = e^1$$

Thus, we have

$$x = (x - 8)(e)$$

$$(1 - e)x = -8e$$

$$x = -\frac{8e}{1 - e} \approx 12.66$$

(b) We begin by taking the natural logarithm of both sides of the given equation, as follows:

$$y = e^{2x-5}$$

$$\ln y = \ln e^{2x-5}$$

Now, applying the inverse property of the exponential and natural logarithmic function, we have

$$\ln y = 2x - 5$$

$$5 + \ln y = 2x$$

$$\frac{5 + \ln y}{2} = x$$

**EXAMPLE 5**

**An Application to Compound Interest**

If $P$ dollars are deposited at 8% interest compounded continuously, how long will it take for the original deposit to double? (See the Introductory Example of this section.)

SOLUTION

To represent a doubling of the deposit we write

$$Pe^{0.08t} = 2P$$

Thus, we have

$$e^{0.08t} = 2$$

$$0.08t = \ln 2$$

$$t = \frac{\ln 2}{0.08} \approx 8.66$$

Therefore, the balance will double by the end of 8 years and 8 months.

**EXAMPLE 6**

**An Application to Radioactive Carbon Dating**

In the Introductory Example to Section 5.1, we looked at the exponential equation

$$10^{-13} = 10^{-12}2^{-t/5700}$$

where $t$ represented the age of a particular fossil in years. Use the properties of logarithms to solve this equation for $t$.

SOLUTION

Dividing both sides of the given equation by $10^{-12}$, we have

$$10^{-1} = 2^{-t/5700}$$

Now, taking the natural logarithm of both sides, we have

$$\ln (10^{-1}) = \ln (2^{-t/5700})$$

$$(-1)(\ln 10) = \left(\frac{-t}{5700}\right)(\ln 2)$$

$$\frac{\ln 10}{\ln 2}(5700) = t$$

$$t \approx 18{,}935 \text{ years}$$

# SECTION EXERCISES 5.3

In Exercises 1–8, write the logarithmic equation as an exponential equation and vice versa.

**1.** $\ln 2 = 0.6931 \ldots$

**2.** $\ln 8.4 = 2.128 \ldots$

**3.** $\ln 0.5 = -0.6931 \ldots$

**4.** $\ln 0.056 = -2.2882 \ldots$

**5.** $e^0 = 1$

**6.** $e^2 = 7.389 \ldots$

**7.** $e^{-2} = 0.1353 \ldots$

**8.** $e^{0.25} = 1.284 \ldots$

In Exercises 9–12, use the graph of $y = \ln x$ to match the given function with the correct graph. [Graphs are labeled (a), (b), (c), and (d).]

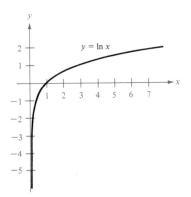

**9.** $f(x) = \ln x + 2$

**10.** $f(x) = -\ln x$

**11.** $f(x) = \ln (x + 2)$

**12.** $f(x) = -\ln (x - 1)$

(a)

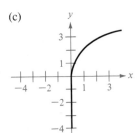

(b)

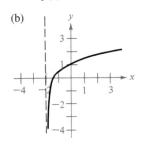

(c)

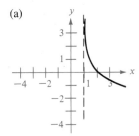

(d)

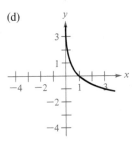

In Exercises 13–18, sketch the graph of the given logarithmic function.

**13.** $y = \ln (x - 1)$

**14.** $y = \ln |x|$

**15.** $y = \ln 2x$

**16.** $y = \ln x + 5$

**17.** $y = 3 \ln x$

**18.** $y = \dfrac{\ln x}{4}$

In Exercises 19–22, show that the given functions are inverses of each other and sketch their graphs on the same coordinate axes.

**19.** $f(x) = e^{2x}, \ g(x) = \ln \sqrt{x}$

**20.** $f(x) = e^x - 1, \ g(x) = \ln (x + 1)$

**21.** $f(x) = e^{x-1}, \ g(x) = 1 + \ln x$

**22.** $f(x) = e^{x/3}, \ g(x) = \ln x^3$

In Exercises 23–28, apply the inverse properties of $\ln x$ and $e^x$ to simplify the given expression.

**23.** $\ln e^{x^2}$

**24.** $\ln e^{2x-1}$

**25.** $e^{\ln (5x+2)}$

**26.** $-1 + \ln e^{2x}$

**27.** $e^{\ln \sqrt{x}}$

**28.** $-8 + e^{\ln x^3}$

In Exercises 29 and 30, use the properties of logarithms and the fact that $\ln 2 \approx 0.6931$ and $\ln 3 \approx 1.0986$ to approximate the given logarithm.

**29.** (a) $\ln 6$   (b) $\ln \dfrac{2}{3}$

   (c) $\ln 81$   (d) $\ln \sqrt{3}$

**30.** (a) $\ln 0.25$   (b) $\ln 24$

   (c) $\ln \sqrt[3]{12}$   (d) $\ln \dfrac{1}{72}$

In Exercises 31–40, use the properties of logarithms to write the given expression as a sum, difference, or multiple of logarithms.

**31.** $\ln \dfrac{2}{3}$

**32.** $\ln \dfrac{1}{5}$

**33.** $\ln xyz$

**34.** $\ln \dfrac{xy}{z}$

**35.** $\ln \sqrt{a - 1}$

**36.** $\ln \sqrt{2^3}$

**37.** $\ln \dfrac{2x}{\sqrt{x^2 - 1}}$

**38.** $\ln \dfrac{1}{e}$

**39.** $\ln \left( \dfrac{x^2 - 1}{x^3} \right)^3$

**40.** $\ln z(z - 1)^2$

In Exercises 41–48, write the given expression as a single logarithm.

**41.** $\ln (x - 2) - \ln (x + 2)$

**42.** $\ln (2x + 1) + \ln (2x - 1)$

**43.** $3 \ln x + 2 \ln y - 4 \ln z$

**44.** $\dfrac{1}{3}[2 \ln (x + 3) + \ln x - \ln (x^2 - 1)]$

**45.** $2[\ln x - \ln (x + 1) - \ln (x - 1)]$

**46.** $2 \ln 3 - \dfrac{1}{2} \ln (x^2 + 1)$

**47.** $\dfrac{3}{2}[\ln x(x^2 + 1) - \ln (x + 1)]$

**48.** $2 \ln x + \dfrac{1}{2} \ln (x + 1)$

In Exercises 49–60, solve for $x$ or $t$ as indicated.

**49.** $e^{\ln x} = 4$

**50.** $e^{\ln x^2} - 9 = 0$

**51.** $\ln x = 0$

**52.** $2 \ln x = 4$

**53.** $e^{x+1} = 4$

**54.** $e^{-0.5x} = 0.075$

**55.** $500e^{-0.11t} = 600$

**56.** $e^{-0.0174t} = 0.5$

**57.** $5^{2x} = 15$

**58.** $2^{1-x} = 6$

**59.** $500(1.07)^t = 1000$

**60.** $1000\left(1 + \dfrac{0.07}{12}\right)^{12t} = 3000$

**61.** A deposit of \$1,000 is made into a fund with an annual interest rate of 11%. Find the time for the investment to double if the interest is compounded
  (a) annually    (b) monthly
  (c) daily    (d) continuously

**62.** A deposit of \$1,000 is made into a fund with an annual interest rate of $10\frac{1}{2}$%. Find the time for the investment to triple if the interest is compounded
  (a) annually    (b) monthly
  (c) daily    (d) continuously

**63.** Complete the following table for the time $t$ necessary for $P$ dollars to triple if interest is compounded continuously at the rate $r$.

| $r$ | 2% | 4% | 6% | 8% | 10% | 12% |
|---|---|---|---|---|---|---|
| $t$ |  |  |  |  |  |  |

**64.** From the Introductory Example of Section 5.1 we know that the formula for the ratio $R$ of carbon isotopes to carbon-14 atoms is given by

$$R = 10^{-12}2^{-t/5700}$$

Estimate the age of a fossil for which the ratio of carbon isotopes to atoms is

$$R = 0.32 \times 10^{-12}$$

**65.** The demand function for a certain product is given by

$$p = 500 - 0.5e^{0.004x}$$

Find the quantity $x$ demanded for a price of
  (a) $p = \$350$    (b) $p = \$300$

**66.** The population $P$ of a city is given by

$$P = 105{,}300e^{0.015t}$$

where $t$ is the time in years, with $t = 0$ corresponding to 1980. According to this model, in what year will the city have a population of 150,000?

**67.** Use a calculator to demonstrate that

$$\dfrac{\ln x}{\ln y} \neq \ln \dfrac{x}{y} = \ln x - \ln y$$

by completing the following table:

| $x$ | $y$ | $\dfrac{\ln x}{\ln y}$ | $\ln \dfrac{x}{y}$ | $\ln x - \ln y$ |
|---|---|---|---|---|
| 1 | 2 |  |  |  |
| 3 | 4 |  |  |  |
| 10 | 5 |  |  |  |
| 4 | 0.5 |  |  |  |

**68.** (a) Use a calculator to complete the following table for the function

$$f(x) = \dfrac{\ln x}{x}$$

| $x$ | 1 | 5 | 10 | $10^2$ | $10^4$ | $10^6$ |
|---|---|---|---|---|---|---|
| $f(x)$ |  |  |  |  |  |  |

(b) Use the table of part (a) to determine $\lim\limits_{x \to \infty} f(x)$.

# Logarithmic Functions: Differentiation and Integration

During the early development of the United States, its population was almost entirely agricultural. As the country became more industrialized, technological advances made it possible for a smaller and smaller percentage of the population to produce enough food and other farm products to sustain the entire population. This phenomenon has continued right up to the present day, with the result that less than 4% of the total labor force worked on farms in 1980. A mathematical model for the number of farm workers $p$ during the 35-year period from 1950 to 1985 is given by

$$p = \frac{1,000,000}{0.0925 + 0.0058t}, \quad 0 \leq t \leq 35$$

where $t$ measures the time in years, with 1950 corresponding to $t = 0$. To estimate the total labor (in worker-years)

required to run U.S. farms over a given time period from $t_1$ to $t_2$, we can integrate this function between $t_1$ and $t_2$, as shown in Figure 5.16. In this section, we will introduce the **log rule for integration.** Using this rule, we have

$$\text{Total labor} = \int_{t_1}^{t_2} \frac{1,000,000}{0.0925 + 0.0058t}\, dt$$

$$= \frac{1,000,000}{0.0058}\Big[\ln (0.0925 + 0.0058t)\Big]_{t_1}^{t_2}$$

$$= \frac{1,000,000}{0.0058}\ln \frac{0.0925 + 0.0058t_2}{0.0925 + 0.0058t_1}$$

With this model we estimate the farm labor over the past three decades as follows.

| Decade | Estimated labor in worker-years |
|---|---|
| 1950–1960 | 83,923,179 |
| 1960–1970 | 56,202,751 |
| 1970–1980 | 42,316,414 |

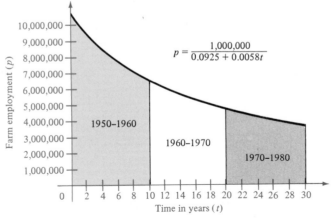

$$p = \frac{1,000,000}{0.0925 + 0.0058t}$$

Farm employment ($p$)

1950–1960

1960–1970

1970–1980

Time in years ($t$)

Number of U.S. Farm Workers

FIGURE 5.16

■ **Differentiation of Logarithmic Functions**
■ **The Log Rule for Integration**

To find the derivative of the natural logarithmic function, we could resort to the definition of the derivative and attempt to evaluate the resulting limit:

$$\frac{d}{dx}[\ln x] = \lim_{\Delta x \to 0} \frac{\ln (x + \Delta x) - \ln x}{\Delta x}$$

However, since we already know the derivative of $e^x$, it is much simpler to use implicit differentiation together with the inverse relationship between $e^x$ and $\ln x$. To do this, we let

$$y = \ln x$$

Then

$$e^y = x$$

and

$$\frac{d}{dx}[e^y] = \frac{d}{dx}[x]$$

$$e^y \frac{dy}{dx} = 1$$

$$\frac{dy}{dx} = \frac{1}{e^y}$$

$$= \frac{1}{x}$$

We summarize this rule, together with its Chain Rule version, as follows. As usual, we assume that $u$ is a differentiable function of $x$.

| | |
|---|---|
| **Derivative of natural logarithmic function** | $\dfrac{d}{dx}[\ln x] = \dfrac{1}{x}$ $\quad$ and $\quad$ $\dfrac{d}{dx}[\ln u] = \dfrac{1}{u}\dfrac{du}{dx}$ |

**EXAMPLE 1**

**Differentiating Logarithmic Functions**

Find the derivative of

$$f(x) = \ln 2x$$

**SOLUTION**

Letting $u = 2x$, we have

$$f'(x) = \frac{1}{u}\frac{du}{dx}$$

$$= \frac{1}{2x}(2)$$

$$= \frac{1}{x}$$

**EXAMPLE 2**

**Differentiating a Logarithmic Function**

Find the derivatives of the following:

(a) $f(x) = \ln(2x^2 + 4)$              (b) $f(x) = x \ln x$

**SOLUTION**

(a) Letting $u = 2x^2 + 4$, we have

$$f'(x) = \frac{1}{u}\frac{du}{dx}$$

$$= \frac{1}{2x^2 + 4}(4x)$$

$$= \frac{2x}{x^2 + 2}$$

(b) Using the Product Rule, we have

$$f'(x) = x\frac{d}{dx}[\ln x] + \ln x\frac{d}{dx}[x]$$

$$= x\left(\frac{1}{x}\right) + (\ln x)(1)$$

$$= 1 + \ln x$$

When logarithms were first introduced by John Napier in the latter part of the sixteenth century, he used logarithmic properties to simplify *calculations* involving products, quotients, and powers. Today, with the availability of calculators, we have little need for this particular application of logarithms. However, we do find a great value in using logarithmic properties to simplify *differentiation* of logarithmic functions involving products, quotients, and powers. The following examples illustrate this technique.

**EXAMPLE 3**

**Logarithmic Properties as an Aid to Differentiation**

Find the derivative of

$$f(x) = \ln \sqrt{x + 1}$$

**SOLUTION**    By rewriting this function as

$$f(x) = \ln \sqrt{x + 1}$$

$$= \frac{1}{2} \ln (x + 1) \qquad \text{Rewrite}$$

we find the derivative to be

$$f'(x) = \frac{d}{dx}[\ln \sqrt{x + 1}]$$

$$= \frac{d}{dx}\left[\frac{1}{2} \ln (x + 1)\right]$$

$$= \frac{1}{2}\left(\frac{1}{x + 1}\right)$$

$$= \frac{1}{2(x + 1)}$$

▉ **Remark:** To see the advantage of rewriting before differentiating, try differentiating $f(x) = \ln\sqrt{x + 1}$ in its given form and compare your result to that obtained in Example 3.

The next example is an even more dramatic illustration of the benefit of rewriting a function before differentiating.

**EXAMPLE 4**    **Rewriting before Differentiating**

Find the derivative of

$$f(x) = \ln \frac{x(x^2 + 1)^2}{\sqrt{2x^3 - 1}}$$

**SOLUTION**    Without rewriting the expression for $f(x)$, we would have

$$f'(x) = \frac{1}{[x(x^2 + 1)^2/\sqrt{2x^3 - 1}]} \frac{d}{dx}\left[\frac{x(x^2 + 1)^2}{\sqrt{2x^3 - 1}}\right]$$

Of course, we would like to avoid differentiating quotients like the one on the right if at all possible. In this case, we can do just that by using the properties of logarithms to rewrite $f(x)$ as

$$f(x) = \ln \frac{x(x^2 + 1)^2}{\sqrt{2x^3 - 1}}$$

$$= \ln x + 2 \ln (x^2 + 1) - \frac{1}{2} \ln (2x^3 - 1)$$

In this form, the derivative is simply

$$f'(x) = \frac{1}{x} + \frac{4x}{x^2 + 1} - \frac{3x^2}{2x^3 - 1}$$

▉

## Integration of Logarithmic Functions

The differentiation formulas

$$\frac{d}{dx}[\ln x] = \frac{1}{x} \quad \text{and} \quad \frac{d}{dx}[\ln u] = \frac{1}{u}\frac{du}{dx}$$

allow us to patch up the hole in our General Power Rule for integration. Recall from Section 4.2 that

$$\int u^n \frac{du}{dx}\, dx = \frac{u^{n+1}}{n+1} + C, \quad n \neq -1$$

Having the derivative formulas for logarithmic functions, we are now in a position to evaluate this integral for $n = -1$, as stated in the following theorem.

**The Log Rule for integration**

$$\int \frac{1}{x}\, dx = \ln|x| + C$$

Moreover, if $u$ is a differentiable function of $x$, then

$$\int \frac{1}{u}\frac{du}{dx}\, dx = \int \frac{1}{u}\, du = \ln|u| + C$$

**Remark:** Be sure to note the use of absolute value in the Log Rule. For those special cases in which $u$ or $x$ is restricted to positive values, we can omit the absolute value sign and write

$$\int \frac{1}{x}\, dx = \ln x + C, \quad x > 0$$

and

$$\int \frac{1}{u}\frac{du}{dx}\, dx = \ln u + C, \quad u > 0$$

**EXAMPLE 5**

**Integrating with the Log Rule**

Evaluate the integral

$$\int \frac{dx}{2x - 1}$$

**SOLUTION**

Letting $u = 2x - 1$, we have $du/dx = 2$. Thus, to apply the Log Rule we multiply and divide by 2 and write

$$\int \frac{dx}{2x-1} = \frac{1}{2} \int \frac{1}{2x-1} (2)\, dx$$

$$= \frac{1}{2} \int \frac{1}{u} \frac{du}{dx}\, dx$$

$$= \frac{1}{2} \ln |u| + C$$

$$= \frac{1}{2} \ln |2x-1| + C$$

**EXAMPLE 6**

**Integrating with the Log Rule**

Evaluate the integral

$$\int \frac{x^2}{1-x^3}\, dx$$

**SOLUTION**

Letting $u = 1 - x^3$, we have $du/dx = -3x^2$. Thus, to apply the Log Rule we multiply and divide by $-3$ and write

$$\int \frac{x^2}{1-x^3}\, dx = -\frac{1}{3} \int \frac{1}{1-x^3} (-3x^2)\, dx$$

$$= -\frac{1}{3} \int \frac{1}{u} \frac{du}{dx}\, dx$$

$$= -\frac{1}{3} \ln |u| + C$$

$$= -\frac{1}{3} \ln |1-x^3| + C$$

**EXAMPLE 7**

**Finding Area with the Log Rule**

Find the area of the region bounded by the graphs of

$$y = \frac{x}{x^2+1}, \quad y = 0, \quad x = 3$$

**SOLUTION**

In Figure 5.17, we see that the area of the specified region is given by the definite integral

$$\text{Area} = \int_0^3 \frac{x}{x^2+1}\, dx$$

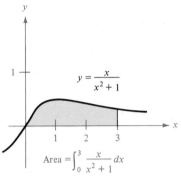

$$y = \frac{x}{x^2 + 1}$$

Area $= \displaystyle\int_0^3 \frac{x}{x^2 + 1}\, dx$

FIGURE 5.17

Letting $u = x^2 + 1$, we have $du/dx = 2x$. Thus, to apply the Log Rule we multiply and divide by 2 and write

$$\int_0^3 \frac{x}{x^2 + 1}\, dx = \frac{1}{2} \int_0^3 \frac{2x}{x^2 + 1}\, dx$$

$$= \frac{1}{2} \left[ \ln\,(x^2 + 1) \right]_0^3$$

(Note that an absolute value sign is unnecessary since $x^2 + 1 > 0$ on the interval under consideration.) Thus, the area is

$$\int_0^3 \frac{x}{x^2 + 1}\, dx = \frac{1}{2}(\ln\,10 - \ln\,1)$$

$$= \frac{1}{2} \ln\,10 \approx 1.15$$

Integrals to which the Log Rule can be applied are often given in disguised form. For instance, if a rational function has a numerator of degree greater than or equal to that of the denominator, division may reveal a form to which we can apply the Log Rule. Our next example is a case in point.

**EXAMPLE 8**

**Rewriting before Integrating**

Evaluate the integral

$$\int \frac{x^2 + x + 1}{x^2 + 1}\, dx$$

**SOLUTION**

First we obtain the following form by dividing:

$$\frac{x^2 + x + 1}{x^2 + 1} = 1 + \frac{x}{x^2 + 1}$$

Thus,
$$\int \frac{x^2 + x + 1}{x^2 + 1}\, dx = \int \left(1 + \frac{x}{x^2 + 1}\right) dx$$

$$= \int (1)\, dx + \frac{1}{2} \int \frac{2x}{x^2 + 1}\, dx$$

$$= x + \frac{1}{2} \ln (x^2 + 1) + C$$

For our last example in this section we summarize some additional situations for which it is helpful to rewrite the integrand in order to recognize the antiderivative. Study this example carefully.

**EXAMPLE 9**

**Disguised Forms of the Log Rule**

Evaluate the following integrals:

(a) $\displaystyle\int \frac{3x^2 + 2x - 1}{x^2}\, dx$

(b) $\displaystyle\int \frac{1}{1 + e^{-x}}\, dx$

(c) $\displaystyle\int \frac{x^3 - 1}{(x - 1)^2}\, dx$

**SOLUTION**

(a) Rewriting the integrand as the sum of three fractions, we have

$$\int \frac{3x^2 + 2x - 1}{x^2}\, dx = \int \left(\frac{3x^2}{x^2} + \frac{2x}{x^2} - \frac{1}{x^2}\right) dx$$

$$= \int \left(3 + \frac{2}{x} - \frac{1}{x^2}\right) dx$$

$$= 3x + 2 \ln |x| + \frac{1}{x} + C$$

(b) In this case, we can rewrite the integrand by multiplying and dividing by $e^x$.

$$\int \frac{1}{1 + e^{-x}}\, dx = \int \left(\frac{e^x}{e^x}\right) \frac{1}{1 + e^{-x}}\, dx$$

$$= \int \frac{e^x}{e^x + 1}\, dx$$

$$= \ln (e^x + 1) + C$$

(c) Dividing the denominator into the numerator, we obtain the following:

$$\int \frac{x^2 + x + 1}{x - 1}\, dx = \int \left(x + 2 + \frac{3}{x - 1}\right) dx$$

$$= \frac{x^2}{2} + 2x + 3 \ln |x - 1| + C$$

## SECTION EXERCISES 5.4

In Exercises 1–6, find the slope of the tangent line to the graph of the given logarithmic function at the point $(1, 0)$.

**1.** $y = \ln x^3$

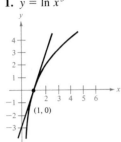
(1, 0)

**2.** $y = \ln x^{5/2}$

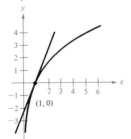
(1, 0)

**3.** $y = \ln x^2$

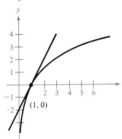
(1, 0)

**4.** $y = \ln x^{1/2}$

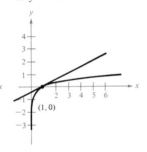
(1, 0)

**5.** $y = \ln x^{3/2}$

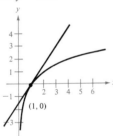
(1, 0)

**6.** $y = \ln x$

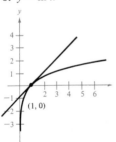
(1, 0)

In Exercises 7–34, find the derivative of the logarithmic function. (Remember that it may be helpful to use logarithmic properties to rewrite the given function *before* differentiating.)

**7.** $y = \ln x^2$

**8.** $y = \ln (x^2 + 3)$

**9.** $f(x) = \ln 2x$

**10.** $f(x) = \ln (1 - x^2)$

**11.** $y = \ln \sqrt{x^4 - 4x}$

**12.** $y = \ln (1 - x)^{3/2}$

**13.** $y = (\ln x)^4$

**14.** $y = (\ln x^2)^2$

**15.** $y = x \ln x$

**16.** $f(x) = x^2 \ln x$

**17.** $y = \ln x\sqrt{x^2 - 1}$

**18.** $g(x) = \ln [(2x + 1)(3x - 2)]$

**19.** $y = \ln \dfrac{x}{x^2 + 1}$

**20.** $y = \ln \dfrac{x}{x + 1}$

**21.** $y = \ln \dfrac{x + 1}{x - 1}$

**22.** $y = \ln \dfrac{x - 1}{x + 1}$

**23.** $y = \ln \sqrt{\dfrac{x + 1}{x - 1}}$

**24.** $y = \ln \sqrt[3]{\dfrac{x - 1}{x + 1}}$

**25.** $y = \dfrac{\ln x}{x^2}$

**26.** $y = \dfrac{\ln x}{x}$

**27.** $y = \ln \dfrac{\sqrt{4 + x^2}}{x}$

**28.** $y = \ln (x + \sqrt{4 + x^2})$

**29.** $y = \ln \sqrt{x^2 - 4}$

**30.** $y = \dfrac{-\sqrt{x^2 + 1}}{x} + \ln (x + \sqrt{x^2 + 1})$

**31.** $g(x) = e^{-x} \ln x$

**32.** $g(x) = \ln \dfrac{e^x + e^{-x}}{2}$

**33.** $f(x) = \ln e^{x^2}$

**34.** $f(x) = \ln \dfrac{1 + e^x}{1 - e^x}$

In Exercises 35–38, find $dx/dp$ for the given demand function.

**35.** $x = \ln \dfrac{1000}{p}$

**36.** $x = 1000 - p \ln p$

**37.** $x = 500 \ln \dfrac{p}{p^2 + 1}$

**38.** $x = 300 - \ln (\ln p)$

In Exercises 39 and 40, find $f''(x)$.

**39.** $f(x) = x \ln x - 4x$

**40.** $f(x) = 2 \ln x + 3$

In Exercises 41–46, find any relative extrema and inflection points and sketch the graph of the function.

**41.** $y = x - \ln x$

**42.** $y = \dfrac{x^2}{2} - \ln x$

**43.** $y = \dfrac{\ln x}{x}$

**44.** $y = x \ln x$

**45.** $y = x^2 \ln x$

**46.** $y = (\ln x)^2$

In Exercises 47–66, evaluate the given integral.

**47.** $\displaystyle \int \dfrac{1}{x + 1} \, dx$

**48.** $\displaystyle \int \dfrac{1}{x - 5} \, dx$

**49.** $\displaystyle \int \dfrac{1}{3 - 2x} \, dx$

**50.** $\displaystyle \int \dfrac{1}{6x + 1} \, dx$

**51.** $\displaystyle \int \dfrac{x}{x^2 + 1} \, dx$

**52.** $\displaystyle \int \dfrac{x^2}{3 - x^3} \, dx$

**53.** $\displaystyle \int_0^1 \dfrac{x^2}{x^3 + 1} \, dx$

**54.** $\displaystyle \int_0^2 \dfrac{x}{x^2 + 4} \, dx$

**55.** $\displaystyle \int \dfrac{x^2 - 4}{x} \, dx$

**56.** $\displaystyle \int \dfrac{x + 5}{x} \, dx$

**57.** $\displaystyle\int_1^e \frac{(1 + \ln x)^2}{x}\,dx$

**58.** $\displaystyle\int_0^1 \frac{x - 1}{x + 1}\,dx$

**59.** $\displaystyle\int_0^2 \frac{x^2 - 2}{x + 1}\,dx$

**60.** $\displaystyle\int \frac{1}{(x + 1)^2}\,dx$

**61.** $\displaystyle\int \frac{1}{\sqrt{x + 1}}\,dx$

**62.** $\displaystyle\int \frac{x + 3}{x^2 + 6x + 7}\,dx$

**63.** $\displaystyle\int \frac{x^2 + 2x + 3}{x^3 + 3x^2 + 9x + 1}\,dx$

**64.** $\displaystyle\int \frac{(\ln x)^2}{x}\,dx$

**65.** $\displaystyle\int \frac{e^x + e^{-x}}{e^x - e^{-x}}\,dx$

**66.** $\displaystyle\int \frac{e^{2x}}{1 + e^{2x}}\,dx$

In Exercises 67 and 68, find the area of the region bounded by the graphs of the given equations.

**67.** $y = \dfrac{x^2 + 4}{x}$

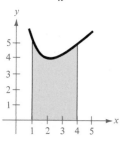

**68.** $y = \dfrac{x + 5}{x}$

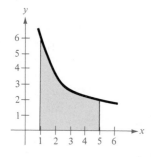

**69.** A population of bacteria is growing at the rate of

$$\frac{dP}{dt} = \frac{3000}{1 + 0.25t}$$

where $t$ is the time in days. Assuming that the initial population (when $t = 0$) is 1000, write an equation that gives the population at any time $t$, and then find the population when $t = 3$ days.

**70.** The demand function for a product is given by

$$p = \frac{90{,}000}{400 + 3x}$$

Find the *average* price $p$ on the interval $40 \le x \le 50$.

# Exponential Growth and Decay

It has been stated that the world's population is growing exponentially. Is this true, and if so what does it mean? Answering the second question first, we say that **exponential growth** implies that the rate of change of the population is proportional to the amount of population at any time. In this section, we will see that the *only* functions having this property are those of the form

$$y = Ce^{kt}$$

To see whether or not the world population is growing exponentially, we used the world population at different times in history, as listed in the *Information Please Almanac* for 1981. With the aid of a computer, we found the best fit of an exponential model to be

$$y = 6164e^{0.00667t}$$

where $y$ is the population and $t$ is the time in years dating from A.D. 1. Table 5.8 lists the estimates for this model, and Figure 5.18 compares the exponential model with the actual growth curve. We can see that the fit is not good. On this basis, we can conclude that world population has *not* followed an exponential growth pattern. It is important to realize that mathematicians use the term *exponential growth* in a technical sense and not simply to describe the growth of a curve whose rate of change is increasing.

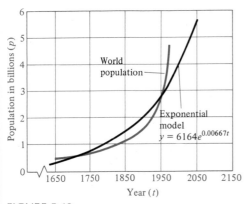

FIGURE 5.18

**TABLE 5.8**

| Year | Actual population | Estimated population |
|------|-------------------|----------------------|
| 1650 | 470,000,000 | 371,000,000 |
| 1750 | 695,000,000 | 723,000,000 |
| 1850 | 1,091,000,000 | 1,409,000,000 |
| 1900 | 1,571,000,000 | 1,966,000,000 |
| 1930 | 2,070,000,000 | 2,402,000,000 |
| 1950 | 2,501,000,000 | 2,745,000,000 |
| 1960 | 2,986,000,000 | 2,934,000,000 |
| 1970 | 3,610,000,000 | 3,137,000,000 |
| 1978 | 4,258,000,000 | 3,308,000,000 |

■ **Exponential Growth and Decay**
■ **Applications**

One of the more common applications of integrals of the form

$$\int \frac{1}{u}\frac{du}{dx}\,dx$$

arises in situations involving exponential growth or decay. In this type of application, we deal with a substance or a population whose *rate of growth (or decay) at a particular time is proportional to the amount of the substance present or the size of the population at that time.* For example, the rate of decomposition of a radioactive substance is proportional to the amount of radioactive substance at a given instant. Or, similarly, the rate of growth of a bacteria culture may be proportional to the number of bacteria in the culture at the time $t$. In its simplest form this relationship is described by the differential equation

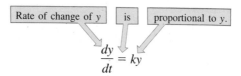

where $k$ is a constant and $y$ is a function of $t$. A **differential equation** is an equation involving derivatives, of first or higher order, of one variable with respect to another. The general solution of this differential equation is given in the following theorem.

---

**Law of exponential growth and decay**

If $y$ is a quantity whose rate of change (with respect to time) is proportional to the quantity present at any time $t$, then $y$ is of the form

$$y = Ce^{kt}$$

where $C$ is called the **initial value** and $k$ is called the **constant of proportionality.** **Exponential growth** is indicated by $k > 0$ and **exponential decay** by $k < 0$.

---

Proof

Since the rate of change of $y$ is proportional to $y$, we have

$$\frac{dy}{dt} = ky$$

An obvious solution to this differential equation is the constant function $y = 0$. To find the other solutions, we assume $y \neq 0$ and divide both sides of the equation by $y$ to obtain

$$\frac{1}{y}\frac{dy}{dt} = k$$

Integrating both sides with respect to $t$, we have

$$\int \frac{1}{y} \frac{dy}{dt} \, dt = \int k \, dt$$

$$\ln |y| = kt + C_1$$

(Note that we need to include only one constant of integration.) Now, solving for $y$, we have

$$|y| = e^{kt+C_1} = e^{C_1} e^{kt}$$

$$y = \pm e^{C_1} e^{kt}$$

Finally, since $y = \pm e^{C_1} e^{kt}$ represents all nonzero solutions and $y = 0$ is already known to be a solution, we can write the general form of the solution as

$$y = Ce^{kt}$$

where $C$ is a real number.

The next example involves radioactive decay, which is measured in terms of **half-life**—the number of years required for half of the atoms in a sample of radioactive material to decay. For example, the half-lives of some common radioactive isotopes are as follows:

| | |
|---|---|
| Uranium ($U^{238}$) | 4,510,000,000 years |
| Plutonium ($Pu^{230}$) | 24,360 years |
| Carbon ($C^{14}$) | 5,730 years |
| Radium ($Ra^{226}$) | 1,620 years |
| Einsteinium ($Es^{254}$) | 276 days |
| Nobelium ($No^{257}$) | 23 seconds |

**EXAMPLE 1**

**Radioactive Decay**

Let $y$ represent the mass of a particular radioactive element whose half-life is 25 years. How much of 1 gram would remain after 15 years?

**SOLUTION**

Assuming that the rate of decay is proportional to $y$, we then have

$$y = Ce^{kt}$$

Since we are given that $y = 1$ when $t = 0$, we know the initial value is

$$1 = Ce^0 = C$$

Furthermore, since $y = \frac{1}{2}$ when $t = 25$, we have

$$\frac{1}{2} = e^{k(25)}$$

$$\ln \frac{1}{2} = 25k$$

$$k = \frac{\ln (1/2)}{25} \approx -0.0277$$

and the equation for $y$ as a function of time is

$$y = Ce^{kt} \approx e^{-0.0277t}$$

Finally, when $t = 15$ years, the amount remaining is

$$y \approx e^{-0.0277(15)} \approx 0.660 \text{ gram}$$

as indicated in Figure 5.19.

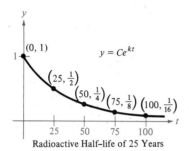

FIGURE 5.19    Radioactive Half–life of 25 Years

Example 1 demonstrates that the problem of finding an exponential growth function basically centers around solving for the constants $C$ and $k$. Furthermore, we can see that it is easy to solve for $C$ when we are given the value of $y$ at $t = 0$. In the next example, we demonstrate a procedure for solving for $C$ and $k$ when we do not know the value of $y$ at $t = 0$.

**EXAMPLE 2**

**Population Growth**

In a research experiment a population of fruit flies increases according to the law of exponential growth. If there are 100 flies after the second day of the experiment and 300 flies after the fourth day, how many flies were in the original population?

**SOLUTION**

Let $y$ be the number of flies at time $t$. Since the population increases according to the law of exponential growth, we know $y$ is of the form $y = Ce^{kt}$, where $y = Ce^{k(0)} = C$ is the number of flies in the original population. Since $y = 100$ when $t = 2$ and $y = 300$ when $t = 4$, we have the equations

$$100 = Ce^{2k} \quad \text{and} \quad 300 = Ce^{4k}$$

Substituting the value $C = 100/e^{2k}$ into the second equation, we have

$$300 = \left(\frac{100}{e^{2k}}\right)e^{4k} = 100e^{2k}$$

$$\frac{300}{100} = e^{2k}$$

$$\ln 3 = 2k$$

$$k = \frac{\ln 3}{2} \approx 0.5493$$

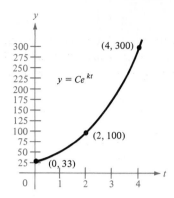

FIGURE 5.20

Thus, the original population is

$$y = \frac{100}{e^{2(0.5493)}} \approx 33 \text{ flies}$$

as shown in Figure 5.20.

**EXAMPLE 3**

**Continuously Compounded Interest**

Money is deposited in a savings account for which the interest is compounded continuously. If the balance in the account doubles in 6 years, what is the annual percentage rate?

**SOLUTION**

The formula for the balance using continuously compounded interest is given by the exponential growth function

$$A = Pe^{rt}$$

When $t = 6$, $A = 2P$, as shown in Figure 5.21. Thus, we have

$$2P = Pe^{6r}$$

$$2 = e^{6r}$$

$$\ln 2 = 6r$$

and we conclude that the annual percentage rate is

$$r = \frac{\ln 2}{6} \approx 0.1155 = 11.55\%$$

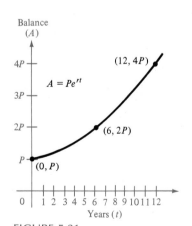

FIGURE 5.21

EXAMPLE 4
**Declining Sales Following Discontinuation of Advertising**

Four months after it discontinued its advertising campaign, a manufacturer noticed that sales had dropped from 100,000 units per month to 80,000 units. If the sales follow an exponential pattern of decline, what will they be after another 4 months?

**SOLUTION**

Using the exponential model

$$y = Ce^{kt}$$

where $t$ is measured in months, we have

$$100,000 = Ce^0 = C \quad \text{and} \quad 80,000 = Ce^{4k}$$

Thus,

$$80,000 = 100,000e^{4k}$$

$$0.8 = e^{4k}$$

$$\ln 0.8 = 4k$$

$$k = \frac{\ln 0.8}{4} \approx -0.0558$$

Finally, after 4 more months ($t = 8$), we can expect sales to drop to

$$y = 100,000e^{-0.0558(8)} \approx 64,000 \text{ units}$$

as shown in Figure 5.22.

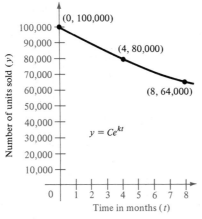

FIGURE 5.22

## SECTION EXERCISES 5.5

In Exercises 1–6, find the exponential function $y = Ce^{kt}$ that passes through the two given points.

**1.** $y = Ce^{kt}$

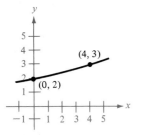

**2.** $y = Ce^{kt}$

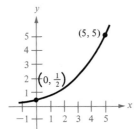

**3.** $y = Ce^{kt}$

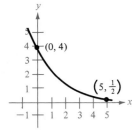

**4.** $y = Ce^{kt}$

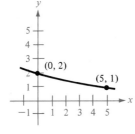

**5.** $y = Ce^{kt}$

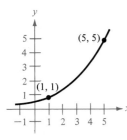

**6.** $y = Ce^{kt}$

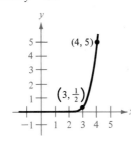

In Exercises 7–12, complete the table for the given radioactive isotope.

| Isotope | Half-life (in years) | Initial quantity | Amount after 1,000 years | Amount after 10,000 years |
|---------|---------------------|------------------|--------------------------|---------------------------|
| **7.** $Ra^{226}$ | 1,620 | 10 grams | | |
| **8.** $Ra^{226}$ | 1,620 | | 1.5 grams | |
| **9.** $C^{14}$ | 5,730 | | | 2 grams |
| **10.** $C^{14}$ | 5,730 | 3 grams | | |
| **11.** $Pu^{230}$ | 24,360 | | 2.1 grams | |
| **12.** $Pu^{230}$ | 24,360 | | | 0.4 gram |

**13.** What percentage of a present amount of radioactive radium ($Ra^{226}$) will remain after 100 years?

**14.** Find the half-life of a radioactive material if after 1 year 99.57% of the initial amount remains.

**15.** The number of a certain type of bacteria increases continuously at a rate proportional to the number present. If there are 100 present at a given time and 300 present 5 hours later, how many will there be 10 hours after the initial time? How long does it take the number of bacteria to double?

**16.** In 1970 the population of a town was 2500, and in 1980 it was 3350. Assuming the population increases continuously at a constant rate proportional to the existing population, estimate the population in 1990. How many years are necessary for the population to double?

In Exercises 17–22, complete the table for a savings account in which interest is compounded continuously.

| Initial investment | Annual rate | Time to double | Amount after 10 years | Amount after 25 years |
|--------------------|-------------|----------------|-----------------------|-----------------------|
| **17.** $1,000 | 12% | | | |
| **18.** $20,000 | $10\frac{1}{2}$% | | | |
| **19.** $750 | | $7\frac{3}{4}$ years | | |
| **20.** $10,000 | | 5 years | | |
| **21.** $500 | | | $1,292.85 | |
| **22.** $2,000 | | | | $6,008.33 |

**23.** During a slump in the economy, a company found that its annual revenues dropped from $742,000 in 1984 to $632,000 in 1986. If the revenue is following an exponential pattern of decline, what is the expected revenue for 1987? (Let $t = 0$ represent 1984.)

**24.** The revenues in 1985 and 1986 from a certain company's products were $650,000 and $708,500, respectively. If the revenue is following an exponential pattern of growth, what is the expected revenue for 1987? (Let $t = 0$ represent 1985.)

**25.** The sales $S$ (in thousands of units) of a new product after it has been on the market $t$ years are given by

$$S(t) = Ce^{k/t}$$

(a) Find $S$ as a function of $t$ if 5000 units have been sold after 1 year and the saturation point for the market is 30,000. That is,

$$S(1) = 5000 \quad \text{and} \quad \lim_{t \to \infty} S(t) = 30,000$$

(b) How many units will be sold after 5 years?

(c) Sketch a graph of this sales function.

**26.** The sales $S$ (in thousands of units) of a new product after it has been on the market $t$ years are given by

$$S(t) = 30(1 - e^{kt})$$

(a) Find $S$ as a function of $t$ if 5000 units have been sold after 1 year.

(b) How many units will saturate this market?

(c) How many units will be sold after 5 years?

(d) Sketch a graph of this sales function.

**27.** The management at a certain factory has found that the maximum number of units a worker can produce in a day is 30. The learning curve for the number of units $N$ produced per day after a new employee has worked $t$ days is given by

$$N = 30(1 - e^{kt})$$

After 20 days on the job, a particular worker produced 19 units.

(a) Find the learning curve for this worker (that is, find the value of $k$).

(b) How many days should pass before this worker is producing 25 units per day?

**28.** If the management in Exercise 27 requires that a new employee be producing at least 20 units per day after 30 days on the job, find

(a) the learning curve describing this minimum requirement

(b) the number of days before a minimal achiever is producing 25 units per day

**29.** A small business assumes that the demand function for one of its products is the exponential function

$$p = Ce^{kx}$$

(a) Find $p$ as a function of $x$ if the quantity demanded is 1000 units and 1200 units when the price is $45 and $40, respectively.

(b) Find the values of $x$ and $p$ that will maximize revenue for this product.

**30.** Repeat Exercise 29, assuming that the quantity demanded is 300 units and 400 units when the price is $5 and $4, respectively.

## Important Terms

| | | | |
|---|---|---|---|
| Exponential function | Compound interest | Exponential decay | Half-life |
| Base of exponential function | Continuous compounding | Differential equation | |
| The number $e \approx 2.71828$ | Natural logarithmic function | Initial value, $C$ | |
| | Exponential growth | Constant of proportionality, $k$ | |

## Important Techniques

Sketching the graph of an exponential function
Differentiating exponential functions
Integrating exponential functions
Sketching the graph of a logarithmic function

Solving equations involving exponential and logarithmic functions
Differentiating logarithmic functions
Integrating with the Log Rule
Solving exponential growth and decay problems

## Important Formulas

Properties of exponents:

1. $a^0 = 1$
2. $a^x a^y = a^{x+y}$
3. $\dfrac{a^x}{a^y} = a^{x-y}$
4. $(a^x)^y = a^{xy}$
5. $(ab)^x = a^x b^x$
6. $\left(\dfrac{a}{b}\right)^x = \dfrac{a^x}{b^x}$
7. $a^{-x} = \dfrac{1}{a^x}$

Limit definition of $e$:

$$e = \lim_{x \to 0} (1 + x)^{1/x}$$

Compound interest:

$P$ = amount of deposit
$t$ = number of years
$A$ = balance after $t$ years
$r$ = annual interest rate (decimal form)

Compounded $n$ times per year,

$$A = P\left(1 + \frac{r}{n}\right)^{nt}$$

Compounded continuously,

$$A = Pe^{rt}$$

Differentiation and integration of exponential functions:

$$\frac{d}{dx}[e^x] = e^x \qquad \frac{d}{dx}[e^u] = e^u \frac{du}{dx}$$

$$\int e^x \, dx = e^x + C$$

$$\int e^u \frac{du}{dx} \, dx = e^u + C$$

Inverse properties of $\ln x$ and $e^x$:

$$\ln(e^x) = x$$

$$e^{\ln x} = x$$

Properties of logarithms:

1. $\ln(xy) = \ln x + \ln y$
2. $\ln \dfrac{x}{y} = \ln x - \ln y$
3. $\ln(x^y) = y \ln x$

Differentiation and integration involving logarithmic functions:

$$\frac{d}{dx}[\ln x] = \frac{1}{x}$$

$$\frac{d}{dx}[\ln u] = \frac{1}{u} \frac{du}{dx}$$

$$\int \frac{1}{x} \, dx = \ln|x| + C$$

$$\int \frac{1}{u} \frac{du}{dx} \, dx = \ln|u| + C$$

Law of exponential growth and decay:

$$y = Ce^{kt}$$

In Exercises 1–4, sketch the graph of the function.

**1.** $f(x) = e^{-2x}$

**2.** $f(t) = e^t - 1$

**3.** $g(x) = \ln(x + 1)$

**4.** $f(x) = \ln \dfrac{x}{4}$

In Exercises 5–8, solve the given equation for $x$.

**5.** $e^{\ln x} = 3$

**6.** $100e^{0.07x} = 125$

**7.** $\ln 4x - \ln(x + 1) = 0$

**8.** $\ln x + \ln(x - 3) = 0$

In Exercises 9–30, find $dy/dx$.

**9.** $y = \ln e^{-x^2}$

**10.** $y = \ln \dfrac{e^x}{1 + e^x}$

**11.** $y = x^2 e^x$

**12.** $y = e^{-x^2/2}$

**13.** $y = \sqrt{e^{2x} + e^{-2x}}$

**14.** $y = \sqrt[3]{xe^{3x}}$

**15.** $y = \dfrac{e^{2x}}{x}$

**16.** $y = \dfrac{x^2}{e^x}$

**17.** $ye^x = xy + 1$

**18.** $y = \ln xy$

**19.** $y = \ln \sqrt{x}$

**20.** $y = \ln \dfrac{x(x - 1)}{x - 2}$

**21.** $y = x\sqrt{\ln x}$

**22.** $y = \ln [x(x^2 - 2)^{2/3}]$

**23.** $y \ln x + y^2 = 1$

**24.** $\ln(x + y) = x$

**25.** $\ln y = x \ln x$

**26.** $\ln y = xy$

**27.** $y = \dfrac{1}{b^2} \left[ \ln(a + bx) + \dfrac{a}{a + bx} \right]$

**28.** $y = \dfrac{1}{b^2}[a + bx - a \ln(a + bx)]$

**29.** $y = -\dfrac{1}{a} \ln \dfrac{a + bx}{x}$

**30.** $y = -\dfrac{1}{ax} + \dfrac{b}{a^2} \ln \dfrac{a + bx}{x}$

In Exercises 31–58, evaluate the given integral.

**31.** $\displaystyle \int xe^{-3x^2}\,dx$

**32.** $\displaystyle \int x^2 e^{x^3+1}\,dx$

**33.** $\displaystyle \int \dfrac{e^{4x} - e^{2x} + 1}{e^x}\,dx$

**34.** $\displaystyle \int \dfrac{e^{1/x}}{x^2}\,dx$

**35.** $\displaystyle \int \dfrac{e^x}{e^x - 1}\,dx$

**36.** $\displaystyle \int \dfrac{e^{2x} - e^{-2x}}{e^{2x} + e^{-2x}}\,dx$

**37.** $\displaystyle \int \dfrac{e^{-2x}}{1 + e^{-2x}}\,dx$

**38.** $\displaystyle \int \dfrac{x - 1}{3x^2 - 6x - 1}\,dx$

**39.** $\displaystyle \int \dfrac{x}{1 - x^2}\,dx$

**40.** $\displaystyle \int \dfrac{x}{16 + x^2}\,dx$

**41.** $\displaystyle \int \dfrac{x}{\sqrt{1 - x^2}}\,dx$

**42.** $\displaystyle \int \dfrac{x}{(16 + x^2)^{3/2}}\,dx$

**43.** $\displaystyle \int \dfrac{1}{7x - 2}\,dx$

**44.** $\displaystyle \int \dfrac{x}{x^2 - 1}\,dx$

**45.** $\displaystyle \int \dfrac{1}{(7x - 2)^3}\,dx$

**46.** $\displaystyle \int \dfrac{x}{(x^2 - 1)^2}\,dx$

**47.** $\displaystyle \int \dfrac{\ln x^2}{x}\,dx$

**48.** $\displaystyle \int \dfrac{\ln 2x}{x}\,dx$

**49.** $\displaystyle \int \dfrac{x^2 + 3}{x}\,dx$

**50.** $\displaystyle \int \dfrac{x^3 + 1}{x^2}\,dx$

**51.** $\displaystyle \int \dfrac{x^2}{x^3 - 1}\,dx$

**52.** $\displaystyle \int \left(x + \dfrac{1}{x}\right)^2\,dx$

**53.** $\displaystyle \int \dfrac{x}{x + 1}\,dx$

**54.** $\displaystyle \int \dfrac{x + 2}{2x + 3}\,dx$

**55.** $\displaystyle \int_1^2 \dfrac{e^{1/x}}{x^2}\,dx$

**56.** $\displaystyle \int_0^2 \dfrac{1}{1 + x}\,dx$

**57.** $\displaystyle \int_1^4 \dfrac{x + 1}{x}\,dx$

**58.** $\displaystyle \int_1^e \dfrac{\ln x}{x}\,dx$

In Exercises 59 and 60, find the area of the region bounded by the graphs of the equations.

**59.** $y = \dfrac{1}{x + 1}$, $y = 0$, $x = 0$, $x = 4$

**60.** $y = e^{-x/4}$, $y = 0$, $x = 0$, $x = 4$

**61.** A deposit of $500 earns interest at the rate of 5% compounded continuously. Find its value after
(a) 1 year     (b) 10 years     (c) 100 years

**62.** A deposit earns interest at the rate $r$ compounded continuously and doubles in value in 10 years. Find $r$.

**63.** How large a deposit, at 7% interest compounded continuously, must be made to obtain a balance of $10,000 in 15 years?

**64.** A deposit of $2,500 is made in a savings account at an annual percentage rate of 12% compounded continuously. Find the average balance in this account during the first 5 years.

**65.** A population is growing continuously at the rate of $2\frac{1}{2}\%$ per year. Find the time necessary for the population to
(a) double in size     (b) triple in size

**66.** Under ideal conditions air pressure decreases continuously with height above sea level at a rate proportional to the pressure at that height. If the barometer reads 30 inches at sea level and 15 inches at 18,000 feet, find the barometric pressure at 35,000 feet.

**67.** A solution of a certain drug contains 500 units per milliliter when it is prepared. After 40 days, it contains 300 units per milliliter. Assuming that the rate of decomposition is proportional to the amount present, find an equation giving the amount $A$ after $t$ days.

In Exercises 68 and 69, the **exponential density function** is given by $f(t) = (1/\mu)e^{-t/\mu}$, where $t$ is the time in minutes and $\mu$ is the average time between successive events. The definite integral

$$P(a \le t \le b) = \int_a^b \frac{1}{\mu} e^{-t/\mu} \, dt$$

gives the probability that the elapsed time before the next occurrence lies between $a$ and $b$ units of time.

**68.** Trucks arrive at a terminal at an average rate of 3 per hour ($\mu = 20$ minutes is the average time between arrivals). If a truck has just arrived, find the probability that the next arrival will be
   (a) within 10 minutes
   (b) within 30 minutes
   (c) between 15 and 30 minutes
   (d) within 60 minutes

**69.** The average time between incoming calls at a switchboard is $\mu = 3$ minutes. Complete the accompanying table to show the probabilities of time elapsed between incoming calls.

| Time elapsed between calls in minutes | 0–2 | 2–4 | 4–6 | 6–8 | 8–10 |
|---|---|---|---|---|---|
| Probability | | | | | |

# Techniques of
# Integration

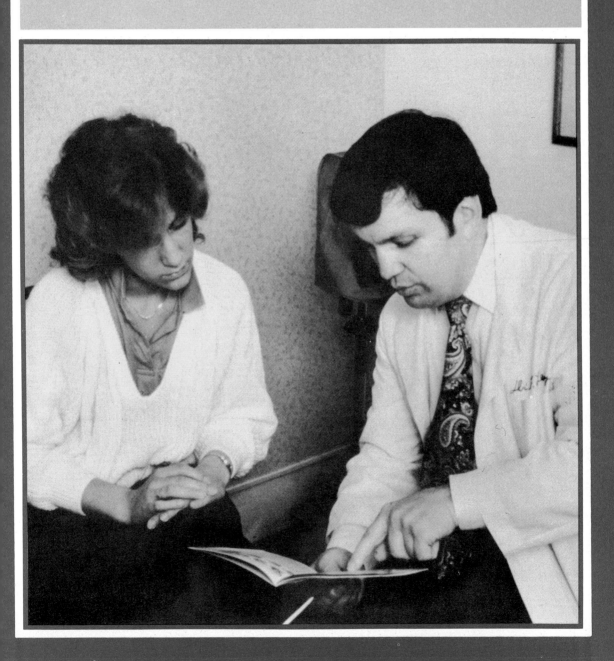

# CHAPTER 6

# Integration by Substitution

Most mathematical models used to predict human responses must allow considerable leeway to account for individual differences. For example, some people have body chemistries that allow them to expel various drugs quickly, whereas others retain the drugs much longer. In Figure 6.1 we have pictured three different frequency distributions showing various drug-retention patterns. Note that each is of the form

$$y = x^r(1 - x)^s, \quad 0 \le r, s$$

Suppose that a drug such as calcium is given to a patient and that the percentage $x$ (in decimal form) of the original dosage remaining in the person's system after 24 hours is measured. Assume the retention pattern is given by

$$y = x^2(1 - x)^{1/2}$$

where $y$ represents the frequency. The graph in Figure 6.1(a) shows that most people would retain from 60% to 90% of the dosage after 24 hours, and very few would retain 25% or less. To find the portion of the population that is expected to retain between 60% and 90% of the drug, we can compute the ratio of the area under the curve between 0.6 and 0.9 to the total area under the curve. That is,

$$\text{Portion of population retaining between 60\% and 90\%} = \frac{\int_{0.6}^{0.9} x^2(1 - x)^{1/2} \, dx}{\int_{0}^{1} x^2(1 - x)^{1/2} \, dx}$$

To find the antiderivatives for these two integrals, we use a procedure called **substitution.** In this particular case the ratio is 0.530, and we conclude that 53.0% of the population would retain between 60% and 90% of the drug.

(a)

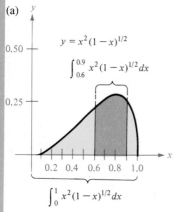

(b)

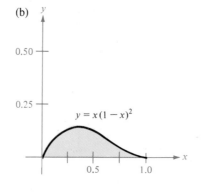

(c)

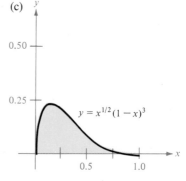

FIGURE 6.1

■ **Review of Basic Integration Formulas**
■ **Integration by Substitution**

In Chapters 2, 4, and 5, we presented a number of rules for differentiating and integrating functions. In general, the rules for integrating were derived from corresponding differentiation formulas. It may surprise you to learn that, although you now have all the necessary tools for *differentiating* algebraic, exponential, and logarithmic functions, your set of tools for *integrating* these functions is by no means complete! The primary objective of this chapter is to develop several techniques that greatly expand the set of integrals to which the basic integration formulas can be applied.

| | | |
|---|---|---|
| **Basic integration formulas** | 1. Constant Rule: | $\displaystyle\int dx = x + C$ |
| | 2. Simple Power Rule: | $\displaystyle\int x^n \, dx = \frac{x^{n+1}}{n+1} + C, \quad n \neq -1$ |
| | 3. General Power Rule: | $\displaystyle\int u^n \frac{du}{dx} \, dx = \int u^n \, du = \frac{u^{n+1}}{n+1} + C, \quad n \neq -1$ |
| | 4. Simple Log Rule: | $\displaystyle\int \frac{1}{x} \, dx = \ln|x| + C$ |
| | 5. General Log Rule: | $\displaystyle\int \frac{1}{u} \frac{du}{dx} \, dx = \int \frac{1}{u} \, du = \ln|u| + C$ |
| | 6. Simple Exponential Rule: | $\displaystyle\int e^x \, dx = e^x + C$ |
| | 7. General Exponential Rule: | $\displaystyle\int e^u \frac{du}{dx} \, dx = \int e^u \, du = e^u + C$ |

You don't have to work many integration problems before you realize that integration is not nearly as straightforward as differentiation is. A major part of any integration problem is the recognition of which basic integration formula (or formulas) to use to solve the problem. This requires memorization of the basic formulas, familiarity with various procedures for rewriting integrands in the basic forms, and a lot of practice.

For example, consider the ingenuity required in the integration of

$$\int \frac{x}{(x+1)^2} \, dx$$

None of the seven basic formulas can be applied directly to this integral. However, if we add and subtract 1 in the numerator, we can apply the Log Rule and the Power Rule as follows:

$$\int \frac{x}{(x+1)^2}\, dx = \int \frac{x+1-1}{(x+1)^2}\, dx$$

$$= \int \left[ \frac{x+1}{(x+1)^2} - \frac{1}{(x+1)^2} \right] dx$$

$$= \int \left[ \frac{1}{x+1} - (x+1)^{-2} \right] dx$$

$$= \ln|x+1| + \frac{1}{x+1} + C$$

How did we know that adding and subtracting 1 would be the key to evaluating this integral? We've had many years of practice—that's how we knew! Since you have had considerably less time to practice there is, fortunately, a nice integration technique for solving this integral in a straightforward manner. This technique is called **substitution,** and it involves a change of variable that permits us to rewrite the integrand in a form to which we can apply the basic integration rules.

One of the critical steps in this procedure is rewriting $dx$ in terms of the new variable. In general, if we use the substitution

$$u = f(x)$$

we have

$$\frac{du}{dx} = f'(x)$$

and

$$du = f'(x)\, dx$$

**EXAMPLE 1**

**Integration by Substitution**

Use the substitution $u = x + 1$ to evaluate the integral

$$\int \frac{x}{(x+1)^2}\, dx$$

**SOLUTION**

From the substitution $u = x + 1$, we have

$$x = u - 1, \qquad \frac{dx}{du} = 1, \qquad dx = du$$

Now, replacing *all* instances of $x$ and $dx$ with the appropriate $u$-variable forms, we have

$$\int \frac{x}{(x+1)^2}\, dx = \int \frac{u-1}{u^2}\, du$$

Then we write
$$\int \frac{u-1}{u^2}\, du = \int \left(\frac{u}{u^2} - \frac{1}{u^2}\right) du$$

$$= \int \left(\frac{1}{u} - \frac{1}{u^2}\right) du$$

$$= \ln |u| + \frac{1}{u} + C$$

Finally, we return to $x$-variable form by resubstituting $u = x + 1$ to obtain
$$\int \frac{x}{(x+1)^2}\, dx = \ln |x+1| + \frac{1}{x+1} + C$$

The basic steps involved in integration by substitution can be outlined as follows.

**Integration by substitution**

1. Let $u$ be some function of $x$ (usually one that appears in the integrand).
2. Solve for $x$ and $dx$ in terms of $u$ and $du$.
3. Convert the entire integral to $u$-variable form and try to fit it to one (or more) of the basic integration formulas. If none seems to fit, consider trying a different substitution.
4. After integrating, rewrite the antiderivative as a function of $x$.

**EXAMPLE 2**

**Integration by Substitution**

Evaluate the integral
$$\int x\sqrt{x^2 - 1}\, dx$$

**SOLUTION**

Consider the substitution $u = x^2 - 1$. Then we have
$$du = 2x\, dx$$

Since $x\, dx$ is already present in the integral, we multiply and divide by 2 to obtain
$$\int x\sqrt{x^2-1}\, dx = \frac{1}{2}\int \underbrace{(x^2-1)^{1/2}}_{u^{1/2}}\ \underbrace{2x\, dx}_{du}$$

$$= \frac{1}{2}\int u^{1/2}\, du$$

$$= \frac{1}{2}\frac{u^{3/2}}{3/2} + C \qquad \text{Power Rule}$$

$$= \frac{1}{3}u^{3/2} + C = \frac{1}{3}(x^2-1)^{3/2} + C$$

## EXAMPLE 3

**Integration by Substitution**

Evaluate the integral

$$\int \frac{e^{2x}}{1 + e^{2x}} \, dx$$

**SOLUTION**

Letting $u$ be the denominator of the integrand, we have

$$u = 1 + e^{2x} \qquad \text{and} \qquad du = 2e^{2x} \, dx$$

Then, multiplying and dividing by 2, we have

$$\int \frac{e^{2x}}{1 + e^{2x}} \, dx = \frac{1}{2} \int \underbrace{\frac{1}{1 + e^{2x}}}_{1/u} \underbrace{2e^{2x} \, dx}_{du}$$

$$= \frac{1}{2} \int \frac{1}{u} \, du$$

$$= \frac{1}{2} \ln |u| + C \qquad \qquad \text{Log Rule}$$

$$= \frac{1}{2} \ln (1 + e^{2x}) + C$$

## EXAMPLE 4

**Integration by Substitution**

Evaluate the integral

$$\int x\sqrt{x - 1} \, dx$$

**SOLUTION**

First, we note that, as it stands, this integral *does not* fit any of the basic integration formulas. However, using the substitution $u = x - 1$, we have

$$du = dx \qquad \text{and} \qquad x = u + 1$$

We now substitute for $x$, $x - 1$, and $dx$ as follows:

$$\int x(x - 1)^{1/2} \, dx = \int (u + 1)u^{1/2} \, du$$

$$= \int (u^{3/2} + u^{1/2}) \, du$$

$$= \frac{u^{5/2}}{5/2} + \frac{u^{3/2}}{3/2} + C$$

$$= \frac{2}{5}(x - 1)^{5/2} + \frac{2}{3}(x - 1)^{3/2} + C$$

$$= \frac{6}{15}(x - 1)^{5/2} + \frac{10}{15}(x - 1)^{3/2} + C$$

$$= \frac{2}{15}(x - 1)^{3/2}[3(x - 1) + 5] + C$$

$$= \frac{2}{15}(x - 1)^{3/2}(3x + 2) + C$$

Example 4 demonstrates one of the characteristics of integration by substitution. That is, the form of the antiderivative as it exists immediately after resubstitution into $x$-variable form can often be simplified. Thus, when working the exercises in this section, don't be too quick to think your answer is incorrect just because it doesn't look exactly like the answer given in the back of the text. You may be able to reconcile the two answers by algebraic simplification. A second way to obtain a different form of an antiderivative is demonstrated in the next example.

**EXAMPLE 5**

**An Extra Constant Term in the Solution**

Evaluate the integral

$$\int \frac{1}{\sqrt{x} + 1} \, dx$$

**SOLUTION**

Since none of the basic formulas apply, we try the substitution $u = \sqrt{x} + 1$. Then, to find $dx$ in terms of $u$ and $du$, we write

$$\sqrt{x} = u - 1, \qquad x = (u - 1)^2, \qquad dx = 2(u - 1) \, du$$

Thus,

$$\int \frac{1}{\sqrt{x} + 1} \, dx = \int \frac{1}{u} 2(u - 1) \, du$$

$$= 2 \int \frac{u - 1}{u} \, du$$

$$= 2 \int \left(1 - \frac{1}{u}\right) du$$

$$= 2(u - \ln |u|) + C_1$$

$$= 2[\sqrt{x} + 1 - \ln (\sqrt{x} + 1)] + C_1$$

$$= 2\sqrt{x} - 2 \ln (\sqrt{x} + 1) + C$$

Note that the constant term 1 can be incorporated into the constant $C_1$ to produce $C$.

**Remark:** You should not conclude that there is only one substitution that will work in a given integral. For instance, the substitution $u = \sqrt{x}$ would have worked just as well in Example 5, and we suggest that you try reworking the example with this substitution.

The fourth step outlined in the guidelines for integration by substitution suggests that we convert back to variable $x$. However, to evaluate *definite* integrals it is often more convenient to determine the limits of integration for variable $u$. This is usually easier than converting back to variable $x$ and evaluating the antiderivative at the original limits. The next example illustrates this procedure.

**EXAMPLE 6**

**Converting the Limits of Integration for Definite Integrals**

Find the area $A$ of the region bounded by

$$y = \frac{x}{\sqrt{2x - 1}}$$

and the $x$-axis, from $x = 1$ to $x = 5$.

**SOLUTION**

From Figure 6.2 we can see that the area of the region is given by

$$A = \int_1^5 \frac{x}{\sqrt{2x - 1}} \, dx$$

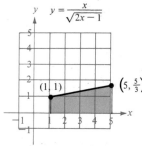

Region Before Substitution

FIGURE 6.2

To evaluate this integral, we let $u = \sqrt{2x - 1}$. Then,

$$u^2 = 2x - 1$$

$$x = \frac{u^2 + 1}{2}$$

$$dx = u \, du$$

Before substituting, we determine the new upper and lower limits of integration. That is, we find the values of $u$ corresponding to the original limits of integration ($x = 1$ and $x = 5$).

| *Changing the lower limit* | *Changing the upper limit* |
|---|---|
| When $x = 1$, $u = \sqrt{2 - 1} = 1$. | When $x = 5$, $u = \sqrt{10 - 1} = 3$. |

Now we substitute to obtain

$$\int_1^5 \frac{x}{\sqrt{2x-1}}\, dx = \int_1^3 \frac{1}{2}\left(\frac{u^2+1}{u}\right) u\, du$$

$$= \frac{1}{2}\int_1^3 (u^2+1)\, du$$

$$= \frac{1}{2}\left[\frac{u^3}{3}+u\right]_1^3$$

$$= \frac{1}{2}\left(9+3-\frac{1}{3}-1\right)$$

$$= \frac{16}{3}$$

**Remark:** In Example 6 we can interpret the equation

$$\int_1^5 \frac{x}{\sqrt{2x-1}}\, dx = \int_1^3 \frac{u^2+1}{2}\, du$$

geometrically to mean that the two *different* regions shown in Figures 6.2 and 6.3 have the *same* area.

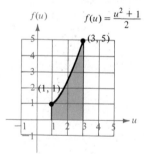

$f(u)$   $f(u) = \dfrac{u^2+1}{2}$

Region After Substitution
FIGURE 6.3

When solving definite integrals by substitution, don't be surprised if the upper limit of integration of the $u$-variable form is smaller than the lower limit. If this happens, don't rearrange the limits. Simply evaluate as usual. For example, after substituting $u = \sqrt{1-x}$ in the integral

$$\int_0^1 x^2(1-x)^{1/2}\, dx$$

we have $u = \sqrt[3]{1 - 1} = 0$ when $x = 1$ and $u = \sqrt[3]{1 - 0} = 1$ when $x = 0$. Thus, the correct $u$-variable form of this integral is

$$-2 \int_1^0 (1 - u^3)^2 u^2 \, du$$

This situation is illustrated in the next example.

**EXAMPLE 7**

**An Application to Probability**

A psychologist finds that the probability of recall in a certain memory experiment is given by the model

$$P(a \le x \le b) = \int_a^b \frac{28}{9} x \sqrt[3]{1 - x} \, dx, \quad 0 \le a \le b \le 1$$

where $x$ represents the percentage of recall, as shown in Figure 6.4. Find the probability that a randomly chosen individual will recall between 0% and $87\frac{1}{2}$% of the material.

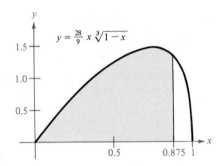

FIGURE 6.4

**SOLUTION**

Since $a = 0$ and $b = 0.875$, the probability is given by

$$P(0 \le x \le 0.875) = \int_0^{0.875} \frac{28}{9} x \sqrt[3]{1 - x} \, dx$$

Now, letting $u = \sqrt[3]{1 - x}$, we have

$$u^3 = 1 - x$$
$$x = 1 - u^3$$
$$dx = -3u^2 \, du$$

and the new upper and lower limits of integration are as follows.

| Lower limit | Upper limit |
|---|---|
| When $x = 0$, $u = \sqrt[3]{1 - 0} = 1$. | When $x = 0.875$, $u = \sqrt[3]{1 - 0.875}$ |
| | $= \sqrt[3]{0.125}$ |
| | $= 0.5 = \dfrac{1}{2}$ |

Thus, the $u$-variable form of the integral is

$$\int_0^{0.875} \frac{28}{9} x \sqrt[3]{1-x}\, dx = \int_1^{1/2} \left[\frac{28}{9}(1-u^3)(u)(-3u^2)\right] du$$

$$= 3\left(\frac{28}{9}\right) \int_1^{1/2} (u^6 - u^3)\, du$$

$$= \frac{28}{3}\left[\frac{u^7}{7} - \frac{u^4}{4}\right]_1^{1/2}$$

$$= \frac{28}{3}\left[\left(\frac{1}{896} - \frac{1}{64}\right) - \left(\frac{1}{7} - \frac{1}{4}\right)\right]$$

$$= \frac{83}{96} \approx 0.8646$$

Finally, we conclude that the probability of recalling between 0% and $87\frac{1}{2}$% of the material is

$$P(0 \le x \le 0.875) \approx 86.46\%$$

## SECTION EXERCISES 6.1

In Exercises 1–40, find the indefinite integral.

**1.** $\displaystyle\int (3x-2)^4\, dx$

**2.** $\displaystyle\int (-2x+5)^{3/2}\, dx$

**3.** $\displaystyle\int \frac{2}{(t-9)^2}\, dt$

**4.** $\displaystyle\int \frac{2t-1}{t^2 - t + 2}\, dt$

**5.** $\displaystyle\int 2x\sqrt{1+x^2}\, dx$

**6.** $\displaystyle\int x^2(3+x^3)^5\, dx$

**7.** $\displaystyle\int \frac{12x+2}{3x^2 + x}\, dx$

**8.** $\displaystyle\int \frac{6x^2+2}{x^3+x}\, dx$

**9.** $\displaystyle\int \frac{1}{(5x+1)^3}\, dx$

**10.** $\displaystyle\int \frac{1}{\sqrt{5x+1}}\, dx$

**11.** $\displaystyle\int \frac{\ln 2x^2}{x}\, dx$

**12.** $\displaystyle\int \frac{x\ln(2x^2+2)}{x^2+1}\, dx$

**13.** $\displaystyle\int \frac{e^{3x}}{1-e^{3x}}\, dx$

**14.** $\displaystyle\int 2x(e^{x^2})^3\, dx$

**15.** $\displaystyle\int \frac{x^2}{x-1}\, dx$

**16.** $\displaystyle\int \frac{2x}{x-4}\, dx$

**17.** $\displaystyle\int x\sqrt{4-2x^2}\, dx$

**18.** $\displaystyle\int \frac{t}{\sqrt{1-t^2}}\, dt$

**19.** $\displaystyle\int e^{5x}\, dx$

**20.** $\displaystyle\int \frac{e^x}{1+e^x}\, dx$

**21.** $\displaystyle\int \frac{x}{(x+1)^3}\, dx$

**22.** $\displaystyle\int \frac{x^2}{(x+1)^3}\, dx$

**23.** $\displaystyle\int \frac{x}{(3x-1)^2}\, dx$

**24.** $\displaystyle\int \frac{5x}{(x-4)^3}\, dx$

**25.** $\displaystyle\int x(1-x)^4\, dx$

**26.** $\displaystyle\int x^2(1-x)^3\, dx$

**27.** $\displaystyle\int \frac{x-1}{x^2-2x}\, dx$

**28.** $\displaystyle\int \frac{x}{(x^2-1)^2}\, dx$

**29.** $\displaystyle\int x\sqrt{x-3}\, dx$

**30.** $\displaystyle\int x\sqrt{2x+1}\, dx$

**31.** $\displaystyle\int x^2\sqrt{1-x}\, dx$

**32.** $\displaystyle\int x^3\sqrt{x+2}\, dx$

**33.** $\displaystyle\int \frac{x^2-1}{\sqrt{2x-1}}\, dx$

**34.** $\displaystyle\int \frac{2x-1}{\sqrt{x+3}}\, dx$

**35.** $\displaystyle\int \frac{1}{1+\sqrt{t}}\, dt$

**36.** $\displaystyle\int \frac{1-\sqrt{x}}{1+\sqrt{x}}\, dx$

**37.** $\displaystyle\int \frac{2\sqrt{t}+1}{t}\, dt$

**38.** $\displaystyle\int t\sqrt[3]{t+1}\, dt$

**39.** $\displaystyle\int \frac{1}{6x+\sqrt{2x}}\, dx$

**40.** $\displaystyle\int (x-1)\sqrt[3]{2-x}\, dx$

In Exercises 41–52, evaluate the definite integral.

**41.** $\displaystyle\int_0^4 \sqrt{2x+1}\, dx$

**42.** $\displaystyle\int_2^4 \frac{1-3x^2}{2(x-x^3)}\, dx$

**43.** $\displaystyle\int_0^1 3xe^{x^2}\,dx$

**44.** $\displaystyle\int_1^{e^2} \frac{3(\ln x)^2}{x}\,dx$

**45.** $\displaystyle\int_0^5 \frac{x}{(x+5)^2}\,dx$

**46.** $\displaystyle\int_0^1 x(x+5)^4\,dx$

**47.** $\displaystyle\int_0^{0.5} x(1-x)^3\,dx$

**48.** $\displaystyle\int_0^{0.5} x^2(1-x)^3\,dx$

**49.** $\displaystyle\int_3^7 x\sqrt{x-3}\,dx$

**50.** $\displaystyle\int_0^4 \frac{x}{\sqrt{2x+1}}\,dx$

**51.** $\displaystyle\int_0^7 x\sqrt[3]{x+1}\,dx$

**52.** $\displaystyle\int_1^2 (x-1)\sqrt{2-x}\,dx$

In Exercises 53–56, find the area of the region bounded by the graphs of the given equations.

**53.** $y = x\sqrt{x+1},\ y = 0$

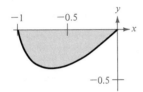

**54.** $y = x\sqrt[3]{1-x},\ y = 0$

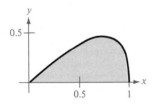

**55.** $y^2 = x^2(1-x^2)$

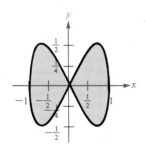

(Hint: Find the area of the region bounded by the graphs of $y = x\sqrt{1-x^2}$ and $y = 0$. Then multiply by 4.)

**56.** $y = \dfrac{1}{1+\sqrt{x}},\ y = 0,\ x = 0,\ x = 4$

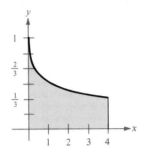

In Exercises 57 and 58, find the volume of the solid generated by revolving the region bounded by the graphs of the given equations about the $x$-axis.

**57.** $y = x\sqrt{1-x^2}$

**58.** $y = \sqrt{x}(1-x)^2,\ y = 0$

In Exercises 59 and 60, find the average amount by which the function $f(x)$ exceeds the function $g(x)$ on the given interval.

**59.** $f(x) = \dfrac{1}{x+1},\ g(x) = \dfrac{x}{(x+1)^2},\ [0, 1]$

**60.** $f(x) = x\sqrt{4x+1},\ g(x) = 2\sqrt{x^3},\ [0, 2]$

**61.** In the Introductory Example of this section, the drug-retention pattern was given by

$$y = x^2(1-x)^{1/2}$$

Find the portion of the population retaining between 0% and 50% of the drug after 24 hours.

**62.** Verify the result in the Introductory Example of this section. That is, show that 53.0% of the population retain between 60% and 90% of the drug after 24 hours by showing that

$$\frac{\displaystyle\int_{0.6}^{0.9} x^2(1-x)^{1/2}\,dx}{\displaystyle\int_0^1 x^2(1-x)^{1/2}\,dx} = 0.530$$

**63.** The probability of recall in a certain experiment is found to be

$$P(a \le x \le b) = \int_a^b \frac{15}{4}x\sqrt{1-x}\,dx$$

where $x$ represents the percentage of recall. (See Figure 6.5.)

(a) What is the probability that a randomly chosen individual will recall between 50% and 75% of the material?

(b) What is the median percentage recall? That is, for what value of $b$ is it true that the probability from 0 to $b$ is 0.5?

**64.** The probability of finding between $a$ and $b$ percent of iron in ore samples taken from a certain region is given by

$$P(a \le x \le b) = \int_a^b \frac{1155}{32} x^3 (1 - x)^{3/2} \, dx$$

(See Figure 6.6.) Find the probability that a sample will contain between

(a) 0% and 25%      (b) 50% and 100%

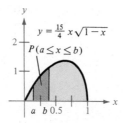

FIGURE 6.5

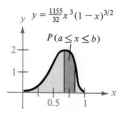

FIGURE 6.6

**65.** A company sells a seasonal product that has a daily revenue approximated by

$$R = 0.06t^2(365 - t)^{1/2} + 1250, \quad 0 \le t \le 365$$

Find the average daily revenue over the period of one year.

# Integration by Parts

An important economic consideration to livestock and poultry growers is the relationship between the total input (feed, utilities, etc.) and the total output (meat). Measured in terms of dollars, the ratio of the total input to the total output is called the *rate of return*. For instance, if the ratio is 2, then every dollar of cost will bring in two dollars of revenue. This ratio is important to livestock and poultry growers because young animals tend to put on more weight relative to food intake than older animals do. Thus, money spent on feeding young animals brings a higher rate of return than does money spent on feeding older animals.

To illustrate this, suppose that a certain breed of fryer gains weight according to the model

$$W = 0.1 + 10.85 \ln (1 + t), \quad 0 \leq t \leq 1$$

where $W$ is the fryer's weight in pounds and $t$ is the time in years, with $t = 0$ corresponding to the time the chick is hatched. Furthermore, suppose that each fryer's daily food intake is proportional to its present weight. Then the total cost of feed necessary to raise a bird to age $t$ is pro-

portional to the area under the weight curve between 0 and $t$, as shown in Figure 6.7. Now let us compare the return ratios for selling a fryer after six months and after one year. For a six-month-old fryer the return ratio is

$$\frac{\text{Revenue}}{\text{Cost}} = \frac{k_2[10.85 \ln (1 + 0.5) + 0.1]}{k_1 \displaystyle\int_0^{0.5} [10.85 \ln (1 + t) + 0.1] \, dt}$$

$$\approx \frac{4.499k_2}{1.224k_1} \approx 3.676\frac{k_2}{k_1}$$

where $k_2$ is the selling price per pound and $k_1$ is determined from the cost of feed. In this section, you will see that the logarithmic integral representing the cost can be solved using an integration technique called **integration by parts.** For the one-year-old fryer the return ratio turns out to be

$$\frac{\text{Revenue}}{\text{Cost}} \approx \frac{7.621k_2}{4.291k_1} \approx 1.776\frac{k_2}{k_1}$$

Thus, even though the older bird can be sold for more, it takes more than three times the feed to raise it, and the return ratio is less than half the return ratio for the younger bird. (If we include the cost of new chicks and daily overhead, the difference will be smaller but will still favor the younger bird.)

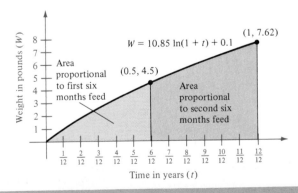

FIGURE 6.7

## Integration by Parts

In Section 6.1, we discussed the technique of integration by substitution. In this section, we introduce a second integration method known as **integration by parts.** This method of integration applies to a wide variety of functions and is particularly useful for integrands involving a product of algebraic and exponential (or logarithmic) functions. For instance, integration by parts works well with products like

$$x \ln x \quad \text{and} \quad x^2 e^x$$

Integration by parts is based on the formula for the derivative of a product:

$$\frac{d}{dx}[uv] = u\frac{dv}{dx} + v\frac{du}{dx} = uv' + vu'$$

where both $u$ and $v$ are differentiable functions of $x$. Integrating both sides of this equation with respect to $x$, we have

$$uv = \int uv'\, dx + \int vu'\, dx = \int u\, dv + \int v\, du$$

Rewriting this equation, we get the following formula for integration by parts. You should memorize this basic integration formula.

| Integration by parts | $$\int u\, dv = uv - \int v\, du$$ |
|---|---|

Note that the integration by parts formula expresses the original integral in terms of another integral. Depending on the choices for $u$ and $dv$, it may be easier to evaluate the second integral than the original one. Since the choices for $u$ and $dv$ are critical in the integration by parts process, we provide the following general guidelines.

| Guidelines for using integration by parts | Try one of the following:<br><br>1. Let $dv$ be the most complicated portion of the integrand that fits a basic integration formula. Then $u$ will be the remaining factor(s) of the integrand.<br>2. Let $u$ be that portion of the integrand whose derivative is a simpler function than $u$ itself. Then $dv$ will be the remaining factor(s) of the integrand. |
|---|---|

**Remark:** These are only suggested guidelines, and they should not be followed blindly. It is usually best to consider the guidelines in the stated order, giving greater consideration to the first one.

**EXAMPLE 1**

**Integration by Parts**

Evaluate the integral

$$\int xe^x \, dx$$

**SOLUTION**

To apply integration by parts, we write the integral in the form $\int u \, dv$. That is, we break $xe^x \, dx$ into two parts, one part representing $u$, the other $dv$. There are several ways to do this.

$$\underbrace{\int (x)(e^x \, dx)}_{u \quad\quad dv} \quad \underbrace{\int (e^x)(x \, dx)}_{u \quad\quad dv} \quad \underbrace{\int (1)(xe^x \, dx)}_{u \quad\quad dv} \quad \underbrace{\int (xe^x)(dx)}_{u \quad\quad dv}$$

Following our guidelines, we choose the first option since $e^x$ is the most complicated portion of the integrand that fits a basic integration formula. Thus, we have

$$dv = e^x \, dx \quad\Longrightarrow\quad v = \int dv = \int e^x \, dx = e^x$$

$$u = x \quad\quad\Longrightarrow\quad du = dx$$

Now, by the integration by parts formula, we have

$$\int u \, dv \;=\; uv \;-\; \int v \, du$$

$$\int xe^x \, dx = xe^x - \int e^x \, dx$$

Finally, we integrate to obtain

$$\int xe^x \, dx = xe^x - e^x + C$$

**Remark:** Note in Example 1 that we do not need to include a constant of integration when solving $v = \int e^x \, dx = e^x + C_1$. To demonstrate this, we replace $v$ by $v + C_1$ in the general formula to obtain

$$\int u \, dv = u(v + C_1) - \int (v + C_1) \, du$$

$$= uv + C_1u - \int C_1 \, du - \int v \, du$$

$$= uv + C_1u - C_1u - \int v \, du$$

$$= uv - \int v \, du$$

Thus, we may drop the first constant of integration when integrating by parts.

**EXAMPLE 2**

**Integration by Parts**

Evaluate

$$\int x^2 \ln x \, dx$$

**SOLUTION**

In this case $x^2$ is more easily integrated than $\ln x$. Furthermore, the derivative of $\ln x$ is simpler than $\ln x$. Therefore, we let $dv = x^2 \, dx$:

$$dv = x^2 \, dx \implies v = \int x^2 \, dx = \frac{x^3}{3}$$

$$u = \ln x \implies du = \frac{1}{x} \, dx$$

Therefore, we have

$$\int x^2 \ln x \, dx = \frac{x^3}{3} \ln x - \int \left(\frac{x^3}{3}\right)\left(\frac{1}{x}\right) dx$$

$$= \frac{x^3}{3} \ln x - \frac{1}{3} \int x^2 \, dx$$

$$= \frac{x^3}{3} \ln x - \frac{x^3}{9} + C$$

It may happen that a particular integral requires repeated application of integration by parts. This is demonstrated in the next example.

**EXAMPLE 3**

**Repeated Application of Integration by Parts**

Evaluate

$$\int x^2 e^x \, dx$$

**SOLUTION**

We may consider $x^2$ and $e^x$ to be equally easy to integrate. However, although the derivative of $x^2$ becomes simpler, the derivative of $e^x$ does not. Therefore, we let $u = x^2$ and write

$$dv = e^x \, dx \implies v = \int e^x \, dx = e^x$$

$$u = x^2 \implies du = 2x \, dx$$

and it follows that

$$\int x^2 e^x \, dx = x^2 e^x - \int 2x e^x \, dx$$

Now we apply integration by parts to the new integral. With the same considerations in mind, we have

$$dv = e^x \, dx \quad \Rightarrow \quad v = \int e^x \, dx = e^x$$

$$u = 2x \quad \Rightarrow \quad du = 2 \, dx$$

and it follows that

$$\int x^2 e^x \, dx = x^2 e^x - \int 2x e^x \, dx$$

$$= x^2 e^x - 2x e^x + \int 2 e^x \, dx$$

$$= x^2 e^x - 2x e^x + 2 e^x + C$$

$$= e^x (x^2 - 2x + 2) + C$$

When making repeated applications of integration by parts, we need to be careful not to interchange the substitutions in successive applications. For instance, in Example 3 our first substitutions were $dv = e^x \, dx$ and $u = x^2$. If in the second application we had switched our substitutions to

$$dv = 2x \, dx \quad \Rightarrow \quad v = \int 2x \, dx = x^2$$

$$u = e^x \quad \Rightarrow \quad du = e^x \, dx$$

we would have obtained

$$\int x^2 e^x \, dx = x^2 e^x - \int 2x e^x \, dx$$

$$= x^2 e^x - x^2 e^x + \int x^2 e^x \, dx$$

$$= \int x^2 e^x \, dx$$

which tells us nothing. By switching substitutions we undid the previous integration and returned to the *original* integral.

**EXAMPLE 4**

**Integration by Parts Applied to a Quotient**

Evaluate

$$\int \frac{x e^x}{(x + 1)^2} \, dx$$

**SOLUTION**   We let

$$dv = \frac{1}{(x + 1)^2} \, dx$$

since this choice is easily integrated using the Power Rule. Thus, we have

$$dv = \frac{1}{(x + 1)^2} \, dx \quad \Longrightarrow \quad v = \int \frac{1}{(x + 1)^2} \, dx = -\frac{1}{x + 1}$$

$$u = xe^x \quad \Longrightarrow \quad du = (xe^x + e^x) \, dx = e^x(x + 1) \, dx$$

Therefore,

$$\int \frac{xe^x}{(x + 1)^2} \, dx = xe^x\left(\frac{-1}{x + 1}\right) - \int (x + 1)e^x\left(\frac{-1}{x + 1}\right) dx$$

$$= -\frac{xe^x}{x + 1} + \int e^x \, dx$$

$$= -\frac{xe^x}{x + 1} + e^x + C$$

$$= \frac{e^x}{x + 1} + C$$

One unusual application of integration by parts involves integrands consisting of a single factor such as $\int \ln x \, dx$. In such cases, we let $dv = dx$, as illustrated in the next example.

**EXAMPLE 5**   **Integration by Parts Applied to a Single Factor**
Evaluate

$$\int_1^e \ln x \, dx$$

**SOLUTION**   At first it may appear that integration by parts does not apply. However, if we let $dv = dx$, we have

$$dv = dx \quad \Longrightarrow \quad v = \int dx = x$$

$$u = \ln x \quad \Longrightarrow \quad du = \frac{1}{x} \, dx$$

Therefore, we have

$$\int \ln x \, dx = x \ln x - \int \frac{1}{x}(x) \, dx = x \ln x - \int 1 \, dx = x \ln x - x + C$$

Now, using this antiderivative to find the definite integral, we have

$$\int_1^e \ln x \, dx = \left[ x \ln x - x \right]_1^e = [e \ln e - e] - [1 \ln 1 - 1] = (e - e) - (0 - 1) = 1$$

(See Figure 6.8.)

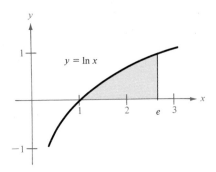

FIGURE 6.8

Before starting the exercises in this section remember that merely knowing *how* to use the various integrating techniques is not enough. You also need to know *when* to use them. Integration is first and foremost a problem of recognition—recognizing which formula or technique to apply to obtain an antiderivative. Frequently the slightest alteration of an integrand will necessitate the use of a different integration technique. For example, consider the integrals

$$\int x \ln x \, dx, \qquad \int \frac{\ln x}{x} \, dx, \qquad \int \frac{dx}{x \ln x}$$

whose antiderivatives are obtained in the manner shown in Table 6.1.

TABLE 6.1

| Integral | Technique | Antiderivative |
|----------|-----------|----------------|
| $\int x \ln x \, dx$ | Integration by parts | $\dfrac{x^2}{2} \ln x - \dfrac{x^2}{4} + C$ |
| $\int \dfrac{\ln x}{x} \, dx$ | Power Rule: $\int u^n \dfrac{du}{dx} \, dx$ | $\dfrac{(\ln x)^2}{2} + C$ |
| $\int \dfrac{1}{x \ln x} \, dx$ | Log Rule: $\int \dfrac{1}{u} \dfrac{du}{dx} \, dx$ | $\ln |\ln x| + C$ |

## An Application

In some types of business applications, the future income for a business is known. For instance, the annual income might be given by $c(t)$. During a period of $t_1$ years, the total income would then be given by

$$\text{Income during } t_1 \text{ years} = \int_0^{t_1} c(t)\, dt$$

To put a present value on such an income, it is useful to find the amount of money that would have to be deposited in an account (paying an annual interest rate of $r$ compounded continuously) so that the total amount obtained from the account would equal the income up to time $t_1$. The formula for this amount is given by

$$\text{Present value} = \int_0^{t_1} c(t)e^{-rt}\, dt$$

and we call this amount the **present value** of the income over $t_1$ years. The solution of this integral often involves integration by parts, as demonstrated in the next example.

---

**EXAMPLE 6**

### Finding the Value of Income

A small company expects its annual income during the next 5 years to be given by the model

$$c(t) = 100{,}000t, \quad 0 \le t \le 5$$

What is the present value of this income over the 5-year period? (Assume an annual interest rate of 10%.)

**SOLUTION**

The present value of the income is given by the definite integral

$$\text{Present value} = \int_0^5 100{,}000te^{-0.1t}\, dt = 100{,}000 \int_0^5 te^{-0.1t}\, dt$$

Using integration by parts, we let $dv = e^{-0.1t}\, dt$ and obtain

$$dv = e^{-0.1t}\, dt \quad \Longrightarrow \quad v = \int e^{-0.1t}\, dt = -10e^{-0.1t}$$

$$u = t \quad \Longrightarrow \quad du = dt$$

---

which implies that

$$\int te^{-0.1t}\,dt = -10te^{-0.1t} + 10\int e^{-0.1t}\,dt$$

$$= -10e^{-0.1t}(t + 10)$$

Therefore, the present value is given by

$$\text{Present value} = 100{,}000\int_0^5 te^{-0.1t}\,dt$$

$$= 100{,}000\left[-10e^{-0.1t}(t + 10)\right]_0^5$$

$$= 1{,}000{,}000[-15e^{-0.5} + 10]$$

$$\approx \$902{,}040$$

In Example 6, note that the actual income during the 5-year period is given by

$$\text{Actual income} = \int_0^5 100{,}000t\,dt$$

$$= 100{,}000\left[\frac{t^2}{2}\right]_0^5$$

$$= \$1{,}250{,}000$$

However, since most of this income is not available during the early years, the present value of the income is less than the actual (or future) value.

## SECTION EXERCISES 6.2

In Exercises 1–24, evaluate the given integral. (Note: Integration by parts is not required for all the integrals.)

**1.** $\displaystyle\int e^{2x}\,dx$

**2.** $\displaystyle\int e^{-2x}\,dx$

**3.** $\displaystyle\int xe^{2x}\,dx$

**4.** $\displaystyle\int xe^{-2x}\,dx$

**5.** $\displaystyle\int xe^{x^2}\,dx$

**6.** $\displaystyle\int x^2e^{x^3}\,dx$

**7.** $\displaystyle\int x^2e^{-2x}\,dx$

**8.** $\displaystyle\int \frac{x}{e^x}\,dx$

**9.** $\displaystyle\int x^3e^x\,dx$

**10.** $\displaystyle\int \frac{e^{1/t}}{t^2}\,dt$

**11.** $\displaystyle\int x^3\ln x\,dx$

**12.** $\displaystyle\int x^2\ln x\,dx$

**13.** $\displaystyle\int t\ln(t + 1)\,dt$

**14.** $\displaystyle\int \frac{1}{x(\ln x)^3}\,dx$

**15.** $\displaystyle\int (\ln x)^2\,dx$

**16.** $\displaystyle\int \ln 3x\,dx$

**17.** $\displaystyle\int \frac{(\ln x)^2}{x}\,dx$

**18.** $\displaystyle\int \frac{\ln x}{x^2}\,dx$

**19.** $\displaystyle\int x\sqrt{x - 1}\,dx$

**20.** $\displaystyle\int x^2\sqrt{x - 1}\,dx$

**21.** $\displaystyle\int (x^2 - 1)e^x\,dx$

**22.** $\displaystyle\int \frac{x}{\sqrt{2 + 3x}}\,dx$

**23.** $\displaystyle\int \frac{xe^{2x}}{(2x + 1)^2}\,dx$

**24.** $\displaystyle\int \frac{x^3e^{x^2}}{(x^2 + 1)^2}\,dx$

In Exercises 25–28, evaluate the definite integral.

**25.** $\displaystyle\int_0^1 x^2e^x\,dx$

**26.** $\displaystyle\int_0^2 \frac{x^2}{e^x}\,dx$

**27.** $\displaystyle\int_1^e x^4\ln x\,dx$

**28.** $\displaystyle\int_0^1 \ln(1 + 2x)\,dx$

**29.** Integrate $\displaystyle\int 2x\sqrt{2x-3}\ dx$

   (a) by parts, letting $dv = \sqrt{2x-3}\ dx$

   (b) by substitution, letting $u = \sqrt{2x-3}$

**30.** Integrate $\displaystyle\int x\sqrt{4+x}\ dx$

   (a) by parts, letting $dv = \sqrt{4+x}\ dx$

   (b) by substitution, letting $u = \sqrt{4+x}$

**31.** Integrate $\displaystyle\int \frac{x}{\sqrt{4+5x}}\ dx$

   (a) by parts, letting $dv = \dfrac{1}{\sqrt{4+5x}}\ dx$

   (b) by substitution, letting $u = \sqrt{4+5x}$

**32.** Integrate $\displaystyle\int x\sqrt{4-x}\ dx$

   (a) by parts, letting $dv = \sqrt{4-x}$

   (b) by substitution, letting $u = \sqrt{4-x}$

In Exercises 33 and 34, use integration by parts to verify the given formula.

**33.** $\displaystyle\int x^n \ln x\ dx = \frac{x^{n+1}}{(n+1)^2}[-1 + (n+1)\ln x] + C$,

   $n \neq -1$

**34.** $\displaystyle\int x^n e^{ax}\ dx = \frac{x^n e^{ax}}{a} - \frac{n}{a}\int x^{n-1}e^{ax}\ dx$

In Exercises 35–38, use the results of Exercises 33 and 34 to evaluate the given integral.

**35.** $\displaystyle\int x^2 e^{5x}\ dx$

**36.** $\displaystyle\int xe^{-3x}\ dx$

**37.** $\displaystyle\int x^5 \ln x\ dx$

**38.** $\displaystyle\int x^3 \ln x\ dx$

In Exercises 39 and 40, find the area of the region bounded by the graphs of the given equations.

**39.** $y = xe^{-x}$, $y = 0$, $x = 4$

**40.** $y = \dfrac{1}{9}xe^{-x/3}$, $y = 0$, $x = 0$, $x = 3$

**41.** Given the region bounded by the graphs of $y = \ln x$, $y = 0$, and $x = e$, find

   (a) the area of the region

   (b) the volume of the solid generated by revolving the region about the $x$-axis

**42.** Find the volume of the solid generated by revolving the region bounded by the graphs of $y = xe^x$, $y = 0$, $x = 0$, and $x = 1$ about the $x$-axis.

**43.** A model for the ability $M$ of a child to memorize, measured on a scale from 0 to 10, is given by

$$M = 1 + 1.6t\ln t, \quad 0 < t \le 4$$

where $t$ is the child's age in years. Find the average value of this function

   (a) between the child's first and second birthdays

   (b) between the child's third and fourth birthdays

**44.** A company sells a seasonal product for which the model for the daily revenue is

$$R = 410.5t^2 e^{-t/30} + 25{,}000, \quad 0 \le t \le 365$$

where $t$ is the time in days.

   (a) Find the average daily receipts during the first quarter, $0 \le t \le 91$.

   (b) Find the average daily receipts during the fourth quarter, $274 \le t \le 365$.

In Exercises 45–50, find the present value of the income given by $c(t)$, measured in dollars, over $t_1$ years at the given annual interest rate $r$.

**45.** $c(t) = 5000$, $r = 11\%$, $t_1 = 4$ years

**46.** $c(t) = 450$, $r = 8\%$, $t_1 = 10$ years

**47.** $c(t) = 100{,}000 + 4000t$, $r = 10\%$, $t_1 = 10$ years

**48.** $c(t) = 30{,}000 + 500t$, $r = 9\%$, $t_1 = 6$ years

**49.** $c(t) = 1000 + 50e^{t/2}$, $r = 6\%$, $t_1 = 4$ years

**50.** $c(t) = 5000 + 25te^{t/10}$, $r = 10\%$, $t_1 = 10$ years

# Partial Fractions

In Section 5.5, the exponential growth function was derived under the assumption that the rate of growth was proportional to the existing quantity. In many situations, there exists some upper limit $L$ past which growth cannot occur. For instance, population growth is limited by the food supply or space, and growth in sales of some new product will stop when the market becomes saturated. Under such conditions, we assume that the rate of growth is proportional not only to the existing amount $y$ but *also* to the difference between the existing amount and the limit $L$. That is,

$$\frac{dy}{dt} = ky(L - y)$$

We can express this relationship in integral form as follows:

$$\int \frac{1}{y(L - y)}\, dy = \int k\, dt$$

To solve the integral on the left, we can use an integration technique called the method of **partial fractions.** The solution turns out to be the *logistics growth function* discussed in Sections 5.1 and 5.2.

For example, if a bacterial culture weighs 1 gram when $t = 0$ and has an upper limit of $L = 10$ grams, then its growth function is of the form

$$y = \frac{10}{1 + 9e^{-kt}}$$

where $y$ is the weight (in grams) and $t$ is the time (in hours). Furthermore, if we know the weight at some other time, then we can solve for $k$. For instance, if $y = 2$ when $t = 1$, we can determine that $k = \ln \frac{9}{4} \approx 0.8109$, as shown in Figure 6.9. On the other hand, if $y = 2$ when $t = 2$, we can determine that $k = \ln \frac{3}{2} \approx 0.4055$, as shown in Figure 6.10.

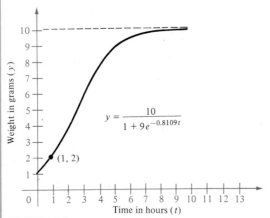

FIGURE 6.9

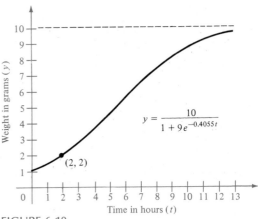

FIGURE 6.10

## SECTION TOPIC

### ■ Partial Fractions

In Sections 6.1 and 6.2, we introduced two techniques that allow us to apply our fundamental integration rules to a wide variety of functions. In this section, we discuss yet a *third* technique that permits us to rewrite rational functions in a form to which we can apply our fundamental integration rules. We call this procedure the method of **partial fractions;** it involves the decomposition of a rational function into the sum of two or more simpler rational functions. The need for this procedure can be seen in the discussion of the following problem.

How can we evaluate the following integral?

$$\int \frac{x + 7}{x^2 - x - 6} \, dx$$

Suppose we knew that

$$\frac{x + 7}{x^2 - x - 6} = \frac{2}{x - 3} - \frac{1}{x + 2}$$

Then we could write

$$\int \frac{x + 7}{x^2 - x - 6} \, dx = \int \left( \frac{2}{x - 3} - \frac{1}{x + 2} \right) dx$$

$$= 2 \int \frac{dx}{x - 3} - \int \frac{dx}{x + 2}$$

$$= 2 \ln |x - 3| - \ln |x + 2| + C$$

The use of this method depends on our ability to factor the denominator, $x^2 - x - 6$, and to find the *partial fractions*

$$\frac{2}{x - 3}$$

and

$$-\frac{1}{x + 2}$$

Recall that one of the basic concerns of algebra is finding the factors of a polynomial. For instance, the polynomial $x^3 + x^2 - x - 1$ can be written as

$$x^3 + x^2 - x - 1 = (x - 1)(x + 1)^2$$

where $(x - 1)$ is a linear factor and $(x + 1)^2$ is a *repeated* linear factor. We can use this factorization to find the partial fraction decomposition of any rational function having $x^3 + x^2 - x - 1$ as its denominator. Specifically, if $N(x)$ is a polynomial of degree less than three, then the partial fraction decomposition of

$$\frac{N(x)}{x^3 + x^2 - x - 1}$$

has the form

$$\frac{N(x)}{x^3 + x^2 - x - 1} = \frac{N(x)}{(x-1)(x+1)^2}$$

$$= \frac{A}{x-1} + \frac{B}{x+1} + \frac{C}{(x+1)^2}$$

where $A$, $B$, and $C$ are constants. Note that the repeated linear factor $(x+1)^2$ results in *two* fractions: one for $(x+1)$ and one for $(x+1)^2$. If $(x+1)^3$ were a factor, then we would use three fractions: one for $(x+1)$, one for $(x+1)^2$, and one for $(x+1)^3$. In general, the number of fractions resulting from a repeated linear factor is equal to the number of times the factor is repeated.

An algebraic technique for determining the value of the constants in the numerators is demonstrated in the examples that follow.

**EXAMPLE 1**

**Distinct Linear Factors**

Write the partial fraction decomposition for the rational function

$$\frac{x+7}{x^2 - x - 6}$$

**SOLUTION**

Since $x^2 - x - 6 = (x-3)(x+2)$, we include one partial fraction for each factor and write

$$\frac{x+7}{x^2 - x - 6} = \frac{A}{x-3} + \frac{B}{x+2}$$

Multiplying both sides of this equation by the lowest common denominator, $(x-3)(x+2)$, leads to the **basic equation**

$$x + 7 = A(x+2) + B(x-3) \qquad \text{\small Basic equation}$$

Since this equation is to be true for all $x$, we can substitute *convenient* values for $x$ to obtain equations in $A$ and $B$, which we then solve. Values of $x$ that are especially convenient are ones that make the factors of the lowest common denominator zero. For instance, if $x = -2$, then

$$-2 + 7 = A(0) + B(-5)$$

$$5 = -5B$$

$$-1 = B$$

If $x = 3$, then

$$3 + 7 = A(5) + B(0)$$

$$10 = 5A$$

$$2 = A$$

The decomposition is, therefore,

$$\frac{x + 7}{x^2 - x - 6} = \frac{2}{x - 3} - \frac{1}{x + 2}$$

as indicated at the beginning of this section.

**Remark:** Be sure you see that the substitutions for $x$ in Example 1 were chosen for their convenience in determining values for $A$ and $B$. We chose $x = -2$ to eliminate the term $A(x + 2)$, and $x = 3$ was chosen to eliminate the term $B(x - 3)$.

In Example 1, each of the factors of the denominator occurs only once. In the next example, one of the factors is repeated.

**EXAMPLE 2**

**Repeated Linear Factors**

Evaluate the integral

$$\int \frac{5x^2 + 20x + 6}{x^3 + 2x^2 + x}\, dx$$

**SOLUTION**

Since

$$x^3 + 2x^2 + x = x(x^2 + 2x + 1)$$
$$= x(x + 1)^2$$

we include one fraction for each power of $x$ and $(x + 1)$, and we write

$$\frac{5x^2 + 20x + 6}{x(x + 1)^2} = \frac{A}{x} + \frac{B}{x + 1} + \frac{C}{(x + 1)^2}$$

Multiplying by the lowest common denominator, $x(x + 1)^2$, leads to the basic equation

$$5x^2 + 20x + 6 = A(x + 1)^2 + Bx(x + 1) + Cx \qquad \text{Basic equation}$$

Substituting $x = -1$ eliminates the $A$ and $B$ terms and yields

$$5 - 20 + 6 = 0 + 0 - C$$
$$C = 9$$

If $x = 0$, then

$$6 = A(1) + 0 + 0$$
$$6 = A$$

At this point we have exhausted the most convenient choices for $x$ and have yet to find the value of $B$. Under such circumstances, we use any other value for $x$ along with the calculated values of $A$ and $C$. [Note that it is necessary to make as many substitutions for $x$ as there are unknowns $(A, B, C, \ldots)$ to be determined.] Thus, for $x = 1$, $A = 6$, and $C = 9$, we have

$$5 + 20 + 6 = A(4) + B(2) + C$$

$$31 = 6(4) + 2B + 9$$

$$-2 = 2B$$

$$-1 = B$$

Therefore,

$$\frac{5x^2 + 20x + 6}{x(x + 1)^2} = \frac{6}{x} - \frac{1}{x + 1} + \frac{9}{(x + 1)^2}$$

and it follows that

$$\int \frac{5x^2 + 20x + 6}{x^3 + 2x^2 + x}\,dx = \int \frac{6}{x}\,dx - \int \frac{dx}{x + 1} + \int 9(x + 1)^{-2}\,dx$$

$$= 6 \ln |x| - \ln |x + 1| + 9\frac{(x + 1)^{-1}}{-1} + C$$

$$= \ln \left| \frac{x^6}{x + 1} \right| - \frac{9}{x + 1} + C$$

We can only apply the partial fraction decomposition outlined in Examples 1 and 2 to a *proper* rational function—that is, to a rational function whose numerator is of lower degree than its denominator. If the numerator is of equal or greater degree, we must divide first. This is demonstrated in the next example.

**EXAMPLE 3**

**Dividing to Obtain a Proper Rational Function**

Evaluate the integral

$$\int \frac{x^5 + x - 1}{x^4 - x^3}\,dx$$

**SOLUTION**

Since the numerator is fifth degree and the denominator is only fourth degree, we divide to obtain

$$\int \frac{x^5 + x - 1}{x^4 - x^3}\,dx = \int \left( x + 1 + \frac{x^3 + x - 1}{x^4 - x^3} \right)\,dx$$

Now, applying partial fractions, we have

$$\frac{x^3 + x - 1}{x^3(x - 1)} = \frac{A}{x} + \frac{B}{x^2} + \frac{C}{x^3} + \frac{D}{x - 1}$$

Using techniques similar to those in the first three examples, we find that

$$A = 0, \quad B = 0, \quad C = 1, \quad D = 1$$

Thus, we have

$$\int \frac{x^5 + x - 1}{x^4 - x^3} \, dx = \int \left( x + 1 + \frac{1}{x^3} + \frac{1}{x - 1} \right) dx$$

$$= \frac{x^2}{2} + x - \frac{1}{2x^2} + \ln |x - 1| + C$$

In integration problems, it often happens that we need to use more than one integration technique to find the solution. For instance, in the next example, we begin the problem by using a $u$-substitution. We then use the method of partial fractions.

**EXAMPLE 4**

**A Substitution Leading to Partial Fractions**

Evaluate the integral

$$\int \frac{1}{x\sqrt{x + 1}} \, dx$$

**SOLUTION**

Substituting $u = \sqrt{x + 1}$, we have

$$x = u^2 - 1 \qquad \text{and} \qquad dx = 2u \, du$$

Thus,

$$\int \frac{1}{x\sqrt{x + 1}} \, dx = \int \frac{1}{(u^2 - 1)u} 2u \, du$$

$$= \int \frac{2}{u^2 - 1} \, du$$

Now, since $u^2 - 1 = (u + 1)(u - 1)$, we can apply partial fractions to obtain

$$\frac{2}{u^2 - 1} = \frac{A}{u + 1} + \frac{B}{u - 1}$$

which leads to the basic equation

$$2 = A(u - 1) + B(u + 1)$$

Letting $u = -1$, we have

$$2 = A(-2) + B(0)$$

$$-1 = A$$

and letting $u = 1$, we have

$$2 = A(0) + B(2)$$

$$1 = B$$

Thus,

$$\int \frac{2}{u^2 - 1} \, du = \int \left( \frac{-1}{u + 1} + \frac{1}{u - 1} \right) du$$

$$= -\ln |u + 1| + \ln |u - 1| + C$$

Finally, resubstituting $u = \sqrt{x + 1}$, we have

$$\int \frac{1}{x\sqrt{x + 1}} \, dx = -\ln |\sqrt{x + 1} + 1| + \ln |\sqrt{x + 1} - 1| + C$$

$$= \ln \left| \frac{\sqrt{x + 1} - 1}{\sqrt{x + 1} + 1} \right| + C$$

Before concluding this section, we make a few observations about the integration of rational functions. First, it is not necessary to use the partial fractions technique on all integrals of the form

$$\int \frac{N(x)}{D(x)} \, dx$$

For instance, even though

$$\int \frac{x^2 + 2x}{x^3 + 3x^2 - 4} \, dx,$$

can be evaluated by using partial fractions and writing

$$\int \frac{x^2 + 2x}{x^3 + 3x^2 - 4} \, dx = \int \frac{x^2 + 2x}{(x - 1)(x + 2)^2} \, dx$$

$$= \int \left( \frac{A}{x - 1} + \frac{B}{x + 2} + \frac{C}{(x + 2)^2} \right) dx$$

it is more easily evaluated by using the Log Rule and writing

$$\int \frac{x^2 + 2x}{x^3 + 3x^2 - 4} \, dx = \frac{1}{3} \int \frac{3x^2 + 6x}{x^3 + 3x^2 - 4} \, dx$$

$$= \frac{1}{3} \ln |x^3 + 3x^2 - 4| + C$$

Second, if the given integral is not in reduced form, reducing it may eliminate the need to use partial fractions. For instance,

$$\int \frac{x^2 - x - 2}{x^3 - 2x - 4} \, dx = \int \frac{(x + 1)(x - 2)}{(x - 2)(x^2 + 2x + 2)} \, dx$$

$$= \int \frac{x + 1}{x^2 + 2x + 2} \, dx$$

The latter integral is a logarithmic form.

Finally, the partial fractions technique can be used with some quotients involving exponential functions. For instance, the substitution $u = e^x$ allows us to write

$$\int \frac{e^x}{e^x(e^x - 1)}\, dx = \int \frac{du}{u(u - 1)}$$

$$= \int \left(\frac{A}{u} + \frac{B}{u - 1}\right) du$$

## SECTION EXERCISES 6.3

In Exercises 1–10, write the given expression as a sum of partial fractions.

**1.** $\dfrac{2(x + 20)}{x^2 - 25}$

**2.** $\dfrac{10x + 3}{x^2 + x}$

**3.** $\dfrac{8x + 3}{x^2 - 3x}$

**4.** $\dfrac{3x + 11}{x^2 - 2x - 3}$

**5.** $\dfrac{4x - 13}{x^2 - 3x - 10}$

**6.** $\dfrac{7x + 5}{6(2x^2 + 3x + 1)}$

**7.** $\dfrac{2x^2 - 2x - 3}{x^3 + x^2}$

**8.** $\dfrac{3x^2 - x + 1}{x(x + 1)^2}$

**9.** $\dfrac{x + 1}{3(x - 2)^2}$

**10.** $\dfrac{8x^2 + 15x + 9}{(x + 1)^3}$

In Exercises 11–30, find the indefinite integral.

**11.** $\displaystyle\int \frac{1}{x^2 - 1}\, dx$

**12.** $\displaystyle\int \frac{9}{x^2 - 9}\, dx$

**13.** $\displaystyle\int \frac{-2}{x^2 - 16}\, dx$

**14.** $\displaystyle\int \frac{-4}{x^2 - 4}\, dx$

**15.** $\displaystyle\int \frac{1}{x^2 + x}\, dx$

**16.** $\displaystyle\int \frac{3}{x^2 - 3x}\, dx$

**17.** $\displaystyle\int \frac{1}{2x^2 + x}\, dx$

**18.** $\displaystyle\int \frac{5}{x^2 + x - 6}\, dx$

**19.** $\displaystyle\int \frac{3}{x^2 + x - 2}\, dx$

**20.** $\displaystyle\int \frac{1}{4x^2 - 9}\, dx$

**21.** $\displaystyle\int \frac{5 - x}{2x^2 + x - 1}\, dx$

**22.** $\displaystyle\int \frac{x + 1}{x^2 + 4x + 3}\, dx$

**23.** $\displaystyle\int \frac{x^2 + 12x + 12}{x^3 - 4x}\, dx$

**24.** $\displaystyle\int \frac{3x^2 - 7x - 2}{x^3 - x}\, dx$

**25.** $\displaystyle\int \frac{x + 2}{x^2 - 4x}\, dx$

**26.** $\displaystyle\int \frac{4x^2 + 2x - 1}{x^3 + x^2}\, dx$

**27.** $\displaystyle\int \frac{2x - 3}{(x - 1)^2}\, dx$

**28.** $\displaystyle\int \frac{x^4}{(x - 1)^3}\, dx$

**29.** $\displaystyle\int \frac{4x^2 - 1}{2x(x^2 + 2x + 1)}\, dx$

**30.** $\displaystyle\int \frac{3x}{x^2 - 6x + 9}\, dx$

In Exercises 31–34, evaluate the definite integral.

**31.** $\displaystyle\int_3^4 \frac{1}{x^2 - 4}\, dx$

**32.** $\displaystyle\int_0^1 \frac{3}{2x^2 + 5x + 2}\, dx$

**33.** $\displaystyle\int_1^5 \frac{x - 1}{x^2(x + 1)}\, dx$

**34.** $\displaystyle\int_0^1 \frac{x^2 - x}{x^2 + x + 1}\, dx$

In Exercises 35–40, evaluate the indefinite integral using the indicated substitution.

| Integral | Substitution |
|---|---|
| **35.** $\displaystyle\int \frac{e^x}{(e^x - 1)(e^x + 4)}\, dx$ | $u = e^x$ |
| **36.** $\displaystyle\int \frac{e^x}{(e^{2x} - 1)(e^x + 1)}\, dx$ | $u = e^x$ |
| **37.** $\displaystyle\int \frac{1}{x\sqrt{4 + x^2}}\, dx$ | $u = \sqrt{4 + x^2}$ |
| **38.** $\displaystyle\int \frac{1}{x\sqrt{4 + x}}\, dx$ | $u = \sqrt{4 + x}$ |
| **39.** $\displaystyle\int \frac{1}{\sqrt{x}(\sqrt{x} + 1)^2}\, dx$ | $u = \sqrt{x}$ |
| **40.** $\displaystyle\int \frac{1}{x[1 - (\ln x)^2]^2}\, dx$ | $u = \ln x$ |

**41.** (a) Write the rational expression $1/(a^2 - x^2)$ as a sum of partial fractions.

(b) Use the result of part (a) to show that

$$\int \frac{1}{a^2 - x^2}\, dx = \frac{1}{2a} \ln \left|\frac{a + x}{a - x}\right| + C$$

**42.** (a) Write the rational expression $1/[x(x + a)]$ as a sum of partial fractions.

(b) Use the result of part (a) to show that

$$\int \frac{1}{x(x + a)}\, dx = \frac{1}{a} \ln \left|\frac{x}{x + a}\right| + C$$

**43.** Find the area of the region bounded by the graphs of

$$y = \frac{7}{16 - x^2} \quad \text{and} \quad y = 1$$

**44.** Find the area of the region bounded by the graphs of

$$y = \frac{-4}{(x + 2)(x - 3)} \quad \text{and} \quad y = 1$$

**45.** Referring to the Introductory Example of this section, use integration by partial fractions to show that

$$y = \frac{10}{1 + 9e^{-kt}}$$

is a solution to the equation

$$\int \frac{1}{y(10 - y)}\, dy = \int k\, dt$$

(Note: To solve for the constant of integration, use the condition that $y = 1$ when $t = 0$.)

**46.** A conservation organization releases 100 animals of an endangered species into a game preserve. The organization believes that the preserve has a carrying capacity of 1000 of these animals and that the growth of the herd will be logistic. That is, the size $y$ of the herd will follow the equation

$$\int \frac{1}{y(1000 - y)}\, dy = \int k\, dt$$

where $t$ is measured in years. Find this logistic curve. (To solve for the constant of integration $C$ and the proportionality constant $k$, assume that $y = 100$ when $t = 0$ and $y = 134$ when $t = 2$.) Sketch the graph of your solution.

**47.** A single infected individual enters a community of $n$ individuals susceptible to the disease. Let $x$ be the number of newly infected individuals after time $t$. The common *epidemic model* assumes that the disease spreads at a rate proportional to the product of the total number infected and the number of susceptible individuals not yet infected. Thus,

$$\frac{dx}{dt} = k(x + 1)(n - x)$$

and we obtain

$$\int \frac{1}{(x + 1)(n - x)}\, dx = \int k\, dt$$

Solve for $x$ as a function of $t$.

# Integration by Tables and Completing the Square

In several applications in this text, we have considered the total revenue function

$$R = xp$$

where $x$ is the number of units sold at a price of $p$ dollars per unit. Another way to look at revenue is as a function of time.

As a case in point, suppose that a company is beginning to market a new type of home care product. Marketing tests indicate that this product should eventually be able to capture 4% of the national market. If the national market is 12,000,000 units per year, this means that the company expects its annual sales to be approaching 480,000 units per year. Since the product is unknown when it is first introduced, initial sales are expected to come in at a lower rate. Suppose that the product sells for $1 per unit and that the projected revenue rate function is

$$\frac{dR}{dt} = 480,000\left(1 - \frac{1}{\sqrt{t^2 + 1}}\right)$$

where $dR/dt$ is measured in dollars per year and $t$ is the time in years. (See Figure 6.11.) What is the total revenue during the first year? During the second year? These questions can be answered by integration. For instance, the total revenue during the first year is given by

$$R = \int_0^1 480,000\left(1 - \frac{1}{\sqrt{t^2 + 1}}\right) dt$$

$$= 480,000\left[t - \ln\left|t + \sqrt{t^2 + 1}\right|\right]_0^1$$

$$\approx \$56,941$$

The antiderivative of

$$\frac{1}{\sqrt{t^2 + 1}}$$

was obtained from Formula 27 of the **integration tables** listed in this section. Table 6.2 lists the projected total revenue for the first 10 years of sales. Note that the annual sales are approaching $480,000.

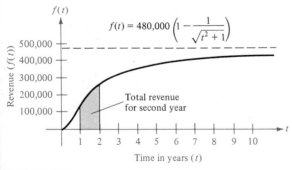

FIGURE 6.11

TABLE 6.2

| Year | Total revenue ($) |
|------|-------------------|
| 1 | 56,900 |
| 2 | 210,100 |
| 3 | 300,100 |
| 4 | 347,400 |
| 5 | 375,500 |
| 6 | 393,900 |
| 7 | 406,900 |
| 8 | 416,900 |
| 9 | 423,900 |
| 10 | 429,700 |

■ **Integration by Tables**
■ **Reduction Formulas**
■ **Completing the Square**

So far in this chapter we have discussed a number of integration techniques to use along with the fundamental integration formulas. Certainly we have not considered every possible method for finding an antiderivative, but we have considered some of the most important ones.

In this section, we expand our list of integration formulas to form a table of integrals. As we add new integration formulas to our basic list, two things occur. On the one hand, it becomes increasingly more difficult to memorize, or even become familiar with, the entire list of formulas. On the other hand, with a longer list we will likely need fewer techniques for fitting an integral to one of the formulas on the list. We call the procedure of integrating by means of a long unmemorized list of formulas **integration by tables.**

Integration by tables is not to be considered a trivial task. It requires considerable thought and insight, and it often involves a substitution procedure. Many people find a table of integrals to be a valuable supplement to the integration techniques discussed in previous sections of this chapter. We encourage you to gain competence in the use of the table of integrals as well as continuing to improve in the use of the various integrating techniques. In doing so you should find that a combination of techniques and tables is the most versatile approach to integration.

To assist you in using the table of integrals, we have grouped the formulas according to the form of the integrand as follows:

|  |  |
|---|---|
| Forms involving $u^n$ | Forms involving $a + bu$ |
| Forms involving $\sqrt{a + bu}$ | Forms involving $(u^2 - a^2)$ |
| Forms involving $\sqrt{u^2 \pm a^2}$ | Forms involving $\sqrt{a^2 - u^2}$ |
| Forms involving $e^u$ | Forms involving $\ln u$ |

**Tables of integrals**    **Forms Involving $u^n$**

1. $\displaystyle\int u^n \, du = \frac{u^{n+1}}{n+1} + C, \quad n \neq -1$

2. $\displaystyle\int \frac{1}{u} \, du = \ln |u| + C$

**Forms Involving $a + bu$**

3. $\displaystyle\int \frac{u}{a + bu} \, du = \frac{1}{b^2}[bu - a \ln |a + bu|] + C$

4. $\displaystyle\int \frac{u}{(a + bu)^2} \, du = \frac{1}{b^2}\left[\frac{a}{a + bu} + \ln |a + bu|\right] + C$

5. $\displaystyle\int \frac{u}{(a+bu)^n}\,du = \frac{1}{b^2}\left[\frac{-1}{(n-2)(a+bu)^{n-2}} + \frac{a}{(n-1)(a+bu)^{n-1}}\right] + C, \quad n \ne 1, 2$

6. $\displaystyle\int \frac{u^2}{a+bu}\,du = \frac{1}{b^3}\left[-\frac{bu}{2}(2a-bu) + a^2\ln|a+bu|\right] + C$

7. $\displaystyle\int \frac{u^2}{(a+bu)^2}\,du = \frac{1}{b^3}\left[bu - \frac{a^2}{a+bu} - 2a\ln|a+bu|\right] + C$

8. $\displaystyle\int \frac{u^2}{(a+bu)^3}\,du = \frac{1}{b^3}\left[\frac{2a}{a+bu} - \frac{a^2}{2(a+bu)^2} + \ln|a+bu|\right] + C$

9. $\displaystyle\int \frac{u^2}{(a+bu)^n}\,du = \frac{1}{b^3}\left[\frac{-1}{(n-3)(a+bu)^{n-3}} + \frac{2a}{(n-2)(a+bu)^{n-2}} - \frac{a^2}{(n-1)(a+bu)^{n-1}}\right] + C, \, n \ne 1, 2, 3$

10. $\displaystyle\int \frac{1}{u(a+bu)}\,du = \frac{1}{a}\ln\left|\frac{u}{a+bu}\right| + C$

11. $\displaystyle\int \frac{1}{u(a+bu)^2}\,du = \frac{1}{a}\left[\frac{1}{a+bu} + \frac{1}{a}\ln\left|\frac{u}{a+bu}\right|\right] + C$

12. $\displaystyle\int \frac{1}{u^2(a+bu)}\,du = -\frac{1}{a}\left[\frac{1}{u} + \frac{b}{a}\ln\left|\frac{u}{a+bu}\right|\right] + C$

13. $\displaystyle\int \frac{1}{u^2(a+bu)^2}\,du = -\frac{1}{a^2}\left[\frac{a+2bu}{u(a+bu)} + \frac{2b}{a}\ln\left|\frac{u}{a+bu}\right|\right] + C$

## Forms Involving $\sqrt{a+bu}$

14. $\displaystyle\int u^n\sqrt{a+bu}\,du = \frac{2}{b(2n+3)}\left[u^n(a+bu)^{3/2} - na\int u^{n-1}\sqrt{a+bu}\,du\right]$

15. $\displaystyle\int \frac{1}{u\sqrt{a+bu}}\,du = \frac{1}{\sqrt{a}}\ln\left|\frac{\sqrt{a+bu}-\sqrt{a}}{\sqrt{a+bu}+\sqrt{a}}\right| + C, \quad 0 < a$

16. $\displaystyle\int \frac{1}{u^n\sqrt{a+bu}}\,du = \frac{-1}{a(n-1)}\left[\frac{\sqrt{a+bu}}{u^{n-1}} + \frac{(2n-3)b}{2}\int \frac{1}{u^{n-1}\sqrt{a+bu}}\,du\right], \quad n \ne 1$

17. $\displaystyle\int \frac{\sqrt{a+bu}}{u}\,du = 2\sqrt{a+bu} + a\int \frac{1}{u\sqrt{a+bu}}\,du$

18. $\displaystyle\int \frac{\sqrt{a+bu}}{u^n}\,du = \frac{-1}{a(n-1)}\left[\frac{(a+bu)^{3/2}}{u^{n-1}} + \frac{(2n-5)b}{2}\int \frac{\sqrt{a+bu}}{u^{n-1}}\,du\right], \quad n \ne 1$

19. $\displaystyle\int \frac{u}{\sqrt{a+bu}}\,du = -\frac{2(2a-bu)}{3b^2}\sqrt{a+bu} + C$

20. $\displaystyle\int \frac{u^n}{\sqrt{a+bu}}\,du = \frac{2}{(2n+1)b}\left[u^n\sqrt{a+bu} - na\int \frac{u^{n-1}}{\sqrt{a+bu}}\,du\right]$

## Forms Involving $u^2 - a^2$, $\quad 0 < a$

21. $\displaystyle\int \frac{1}{u^2-a^2}\,du = -\int \frac{1}{a^2-u^2}\,du = \frac{1}{2a}\ln\left|\frac{u-a}{u+a}\right| + C$

22. $\displaystyle\int \frac{1}{(u^2-a^2)^n}\,du = \frac{-1}{2a^2(n-1)}\left[\frac{u}{(u^2-a^2)^{n-1}} + (2n-3)\int \frac{1}{(u^2-a^2)^{n-1}}\,du\right], \quad n \ne 1$

**Forms Involving $\sqrt{u^2 \pm a^2}$, $0 < a$**

23. $\displaystyle\int \sqrt{u^2 \pm a^2}\, du = \frac{1}{2}[u\sqrt{u^2 \pm a^2} \pm a^2 \ln | u + \sqrt{u^2 \pm a^2} |] + C$

24. $\displaystyle\int u^2\sqrt{u^2 \pm a^2}\, du = \frac{1}{8}[u(2u^2 \pm a^2)\sqrt{u^2 \pm a^2} - a^4 \ln | u + \sqrt{u^2 \pm a^2} |] + C$

25. $\displaystyle\int \frac{\sqrt{u^2 + a^2}}{u}\, du = \sqrt{u^2 + a^2} - a \ln \left| \frac{a + \sqrt{u^2 + a^2}}{u} \right| + C$

26. $\displaystyle\int \frac{\sqrt{u^2 \pm a^2}}{u^2}\, du = \frac{-\sqrt{u^2 \pm a^2}}{u} + \ln | u + \sqrt{u^2 \pm a^2} | + C$

27. $\displaystyle\int \frac{1}{\sqrt{u^2 \pm a^2}}\, du = \ln | u + \sqrt{u^2 \pm a^2} | + C$

28. $\displaystyle\int \frac{1}{u\sqrt{u^2 + a^2}}\, du = -\frac{1}{a} \ln \left| \frac{a + \sqrt{u^2 + a^2}}{u} \right| + C$

29. $\displaystyle\int \frac{u^2}{\sqrt{u^2 \pm a^2}}\, du = \frac{1}{2}[u\sqrt{u^2 \pm a^2} \mp a^2 \ln | u + \sqrt{u^2 \pm a^2} |] + C$

30. $\displaystyle\int \frac{1}{u^2\sqrt{u^2 \pm a^2}}\, du = \mp \frac{\sqrt{u^2 \pm a^2}}{a^2 u} + C$

31. $\displaystyle\int \frac{1}{(u^2 \pm a^2)^{3/2}}\, du = \frac{\pm u}{a^2\sqrt{u^2 \pm a^2}} + C$

**Forms Involving $\sqrt{a^2 - u^2}$, $0 < a$**

32. $\displaystyle\int \frac{\sqrt{a^2 - u^2}}{u}\, du = \sqrt{a^2 - u^2} - a \ln \left| \frac{a + \sqrt{a^2 - u^2}}{u} \right| + C$

33. $\displaystyle\int \frac{1}{u\sqrt{a^2 - u^2}}\, du = -\frac{1}{a} \ln \left| \frac{a + \sqrt{a^2 - u^2}}{u} \right| + C$

34. $\displaystyle\int \frac{1}{u^2\sqrt{a^2 - u^2}}\, du = \frac{-\sqrt{a^2 - u^2}}{a^2 u} + C$

35. $\displaystyle\int \frac{1}{(a^2 - u^2)^{3/2}}\, du = \frac{u}{a^2\sqrt{a^2 - u^2}} + C$

**Forms Involving $e^u$**

36. $\displaystyle\int e^u\, du = e^u + C$

37. $\displaystyle\int ue^u\, du = (u - 1)e^u + C$

38. $\displaystyle\int u^n e^u\, du = u^n e^u - n \int u^{n-1} e^u\, du$

39. $\displaystyle\int \frac{1}{1 + e^u}\, du = u - \ln (1 + e^u) + C$

40. $\displaystyle\int \frac{1}{1 + e^{nu}}\, du = u - \frac{1}{n} \ln (1 + e^{nu}) + C$

## Forms Involving ln *u*

41. $\displaystyle\int \ln u \, du = u[-1 + \ln u] + C$

42. $\displaystyle\int u \ln u \, du = \frac{u^2}{4}[-1 + 2 \ln u] + C$

43. $\displaystyle\int u^n \ln u \, du = \frac{u^{n+1}}{(n+1)^2}[-1 + (n+1) \ln u] + C, \quad n \neq -1$

44. $\displaystyle\int (\ln u)^2 \, du = u[2 - 2 \ln u + (\ln u)^2] + C$

45. $\displaystyle\int (\ln u)^n \, du = u(\ln u)^n - n \int (\ln u)^{n-1} \, du$

In the examples in this section, we demonstrate the use of this table of integrals to find antiderivatives.

**EXAMPLE 1**

**Integration by Tables**

Evaluate the integral

$$\int \frac{x}{\sqrt{x-1}} \, dx$$

**SOLUTION**

Since the expression inside the radical is linear, we consider forms involving $\sqrt{a + bu}$, as in Formula 19.

$$\int \frac{u}{\sqrt{a+bu}} \, du = -\frac{2(2a - bu)}{3b^2}\sqrt{a+bu} + C$$

We let $a = -1$, $b = 1$, and $u = x$. Then $du = dx$, and we write

$$\int \frac{x}{\sqrt{x-1}} \, dx = -\frac{2(-2-x)}{3}\sqrt{x-1} + C$$

$$= \frac{2}{3}(2+x)\sqrt{x-1} + C$$

**EXAMPLE 2**

**Integration by Tables**

Evaluate the integral

$$\int x\sqrt{x^4 - 9} \, dx$$

**SOLUTION**    If we let $u = x^2$ and $a = 3$, then $du = 2x\,dx$, and we can use Formula 23.

$$\int \sqrt{u^2 - a^2}\,du = \frac{1}{2}[u\sqrt{u^2 - a^2} - a^2 \ln |u + \sqrt{u^2 - a^2}|] + C$$

Then we can write

$$\int x\sqrt{x^4 - 9}\,dx = \frac{1}{2}\int \sqrt{(x^2)^2 - (3)^2}(2x)\,dx$$

$$= \frac{1}{4}[x^2\sqrt{x^4 - 9} - 9 \ln |x^2 + \sqrt{x^4 - 9}|] + C$$

---

**EXAMPLE 3**    **Integration by Tables**

Evaluate the integral

$$\int \frac{dx}{x\sqrt{x + 1}}$$

**SOLUTION**    Considering forms involving $\sqrt{a + bu}$, where $a = 1$, $b = 1$, and $u = x$, we choose Formula 15, which states that

$$\int \frac{1}{u\sqrt{a + bu}}\,du = \frac{1}{\sqrt{a}} \ln \left| \frac{\sqrt{a + bu} - \sqrt{a}}{\sqrt{a + bu} + \sqrt{a}} \right| + C, \quad 0 < a$$

Therefore,

$$\int \frac{dx}{x\sqrt{x + 1}} = \ln \left| \frac{\sqrt{x + 1} - 1}{\sqrt{x + 1} + 1} \right| + C$$

---

**EXAMPLE 4**    **Integration by Tables**

Evaluate the integral

$$\int \frac{x}{1 + e^{-x^2}}\,dx$$

**SOLUTION**    Of the forms involving $e^u$, we use Formula 39,

$$\int \frac{1}{1 + e^u}\,du = u - \ln (1 + e^u) + C$$

with $u = -x^2$ and $du = -2x\,dx$. Thus, we obtain

$$\int \frac{x}{1 + e^{-x^2}}\,dx = -\frac{1}{2}\int \frac{-2x}{1 + e^{-x^2}}\,dx$$

$$= -\frac{1}{2}[-x^2 - \ln (1 + e^{-x^2})] + C$$

$$= \frac{1}{2}[x^2 + \ln (1 + e^{-x^2})] + C$$

Notice that a number of the formulas in our table of integrals have the form

$$\int f(x)\, dx = g(x) + \int h(x)\, dx$$

where the right-hand member of the formula contains another integral. Such integration formulas are referred to as **reduction formulas,** since they reduce a given integral to the sum of a function and a simpler integral. We demonstrate the use of reduction formulas in the next two examples.

**EXAMPLE 5**

**Using Reduction Formulas**

Evaluate the integral

$$\int x^2 e^x\, dx$$

SOLUTION

Using Formula 38,

$$\int u^n e^u\, du = u^n e^u - n \int u^{n-1} e^u\, du$$

we let $u = x$ and $n = 2$. Then $du = dx$, and we have

$$\int x^2 e^x\, dx = x^2 e^x - 2 \int x e^x\, dx$$

Then, from Formula 37,

$$\int u e^u\, du = (u - 1)e^u + C$$

and we have

$$\int x^2 e^x\, dx = x^2 e^x - 2(x - 1)e^x + C$$

$$= e^x(x^2 - 2x + 2) + C$$

**EXAMPLE 6**

**Using Reduction Formulas**

Evaluate the integral

$$\int \frac{\sqrt{3 - 5x}}{2x}\, dx$$

SOLUTION

By Formula 17,

$$\int \frac{\sqrt{a + bu}}{u}\, du = 2\sqrt{a + bu} + a \int \frac{1}{u\sqrt{a + bu}}\, du$$

With $a = 3$, $b = -5$, and $u = x$, it follows that

$$\frac{1}{2} \int \frac{\sqrt{3 - 5x}}{x} \, dx = \frac{1}{2} \left[ 2\sqrt{3 - 5x} + 3 \int \frac{dx}{x\sqrt{3 - 5x}} \right]$$

$$= \sqrt{3 - 5x} + \frac{3}{2} \int \frac{dx}{x\sqrt{3 - 5x}}$$

Now, by Formula 15, with $u = x$, $a = 3$, and $b = -5$, it follows that

$$\int \frac{\sqrt{3 - 5x}}{2x} \, dx = \sqrt{3 - 5x} + \frac{3}{2} \left[ \frac{1}{\sqrt{3}} \ln \left| \frac{\sqrt{3 - 5x} - \sqrt{3}}{\sqrt{3 - 5x} + \sqrt{3}} \right| \right] + C$$

$$= \sqrt{3 - 5x} + \frac{\sqrt{3}}{2} \ln \left| \frac{\sqrt{3 - 5x} - \sqrt{3}}{\sqrt{3 - 5x} + \sqrt{3}} \right| + C$$

## Completing the Square

Several of the integration formulas listed in our table involve the sum or difference of two squares. We can extend the application of these formulas through the use of an algebraic technique called **completing the square.** This technique provides a means for writing any quadratic polynomial as the sum or difference of two squares. For instance, the polynomial $x^2 + bx + c$ can be written as follows:

$$x^2 + bx + c = x^2 + bx + \left( \frac{b}{2} \right)^2 - \left( \frac{b}{2} \right)^2 + c$$

$$= \left( x + \frac{b}{2} \right)^2 + \left[ c - \left( \frac{b}{2} \right)^2 \right]$$

Thus, we have written $x^2 + bx + c$ in the form $u^2 \pm a^2$, where

$$u = x + \frac{b}{2} \qquad \text{and} \qquad a^2 = \left| c - \left( \frac{b}{2} \right)^2 \right|$$

**EXAMPLE 7**

**Completing the Square**

Evaluate the integral

$$\int \frac{dx}{x^2 - 4x + 1}$$

**SOLUTION**

By completing the square, we obtain

$$x^2 - 4x + 1 = (x^2 - 4x + 4) - 4 + 1 = (x - 2)^2 - 3$$

Therefore,

$$\int \frac{dx}{x^2 - 4x + 1} = \int \frac{dx}{(x - 2)^2 - 3}$$

Considering $u = x - 2$ and $a = \sqrt{3}$, we apply Formula 21,

$$\int \frac{1}{u^2 - a^2}\, du = \frac{1}{2a} \ln \left| \frac{u - a}{u + a} \right| + C$$

to conclude that

$$\int \frac{dx}{x^2 - 4x + 1} = \frac{1}{2\sqrt{3}} \ln \left| \frac{x - 2 - \sqrt{3}}{x - 2 + \sqrt{3}} \right| + C$$

## EXAMPLE 8

**Completing the Square**

Evaluate the integral

$$\int \frac{dx}{\sqrt{x^2 + 2x}}$$

**SOLUTION**

By completing the square within the radical, we obtain

$$\int \frac{dx}{\sqrt{x^2 + 2x}} = \int \frac{dx}{\sqrt{(x^2 + 2x + 1) - 1}}$$

$$= \int \frac{dx}{\sqrt{(x + 1)^2 - 1}}$$

Now, letting $u = x + 1$ and $a = 1$, we apply Formula 27,

$$\int \frac{1}{\sqrt{u^2 - a^2}}\, du = \ln \left| u + \sqrt{u^2 - a^2} \right| + C$$

to obtain

$$\int \frac{1}{\sqrt{x^2 + 2x}}\, dx = \ln \left| (x + 1) + \sqrt{(x + 1)^2 - 1} \right| + C$$

$$= \ln \left| (x + 1) + \sqrt{x^2 + 2x} \right| + C$$

## SECTION EXERCISES 6.4

In Exercises 1–10, use the indicated formula from the table of integrals in this section to find the indefinite integral.

1. $\displaystyle \int \frac{x}{(2 + 3x)^2}\, dx$, Formula 4

2. $\displaystyle \int \frac{1}{x(2 + 3x)^2}\, dx$, Formula 11

3. $\displaystyle \int \frac{x}{\sqrt{2 + 3x}}\, dx$, Formula 19

4. $\displaystyle \int \frac{4}{x^2 - 9}\, dx$, Formula 21

5. $\displaystyle \int \frac{2x}{\sqrt{x^4 - 9}}\, dx$, Formula 27

6. $\displaystyle \int x^2 \sqrt{x^2 + 9}\, dx$, Formula 24

**7.** $\int x^3 e^{x^2}\, dx$, Formula 37

**8.** $\int \dfrac{x}{1 + e^{x^2}}\, dx$, Formula 39

**9.** $\int x \ln (x^2 + 1)\, dx$, Formula 41

**10.** $\int x^3 \ln x\, dx$, Formula 43

In Exercises 11–40, use the table of integrals in this section to find the indefinite integral.

**11.** $\int \dfrac{1}{x(1 + x)}\, dx$

**12.** $\int \dfrac{1}{x(1 + x)^2}\, dx$

**13.** $\int \dfrac{1}{x\sqrt{x^2 + 1}}\, dx$

**14.** $\int \dfrac{1}{\sqrt{x^2 - 1}}\, dx$

**15.** $\int \dfrac{1}{x\sqrt{4 - x^2}}\, dx$

**16.** $\int \dfrac{\sqrt{x^2 - 9}}{x^2}\, dx$

**17.** $\int x \ln x\, dx$

**18.** $\int x^2 (\ln x^3)^2\, dx$

**19.** $\int \dfrac{e^x}{e^{2x}(1 + e^x)}\, dx$

**20.** $\int \dfrac{1}{1 + e^x}\, dx$

**21.** $\int x\sqrt{x^4 - 9}\, dx$

**22.** $\int \dfrac{x}{x^4 - 9}\, dx$

**23.** $\int \dfrac{t^2}{(2 + 3t)^3}\, dt$

**24.** $\int \dfrac{\sqrt{3 + 4t}}{t}\, dt$

**25.** $\int \dfrac{s}{s^2\sqrt{3 + s}}\, ds$

**26.** $\int \sqrt{3 + x^2}\, dx$

**27.** $\int \dfrac{x^2}{1 + x}\, dx$

**28.** $\int \dfrac{1}{x^2\sqrt{x^2 - 4}}\, dx$

**29.** $\int \dfrac{1}{x^2\sqrt{1 - x^2}}\, dx$

**30.** $\int \dfrac{2x}{(1 - 3x)^2}\, dx$

**31.** $\int x^2 \ln x\, dx$

**32.** $\int xe^{x^2}\, dx$

**33.** $\int \dfrac{x^2}{(3x - 5)^2}\, dx$

**34.** $\int \dfrac{1}{2x^2(2x - 1)^2}\, dx$

**35.** $\int x^2\sqrt{x^2 + 4}\, dx$

**36.** $\int \dfrac{1}{\sqrt{x}(1 + 2\sqrt{x})}\, dx$

**37.** $\int \dfrac{1}{1 + e^{2x}}\, dx$

**38.** $\int \dfrac{e^x}{(1 - e^{2x})^{3/2}}\, dx$

**39.** $\int \dfrac{\ln x}{x(3 + 2 \ln x)}\, dx$

**40.** $\int (\ln x)^3\, dx$

In Exercises 41 and 42, complete the square to express each polynomial as the sum or difference of squares.

**41.** (a) $x^2 + 6x$      (b) $x^2 - 8x + 9$
(c) $x^4 + 2x^2 - 5$      (d) $3 - 2x - x^2$

**42.** (a) $2x^2 + 12x + 14$      (b) $3x^2 - 12x - 9$
(c) $x^2 - 2x$      (d) $9 + 8x - x^2$

In Exercises 43–50, complete the square and then use the table of integrals in this section to find the indefinite integral.

**43.** $\int \dfrac{1}{x^2 - 2x - 3}\, dx$

**44.** $\int \dfrac{1}{(x^2 + 4x - 5)^{3/2}}\, dx$

**45.** $\int \dfrac{1}{(x - 1)\sqrt{x^2 - 2x + 2}}\, dx$

**46.** $\int \sqrt{x^2 - 6x}\, dx$

**47.** $\int \dfrac{1}{2x^2 - 4x - 6}\, dx$

**48.** $\int \dfrac{\sqrt{7 - 6x - x^2}}{x + 3}\, dx$

**49.** $\int \dfrac{x}{\sqrt{x^4 + 2x^2 + 2}}\, dx$

**50.** $\int \dfrac{x\sqrt{x^4 + 4x^2 + 5}}{x^2 + 2}\, dx$

In Exercises 51 and 52, find the area of the region bounded by the graphs of the given equations.

**51.** $y = \dfrac{x}{\sqrt{x + 1}}$, $y = 0$, $x = 8$

**52.** $y = \dfrac{x}{1 + e^{x^2}}$, $y = 0$, $x = 2$

In Exercises 53 and 54, find the average value of the growth function over the given interval, where $N$ is the size of a population and $t$ is the time in days.

**53.** $N = \dfrac{50}{1 + e^{4.8 - 1.9t}}$, [3, 4]

**54.** $N = \dfrac{375}{1 + e^{4.20 - 0.25t}}$, [21, 28]

**55.** Find the consumer surplus and producer surplus for a given product if the demand and supply functions are

$$\text{Demand: } p = \dfrac{60}{\sqrt{x^2 + 81}} \qquad \text{Supply: } p = \dfrac{x}{3}$$

(Note: For a definition of consumer and producer surplus, refer to Exercises 37–46 in Section 4.4.)

# Numerical Integration

In Section 5.2, we looked at the normal probability density function

$$f(x) = \frac{1}{\sigma\sqrt{2\pi}} e^{-x^2/2\sigma^2}$$

where $x = 0$ represents the *mean* of the distribution and $\sigma$ is the *standard deviation* from the mean. ($\sigma$ is the lowercase Greek letter sigma.) In such a normal distribution, approximately 68% of the area under the curve lies within one standard deviation from the mean, as shown in Figure 6.12. This area is represented by the integral

$$\frac{1}{\sigma\sqrt{2\pi}} \int_{-\sigma}^{\sigma} e^{-x^2/2\sigma^2} \, dx$$

However, at this stage we know of no way to find an antiderivative for $e^{-x^2/2\sigma^2}$. In this section, we will see that, when we cannot find an antiderivative for a certain definite integral, we can approximate its value through a procedure called **numerical integration.** Specifically, using **Simpson's Rule** (with $n = 10$), we have

$$\frac{1}{\sigma\sqrt{2\pi}} \int_{-\sigma}^{\sigma} e^{-x^2/2\sigma^2} \, dx \approx \frac{2\sigma}{30} \left( \frac{1}{\sigma\sqrt{2\pi}} \right) [e^{-0.5} + 4e^{-0.32}$$

$$+ 2e^{-0.18} + 4e^{-0.08}$$

$$+ 2e^{-0.02} + 4e^0 + 2e^{-0.02}$$

$$+ 4e^{-0.08} + 2e^{-0.18}$$

$$+ 4e^{-0.32} + e^{-0.5}]$$

$$\approx 0.6827 = 68.27\%$$

Extending this procedure to find the area of the region lying within two, three, and four standard deviations from the mean, we have the results shown in Table 6.3.

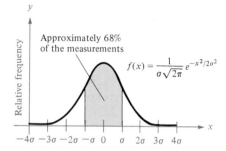

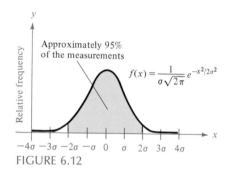

FIGURE 6.12

TABLE 6.3

| $n$ | $\dfrac{1}{\sigma\sqrt{2\pi}} \displaystyle\int_{-n\sigma}^{n\sigma} e^{-x^2/2\sigma^2} \, dx$ |
| --- | --- |
| 1 | 68.27% |
| 2 | 95.45% |
| 3 | 99.73% |
| 4 | 99.99% |
| Rounded to the nearest 0.01% | |

- **Trapezoidal Rule**
- **Simpson's Rule**

When we began our discussion of integration techniques, we mentioned that the search for an antiderivative is not nearly as straightforward a process as is differentiation. Occasionally we encounter functions for which we cannot find antiderivatives. Of course, inability to find an antiderivative for a particular function may be due to a lack of cleverness on our part. On the other hand, some elementary functions* simply do not possess antiderivatives that are elementary functions. For example, there are no elementary functions that have any one of the following functions as their derivative:

$$\sqrt[3]{x}\sqrt{1-x}, \qquad \sqrt{1-x^3}, \qquad e^{-x^2}$$

$$\frac{1}{\sqrt{x}e^x}, \qquad \frac{e^x}{x}, \qquad \frac{1}{\ln x}$$

Up to this point we have been evaluating definite integrals by means of antiderivatives and the Fundamental Theorem of Calculus. However, if we wish to evaluate a definite integral involving a function whose antiderivative we cannot find, then the Fundamental Theorem cannot be applied, and we must resort to an approximation technique. We describe two such techniques in this section.

### The Trapezoidal Rule

One way we can approximate the definite integral $\int_a^b f(x)\,dx$ is by the use of $n$ trapezoids, as shown in Figure 6.13. In the development of this method we assume that $f$ is continuous and positive on the interval $[a, b]$ and that this definite integral represents the area of a region bounded by $f$ and the $x$-axis, from $x = a$ to $x = b$.

First we partition the interval $[a, b]$ into $n$ equal subintervals, each of width

$$\Delta x = \frac{b - a}{n}$$

such that

$$a = x_0 < x_1 < x_2 < \cdots < x_{n-1} < x_n = b$$

We then form trapezoids for each subinterval, as shown in Figure 6.13. Now the first trapezoid (standing on end as shown in Figure 6.14) has an area given by

Area of first trapezoid = (average of two bases)(height)

$$= \left[\frac{f(x_0) + f(x_1)}{2}\right]\left(\frac{b - a}{n}\right)$$

---

*An *elementary function* is one that is formed from algebraic, exponential, logarithmic, trigonometric, or inverse trigonometric functions.

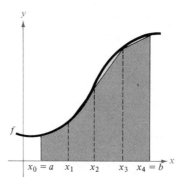

FIGURE 6.13

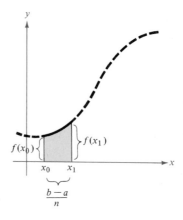

$$\frac{b-a}{n}$$

FIGURE 6.14

Extending this procedure to all $n$ trapezoids, we have

$$\text{Area of first trapezoid} = \left[\frac{f(x_0) + f(x_1)}{2}\right]\left(\frac{b-a}{n}\right)$$

$$\text{Area of second trapezoid} = \left[\frac{f(x_1) + f(x_2)}{2}\right]\left(\frac{b-a}{n}\right)$$

$$\vdots$$

$$\text{Area of } n\text{th trapezoid} = \left[\frac{f(x_{n-1}) + f(x_n)}{2}\right]\left(\frac{b-a}{n}\right)$$

Finally, the sum of the areas of the $n$ trapezoids is

$$\left(\frac{b-a}{n}\right)\left[\frac{f(x_0) + f(x_1)}{2} + \frac{f(x_1) + f(x_2)}{2} + \cdots + \frac{f(x_{n-1}) + f(x_n)}{2}\right]$$

$$= \left(\frac{b-a}{2n}\right)[f(x_0) + f(x_1) + f(x_1) + f(x_2) + \cdots + f(x_{n-1}) + f(x_n)]$$

$$= \left(\frac{b-a}{2n}\right)[f(x_0) + 2f(x_1) + 2f(x_2) + \cdots + 2f(x_{n-1}) + f(x_n)]$$

Thus, we arrive at the following approximation, which we call the Trapezoidal Rule.

**Trapezoidal Rule**    If $f$ is continuous on the interval $[a, b]$, then

$$\int_a^b f(x)\, dx \approx \frac{b-a}{2n}[f(x_0) + 2f(x_1) + 2f(x_2) + \cdots + 2f(x_{n-1}) + f(x_n)]$$

**EXAMPLE 1**

**Approximation with the Trapezoidal Rule**

Use the Trapezoidal Rule to approximate the definite integral

$$\int_0^1 e^x \, dx$$

Compare the results for $n = 4$ and $n = 8$.

**SOLUTION**

When $n = 4$, we have $\Delta x = \frac{1}{4}$ and

$$x_0 = 0, \quad x_1 = \frac{1}{4}, \quad x_2 = \frac{2}{4}, \quad x_3 = \frac{3}{4}, \quad x_4 = 1$$

Therefore, by the Trapezoidal Rule, we have

$$\int_0^1 e^x \, dx \approx \frac{1}{8}[e^0 + 2e^{0.25} + 2e^{0.5} + 2e^{0.75} + e^1] \approx 1.7272$$

(See Figure 6.15.)

When $n = 8$, we have $\Delta x = \frac{1}{8}$ and

$$x_0 = 0, \quad x_1 = \frac{1}{8}, \quad x_2 = \frac{2}{8}, \quad x_3 = \frac{3}{8}, \quad x_4 = \frac{4}{8}$$

$$x_5 = \frac{5}{8}, \quad x_6 = \frac{6}{8}, \quad x_7 = \frac{7}{8}, \quad x_8 = 1$$

and it follows that

$$\int_0^1 e^x \, dx \approx \frac{1}{16}[e^0 + 2e^{0.125} + 2e^{0.25} + 2e^{0.375} + 2e^{0.5} + 2e^{0.625}$$

$$+ 2e^{0.75} + 2e^{0.875} + e^1]$$

$$\approx 1.7205$$

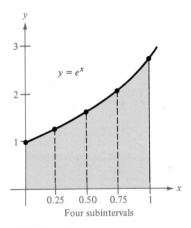

Four subintervals

FIGURE 6.15

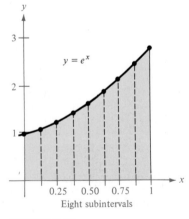

Eight subintervals

FIGURE 6.16

(See Figure 6.16.) Of course, *for this particular example* we could have found an antiderivative and determined that the exact area of the region is

$$e - 1 \approx 1.7183$$

Although you may not yet be excited about using a rather lengthy approximation method to evaluate the integral $\int_0^1 e^x \, dx$, two important points must be brought to your attention. First, this approximation method involving trapezoids becomes more accurate as $n$ increases. (For $n = 16$ in Example 1, the Trapezoidal Rule yields an approximation of 1.7188.) Second, though we could have used the Fundamental Theorem to evaluate the integral in Example 1, this theorem cannot be used to evaluate an integral so simple looking as $\int_0^1 e^{x^2} \, dx$, because $e^{x^2}$ has no elementary antiderivative. Yet the Trapezoidal Rule can be readily applied to this integral.

## Simpson's Rule

One way to view the trapezoidal approximation of a definite integral is to say that on each subinterval we approximate $f$ by a first-degree polynomial. In Simpson's Rule we carry this procedure one step further and approximate $f$ by second-degree polynomials.

Before presenting Simpson's Rule, we give the following theorem for evaluating integrals of second-degree polynomials.

**Integral of quadratic function**

If $p(x) = Ax^2 + Bx + C$, then

$$\int_a^b p(x) \, dx = \left( \frac{b - a}{6} \right) \left[ p(a) + 4p\left( \frac{a + b}{2} \right) + p(b) \right]$$

Proof

Although the proof of this theorem is logically straightforward, it is quite messy in terms of the algebra involved. Thus, for the sake of brevity, we merely give an outline for the proof. We encourage you to supply the missing algebraic steps.

$$\int_a^b (Ax^2 + Bx + C) \, dx = \left[ \frac{Ax^3}{3} + \frac{Bx^2}{2} + Cx \right]_a^b$$

$$= \frac{A(b^3 - a^3)}{3} + \frac{B(b^2 - a^2)}{2} + C(b - a)$$

$$= \left( \frac{b - a}{6} \right) [2A(a^2 + ab + b^2) + 3B(b + a) + 6C]$$

$$= \frac{b - a}{6} \left[ p(a) + 4p\left( \frac{a + b}{2} \right) + p(b) \right]$$

To develop Simpson's Rule for approximating the value of the definite integral $\int_a^b f(x)\,dx$, we again partition the interval $[a, b]$ into $n$ equal parts, each of width

$$\frac{b - a}{n}$$

However, this time we require $n$ to be even and then group the subintervals into pairs such that

$$a = \underbrace{x_0 < x_1 < x_2}_{[x_0,\, x_2]} \underbrace{< x_3 < x_4}_{[x_2,\, x_4]} < \cdots < \underbrace{x_{n-2} < x_{n-1} < x_n}_{[x_{n-2},\, x_n]} = b$$

Then on the subinterval $[x_0, x_2]$ we approximate $y = f(x)$ by the second-degree polynomial $p$ that passes through the points

$$(x_0, y_0), \qquad (x_1, y_1), \qquad (x_2, y_2)$$

as shown in Figure 6.17. Now, using $p$ as an approximation for $f$, we have

$$\int_{x_0}^{x_2} f(x)\,dx \approx \int_{x_0}^{x_2} p(x)\,dx$$

$$= \frac{x_2 - x_0}{6}\left[p(x_0) + 4p\left(\frac{x_2 + x_0}{2}\right) + p(x_2)\right]$$

$$= \frac{2[(b - a)/n]}{6}[p(x_0) + 4p(x_1) + p(x_2)]$$

$$= \frac{b - a}{3n}[f(x_0) + 4f(x_1) + f(x_2)]$$

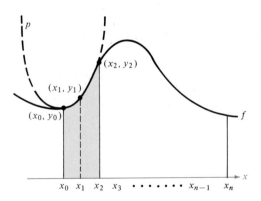

$$\int_{x_0}^{x_2} p(x)\,dx \approx \int_{x_0}^{x_2} f(x)\,dx$$

FIGURE 6.17

Repeating this procedure on each subinterval $[x_{i-2}, x_i]$ of interval $[a, b]$ results in the formula

$$\int_a^b f(x)\, dx \approx \frac{b-a}{3n}\{[f(x_0) + 4f(x_1) + f(x_2)] + [f(x_2) + 4f(x_3) + f(x_4)]$$

$$+ \cdots + [f(x_{n-2}) + 4f(x_{n-1}) + f(x_n)]\}$$

Grouping like terms, we obtain the following approximation formula, known as *Simpson's Rule*.

---

**Simpson's Rule**
**($n$ is even)**

If $f$ is continuous on the interval $[a, b]$, then

$$\int_a^b f(x)\, dx \approx \frac{b-a}{3n}[f(x_0) + 4f(x_1) + 2f(x_2) + 4f(x_3) + \cdots + 4f(x_{n-1}) + f(x_n)]$$

---

In Example 1, we used the Trapezoidal Rule to estimate $\int_0^1 e^x\, dx$. In the next example, we see how well Simpson's Rule works for the same integral.

**EXAMPLE 2**

**Approximation with Simpson's Rule**

Use Simpson's Rule to approximate the integral

$$\int_0^1 e^x\, dx$$

Compare the results for $n = 4$ and $n = 8$.

**SOLUTION**

When $n = 4$, we have

$$\int_0^1 e^x\, dx \approx \frac{1}{12}[e^0 + 4e^{0.25} + 2e^{0.5} + 4e^{0.75} + e^1]$$

$$\approx 1.718319$$

When $n = 8$, we have

$$\int_0^1 e^x\, dx \approx \frac{1}{24}[e^0 + 4e^{0.125} + 2e^{0.25} + 4e^{0.375} + 2e^{0.5} + 4e^{0.625}$$

$$+ 2e^{0.75} + 4e^{0.875} + e^1]$$

$$\approx 1.718284$$

Recall that the actual value of this integral is

$$e - 1 \approx 1.718282$$

Thus, with only eight subintervals we have obtained an approximation that is correct to the nearest 0.00001—an impressive result!

In Examples 1 and 2, we were able to calculate the exact value of the integral and compare that to our approximations to see how good they were. Of course, in practice we would not bother with an approximation if it were possible to evaluate the integral exactly. However, if it is necessary to use an approximation technique to determine the value of a definite integral, then it is important to know how good we can expect our approximation to be. The following theorem, which we list without proof, gives the formulas for estimating the error involved in the use of the Trapezoidal and Simpson's Rules.

**Error in Trapezoidal and Simpson's Rules**

The error $E$ in approximating $\int_a^b f(x)\,dx$ is

$$|E| \leq \frac{(b-a)^3}{12n^2}[\max |f''(x)|], \quad a \leq x \leq b$$

for the Trapezoidal Rule and

$$|E| \leq \frac{(b-a)^5}{180n^4}[\max |f^{(4)}(x)|], \quad a \leq x \leq b$$

for Simpson's Rule.

This theorem indicates that the errors generated by the Trapezoidal and Simpson's Rules have upper bounds dependent upon the extreme values of $f''(x)$ and $f^{(4)}(x)$ respectively, in the interval $[a, b]$. Furthermore, it is evident that the bounds for these errors can be made arbitrarily small by *increasing n* provided, of course, that $f''$ and $f^{(4)}$ exist and are bounded in $[a, b]$. The next examples show how to find a value of $n$ that will bound the error within a predetermined tolerance interval. To solve this type of problem, we suggest the following steps.

| *Trapezoidal Rule* | *Simpson's Rule* |
|---|---|
| 1. Find $f''(x)$. | 1. Find $f^{(4)}(x)$. |
| 2. Find the maximum of $|f''(x)|$ on the interval $[a, b]$. | 2. Find the maximum of $|f^{(4)}(x)|$ on the interval $[a, b]$. |
| 3. Set up the inequality | 3. Set up the inequality |
| $$|E| \leq \frac{(b-a)^3}{12n^2}[\max |f''(x)|]$$ | $$|E| \leq \frac{(b-a)^5}{180n^4}[\max |f^{(4)}(x)|]$$ |
| 4. For an error less than $\varepsilon$, solve for $n$ in the inequality | 4. For an error less than $\varepsilon$, solve for $n$ in the inequality |
| $$\frac{(b-a)^3}{12n^2}[\max |f''(x)|] < \varepsilon$$ | $$\frac{(b-a)^5}{180n^4}[\max |f^{(4)}(x)|] < \varepsilon$$ |
| 5. Partition $[a, b]$ into $n$ subintervals and apply the Trapezoidal Rule. | 5. Partition $[a, b]$ into $n$ subintervals and apply Simpson's Rule. |

**EXAMPLE 3**

**Approximating the Error in the Trapezoidal Rule**

Using the Trapezoidal Rule, estimate the value of

$$\int_0^1 e^{-x^2}\, dx$$

choosing a value of $n$ so that the approximation error is less than 0.01.

**SOLUTION**

1. We begin by finding the second derivative of $f(x) = e^{-x^2}$.

$$f(x) = e^{-x^2}$$
$$f'(x) = -2xe^{-x^2}$$
$$f''(x) = 4x^2e^{-x^2} - 2e^{-x^2}$$
$$= 2e^{-x^2}(2x^2 - 1)$$

2. Using the methods of differential calculus, we can determine that $f''$ has only one critical value ($x = 0$) in the interval [0, 1] and that the maximum value of $|f''(x)|$ on this interval is $|f''(0)| = 2$.

3. The approximation error using the Trapezoidal Rule is bounded as follows:

$$|E| \le \frac{(b-a)^3}{12n^2} \quad (2)$$

$$= \frac{1}{12n^2}(2) = \frac{1}{6n^2}$$

4. To ensure that the approximation has an error less than 0.01, we must choose $n$ so that

$$|E| \le \frac{1}{6n^2} < 0.01 = \frac{1}{100}$$

$$100 < 6n^2$$

$$\frac{50}{3} < n^2$$

$$4.08 \approx \sqrt{\frac{50}{3}} < n$$

Therefore, we choose $n = 5$ (since $n$ must be greater than 4.08).

5. Finally, we partition the interval [0, 1] into five subintervals as shown in Figure 6.18, and we apply the Trapezoidal Rule to obtain

$$\int_0^1 e^{-x^2}\, dx \approx \frac{1}{10}\left(\frac{1}{e^0} + \frac{2}{e^{0.04}} + \frac{2}{e^{0.16}} + \frac{2}{e^{0.36}} + \frac{2}{e^{0.64}} + \frac{1}{e^1}\right)$$

$$\approx 0.744$$

Therefore, with an error less than 0.01, we know that

$$0.734 \le \int_0^1 e^{-x^2}\, dx \le 0.754$$

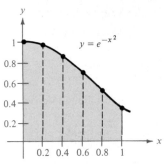

$y = e^{-x^2}$

FIGURE 6.18

## EXAMPLE 4

**Approximating the Error in Simpson's Rule**

Use Simpson's Rule to estimate the definite integral

$$\int_0^1 \ln (x^2 + 1) \, dx$$

Determine $n$ so that the approximation error is less than 0.0001.

**SOLUTION**

1. Since the error in Simpson's Rule involves the fourth derivative, we use repeated differentiation to obtain

$$f(x) = \ln (x^2 + 1)$$

$$f'(x) = \frac{2x}{x^2 + 1}$$

$$f''(x) = \frac{-2(x^2 - 1)}{(x^2 + 1)^2}$$

$$f'''(x) = \frac{4x(x^2 - 3)}{(x^2 + 1)^3}$$

$$f^{(4)}(x) = \frac{-12(x^4 - 6x^2 + 1)}{(x^2 + 1)^4}$$

2. On the interval $[0, 1]$, $|f^{(4)}(x)|$ has a maximum value of 12.
3. The approximation error using Simpson's Rule is bounded as follows:

$$|E| \le \frac{(b - a)^5}{180n^4} (12) = \frac{1}{15n^4}$$

4. To ensure that the approximation has an error less than 0.0001, we must choose an *even* integer $n$ so that

$$|E| \le \frac{1}{15n^4} < 0.0001$$

$$\frac{10,000}{15} < n^4$$

$$666.67 < n^4$$

$$5.08 < n$$

Therefore, we choose $n = 6$.

5. Finally, we partition the interval $[0, 1]$ into six subintervals as shown in Figure 6.19, and apply Simpson's Rule to obtain

$$\int_0^1 \ln (x^2 + 1) \, dx \approx \frac{1}{18}\left[ \ln (1) + 4 \ln \left(\frac{1}{36} + 1\right) + 2 \ln \left(\frac{4}{36} + 1\right) \right.$$

$$+ 4 \ln \left(\frac{9}{36} + 1\right) + 2 \ln \left(\frac{16}{36} + 1\right)$$

$$\left. + 4 \ln \left(\frac{25}{36} + 1\right) + \ln (2) \right]$$

$$\approx 0.26394$$

Therefore, with an error less than $0.0001$, we know that

$$0.26384 \le \int_0^1 \ln (x^2 + 1) \, dx \le 0.26404$$

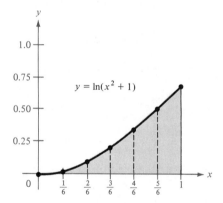

$$y = \ln(x^2 + 1)$$

FIGURE 6.19

Looking back at the examples in this section, you may wonder why we introduced the Trapezoidal Rule, since for a fixed $n$ Simpson's Rule usually gives a better approximation. The main reason for including the Trapezoidal Rule is that its error can be more easily estimated than can the error involved in Simpson's Rule. Certainly an approximation method is of little benefit if we have no idea of the potential error in the approximation. For instance, if $f(x) = \sqrt{x} \ln (x + 1)$, then to estimate the error in Simpson's Rule we would need to determine the fourth derivative of $f$, which is a monumental task. Because the estimation of the error in Simpson's Rule involves the fourth derivative, we sometimes prefer to use the Trapezoidal Rule even though we may have to use a larger $n$ to obtain the desired accuracy.

## SECTION EXERCISES 6.5

In Exercises 1–10, use the Trapezoidal Rule and Simpson's Rule to approximate the value of the definite integral for the indicated value of $n$. Compare these results with the exact value of the definite integral. Round your answers to four decimal places.

**1.** $\int_0^2 x^2 \, dx, \; n = 4$

**2.** $\int_0^1 \left( \frac{x^2}{2} + 1 \right) dx, \; n = 4$

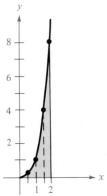

**3.** $\int_0^2 x^3 \, dx, \; n = 4$

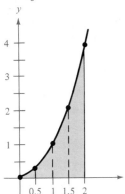

**4.** $\int_1^2 \frac{1}{x} \, dx, \; n = 4$

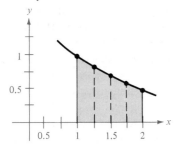

**5.** $\int_0^2 x^3 \, dx, \; n = 8$

**6.** $\int_1^2 \frac{1}{x} \, dx, \; n = 8$

**7.** $\int_1^2 \frac{1}{x^2} \, dx, \; n = 4$

**8.** $\int_0^4 \sqrt{x} \, dx, \; n = 8$

**9.** $\int_0^1 \frac{1}{1 + x} \, dx, \; n = 4$

**10.** $\int_0^2 x\sqrt{x^2 + 1} \, dx, \; n = 4$

In Exercises 11–20, approximate the given integral using (a) the Trapezoidal Rule and (b) Simpson's Rule. (Round your answers to three significant digits.)

**11.** $\int_0^2 \sqrt{1 + x^3} \, dx, \; n = 2$

**12.** $\int_0^2 \frac{1}{\sqrt{1 + x^3}} \, dx, \; n = 4$

**13.** $\int_0^1 \sqrt{x}\sqrt{1 - x} \, dx, \; n = 4$

**14.** $\int_0^1 \frac{1}{1 + x^2} \, dx, \; n = 4$

**15.** $\int_0^1 \sqrt{1 - x^2} \, dx, \; n = 4$

**16.** $\int_0^2 e^{-x^2} \, dx, \; n = 2$

**17.** $\int_0^1 \sqrt{1 - x^2} \, dx, \; n = 8$

**18.** $\int_0^2 e^{-x^2} \, dx, \; n = 4$

**19.** $\int_1^3 \frac{1}{2 + x + x^2} \, dx, \; n = 2$

**20.** $\int_0^3 \frac{x}{2 + x + x^2} \, dx, \; n = 6$

In Exercises 21 and 22, use Simpson's Rule with $n = 8$ to approximate the present value of the income $c(t)$ over $t_1$ years at the given annual interest rate $r$. (Present value is defined in Section 6.2.)

**21.** $c(t) = 6000 + 200\sqrt{t}, \; r = 7\%, \; t_1 = 4$

**22.** $c(t) = 200{,}000 + 15{,}000\sqrt[3]{t}, \; r = 10\%, \; t_1 = 8$

In Exercises 23 and 24, use Simpson's Rule with $n = 6$ to approximate the indicated normal probability. The standard normal probability density function is

$$f(x) = \frac{1}{\sqrt{2\pi}} e^{-x^2/2}$$

If $x$ is chosen at random from a population with this density, then the probability that $x$ lies in the interval $[a, b]$ is given by $P(a \le x \le b) = \int_a^b f(x)\, dx$.

**23.** $P(0 \le x \le 1)$

**24.** $P(0 \le x \le 2)$

In Exercises 25 and 26, use Simpson's Rule to estimate the number of square feet of land in the given lot where $x$ and $y$ are measured in feet, as shown in Figures 6.20 and 6.21. In each case the land is bounded by a stream and two straight roads.

**25.** See Figure 6.20.

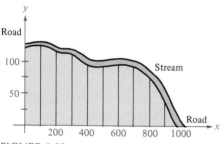

FIGURE 6.20

| $x$ | 0 | 100 | 200 | 300 | 400 | 500 | 600 | 700 | 800 | 900 | 1000 |
|---|---|---|---|---|---|---|---|---|---|---|---|
| $y$ | 125 | 125 | 120 | 112 | 90 | 90 | 95 | 88 | 75 | 35 | 0 |

**26.** See Figure 6.21.

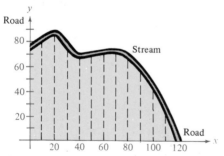

FIGURE 6.21

| $x$ | 0 | 10 | 20 | 30 | 40 | 50 | 60 | 70 | 80 | 90 | 100 | 110 | 120 |
|---|---|---|---|---|---|---|---|---|---|---|---|---|---|
| $y$ | 75 | 81 | 84 | 76 | 67 | 68 | 69 | 72 | 68 | 56 | 42 | 23 | 0 |

In Exercises 27–30, find the maximum possible error in approximating the given integral using (a) the Trapezoidal Rule and (b) Simpson's Rule. (Let $n = 4$.)

**27.** $\int_0^2 x^3\, dx$

**28.** $\int_0^1 \frac{1}{x+1}\, dx$

**29.** $\int_0^1 e^{x^3}\, dx$

**30.** $\int_0^1 e^{-x^2}\, dx$

In Exercises 31 and 32, find $n$ so that the error in the approximation of the definite integral is less than 0.00001 using (a) the Trapezoidal Rule and (b) Simpson's Rule.

**31.** $\int_0^1 e^{-x^2}\, dx$

**32.** $\int_1^3 \frac{1}{x}\, dx$

**33.** Prove that Simpson's Rule is exact when used to approximate the integral of a cubic polynomial function, and demonstrate the result for

$$\int_0^1 x^3\, dx, \quad n = 2$$

**34.** The suspension cable on a bridge that is 400 feet long is in the shape of a parabola whose equation is given by

$$y = \frac{x^2}{800}$$

(See Figure 6.22.) The length of the cable is given by the integral

$$\int_{-200}^{200} \sqrt{1 + (y')^2}\, dx = 2\int_0^{200} \sqrt{1 + \left(\frac{x}{400}\right)^2}\, dx$$

(a) Use the table of integrals in Section 6.4 to find the length of the cable.

(b) Use Simpson's Rule with $n = 10$ to approximate the length of the cable.

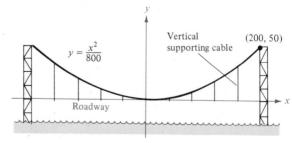

FIGURE 6.22

# Improper Integrals

### INTRODUCTORY EXAMPLE
### More About the Normal Distribution

In the Introductory Example to the preceding section, we discussed the normal probability density function. In that discussion we mentioned that approximately 99.73% of the measurements in a normal distribution fall within three standard deviations of the mean. Furthermore, we saw that nearly 100% (but *not* all) of the measurements fall within four standard deviations of the mean. Since even four standard deviations do not guarantee the inclusion of all measurements, we are led to ask: How many standard deviations away from the mean must we move before we can be sure to capture *all* measurements in a normal distribution? For example, if the mean height of a normally distributed population is 70 inches with a standard deviation of 3 inches, can we be sure that no one in the population has a height that differs from the mean by more than

five, six, or even seven standard deviations? The answer (at least theoretically) is that we can never be sure that *all* measurements in a truly normal distribution lie within $n$ standard deviations of the mean, no matter how large we take $n$ to be. Thus, we say that only in the limit does the area under the normal curve equal 1. That is,

$$\lim_{n \to \infty} \frac{1}{\sigma\sqrt{2\pi}} \int_{-n\sigma}^{n\sigma} e^{-x^2/2\sigma^2} \, dx = 1$$

We indicate this limit by the **improper integral**

$$\frac{1}{\sigma\sqrt{2\pi}} \int_{-\infty}^{\infty} e^{-x^2/2\sigma^2} \, dx$$

which represents the area bounded above by the normal curve and below by the x-axis and extending without bound to the right and left, as shown in Figure 6.23.

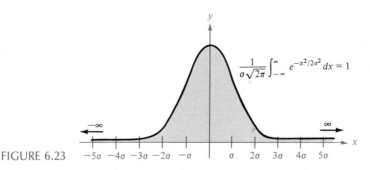

$$\frac{1}{\sigma\sqrt{2\pi}} \int_{-\infty}^{\infty} e^{-x^2/2\sigma^2} \, dx = 1$$

FIGURE 6.23

- **Improper Integrals**
- **Convergence and Divergence of Improper Integrals**

The definition of the definite integral $\int_a^b f(x)\,dx$ included the requirements that the interval $[a, b]$ be finite and that $f$ be bounded on $[a, b]$. Furthermore, the Fundamental Theorem of Calculus, by which we have been evaluating definite integrals, requires that $f$ be continuous on $[a, b]$. In this section, we discuss a limit procedure for evaluating integrals that do not satisfy these requirements because either (1) one or both of the limits of integration are infinite or (2) $f$ has an infinite discontinuity on the interval $[a, b]$.

Integrals possessing either of these characteristics are called **improper integrals.** For instance, the integrals

$$\int_0^\infty e^{-x}\,dx \qquad \text{and} \qquad \int_{-\infty}^\infty \frac{dx}{x^2 + 1}$$

are improper because one or both of their limits of integration are infinite. (See Figures 6.24 and 6.25.)

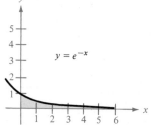

$y = e^{-x}$

FIGURE 6.24

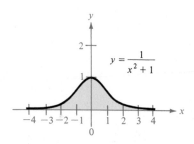

$y = \dfrac{1}{x^2 + 1}$

FIGURE 6.25

Similarly, the integrals

$$\int_1^5 \frac{dx}{\sqrt{x - 1}} \qquad \text{and} \qquad \int_{-2}^2 \frac{dx}{(x + 1)^2}$$

are improper because their integrands approach infinity somewhere in the interval of integration. (See Figures 6.26 and 6.27.)

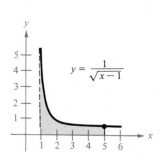

$y = \dfrac{1}{\sqrt{x - 1}}$

FIGURE 6.26

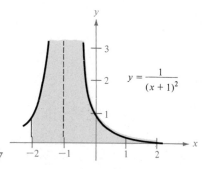

$y = \dfrac{1}{(x + 1)^2}$

FIGURE 6.27

### Development of Improper Integrals

To get an idea how we might evaluate an improper integral, consider the integral

$$\int_1^b \frac{1}{x^2}\,dx$$

Remember that as long as $b$ is a real number (no matter how large), this is a definite integral (not an improper integral) and we can evaluate it as follows:

$$\int_1^b \frac{1}{x^2}\,dx = -\frac{1}{x}\Big]_1^b = -\frac{1}{b} + 1 = 1 - \frac{1}{b}$$

Figure 6.28 shows the region whose area is represented by this integral, and Table 6.4 lists the region's area for several values of $b$.

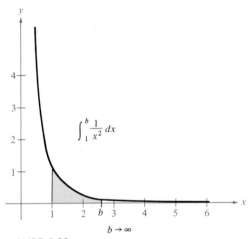

$$\int_1^b \frac{1}{x^2}\,dx$$

$$b \to \infty$$

FIGURE 6.28

TABLE 6.4

| $b$ | 2 | 5 | 10 | 100 | 1000 | 10,000 |
|---|---|---|---|---|---|---|
| $\int_1^b \frac{1}{x^2}\,dx$ | 0.5 | 0.8 | 0.9 | 0.99 | 0.999 | 0.9999 |

Now, if we let $b \to \infty$, we have

$$\lim_{b \to \infty} \int_1^b \frac{1}{x^2}\,dx = \lim_{b \to \infty}\left[\frac{-1}{x}\right]_1^b = \lim_{b \to \infty}\left(1 - \frac{1}{b}\right) = 1$$

We denote this limit by the improper integral

$$\int_1^\infty \frac{1}{x^2}\,dx = 1$$

and we can interpret it as the area of the unbounded region under $y = 1/x^2$, above the $x$-axis, and to the right of $x = 1$.

More generally, we define improper integrals having infinite limits of integration as follows.

**Improper integrals (infinite limits of integration)**

1. If $f$ is continuous on the interval $[a, \infty)$, then

$$\int_a^{\infty} f(x)\, dx = \lim_{b \to \infty} \int_a^b f(x)\, dx$$

2. If $f$ is continuous on the interval $(-\infty, b]$, then

$$\int_{-\infty}^b f(x)\, dx = \lim_{a \to -\infty} \int_a^b f(x)\, dx$$

3. If $f$ is continuous on the interval $(-\infty, \infty)$, then

$$\int_{-\infty}^{\infty} f(x)\, dx = \int_{-\infty}^c f(x)\, dx + \int_c^{\infty} f(x)\, dx$$

where $c$ is any real number.

In each case above, if the limit exists then the improper integral is said to **converge;** otherwise the improper integral **diverges.** This means that in the third case the integral will diverge if either one of the integrals on the right diverges.

**EXAMPLE 1**

**A Divergent Improper Integral**

Determine the convergence or divergence of the integral

$$\int_1^{\infty} \frac{1}{x}\, dx$$

**SOLUTION**

$$\int_1^{\infty} \frac{1}{x}\, dx = \lim_{b \to \infty} \int_1^b \frac{1}{x}\, dx = \lim_{b \to \infty} \left[ \ln x \right]_1^b$$

$$= \lim_{b \to \infty} (\ln b - 0) = \infty$$

Thus, the integral diverges since the limit does not exist.

A comparison of the improper integral

$$\int_1^{\infty} \frac{1}{x^2}\, dx$$

which converges to 1, and

$$\int_1^{\infty} \frac{1}{x}\, dx$$

which diverges, suggests the somewhat unpredictable nature of improper integrals.

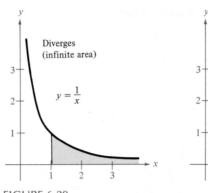

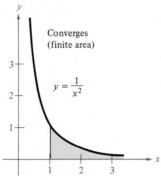

FIGURE 6.29

The functions being integrated have similar-looking graphs, as shown in Figure 6.29, yet the shaded region under $y = 1/x^2$ to the right of $x = 1$ has a finite area, whereas the corresponding region under $y = 1/x$ has an infinite area (see Example 1).

**EXAMPLE 2**

**A Convergent Improper Integral**

Evaluate the improper integral

$$\int_{-\infty}^{0} \frac{dx}{(1 - 2x)^{3/2}}$$

**SOLUTION**

$$\int_{-\infty}^{0} \frac{dx}{(1 - 2x)^{3/2}} = \lim_{b \to -\infty} \int_{b}^{0} \frac{dx}{(1 - 2x)^{3/2}}$$

$$= \lim_{b \to -\infty} \left[ \frac{1}{\sqrt{1 - 2x}} \right]_{b}^{0}$$

$$= \lim_{b \to -\infty} \left[ 1 - \frac{1}{\sqrt{1 - 2b}} \right]$$

$$= 1 - 0 = 1$$

Consequently, this improper integral converges. (See Figure 6.30.)

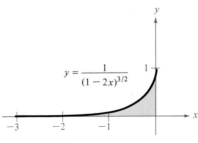

FIGURE 6.30

EXAMPLE 3

**A Convergent Improper Integral**

Evaluate the improper integral

$$\int_0^\infty 2xe^{-x^2}\, dx$$

SOLUTION

$$\int_0^\infty 2xe^{-x^2}\, dx = \lim_{b\to\infty} \int_0^b 2xe^{-x^2}\, dx$$

$$= \lim_{b\to\infty}\left[-e^{-x^2}\right]_0^b$$

$$= \lim_{b\to\infty}\left[-e^{-b^2} + 1\right]$$

$$= 0 + 1 = 1$$

(See Figure 6.31.)

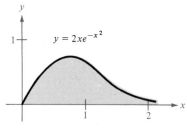

FIGURE 6.31

A second general type of improper integral involves integrands that approach infinity at one of the limits of integration. Such integrals are described in the following definitions.

**Improper integrals (infinite integrands)**

1. If $f$ is continuous on the interval $[a, b)$ and approaches infinity at $b$, then

$$\int_a^b f(x)\, dx = \lim_{c\to b^-} \int_a^c f(x)\, dx$$

2. If $f$ is continuous on the interval $(a, b]$ and approaches infinity at $a$, then

$$\int_a^b f(x)\, dx = \lim_{c\to a^+} \int_c^b f(x)\, dx$$

3. If $f$ is continuous on the interval $[a, b]$, except for some $c$ in $(a, b)$ at which $f$ approaches infinity, then

$$\int_a^b f(x)\, dx = \int_a^c f(x)\, dx + \int_c^b f(x)\, dx$$

**EXAMPLE 4**

**A Convergent Improper Integral**

Evaluate the improper integral

$$\int_1^2 \frac{dx}{\sqrt[3]{x-1}}$$

**SOLUTION**

Since the integrand approaches infinity at $x = 1$, we write

$$\int_1^2 \frac{dx}{\sqrt[3]{x-1}} = \lim_{b \to 1^+} \int_b^2 \frac{dx}{\sqrt[3]{x-1}}$$

$$= \lim_{b \to 1^+} \left[ \frac{3}{2}(x-1)^{2/3} \right]_b^2$$

$$= \lim_{b \to 1^+} \left[ \frac{3}{2} - \frac{3}{2}(b-1)^{2/3} \right]$$

$$= \frac{3}{2} - 0 = \frac{3}{2}$$

and the integral converges. (See Figure 6.32.)

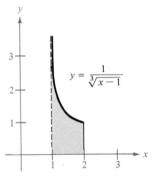

$$y = \frac{1}{\sqrt[3]{x-1}}$$

FIGURE 6.32

**EXAMPLE 5**

**A Divergent Improper Integral**

Evaluate the improper integral

$$\int_1^2 \frac{2\,dx}{x^2 - 2x}$$

**SOLUTION**

Separating the integrand into its partial fractions, we obtain

$$\int \frac{2\,dx}{x^2 - 2x} = \int \left( \frac{1}{x-2} - \frac{1}{x} \right) dx$$

Now we evaluate the improper integral as follows:

$$\int_1^2 \frac{2\,dx}{x^2 - 2x} = \int_1^2 \left( \frac{1}{x-2} - \frac{1}{x} \right) dx$$

$$= \lim_{b \to 2^-} \int_1^b \left( \frac{1}{x-2} - \frac{1}{x} \right) dx$$

$$= \lim_{b \to 2^-} \left[ \ln|x-2| - \ln|x| \right]_1^b$$

$$= -\infty - 0 = -\infty$$

Therefore, we conclude that the integral diverges. (See Figure 6.33.)

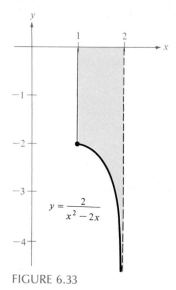

$$y = \frac{2}{x^2 - 2x}$$

FIGURE 6.33

**EXAMPLE 6**

**A Divergent Improper Integral**

Evaluate the integral

$$\int_0^2 \frac{1}{x^3} \, dx$$

**SOLUTION**

Since the integrand has an infinite discontinuity at $x = 0$, we write

$$\int_0^2 \frac{1}{x^3} \, dx = \lim_{b \to 0^+} \int_b^2 \frac{1}{x^3} \, dx$$

$$= \lim_{b \to 0^+} \left[ -\frac{1}{2x^2} \right]_b^2$$

$$= \lim_{b \to 0^+} \left[ -\frac{1}{8} + \frac{1}{2b^2} \right]$$

$$= -\frac{1}{8} + \infty = \infty$$

and conclude that the integral diverges.

**EXAMPLE 7**

**An Improper Integral with an Interior Discontinuity**

Evaluate the integral

$$\int_{-1}^{2} \frac{1}{x^3} \, dx$$

**SOLUTION**

This integral is improper because the integrand has an infinite discontinuity at the interior point $x = 0$, as shown in Figure 6.34. Thus, we write

$$\int_{-1}^{2} \frac{1}{x^3} \, dx = \int_{-1}^{0} \frac{1}{x^3} \, dx + \int_{0}^{2} \frac{1}{x^3} \, dx$$

From Example 6 we know that the second integral diverges. Therefore, the original improper integral also diverges.

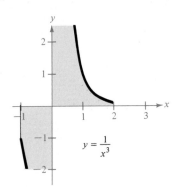

FIGURE 6.34

▬▬ **Remark:** Had we not recognized that the integral in Example 7 was improper, we would have obtained the incorrect result

$$\int_{-1}^{2} \frac{1}{x^3} \, dx = \frac{-1}{2x^2} \Big]_{-1}^{2} = -\frac{1}{8} + \frac{1}{2} = \frac{3}{8} \qquad \text{Incorrect}$$

Improper integrals in which the integrand approaches infinity at some point *between* the limits of integration are quite often overlooked, so keep alert for such possibilities.

## SECTION EXERCISES 6.6 ▬▬▬▬▬▬▬▬▬▬▬▬▬▬▬▬▬▬▬▬▬

In Exercises 1–24, determine the divergence or convergence of the given improper integral. Evaluate the integral if it converges. Use the table of integrals when necessary.

**1.** $\int_{0}^{4} \frac{1}{\sqrt{x}} \, dx$

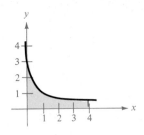

**2.** $\int_{3}^{4} \frac{1}{\sqrt{x-3}} \, dx$

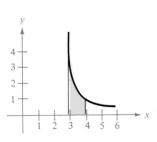

**3.** $\displaystyle\int_0^2 \frac{1}{(x-1)^{2/3}}\,dx$

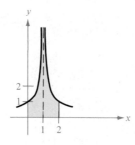

**4.** $\displaystyle\int_0^2 \frac{1}{(x-1)^2}\,dx$

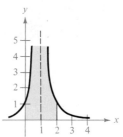

**5.** $\displaystyle\int_0^\infty e^{-x}\,dx$

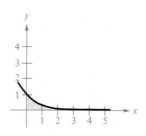

**6.** $\displaystyle\int_{-\infty}^0 e^{2x}\,dx$

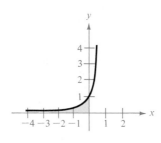

**7.** $\displaystyle\int_0^1 \frac{1}{1-x}\,dx$

**8.** $\displaystyle\int_0^{27} \frac{5}{\sqrt[3]{x}}\,dx$

**9.** $\displaystyle\int_0^8 \frac{1}{\sqrt[3]{8-x}}\,dx$

**10.** $\displaystyle\int_0^2 \frac{x}{\sqrt{4-x^2}}\,dx$

**11.** $\displaystyle\int_0^1 \frac{1}{x^2}\,dx$

**12.** $\displaystyle\int_0^1 \frac{1}{x}\,dx$

**13.** $\displaystyle\int_0^2 \frac{1}{\sqrt[3]{x-1}}\,dx$

**14.** $\displaystyle\int_0^2 \frac{1}{(x-1)^{4/3}}\,dx$

**15.** $\displaystyle\int_2^4 \frac{1}{\sqrt{x^2-4}}\,dx$

**16.** $\displaystyle\int_0^e \ln x\,dx$

**17.** $\displaystyle\int_1^\infty \frac{1}{x^2}\,dx$

**18.** $\displaystyle\int_1^\infty \frac{1}{\sqrt{x}}\,dx$

**19.** $\displaystyle\int_0^\infty e^{x/2}\,dx$

**20.** $\displaystyle\int_0^\infty \frac{5}{e^{2x}}\,dx$

**21.** $\displaystyle\int_{-\infty}^\infty x^2 e^{-x^3}\,dx$

**22.** $\displaystyle\int_0^\infty x^2 e^{-x^3}\,dx$

**23.** $\displaystyle\int_5^\infty \frac{x}{\sqrt{x^2-16}}\,dx$

**24.** $\displaystyle\int_{1/2}^\infty \frac{1}{\sqrt{2x-1}}\,dx$

In Exercises 25–28, evaluate the convergent improper integral by using the fact that

$$\lim_{x\to\infty} x^n e^{-ax} = 0, \quad a>0, \; n>0$$

**25.** $\displaystyle\int_0^\infty x^2 e^{-x}\,dx$

**26.** $\displaystyle\int_0^\infty (x-1)e^{-x}\,dx$

**27.** $\displaystyle\int_0^\infty xe^{-2x}\,dx$

**28.** $\displaystyle\int_0^\infty xe^{-x}\,dx$

In Exercises 29 and 30, (a) find the area of the region bounded by the graphs of the given equations and (b) find the volume of the solid generated by revolving the region about the $x$-axis.

**29.** $y = \dfrac{1}{x^2}$, $y = 0$, $x \geq 1$

**30.** $y = e^{-x}$, $y = 0$, $x \geq 0$

In Exercises 31 and 32, find the capitalized cost $C$ of an asset (a) for $n = 5$ years, (b) for $n = 10$ years, and (c) forever. The **capitalized cost** is given by

$$C = C_0 + \int_0^n c(t)e^{-rt}\,dt$$

where $C_0$ is the original investment, $t$ is the time in years, $r$ is the annual rate compounded continuously, and $c(t)$ is the annual cost of maintenance.

**31.** $C_0 = \$650{,}000$, $c(t) = \$25{,}000$, $r = 10\%$

**32.** $C_0 = \$650{,}000$, $c(t) = \$25{,}000(1 + .08t)$, $r = 12\%$ (See Exercise 33.)

**33.** Demonstrate that

$$\lim_{x\to\infty} x^n e^{-x} = 0$$

by completing the following table for
(a) $n = 1$    (b) $n = 2$    (c) $n = 5$

| $x$ | 1 | 10 | 25 | 50 |
|---|---|---|---|---|
| $x^n e^{-x}$ | | | | |

# Random Variables and Probability

In the study of probability problems are usually divided into two separate categories depending on whether the number of possible outcomes is finite or infinite. In the finite case (called discrete probability), one of the most common types of distributions of possible outcomes is the **binomial distribution.** The term binomial refers to situations involving *two* basic types of outcomes—for example, a coin may land head up or tail up, or a newborn infant may be a boy or a girl.

For example, suppose that a couple has 8 children. Assuming that it is equally likely that each child will be a girl or a boy, what is the likelihood that all 8 children will be girls? To answer this question we count the number of different outcomes that could occur. There is only one way in which all children could be girls. However, there are 8 ways to have 7 girls and 1 boy—that is, the first could be a boy and the other seven could be girls, or the

second could be a boy and the other seven girls, and so on. Using the binomial distribution, we obtain the following numbers of ways to obtain 8, 7, 6, 5, 4, 3, 2, 1, or 0 girls.

TABLE 6.5

| Number of girls | 8 | 7 | 6 | 5 | 4 | 3 | 2 | 1 | 0 |
|---|---|---|---|---|---|---|---|---|---|
| Number of ways this can occur | 1 | 8 | 28 | 56 | 70 | 56 | 28 | 8 | 1 |

The total number of ways of obtaining different combinations of girls and boys out of 8 children is 256—the sum of the numbers in the second row of Table 6.5. Thus we can determine that the probability that a family with 8 children will have all girls is only 1 out of 256. That is,

$$P(8) = \frac{1}{256}$$

The probability distribution for all 9 possibilities is shown in Figure 6.35.

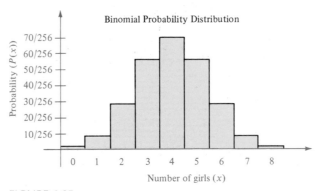

FIGURE 6.35

- **Discrete Random Variable**
- **Continuous Random Variable**
- **Probability**

In everyday life we often use ambiguous terminology such as "fairly certain," "probable," or "highly unlikely" to assign measurements to uncertainties. In mathematics we attempt to remove this ambiguity by assigning a number to the likelihood of the occurrence of an event. We call this numerical measurement the *probability* that the event will occur. For example, if we toss a fair coin, we say the probability that it will land head up is $\frac{1}{2}$.

In the study of probability we call any happening whose result is uncertain an *experiment*. The collection of all possible outcomes of an experiment is called the **sample space,** and a subcollection of the sample space is called an **event.** For example, if three coins are tossed, the sample space can be represented as follows:

$$S = \{HHH, HHT, HTH, HTT, THH, THT, TTH, TTT\}$$

Often it is convenient to assign a numerical value to each outcome in a sample space. This can be done in a variety of ways. For instance, in the sample space listed above, the numerical value $x$ of each outcome could be the number of heads:

$$S = \{HHH, HHT, HTH, HTT, THH, THT, TTH, TTT\}$$

|   |   |   |   |   |   |   |   |
|---|---|---|---|---|---|---|---|
| ↓ | ↓ | ↓ | ↓ | ↓ | ↓ | ↓ | ↓ |
| 3 | 2 | 2 | 1 | 2 | 1 | 1 | 0 |

A function that assigns a numerical value to each outcome in the sample space is called a **random variable.** When the values of the random variable consist of a finite set of numbers, we call the random variable **discrete.** The number of times the value $x$ occurs is called the **frequency** of $x$ and is denoted by $n(x)$. If each outcome in the sample space is *equally likely* (as it is in this example), then the **probability** of $x$ is given by

$$P(x) = \frac{n(x)}{n(S)}$$

where $n(S)$ is the number of outcomes in the sample space. The collection of all probabilities corresponding to the values of the random variable is called a **probability distribution.** For instance, the probability distribution for the random variable representing the number of heads that turn up when three coins are tossed is shown in Table 6.6.

TABLE 6.6

| Random variable, $x$ | 0 | 1 | 2 | 3 |
|---|---|---|---|---|
| Probability of $x$, $P(x)$ | $\frac{1}{8}$ | $\frac{3}{8}$ | $\frac{3}{8}$ | $\frac{1}{8}$ |

There are two important observations we can make about the probability distribution shown in Table 6.6. First, the probability of each value of the random variable is a number between 0 and 1. Second, the sum of the probabilities of the values of the random variable is 1. That is,

$$P(0) + P(1) + P(2) + P(3) = \frac{1}{8} + \frac{3}{8} + \frac{3}{8} + \frac{1}{8} = 1$$

These two properties are true of probability distributions for any discrete random variable.

| Probability distribution for a discrete random variable | If the range of a discrete random variable consists of $m$ different values $\{x_1, x_2, x_3, \ldots, x_m\}$, then the following properties are true. <br><br> 1. $0 \le P(x_i) \le 1$, $i = 1, 2, 3, \ldots, m$ <br> 2. $P(x_1) + P(x_2) + P(x_3) + \cdots + P(x_m) = 1$ |
|---|---|

**EXAMPLE 1**

**Finding a Probability Distribution**

Five coins are tossed. Graph the probability distribution for the random variable giving the number of heads that turn up.

**SOLUTION**

Since 0, 1, 2, 3, 4, or 5 heads can turn up, the values of this random variable correspond to the following events:

| $x$ | Event | $n(x)$ |
|---|---|---|
| 0 | TTTTT | 1 |
| 1 | HTTTT, THTTT, TTHTT, TTTHT, TTTTH | 5 |
| 2 | HHTTT, HTHTT, HTTHT, HTTTH, THHTT <br> THTHT, THTTH, TTHHT, TTHTH, TTTHH | 10 |
| 3 | HHHTT, HHTHT, HHTTH, HTHHT, HTHTH <br> HTTHH, THHHT, THHTH, THTHH, TTHHH | 10 |
| 4 | HHHHT, HHHTH, HHTHH, HTHHH, THHHH | 5 |
| 5 | HHHHH | 1 |

The number of outcomes in the sample space is $n(S) = 32$. Therefore the probability of each value of the random variable is as shown in Table 6.7.

TABLE 6.7

| Random variable, $x$ | 0 | 1 | 2 | 3 | 4 | 5 |
|---|---|---|---|---|---|---|
| Probability, $P(x)$ | $\frac{1}{32}$ | $\frac{5}{32}$ | $\frac{10}{32}$ | $\frac{10}{32}$ | $\frac{5}{32}$ | $\frac{1}{32}$ |

A graph of this probability distribution is shown in Figure 6.36.

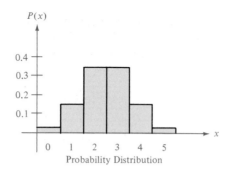

FIGURE 6.36      Probability Distribution

■■■ **Remark:** Note in Figure 6.36 that we represent each value of the random variable by an *interval* on the *x*-axis.

## Continuous Random Variables

The random variable in Example 1 is discrete (since its range consists of a finite set of values). In many applications of probability, it is useful to consider a random variable whose range consists of an interval on the real line. We call such a random variable **continuous.** For example, the random variable measuring the height of a person in a population is continuous.

---

**Definition of continuous random variable**

A random variable whose range is an interval $[a, b]$ on the real number line is called **continuous.**

---

■■■ **Remark:** The interval on the real line may be open or closed, bounded or unbounded. For instance, the interval could be $(0, \infty)$ or $(-\infty, \infty)$.

To define the probability of an event involving a continuous random variable, we cannot simply count the number of ways the event can occur (as we can with a discrete random variable). Rather, we introduce a new function $f$ called a **probability density function.** This function must be nonnegative and have the property that the area of the region bounded by the graph of $f$ and the *x*-axis, $a \le x \le b$, is 1.

---

**Probability density function**

A function $f$ is called a **probability density function** for a continuous random variable with a range of $a \le x \le b$ if the following properties are true:

$$1.\ f(x) \ge 0 \text{ for all } x \quad \text{and} \quad 2.\ \int_a^b f(x)\, dx = 1$$

The probability that $x$ lies in the interval $[c, d]$ is given by

$$P(c \le x \le d) = \int_c^d f(x)\, dx$$

---

This definition is illustrated graphically in Figure 6.37.

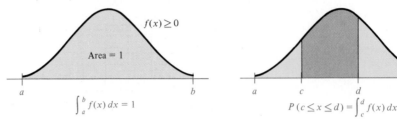

FIGURE 6.37

EXAMPLE 2

**Verifying Probability Density Functions**

Show that each of the following functions is a probability density function over the interval $[0, 1]$:

(a) $f(x) = 12x(1 - x)^2$

(b) $f(x) = \dfrac{105}{16}x^2(1 - x)^{1/2}$

**SOLUTION**

We begin by observing that both functions are nonnegative in the interval $[0, 1]$. Therefore, we can conclude that they are probability density functions by showing that

$$\int_0^1 f(x)\, dx = 1$$

(a) Expanding to polynomial form, we can evaluate this integral as follows:

$$\int_0^1 12x(1 - x)^2\, dx = 12\int_0^1 (x^3 - 2x^2 + x)\, dx$$

$$= 12\left[\frac{x^4}{4} - \frac{2x^3}{3} + \frac{x^2}{2}\right]_0^1$$

$$= 12\left[\frac{1}{4} - \frac{2}{3} + \frac{1}{2}\right]$$

$$= 12\left(\frac{1}{12}\right) = 1$$

(b) Using the substitution $u = 1 - x$, we have $dx = -du$ and we write

$$\int_0^1 \frac{105}{16}x^2(1 - x)^{1/2} = -\frac{105}{16}\int_1^0 (1 - u)^2 u^{1/2}\, du$$

$$= \frac{105}{16}\int_0^1 (u^{5/2} - 2u^{3/2} + u^{1/2})\, du$$

$$= \frac{105}{16}\left[\frac{u^{7/2}}{7/2} - \frac{2u^{5/2}}{5/2} + \frac{u^{3/2}}{3/2}\right]_0^1$$

$$= \frac{105}{16}\left[\frac{2}{7} - \frac{4}{5} + \frac{2}{3}\right]$$

$$= \frac{105}{16}\left(\frac{16}{105}\right) = 1$$

The graphs of both functions are shown in Figure 6.38.

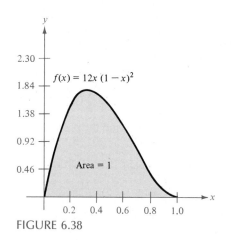

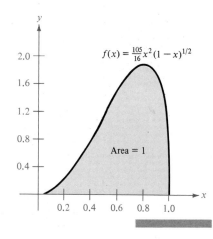

FIGURE 6.38

## EXAMPLE 3

### Finding Probability

For the probability density function given in part (a) of Example 2,

$$f(x) = 12x(1 - x)^2$$

find the probability that $x$ lies in the interval $0.5 \le x \le 0.75$.

**SOLUTION**

From the definition of probability we have

$$P(0.5 \le x \le 0.75) = 12\int_{0.5}^{0.75} x(1 - x)^2 \, dx$$

$$= 12\int_{0.5}^{0.75} (x^3 - 2x^2 + x) \, dx$$

$$= 12\left[\frac{x^4}{4} - \frac{2x^3}{3} + \frac{x^2}{2}\right]_{0.5}^{0.75}$$

$$= 12\left[\frac{0.75^4}{4} - \frac{2(0.75^3)}{3} + \frac{0.75^2}{2} - \frac{0.5^4}{4} + \frac{2(0.5^3)}{3} - \frac{0.5^2}{2}\right]$$

$$\approx 0.262$$

Therefore, the probability that $x$ lies in the interval $[0.5, 0.75]$ is approximately 26.2%, as indicated in Figure 6.39.

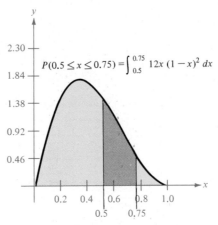

FIGURE 6.39

Note that had we been asked to find the probability that $x$ lies in the any of the intervals

$$0.5 < x < 0.75, \qquad 0.5 \le x < 0.75, \qquad 0.5 < x \le 0.75$$

we would have obtained the same solution as that given in Example 3, since the omission of either endpoint adds nothing to the probability. This demonstrates an important difference between discrete and continuous random variables. For a continuous random variable the probability that $x$ will be precisely one value (such as 0.5) is considered to be zero, since

$$P(0.5 \le x \le 0.5) = \int_{0.5}^{0.5} f(x) \, dx = 0$$

You should not interpret this result to mean that it is impossible for a continuous random variable $x$ to have the value 0.5. It simply means that the probability that $x$ will have this exact value is insignificant.

## Applications

**EXAMPLE 4**

**The Useable Lifetime of a Product**

The useable lifetime (in years) of a certain product is estimated by the following probability density function:

$$f(t) = \frac{1}{10} e^{-t/10}, \quad 0 \le t$$

Find the probability that a randomly selected unit will have a lifetime falling in the following intervals:
(a) Less than 2 years
(b) More than 2 years but less than 4 years
(c) More than 4 years

**SOLUTION**

(a) The probability that the unit will last less than 2 years is given by the integral

$$P(0 \leq t \leq 2) = \frac{1}{10} \int_0^2 e^{-t/10} \, dt$$

$$= \left[ -e^{-t/10} \right]_0^2$$

$$= -e^{-0.2} + 1 \approx 0.181$$

(b) The probability that the unit will last more than 2 years but less than 4 years is given by the integral

$$P(2 \leq t \leq 4) = \frac{1}{10} \int_2^4 e^{-t/10} \, dt$$

$$= \left[ -e^{-t/10} \right]_2^4$$

$$= -e^{-0.4} + e^{-0.2} \approx 0.148$$

(c) The probability that the unit will last more than 4 years is given by the *improper* integral

$$P(4 \leq t < \infty) = \frac{1}{10} \int_4^\infty e^{-t/10} \, dt$$

$$= \lim_{b \to \infty} \left[ -e^{-t/10} \right]_4^b$$

$$= \lim_{b \to \infty} \left[ -e^{-b/10} + e^{-0.4} \right]$$

$$= e^{-0.4} \approx 0.670$$

These three probabilities are illustrated graphically in Figure 6.40.

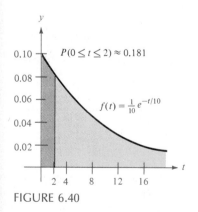

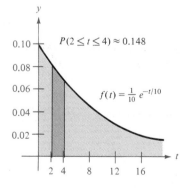

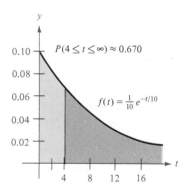

FIGURE 6.40

## EXAMPLE 5

### The Weekly Demand for a Product

The weekly demand for a certain product is estimated by the probability density function

$$f(x) = \frac{1}{36}(-x^2 + 6x), \quad 0 \le x \le 6$$

where $x$ is measured in thousands of units. Find the probability that the sales for a randomly chosen week will be between 2000 and 4000 units.

**SOLUTION**

The probability is given by the integral

$$P(2 \le x \le 4) = \frac{1}{36}\int_2^4 (-x^2 + 6x)\, dx$$

$$= \frac{1}{36}\left[ -\frac{x^3}{3} + 3x^2 \right]_2^4$$

$$= \frac{1}{36}\left[ -\frac{64}{3} + 48 + \frac{8}{3} - 12 \right]$$

$$= \frac{13}{27} \approx 0.481$$

Thus, the probability that the weekly sales will be between 2000 and 4000 units is approximately 48.1%, as indicated in Figure 6.41.

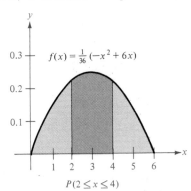

FIGURE 6.41       $P(2 \le x \le 4)$

## SECTION EXERCISES 6.7

1. Consider a couple that has 4 children. Assume that it is equally likely that each child will be a girl or a boy.

   (a) Complete the following set to form the sample space consisting of 16 elements:

   $$S = \{gggg, gggb, ggbg, \ldots\}$$

   (b) Complete the following table to form the probability distribution if the random variable $x$ is the number of girls in the family.

| Random variable, $x$ | 0 | 1 | 2 | 3 | 4 |
|---|---|---|---|---|---|
| Probability of $x$, $P(x)$ | | | | | |

   (c) Use the table of part (b) to graph the probability distribution.

   (d) Use the table of part (b) to find the probability of having at least 1 boy.

2. Consider the experiment of tossing a 6-sided die twice.

(a) Complete the following set to form the sample space of 36 elements. Note that each element is an ordered pair in which the entries are the number of points on the first and second toss, respectively.

$$S = \{(1, 1), (1, 2), \ldots, (2, 1), (2, 2), \ldots\}$$

(b) Complete the following table to form the probability distribution if the random variable $x$ is the sum of the number of points.

| Random variable, $x$ | 2 | 3 | 4 | 5 | 6 | 7 | 8 | 9 | 10 | 11 | 12 |
|---|---|---|---|---|---|---|---|---|---|---|---|
| Probability of $x$, $P(x)$ | | | | | | | | | | | |

(c) Use the table of part (b) to graph the probability distribution.

(d) Use the table of part (b) to find $P(5 \le x \le 9)$.

In Exercises 3–8, verify that the function $f$ is a probability density function over the given interval.

3. $f(x) = 6x(1 - x)$, $[0, 1]$

4. $f(x) = 12x^2(1 - x)$, $[0, 1]$

5. $f(x) = \dfrac{4}{27}x^2(3 - x)$, $[0, 3]$

6. $f(x) = \dfrac{2}{9}x(3 - x)$, $[0, 3]$

7. $f(x) = \dfrac{1}{3}e^{-x/3}$, $[0, \infty)$

8. $f(x) = \dfrac{1}{4}$, $[8, 12]$

In Exercises 9–14, find the constant $k$ so that the function $f$ is a probability density function over the given interval.

9. $f(x) = kx$, $[1, 5]$

10. $f(x) = kx^3$, $[0, 4]$

11. $f(x) = k(4 - x^2)$, $[-2, 2]$

12. $f(x) = k\sqrt{x}(1 - x)$, $[0, 1]$

13. $f(x) = ke^{-x/2}$, $[0, \infty)$

14. $f(x) = \dfrac{k}{b - a}$, $[a, b]$

In Exercises 15–20, sketch the graph of the given probability density function over the indicated interval and find the indicated probabilities.

15. $f(x) = \dfrac{1}{10}$, $[0, 10]$

   (a) $P(0 < x < 6)$         (b) $P(4 < x < 6)$
   (c) $P(8 < x < 10)$        (d) $P(x \ge 2)$

16. $f(x) = \dfrac{x}{50}$, $[0, 10]$

   (a) $P(0 < x < 6)$         (b) $P(4 < x < 6)$
   (c) $P(8 < x < 10)$        (d) $P(x \ge 2)$

17. $f(x) = \dfrac{3}{16}\sqrt{x}$, $[0, 4]$

   (a) $P(0 < x < 2)$         (b) $P(2 < x < 4)$
   (c) $P(1 < x < 3)$         (d) $P(x \le 3)$

18. $f(x) = \dfrac{4}{5(x + 1)^2}$, $[0, 4]$

   (a) $P(0 < x < 2)$         (b) $P(2 < x < 4)$
   (c) $P(1 < x < 3)$         (d) $P(x \le 3)$

19. $f(t) = \dfrac{1}{2}e^{-t/2}$, $[0, \infty)$

   (a) $P(t < 2)$             (b) $P(t \ge 2)$
   (c) $P(1 < t < 4)$         (d) $P(t = 3)$

20. $f(t) = \dfrac{3}{256}(16 - t^2)$, $[-4, 4]$

   (a) $P(t < -2)$            (b) $P(t > 2)$
   (c) $P(-1 < t < 1)$        (d) $P(t > -2)$

21. Buses arrive and depart from a college every 20 minutes. The probability density function for the waiting time $t$ (in minutes) for a person arriving at random at the bus stop is

$$f(t) = \dfrac{1}{20}, \quad [0, 20]$$

Find the probability that the person will wait
(a) no more than 5 minutes
(b) at least 12 minutes

22. The time $t$ (in hours) required for a new employee to successfully learn to operate a machine in a manufacturing process is described by the probability density function

$$f(t) = \dfrac{2}{81}t\sqrt{9 - t}, \quad [0, 9]$$

Find the probability that a given new employee will learn to operate the machine
(a) in less than 3 hours
(b) in more than 4 hours but less than 8 hours

In Exercises 23–26, find the required probabilities using the **exponential density function**

$$f(t) = \frac{1}{\lambda} e^{-t/\lambda}, \quad [0, \infty)$$

**23.** The waiting time (in minutes) for service at the checkout at a certain grocery store is exponentially distributed with $\lambda = 3$. Find the probability of waiting
(a) less than 2 minutes
(b) more than 2 minutes but less than 4 minutes
(c) at least 2 minutes
[Note that the last probability can be obtained from an improper integral or from part (a), since $P(t \geq 2) = 1 - P(t < 2)$.]

**24.** The lifetime (in years) of a battery is exponentially distributed with $\lambda = 5$. Find the probability that the lifetime of a given battery will be
(a) less than 6 years
(b) more than 2 years but less than 6 years
(c) more than 8 years

**25.** The length of time (in hours) required to unload trucks at a depot is exponentially distributed with $\lambda = \frac{3}{4}$. What proportion of the trucks can be unloaded in less than 1 hour?

**26.** The time (in years) until failure of a component in a machine is exponentially distributed with $\lambda = 3.5$. A certain manufacturer has a large number of these machines and plans to replace the components in all the machines during regularly scheduled maintenance periods. How much time should elapse between maintenance periods if at least 90% of the components are to remain working throughout the period?

**27.** The weekly demand $x$ (in tons) for a certain product is a continuous random variable with the density function

$$f(x) = \frac{1}{25} x e^{-x/5}, \quad [0, \infty)$$

Find
(a) $P(x < 5)$    (b) $P(5 < x < 10)$
(c) $P(x > 10) = 1 - P(x \leq 10)$

**28.** Given the conditions of Exercise 27, determine the number of tons that should be ordered each week so that the demand can be met for 90% of the weeks.

**29.** The probability density function for the percentage of recall in a certain learning experiment is found to be

$$f(x) = \frac{15}{4} x \sqrt{1 - x}, \quad [0, 1]$$

What is the probability that a randomly chosen individual in the experiment will recall between
(a) 0% and 25% of the material
(b) 50% and 75% of the material

**30.** The probability density function for the percentage of iron in ore samples taken from a certain region is

$$f(x) = \frac{1155}{32} x^3 (1 - x)^{3/2}, \quad [0, 1]$$

Find the probability that a sample will contain between
(a) 0% and 25% iron    (b) 50% and 100% iron

# Expected Value, Standard Deviation, and Median

In Example 4 in Section 6.7, we used the exponential probability density function

$$f(t) = \frac{1}{10}e^{-t/10}, \quad 0 \le t < \infty$$

to represent the useable lifetime of a product. In that example, we found the probability that a randomly chosen unit would last between 0 and 2 years, between 2 and 4 years, and more than 4 years.

We now look at a different question. If you purchased one of these products, how long would you expect it to last? One way to answer this question would be to sample several units and find their average lifetime.* We could do this by finding the usual arithmetic average—

that is, we could divide the sum of the lifetimes by the number of units in the sample to obtain

$$\frac{\text{Average lifetime}}{\text{of sample}} = \frac{t_1 + t_2 + t_3 + \cdots + t_n}{n}$$

From the density function shown in Figure 6.42 we would expect most of the lifetimes in this sample to be 20 years or less, and we would expect very few (if any) lifetimes to be more than 50 years.

As we increase the size of our sample, we find that the average lifetime of the sample approaches a value that we call the **expected value** or the **mean.** It is possible to show that this limiting value is given by the integral

$$\text{Expected value} = \int_0^\infty t\, f(t)\, dt$$

$$= \int_0^\infty t\left(\frac{1}{10}e^{-t/10}\right) dt = 10$$

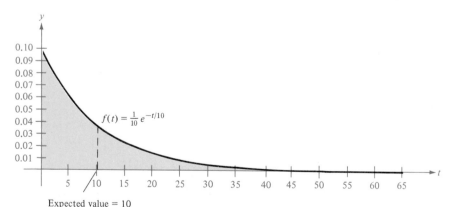

Expected value = 10

FIGURE 6.42

---

*Another way to answer this question is to use the concept of median as discussed in Example 5 in this section.

■ **Expected Value**
■ **Variance**
■ **Standard Deviation**
■ **Median**

In Example 5 of the previous section, we looked at the probability density function

$$f(x) = \frac{1}{36}(-x^2 + 6x), \quad 0 \le x \le 6$$

where $x$ is measured in thousands of units. We used this function as a model to predict the weekly demand for a product. From this model how large would you expect the demand to be for a typical week? One way to answer this question is to use the concept of **expected value,** which is defined as follows.

| | |
|---|---|
| **Definition of expected value** | If $f$ is a probability density function for a continuous random variable $x$ over the interval $[a, b]$, then the **expected value** of $x$ is given by $$\mu = E(x) = \int_a^b xf(x)\, dx$$ ($\mu$ is the lowercase Greek letter mu.) |

▬▬ **Remark:** The expected value of $x$ is also called the **mean** of $x$.

**EXAMPLE 1**
▬▬▬▬

**Finding the Expected Value**

Find the expected value of the probability density function

$$f(x) = \frac{1}{36}(-x^2 + 6x), \quad 0 \le x \le 6$$

**SOLUTION**

Using the definition of expected value, we have

$$E(x) = \frac{1}{36}\int_0^6 x(-x^2 + 6x)\, dx$$

$$= \frac{1}{36}\int_0^6 (-x^3 + 6x^2)\, dx$$

$$= \frac{1}{36}\left[\frac{-x^4}{4} + 2x^3\right]_0^6$$

$$= \frac{1}{36}[-324 + 432]$$

$$= \frac{108}{36} = 3$$

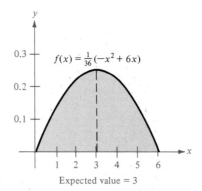

$f(x) = \frac{1}{36}(-x^2 + 6x)$

FIGURE 6.43

Expected value = 3

■■■■ **Remark:** Note from Figure 6.43 that an expected value of 3 seems reasonable since the region is symmetrical about the line $x = 3$.

One way to interpret the result of Example 1 is to say that if we sampled the demand for several randomly selected weeks,

$$x_1, x_2, x_3, \ldots, x_n$$

the average of these values would approximate the expected value. That is,

$$E(x) = \frac{x_1 + x_2 + x_3 + \cdots + x_n}{n} \approx 3$$

The expected value found in Example 1 happens to lie at the midpoint of the interval corresponding to the range of the continuous random variable. This occurs because the region lying below the probability density function is symmetrical (about the midpoint of the interval). In the next example, we look at the expected value for a nonsymmetrical region.

**EXAMPLE 2**

**Finding the Expected Value**

Find the expected value of the probability density function

$$f(x) = 2 - 2x, \quad 0 \le x \le 1$$

**SOLUTION**

The expected value is given by the integral

$$E(x) = \int_0^1 x(2 - 2x)\, dx$$

$$= \int_0^1 (2x - 2x^2)\, dx$$

$$= \left[ x^2 - \frac{2x^3}{3} \right]_0^1$$

$$= 1 - \frac{2}{3} = \frac{1}{3}$$

Therefore, the expected value is $\frac{1}{3}$, as indicated in Figure 6.44.

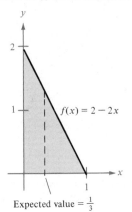

$f(x) = 2 - 2x$

FIGURE 6.44

Expected value $= \frac{1}{3}$

## Variance and Standard Deviation

An important question in the study of probability is ''How much does a random variable *vary* from the mean?'' For instance, each of the probability distribution functions shown in Figure 6.45 has a mean of 0 on the real line $(-\infty, \infty)$, and yet the ranges of the random variables differ considerably. As a measure of the amount that a continuous random variable varies (from the mean or expected value) we use the concepts of **variance** and **standard deviation.**

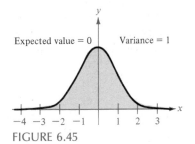

Expected value = 0     Variance = 1

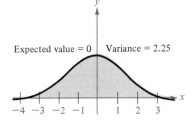

Expected value = 0     Variance = 2.25

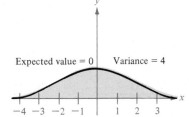

Expected value = 0     Variance = 4

FIGURE 6.45

| **Definition of variance and standard deviation** | If $f$ is a probability density function for a continuous random variable $x$ over the interval $[a, b]$, then the **variance** of $x$ is given by $$V(x) = \int_a^b (x - \mu)^2 f(x)\ dx$$ where $\mu$ is the expected value of $x$. Moreover, the **standard deviation** of $x$ is given by $$\sigma = \sqrt{V(x)}$$ ($\sigma$ is the lowercase Greek letter sigma.) |
| --- | --- |

**EXAMPLE 3**

**Finding Variance and Standard Deviation**

Find the variance and standard deviation for the probability density function

$$f(x) = 2 - 2x, \quad 0 \le x \le 1$$

**SOLUTION**

From Example 2 we know that the expected value is $\mu = \frac{1}{3}$. Therefore, the variance is given by

$$V(x) = \int_0^1 \left(x - \frac{1}{3}\right)^2 (2 - 2x)\, dx$$

$$= \int_0^1 \left(-2x^3 + \frac{10x^2}{3} - \frac{14x}{9} + \frac{2}{9}\right) dx$$

$$= \left[\frac{-x^4}{2} + \frac{10x^3}{9} - \frac{7x^2}{9} + \frac{2x}{9}\right]_0^1$$

$$= -\frac{1}{2} + \frac{10}{9} - \frac{7}{9} + \frac{2}{9}$$

$$= \frac{1}{18}$$

$$\approx 0.056$$

Taking the square root of this number, we find the standard deviation to be

$$\sigma = \sqrt{\frac{1}{18}}$$

$$= \frac{\sqrt{2}}{6}$$

$$\approx 0.236$$

The integral for variance is often difficult to evaluate. The following alternative formula for variance will usually help simplify the task.

---

**Alternative formula for variance**

If $f$ is a probability density function for a continuous random variable $x$ over the interval $[a, b]$, then the variance of $x$ is given by

$$V(x) = \int_a^b x^2 f(x)\, dx - \mu^2$$

where $\mu$ is the expected value of $x$.

---

**Proof**    Using properties of definite integrals, we have

$$V(x) = \int_a^b (x - \mu)^2 f(x)\, dx$$

$$= \int_a^b (x^2 - 2x\mu + \mu^2) f(x)\, dx$$

$$= \int_a^b [x^2 f(x) - 2x\mu f(x) + \mu^2 f(x)]\, dx$$

$$= \int_a^b x^2 f(x)\, dx - 2\mu \underbrace{\int_a^b x f(x)\, dx}_{\mu} + \mu^2 \underbrace{\int_a^b f(x)\, dx}_{1}$$

$$= \int_a^b x^2 f(x)\, dx - 2\mu^2 + \mu^2$$

$$= \int_a^b x^2 f(x)\, dx - \mu^2$$

**EXAMPLE 4**

**Finding Variance by the Alternative Formula**

Use the alternative formula to find the variance for the probability density function given in Example 2.

**SOLUTION**    Using an expected value of $\mu = \frac{1}{3}$, we have

$$V(x) = \int_0^1 x^2 (2 - 2x)\, dx - \left(\frac{1}{3}\right)^2$$

$$= \int_0^1 (2x^2 - 2x^3)\, dx - \frac{1}{9}$$

$$= \left[\frac{2x^3}{3} - \frac{x^4}{2}\right]_0^1 - \frac{1}{9}$$

$$= \frac{2}{3} - \frac{1}{2} - \frac{1}{9} = \frac{1}{18}$$

### Median

At the beginning of this section, we looked at the model

$$f(x) = \frac{1}{36}(-x^2 + 6x), \quad 0 \le x \le 6$$

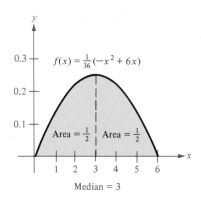

FIGURE 6.46

Median = 3

and asked the question "From this model how large would you expect the demand to be for a typical week?" We answered this question using the concept of *expected value* or *mean*. In the study of probability, we call the mean a **measure of central tendency.** Another useful measure of central tendency is the **median.** We define the median to be the number $m$ such that precisely half of the $x$-values lie below $m$ and the other half of the $x$-values lie above $m$. That is,

$$P(a \leq x \leq m) = 0.5$$

Because of the symmetry in the graph of the probability density function shown in Figure 6.46, it is easy to see that the median coincides with the expected value. That is, the median is $m = 3$. However, it is not always true that these two measures of central tendency are the same. This is demonstrated in the next example.

**EXAMPLE 5**

**Finding the Median**

In Example 4 of Section 6.7, we used the probability density function

$$f(t) = \frac{1}{10} e^{-t/10}, \quad 0 \leq t < \infty$$

to predict the useable lifetime of a product. It is possible to show that the expected value for this model is $\mu = 10$. Find the median.

**SOLUTION**

Using the definition of median, we have

$$P(0 \leq t \leq m) = \frac{1}{10} \int_0^m e^{-t/10} \, dt = \left[ -e^{-t/10} \right]_0^m = -e^{-m/10} + 1 = 0.5$$

Now, to solve for $m$, we write

$$-e^{-m/10} + 1 = 0.5$$

$$e^{-m/10} = 0.5$$

$$-\frac{m}{10} = \ln 0.5$$

$$m = -10 \ln 0.5 \approx 6.93 \text{ years}$$

This means that the median is approximately 6.93 years. In other words, using the median as a measure of central tendency we can predict that a typical unit will last for 6.93 years. (See Figure 6.47.)

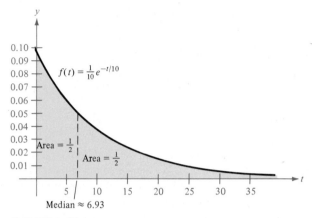

FIGURE 6.47

**EXAMPLE 6**

**Comparing the Expected Value and the Median**

Find the expected value and the median for the probability density function given by

$$f(x) = \frac{2}{x^3}, \quad 1 \leq x < \infty$$

**SOLUTION**

The expected value is given by

$$\mu = \int_1^\infty x\left(\frac{2}{x^3}\right) dx$$

$$= \int_1^\infty \frac{2}{x^2} dx$$

$$= \lim_{b \to \infty} \left[ -\frac{2}{x} \right]_1^b$$

$$= \lim_{b \to \infty} \left( -\frac{2}{b} + 2 \right)$$

$$= 2$$

We can find the median $m$ by solving for $m$ in the equation

$$\int_1^m \frac{2}{x^3} dx = \frac{1}{2}$$

$$\left[ -\frac{1}{x^2} \right]_1^m = \frac{1}{2}$$

$$-\frac{1}{m^2} + 1 = \frac{1}{2}$$

$$\frac{1}{2} = \frac{1}{m^2}$$

$$m^2 = 2$$

$$m = \sqrt{2} \approx 1.414$$

Figure 6.48 compares the expected value and median for this probability density function.

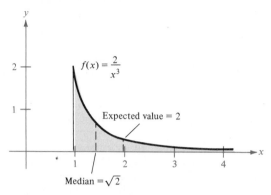

FIGURE 6.48

In Sections 6.7 and 6.8, we have given only a brief introduction to some of the important applications of calculus to the theory of probability. Unfortunately, some of the most useful probability density functions result in very complicated integrals whose evaluation is beyond the scope of this text. For instance, the most widely used probability density function is the **normal probability density function** described in the introductory examples for Sections 6.5 and 6.6. This function is given by

$$f(x) = \frac{1}{\sigma\sqrt{2\pi}} e^{-(x-\mu)^2/2\sigma^2}, \quad -\infty < x < \infty$$

This probability density function has the following properties:

$$\text{Expected value} = \mu$$

$$\text{Median} = \mu$$

$$\text{Variance} = \sigma^2$$

$$\text{Standard deviation} = \sigma$$

# SECTION EXERCISES 6.8

In Exercises 1–10, use the given probability density function over the indicated interval to find (a) the mean, (b) the variance, and (c) the median of the random variable. Locate the mean and median on the graph of the density function.

**1.** $f(x) = \dfrac{1}{8}$, $[0, 8]$

**2.** $f(x) = \dfrac{1}{4}$, $[0, 4]$

**3.** $f(t) = \dfrac{t}{32}$, $[0, 8]$

**4.** $f(x) = \dfrac{4}{3x^2}$, $[1, 4]$

**5.** $f(x) = 6x(1 - x)$, $[0, 1]$

**6.** $f(x) = \dfrac{3}{32}x(4 - x)$, $[0, 4]$

**7.** $f(x) = \dfrac{5}{2}x^{3/2}$, $[0, 1]$

**8.** $f(x) = \dfrac{3}{16}\sqrt{4 - x}$, $[0, 4]$

**9.** $f(x) = \dfrac{4}{3(x + 1)^2}$, $[0, 3]$

**10.** $f(x) = \dfrac{1}{18}\sqrt{9 - x}$, $[0, 9]$

In Exercises 11 and 12, find the median of the exponential probability density function.

**11.** $f(t) = \dfrac{1}{7}e^{-t/7}$, $[0, \infty)$

**12.** $f(t) = \dfrac{2}{5}e^{-2t/5}$, $[0, \infty)$

In Exercises 13 and 14, find the mean and standard deviation of the probability density function using the limit

$$\lim_{x \to \infty} x^n e^{-x} = 0$$

**13.** $f(x) = \dfrac{1}{\lambda}e^{-x/\lambda}$, $[0, \infty)$

(exponential probability density function)

**14.** $f(x) = \dfrac{1}{\lambda^2}xe^{-x/\lambda}$, $[0, \infty)$

In Exercises 15–20, use the result of Exercise 13.

**15.** Determine the mean, variance, and standard deviation of the probability density function

$$f(x) = \dfrac{1}{8}e^{-x/8}, \quad [0, \infty)$$

**16.** Determine the mean, variance, and standard deviation of the probability density function

$$f(x) = \dfrac{5}{3}e^{-5x/3}, \quad [0, \infty)$$

**17.** The time $t$ until failure of a machine component is exponentially distributed, with a mean of 4 years.
 (a) Find the probability density function for the random variable $t$.
 (b) Find the probability that a given component will fail in less than 3 years.

**18.** The time $t$ between arrivals of customers at a gasoline station is exponentially distributed, with a mean of 2 minutes.
 (a) Find the probability density function for the random variable $t$.
 (b) If a customer has just arrived, determine the probability that it will be more than 3 minutes before the next arrival.

**19.** The waiting time $t$ for service at a customer service desk in a department store is exponentially distributed, with a mean of 5 minutes.
 (a) Find the probability density function for the random variable $t$.
 (b) Find $P(\mu - \sigma < t < \mu + \sigma)$.

**20.** The service time $t$ for a customer at the service desk in a department store is exponentially distributed, with a mean of 3.5 minutes.
 (a) Find the probability density function for the random variable $t$.
 (b) Find $P(\mu - \sigma < t < \mu + \sigma)$.

In Exercises 21 and 22, use the result of Exercise 14.

**21.** The daily demand $x$ for a certain product (in tons) is a random variable with the probability density function

$$f(x) = \dfrac{1}{36}xe^{-x/6}, \quad [0, \infty)$$

 (a) Determine the expected daily demand.
 (b) Find $P(x \le 5)$.

**22.** The daily demand $x$ for water (in millions of gallons) in a certain town is a random variable with the probability density function

$$f(x) = \dfrac{1}{9}xe^{-x/3}, \quad [0, \infty)$$

 (a) Determine the expected value and the standard deviation of demand.

(b) Find the probability that demand is greater than 4 million gallons on a given day. [Note that $P(x > 4) = 1 - P(x < 4)$.]

**23.** The daily demand $x$ for a certain product (in hundreds of pounds) is a random variable with the probability density function

$$f(x) = \frac{1}{36} x(6 - x), \quad [0, 6]$$

(a) Determine the expected value and the standard deviation of demand.
(b) Determine the median of the random variable.
(c) Find $P(\mu - \sigma < x < \mu + \sigma)$.

**24.** Repeat Exercise 23 if the probability density function is given by

$$f(x) = \frac{3}{256}(x - 2)(10 - x), \quad [2, 10]$$

**25.** The percentage recall $x$ in a learning experiment is a random variable with the probability density function

$$f(x) = \frac{15}{4}x\sqrt{1 - x}, \quad [0, 1]$$

Determine the mean and variance of the random variable $x$.

**26.** The percentage of iron $x$ in samples of ore is a random variable with the probability density function

$$f(x) = \frac{1155}{32}x^3(1 - x)^{3/2}, \quad [0, 1]$$

Determine the expected percentage of iron in each ore sample.

In Exercises 27–30, use the normal probability density function

$$f(x) = \frac{1}{\sigma\sqrt{2\pi}}e^{-(x-\mu)^2/2\sigma^2}, \quad -\infty < x < \infty$$

**27.** Sketch the graph of the normal density function if $\mu = 0$ and $\sigma = 1$. Show the maximum and the points of inflection on the graph.

**28.** For the density function shown in Exercise 27, use integration to verify that the mean is zero.

**29.** The weekly profit $x$ (in thousands of dollars) of a given store is normally distributed with $\mu = 8$ and $\sigma = 3$. Use Simpson's Rule with $n = 6$ to approximate $P(8 < x < 11)$.

**30.** The IQs of students in a certain school are normally distributed, with mean 110 and standard deviation 10. Use Simpson's Rule with $n = 6$ to approximate the probability that a student selected at random will have an IQ within one standard deviation of the mean.

## Important Terms

Integration by substitution
Integration by parts
Partial fraction
Basic equation
Proper rational function
Integration by tables
Reduction formula
Completing the square
Trapezoidal Rule
Simpson's Rule
Improper integral
Convergence of an improper integral
Divergence of an improper integral

Sample space, $S$
Event
Random variable, $x$
Discrete random variable
Frequency, $n(x)$
Probability, $P(x)$
Probability distribution
Continuous random variable
Probability density function
Expected value (or mean), $E(x)$, $\mu$
Variance, $V(x)$
Standard deviation, $\sigma$
Median, $m$
Normal probability density function

## Important Techniques

Using $u$-substitution to evaluate an indefinite integral
Using $u$-substitution to evaluate a definite integral
Using integration by parts to evaluate an indefinite integral
Using the method of partial fractions to evaluate an indefinite integral
Using an integration table to evaluate an indefinite integral

Rewriting an indefinite integral by completing the square
Evaluating an improper integral
Finding the probability of an event for a discrete random variable
Finding the probability of an event for a continuous random variable
Finding the expected value, variance, standard deviation, and median

## Important Formulas

Constant Rule:

$$\int dx = x + C$$

Simple Power Rule:

$$\int x^n \, dx = \frac{x^{n+1}}{n+1} + C, \quad n \neq -1$$

General Power Rule:

$$\int u^n \frac{du}{dx} \, dx = \int u^n \, du = \frac{u^{n+1}}{n+1} + C, \quad n \neq -1$$

Simple Log Rule:

$$\int \frac{1}{x} \, dx = \ln |x| + C$$

General Log Rule:

$$\int \frac{1}{u} \frac{du}{dx} \, dx = \int \frac{1}{u} \, du = \ln |u| + C$$

Simple Exponential Rule:

$$\int e^x \, dx = e^x + C$$

General Exponential Rule:

$$\int e^u \frac{du}{dx} \, dx = \int e^u \, du = e^u + C$$

Integration by parts:

$$\int u \, dv = uv - \int v \, du$$

Trapezoidal Rule:

$$\int_a^b f(x) \, dx \approx \frac{b-a}{2n}[f(x_0) + 2f(x_1) + 2f(x_2) + \cdots + 2f(x_{n-1}) + f(x_n)]$$

Simpson's Rule ($n$ is even):

$$\int_a^b f(x) \, dx \approx \frac{b-a}{3n}[f(x_0) + 4f(x_1) + 2f(x_2) + 4f(x_3) + \cdots + 4f(x_{n-1}) + f(x_n)]$$

Error in Trapezoidal Rule:

$$|E| \leq \frac{(b-a)^3}{12n^2}[\max |f''(x)|], \quad a \leq x \leq b$$

Error in Simpson's Rule:

$$|E| \le \frac{(b-a)^5}{180n^4}[\max |f^{(4)}(x)|], \quad a \le x \le b$$

Improper integrals (infinite limits of integration):

1. If $f$ is continuous on the interval $[a, \infty)$, then

$$\int_a^\infty f(x)\, dx = \lim_{b \to \infty} \int_a^b f(x)\, dx$$

2. If $f$ is continuous on the interval $(-\infty, b]$, then

$$\int_{-\infty}^b f(x)\, dx = \lim_{a \to -\infty} \int_a^b f(x)\, dx$$

3. If $f$ is continuous on the interval $(-\infty, \infty)$, then

$$\int_{-\infty}^\infty f(x)\, dx = \int_{-\infty}^c f(x)\, dx + \int_c^\infty f(x)\, dx$$

   where $c$ is any real number.

Improper integrals (infinite integrands):

1. If $f$ is continuous on the interval $[a, b)$ and approaches infinity at $b$, then

$$\int_a^b f(x)\, dx = \lim_{c \to b^-} \int_a^c f(x)\, dx$$

2. If $f$ is continuous on the interval $(a, b]$ and approaches infinity at $a$, then

$$\int_a^b f(x)\, dx = \lim_{c \to a^+} \int_c^b f(x)\, dx$$

3. If $f$ is continuous on the interval $[a, b]$, except for some $c$ in $(a, b)$ at which $f$ approaches infinity, then

$$\int_a^b f(x)\, dx = \int_a^c f(x)\, dx + \int_c^b f(x)\, dx$$

Discrete random variable:

$$P(x) = \frac{n(x)}{n(S)}$$

1. $0 \le P(x_i) \le 1,\ i = 1, 2, 3, \ldots, m$
2. $P(x_1) + P(x_2) + P(x_3) + \cdots + P(x_m) = 1$

Continuous random variable:

$$P(c \le x \le d) = \int_c^d f(x)\, dx$$

1. $f(x) \ge 0$ for all $x$ in $[a, b]$

2. $\displaystyle\int_a^b f(x)\, dx = 1$

$$\mu = E(x) = \int_a^b xf(x)\, dx$$

$$V(x) = \int_a^b (x - \mu)^2 f(x)\, dx = \int_a^b x^2 f(x)\, dx - \mu^2$$

$$\sigma = \sqrt{V(x)}$$

# REVIEW EXERCISES FOR CHAPTER 6

In Exercises 1–24, find the indefinite integral.

**1.** $\int x(2-x)^3 \, dx$

**2.** $\int x^2(1-x)^2 \, dx$

**3.** $\int x(x+1)^{3/2} \, dx$

**4.** $\int 2x\sqrt{x-3} \, dx$

**5.** $\int x\sqrt{x+3} \, dx$

**6.** $\int x^2\sqrt{1+x} \, dx$

**7.** $\int x^2\sqrt{1-x} \, dx$

**8.** $\int (y+1)\sqrt{1-y} \, dy$

**9.** $\int \dfrac{\sqrt{x}}{1+\sqrt{x}} \, dx$

**10.** $\int \dfrac{1}{\sqrt{x}(1+\sqrt{x})} \, dx$

**11.** $\int \dfrac{\ln 2x}{\sqrt{x}} \, dx$

**12.** $\int \dfrac{\ln x}{x} \, dx$

**13.** $\int (x-1)e^x \, dx$

**14.** $\int \dfrac{e^x}{1+e^x} \, dx$

**15.** $\int \sqrt{x}\ln x \, dx$

**16.** $\int \ln \dfrac{x}{x+1} \, dx$

**17.** $\int \dfrac{9}{x^2-9} \, dx$

**18.** $\int \dfrac{x^2}{x^2+2x-15} \, dx$

**19.** $\int \dfrac{1}{x(x+5)} \, dx$

**20.** $\int \dfrac{4x^2-x-5}{x^2(x+5)} \, dx$

**21.** $\int \dfrac{4x-2}{3(x-1)^2} \, dx$

**22.** $\int \dfrac{4x^3+4x}{(x^2+1)^2} \, dx$

**23.** $\int \dfrac{\sqrt{x^2+9}}{x} \, dx$ (Use tables)

**24.** $\int \dfrac{1}{(9-x^2)^{3/2}} \, dx$ (Use tables)

In Exercises 25–28, use the Trapezoidal Rule and Simpson's Rule to approximate the definite integral.

**25.** $\int_0^1 \dfrac{x^{3/2}}{2-x^2} \, dx, \; n=4$

**26.** $\int_0^2 \dfrac{1}{\sqrt{1+x^3}} \, dx, \; n=8$

**27.** $\int_1^2 \dfrac{1}{1+\ln x} \, dx, \; n=4$

**28.** $\int_0^1 e^{x^2} \, dx, \; n=6$

In Exercises 29–34, determine the convergence or divergence of the improper integral and evaluate those that converge.

**29.** $\int_0^{16} \dfrac{1}{\sqrt[4]{x}} \, dx$

**30.** $\int_0^1 \dfrac{6}{x-1} \, dx$

**31.** $\int_1^\infty x^2 \ln x \, dx$

**32.** $\int_1^\infty \dfrac{\ln x}{x} \, dx$

**33.** $\int_0^4 \dfrac{x}{\sqrt{16-x^2}} \, dx$

**34.** $\int_0^\infty \dfrac{e^{-1/x}}{x^2} \, dx$

In Exercises 35–38, find the area of the region enclosed by the graphs of the equations.

**35.** $y = x\sqrt{4-x}, \; y=0$

**36.** $y = x\sqrt{x+1}, \; y=0$

**37.** $y = \dfrac{1}{8}e^{-x/8}, \; y=0, \; x \geq 0$

**38.** $y = \dfrac{1}{25-x^2}, \; y=0, \; x=0, \; x=4$

In Exercises 39 and 40, find the volume of the solid generated by revolving the region bounded by the graphs of the given equations about the $x$-axis.

**39.** $y = \sqrt{x}(1-x)^2, \; y=0$

**40.** $y = xe^{-x}, \; y=0, \; x \geq 0$

In Exercises 41–44, find the constant $k$ so that the function $f$ is a probability density function over the given interval.

**41.** $f(x) = k(4-x), \; [0, 4]$

**42.** $f(x) = kx^2(2-x), \; [0, 2]$

**43.** $f(x) = \dfrac{k}{\sqrt{x}}, \; [1, 9]$

**44.** $f(x) = kx^{3/2}(4-x), \; [0, 4]$

In Exercises 45–48, find the mean, median, standard deviation, and required probability for the given probability density function.

**45.** $f(x) = \dfrac{1}{50}(10-x), \; [0, 10]$

    (a) $P(0 < x < 2)$

    (b) $P(8 < x < 10)$

**46.** $f(x) = \dfrac{1}{36}(9-x^2), \; [-3, 3]$

    (a) $P(-1 < x < 2)$

    (b) $P(-2 < x < 1)$

**47.** $f(x) = \dfrac{2}{(x+1)^2}, \; [0, 1]$

    (a) $P\left(0 < x < \dfrac{1}{2}\right)$

    (b) $P\left(\dfrac{1}{2} < x < 1\right)$

**48.** $f(x) = \dfrac{3}{128}\sqrt{x}$, $[0, 16]$

 (a) $P(4 < x < 9)$

 (b) $P(9 < x < 16)$

**49.** Find the median of the exponential probability density function

$$f(x) = \frac{1}{12}e^{-x/12}, \quad [0, \infty)$$

**50.** Find the median of the probability density function

$$f(x) = \frac{1}{4}xe^{-x/2}, \quad [0, \infty)$$

**51.** The time $t$ (in days) until recovery after a certain medical procedure is a random variable with the probability density function

$$f(t) = \frac{1}{2\sqrt{t-2}}, \quad [3, 6]$$

 (a) Find the probability that a patient selected at random will take more than 4 days for recovery.

 (b) Determine the expected time for recovery.

**52.** The daily demand $x$ (in hundreds of pounds) for a perishable product is a random variable with the probability density function

$$f(x) = \frac{20x}{9(x^2 + 1)^2}, \quad [0, 3]$$

Determine the number of pounds that should be in stock at the beginning of each day if management wants an adequate supply to meet demand on three-fourths of the days.

# Differential
# Equations

# CHAPTER 7

# Solutions of Differential Equations

Economists refer to the value of the total annual production of goods and services of a nation's economy as the *national income*. This national income is commonly assumed to be proportional to the total stock of *capital* (such as factories and equipment) in the country. Thus, we have

$$y = kx$$

where $y$ is the national income, $x$ is the capital, and $k$ is the proportionality constant. That portion of the national income that is fed back into the economy as new capital is called *capital investment*. The total capital investment is actually a measure of the rate of change of the capital, and we have

$$\text{Investment} = \frac{dx}{dt} = \frac{1}{k}\frac{dy}{dt}$$

That portion of the national income that is consumed is called the *national consumption*. Hence, income, investment, and consumption are related by the equation

Income = consumption + investment

One useful economic model for this relationship assumes that consumption is linearly related to income and that it has two parts, necessities and luxuries. A country must consume a certain amount to survive, and this model assumes that after the survival level is reached people are willing to spend a fixed percentage of their extra income on luxuries and save the rest. Thus, for an income of $y$, we have

Income = [necessities + luxuries] + investment

$$y = [a + b(y - a)] \qquad + \frac{1}{k}\frac{dy}{dt}$$

which yields the following **differential equation:**

$$y = a(1 - b) + by + \frac{1}{k}\frac{dy}{dt}$$

$$\frac{dy}{dt} = k(1 - b)(y - a)$$

The **solution** to this equation is

$$y = a + Ce^{k(1-b)t}$$

where $C$ represents the income above the necessity level at the time $t = 0$. Figure 7.1 shows a graph for national income in which the factor $b$ is 0.75. In other words, this model assumes people will consume 75% of their extra income on luxuries.

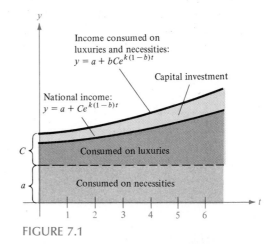

FIGURE 7.1

■ **Differential Equation**
■ **General Solution of a Differential Equation**
■ **Particular Solution of a Differential Equation**
■ **Initial Conditions for a Differential Equation**

A **differential equation** is an equation involving a function and one or more of its derivatives. For instance,

$$3\frac{dy}{dx} - 2xy = 0$$

is a differential equation in which $y = f(x)$ is a differentiable function of $x$.

A function $y = f(x)$ is called a **solution** of a given differential equation if the equation is satisfied when $y$ and its derivatives are replaced by $f(x)$ and its derivatives. For example,

$$y = e^{-2x}$$

is a solution to the differential equation

$$y' + 2y = 0$$

because $y' = -2e^{-2x}$, and by substitution we have

$$\overbrace{-2e^{-2x}}^{y'} + \overbrace{2e^{-2x}}^{2y} = 0$$

Furthermore, we can readily see that

$$y = 2e^{-2x}, \qquad y = 3e^{-2x}, \qquad y = \frac{1}{2}e^{-2x}$$

are also solutions to this same differential equation. In fact, the functions

$$y = Ce^{-2x}$$

where $C$ is any real number, are all solutions. We call this family of solutions the **general solution** of the differential equation $y' + 2y = 0$.

Recall from Example 7 in Section 4.1 that the differential equation

$$s''(t) = -32$$

has the general solution

$$s(t) = -16t^2 + C_1 t + C_2$$

Note that this general solution has *two* arbitrary constants. A differential equation usually has a general solution in which the number of arbitrary constants is the same as the order of the highest derivative appearing in the equation.

A **particular solution** of a differential equation is any solution that is obtained by assigning specific values to the constants in the general solution. (Occasionally a differential equation has other solutions, not obtainable from the general solution by

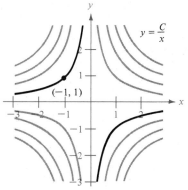

Solution curves for $xy' + y = 0$

FIGURE 7.2

assigning values to the arbitrary constants; such solutions are called *singular solutions*. In our brief discussion of differential equations, we will omit any further reference to singular solutions.)

Geometrically, the general solution of a given differential equation represents a family of curves, called **solution curves,** one for each value assigned to the arbitrary constants. For instance, we can easily verify that

$$y = \frac{C}{x}$$

is the general solution to the differential equation

$$xy' + y = 0$$

Figure 7.2 shows several of the solution curves corresponding to different values of $C$.

In practice, particular solutions of a differential equation are usually obtained from **initial conditions** placed on the unknown function and its derivatives. For instance, for Figure 7.2, suppose we want to obtain the particular solution whose curve passes through the point $(-1, 1)$. We have the initial condition

$$f(-1) = 1 \quad \text{or} \quad y = 1 \quad \text{when} \quad x = -1$$

Substituting this condition into the general solution yields

$$y = \frac{C}{x}$$

$$1 = \frac{C}{-1}$$

$$-1 = C$$

and the particular solution is

$$y = -\frac{1}{x}$$

**EXAMPLE 1**

**Checking Possible Solutions of a Differential Equation**

Determine whether or not the given functions are solutions of the differential equation

$$y'' - y = 0$$

(a) $y = e^x$            (b) $y = 4e^{-x}$
(c) $y = 0$            (d) $y = e^{2x}$
(e) $y = Ce^x$

**SOLUTION**

(a) Since $y = e^x$, $y' = e^x$, and $y'' = e^x$, we have

$$y'' - y = e^x - e^x = 0$$

Hence, $y = e^x$ *is* a solution.

(b) Since $y = 4e^{-x}$, $y' = -4e^{-x}$, and $y'' = 4e^{-x}$, we have

$$y'' - y = 4e^{-x} - 4e^{-x} = 0$$

Hence, $y = 4e^{-x}$ *is* a solution.

(c) Since $y = 0$, $y' = 0$, and $y'' = 0$, we have

$$y'' - y = 0 - 0 = 0$$

Hence, $y = 0$ *is* a solution.

(d) Since $y = e^{2x}$, $y' = 2e^{2x}$, and $y'' = 4e^{2x}$, we have

$$y'' - y = 4e^{2x} - e^{2x} = 3e^{2x} \neq 0$$

Hence, $y = e^{2x}$ *is not* a solution.

(e) Since $y = Ce^x$, $y' = Ce^x$, and $y'' = Ce^x$, we have

$$y'' - y = Ce^x - Ce^x = 0$$

Hence, $y = Ce^x$ *is* a solution for any value of $C$.

**EXAMPLE 2**

**Finding a Particular Solution of a Differential Equation**

Verify that $y = Cx^3$ is a solution of the differential equation

$$xy' - 3y = 0$$

for any value of $C$. Then find the particular solution determined by the initial condition $y = 2$ when $x = -3$.

**SOLUTION**

Since $y = Cx^3$ and $y' = 3Cx^2$, we have

$$xy' - 3y = x(3Cx^2) - 3(Cx^3) = 0$$

Hence, $y = Cx^3$ is a solution for any value of $C$. Furthermore, from the initial condition we determine that

$$y = Cx^3$$

$$2 = C(-3)^3$$

$$-\frac{2}{27} = C$$

Therefore, the particular solution is

$$y = -\frac{2}{27}x^3$$

**EXAMPLE 3**

**Finding a Particular Solution of a Differential Equation**

Given the general solution

$$s(t) = -16t^2 + C_1 t + C_2$$

of the differential equation

$$s''(t) = -32$$

find the particular solution satisfying the initial conditions

$$s(0) = 80 \quad \text{and} \quad s'(0) = 64$$

**SOLUTION**

From the initial condition $s(0) = 80$ we have

$$80 = -16(0)^2 + C_1(0) + C_2$$
$$80 = C_2$$

Now, differentiating the general solution, we have

$$s'(t) = -32t + C_1$$

and from the initial condition $s'(0) = 64$ we have

$$64 = -32(0) + C_1$$
$$64 = C_1$$

Finally, we conclude that the particular solution satisfying the given initial conditions is

$$s(t) = -16t^2 + 64t + 80$$

**SECTION EXERCISES 7.1**

In Exercises 1–20, verify that the given equation represents a solution of the differential equation for any value of C.

| Solution | Differential equation |
|----------|----------------------|
| **1.** $y = \dfrac{1}{x} + C$ | $\dfrac{dy}{dx} = -\dfrac{1}{x^2}$ |

| Solution | Differential equation |
|----------|----------------------|
| **2.** $y = \sqrt{4 - x^2} + C$ | $\dfrac{dy}{dx} = -\dfrac{x}{\sqrt{4 - x^2}}$ |
| **3.** $y = Ce^{4x}$ | $\dfrac{dy}{dx} = 4y$ |

| Solution | Differential equation |
|---|---|
| **4.** $y = Ce^{-2x}$ | $\dfrac{dy}{dx} = -2y$ |
| **5.** $y = Ce^{-t/2} + 5$ | $2\dfrac{dy}{dt} + y - 5 = 0$ |
| **6.** $y = Ce^{-t} + 10$ | $y' + y - 10 = 0$ |
| **7.** $y = Cx^2 - 3x$ | $xy' - 3x - 2y = 0$ |
| **8.** $y^2 + 2xy - x^2 = C$ | $(x + y)y' - x + y = 0$ |
| **9.** $y = x \ln x + Cx - 2$ | $x(y' - 1) - (y + 2) = 0$ |
| **10.** $y = x \ln x^2 + 2x^{3/2} + Cx$ | $y' - \dfrac{y}{x} = 2 + \sqrt{x}$ |
| **11.** $y = x^2 + 2x + \dfrac{C}{x}$ | $xy' + y = x(3x + 4)$ |
| **12.** $y = C_1 + C_2 e^x$ | $y'' - y' = 0$ |
| **13.** $y = C_1 e^{x/2} + C_2 e^{-2x}$ | $2y'' + 3y' - 2y = 0$ |
| **14.** $y = C_1 e^{4x} + C_2 e^{-x}$ | $y'' - 3y' - 4y = 0$ |
| **15.** $y = \dfrac{bx^4}{4 - a} + Cx^a$ | $y' - \dfrac{ay}{x} = bx^3$ |
| **16.** $y = \dfrac{x^3}{5} - x + C\sqrt{x}$ | $2xy' - y = x^3 - x$ |
| **17.** $x^2 + y^2 = Cy$ | $y' = \dfrac{2xy}{x^2 - y^2}$ |
| **18.** $y = Ce^{x-x^2}$ | $y' + (2x - 1)y = 0$ |
| **19.** $y = \dfrac{2}{1 + Ce^{x^2}}$ | $y' + 2xy = xy^2$ |
| **20.** $y = x(\ln x + C)$ | $x + y - xy' = 0$ |

In Exercises 21–24, determine whether the given function is a solution to the differential equation $y^{(4)} - 16y = 0$.

**21.** $y = e^{-2x}$      **22.** $y = 5 \ln x$

**23.** $y = \dfrac{4}{x}$      **24.** $y = 4e^{2x}$

In Exercises 25–32, verify that the given general solutions satisfy the differential equation. Then find the particular solution satisfying the given initial conditions.

**25.** $y = Ce^{-2x}$, $y' + 2y = 0$
    $y = 3$ when $x = 0$

**26.** $2x^2 + 3y^2 = C$, $2x + 3yy' = 0$
    $y = 2$ when $x = 1$

**27.** $y = C_1 + C_2 \ln |x|$, $xy'' + y' = 0$
    $y = 5$ and $y' = \dfrac{1}{2}$ when $x = 1$

**28.** $y = C_1 x + C_2 x^3$, $x^2 y'' - 3xy' + 3y = 0$
    $y = 0$ and $y' = 4$ when $x = 2$

**29.** $y = C_1 e^{6x} + C_2 e^{-5x}$, $y'' - y' - 30y = 0$
    $y = 0$ and $y' = -4$ when $x = 0$

**30.** $y = Ce^{x-x^2}$, $y' + (2x - 1)y = 0$
    $y = 2$ when $x = 1$

**31.** $y = e^{2x/3}(C_1 + C_2 x)$, $9y'' - 12y' + 4y = 0$
    $y = 4$ when $x = 0$ and $y = 0$ when $x = 3$

**32.** $y = \left(C_1 + C_2 x + \dfrac{1}{12}x^4\right)e^{2x}$, $y'' - 4y' + 4y = x^2 e^{2x}$
    $y = 2$ and $y' = 1$ when $x = 0$

In Exercises 33–36, some of the curves corresponding to different values of $C$ in the general solution to the differential equation are given. Find the particular solution that passes through the point indicated on each graph.

**33.** $y^2 = Cx^3$, $2xy' - 3y = 0$

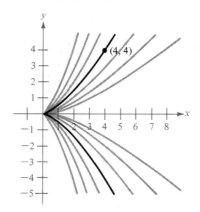

**34.** $2x^2 - y^2 = C$, $yy' - 2x = 0$

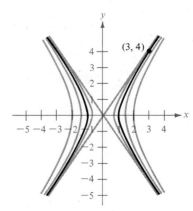

**35.** $y = Ce^x$, $y' - y = 0$

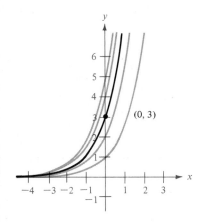

**36.** $y^2 = 2Cx$, $2xy' - y = 0$

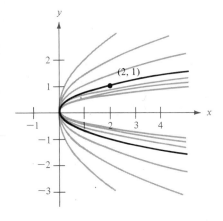

In Exercises 37–40, the general solution to the differential equation is given. Sketch the graph of the particular solutions for the given values of $C$.

**37.** $x^2 + y^2 = C$, $yy' + x = 0$
   $C = 0$, $C = 1$, $C = 4$
**38.** $4y^2 - x^2 = C$, $2yy' - x = 0$
   $C = 0$, $C = \pm1$, $C = \pm4$
**39.** $y = C(x + 2)^2$, $(x + 2)y' - 2y = 0$
   $C = 0$, $C = \pm1$, $C = \pm2$
**40.** $y = Ce^{-x}$, $y' + y = 0$
   $C = 0$, $C = \pm1$, $C = \pm2$

In Exercises 41–46, use integration to find a general solution to the differential equation.

**41.** $\dfrac{dy}{dx} = 3x^2$

**42.** $\dfrac{dy}{dx} = \dfrac{1}{1 + x}$

**43.** $\dfrac{dy}{dx} = \dfrac{x - 2}{x}$

**44.** $\dfrac{dy}{dx} = \dfrac{1}{x^2 - 1}$

**45.** $\dfrac{dy}{dx} = x\sqrt{x - 3}$

**46.** $\dfrac{dy}{dx} = xe^x$

**47.** The limiting capacity of the habitat for a particular wildlife herd is $L$. The growth rate $dN/dt$ of the herd is proportional to the unutilized opportunity for growth, as described by the differential equation

$$\frac{dN}{dt} = k(L - N)$$

The general solution to this differential equation is

$$N = L - Ce^{-kt}$$

Suppose 100 animals are released into a tract of land that can support 750 of these animals. After 2 years the herd has grown to 160 animals.
   (a) Find the population function in terms of the time $t$ in years.
   (b) Sketch the graph of this population function.

**48.** The rate of growth of an investment is proportional to the amount in the investment at any time $t$. That is,

$$\frac{dA}{dt} = kA$$

   (a) Show that $A = Ce^{kt}$ is a solution to this differential equation.
   (b) Find the particular solution to this differential equation if the initial investment is $1,000 and 10 years later the amount is $3,320.12.

# Separation of Variables

Newton's Law of Cooling describes the rate at which the temperature of a small object will change when the object is placed in surroundings of a different temperature. Specifically, if we let $T$ be the temperature (in degrees Fahrenheit) of the object at time $t$ (in minutes), then the rate of change of $T$ is proportional to the difference between $T$ and the temperature of the surroundings.

As an example, consider a cold-blooded snake that has been lying in the sun and has reached a body temperature of $120°$. Suppose the snake is disturbed by a predator and seeks refuge in a cave where the air temperature is $70°$. The approximate rate of change in the snake's body temperature is given by the differential equation

$$\frac{dT}{dt} = k(T - 70), \quad 70 \le T \le 120$$

If the snake's body temperature drops to $100°$ after 10 minutes, what will its body temperature be after 20 minutes?

To solve this differential equation, we can use a technique called **separation of variables** to obtain

$$\frac{dT}{T - 70} = k \, dt$$

$$\int \frac{dT}{T - 70} = \int k \, dt$$

which yields

$$T = 70 + Ce^{kt}$$

Using the conditions $T(0) = 120$ and $T(10) = 100$, we solve for the constants $C$ and $k$ to obtain the particular solution

$$T = 70 + 50e^{-0.0511t}$$

Finally, when $t = 20$ minutes, we have $T = 88°$. Figure 7.3 shows a graph of the snake's body temperature as it asymptotically approaches $70°$.

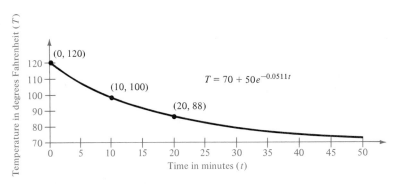

FIGURE 7.3

■ **Separation of Variables**

The simplest types of differential equations are those of the form

$$\frac{dy}{dx} = f(x)$$

In Chapter 4, we saw that we can solve such an equation by means of antidifferentiation (integration) to obtain

$$y = \int f(x)\, dx$$

In this section, we will see that integration can be used to solve another important class of differential equations—those in which the variables can be separated. This method is called **separation of variables** and is outlined as follows.

---

**Separation of variables**

If $f$ and $g$ are continuous functions, then the differential equation

$$\frac{dy}{dx} = f(x)g(y)$$

has a general solution of the form

$$\int \frac{1}{g(y)}\, dy = \int f(x)\, dx + C$$

where $C$ is an arbitrary constant.

---

**EXAMPLE 1**

**Separation of Variables**

Find the general solution of the equation

$$\frac{dy}{dx} = \frac{x}{y}$$

by separation of variables.

**SOLUTION**

We begin by separating variables to obtain

$$y\, dy = x\, dx$$

Then, integrating both sides, we obtain

$$\int y\, dy = \int x\, dx$$

$$\frac{y^2}{2} = \frac{x^2}{2} + C_1$$

---

Now, letting $C = 2C_1$, we find the general solution to be

$$y^2 = x^2 + C$$

■■■ **Remark:** We encourage you to verify each solution obtained in the examples in this section. For instance, in Example 1 you can use implicit differentiation to verify that the general solution

$$y^2 = x^2 + C$$

satisfies the differential equation $dy/dx = x/y$.

**EXAMPLE 2**

**Separation of Variables**

Find the general solution of the equation

$$\frac{dy}{dx}e^y = 2x$$

by separation of variables.

**SOLUTION**

Separating variables, we have

$$\int e^y \, dy = \int 2x \, dx$$

$$e^y = x^2 + C$$

Taking the natural logarithm of both sides, we can write this solution as

$$y = \ln{(x^2 + C)}$$

**EXAMPLE 3**

**Separation of Variables**

Find the general solution of the equation

$$(x^2 + 4)\frac{dy}{dx} = xy$$

by separation of variables.

**SOLUTION**

The given equation has the form

$$\frac{dy}{dx} = \frac{xy}{x^2 + 4}$$

$$= \left(\frac{x}{x^2 + 4}\right)(y)$$

Thus, we can separate the variables to obtain

$$\int \frac{1}{y} \, dy = \int \frac{x}{x^2 + 4} \, dx$$

$$\ln |y| = \frac{1}{2} \ln (x^2 + 4) + C_1$$

For convenience, we let $C_1 = \ln |C|$ and write

$$\ln |y| = \ln \sqrt{x^2 + 4} + \ln |C|$$
$$= \ln |C\sqrt{x^2 + 4}|$$

Therefore, the general solution is

$$y = C\sqrt{x^2 + 4}$$

In Examples 1, 2, and 3, we found the *general* solution of the given differential equation. In the next example, we demonstrate how to use an initial condition to find a *particular* solution.

**EXAMPLE 4**

**Finding a Particular Solution**

Solve the differential equation

$$e^{x^2} + \frac{y}{x}y' = 0$$

subject to the initial condition $y(0) = 1$.

**SOLUTION**

To separate the variables, we write

$$\frac{y}{x}\frac{dy}{dx} = -e^{x^2}$$

$$y \, dy = -xe^{x^2} \, dx$$

$$\int y \, dy = -\int xe^{x^2} \, dx$$

$$\int y \, dy = -\frac{1}{2} \int 2xe^{x^2} \, dx$$

$$\frac{y^2}{2} = -\frac{1}{2}e^{x^2} + C_1$$

$$y^2 = -e^{x^2} + C$$

Since $y = 1$ when $x = 0$, we find $C$ to be

$$1 = -1 + C$$

$$2 = C$$

Finally, the particular solution satisfying the given initial condition is

$$y^2 = -e^{x^2} + 2$$

In our description of the technique of separation of variables, we assumed that we begin with an equation of the form

$$\frac{dy}{dx} = f(x)g(y)$$

In practice it often happens that $f$ is a constant function. Don't let this bother you. You can still separate the variables as demonstrated in the next example.

**EXAMPLE 5**

**Separating Variables When $f$ Is a Constant Function**

In the Introductory Example of Section 7.1, we discussed the differential equation

$$\frac{dy}{dt} = k(1 - b)(y - a)$$

where $y$ is the national income, $t$ is the time in years, and $a$, $b$, and $k$ are constants. Use the method of separation of variables to solve this equation.

**SOLUTION**

Considering $k(1 - b)$ to be a *constant* function of $t$, we separate variables as follows:

$$\int \frac{1}{y - a}\, dy = \int k(1 - b)\, dt$$

$$\ln|y - a| = k(1 - b)t + C_1$$

$$y - a = Ce^{k(1-b)t}$$

$$y = a + Ce^{k(1-b)t}$$

**EXAMPLE 6**

**An Application**

Find the equation of the curve having the following characteristics:

1. At *each* point $(x, y)$ on the curve, the slope of the curve is $-x/2y$.
2. The curve passes through the point $(2, 1)$.

**SOLUTION**

Since the slope of the curve is given, we have the differential equation

$$\frac{dy}{dx} = -\frac{x}{2y}$$

with the initial condition $y(2) = 1$. Separating the variables and integrating, we get

$$2y\, dy = -x\, dx$$

$$\int 2y\, dy = \int -x\, dx$$

$$y^2 = -\frac{x^2}{2} + C_1$$

$$2y^2 = -x^2 + C$$

$$x^2 + 2y^2 = C$$

Now, since $y = 1$ when $x = 2$, it follows that

$$2^2 + 2(1^2) = C$$

$$6 = C$$

Therefore, the equation of the specified curve is

$$x^2 + 2y^2 = 6$$

as shown in Figure 7.4.

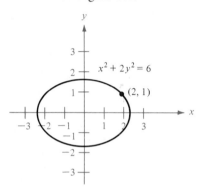

FIGURE 7.4

## SECTION EXERCISES 7.2

In Exercises 1–20, use separation of variables to find the general solution to the differential equation.

**1.** $\dfrac{dy}{dx} = 2x$

**2.** $\dfrac{dy}{dx} = \dfrac{1}{x}$

**3.** $3y^2 \dfrac{dy}{dx} = 1$

**4.** $(y + 1)\dfrac{dy}{dx} = 2x$

**5.** $y' - xy = 0$

**6.** $(1 + y)y' - 4x = 0$

**7.** $(1 + x)y' - 2y = 0$

**8.** $y' - y = 5$

**9.** $e^y \dfrac{dy}{dt} = 3t^2 + 1$

**10.** $\dfrac{dy}{dt} = \sqrt{\dfrac{t}{y}}$

**11.** $\dfrac{dy}{dx} = \dfrac{x}{y}$

**12.** $\dfrac{dy}{dx} = \dfrac{x^2 + 2}{3y^2}$

**13.** $(2 + x)y' = 2y$

**14.** $xy' = y$

**15.** $y \ln x - xy' = 0$

**16.** $y' - y(x + 1) = 0$

**17.** $y' - (x + 1)(y + 1) = 0$

**18.** $yy' - 2xe^x = 0$

**19.** $xyy' + y^2 = 1$

**20.** $y' \ln y = x^2 e^x$

In Exercises 21–26, find the particular solution of the differential equation that satisfies the given initial condition.

**21.** $yy' - e^x = 0$, $y(0) = 4$

**22.** $\sqrt{x} + \sqrt{y}y' = 0$, $y(1) = 4$

**23.** $y(x + 1) + y' = 0$, $y(-2) = 1$

**24.** $xyy' - \ln x = 0$, $y(1) = 0$

**25.** $dP - kP\,dt = 0$, $P(0) = P_0$

**26.** $dT + k(T - 70)\,dt = 0$, $T(0) = 140$

In Exercises 27 and 28, find an equation for the curve that passes through the given point and has the specified slope.

**27.** Point: $(1, 1)$; slope: $y' = -\dfrac{9x}{16y}$

**28.** Point: $(8, 2)$; slope: $y' = \dfrac{2y}{3x}$

In Exercises 29 and 30, solve the given differential equation to find velocity $v$ as a function of time if $v = 0$ when $t = 0$. The differential equation was derived to describe the motion of two people on a toboggan after consideration of the force of gravity, friction, and air resistance.

**29.** $12.5\dfrac{dv}{dt} = 43.2 - 1.25v$

**30.** $12.5\dfrac{dv}{dt} = 43.2 - 1.75v$

In Exercises 31 and 32, use Newton's Law of Cooling given in the Introductory Example.

**31.** A room is kept at a constant temperature of 70°. An object placed in the room cools from 350° to 150° in 45 minutes. How long will it take for the object to cool to a temperature of 80°?

**32.** Food at a temperature of 70° is placed in a freezer that is set at 0°. After 1 hour, the temperature of the food is 48°. Find the temperature of the food after it has been in the freezer 6 hours, and find how long it will take for the food to reach a temperature of 10°.

# First-Order Linear Differential Equations

The price $p$ per unit of a product tends to be a function of time $t$. Often $p$ increases with time, but occasionally, with new products such as hand-held calculators, personal computers, or video recorders, the price decreases as the initial cost of technology is offset by increased demand. By considering the price to be a function of time, economists can describe the supply and the demand for a product as functions of both the price *and* the rate of change of the price.

For instance, suppose that the supply and demand functions for a particular product are given by

$$D(t) = 100 + 4p(t) - 3p'(t) \qquad \text{Demand function}$$

and

$$S(t) = 60 + 6p(t) + p'(t) \qquad \text{Supply function}$$

Moreover, suppose that the price at time $t = 0$ is $25.00. We know that the equilibrium point for these two functions is found by setting $D(t)$ equal to $S(t)$. Thus, the equi-

librium point for this product is given by

$$100 + 4p(t) - 3p'(t) = 60 + 6p(t) + p'(t)$$

$$40 - 2p(t) - 4p'(t) = 0$$

or $\qquad p'(t) + \dfrac{1}{2}p(t) = 10$

We call this equation a **first-order linear differential equation** in $p(t)$. In this section, we will see that the solution to this equation is given by

$$p(t) = \frac{1}{e^{t/2}} \int 10e^{t/2}\, dt$$

$$= e^{-t/2}(20e^{t/2} + C)$$

$$= 20 + Ce^{-t/2}$$

Using the initial condition $p(0) = 25$, we can solve for $C$ to obtain $C = 5$, and we conclude that the price as a function of time is given by

$$p(t) = 20 + 5e^{-t/2}$$

The graph of this function is shown in Figure 7.5.

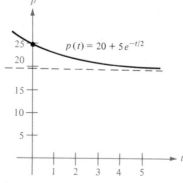

FIGURE 7.5

## ■ Linear Differential Equations

In this section, we will look at a technique for solving one of the most important classes of differential equations—*linear* differential equations.

| | |
|---|---|
| **Definition of linear differential equation** | A first-order linear differential equation is an equation of the form $$y' + P(x)y = Q(x)$$ where $P$ and $Q$ are functions of $x$. |

■■■■ **Remark:** The term "first-order" refers to the fact that the highest order derivative of $y$ in the equation is the first derivative.

To solve the first-order linear differential equation

$$y' + P(x)y = Q(x)$$

we use an **integrating factor** $u(x)$ which converts the left side into the derivative of the product $u(x)y$. One such factor is

$$u(x) = Ce^{\int P(x)\, dx}$$

Multiplying both sides of the original equation by this factor, we obtain

$$y'e^{\int P(x)\, dx} + P(x)e^{\int P(x)\, dx}y = Q(x)e^{\int P(x)\, dx}$$

Now the left side of the equation is the derivative of

$$ye^{\int P(x)\, dx}$$

(Try using implicit differentiation to verify this.) In other words, integrating both sides of the equation with respect to $x$, we obtain the general solution

$$ye^{\int P(x)\, dx} = \int Q(x)e^{\int P(x)\, dx}\, dx$$

| | |
|---|---|
| **Solution of a first-order linear differential equation** | 1. Write the equation in standard form $y' + P(x)y = Q(x)$. <br> 2. Find the integrating factor $u(x) = e^{\int P(x)\, dx}$. <br> 3. Multiply both sides of the given equation by $u(x)$ and integrate to obtain $$yu(x) = \int Q(x)u(x)\, dx$$ <br> 4. Solve for $y$ to obtain the general solution $$y = \frac{1}{u(x)} \int Q(x)u(x)\, dx$$ |

EXAMPLE 1

**Solving a First-Order Linear Differential Equation**

Find the general solution of

$$y' + y = e^x \qquad \text{Standard form}$$

**SOLUTION**

For this equation, we have $P(x) = 1$ and $Q(x) = e^x$. Therefore,

$$\int P(x)\, dx = \int dx = x$$

and the integrating factor $u(x)$ is

$$u(x) = e^{\int P(x)\, dx} = e^x \qquad \text{Integrating factor}$$

Multiplying the original equation by this integrating factor produces

$$y'e^x + ye^x = e^{2x}$$

Now, since the left side of the equation is the derivative of $ye^x$, we have

$$ye^x = \int e^{2x}\, dx$$

$$= \frac{1}{2}e^{2x} + C$$

which implies that the solution is

$$y = \frac{1}{2}e^x + Ce^{-x} \qquad \text{General solution}$$

In Example 1, the differential equation was given in standard form $y' + P(x)y = Q(x)$. For equations that are not written in standard form, we first convert to standard form in order to determine the functions $P(x)$ and $Q(x)$, as demonstrated in the next example.

**EXAMPLE 2**

**Solving a First-Order Linear Differential Equation**

Find the general solution to

$$xy' - 2y = x^2$$

**SOLUTION**

The *standard form* of the given equation is

$$y' + P(x)y = Q(x)$$

$$y' - \left(\frac{2}{x}\right)y = x \qquad \text{Standard form}$$

Thus, $P(x) = -2/x$, $Q(x) = x$, and we have

$$\int P(x)\, dx = -\int \frac{2}{x}\, dx = -\ln x^2$$

This implies that the integrating factor $u(x)$ is

$$u(x) = e^{\int P(x)\,dx} = e^{-\ln x^2} = \frac{1}{x^2} \qquad \text{Integrating factor}$$

Therefore, the general solution is given by

$$y = \frac{1}{u(x)} \int Q(x)u(x)\,dx$$

$$= \frac{1}{1/x^2} \int \frac{1}{x^2}x\,dx$$

$$= x^2 \int \frac{1}{x}\,dx$$

$$= x^2(\ln|x| + C)$$

$$= x^2 \ln|x| + Cx^2 \qquad \text{General solution} \qquad \rule{3cm}{0.3cm}$$

In the next example, we demonstrate the use of an initial condition to find a particular solution of a first-order linear differential equation.

**EXAMPLE 3**

**Finding a Particular Solution**

Find the particular solution of the differential equation

$$x^3 y' + 2y = e^{1/x^2}$$

that satisfies the initial condition $y = e$ when $x = 1$.

**SOLUTION**

We begin by finding the general solution. In standard form, this equation is

$$y' + \overbrace{\frac{2}{x^3}}^{P(x)} y = \overbrace{\frac{1}{x^3}}^{Q(x)} e^{1/x^2} \qquad \text{Standard form}$$

and we have

$$\int P(x)\,dx = \int \frac{2}{x^3}\,dx = -\frac{1}{x^2}$$

Thus, the integrating factor is

$$u(x) = e^{\int P(x)\,dx}$$

$$= e^{-1/x^2}$$

Therefore, the general solution is given by

$$y = \frac{1}{u(x)} \int Q(x)u(x)\,dx$$

$$= e^{1/x^2} \int \frac{1}{x^3} e^{1/x^2} e^{-1/x^2}\,dx$$

$$= e^{1/x^2} \int \frac{1}{x^3}\, dx$$

$$= e^{1/x^2}\left(-\frac{1}{2x^2} + C\right) \qquad \text{General solution}$$

Finally, using the initial condition $y = e$ when $x = 1$, we have

$$e = e\left(-\frac{1}{2} + C\right)$$

$$1 = -\frac{1}{2} + C$$

$$\frac{3}{2} = C$$

and the particular solution is

$$y = e^{1/x^2}\left(-\frac{1}{2x^2} + \frac{3}{2}\right) \qquad \text{Particular solution}$$

## SECTION EXERCISES 7.3

In Exercises 1–12, solve the first-order linear differential equation.

**1.** $\dfrac{dy}{dx} + 3y = 6$

**2.** $\dfrac{dy}{dx} + 5y = 15$

**3.** $\dfrac{dy}{dx} + y = e^{-x}$

**4.** $\dfrac{dy}{dx} + 3y = e^{-3x}$

**5.** $\dfrac{dy}{dx} + \dfrac{y}{x} = 3x + 4$

**6.** $\dfrac{dy}{dx} + \dfrac{2y}{x} = 3x + 1$

**7.** $y' + 2xy = 2x$

**8.** $y' + 5y = e^{5x}$

**9.** $(x - 1)y' + y = x^2 - 1$

**10.** $x^3 y' + 2y = e^{1/x^2}$

**11.** $xy' + y = x \ln x$

**12.** $xy' + y = x^2 \ln x$

In Exercises 13–20, find the particular solution that satisfies the given initial condition.

**13.** $y' + y = e^x$, $y = 2$ when $x = 0$

**14.** $y' + 3x^2 y = 3x^2$, $y = 6$ when $x = 0$

**15.** $xy' + y = 0$, $y = 2$ when $x = 2$

**16.** $y' + (2x - 1)y = 0$, $y = 2$ when $x = 1$

**17.** $y' + y = x$, $y = 4$ when $x = 0$

**18.** $y' + 2y = e^{-2x}$, $y = 4$ when $x = 1$

**19.** $xy' - 2y = -x^2$, $y = 5$ when $x = 1$

**20.** $x^2 y' - 4xy = 10$, $y = 10$ when $x = 1$

**21.** A brokerage firm opens a new real estate investment plan for which the earnings are equivalent to continuous compounding at the rate of $r$. The firm estimates that deposits from investors will create a net cash flow of $Pt$ dollars, where $t$ is time in years. The rate of increase in the amount $A$ in the real estate investment is

$$\frac{dA}{dt} = rA + Pt$$

Solve the differential equation and find the amount $A$ as a function of $t$. Assume that $A = 0$ when $t = 0$.

**22.** Use the result of Exercise 21 with $P = \$500{,}000$ and $r = 12\%$ to find
(a) the amount $A$ in the fund after 10 years
(b) the amount invested and the total earnings over the 10 years

**23.** The rate of increase (in thousands of units) in sales $S$ is estimated to be

$$\frac{dS}{dt} = 0.2(100 - S) + 0.2t$$

where $t$ is time in years. Solve the differential equation and complete the following table to estimate sales of a new product for the first 10 years.

| $t$ | 0 | 1 | 2 | 3 | 4 | 5 | 6 | 7 | 8 | 9 | 10 |
|---|---|---|---|---|---|---|---|---|---|---|---|
| $S$ | 0 | | | | | | | | | | |

**24.** The rate of increase in sales $S$ is given by

$$\frac{dS}{dt} = k_1(L - S) + k_2 t$$

where $t$ is time in years and $S = 0$ when $t = 0$. Solve this differential equation for $S$ as a function of $t$.

# Applications of Differential Equations

In this text we have made many references to mathematical models. By now you should be familiar with the two major goals in developing a model: accuracy and simplicity. In practice, we often find a model by making various simplifying assumptions and then checking the accuracy produced by these assumptions.

As a case in point, let us consider three models of differential equations to predict the yield of a crop as a function of time. For our basic model we assume that $w$ is the *current* dry weight of the crop, 1 is the *maximum* dry weight, and the rate of growth of the crop is

$$\frac{dw}{dt} = f(t)[(1 - w)w]$$

where $f(t)$ is the coefficient of growth as a function of time. For simplicity we measure the time in growing sea-sons. That is, $t = 0$ represents the time of germination of the crop seeds and $t = 1$ represents the harvest time.

Now we consider possibilities for the function $f(t)$:

(a) $f_1(t) = k_1$
(b) $f_2(t) = k_2 t$
(c) $f_3(t) = k_3 t(1 - t)$

The solutions to the three resulting differential equations are as follows:

(a) $w = \dfrac{1}{1 + C_1 e^{-k_1 t}}$

(b) $w = \dfrac{1}{1 + C_2 e^{-k_2 t^2/2}}$

(c) $w = \dfrac{1}{1 + C_3 e^{-k_3 t^2(3 - 2t)/6}}$

Figure 7.6 compares the different growth patterns of these three models. Of course, to decide which model is best we would need to compare each model's values with the actual measurements for crop weight.

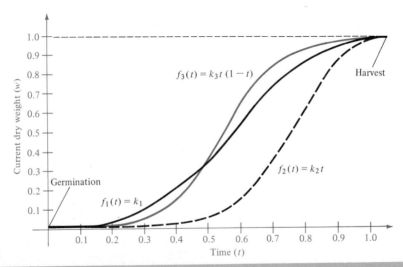

FIGURE 7.6

■ **Applications of Differential Equations**

In this concluding section on differential equations, we look at several types of applications involving differential equations.

Recall from Section 5.5 that the mathematical model for *exponential growth (or decay)* resulted from the differential equation

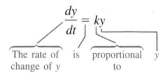

The *logistics growth curve* (described in the Introductory Example of Section 6.3) was developed from the differential equation

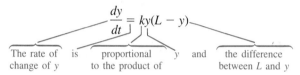

The model for *Newton's Law of Cooling* (Section 7.2) was developed from the differential equation

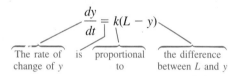

These three instances suggest that the key terms to look for in constructing a differential equation for a given relationship are "rate of change" and "proportional."

**EXAMPLE 1**

**A Model for Advertising**

A new product is introduced through an advertising campaign to a population of 1,000,000 potential consumers. Assume that the rate at which the population hears about the product is proportional to the number of people who are not yet aware of the product. If by the end of 1 year half of the population is aware of the product, how many will have heard by the end of 2 years?

**SOLUTION**

Let $y$ be the number of people in the population at time $t$ who have heard of the product. This means that $1,000,000 - y$ represents the number of people at time $t$ who have not yet heard of the product. That is,

$$y = \text{number of people who have heard}$$

$$1,000,000 - y = \text{number of people who have not heard}$$

$$\frac{dy}{dt} = \text{rate of change of } y$$

Since the rate of change of $y$ is proportional to the number of people who have not heard, we can write the differential equation

$$\frac{dy}{dt} = k(1,000,000 - y)$$

The rate of    is    proportional    the number who
change of $y$          to        haven't heard

Using separation of variables, we find the solution of this equation to be

$$\int \frac{1}{1,000,000 - y} \, dy = \int k \, dt$$

$$-\ln (1,000,000 - y) = kt + C_1$$

$$1,000,000 - y = Ce^{-kt} \qquad\qquad \text{Let } C = e^{-C_1}$$

$$y = 1,000,000 - Ce^{-kt}$$

Now, since $y = 0$ when $t = 0$, we find $C$ as follows:

$$0 = 1,000,000 - C(1)$$

$$C = 1,000,000$$

Moreover, since $y = 500,000$ when $t = 1$, we have

$$500,000 = 1,000,000(1 - e^{-k})$$

$$\frac{1}{2} = 1 - e^{-k}$$

$$e^{-k} = \frac{1}{2}$$

$$-k = \ln \frac{1}{2}$$

$$k = \ln 2$$

Therefore, the model giving $y$ as a function of time is

$$y = 1,000,000(1 - e^{-t \ln 2})$$

Finally, when $t = 2$, the number of people who have heard of the product is

$$y = 1,000,000(1 - e^{-2 \ln 2})$$

$$= 1,000,000[1 - e^{\ln (1/4)}]$$

$$= 1,000,000\left[1 - \frac{1}{4}\right]$$

$$= 750,000$$

The graph of $y$ is shown in Figure 7.7.

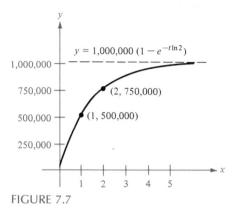

$$y = 1{,}000{,}000\,(1 - e^{-t\ln 2})$$

(2, 750,000)

(1, 500,000)

FIGURE 7.7

## EXAMPLE 2

### A Chemical Reaction Model

During a certain chemical reaction, substance $A$ is converted into substance $B$ at a rate proportional to the square of the amount of $A$. When $t = 0$, 60 grams of $A$ are present, and after 1 hour ($t = 1$), only 10 grams of $A$ remain unconverted. How much of $A$ is present after 2 hours?

## SOLUTION

Letting $y$ be the amount of unconverted $A$ at any time $t$, we have the differential equation

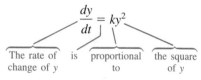

$$\frac{dy}{dt} = ky^2$$

The rate of change of $y$   is   proportional to   the square of $y$

Separating variables, we have

$$\int \frac{dy}{y^2} = \int k\,dt$$

$$-\frac{1}{y} = kt + C$$

$$y = \frac{-1}{kt + C}$$

Since $y = 60$ when $t = 0$, we have

$$60 = -\frac{1}{C}$$

$$C = -\frac{1}{60}$$

Furthermore, since $y = 10$ when $t = 1$, we have

$$10 = \frac{-1}{k - (1/60)}$$

$$10 = \frac{-60}{60k - 1}$$

$$k = -\frac{1}{12}$$

Thus, the model describing this chemical reaction is

$$y = \frac{-1}{(-1/12)t - (1/60)}$$

$$= \frac{60}{5t + 1}$$

Finally, when $t = 2$,

$$y = \frac{60}{11} \approx 5.45 \text{ grams}$$

Figure 7.8 shows the graph of this function.

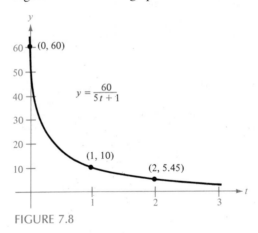

FIGURE 7.8

## EXAMPLE 3

### The Gompertz Growth Equation

The *Gompertz growth equation* is a mathematical model describing a limited growth situation in which the rate of growth of $y$ is proportional to the product of $y$ and to the natural log of the ratio of $L$ to $y$, where $L$ is the maximum population size. Set up a differential equation for this model and sketch its graph. Assume that $L = 100$ and use the conditions $y = 10$ when $t = 0$ and $y = 50$ when $t = 10$, where $t$ is measured in days.

**SOLUTION**  Since $L = 100$, the natural log of the ratio of $L$ to $y$ is $\ln(100/y)$, and we have the differential equation

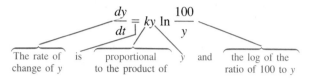

$$\frac{dy}{dt} = ky \ln \frac{100}{y}$$

The rate of change of $y$   is   proportional to the product of   $y$   and   the log of the ratio of 100 to $y$

Separating variables gives us

$$\int \frac{dy}{y \ln(100/y)} = \int k \, dt$$

Letting $u = \ln(100/y)$, we have $du = -(1/y) \, dy$, which yields

$$-\int \frac{-(1/y)}{\ln(100/y)} \, dy = \int k \, dt$$

$$-\ln \left| \ln \frac{100}{y} \right| = kt + C_1$$

$$\ln \frac{100}{y} = Ce^{-kt}$$

$$\frac{100}{y} = e^{Ce^{-kt}}$$

$$y = \frac{100}{e^{Ce^{-kt}}} = 100e^{-Ce^{-kt}}$$

Now since $y = 10$ when $t = 0$, we have

$$10 = 100e^{-C}$$

$$C = -\ln \frac{1}{10} \approx 2.3026$$

Furthermore, since $y = 50$ when $t = 10$, we have

$$50 = 100e^{-2.3026e^{-10k}}$$

$$\ln \frac{1}{2} = -2.3026e^{-10k}$$

$$\frac{-0.6931}{-2.3026} = e^{-10k}$$

$$k = -\frac{1}{10} \ln(0.3010) \approx 0.12005$$

Thus, the model for this particular growth pattern is

$$y = 100e^{-2.3026e^{-0.12005t}}$$

Figure 7.9 shows the graph of this equation.

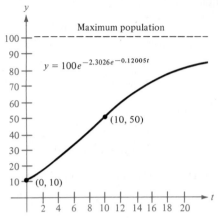

FIGURE 7.9

**EXAMPLE 4**

**Hybrid Selection Model**

In the study of genetics, one encounters a family of differential equations of the form

$$\frac{dy}{dt} = ky(1 - y)(a - by)$$

where $y$ represents the portion of the population having a certain characteristic and $t$ represents the time measured in generations. If $y = 0.5$ when $t = 0$, $y = 0.8$ when $t = 4$, $a = 2$, and $b = 1$, find the model relating $y$ and $t$.

**SOLUTION**

Separating variables, we have

$$\int \frac{1}{y(1 - y)(2 - y)} \, dy = \int k \, dt$$

Now, integrating by partial fractions, we obtain

$$\int \left( \frac{1/2}{y} + \frac{1}{1 - y} - \frac{1/2}{2 - y} \right) dy = \int k \, dt$$

$$\frac{1}{2} \ln |y| - \ln |1 - y| + \frac{1}{2} \ln |2 - y| = kt + C_2$$

$$\ln \left| \frac{\sqrt{y}\sqrt{2 - y}}{1 - y} \right| = kt + C_2$$

$$\left| \frac{\sqrt{y}\sqrt{2-y}}{1-y} \right| = C_1 e^{kt}$$

$$\frac{y(2-y)}{(1-y)^2} = Ce^{2kt}$$

Since $y = 0.5$ when $t = 0$, we have

$$\frac{0.5(1.5)}{(0.5)^2} = C$$

$$3 = C$$

Also, since $y = 0.8$ when $t = 4$, we have

$$\frac{0.8(1.2)}{(0.2)^2} = 3e^{8k}$$

$$8 = e^{8k}$$

$$\frac{1}{8}\ln 8 = k$$

$$0.2599t \approx k$$

Thus, the model relating $y$ and $t$ is

$$\frac{y(2-y)}{(1-y)^2} = 3e^{0.5198t}$$

Figure 7.10 shows the graph of this model.

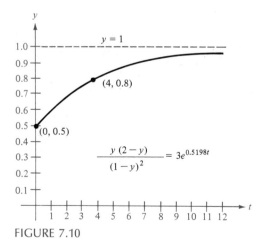

FIGURE 7.10

**EXAMPLE 5**

**A Mixture Problem**

A tank contains 40 gallons of a solution composed of 90 percent water and 10 percent alcohol. A second solution containing 50 percent water and 50 percent alcohol is added to the tank at the rate of 4 gal/min. As the second solution is being added, the tank is being drained at the rate of 4 gal/min, as shown in Figure 7.11. Assuming the solution in the tank is stirred constantly, how much alcohol is in the tank after 10 minutes?

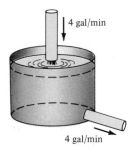

4 gal/min

4 gal/min

FIGURE 7.11

**SOLUTION**

Let $y$ be the number of gallons of alcohol in the tank at any time $t$. This means that the percentage of alcohol in the tank at any time is $y/40$, and since 4 gallons of solution is being drained every minute, the rate of change of $y$ is given by

$$\frac{dy}{dt} = \underbrace{-4\left(\frac{y}{40}\right)}_{\substack{\text{alcohol} \\ \text{draining}}} + \underbrace{2}_{\substack{\text{alcohol} \\ \text{entering}}}$$

where 2 represents the 2 gallons of alcohol entering each minute in the 50% solution. In standard form, this first-order linear differential equation is

$$y' + \frac{1}{10}y = 2$$

To solve this linear equation, we let $P(x) = \frac{1}{10}$ and obtain

$$\int P(t)\, dt = \int \frac{1}{10}\, dt = \frac{t}{10}$$

Thus, the integrating factor is

$$u(t) = e^{\int P(t)\, dt} = e^{t/10}$$

and the general solution is

$$y = e^{-t/10} \int 2e^{t/10}\, dt$$

$$= e^{-t/10}(20e^{t/10} + C)$$

$$= 20 + Ce^{-t/10}$$

Since $y = 4$ when $t = 0$, we have

$$4 = 20 + C \quad \Rightarrow \quad C = -16$$

which means that the particular solution is

$$y = 20 - 16e^{-t/10}$$

Finally, when $t = 10$, the amount of alcohol in the tank is

$$y = 20 - 16e^{-1} \approx 14.11 \text{ gallons}$$

**Remark:** From the examples in this section you can see that a great deal of the work in applying differential equations involves solving for the constants $k$ and $C$. Because of the many opportunities for error, it is a good idea to check to see if the initial conditions satisfy your final equation.

## SECTION EXERCISES 7.4

In Exercises 1–6, assume that the rate of change of $y$ is proportional to $y$. Solve the resulting differential equation $dy/dx = ky$ and find the particular solution that passes through the given points.

**1.** $(0, 1)$, $(3, 2)$

**2.** $(0, 4)$, $(1, 6)$

**3.** $(0, 4)$, $(4, 1)$

**4.** $(0, 60)$, $(5, 30)$

**5.** $(2, 2)$, $(3, 4)$

**6.** $(1, 4)$, $(2, 1)$

In Exercises 7–10, assume that the rate of change of $y$ is proportional to the product of $y$ and the difference of $L$ and $y$. Solve the resulting differential equation $dy/dx = ky(L - y)$ and find the particular solution that passes through the given points for the indicated value of $L$.

**7.** $L = 20$; $(0, 1)$, $(5, 10)$

**8.** $L = 100$; $(0, 10)$, $(5, 30)$

**9.** $L = 5000$; $(0, 250)$, $(25, 2000)$

**10.** $L = 1000$; $(0, 100)$, $(4, 750)$

In Exercises 11 and 12, use the chemical reaction model of Example 2 to find the amount $y$ as a function of $t$ and sketch the graph of the function.

**11.** $y = 100$ grams when $t = 0$; $y = 37$ grams when $t = 2$

**12.** $y = 75$ grams when $t = 0$; $y = 12$ grams when $t = 1$

In Exercises 13 and 14, use the Gompertz growth model described in Example 3 to find the growth function and sketch its graph.

**13.** $L = 500$; $y = 100$ when $t = 0$; $y = 150$ when $t = 2$

**14.** $L = 5000$; $y = 500$ when $t = 0$; $y = 625$ when $t = 1$

In Exercises 15 and 16, assume that the rate of change in the proportion $P$ of correct responses after $n$ trials is pro-

portional to the product of $P$ and $L - P$ where $L$ is the limiting proportion of correct responses.

**15.** Write and solve the differential equation for this learning theory model.

**16.** Use the solution of Exercise 15 to write $P$ as a function of $n$ and then sketch the graph of the solution.

(a) $L = 1.00$; $P = 0.50$ when $n = 0$; $P = 0.85$ when $n = 4$

(b) $L = 0.80$; $P = 0.25$ when $n = 0$; $P = 0.60$ when $n = 10$

**17.** Assume that the rate of change in the number of miles $s$ of road cleared per hour by a snowplow is inversely proportional to the height $h$ of snow. That is,

$$\frac{ds}{dh} = \frac{k}{h}$$

Find $s$ as a function of $h$ if $s = 25$ miles when $h = 2$ inches and $s = 12$ miles when $h = 6$ inches ($2 \le h \le 15$).

**18.** A wet towel hung from a clothesline to dry loses moisture through evaporation at a rate proportional to its moisture content. If after 1 hour the towel has lost 40% of its original moisture content, after how long will it have lost 80%?

**19.** Let $x$ and $y$ be the sizes of two organs of a particular mammal at time $t$. Empirical data indicate that the relative growth rates of these two organs are equal, and hence we have

$$\frac{1}{x}\frac{dx}{dt} = \frac{1}{y}\frac{dy}{dt}$$

Solve this differential equation, writing $y$ as a function of $x$.

**20.** When predicting population growth, demographers must consider birth and death rates as well as the net change caused by the difference between the rates of immigration and emigration. Let $P$ be the population at time $t$ and $N$ be the net increase per unit time due to the difference between immigration and emigration. Thus, the rate of growth of the population is given by

$$\frac{dP}{dt} = kP + N, \quad N \text{ is constant}$$

Solve this differential equation to find $P$ as a function of time.

**21.** A large corporation starts at time $t = 0$ to invest part of its receipts at a rate of $P$ dollars per year in a fund for future corporate expansion. Assume that the fund earns $r$ percent interest per year compounded continuously. Thus, the rate of growth of the amount $A$ in the fund is given by

$$\frac{dA}{dt} = rA + P$$

where $A = 0$ when $t = 0$. Solve this differential equation for $A$ as a function of $t$.

In Exercises 22–24, use the result of Exercise 21.

**22.** Find $A$ if
(a) $P = \$100,000$, $r = 12\%$, and $t = 5$ years
(b) $P = \$250,000$, $r = 15\%$, and $t = 10$ years

**23.** Find $P$ if the corporation needs $\$120,000,000$ in 8 years and the fund earns $16\frac{1}{4}\%$ interest compounded continuously.

**24.** Find $t$ if the corporation needs $\$800,000$ and it can invest $\$75,000$ per year in a fund earning $13\%$ interest compounded continuously.

In Exercises 25–28, suppose that a medical researcher wants to determine the concentration $C$ (in moles per liter) of a tracer drug injected into a moving fluid. We can start solving this problem by considering a single-compartment dilution model. (See Figure 7.12.) We assume that the

fluid in the compartment is continuously mixed and that the volume of fluid in the compartment is constant.

**25.** If the tracer is injected instantaneously at time $t = 0$, then the concentration of the fluid in the compartment begins diluting according to the differential equation

$$\frac{dC}{dt} = \left(-\frac{R}{V}\right)C, \quad C = C_0 \text{ when } t = 0$$

(a) Solve this differential equation to find the concentration as a function of time.
(b) Find the limit of $C$ as $t \to \infty$.

**26.** Use the solution of the differential equation in Exercise 25 to find the concentration as a function of time and sketch its graph if
(a) $V = 2$ liters, $R = 0.5$ L/min, and $C_0 = 0.6$ mol/L
(b) $V = 2$ liters, $R = 1.5$ L/min, and $C_0 = 0.6$ mol/L

**27.** In Exercises 25 and 26, we assumed that there was a single initial injection of the tracer drug into the compartment. Now let us consider the case in which the tracer is continuously injected (beginning at $t = 0$) at the rate of $Q$ mol/min. Considering $Q$ to be negligible compared with $R$, we have the differential equation

$$\frac{dC}{dt} = \frac{Q}{V} - \left(\frac{R}{V}\right)C, \quad C = 0 \text{ when } t = 0$$

(a) Solve this differential equation to find the concentration as a function of time.
(b) Find the limit of $C$ as $t \to \infty$.

**28.** Use the solution of Exercise 27 to find the concentration as a function of time if $V = 2$ liters, $R = 1$ L/min, and $Q = 0.75$ mol/min.

**29.** Glucose is added intraveneously to the bloodstream at the rate of $q$ units per minute, and the body removes glucose from the bloodstream at a rate proportional to the amount present. Assume $Q(t)$ is the amount of glucose in the bloodstream at time $t$.
(a) Determine the differential equation describing the rate of change with respect to time of glucose in the bloodstream.
(b) Solve the differential equation, letting $Q = Q_0$ when $t = 0$.
(c) Find the limit of $Q(t)$ as $t \to \infty$.

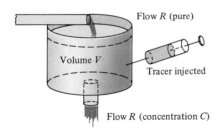

Flow $R$ (pure)

Volume $V$

Tracer injected

Flow $R$ (concentration $C$)

FIGURE 7.12

## Important Terms

Differential equation

General solution

Particular solution

Solution curve

Initial condition

Separation of variables

First-order linear differential equation

Integrating factor

## Important Techniques

Verifying solutions of differential equations

Finding a particular solution of a differential equation

Using separation of variables to solve a differential equation

Solving a first-order linear differential equation

Solving applied problems involving differential equations

## Important Formulas

The differential equation

$$\frac{dy}{dx} = f(x)g(y)$$

has a general solution of the form

$$\int \frac{1}{g(y)} \, dy = \int f(x) \, dx + C$$

where $C$ is an arbitrary constant.

The general solution of $y' + P(x)y = Q(x)$ is

$$y = \frac{1}{u(x)} \int Q(x)u(x) \, dx$$

where $u(x) = e^{\int P(x) \, dx}$

# REVIEW EXERCISES FOR CHAPTER 7

In Exercises 1–10, verify that $y$ is a solution of the given differential equation.

| Solution | Equation |
|---|---|
| **1.** $y = x \ln x^2 + 2x^{3/2} + Cx$ | $xy' - y = x(2 + \sqrt{x})$ |
| **2.** $y = Ce^{-(x-2)^2/2}$ | $y' + xy = 2y$ |
| **3.** $y = C(x - 1)^2$ | $\dfrac{dy}{dx} - \dfrac{2y}{x} = \dfrac{1}{x}\dfrac{dy}{dx}$ |
| **4.** $y = e^{x^3}(x + C)$ | $\dfrac{dy}{dx} - 3x^2y = e^{x^3}$ |
| **5.** $y = -\dfrac{1}{3} + Ce^{-3/x}$ | $y' - \dfrac{3y}{x^2} = \dfrac{1}{x^2}$ |
| **6.** $y = \dfrac{bx^4}{4 - a} + Cx^a$ | $y' - \left(\dfrac{a}{x}\right)y = bx^3$ |
| **7.** $y = \left(C_1 + C_2x + \dfrac{1}{3}x^3\right)e^x$ | $y'' - 2y' + y = 2xe^x$ |
| **8.** $y = (C_1 + C_2x - \ln|x|)e^{-x}$ | $y'' + 2y' + y = \dfrac{1}{e^xx^2}$ |
| **9.** $y = \dfrac{1}{5}x^3 - x + C\sqrt{x}$ | $2xy' - y = x^3 - x$ |
| **10.** $y = x \ln|x| - 2 + Cx$ | $xy' = x + y + 2$ |

In Exercises 11–14, verify that the general solution satisfies the given differential equation. Then find the particular solution satisfying the given initial conditions.

**11.** $y = x^2 + 2x + \dfrac{C}{x}, \ \dfrac{dy}{dx} + \dfrac{y}{x} = 3x + 4$

  $y = 3$ when $x = 1$

**12.** $y = \dfrac{1}{2}e^x + Ce^{-x}, \ y' + y = e^x$

  $y = \dfrac{3}{2}$ when $x = 0$

**13.** $y = 1 + Ce^{-x^2}, \ y' + 2xy = 2x$

  $y = -1$ when $x = 0$

**14.** $y = \dfrac{x^3 - 3x + C}{3(x - 1)}, \ (x - 1)y' + y = x^2 - 1$

  $y = 4$ when $x = 2$

In Exercises 15–20, use separation of variables to find the general solution of the differential equation.

**15.** $yy' - 3x^2 = 0$

**16.** $y' + xy = 0$

**17.** $y' = x^2y^2 - 9x^2$

**18.** $(1 + y)\ln(1 + y) + y' = 0$

**19.** $(1 + x)y' = 1 + y$

**20.** $xyy' - (1 + 2y^2 + y^4) = 0$

In Exercises 21–26, solve the first-order linear differential equation.

**21.** $\dfrac{dy}{dx} - \dfrac{y}{x} = 2 + \sqrt{x}$

**22.** $\dfrac{dy}{dx} - 3x^2y = e^{x^3}$

**23.** $\dfrac{dy}{dx} - \dfrac{3y}{x^2} = \dfrac{1}{x^2}$

**24.** $\dfrac{dy}{dx} - \dfrac{y}{x} = \dfrac{x + 2}{x}$

**25.** $2xy' - y = x^3 - x$

**26.** $xy' - ay = bx^4$

**27.** In a chemical reaction, a certain compound changes into another compound at a rate proportional to the unchanged amount. If initially there is 20 grams of the original compound and 16 grams remains after 1 hour, when will 75% of the compound be changed?

**28.** The absorption of x-rays through a body is defined to be the rate of change of the intensity $I$ of the x-ray with respect to the depth of penetration $r$. The absorption is proportional to the density $\rho$ of the body and to the intensity. This is described by the differential equation

$$\text{Absorption} = \dfrac{dI}{dr} = -k\rho I$$

Solve this differential equation to find the intensity as a function of the depth of penetration. (Assume that the intensity is $I_0$ when $r = 0$.)

**29.** Let $A(t)$ be the amount in a fund earning interest at the annual rate of $r$ compounded continuously. If a continuous cash flow of $P$ dollars per year is withdrawn from the fund, then the rate of decrease of $A$ is given by the differential equation

$$\dfrac{dA}{dt} = rA - P$$

where $A = A_0$ when $t = 0$. Solve this differential equation for $A$ as a function of $t$.

In Exercises 30–32, use the result of Exercise 29.

**30.** Find $A$ if

  (a) $A_0 = \$2,000,000, \quad r = 12\%, \quad P = \$250,000,$
  $t = 5$ years
  (b) $A_0 = \$100,000, \quad r = 10\%, \quad P = \$30,000,$
  $t = 2$ years

**31.** Find $A_0$ if a retired couple want a continuous cash flow of \$40,000 per year for 20 years. (That is, $A = 0$ when $t = 20$.) Assume that their investment will earn 13% interest compounded continuously.

**32.** Find the time necessary to deplete a fund earning 14% interest compounded continuously if $A_0 = \$1,000,000$ and $P = \$200,000$.

**33.** A company introduces a new product for which it is believed that the limit on yearly sales is 500,000 units. Sales are expected to increase at a rate propor- tional to the product of sales and the difference be- tween sales and the upper limit. Initial and second- year sales are 10,000 and 100,000 units, respec- tively.

(a) Estimate third-year sales.

(b) During what year will sales surpass 400,000 units?

# Functions of Several Variables

# CHAPTER 8

# The Three-Dimensional Coordinate System

We have already considered the total revenue derived from the sale of $x$ units of a single product at a price of $p$ dollars per unit,

$$R = xp$$

This formula can be generalized as follows to cover the revenue of several different products:

$$R = x_1 p_1 + x_2 p_2 + \cdots + x_n p_n$$

where $x_i$ represents the number of units and $p_i$ represents the price per unit of the $i$th product. Although this multi-product formula is the result of a fairly simple extension of the single-product case, a graphical interpretation of the multiproduct case is not so easy to come by. The problem lies in the fact that *each* of the variables $x_1, x_2, \ldots, x_n$ requires its own dimension in the graph in addition to $R$. Since we are visually limited to a maximum of three dimensions, our graphical interpretation of revenue functions is limited to one- and two-product cases.

As an example of a two-product situation, consider a manufacturer who sells $x$ units of one product at \$2.00 per unit and $y$ units of another product at \$0.50 per unit. The total revenue for these two products is given by

$$R = (2)(x) + (0.5)(y)$$

$$= 2x + \frac{y}{2}$$

Graphically, we can represent this equation as a **surface in space.** (In this particular case the surface happens to be a plane.) The values of $x$ and $y$ are represented in the horizontal $xy$-plane, and the revenue $R$ is represented by the height of the surface. (See Figure 8.1.)

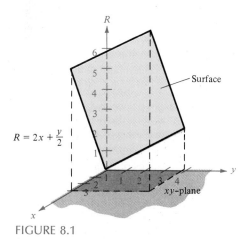

FIGURE 8.1

In Chapter 1, we described the Cartesian plane as the plane determined by two mutually perpendicular number lines called the $x$- and $y$-axes. These axes together with their point of intersection (the origin) allowed us to develop a two-dimensional coordinate system for identifying points in a plane and for discussing topics in plane analytic geometry. To identify a point in space, we need to introduce a third dimension to our model. The geometry of this three-dimensional model is referred to as **solid analytic geometry.**

To construct a **three-dimensional coordinate system,** we begin with the two-dimensional system (the $xy$-plane in a horizontal position), and through the origin, 0, we pass a vertical $z$-axis that is perpendicular to both the $x$- and $y$-axes, as shown in Figure 8.2.

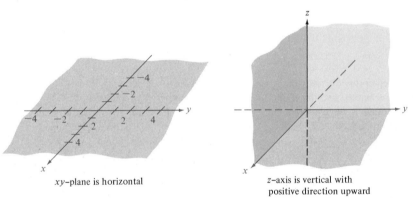

xy–plane is horizontal          z–axis is vertical with
                                positive direction upward

FIGURE 8.2

This particular orientation of the $x$-, $y$-, and $z$-axes is called a **right-handed system.** As an aid to remembering the relationship between the three axes in a right-handed system, imagine that you are standing at the origin with your arms in the direction of the positive $x$- and $y$-axes. The system is right-handed if your *right* hand points in the direction of the $x$-axis, and it is left-handed if your *left* hand points in the direction of the $x$-axis, as shown in Figure 8.3. In this text we will work exclusively with the right-handed system.

Pairwise, the three axes of the three-dimensional system form three **coordinate planes,** as shown in Figure 8.2. The **$xy$-plane** is determined by the $x$- and $y$-axes, the **$xz$-plane** by the $x$- and $z$-axes, and the **$yz$-plane** by the $y$- and $z$-axes. The three coordinate planes separate space into eight **octants,** and we refer to the *first octant* as the one in which all three coordinates are positive.

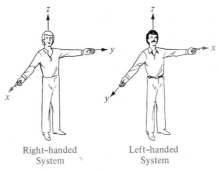

Right-handed
System

Left-handed
System

FIGURE 8.3

A point $P$ in three-dimensional space (3-space) is determined by an ordered triple $(x, y, z)$, where

$$x = \text{the directed distance from the } yz\text{-plane to } P$$

$$y = \text{the directed distance from the } xz\text{-plane to } P$$

$$z = \text{the directed distance from the } xy\text{-plane to } P$$

**EXAMPLE 1**

**Plotting Points in Space**

Plot the following points in space:
(a) $(-2, 5, 4)$      (b) $(1, 6, 0)$      (c) $(2, -5, 3)$      (d) $(3, 3, -2)$

**SOLUTION**

To plot the point $(-2, 5, 4)$, we note that

$$x = -2, \qquad y = 5, \qquad z = 4$$

To help visualize this point $(-2, 5, 4)$, we locate the point $(-2, 5)$ in the $xy$-plane (denoted by a cross). The point $(-2, 5, 4)$ lies 4 units above the cross, as shown in Figure 8.4.

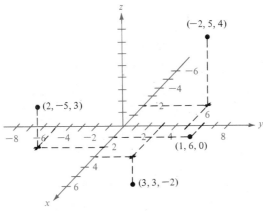

FIGURE 8.4

### The Distance Formula and the Midpoint Formula

Many of the geometric properties of space are simple extensions of those established for the plane in Chapter 1. For instance, by using the Pythagorean Theorem twice (see Figure 8.5), we can establish the *Distance Formula* for the distance $d$ between two points.

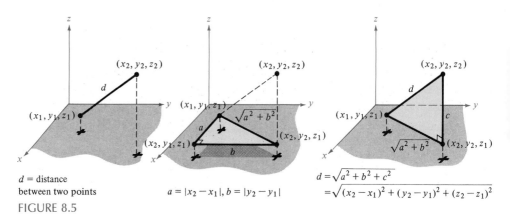

$d$ = distance between two points

$a = |x_2 - x_1|, b = |y_2 - y_1|$

$d = \sqrt{a^2 + b^2 + c^2}$
$= \sqrt{(x_2 - x_1)^2 + (y_2 - y_1)^2 + (z_2 - z_1)^2}$

FIGURE 8.5

---

**The Distance Formula**

The distance between $(x_1, y_1, z_1)$ and $(x_2, y_2, z_2)$ in space is

$$d = \sqrt{(x_2 - x_1)^2 + (y_2 - y_1)^2 + (z_2 - z_1)^2}$$

---

The *Midpoint Formula* is also a straightforward extension of the one for plane geometry. If $(x_1, y_1, z_1)$ and $(x_2, y_2, z_2)$ are two points in space, then the midpoint of the line segment connecting the two points is as follows.

---

**The Midpoint Formula**

The **midpoint** of the line segment connecting the two points $(x_1, y_1, z_1)$ and $(x_2, y_2, z_2)$ is given by

$$\text{Midpoint} = \left( \frac{x_1 + x_2}{2}, \frac{y_1 + y_2}{2}, \frac{z_1 + z_2}{2} \right)$$

---

**EXAMPLE 2**

**The Distance and Midpoint Between Two Points**

Find the distance between $(5, -2, 3)$ and $(0, 4, -3)$. What is the midpoint of the line segment connecting these two points?

---

**SOLUTION**

By the Distance Formula, we have

$$d = \sqrt{(0 - 5)^2 + (4 + 2)^2 + (-3 - 3)^2}$$
$$= \sqrt{25 + 36 + 36}$$
$$= \sqrt{97}$$

The midpoint is

$$\left(\frac{5 + 0}{2}, \frac{-2 + 4}{2}, \frac{3 - 3}{2}\right) = \left(\frac{5}{2}, 1, 0\right)$$

as shown in Figure 8.6.

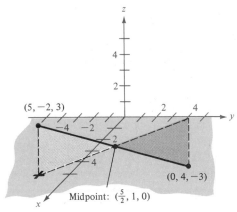

FIGURE 8.6

## The Equation of a Sphere in Space

Since a *sphere* consists of the set of all points in space that lie at a fixed distance from a given point, we can use the Distance Formula to obtain the following **standard equation of a sphere.**

**Standard equation of a sphere in space**

The **standard form of the equation of a sphere** with radius $r$ and center at $(h, k, l)$ is given by

$$(x - h)^2 + (y - k)^2 + (z - l)^2 = r^2$$

**EXAMPLE 3**

**Finding the Equation of a Sphere**

Find the standard form of the equation of the sphere whose center is $(0, 2, -1)$ and whose radius is 3.

**SOLUTION**

Using the standard form of the equation of a sphere,

$$(x - h)^2 + (y - k)^2 + (z - l)^2 = r^2$$

we have

$$(x - 0)^2 + (y - 2)^2 + [z - (-1)]^2 = 3^2$$
$$x^2 + (y - 2)^2 + (z + 1)^2 = 9$$

**EXAMPLE 4**

**Finding the Center and Radius of a Sphere**

Find the center and radius of the sphere whose equation is

$$x^2 + y^2 + z^2 - 2x + 4y - 6z + 8 = 0$$

**SOLUTION**

We can obtain the standard equation of this sphere by completing the square with each variable, as follows:

$$(x^2 - 2x \quad) + (y^2 + 4y \quad) + (z^2 - 6z \quad) = -8$$
$$(x^2 - 2x + 1) + (y^2 + 4y + 4) + (z^2 - 6z + 9) = -8 + 1 + 4 + 9$$
$$(x - 1)^2 + (y + 2)^2 + (z - 3)^2 = 6$$

Therefore, the center of the sphere is at $(1, -2, 3)$, and its radius is $\sqrt{6}$. (See Figure 8.7.)

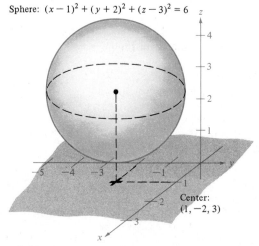

Sphere: $(x - 1)^2 + (y + 2)^2 + (z - 3)^2 = 6$

Center: (1, −2, 3)

FIGURE 8.7

Note in Example 4 that the points satisfying the equation of the given sphere are the points on the sphere's *surface* (not points inside the sphere). In general, we call the collection of points satisfying an equation involving $x$, $y$, and $z$ a **surface in**

**space.** We will discuss the idea of a surface in space in greater detail in the next section.

Finding the intersection of the surface with one (or more) of the three coordinate planes aids greatly in the visualization of a general surface in space. We call such intersections **traces.** For example, the $xy$-trace of a surface consists of all points that are common to both the surface *and* the $xy$-plane. Similarly, the $xz$-trace of a surface consists of all points that are common to both the surface *and* the $xz$-plane. The next example demonstrates the use of a trace in analyzing a surface.

**EXAMPLE 5**

**Finding a Trace of a Surface**

Sketch the $xy$-trace of the sphere whose equation is

$$(x - 3)^2 + (y - 2)^2 + (z + 4)^2 = 5^2$$

**SOLUTION**

To find the $xy$-trace of this surface, we use the fact that every point in the $xy$-plane has a $z$-coordinate of zero. This means that if we substitute $z = 0$ into the given equation, the resulting equation (involving the two variables $x$ and $y$) will represent the intersection of the surface with the $xy$-plane. In other words, by letting $z = 0$, we have

$$(x - 3)^2 + (y - 2)^2 + (0 + 4)^2 = 25$$
$$(x - 3)^2 + (y - 2)^2 + 16 = 25$$
$$(x - 3)^2 + (y - 2)^2 = 9$$
$$(x - 3)^2 + (y - 2)^2 = 3^2$$

Now, since this equation represents a circle (of radius 3) in the $xy$-plane, we can sketch the $xy$-trace as shown in Figure 8.8.

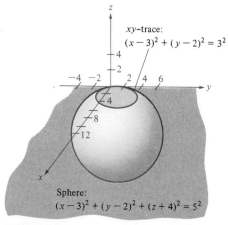

FIGURE 8.8

## SECTION EXERCISES 8.1

In Exercises 1–4, plot the points on the same 3-dimensional coordinate system.

**1.** (a) (2, 1, 3)  (b) (−1, 2, 1)

**2.** (a) (3, −2, 5)  (b) $\left(\frac{3}{2}, 4, -2\right)$

**3.** (a) (5, −2, 2)  (b) (5, −2, −2)

**4.** (a) (0, 4, −5)  (b) (4, 0, 5)

In Exercises 5–8, find the coordinates of the midpoint of the line segment joining the given points.

**5.** (5, −9, 7), (−2, 3, 3)

**6.** (4, 0, −6), (8, 8, 20)

**7.** (−5, −2, 5.5), (6.3, 4.2, −7.1)

**8.** (0, −2, 5), (4, 2, 7)

In Exercises 9 and 10, the coordinates of one endpoint and the midpoint of a line segment are given. Find the coordinates of the other endpoint.

**9.** Endpoint: (−2, 1, 1); midpoint: (0, 2, 5)

**10.** Endpoint: (3, −2, 6); midpoint: (−1, 5, −1)

In Exercises 11–14, find the distance between the two points.

**11.** (4, 1, 5), (8, 2, 6)

**12.** (−4, −1, 1), (2, −1, 5)

**13.** (−1, −5, 7), (−3, 4, −4)

**14.** (8, −2, 2), (8, −2, 4)

In Exercises 15–18, find the lengths of the sides of the triangle with the indicated vertices, and determine whether the triangle is a right triangle, an isosceles triangle, or neither of these.

**15.** (0, 0, 0), (2, 2, 1), (2, −4, 4)

**16.** (5, 3, 4), (7, 1, 3), (3, 5, 3)

**17.** (1, −3, −2), (5, −1, 2), (−1, 1, 2)

**18.** (5, 0, 0), (0, 2, 0), (0, 0, −3)

In Exercises 19–22, find the standard form of the equation of the sphere.

**19.** Center: (0, 2, 5); radius: 2

**20.** Center: (4, −1, 1); radius: 5

**21.** Endpoints of a diameter are (2, 0, 0) and (0, 6, 0)

**22.** Center: (−2, 1, 1); tangent to the $xy$-coordinate plane

In Exercises 23–30, find the center and radius of the sphere.

**23.** $x^2 + y^2 + z^2 - 2x + 6y + 8z + 1 = 0$

**24.** $x^2 + y^2 + z^2 - 5x = 0$

**25.** $x^2 + y^2 + z^2 - 8y = 0$

**26.** $x^2 + y^2 + z^2 - 4y + 6z + 4 = 0$

**27.** $2x^2 + 2y^2 + 2z^2 - 2x + 2y - 4z + 1 = 0$

**28.** $x^2 + y^2 + z^2 = 36$

**29.** $9x^2 + 9y^2 + 9z^2 - 6x + 18y + 1 = 0$

**30.** $4x^2 + 4y^2 + 4z^2 - 4x - 32y + 8z + 33 = 0$

In Exercises 31 and 32, sketch the $xy$-trace of the sphere.

**31.** $(x - 2)^2 + (y - 2)^2 + (z - 3)^2 = 25$

**32.** $x^2 + y^2 + z^2 - 6x - 10y + 6z + 30 = 0$

In Exercises 33 and 34, sketch the $yz$-trace of the sphere.

**33.** $x^2 + y^2 + z^2 - 4x - 4y - 6z - 12 = 0$

**34.** $x^2 + y^2 + z^2 - 6x - 10y + 6z + 30 = 0$

# Surfaces in Space

Sketching the graph of an equation involving three variables can be quite difficult. However, today the problem
can be simplified through the use of a computer. For instance, suppose we want to sketch the graph of the equation

$$z = (x^2 + y^2)e^{1-x^2-y^2}$$

Although there are several types of computer graphics
plotting routines, most use some form of trace analysis to
give the illusion of three dimensions. To use such a program, we need to enter the equation, the region in the
xy-plane over which the graph should be sketched, the
scale for the z-axis, and the number of traces to be taken.
For instance, suppose we choose to sketch the above
equation over the square region bounded by

$$-3 \leq x \leq 3 \qquad \text{Bounds for } x$$

$$-3 \leq y \leq 3 \qquad \text{Bounds for } y$$

Figure 8.9 shows a computer-generated sketch of this
equation using 26 traces taken parallel to the y-axis. To
heighten the three-dimensional effect, the program uses a
"hidden line" routine. That is, it begins by sketching the
traces in the foreground (those corresponding to the largest x-values). Then as each new trace is sketched, the program determines whether all or only part of the next trace
should be shown.

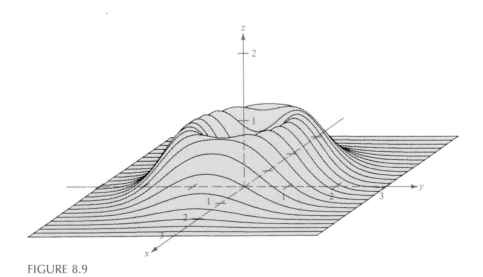

FIGURE 8.9

■ **Equation of a Plane in Space**
■ **Quadric Surfaces in Space**

In Section 8.1, we looked at the equation of one type of surface in space—a sphere. A second type of surface in space is a plane. The **general equation of a plane** is

$$ax + by + cz = d \qquad \text{General equation of a plane}$$

Note the similarity of this equation to the general equation of a line in the plane. As a matter of fact, if we form the intersection of the plane represented by this equation with each of the three coordinate planes, we see that the resulting traces are lines. (See Figure 8.10.) This is what we would expect, since we know from geometry that two nonparallel planes intersect in a line.

In Figure 8.10, the points where the given plane intersects the three coordinate axes are called the $x$-, $y$-, and $z$-intercepts. By connecting these three points, we can form a triangular region, which helps us to visualize the plane in space. Example 1 demonstrates this procedure.

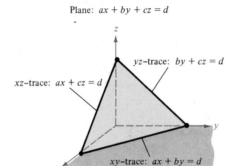

Plane: $ax + by + cz = d$

$xz$-trace: $ax + cz = d$

$yz$-trace: $by + cz = d$

$xy$-trace: $ax + by = d$

FIGURE 8.10

**EXAMPLE 1**

**Sketching a Plane in Space**

Find the $x$-, $y$-, and $z$-intercepts of the plane given by

$$3x + 2y + 4z = 12$$

and then sketch the triangular region formed by connecting these three intercepts.

**SOLUTION**

To find the $x$-intercept, we let both $y$ and $z$ be zero:

$$3x + 2(0) + 4(0) = 12$$
$$3x = 12$$
$$x = 4$$

Thus, the $x$-intercept is $(4, 0, 0)$. To find the $y$-intercept, we let $x$ and $z$ be zero and conclude that $y = 6$, which means that the $y$-intercept is $(0, 6, 0)$. Similarly, by letting $x$ and $y$ be zero, we can determine that $z = 3$ and that the $z$-intercept is $(0, 0, 3)$. Figure 8.11 shows the triangular portion of the given plane formed by connecting these three intercepts.

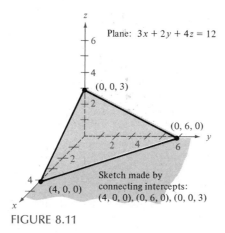

Plane: $3x + 2y + 4z = 12$

$(0, 0, 3)$

$(0, 6, 0)$

$(4, 0, 0)$

Sketch made by connecting intercepts: $(4, 0, 0), (0, 6, 0), (0, 0, 3)$

FIGURE 8.11

In Figures 8.10 and 8.11, we pictured planes that have three intercepts. It is possible for a plane in space to have less than three intercepts. In particular, this occurs when one or more of the coefficients in $ax + by + cz = d$ is zero. Figure 8.12 shows some planes in space that have only one intercept, and Figure 8.13 shows some that have only two intercepts.

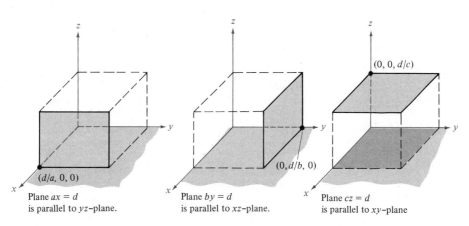

$(d/a, 0, 0)$

Plane $ax = d$
is parallel to $yz$-plane.

$(0, d/b, 0)$

Plane $by = d$
is parallel to $xz$-plane.

$(0, 0, d/c)$

Plane $cz = d$
is parallel to $xy$-plane

Planes Parallel to Coordinate Planes

FIGURE 8.12

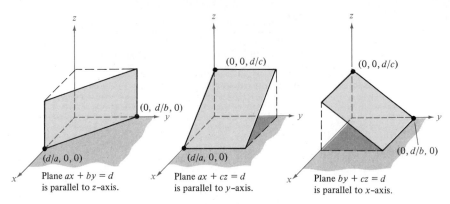

Plane $ax + by = d$ is parallel to $z$-axis.

Plane $ax + cz = d$ is parallel to $y$-axis.

Plane $by + cz = d$ is parallel to $x$-axis.

Planes Parallel to Coordinate Axes

FIGURE 8.13

## Quadric Surfaces

So far we have looked at two types of surfaces in space: spheres and planes. A third common type of surface in space is the one whose equation is of the form

$$Ax^2 + By^2 + Cz^2 + Dx + Ey + Fz + G = 0$$

We call the graph of such an equation a **quadric surface.** We confine our discussion of quadric surfaces to those having their center at the origin and axes along the coordinate axes. The six basic quadric surfaces are as follows:

1. Elliptic cone
2. Elliptic paraboloid
3. Hyperbolic paraboloid
4. Ellipsoid
5. Hyperboloid of one sheet
6. Hyperboloid of two sheets

In this text you will *not* be asked to sketch quadric surfaces in space. However, familiarity with some of the common surfaces will prove to be a great aid in visualizing some of the concepts in this chapter. For this reason, we suggest that you study the equations and quadric surfaces shown in Table 8.1 carefully.

▬▬ **Remark:** Table 8.1 includes several computer-generated drawings of quadric surfaces in space. Note that these drawings make use of trace analysis to enhance the three-dimensional perspective. For these particular surfaces, the computer sketched the traces in several evenly spaced planes (all of which were taken parallel to the $yz$-plane).

In Table 8.1, only one of several orientations of each quadric surface is shown. If the surface is oriented along a different axis, its standard equation will change accordingly, as illustrated in the next examples. In classifying quadric surfaces it is helpful to remember that the two types of paraboloids have one variable raised to the first power. The other four types of basic quadric surfaces have equations that are of *second degree* in all three variables.

TABLE 8.1   Quadric Surfaces

### Elliptic Cone

$$\frac{x^2}{a^2} + \frac{y^2}{b^2} - \frac{z^2}{c^2} = 0$$

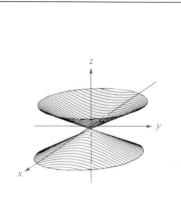

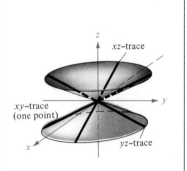

| Trace | Plane |
|---|---|
| Ellipse | Parallel to $xy$-plane |
| Hyperbola | Parallel to $xz$-plane |
| Hyperbola | Parallel to $yz$-plane |

The axis of the cone corresponds to the variable whose coefficient is negative. The traces in the coordinate planes parallel to this axis are intersecting lines.

### Elliptic Paraboloid

$$z = \frac{x^2}{a^2} + \frac{y^2}{b^2}$$

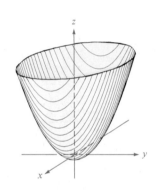

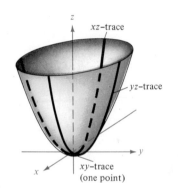

| Trace | Plane |
|---|---|
| Ellipse | Parallel to $xy$-plane |
| Parabola | Parallel to $xz$-plane |
| Parabola | Parallel to $yz$-plane |

The axis of the paraboloid corresponds to the variable raised to the first power.

### Hyperbolic Paraboloid

$$z = \frac{y^2}{b^2} - \frac{x^2}{a^2}$$

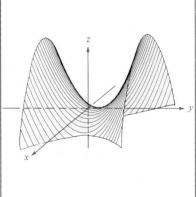

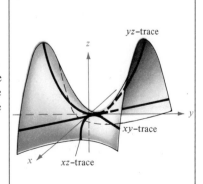

| Trace | Plane |
|---|---|
| Hyperbola | Parallel to $xy$-plane |
| Parabola | Parallel to $xz$-plane |
| Parabola | Parallel to $yz$-plane |

The axis of the paraboloid corresponds to the variable raised to the first power.

TABLE 8.1   Quadric Surfaces (Continued)

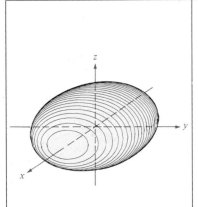

### Ellipsoid

$$\frac{x^2}{a^2} + \frac{y^2}{b^2} + \frac{z^2}{c^2} = 1$$

| *Trace* | *Plane* |
|---------|---------|
| Ellipse | Parallel to $xy$-plane |
| Ellipse | Parallel to $xz$-plane |
| Ellipse | Parallel to $yz$-plane |

The surface is a sphere if
$a = b = c \neq 0$.

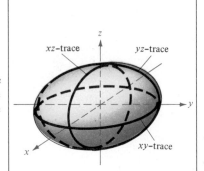

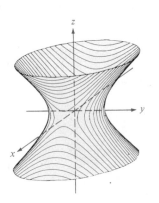

### Hyperboloid of One Sheet

$$\frac{x^2}{a^2} + \frac{y^2}{b^2} - \frac{z^2}{c^2} = 1$$

| *Trace* | *Plane* |
|---------|---------|
| Ellipse | Parallel to $xy$-plane |
| Hyperbola | Parallel to $xz$-plane |
| Hyperbola | Parallel to $yz$-plane |

The axis of the hyperboloid
corresponds to the variable whose
coefficient is negative.

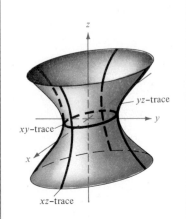

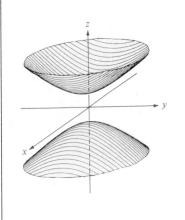

### Hyperboloid of Two Sheets

$$\frac{z^2}{c^2} - \frac{x^2}{a^2} - \frac{y^2}{b^2} = 1$$

| *Trace* | *Plane* |
|---------|---------|
| Ellipse | Parallel to $xy$-plane |
| Hyperbola | Parallel to $xz$-plane |
| Hyperbola | Parallel to $yz$-plane |

The axis of the hyperboloid
corresponds to the variable whose
coefficient is positive. There is no
trace in the coordinate plane
perpendicular to this axis.

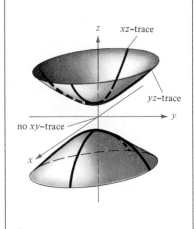

**EXAMPLE 2**

**Classifying a Quadric Surface**

Classify the surface given by

$$x - y^2 - z^2 = 0$$

**SOLUTION**

Since $x$ is raised only to the first power the surface will be a paraboloid, and in this case its axis is the $x$-axis. In the standard form, the equation is

$$x = y^2 + z^2$$

Some convenient traces are

$$xy\text{-trace:} \quad x = y^2 \qquad \text{Parabola}$$
$$(z = 0)$$

$$xz\text{-trace:} \quad x = z^2 \qquad \text{Parabola}$$
$$(y = 0)$$

$$\text{Parallel to } yz\text{-plane:} \quad y^2 + z^2 = 1 \qquad \text{Circle}$$
$$(x = 1)$$

Thus the surface is an elliptic (or circular) paraboloid, as shown in Figure 8.14.

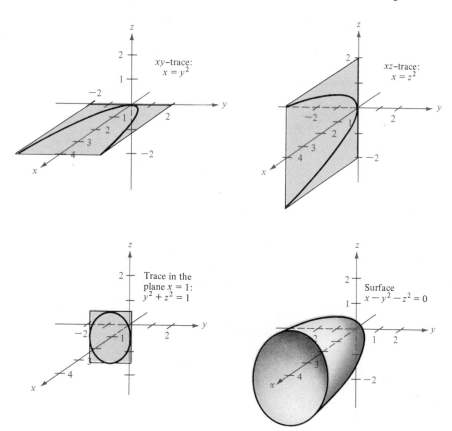

FIGURE 8.14

**EXAMPLE 3**　　　**Classifying a Quadric Surface**

Classify the surface given by

$$x^2 - 4y^2 - 4z^2 - 4 = 0$$

**SOLUTION**　　　We express the given equation in standard form as follows:

$$x^2 - 4y^2 - 4z^2 - 4 = 0 \qquad \text{Given equation}$$

$$\frac{x^2}{4} - y^2 - z^2 - 1 = 0 \qquad \text{Divide by 4}$$

$$\frac{x^2}{4} - y^2 - z^2 = 1 \qquad \text{Standard form}$$

From Table 8.1 we conclude that the surface is a hyperboloid of two sheets with the $x$-axis as its axis. Note in Figure 8.15 that the graph has the following traces:

$$xy\text{-trace:} \quad \frac{x^2}{4} - y^2 = 1 \qquad \text{Hyperbola}$$
$$(z = 0)$$

$$xz\text{-trace:} \quad \frac{x^2}{4} - z^2 = 1 \qquad \text{Hyperbola}$$
$$(y = 0)$$

$$\text{Parallel to } yz\text{-plane:} \quad y^2 + z^2 = 8 \qquad \text{Circle}$$
$$(x = 6)$$

The graph is shown in Figure 8.15.

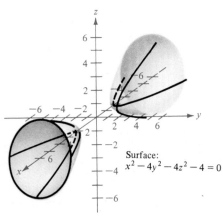

Surface:
$x^2 - 4y^2 - 4z^2 - 4 = 0$

FIGURE 8.15

**EXAMPLE 4**

**Classifying a Quadric Surface**

Classify the surface given by

$$x^2 + 4y^2 + z^2 = 4$$

**SOLUTION**

To write this equation in standard form, we divide by 4 to obtain

$$\frac{x^2}{4} + y^2 + \frac{z^2}{4} = 1 \qquad \text{Standard form}$$

The traces in the three coordinate planes are as follows:

$$xy\text{-trace:} \quad \frac{x^2}{4} + y^2 = 1 \qquad \text{Ellipse}$$
$$(z = 0)$$

$$xz\text{-trace:} \quad x^2 + z^2 = 4 \qquad \text{Circle}$$
$$(y = 0)$$

$$yz\text{-trace:} \quad y^2 + \frac{z^2}{4} = 1 \qquad \text{Ellipse}$$
$$(x = 0)$$

Thus the surface is an ellipsoid, as shown in Figure 8.16.

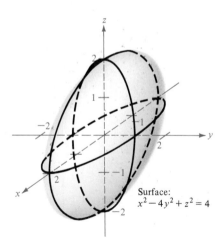

Surface:
$$x^2 - 4y^2 + z^2 = 4$$

FIGURE 8.16

When working the exercises in this section, remember that for the purpose of this text our main goal is to be able to *read* three-dimensional graphs, not to be able to *create* them. In the following exercise set, the only graphing you are asked to do is that involving planes in space.

## SECTION EXERCISES 8.2

In Exercises 1–12, find the intercepts and sketch the graph of the plane.

**1.** $4x + 2y + 6z = 12$

**2.** $3x + 6y + 2z = 6$

**3.** $3x + 3y + 5z = 15$

**4.** $x + y + z = 3$

**5.** $2x - y + 3z = 4$

**6.** $2x - y + z = 4$

**7.** $z = 3$

**8.** $y = -4$

**9.** $y + z = 5$

**10.** $x + 2y = 8$

**11.** $x + y - z = 0$

**12.** $x - 3z = 3$

In Exercises 13–20, determine whether the planes $a_1 x + b_1 y + c_1 z = d_1$ and $a_2 x + b_2 y + c_2 z = d_2$ are parallel, perpendicular, or neither. The planes are parallel if there exists a nonzero constant $k$ such that $a_1 = ka_2$, $b_1 = kb_2$, $c_1 = kc_2$, and perpendicular if $a_1 a_2 + b_1 b_2 + c_1 c_2 = 0$.

**13.** $5x - 3y + z = 4$, $x + 4y + 7z = 1$

**14.** $3x + y - 4z = 3$, $-9x - 3y + 12z = 4$

**15.** $x - 5y - z = 1$, $5x - 25y - 5z = -3$

**16.** $x = 6$, $y = -1$

**17.** $x + 2y = 3$, $4x + 8y = 5$

**18.** $x + 3y + z = 7$, $x - 5z = 0$

**19.** $2x + y = 3$, $x - 5z = 0$

**20.** $2x - z = 1$, $4x + y + 8z = 10$

In Exercises 21–24, find the distance between the point and the plane. The distance $D$ between a point $(x_0, y_0, z_0)$ and the plane $ax + by + cz + d = 0$ is

$$D = \frac{|ax_0 + by_0 + cz_0 + d|}{\sqrt{a^2 + b^2 + c^2}}$$

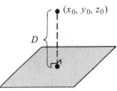

Plane:
$ax + by + cz + d = 0$

**21.** $(0, 0, 0)$, $2x + 3y + z = 12$

**22.** $(1, 5, -4)$, $3x - y + 2z = 6$

**23.** $(1, 2, 3)$, $2x - y + z = 4$

**24.** $(10, 0, 0)$, $x - 3y + 4z = 6$

In Exercises 25–32, match the given equation with the correct graph. [Graphs are labeled (a), (b), (c), (d), (e), (f), (g), and (h).]

**25.** $\dfrac{x^2}{9} + \dfrac{y^2}{16} + \dfrac{z^2}{9} = 1$

**26.** $15x^2 - 4y^2 + 15z^2 = -4$

**27.** $4x^2 - y^2 + 4z^2 = 4$

**28.** $y^2 = 4x^2 + 9z^2$

**29.** $4x^2 - 4y + z^2 = 0$

**30.** $12z = -3y^2 + 4x^2$

**31.** $4x^2 - y^2 + 4z = 0$

**32.** $x^2 + y^2 + z^2 = 9$

(a)

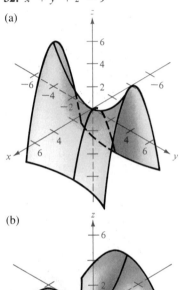

(b)

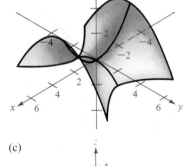

(c)

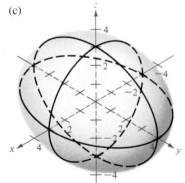

(d)

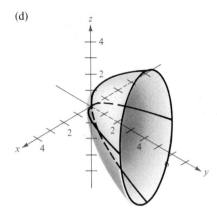

(e)

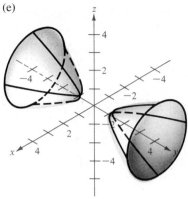

(f)

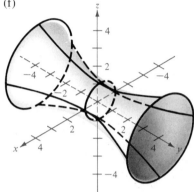

(g)

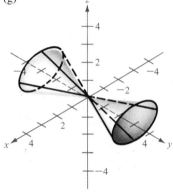

(h)

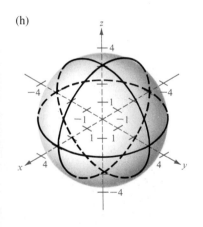

In Exercises 33–48, identify the quadric surface.

**33.** $x^2 + \dfrac{y^2}{4} + z^2 = 1$

**34.** $\dfrac{x^2}{9} + \dfrac{y^2}{16} + \dfrac{z^2}{16} = 1$

**35.** $16x^2 - y^2 + 16z^2 = 4$

**36.** $9x^2 + 4y^2 - 8z^2 = 72$

**37.** $x^2 - y + z^2 = 0$

**38.** $z = 4x^2 + y^2$

**39.** $x^2 - y^2 + z = 0$

**40.** $z^2 - x^2 - \dfrac{y^2}{4} = 1$

**41.** $4x^2 - y^2 + 4z^2 = -16$

**42.** $z^2 = x^2 + \dfrac{y^2}{4}$

**43.** $z^2 = x^2 + 4y^2$

**44.** $4y = x^2 + z^2$

**45.** $3z = -y^2 + x^2$

**46.** $z^2 = 2x^2 + 2y^2$

**47.** $2x^2 + 2y^2 + 2z^2 - 3x + 4z = 10$

**48.** $4x^2 + y^2 - 4z^2 - 16x - 6y - 16z + 9 = 0$

# Functions of Several Variables

The surface area $A$ (in square feet) of a person's skin depends on both the weight and the height of the person. One model for estimating skin surface area is given by the **function of two variables**

$$A = f(x, y) = 0.6416x^{0.425}y^{0.725}$$

where $x$ is the person's weight (in pounds) and $y$ is the person's height (in feet). For example, someone whose weight is 160 pounds and whose height is 6 feet has an estimated surface area of

$$A = f(160, 6) = 0.6416(160^{0.425})(6^{0.725})$$

$$\approx 20.33 \text{ square feet}$$

Table 8.2 lists the surface areas (in square feet) given by this model for several different heights and weights.

Figure 8.17 shows the surface in space given by this function. Note that some of the $xy$-values pictured in Figure 8.17 make no sense for this particular model. For example, a person could not weigh 200 pounds if he were only 2 feet high! To compensate for this problem, we restrict this function to an appropriate **domain** in the $xy$-plane for which the model's estimate is reasonable.

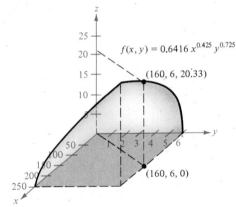

FIGURE 8.17

TABLE 8.2

| Weight (lb) | Height | | | | |
|---|---|---|---|---|---|
| | 5' | 5'4" | 5'8" | 6' | 6'4" |
| 100 | 14.59 | 15.29 | 15.97 | 16.65 | 17.32 |
| 150 | 17.33 | 18.16 | 18.98 | 19.78 | 20.57 |
| 200 | 19.59 | 20.52 | 21.45 | 22.35 | 23.25 |
| 250 | 21.53 | 22.57 | 23.58 | 24.58 | 25.56 |

■ **Functions of Several Variables**
■ **Graphs of Functions of Two Variables**
■ **Applications**

In the first seven chapters of this book, we dealt only with functions of a single independent variable. Many familiar quantities are functions not of one but of two or more variables. For instance, it may be more realistic to consider the demand for a particular product to be a function of the two variables price and amount spent on advertising, rather than a function of price alone. Similarly, the growth of a plant may be considered to be a function of the three variables rainfall, hours of sunshine, and the amount of fertilizer.

We denote functions of two or more variables by a notation similar to that used for functions of a single variable. For example,

$$f(x, y) = x^2 + xy \qquad \text{and} \qquad g(x, y) = e^{x+y}$$

$$\underbrace{\phantom{f(x, y)}}_{\text{2 variables}} \qquad\qquad\qquad \underbrace{\phantom{g(x, y)}}_{\text{2 variables}}$$

are functions of two variables, and

$$f(x, y, z) = x + 2y - 3z$$

$$\underbrace{\phantom{f(x, y, z)}}_{\text{3 variables}}$$

is a function of three variables. We give the following definition of a function of two variables. There are similar definitions for functions of three, four, or $n$ variables.

| | |
|---|---|
| **Definition of a function of two variables** | If to each ordered pair $(x, y)$ in some set $D$ there corresponds a unique real number $f(x, y)$, then $f$ is called a **function of $x$ and $y$.** The set $D$ is the **domain** of $f$, and the corresponding set of values for $f(x, y)$ is the **range** of $f$. |

A function of two variables can be represented geometrically as a surface in space by letting $z = f(x, y)$. Note in Figure 8.18 that although the surface is three-

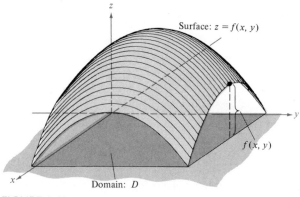

FIGURE 8.18

dimensional, the domain of the function is two-dimensional, since it consists of the points in the *xy*-plane for which the function is defined. In this chapter, when we refer to the function given by the equation $z = f(x, y)$, we will assume (unless it is specifically restricted) that the domain is the set of all points $(x, y)$ for which the equation has meaning.

**EXAMPLE 1**

**Finding the Domain and Range of a Function**

Determine the domain and range of the function defined by

$$f(x, y) = \sqrt{64 - x^2 - y^2}$$

**SOLUTION**

Since $D$ is not otherwise specified, we assume the domain of $f$ to be the set of all points $(x, y)$ such that

$$64 - x^2 - y^2 \geq 0$$

$$x^2 + y^2 \leq 64$$

In other words, $D$ is the set of all points lying on or inside the circle

$$x^2 + y^2 = 8^2$$

The range of $f$ is all values $z = f(x, y)$ such that

$$0 \leq z \leq 8$$

The graph of $f$ is a hemisphere, as shown in Figure 8.19.

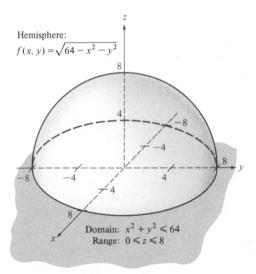

FIGURE 8.19

In Table 8.1 in the preceding section, we saw that it is often possible to obtain a good picture of a surface in space by sketching traces of the surface taken in several

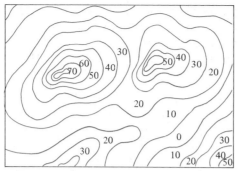

Topographical Map of Two Peaks.
One is over 70 units high and
the other is over 50 units high.

FIGURE 8.20

evenly spaced parallel planes. A valuable adaptation of this method involves the use of horizontal planes whose traces are *projected* onto the $xy$-plane. We call these projections in the $xy$-plane **level curves.** A graph constructed by sketching several level curves on the same $xy$-plane is called a **contour map.** This process is commonly used in constructing topographical maps, such as the one shown in Figure 8.20. Each of the level curves in this map represents the intersection of the surface $z = f(x, y)$ with the plane $z = c$, where $c = 0, 10, 20, \ldots , 70$. We demonstrate the construction of a contour map in the next example.

**EXAMPLE 2**

**Sketching a Contour Map**

Sketch a contour map for the function $f(x, y) = \sqrt{64 - x^2 - y^2}$ using $c_1 = 0$, $c_2 = 1$, $c_3 = 2, \ldots , c_9 = 8$.

**SOLUTION**

For each of these nine values of $c$, we obtain a circular level curve.

$$c_1 = 0: 0 = \sqrt{64 - x^2 - y^2} \implies x^2 + y^2 = 64 = 8^2$$

$$c_2 = 1: 1 = \sqrt{64 - x^2 - y^2} \implies x^2 + y^2 = 63 \approx 7.94^2$$

$$c_3 = 2: 2 = \sqrt{64 - x^2 - y^2} \implies x^2 + y^2 = 60 \approx 7.75^2$$

$$c_4 = 3: 3 = \sqrt{64 - x^2 - y^2} \implies x^2 + y^2 = 55 \approx 7.42^2$$

$$c_5 = 4: 4 = \sqrt{64 - x^2 - y^2} \implies x^2 + y^2 = 48 \approx 6.93^2$$

$$c_6 = 5: 5 = \sqrt{64 - x^2 - y^2} \implies x^2 + y^2 = 39 \approx 6.24^2$$

$$c_7 = 6: 6 = \sqrt{64 - x^2 - y^2} \implies x^2 + y^2 = 28 \approx 5.29^2$$

$$c_8 = 7: 7 = \sqrt{64 - x^2 - y^2} \implies x^2 + y^2 = 15 \approx 3.87^2$$

$$c_9 = 8: 8 = \sqrt{64 - x^2 - y^2} \implies x^2 + y^2 = 0 = 0^2$$

The nine level curves representing this hemisphere are shown in Figure 8.21.

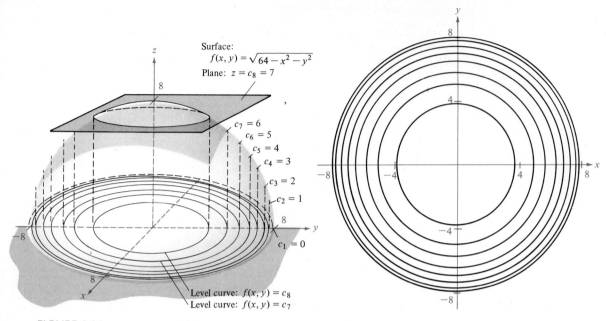

**Surface:**
$$f(x, y) = \sqrt{64 - x^2 - y^2}$$
**Plane:** $z = c_8 = 7$

$c_7 = 6$
$c_6 = 5$
$c_5 = 4$
$c_4 = 3$
$c_3 = 2$
$c_2 = 1$
$c_1 = 0$

Level curve: $f(x, y) = c_8$
Level curve: $f(x, y) = c_7$

FIGURE 8.21

■■ **Remark:** Note in Example 2 that a contour map depicts the variation of $z$ with respect to $x$ and $y$ by the change in the spacing of the level curves. Much space between level curves indicates that $z$ is changing slowly, and little space between level curves indicates a rapid change in $z$. Furthermore, remember that in order to give the correct impression in a contour map it is important to choose $c$-values that are *evenly spaced*.

**EXAMPLE 3**

**Evaluating a Function of Several Variables**

Evaluate each of the following functions at the indicated point:
(a) Given $f(x, y) = 2x^2 - y^2$, find $f(2, 3)$.
(b) Given $f(x, y, z) = e^x(y + z)$, find $f(0, -1, 4)$.

**SOLUTION**

(a) Since $f(x, y) = 2x^2 - y^2$, we have

$$f(2, 3) = 2(2^2) - 3^2$$
$$= 2(4) - 9$$
$$= 8 - 9 = -1$$

(b) Since $f(x, y, z) = e^x(y + z)$, we have

$$f(0, -1, 4) = e^0(-1 + 4)$$
$$= (1)(3) = 3$$

**EXAMPLE 4**

**An Application in Economics**

The *Cobb-Douglas production function* is used in economics to represent the number of units produced by varying amounts of labor and capital. If $x$ measures the units of labor and $y$ measures the units of capital, then the number of units produced is given by

$$f(x, y) = Cx^a y^{1-a}$$

where $C$ is constant and $0 < a < 1$. Suppose that a manufacturer estimates a particular production function to be

$$f(x, y) = 100x^{0.6}y^{0.4}$$

(a) What is the production level when $x = 1000$ and $y = 500$?
(b) What is the production level when $x = 2000$ and $y = 1000$?

**SOLUTION**

(a) When $x = 1000$ and $y = 500$, we have a production level of

$$f(1000, 500) = 100(1000^{0.6})(500^{0.4})$$

$$\approx 100(63.10)(12.01)$$

$$\approx 75,786$$

(b) When $x = 2000$ and $y = 1000$, we have a production level of

$$f(2000, 1000) = 100(2000^{0.6})(1000^{0.4})$$

$$\approx 100(95.64)(15.85)$$

$$\approx 151,572$$

Note that by doubling *both* $x$ and $y$ we doubled the production level. In Exercise 32 at the end of this section, you are asked to show that this result is characteristic of the Cobb-Douglas production function.

**EXAMPLE 5**

**An Application in Banking**

The monthly payment for an installment loan of $P$ dollars taken out over $t$ years at an annual percentage rate of $r$ is given by

$$\text{Monthly payment} = f(P, r, t) = \frac{Pr}{12}\left[\frac{1}{1 - \left(\dfrac{1}{1 + [r/12]}\right)^{12t}}\right]$$

(a) Find the monthly payment for a home mortgage of $75,000 taken out for 30 years at 16% interest. (See the Introductory Example to Section 2.2.)
(b) Find the monthly payment for a home mortgage of $75,000 taken out for 30 years at 8% interest.

**SOLUTION**

(a) If $P = \$75{,}000$, $r = 0.16$, and $t = 30$, the monthly payment is

$$f(75{,}000, 0.16, 30) = \frac{(75{,}000)(0.16)}{12}\left[\frac{1}{1 - \left(\dfrac{1}{1 + [0.16/12]}\right)^{360}}\right]$$

$$= \$1{,}008.57$$

(b) If $P = \$75{,}000$, $r = 0.08$, and $t = 30$, the monthly payment is

$$f(75{,}000, 0.08, 30) = \frac{(75{,}000)(0.08)}{12}\left[\frac{1}{1 - \left(\dfrac{1}{1 + [0.08/12]}\right)^{360}}\right]$$

$$= \$550.32$$

## SECTION EXERCISES 8.3

In Exercises 1–14, find the functional values.

**1.** $f(x, y) = \dfrac{x}{y}$

  (a) $f(3, 2)$           (b) $f(-1, 4)$
  (c) $f(30, 5)$         (d) $f(5, y)$
  (e) $f(x, 2)$          (f) $f(5, t)$

**2.** $f(x, y) = 4 - x^2 - 4y^2$

  (a) $f(0, 0)$          (b) $f(0, 1)$
  (c) $f(2, 3)$          (d) $f(1, y)$
  (e) $f(x, 0)$          (f) $f(t, 1)$

**3.** $f(x, y) = xe^y$

  (a) $f(5, 0)$          (b) $f(3, 2)$
  (c) $f(2, -1)$       (d) $f(5, y)$
  (e) $f(x, 2)$          (f) $f(t, t)$

**4.** $g(x, y) = \ln |x + y|$

  (a) $g(2, 3)$          (b) $g(5, 6)$
  (c) $g(e, 0)$          (d) $g(0, 1)$
  (e) $g(2, -3)$      (f) $g(e, e)$

**5.** $h(x, y, z) = \dfrac{xy}{z}$

  (a) $h(2, 3, 9)$      (b) $h(1, 0, 1)$

**6.** $f(x, y, z) = \sqrt{x + y + z}$

  (a) $f(0, 5, 4)$      (b) $f(6, 8, -3)$

**7.** $V(r, h) = \pi r^2 h$

  (a) $V(3, 10)$       (b) $V(5, 2)$

**8.** $F(r, N) = 500\left[1 + \dfrac{r}{12}\right]^N$

  (a) $F(0.09, 60)$    (b) $F(0.14, 240)$

**9.** $A(P, r, t) = P\left[\left(1 + \dfrac{r}{12}\right)^{12t} - 1\right]\left(1 + \dfrac{12}{r}\right)$

  (a) $A(100, 0.10, 10)$    (b) $A(275, 0.0925, 40)$

**10.** $A(P, r, t) = Pe^{rt}$

  (a) $A(500, 0.10, 5)$    (b) $A(1500, 0.12, 20)$

**11.** $f(x, y) = \displaystyle\int_x^y (2t - 3)\, dt$

  (a) $f(0, 4)$          (b) $f(1, 4)$

**12.** $g(x, y) = \displaystyle\int_x^y \frac{1}{t}\, dt$

  (a) $g(4, 1)$          (b) $g(6, 3)$

**13.** $f(x, y) = x^2 - 2y$

  (a) $\dfrac{f(x + \Delta x, y) - f(x, y)}{\Delta x}$

  (b) $\dfrac{f(x, y + \Delta y) - f(x, y)}{\Delta y}$

**14.** $f(x, y) = 3xy + y^2$

  (a) $\dfrac{f(x + \Delta x, y) - f(x, y)}{\Delta x}$

  (b) $\dfrac{f(x, y + \Delta y) - f(x, y)}{\Delta y}$

In Exercises 15–24, describe the region $R$ in the $xy$-coordinate plane that corresponds to the domain of the given function, and find the range of the function.

**15.** $f(x, y) = \sqrt{4 - x^2 - 4y^2}$

**16.** $f(x, y) = \sqrt{x^2 + y^2 - 1}$

**17.** $f(x, y) = 4 - x^2 - y^2$

**18.** $h(x, y) = e^{x/y}$

**19.** $f(x, y) = \ln (4 - x - y)$

**20.** $f(x, y) = \ln (4 - xy)$

**21.** $h(x, y) = \dfrac{1}{xy}$

**22.** $g(x, y) = \dfrac{1}{x - y}$

**23.** $g(x, y) = x\sqrt{y}$

**24.** $f(x, y) = x^2 + y^2$

In Exercises 25–30, sketch a contour map for the given function using the indicated $c$-values.

| Function | $c$-values |
|---|---|
| **25.** $f(x, y) = \sqrt{25 - x^2 - y^2}$ | $c = 0, 1, 2, 3, 4, 5$ |
| **26.** $f(x, y) = x^2 + y^2$ | $c = 0, 2, 4, 6, 8$ |
| **27.** $f(x, y) = xy$ | $c = 1, -1, 3, -3$ |
| **28.** $f(x, y) = 6 - 2x - 3y$ | $c = 0, 2, 4, 6, 8, 10$ |
| **29.** $f(x, y) = \ln (x - y)$ | $c = 0, \pm\dfrac{1}{2}, \pm1, \pm\dfrac{3}{2},$ $\pm2$ |
| **30.** $f(x, y) = \dfrac{x + y}{x - y}$ | $c = 0, \pm1, \pm2, \pm3$ |

**31.** A manufacturer estimates the Cobb-Douglas production function to be

$$f(x, y) = 100x^{0.75}y^{0.25}$$

Estimate the production level when $x = 1500$ and $y = 1000$.

**32.** Use the Cobb-Douglas production function (Example 4) to show that, if both the number of units of labor and the number of units of capital are doubled, the production level is also doubled.

**33.** A company manufactures two types of woodburning stoves: a free-standing model and a fireplace-insert model. The cost function for producing $x$ free-standing stoves and $y$ fireplace-insert stoves is

$$C(x, y) = 32\sqrt{xy} + 175x + 205y + 1050$$

Find the cost when $x = 80$ and $y = 20$.

**34.** A corporation manufactures a product at two locations. The costs of producing $x_1$ units at location 1 and $x_2$ units at location 2 are

$$C_1(x_1) = 0.02x_1{}^2 + 4x_1 + 500$$

and

$$C_2(x_2) = 0.05x_2{}^2 + 4x_2 + 275$$

respectively. If the product sells for $15 per unit, then the profit function for the product is

$$P(x_1, x_2) = 15(x_1 + x_2) - C_1(x_1) - C_2(x_2)$$

Find

(a) $P(250, 150)$    (b) $P(300, 200)$

**35.** The **Doyle Log Rule** is one of several methods used to determine the lumber yield of a log in board feet in terms of its diameter $d$ in inches and its length $L$ in feet. The number of board feet is given by

$$N(d, L) = \left(\frac{d - 4}{4}\right)^2 L$$

(a) Find the number of board feet of lumber in a log with a diameter of 22 inches and a length of 12 feet.

(b) Find $N(30, 12)$.

**36.** The sum of $1,000 is deposited in a savings account earning $r$ percent interest compounded continuously. The amount $A(r, t)$ after $t$ years is given by

$$A(r, t) = 1000e^{rt}$$

Use this function of two variables to complete the following table.

| Rate | Number of years | | | |
|---|---|---|---|---|
| | 5 | 10 | 15 | 20 |
| 0.08 | | | | |
| 0.10 | | | | |
| 0.12 | | | | |
| 0.14 | | | | |

**37.** (Queuing Model) The average length of time that a customer waits in line for service is given by

$$W(x, y) = \frac{1}{x - y}, \quad y < x$$

where $y$ is the average arrival rate expressed in the number of customers per unit time (hours) and $x$ is the average service rate expressed in the same units. Evaluate $W$ at the following points:

(a) $(15, 10)$    (b) $(12, 9)$

(c) $(12, 6)$    (d) $(4, 2)$

# SECTION 8.4

# Partial Derivatives

## INTRODUCTORY EXAMPLE
## Complementary and Substitute Products

Consumer products in the same or related markets are often classified as *complementary products* or *substitute products*. If two products have a complementary relationship, an increase in the sale of one product will be accompanied by an increase in the sale of the other product. For example, video cassette recorders and video cassettes have a complementary relationship. Thus, we would expect a drop in the price of video cassette recorders to stimulate the sale of video cassettes even if the price of video cassettes remains constant. Similarly, we would expect a drop in the price of video cassettes to stimulate the sale of video cassette recorders even if the price of the recorders remains constant.

If two products have a substitute relationship, an increase in the sale of one product will be accompanied by a decrease in the sale of the other product. For instance, video cassette recorders and video disc recorders both compete for the same home entertainment market, and we would expect a drop in the price of video cassette recorders to serve as a deterrent to the sale of video disc recorders.

Recall from earlier work that the *demand function* for a given product relates the number of units sold to the price per unit. For substitute or complementary products, the demand for one of the products is a function not only of its own price but also of the price of the other product(s). Thus, for two products,

$$\text{Demand for product } 1 = x_1 = f(p_1, p_2)$$

where $p_1$ and $p_2$ are the prices per unit for products 1 and 2, respectively. If these two products are substitutes, an increase in the price of product 2 results in an increase in the demand for product 1. We describe this phenomenon mathematically by saying that the **partial derivative** of $f$ with respect to $p_2$ is positive, and we write

$$\frac{\partial f}{\partial p_2} > 0$$

Figure 8.22 illustrates a substitute demand situation. For complementary products, this partial derivative is negative, as shown in Figure 8.23.

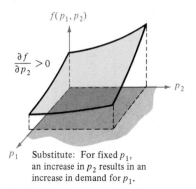

Substitute: For fixed $p_1$, an increase in $p_2$ results in an increase in demand for $p_1$.

FIGURE 8.22

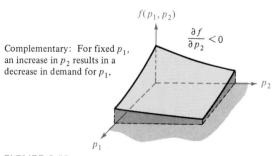

Complementary: For fixed $p_1$, an increase in $p_2$ results in a decrease in demand for $p_1$.

FIGURE 8.23

- **Partial Derivatives of a Function of Two Variables**
- **Geometric Interpretation of Partial Derivatives**
- **Partial Derivatives of Higher Order**

In applications using functions of several variables, the question often arises, "How will the function be affected if we change one or some or all of its independent variables?" We can answer this question, at least in part, by considering the independent variables one at a time. For instance, the gross national product is a function of many variables such as tax rates, unemployment, and wars. Thus, an economist who wants to determine the effect of a tax increase makes calculations using different tax rates holding all other variables constant. Or, to determine the effect of a certain catalyst in an experiment, a chemist conducts the experiment several times using varying amounts of the catalyst while keeping constant other variables such as temperature and pressure.

Mathematically, we follow a similar procedure to determine the rate of change of a function $f$ with respect to one of its several independent variables. In this procedure, we find the derivative of $f$ with respect to one independent variable at a time, while holding the others constant. This process is called **partial differentiation,** and the result is referred to as the **partial derivative** of $f$ with respect to the chosen independent variable.

For functions of two variables, we give the following definition.

| | |
|---|---|
| **Definition of first partial derivatives** | If $z = f(x, y)$, then the **first partial derivatives of $f$ with respect to $x$ and $y$** are the functions $\partial z/\partial x$ and $\partial z/\partial y$ defined as $$\frac{\partial z}{\partial x} = \lim_{\Delta x \to 0} \frac{f(x + \Delta x, y) - f(x, y)}{\Delta x} \qquad y \text{ is held constant}$$ $$\frac{\partial z}{\partial y} = \lim_{\Delta y \to 0} \frac{f(x, y + \Delta y) - f(x, y)}{\Delta y} \qquad x \text{ is held constant}$$ |

▬▬ **Remark:** Note that this definition indicates that partial derivatives of a function of two variables are determined by temporarily considering one variable to be fixed. For instance, if $z = f(x, y)$, then to find $\partial z/\partial x$ we consider $y$ to be a constant and differentiate with respect to $x$. Similarly, to find $\partial z/\partial y$ we consider $x$ fixed and differentiate with respect to $y$.

**EXAMPLE 1**
▬▬▬

**Finding Partial Derivatives**

Find $\partial z/\partial x$ and $\partial z/\partial y$ if

$$z = 3x - x^2y^2 + 2x^3y$$

**SOLUTION**

Considering $y$ to be constant and differentiating with respect to $x$, we have

$$\frac{\partial z}{\partial x} = 3 - 2xy^2 + 6x^2y$$

Similarly, considering $x$ to be constant and differentiating with respect to $y$, we have

$$\frac{\partial z}{\partial y} = -2x^2y + 2x^3$$

There are several notations for first partial derivatives. We list the common ones here, along with the notation for a partial derivative evaluated at some point $(a, b)$.

**Notation for first partial derivatives**

If $z = f(x, y)$, then

$$\frac{\partial z}{\partial x} = f_x(x, y) = z_x = \frac{\partial}{\partial x}f(x, y)$$

and

$$\frac{\partial z}{\partial y} = f_y(x, y) = z_y = \frac{\partial}{\partial y}f(x, y)$$

The first partial derivatives evaluated at the point $(a, b)$ are denoted by

$$\left.\frac{\partial z}{\partial x}\right|_{(a, b)} \qquad \text{or} \qquad f_x(a, b)$$

and

$$\left.\frac{\partial z}{\partial y}\right|_{(a, b)} \qquad \text{or} \qquad f_y(a, b)$$

**EXAMPLE 2**

**Finding and Evaluating Partial Derivatives**

Find the first partial derivatives of

$$f(x, y) = xe^{x^2y}$$

and evaluate each at the point $(1, \ln 2)$.

**SOLUTION**

To find the first partial derivative with respect to $x$, we hold $y$ constant and differentiate using the Product Rule to obtain

$$f_x(x, y) = x\frac{\partial}{\partial x}[e^{x^2y}] + e^{x^2y}\frac{\partial}{\partial x}[x]$$

$$= x(2xy)e^{x^2y} + e^{x^2y}$$

$$= e^{x^2y}(2x^2y + 1)$$

Therefore, at the point $(1, \ln 2)$, we have

$$f_x(1, \ln 2) = e^{1^2 \ln 2}[2(1^2) \ln 2 + 1]$$

$$= 2[2 \ln 2 + 1] \approx 4.773$$

To find the first partial derivative with respect to $y$, we hold $x$ constant and differentiate to obtain

$$f_y(x, y) = x(x^2)e^{x^2 y}$$

$$= x^3 e^{x^2 y}$$

Therefore, at the point $(1, \ln 2)$, we have

$$f_y(1, \ln 2) = 1^3 e^{1^2 \ln 2}$$

$$= 2$$

The partial derivatives of a function of two variables, $z = f(x, y)$, have useful geometric interpretations. If $y$ is fixed, say $y = y_0$, then $z = f(x, y_0)$ represents a curve that is the intersection of the plane $y = y_0$ and the surface $z = f(x, y)$. (See Figure 8.24.) Thus

$$f_x(x_0, y_0)$$

represents the slope of this curve at the point $(x_0, y_0, z_0)$. (Note that both the curve and the tangent line lie in the plane $y = y_0$.) Similarly,

$$f_y(x_0, y_0)$$

represents the slope of the intersection of $z = f(x, y)$ and the plane $x = x_0$ at $(x_0, y_0, z_0)$. (See Figure 8.25.)

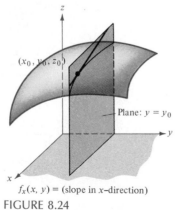

$f_x(x, y) = $ (slope in $x$–direction)

FIGURE 8.24

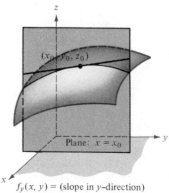

$f_y(x, y) = $ (slope in $y$–direction)

FIGURE 8.25

**EXAMPLE 3**

**Finding the Slope of a Surface in the x- and y-Directions**

Find the slope of the surface given by

$$f(x, y) = -\frac{x^2}{2} - y^2 + \frac{25}{8}$$

at the point $(\frac{1}{2}, 1, 2)$

(a) in the x-direction          (b) in the y-direction

**SOLUTION**

(a) To find the slope in the x-direction, we hold y constant and differentiate with respect to x to obtain

$$f_x(x, y) = -x$$

Thus, at the point $(\frac{1}{2}, 1, 2)$, the slope in the x-direction is

$$f_x\left(\frac{1}{2}, 1\right) = -\frac{1}{2}$$

as shown in Figure 8.26.

(b) To find the slope in the y-direction, we hold x constant and differentiate with respect to y to obtain

$$f_y(x, y) = -2y$$

Thus, at the point $(\frac{1}{2}, 1, 2)$, the slope in the y direction is

$$f_y\left(\frac{1}{2}, 1\right) = -2$$

as shown in Figure 8.27.

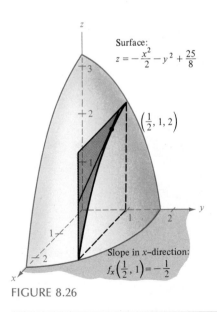

Surface:
$z = -\dfrac{x^2}{2} - y^2 + \dfrac{25}{8}$

$\left(\dfrac{1}{2}, 1, 2\right)$

Slope in x-direction:
$f_x\left(\dfrac{1}{2}, 1\right) = -\dfrac{1}{2}$

FIGURE 8.26

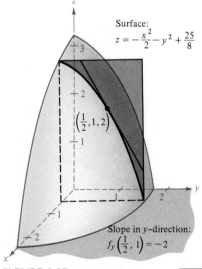

Surface:
$z = -\dfrac{x^2}{2} - y^2 + \dfrac{25}{8}$

$\left(\dfrac{1}{2}, 1, 2\right)$

Slope in y-direction:
$f_y\left(\dfrac{1}{2}, 1\right) = -2$

FIGURE 8.27

The concept of a partial derivative can be extended in a natural way to functions of three or more variables. For instance, if $w = f(x, y, z)$, there are three first partial derivatives, each of which is formed by considering two of the variables to be constant. That is, to define the partial derivative of $w$ with respect to $x$, we consider $y$ and $z$ to be constant and write

$$\frac{\partial w}{\partial x} = f_x(x, y, z) = \lim_{\Delta x \to 0} \frac{f(x + \Delta x, y, z) - f(x, y, z)}{\Delta x}$$

To define the partial derivative of $w$ with respect to $y$, we consider $x$ and $z$ to be constant and write

$$\frac{\partial w}{\partial y} = f_y(x, y, z) = \lim_{\Delta y \to 0} \frac{f(x, y + \Delta y, z) - f(x, y, z)}{\Delta y}$$

To define the partial derivative of $w$ with respect to $z$, we consider $x$ and $y$ to be constant and write

$$\frac{\partial w}{\partial z} = f_z(x, y, z) = \lim_{\Delta z \to 0} \frac{f(x, y, z + \Delta z) - f(x, y, z)}{\Delta z}$$

**EXAMPLE 4**

**Finding Partial Derivatives for a Function of Three Variables**

Find $\partial w/\partial x$, $\partial w/\partial y$, and $\partial w/\partial z$ if

$$w = xe^{xy+2z}$$

**SOLUTION**

Holding $y$ and $z$ constant, we have

$$\frac{\partial w}{\partial x} = x\frac{\partial}{\partial x}[e^{xy+2z}] + e^{xy+2z}\frac{\partial}{\partial x}[x]$$

$$= x(ye^{xy+2z}) + e^{xy+2z}(1)$$

$$= (xy + 1)e^{xy+2z}$$

Holding $x$ and $z$ constant, we have

$$\frac{\partial w}{\partial y} = x(x)e^{xy+2z} = x^2 e^{xy+2z}$$

Holding $x$ and $y$ constant, we have

$$\frac{\partial w}{\partial z} = x(2)e^{xy+2z} = 2xe^{xy+2z}$$

(Note that for this function it was necessary to apply the Product Rule only when finding the partial derivative with respect to $x$.)

There is no simple geometric interpretation for partial derivatives of functions of three or more variables. However, we can interpret the partial derivatives of functions of several variables as *rates of change*, no matter how many variables are involved. This is illustrated in Example 5.

**EXAMPLE 5**

**Using Partial Derivatives to Find Rates of Change**

The volume of a frustum of a cone, Figure 8.28, is given by

$$V = f(R, r, h) = \frac{1}{3}\pi h(R^2 + Rr + r^2)$$

When $R = 10$, $r = 4$, and $h = 6$, what is the rate of change in volume with respect to the upper radius $r$? With respect to the height $h$?

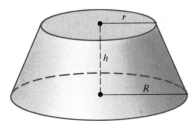

FIGURE 8.28          Frustum of a Cone

**SOLUTION**

Since the partial derivative of the volume with respect to $r$ is

$$\frac{\partial V}{\partial r} = f_r(R, r, h) = \frac{1}{3}\pi h(0 + R + 2r)$$

the rate of change of the volume with respect to the upper radius when $R = 10$, $r = 4$, and $h = 6$ is

$$f_r(10, 4, 6) = \frac{1}{3}\pi(6)(18) = 36\pi$$

Furthermore, since the partial derivative of the volume with respect to $h$ is

$$\frac{\partial V}{\partial h} = f_h(R, r, h) = \frac{1}{3}\pi(R^2 + Rr + r^2)$$

the rate of change of the volume with respect to the height when $R = 10$, $r = 4$, and $h = 6$ is

$$f_h(10, 4, 6) = \frac{1}{3}\pi(100 + 40 + 16) = 52\pi$$

**Higher-Order Partial Derivatives**

As with ordinary derivatives, it is possible to take a partial derivative of second, third, or higher order. For instance, there are four different ways to find a second partial derivative of $z = f(x, y)$.

1. Differentiate twice with respect to $x$:

$$\frac{\partial}{\partial x}\left(\frac{\partial f}{\partial x}\right) = \frac{\partial^2 f}{\partial x^2} = f_{xx}$$

2. Differentiate twice with respect to $y$:

$$\frac{\partial}{\partial y}\left(\frac{\partial f}{\partial y}\right) = \frac{\partial^2 f}{\partial y^2} = f_{yy}$$

3. Differentiate first with respect to $x$ and then with respect to $y$:

$$\frac{\partial}{\partial y}\left(\frac{\partial f}{\partial x}\right) = \frac{\partial^2 f}{\partial y\, \partial x} = f_{xy}$$

4. Differentiate first with respect to $y$ and then with respect to $x$:

$$\frac{\partial}{\partial x}\left(\frac{\partial f}{\partial y}\right) = \frac{\partial^2 f}{\partial x\, \partial y} = f_{yx}$$

The third and fourth cases are called "mixed" partial derivatives. You should notice that with the two types of notation for mixed partials different conventions are used for indicating the order of differentiation. For instance, the partial derivative

$$\frac{\partial}{\partial y}\left(\frac{\partial f}{\partial x}\right) = \frac{\partial^2 f}{\partial y\, \partial x} \qquad \text{Right-to-left order}$$

indicates differentiation with respect to $x$ first, but the partial

$$(f_y)_x = f_{yx} \qquad \text{Left-to-right order}$$

indicates differentiation with respect to $y$ first. You can learn this by remembering that in both notations you differentiate first with respect to the variable "nearest" $f$.

## EXAMPLE 6

**Finding Second Partial Derivatives**

Find the second partial derivatives of

$$f(x, y) = 3xy^2 - 2y + 5x^2y^2$$

and determine the value of $f_{xy}(-1, 2)$.

## SOLUTION

We begin by finding the first partial derivatives

$$f_x(x, y) = 3y^2 + 10xy^2 \qquad \text{and} \qquad f_y(x, y) = 6xy - 2 + 10x^2y$$

Then, differentiating each of these with respect to $x$ and $y$, we have

$$f_{xy}(x, y) = 6y + 20xy \qquad \text{and} \qquad f_{yx}(x, y) = 6y + 20xy$$

$$f_{xx}(x, y) = 10y^2 \qquad \text{and} \qquad f_{yy}(x, y) = 6x + 10x^2$$

Finally,

$$f_{xy}(-1, 2) = 12 - 40 = -28$$

Notice in Example 6 that the two mixed partials are equal. This is often the case. In fact, it can be shown that if a function $f$ has continuous second partial derivatives, then the order in which the partial derivatives are taken is immaterial.

For a function of two variables, there are two first partials and four second partials. For a function of three variables, there are three first partials, $f_x$, $f_y$, $f_z$, and nine second partials,

$$f_{xx}, \; f_{xy}, \; f_{xz}, \; f_{yx}, \; f_{yy}, \; f_{yz}, \; f_{zx}, \; f_{zy}, \; f_{zz}$$

six of which are mixed partials. To take partial derivatives of order three and higher, we follow the same pattern used in taking second partial derivatives. For instance, if $z = f(x, y)$, we have

$$z_{xxx} = \frac{\partial}{\partial x}\left(\frac{\partial^2 f}{\partial x^2}\right) = \frac{\partial^3 f}{\partial x^3} \qquad \text{and} \qquad z_{xxy} = \frac{\partial}{\partial y}\left(\frac{\partial^2 f}{\partial x^2}\right) = \frac{\partial^3 f}{\partial y \, \partial x^2}$$

**EXAMPLE 7**

**Finding Second Partial Derivatives**

Find the second partial derivatives of

$$f(x, y, z) = ye^x + x \ln z$$

**SOLUTION**

Since the first partial derivatives are

$$f_x(x, y, z) = ye^x + \ln z, \qquad f_y(x, y, z) = e^x, \qquad f_z(x, y, z) = x\left(\frac{1}{z}\right) = \frac{x}{z}$$

it follows that the second partial derivatives are

$$f_{xx}(x, y, z) = ye^x, \qquad f_{xy}(x, y, z) = e^x, \qquad f_{xz}(x, y, z) = \frac{1}{z}$$

$$f_{yx}(x, y, z) = e^x, \qquad f_{yy}(x, y, z) = 0, \qquad f_{yz}(x, y, z) = 0$$

$$f_{zx}(x, y, z) = \frac{1}{z}, \qquad f_{zy}(x, y, z) = 0, \qquad f_{zz}(x, y, z) = -\frac{x}{z^2}$$

## SECTION EXERCISES 8.4

In Exercises 1–20, find the first partial derivatives with respect to $x$ and with respect to $y$.

**1.** $f(x, y) = 2x - 3y + 5$

**2.** $f(x, y) = x^2 - 3y^2 + 7$

**3.** $f(x, y) = 5\sqrt{x} - 6y^2 + 4$

**4.** $f(x, y) = x^{-1/2} + 4y^{3/2}$

**5.** $f(x, y) = xy$

**6.** $f(x, y) = \dfrac{x}{y}$

**7.** $z = x\sqrt{y}$

**8.** $z = x^2 - 3xy + y^2$

**9.** $f(x, y) = \sqrt{x^2 + y^2}$

**10.** $f(x, y) = \dfrac{xy}{x^2 + y^2}$

**11.** $z = x^2 e^{2y}$

**12.** $z = xe^{x+y}$

**13.** $h(x, y) = e^{-(x^2+y^2)}$

**14.** $g(x, y) = e^{x/y}$

**15.** $z = \ln(x^2 + y^2)$

**16.** $z = \ln\sqrt{xy}$

**17.** $z = \ln \dfrac{x + y}{x - y}$

**18.** $g(x, y) = \ln\sqrt{x^2 + y^2}$

**19.** $f(x, y) = \displaystyle\int_x^y (t^2 - 1)\, dt$

**20.** $f(x, y) = \displaystyle\int_x^y (2t + 1)\, dt$

In Exercises 21–24, evaluate $f_x$ and $f_y$ at the indicated point.

| Function | Point |
|---|---|
| **21.** $f(x, y) = 3x^2 + xy - y^2$ | $(2, 1)$ |
| **22.** $f(x, y) = e^{3xy}$ | $(0, 4)$ |
| **23.** $f(x, y) = \dfrac{xy}{x - y}$ | $(2, -2)$ |
| **24.** $f(x, y) = \dfrac{4xy}{\sqrt{x^2 + y^2}}$ | $(1, 0)$ |

In Exercises 25–30, find the first partial derivatives with respect to $x$, $y$, and $z$, and evaluate these partial derivatives at the indicated point.

| Function | Point |
|---|---|
| **25.** $w = \sqrt{x^2 + y^2 + z^2}$ | $(2, -1, 2)$ |
| **26.** $w = \dfrac{xy}{x + y + z}$ | $(1, 2, 0)$ |
| **27.** $F(x, y, z) = \ln \sqrt{x^2 + y^2 + z^2}$ | $(3, 0, 4)$ |
| **28.** $G(x, y, z) = \dfrac{1}{\sqrt{1 - x^2 - y^2 - z^2}}$ | $(0, 0, 0)$ |
| **29.** $f(x, y, z) = 3x^2y - 5xyz + 10yz^2$ | $(1, -1, 2)$ |
| **30.** $f(x, y, z) = xye^{z^2}$ | $(2, 1, 0)$ |

In Exercises 31–42, find the second partial derivatives

$$\frac{\partial^2 z}{\partial x^2}, \quad \frac{\partial^2 z}{\partial y^2}, \quad \frac{\partial^2 z}{\partial y\, \partial x}, \quad \frac{\partial^2 z}{\partial x\, \partial y}$$

**31.** $z = x^3 - 4y^2$

**32.** $z = 3x^2 - xy + 2y^3$

**33.** $z = x^3 + 3x^2y - 5y^2$

**34.** $z = x^4 - 3x^2y^2 + y^4$

**35.** $z = 9 + 4x - 6y - x^2 - y^2$

**36.** $z = \sqrt{9 - x^2 - y^2}$

**37.** $z = \dfrac{xy}{x - y}$      **38.** $z = \dfrac{x}{x + y}$

**39.** $z = \sqrt{x^2 + y^2}$      **40.** $z = \ln (x - y)$

**41.** $z = xe^{-y^2}$      **42.** $z = xe^y + ye^x$

In Exercises 43–46, show that each function satisfies **Laplace's equation**

$$\frac{\partial^2 z}{\partial x^2} + \frac{\partial^2 z}{\partial y^2} = 0$$

**43.** $z = 5xy$      **44.** $z = x^2 - y^2$

**45.** $z = x^3 - 3xy^2$      **46.** $z = \dfrac{y}{x^2 + y^2}$

In Exercises 47–52, find the slope of the surface at the indicated point in (a) the $x$-direction and (b) the $y$-direction.

| Function | Point |
|---|---|
| **47.** $z = 2x - 3y + 5$ | $(2, 1, 6)$ |
| **48.** $z = xy$ | $(1, 2, 2)$ |
| **49.** $z = 9x^2 - y^2$ | $(1, 3, 0)$ |
| **50.** $z = x^2 + 4y^2$ | $(2, 1, 8)$ |
| **51.** $z = \sqrt{25 - x^2 - y^2}$ | $(3, 0, 4)$ |
| **52.** $z = \dfrac{x}{y}$ | $(3, 1, 3)$ |

**53.** A company manufactures two types of wood-burning stoves: a free-standing model and a fireplace-insert model. The cost function for producing $x$ free-standing and $y$ fireplace-insert stoves is

$$C = 32\sqrt{xy} + 175x + 205y + 1050$$

Find the marginal costs ($\partial C/\partial x$ and $\partial C/\partial y$) when $x = 80$ and $y = 20$.

**54.** A corporation has two plants that produce the same product. If $x_1$ and $x_2$ are the number of units produced at plant 1 and plant 2, respectively, then the total revenue for the product is given by

$$R = 200x_1 + 200x_2 - 4x_1^2 - 8x_1x_2 - 4x_2^2$$

If $x_1 = 4$ and $x_2 = 12$, find

(a) the marginal revenue for plant 1, $\partial R/\partial x_1$

(b) the marginal revenue for plant 2, $\partial R/\partial x_2$

**55.** Let $x = 1000$ and $y = 500$ in the Cobb-Douglas production function

$$f(x, y) = 100x^{0.6}y^{0.4}$$

(a) Find the marginal productivity of labor, $\partial f/\partial x$.

(b) Find the marginal productivity of capital, $\partial f/\partial y$.

**56.** Repeat Exercise 55 for the production function

$$f(x, y) = 100x^{0.75}y^{0.25}$$

**57.** Let $N$ be the number of applicants to a university, $p$ the charge for food and housing at the university, and $t$ the tuition. Suppose that $N$ is a function of $p$ and $t$ such that $\partial N/\partial p < 0$ and $\partial N/\partial t < 0$. How would you interpret the fact that both partials are negative?

**58.** The temperature $T$ at any point $(x, y)$ in a steel plate is given by

$$T(x, y) = 500 - 0.6x^2 - 1.5y^2$$

where $x$ and $y$ are measured in feet. At the point $(2, 3)$, find the rate of change of the temperature with respect to the distance moved along the plate in the directions of the $x$- and $y$-axes, respectively.

**59.** Using the notation of the Introductory Example for this section, we let $x_1$ and $x_2$ be the demands for products 1 and 2, respectively, and $p_1$ and $p_2$ the prices of products 1 and 2, respectively. Determine if the following demand functions describe complementary or substitute product relationships:

(a) $x_1 = 150 - 2p_1 - \dfrac{5}{2}p_2$, $x_2 = 350 - \dfrac{3}{2}p_1 - 3p_2$

(b) $x_1 = 150 - 2p_1 + 1.8p_2$,
$x_2 = 350 + 0.75p_1 - 1.9p_2$

(c) $x_1 = \dfrac{1000}{\sqrt{p_1 p_2}}$, $x_2 = \dfrac{750}{p_2 \sqrt{p_1}}$

# Extrema of Functions of Two Variables

A certain company markets two substitute products whose demand equations are

Demand for product 1 = $x_1 = 200(p_2 - p_1)$

Demand for product 2 = $x_2 = 500 + 100p_1 - 180p_2$

where $p_1$ and $p_2$ are the product prices in dollars. (Note that $x_1$ increases as $p_2$ increases and $x_2$ increases as $p_1$ increases. This is consistent with our claim that the two products have a substitute demand relationship.)

The costs of producing the two products are $0.50 and $0.75 per unit, respectively. Hence the total cost, the total revenue, and the profit are as follows:

$$C = 0.5x_1 + 0.75x_2$$

$$R = p_1x_1 + p_2x_2$$

$$P = R - C = p_1x_1 + p_2x_2 - 0.5x_1 - 0.75x_2$$

$$= (p_1 - 0.5)x_1 + (p_2 - 0.75)x_2$$

$$= (p_1 - 0.5)200(p_2 - p_1)$$

$$\quad + (p_2 - 0.75)(500 + 100p_1 - 180p_2)$$

$$= -200p_1{}^2 + 300p_1p_2 - 180p_2{}^2 + 25p_1 + 535p_2$$
$$\quad - 375$$

The **maximum profit** occurs when the two first partial derivatives of $P$ are zero:

$$\frac{\partial P}{\partial p_1} = -400p_1 + 300p_2 + 25 = 0$$

$$\frac{\partial P}{\partial p_2} = 300p_1 - 360p_2 + 535 = 0$$

The solutions to this system of equations are $p_1 = \$3.14$ and $p_2 = \$4.10$. Table 8.3 lists the profit for various pairs of prices, and Figure 8.29 reinforces our conclusion that there actually exists a pair of values for $p_1$ and $p_2$ that produces a maximum profit.

TABLE 8.3

| $p_1$ (price for product 1) | $p_2$ (price for product 2) | | | | |
|---|---|---|---|---|---|
| | $3.00 | $3.50 | $4.00 | $4.50 | $5.00 |
| $2.00 | $660.00 | $642.50 | $535.00 | $337.50 | $50.00 |
| $2.50 | $672.50 | $730.00 | $697.50 | $575.00 | $362.50 |
| $3.00 | $585.00 | $717.50 | $760.00 | $712.50 | $575.00 |
| $3.50 | $397.50 | $605.00 | $722.50 | $750.00 | $687.50 |
| $4.00 | $110.00 | $392.50 | $585.00 | $687.50 | $700.00 |

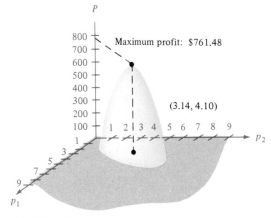

FIGURE 8.29

■ **Extrema**
■ **First-Partials Test**
■ **Second-Partials Test**
■ **Applications of Extrema**

A considerable portion of our study of the derivative of a function of one variable dealt with the finding and testing of the extreme values of a function. In this section, we take up this subject for functions of several variables, and you will find that the ideas are much the same. Although most theorems and definitions will be given in terms of functions of two variables, analogous ones may be stated for functions of three or more variables.

We define first the relative extrema of a function of two variables.

| | |
|---|---|
| **Definition of relative extrema** | Let $f$ be a function defined on a region containing $(x_0, y_0)$. |

1. $f(x_0, y_0)$ is called a **relative maximum** of $f$ if there is a circular region $R$ centered at $(x_0, y_0)$ such that

$$f(x, y) \leq f(x_0, y_0)$$

for all $(x, y)$ in $R$.

2. $f(x_0, y_0)$ is called a **relative minimum** of $f$ if there is a circular region $R$ centered at $(x_0, y_0)$ such that

$$f(x, y) \geq f(x_0, y_0)$$

for all $(x, y)$ in $R$.

Informally, we say that a point is a relative maximum of a surface if it is at least as high as all "nearby" points on the surface. Similarly, we say that a point is a relative minimum if it is at least as low as all "nearby" points on the surface. Several relative maximum and relative minimum points are pictured in Figure 8.30.

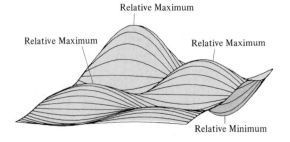

Relative Maximum
Relative Maximum
Relative Maximum
Relative Minimum
Relative Extrema

FIGURE 8.30

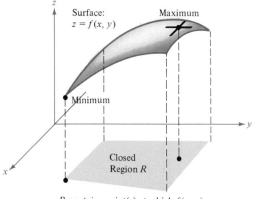

Surface:
$z = f(x, y)$

Maximum

Minimum

Closed
Region $R$

$R$ contains point(s) at which $f(x, y)$
is minimum and point(s) at which
$f(x, y)$ is maximum.

Absolute Extrema

FIGURE 8.31

We must be careful to distinguish between relative and absolute extrema as we did for functions of a single variable. For instance, in Figure 8.31, the minimum value of $f(x, y)$ over the region $R$ is not a relative minimum, whereas the maximum value of $f(x, y)$ over the region $R$ happens also to be a relative maximum.

To locate the relative extrema of a function of two variables, we can use a procedure that is similar to the First-Derivative Test used for functions of a single variable. For functions of two variables, we call this test the **First-Partials Test,** and describe it as follows. (See Figure 8.32.)

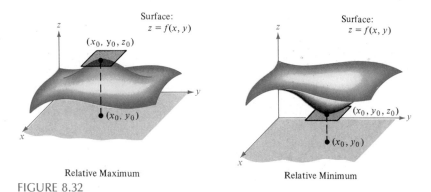

Surface:
$z = f(x, y)$

$(x_0, y_0, z_0)$

$(x_0, y_0)$

Relative Maximum

Surface:
$z = f(x, y)$

$(x_0, y_0, z_0)$

$(x_0, y_0)$

Relative Minimum

FIGURE 8.32

**First-Partials Test for relative extrema**

If $f(x_0, y_0)$ is a relative extremum of $f$ on an open region $R$ in the $xy$-plane, and the first partial derivatives of $f$ exist in $R$, then

$$f_x(x_0, y_0) = 0 = f_y(x_0, y_0)$$

**Remark:** By a *closed region R* in the *xy*-plane we mean all points within *R* as well as all points on the boundary of *R*. A closed region is comparable to a closed interval [*a*, *b*] of real numbers, which includes the endpoints *a* and *b*. Similarly, an *open region* is comparable to an open interval and does *not* include the points on the boundary.

A point $(x_0, y_0)$ for which both $f_x(x_0, y_0) = 0$ and $f_y(x_0, y_0) = 0$, or for which $f_x(x_0, y_0)$ or $f_y(x_0, y_0)$ is undefined, is called a **critical point** of $f$. The First-Partials Test tells us that if the first partial derivatives exist, we need only examine values of $f(x, y)$ at critical points to find local extrema. However, as is true for a function of one variable, the critical points of a function of two variables do not always yield relative maxima or minima. For example, some critical points yield **saddle points** (see Figure 8.33), which are neither relative maxima nor relative minima.

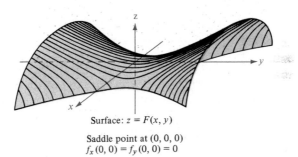

Surface: $z = F(x, y)$

Saddle point at $(0, 0, 0)$
$f_x(0, 0) = f_y(0, 0) = 0$

FIGURE 8.33

## EXAMPLE 1

**Finding Relative Extrema**

Determine the relative extrema of

$$f(x, y) = 2x^2 + y^2 + 8x - 6y + 20$$

**SOLUTION**

We begin by finding the first partial derivatives of $f$:

$$f_x(x, y) = 4x + 8 \quad \text{and} \quad f_y(x, y) = 2y - 6$$

Since these partial derivatives are defined for all $x$, the only critical values are those for which both first partial derivatives are zero. To locate these points, we let $f_x(x, y)$ and $f_y(x, y)$ equal zero and solve the system of equations

$$4x + 8 = 0 \quad \text{and} \quad 2y - 6 = 0$$

to obtain the solution $(-2, 3)$. Now, for all $(x, y) \neq (-2, 3)$, completing the square shows us that

$$f(x, y) = 2(x + 2)^2 + (y - 3)^2 + 3 > 3$$

Hence a relative *minimum* of $f$ occurs at $(-2, 3)$, and its value is

$$f(-2, 3) = 3$$

as shown in Figure 8.34.

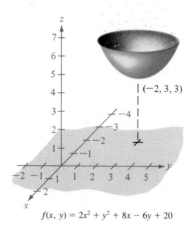

FIGURE 8.34

$f(x, y) = 2x^2 + y^2 + 8x - 6y + 20$

Example 1 shows a relative minimum occurring at one type of critical point—the type for which both $f_x(x, y)$ and $f_y(x, y)$ are zero. In the next example, we look at a relative maximum that occurs at the other type of critical point—the type for which either $f_x(x, y)$ or $f_y(x, y)$ is undefined.

**EXAMPLE 2**

**Finding Relative Extrema**

Determine the relative extrema of

$$f(x, y) = 1 - (x^2 + y^2)^{1/3}$$

**SOLUTION**

Since

$$f_x(x, y) = -\frac{2x}{3(x^2 + y^2)^{2/3}} \quad \text{and} \quad f_y(x, y) = -\frac{2y}{3(x^2 + y^2)^{2/3}}$$

we see that both partial derivatives are defined for all points in the $xy$-plane except for $(0, 0)$. Moreover, $(0, 0)$ is the only critical point, since for $(x, y) \neq (0, 0)$ the partial derivatives cannot both be zero. In Figure 8.35 we see that $f(0, 0)$ is 1. For all other $(x, y)$, it is clear that

$$f(x, y) = 1 - (x^2 + y^2)^{1/3} < 1$$

Therefore, $f(0, 0)$ is a relative *maximum* of $f$.

Surface:
$f(x, y) = 1 - (x^2 + y^2)^{1/3}$

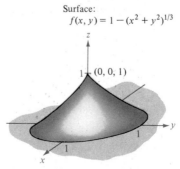

FIGURE 8.35

$f_x(x, y)$ and $f_y(x, y)$ are undefined at $(0, 0)$.

For functions like those in Examples 1 and 2, it is relatively easy to determine the *type* of extrema at the critical points. This can be done by algebraic arguments or by making a rough sketch of the graph of the function. For more complicated functions, such procedures are generally not so fruitful, and hence we seek a test that gives conditions under which a critical point will yield a relative maximum, a relative minimum, or neither. The basic test for determining the relative maxima and minima of a function of two variables is the Second-Partials Test, which is the counterpart of the Second-Derivative Test for functions of one variable.

**Second-Partials Test**  If $f$ has continuous first and second partial derivatives on an open region and there exists a point $(a, b)$ in the region such that

$$f_x(a, b) = 0 \quad \text{and} \quad f_y(a, b) = 0$$

then the quantity

$$d = f_{xx}(a, b)f_{yy}(a, b) - [f_{xy}(a, b)]^2$$

can be used as follows:

1. $f(a, b)$ is a **relative minimum** if $d > 0$ *and* $f_{xx}(a, b) > 0$.
2. $f(a, b)$ is a **relative maximum** if $d > 0$ *and* $f_{xx}(a, b) < 0$.
3. $(a, b, f(a, b))$ is a **saddle point** if $d < 0$.
4. The test gives no information if $d = 0$.

▰▰▰ **Remark:** If $d > 0$, then $f_{xx}(a, b)$ and $f_{yy}(a, b)$ must have the same sign. This means that you can replace $f_{xx}(a, b)$ with $f_{yy}(a, b)$ in the first two parts of the test.

**EXAMPLE 3**

**Applying the Second-Partials Test**

Find the relative extrema of

$$f(x, y) = x^3 - 4xy + 2y^2$$

**SOLUTION**

We begin by finding the critical points of $f$. Since

$$f_x(x, y) = 3x^2 - 4y \quad \text{and} \quad f_y(x, y) = -4x + 4y$$

are defined for all $x$ and $y$, the only critical points are those for which both first partial derivatives are zero. To locate these points, we let $f_x(x, y)$ and $f_y(x, y)$ be zero and obtain the following system of equations:

$$3x^2 - 4y = 0$$

$$-4x + 4y = 0$$

From the second equation we see that $x = y$, and by substitution into the first equation we obtain

$$3(y)^2 - 4y = 0$$

$$y(3y - 4) = 0$$

which implies that $y = x = 0$ or $y = x = \frac{4}{3}$. Therefore, the critical points are $(0, 0)$ and $(\frac{4}{3}, \frac{4}{3})$. Since

$$f_{xx}(x, y) = 6x, \qquad f_{yy}(x, y) = 4, \qquad f_{xy}(x, y) = -4$$

it follows that

$$f_{xx}(0, 0)f_{yy}(0, 0) - [f_{xy}(0, 0)]^2 = 0 - 16 < 0$$

and by part 3 of the Second-Partials Test we conclude that $(0, 0)$ yields a saddle point of $f$. Furthermore, since

$$f_{xx}\left(\frac{4}{3}, \frac{4}{3}\right) = 8 > 0$$

and

$$f_{xx}\left(\frac{4}{3}, \frac{4}{3}\right)f_{yy}\left(\frac{4}{3}, \frac{4}{3}\right) - \left[f_{xy}\left(\frac{4}{3}, \frac{4}{3}\right)\right]^2 = 8(4) - 16 = 16 > 0$$

we conclude by part 1 of the Second-Partials Test that

$$f\left(\frac{4}{3}, \frac{4}{3}\right) = -\frac{32}{27}$$

is a relative minimum of $f$. (See Figure 8.36.)

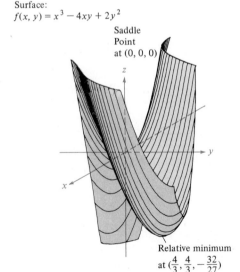

Surface:
$f(x, y) = x^3 - 4xy + 2y^2$

Saddle
Point
at $(0, 0, 0)$

Relative minimum
at $(\frac{4}{3}, \frac{4}{3}, -\frac{32}{27})$

FIGURE 8.36

## Applications of Extrema of Functions of Two Variables

We conclude this section with two examples involving applications of extrema of functions of two variables.

**EXAMPLE 4**

**Finding the Maximum Volume of a Box**

A rectangular box is resting on the $xy$-plane with one vertex at the origin. Find the maximum volume of the box if its vertex opposite the origin lies in the plane $6x + 4y + 3z = 24$, as shown in Figure 8.37.

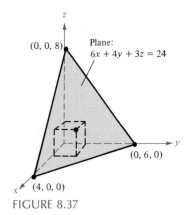

FIGURE 8.37

**SOLUTION**

Since one vertex of the box lies in the plane

$$6x + 4y + 3z = 24 \qquad \text{or} \qquad z = \frac{1}{3}(24 - 6x - 4y)$$

we can denote the volume of the box by

$$V = xyz = \frac{1}{3}xy(24 - 6x - 4y)$$

$$= \frac{1}{3}(24xy - 6x^2y - 4xy^2)$$

Now, from the system

$$V_x = \frac{1}{3}(24y - 12xy - 4y^2) = \frac{y}{3}(24 - 12x - 4y) = 0$$

$$V_y = \frac{1}{3}(24x - 6x^2 - 8xy) = \frac{x}{3}(24 - 6x - 8y) = 0$$

we obtain four solutions $x = 0$, $y = 0$; $x = 0$, $y = 6$; $x = 4$, $y = 0$; and $x = \frac{4}{3}$, $y = 2$. Since the volume is zero for $x = 0$ or $y = 0$, we test the values $x = \frac{4}{3}$ and $y = 2$ to see if they yield a maximum volume. From

$$V_{xx}(x, y) = -4y, \qquad V_{yy}(x, y) = \frac{-8x}{3}, \qquad V_{xy}(x, y) = \frac{1}{3}(24 - 12x - 8y)$$

and

$$V_{xx}\left(\frac{4}{3}, 2\right)V_{yy}\left(\frac{4}{3}, 2\right) - \left[V_{xy}\left(\frac{4}{3}, 2\right)\right]^2 = (-8)\left(\frac{-32}{9}\right) - \left[\frac{-8}{3}\right]^2 = \frac{64}{3} > 0$$

we conclude by the Second-Partials Test that $V$ is maximum for $x = \frac{4}{3}$ and $y = 2$, and the maximum volume is

$$V = xyz = \frac{4}{3}(2)\left(\frac{1}{3}\right)(24 - 8 - 8)$$

$$= \frac{64}{9} \text{ cubic units}$$

In Example 4, we were able to convert a function of three variables to a function of two variables to obtain a solution. In general this is not necessary, since the definitions of local extrema and critical points can be extended to functions of three or more variables. Specifically, if all first partials of $w = f(x, y, z, \ldots)$ exist, then it can be shown that a local maximum or minimum can occur at $(x, y, z, \ldots)$ only if

$$f_x(x, y, z, \ldots) = 0$$
$$f_y(x, y, z, \ldots) = 0$$
$$f_z(x, y, z, \ldots) = 0$$
$$\cdot$$
$$\cdot$$
$$\cdot$$

which means the critical points are obtained by solving this system of equations. The extension of the Second-Partials Test to three or more variables is also possible, though we will not consider such an extension in this text.

**EXAMPLE 5**

**Finding the Maximum Profit**

The profit obtained by producing $x$ units of product 1 and $y$ units of product 2 is given by

$$P = 8x + 10y - (0.001)(x^2 + xy + y^2) - 10{,}000$$

Find the production level that produces a maximum profit.

**SOLUTION**

Since

$$P_x = 8 - (0.001)(2x + y)$$

and
$$P_y = 10 - (0.001)(x + 2y)$$

we conclude that $P_x = 0$ and $P_y = 0$ if

$$8 - (0.001)(2x + y) = 0 \implies 2x + y = 8000$$
$$10 - (0.001)(x + 2y) = 0 \implies x + 2y = 10{,}000$$

Solving this system, we have

$$x = 2000 \quad \text{and} \quad y = 4000$$

Since

$$P_{xx}P_{yy} - (P_{xy})^2 = (-0.002)(-0.002) - (-0.001)^2 > 0$$

and

$$P_{xx} < 0$$

the production level of $x = 2000$ units and $y = 4000$ units yields a maximum profit.

## SECTION EXERCISES 8.5

In Exercises 1–20, examine each function for relative extrema and saddle points.

**1.** $f(x, y) = (x - 1)^2 + (y - 3)^2$

**2.** $f(x, y) = 9 - (x - 3)^2 - (y + 2)^2$

**3.** $f(x, y) = 2x^2 + 2xy + y^2 + 2x - 3$

**4.** $f(x, y) = -x^2 - 5y^2 + 8x - 10y - 13$

**5.** $f(x, y) = -5x^2 + 4xy - y^2 + 16x + 10$

**6.** $f(x, y) = x^2 + 6xy + 10y^2 - 4y + 4$

**7.** $f(x, y) = 2x^2 + 3y^2 - 4x - 12y + 13$

**8.** $f(x, y) = -3x^2 - 2y^2 + 3x - 4y + 5$

**9.** $f(x, y) = x^2 - y^2 - 2x - 4y - 4$

**10.** $f(x, y) = x^2 - 3xy - y^2$

**11.** $f(x, y) = xy$

**12.** $f(x, y) = 120x + 120y - xy - x^2 - y^2$

**13.** $f(x, y) = x^3 - 3xy + y^3$

**14.** $f(x, y) = 4xy - x^4 - y^4$

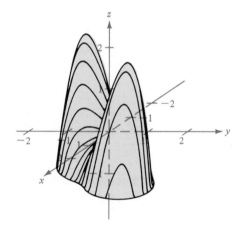

**15.** $f(x, y) = \frac{1}{2}(3x^2 + 1 - 2x^3 - 2xy^2)$

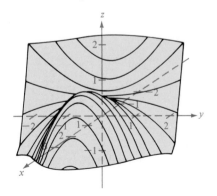

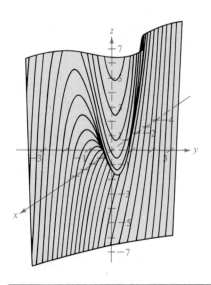

**16.** $f(x, y) = y^3 - 3yx^2 - 3y^2 - 3x^2 + 1$

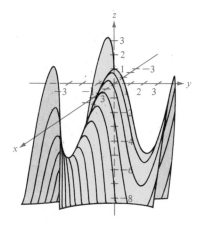

**17.** $f(x, y) = (x^2 + 4y^2)e^{1-x^2-y^2}$

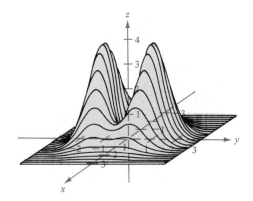

**18.** $f(x, y) = e^{-(x^2+y^2)}$

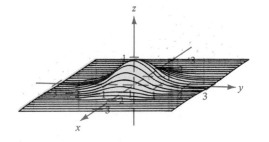

**19.** $f(x, y) = e^{xy}$

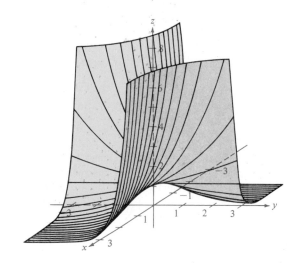

**20.** $f(x, y) = -\dfrac{4x}{x^2 + y^2 + 1}$

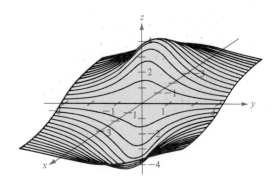

**21.** A company manufactures two products. The total revenue from $x_1$ units of product 1 and $x_2$ units of product 2 is

$$R = -5x_1^2 - 8x_2^2 - 2x_1x_2 + 42x_1 + 102x_2$$

Find $x_1$ and $x_2$ so as to maximize the revenue.

**22.** A retail outlet sells two competitive products, the prices of which are $p_1$ and $p_2$. Find $p_1$ and $p_2$ so as to maximize the total revenue

$$R = 500p_1 + 800p_2 + 1.5p_1p_2 - 1.5p_1^2 - p_2^2$$

In Exercises 23 and 24, find $p_1$ and $p_2$ so as to maximize the total revenue $R = x_1p_1 + x_2p_2$ for a retail outlet that sells two competitive products with the given demand functions.

**23.** $x_1 = 1000 - 2p_1 + p_2$, $x_2 = 1500 + 2p_1 - 1.5p_2$

**24.** $x_1 = 1000 - 4p_1 + 2p_2$, $x_2 = 900 + 4p_1 - 3p_2$

**25.** A corporation manufactures a product at two locations. The costs of producing $x_1$ units at location 1 and $x_2$ units at location 2 are

$$C_1 = 0.02x_1^2 + 4x_1 + 500$$

and

$$C_2 = 0.05x_2^2 + 4x_2 + 275$$

respectively. If the product sells for $15 per unit, find the quantity that must be produced at each location to maximize the profit,

$$P = 15(x_1 + x_2) - C_1 - C_2$$

**26.** Repeat Exercise 25 for a product that sells for $50 per unit with cost functions given by

$$C_1 = 0.1x_1^3 + 20x_1 + 150$$

and

$$C_2 = 0.05x_2^3 + 20.6x_2 + 125$$

**27.** A corporation manufactures a product at two locations. The cost functions for producing $x_1$ units at location 1 and $x_2$ units at location 2 are given by

$$C_1 = 0.05x_1^2 + 15x_1 + 5400$$

and

$$C_2 = 0.03x_2^2 + 15x_2 + 6100$$

respectively. The demand function for the product is given by

$$p = 225 - 0.4(x_1 + x_2)$$

and therefore the total revenue function is

$$R = [225 - 0.4(x_1 + x_2)](x_1 + x_2)$$

Find the production levels at the two locations that will maximize the profit

$$P = R - C_1 - C_2$$

**28.** The material for constructing the base of an open box costs 1.5 times as much as the material for constructing the sides. Find the dimensions of the box of largest volume that can be made for a fixed amount of money $C$.

**29.** Find the dimensions of a rectangular package of largest volume that may be sent by parcel post assuming that the sum of the length and the girth (perimeter of a cross section) cannot exceed 108 inches.

**30.** Show that the rectangular box of given volume and minimum surface area is a cube.

In Exercises 31–34, find three positive numbers $x$, $y$, and $z$ satisfying the given conditions.

**31.** The sum is 30 and the product is maximum.

**32.** The sum is 32 and $P = xy^2z$ is maximum.

**33.** The sum is 30 and the sum of the squares is minimum.

**34.** The sum is 1 and the sum of the squares is minimum.

# Lagrange Multipliers and Constrained Optimization

In the Introductory Example for Section 8.5, we looked at the maximum profit for two substitute products with the demand equations

$$x_1 = 200(p_2 - p_1)$$

and

$$x_2 = 500 + 100p_1 - 180p_2$$

The total demand for the two products is

$$x_1 + x_2 = 200(p_2 - p_1) + 500 + 100p_1 - 180p_2$$

$$= -100p_1 + 20p_2 + 500$$

Note that the total demand is determined solely by the prices $p_1$ and $p_2$. In many situations this assumption is overly simplistic. For example, regardless of the prices of the substitute brands, the annual total demand for some products (such as toothpaste) is relatively constant. In such situations we say that the total market is *limited*, and variations in price do not affect the total market as much as they affect the market share of the substitute brands.

Recall that in the Introductory Example of Section 8.5 the maximum profit for two products in an unlimited market was realized when $p_1 = \$3.14$ and $p_2 = \$4.10$. This situation corresponds to a total demand of

$$x_1 + x_2 = -100p_1 + 20p_2 + 500 = 268 \text{ units}$$

Now let us suppose that the market for these two products is limited to 200 units per year. Thus, we have

$$x_1 + x_2 = -100p_1 + 20p_2 + 500 = 200$$

which becomes

$$-5p_1 + p_2 + 15 = 0$$

We call this additional condition on $p_1$ and $p_2$ a **constraint.** In this particular case the graph of the constraint equation is a vertical plane, as shown in Figure 8.38. Now, to find the maximum profit,

$$P = -200p_1{}^2 + 300p_1p_2 - 180p_2{}^2 + 25p_1$$

$$+ 535p_2 - 375$$

subject to our limited market constraint, we consider only those points that lie on *both* the surface representing the profit and the plane representing the constraint. In this section, we will see that by using a **Lagrange multiplier** we can determine that the maximum profit occurs when $p_1 = \$3.94$ and $p_2 = \$4.70$. (This corresponds to an annual profit of \$712.48.) Recall that the maximum profit in an unlimited market was $P = \$761.48$. Hence, just as the name implies, the addition of a constraint equation *reduced* the maximum profit available for the two products.

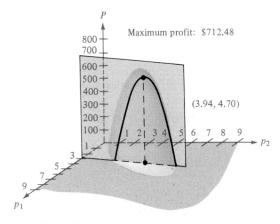

FIGURE 8.38

■ **Constrained Optimization Problems**
■ **Lagrange Multipliers**

The maximum and minimum problems in the preceding section are examples of a general type of problem called an **optimization problem.** In Example 4 of that section, we looked at an important type of optimization called **constrained optimization.** Recall that in Example 4 we wanted to find the dimensions of the rectangular box of maximum volume that would fit in the first octant beneath the plane $6x + 4y + 3z = 24$. (See Figure 8.39.) That is, we wanted to find the maximum of

$$V = xyz \qquad \text{Objective function}$$

subject to the constraint

$$6x + 4y + 3z = 24 \qquad \text{Constraint}$$

In Section 8.5, we solved this problem by solving for $z$ in the constraint equation and then rewriting $V$ as a function of two variables.

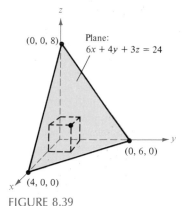

FIGURE 8.39

In this section, we look at a different (and often better) way to solve constrained optimization problems. This new method involves the use of variables called **Lagrange multipliers,** named after the French mathematician Joseph Louis Lagrange (1736–1813). We first outline this procedure and then demonstrate its use in the examples in this section.

| **Method of Lagrange multipliers** | If $f(x, y)$ has a minimum or maximum subject to the constraint $g(x, y) = 0$, then it will occur at one of the critical numbers of the function $F$ defined by $$F(x, y, \lambda) = f(x, y) - \lambda g(x, y)$$ The variable $\lambda$ (the lowercase Greek letter lambda) is called a **Lagrange multiplier.** |
| --- | --- |

For functions of three variables, $F$ has the form

$$F(x, y, z, \lambda) = f(x, y, z) - \lambda g(x, y, z)$$

As we present it, the method of Lagrange multipliers gives us a way of finding critical points but does not tell us whether these points yield minimums, maximums, or neither. To make this distinction, we rely on the context of the problem.

In the first example, we demonstrate the use of Lagrange multipliers to solve the constrained optimization problem discussed in Example 4 of the previous section.

**EXAMPLE 1**

**Using Lagrange Multipliers with One Constraint**

Find the maximum value of

$$V = xyz, \quad x > 0, \, y > 0, \, z > 0 \qquad \text{Objective function}$$

subject to the constraint

$$6x + 4y + 3z = 24 \qquad\qquad \text{Constraint}$$

**SOLUTION**

First we let

$$f(x, y, z) = xyz \qquad \text{and} \qquad g(x, y, z) = 6x + 4y + 3z - 24$$

Then we define a new function $F$ by

$$F(x, y, z, \lambda) = f(x, y, z) - \lambda g(x, y, z)$$

$$= xyz - \lambda(6x + 4y + 3z - 24)$$

To find the critical numbers of $F$, we set the partial derivatives of $F$ with respect to $x$, $y$, $z$, and $\lambda$ equal to zero and obtain the system

$$F_x(x, y, z, \lambda) = yz - 6\lambda = 0$$

$$F_y(x, y, z, \lambda) = xz - 4\lambda = 0$$

$$F_z(x, y, z, \lambda) = xy - 3\lambda = 0$$

$$F_\lambda(x, y, z, \lambda) = -6x - 4y - 3z + 24 = 0$$

Solving for $\lambda$ in the first equation and substituting into the second equation, we obtain

$$xz - 4\left(\frac{yz}{6}\right) = 0 \quad \Longrightarrow \quad y = \frac{3}{2}x$$

Similarly, solving for $\lambda$ in the first equation and substituting into the third equation, we obtain

$$xy - 3\left(\frac{yz}{6}\right) = 0 \quad \Longrightarrow \quad z = 2x$$

Now, substituting these values for $y$ and $z$ into the equation $F_\lambda(x, y, z, \lambda) = 0$, we obtain

$$F_\lambda(x, y, z, \lambda) = -6x - 4\left(\frac{3}{2}x\right) - 3(2x) + 24 = 0$$

$$-18x = -24$$

$$x = \frac{4}{3}$$

Finally, we conclude that the critical values are

$$x = \frac{4}{3}, \quad y = 2, \quad z = \frac{8}{3}$$

which implies that the maximum is

$$V = xyz = \left(\frac{4}{3}\right)(2)\left(\frac{8}{3}\right) = \frac{64}{9}$$

Note that in Example 1 we gave no particular interpretation to the value of $\lambda$ that produced a maximum. In Example 2, we look at a problem in which it is useful to assign a meaning to the value of $\lambda$.

**EXAMPLE 2**

**A Business Application**

The Cobb-Douglas production function (see Example 4, Section 8.3) for a particular manufacturer is given by

$$f(x, y) = 100x^{3/4}y^{1/4} \qquad \text{Objective function}$$

where $x$ represents the units of labor and $y$ represents the units of capital. If units of labor and capital cost $150 and $250 each, respectively, and the total expense for labor and capital is limited to $50,000, then we have the constraint

$$150x + 250y = 50,000 \qquad \text{Constraint}$$

Find the maximum production level for this manufacturer.

**SOLUTION**

Let

$$F(x, y, \lambda) = 100x^{3/4}y^{1/4} - \lambda(150x + 250y - 50,000)$$

Then

$$F_x(x, y, \lambda) = 75x^{-1/4}y^{1/4} - 150\lambda = 0$$

$$F_y(x, y, \lambda) = 25x^{3/4}y^{-3/4} - 250\lambda = 0$$

$$F_\lambda(x, y, \lambda) = -150x - 250y + 50,000 = 0$$

Solving for $\lambda$ in the first equation,

$$\lambda = \frac{75x^{-1/4}y^{1/4}}{150} = \frac{x^{-1/4}y^{1/4}}{2}$$

and substituting into the second equation, we have

$$25x^{3/4}y^{-3/4} - 250\left(\frac{x^{-1/4}y^{1/4}}{2}\right) = 0$$

Multiplying by $x^{1/4}y^{3/4}$, we have

$$25x - 125y = 0$$

$$x = \frac{125y}{25} = 5y$$

Finally, substituting this value of $x$ into the third equation, we have

$$50,000 - 150(5y) - 250y = 0$$

$$1000y = 50,000$$

$$y = 50 \text{ units of capital}$$

$$x = 250 \text{ units of labor}$$

Thus, the maximum production is

$$f(250, 50) = 100(250^{3/4})(50)^{1/4} \approx 16,719 \text{ units}$$

Economists call the Lagrange multiplier obtained in a production function the **marginal productivity of money.** For instance, in Example 2, the marginal productivity of money at $x = 250$ and $y = 50$ is

$$\lambda = \frac{x^{-1/4}y^{1/4}}{2} = \frac{(250^{-1/4})(50^{1/4})}{2} \approx 0.334$$

This means that if one additional dollar is spent on production, approximately 0.334 additional unit of the product can be produced. We use this concept in the next example.

**EXAMPLE 3**

**Using the Marginal Productivity of Money**

Suppose that $70,000 is available for labor and capital for the Cobb-Douglas production function in Example 2. What is the maximum number of units that can be produced?

**SOLUTION**

We could rework the entire problem following the procedure used in Example 2. However, since the only change in the problem is the availability of additional money to spend on labor and capital, we use the fact that the marginal productivity of money is

$$\lambda \approx 0.334$$

Since an additional $20,000 is available and since the maximum production in Example 2 was 16,719 units, we conclude that the maximum production is now

$$16{,}719 + (0.334)(20{,}000) \approx 23{,}400 \text{ units}$$

We suggest that you test this result by working out this example using the procedure described in Example 2 to see if you arrive at the same production level.

For optimization problems with two constraints, we introduce a second Lagrange multiplier, as is illustrated in the next example.

**EXAMPLE 4**

**Optimization with Two Constraints**

If

$$T(x,\ y,\ z) = 20 + 2x + 2y + z^2 \qquad \text{Objective function}$$

represents the temperature at each point on the sphere given by $x^2 + y^2 + z^2 = 11$, find the extreme temperatures on the curve of intersection of the plane $x + y + z - 3 = 0$ and the sphere.

**SOLUTION**

In this case we have two constraints,

$$g(x,\ y,\ z) = x^2 + y^2 + z^2 - 11 = 0 \qquad \text{Constraint \#1}$$

and

$$h(x,\ y,\ z) = x + y + z - 3 = 0 \qquad \text{Constraint \#2}$$

We let the function $F$ be defined by

$$F(x,\ y,\ z,\ \lambda,\ \mu)$$

$$= T(x,\ y,\ z) - \lambda g(x,\ y,\ z) - \mu h(x,\ y,\ z)$$

$$= 20 + 2x + 2y + z^2 - \lambda(x^2 + y^2 + z^2 - 11) - \mu(x + y + z - 3)$$

($\mu$ is the lowercase Greek letter mu). Then, differentiating, we have the system

$$F_x(x,\ y,\ z,\ \lambda,\ \mu) = 2 - 2x\lambda - \mu = 0$$

$$F_y(x,\ y,\ z,\ \lambda,\ \mu) = 2 - 2y\lambda - \mu = 0$$

$$F_z(x,\ y,\ z,\ \lambda,\ \mu) = 2z - 2z\lambda - \mu = 0$$

$$F_\lambda(x,\ y,\ z,\ \lambda,\ \mu) = -x^2 - y^2 - z^2 + 11 = 0$$

$$F_\mu(x,\ y,\ z,\ \lambda,\ \mu) = -x - y - z + 3 = 0$$

Subtracting the first equation from the second yields

$$2\lambda(x - y) = 0$$

which means that $\lambda = 0$ or $x = y$. We consider these two cases separately.

**Case 1:** If $\lambda = 0$ the first equation yields $\mu = 2$, and by substituting into the third equation we have $z = 1$. The fourth and fifth equations now become

$$x^2 + y^2 - 10 = 0 \quad \text{and} \quad x + y - 2 = 0$$

and by substitution we obtain

$$x^2 + (2 - x)^2 - 10 = 0$$
$$2x^2 - 4x - 6 = 0$$
$$x^2 - 2x - 3 = 0$$

with solutions $x = 3$ and $x = -1$. The corresponding $y$-values are $y = -1$ and $y = 3$, respectively. Thus, if $\lambda = 0$ the critical points are $(3, -1, 1)$ and $(-1, 3, 1)$.

**Case 2:** If $x = y$ the fourth and fifth equations become

$$2x^2 + z^2 = 11 \quad \text{and} \quad 2x + z = 3$$

and by substitution we obtain

$$2x^2 + (3 - 2x)^2 = 11$$
$$6x^2 - 12x - 2 = 0$$
$$3x^2 - 6x - 1 = 0$$
$$x = \frac{3 \pm 2\sqrt{3}}{3}$$

and the corresponding values of $z$ are $z = (3 \mp 4\sqrt{3})/3$, which yield two additional critical points.

Finally, to find the optimal solutions, we compare the temperatures at the four critical points.

$$T(3, -1, 1) = T(-1, 3, 1) = 25$$

$$T\left(\frac{3 - 2\sqrt{3}}{3}, \frac{3 - 2\sqrt{3}}{3}, \frac{3 + 4\sqrt{3}}{3}\right) = \frac{91}{3} \approx 30.33$$

$$T\left(\frac{3 + 2\sqrt{3}}{3}, \frac{3 + 2\sqrt{3}}{3}, \frac{3 - 4\sqrt{3}}{3}\right) = \frac{91}{3} \approx 30.33$$

Thus, $T = 25$ is the minimum temperature and $T = \frac{91}{3}$ is the maximum temperature.

We can see from Examples 1, 2, and 4 that the system of equations that arises in the method of Lagrange multipliers is not, in general, a linear system. Because of this nonlinearity the solution of the system often requires some ingenuity, and we encourage you to tailor your method of solution to each individual system. Of course, if the system happens to be linear, we can resort to familiar methods, as is demonstrated in the next example.

**EXAMPLE 5**

**Optimization Involving Four Variables**

Find the minimum of the function

$$f(x, y, z, w) = x^2 + y^2 + z^2 + w^2 \qquad \text{Objective function}$$

subject to the constraint

$$3x + 2y - 4z + w = 3 \qquad \text{Constraint}$$

**SOLUTION**

Let

$$F(x, y, z, w, \lambda) = x^2 + y^2 + z^2 + w^2 - \lambda(3x + 2y - 4z + w - 3)$$

Then we obtain the linear system

$$F_x(x, y, z, w, \lambda) = 2x - 3\lambda = 0, \qquad\qquad x = \frac{3}{2}\lambda$$

$$F_y(x, y, z, w, \lambda) = 2y - 2\lambda = 0, \qquad\qquad y = \lambda$$

$$F_z(x, y, z, w, \lambda) = 2z + 4\lambda = 0, \qquad\qquad z = -2\lambda$$

$$F_w(x, y, z, w, \lambda) = 2w - \lambda = 0, \qquad\qquad w = \frac{1}{2}\lambda$$

$$F_\lambda(x, y, z, w, \lambda) = -3x - 2y + 4z - w + 3 = 0$$

Substituting for $x$, $y$, $z$, and $w$ into the last equation yields

$$3\left(\frac{3}{2}\lambda\right) + 2(\lambda) - 4(-2\lambda) + \left(\frac{1}{2}\lambda\right) = 3$$

$$15\lambda = 3$$

$$\lambda = \frac{1}{5}$$

Therefore, the critical numbers are

$$x = \frac{3}{10}, \qquad y = \frac{1}{5}, \qquad z = -\frac{2}{5}, \qquad w = \frac{1}{10}$$

and the minimum value of $f$ subject to the given constraint is

$$f\left(\frac{3}{10}, \frac{1}{5}, -\frac{2}{5}, \frac{1}{10}\right) = \frac{9}{100} + \frac{1}{25} + \frac{4}{25} + \frac{1}{100} = \frac{3}{10}$$

For our last example, we return to the concepts discussed in the Introductory Example of this section. Note that the system of equations obtained in this problem is also linear.

**EXAMPLE 6**

**Finding the Maximum Profit in a Limited Market**

Referring to the Introductory Example of this section, find the prices yielding a maximum profit

$$P = -200p_1{}^2 + 300p_1p_2 - 180p_2{}^2 + 25p_1 + 535p_2 - 375$$

subject to the constraint

$$-5p_1 + p_2 + 15 = 0$$

**SOLUTION**

Let $F$ be defined by

$$F(p_1, p_2, \lambda) = -200p_1{}^2 + 300p_1p_2 - 180p_2{}^2 + 25p_1 + 535p_2 - 375$$
$$- \lambda(-5p_1 + p_2 + 15)$$

Then we obtain the linear system

$$F_{p_1}(p_1, p_2, \lambda) = -400p_1 + 300p_2 + 25 + 5\lambda = 0$$
$$F_{p_2}(p_1, p_2, \lambda) = 300p_1 - 360p_2 + 535 - \lambda = 0$$
$$F_{\lambda}(p_1, p_2, \lambda) = 5p_1 - p_2 - 15 = 0$$

Now, to solve this system, we divide the first equation by 5 and add the result to the second equation to obtain

$$
\begin{array}{r}
-80p_1 + 60p_2 + 5 + \lambda = 0 \\
\underline{300p_1 - 360p_2 + 535 - \lambda = 0} \\
220p_1 - 300p_2 + 540 \qquad = 0
\end{array}
$$

Finally, dividing this result by 20 and multiplying the third equation by $-15$, we have

$$
\begin{array}{r}
11p_1 - 15p_2 + 27 = 0 \\
\underline{-75p_1 + 15p_2 + 225 = 0} \\
-64p_1 \qquad\quad + 252 = 0
\end{array}
$$

$$p_1 = \frac{252}{64} = \$3.94$$

and

$$p_2 = 5p_1 - 15 = \$4.70$$

## SECTION EXERCISES 8.6

In Exercises 1–20, use Lagrange multipliers to find the indicated extrema. In each case, assume that $x$, $y$, and $z$ are positive.

**1.** Maximize $f(x, y) = xy$
   Constraint: $x + y = 10$
**2.** Maximize $f(x, y) = xy$
   Constraint: $2x + y = 4$
**3.** Minimize $f(x, y) = x^2 + y^2$
   Constraint: $x + y - 4 = 0$
**4.** Minimize $f(x, y) = x^2 + y^2$
   Constraint: $2x - 4y + 5 = 0$
**5.** Maximize $f(x, y) = x^2 - y^2$
   Constraint: $y - x^2 = 0$
**6.** Maximize $f(x, y) = x^2 - y^2$
   Constraint: $x - 2y + 6 = 0$
**7.** Maximize $f(x, y) = 2x + 2xy + y$
   Constraint: $2x + y = 100$
**8.** Maximize $f(x, y) = 3x + y + 10$
   Constraint: $x^2 y = 6$
**9.** Maximize $f(x, y) = \sqrt{6 - x^2 - y^2}$
   Constraint: $x + y - 2 = 0$
**10.** Minimize $f(x, y) = \sqrt{x^2 + y^2}$
   Constraint: $2x + 4y - 15 = 0$
**11.** Maximize $f(x, y) = e^{xy}$
   Constraint: $x^2 + y^2 - 8 = 0$
**12.** Minimize $f(x, y) = 2x + y$
   Constraint: $xy = 32$
**13.** Minimize $f(x, y, z) = x^2 + y^2 + z^2$
   Constraint: $x + y + z - 6 = 0$
**14.** Maximize $f(x, y, z) = xyz$
   Constraint: $x + y + z - 6 = 0$
**15.** Maximize $f(x, y, z) = xyz$
   Constraints: $x + y + z = 32$
   $\qquad\qquad x - y + z = 0$
**16.** Minimize $f(x, y, z) = x^2 + y^2 + z^2$
   Constraints: $x + 2z = 4$
   $\qquad\qquad x + y = 8$
**17.** Maximize $f(x, y, z) = xyz$
   Constraints: $x^2 + z^2 = 5$
   $\qquad\qquad x - 2y = 0$
**18.** Maximize $f(x, y, z) = xy + yz$
   Constraints: $x + 2y = 6$
   $\qquad\qquad x - 3z = 0$
**19.** Minimize $f(x, y, z) = x^2 + y^2 + z^2$
   Constraint: $x + y + z = 1$
**20.** Minimize $f(x, y) = x^2 - 8x + y^2 - 12y + 48$
   Constraint: $x + y = 8$

**21.** Find the dimensions of the rectangular package of largest volume subject to the constraint that the sum of the length and the girth cannot exceed 108 inches. (Maximize $V = xyz$ subject to the constraint $x + 2y + 2z = 108$.) (See Figure 8.40.)

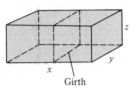

Girth

FIGURE 8.40          FIGURE 8.41

**22.** The material for the base of an open box costs 1.5 times as much as the material for the sides. Find the dimensions of the box of largest volume that can be made for a fixed cost $C$. (Maximize $V = xyz$ subject to $1.5xy + 2xz + 2yz = C$.) (See Figure 8.41.)

**23.** A manufacturer has an order for 1000 units that can be produced at two locations. Let $x_1$ and $x_2$ be the number of units produced at the two plants. Find the number of units that should be produced at each plant to minimize the cost if the cost function is given by

$$C = 0.25x_1^2 + 10x_1 + 0.15x_2^2 + 12x_2$$

**24.** Prove that the product of three positive numbers whose sum is $S$ is maximum if the three numbers are equal.

In Exercises 25–28, find the minimum distance from the curve or surface to the specified point. (Note: Start by minimizing the square of the distance.)

**25.** Line: $2x + 3y = -1$, $(0, 0)$
   [Minimize $d^2 = x^2 + y^2$.]
**26.** Circle: $(x - 4)^2 + y^2 = 4$, $(0, 10)$
   [Minimize $d^2 = x^2 + (y - 10)^2$.]
**27.** Plane: $x + y + z = 1$, $(2, 1, 1)$
   [Minimize $d^2 = (x - 2)^2 + (y - 1)^2 + (z - 1)^2$.]
**28.** Cone: $z = \sqrt{x^2 + y^2}$, $(4, 0, 0)$
   [Minimize $d^2 = (x - 4)^2 + y^2 + z^2$.]

**29.** The production function for a company is

$$f(x, y) = 100x^{0.25}y^{0.75}$$

where $x$ is the number of units of labor and $y$ is the number of units of capital. Suppose that labor costs $48 per unit, capital costs $36 per unit, and management sets a production goal of 20,000 units.

(a) Find the number of units of labor and capital needed to meet the production goal while minimizing the cost.

(b) Show that the conditions of part (a) are met when

$$\frac{\text{Marginal productivity of labor}}{\text{Marginal productivity of capital}} = \frac{\text{unit price of labor}}{\text{unit price of capital}}$$

This proportion is called the **Least-Cost Rule** (or Equimarginal Rule).

**30.** Repeat Exercise 29 for the production function

$$f(x, y) = 100x^{0.6}y^{0.4}$$

**31.** The production function for a company is

$$f(x, y) = 100x^{0.25}y^{0.75}$$

where $x$ is the number of units of labor and $y$ is the number of units of capital. Suppose that labor costs $48 per unit and capital costs $36 per unit. The total cost of labor and capital is limited to $100,000.

(a) Find the maximum production level for this manufacturer.

(b) Find the marginal productivity of money.

(c) Use the marginal productivity of money to find the maximum number of units that can be produced if $125,000 is available for labor and capital.

**32.** Repeat Exercise 31 for the production function

$$f(x, y) = 100x^{0.6}y^{0.4}$$

# The Method of Least Squares

### Egg Consumption in the United States

Although eating patterns in the United States follow fairly stable patterns, the past 20 years have witnessed some gradual shifts away from certain types of foods to other types. For example, two protein foods that have shown significant gains in popularity are chicken and cheese. On the other hand, eggs (another source of protein) have decreased in popularity. Figure 8.42 shows the average annual consumption per person in the United States for these three protein products.

Part of the reason for the loss of popularity of eggs was the cholesterol scare of the 1960s and 1970s. (However, as Figure 8.42 shows, cheese consumption continued to climb during that period, even though cheese is also high in cholesterol.) To combat the decline in consumption, the nation's egg producers decided to invest in a national advertising campaign to push "the incredible edible egg" as an inexpensive and nutritious source of protein. By 1978 this advertising effort seemed to be paying off.

Using the **method of least squares,** we can determine that until 1977 the average egg consumption roughly followed the linear model

$$y = -3.68t + 299.67$$

where $t = 0$ represents 1970. Had this pattern continued through 1979, the average consumption would have dropped to 267. (See Figure 8.42.) Since consumption in 1979 was actually 283, we can conclude that (probably as a result of effective advertising) the average American consumed 16 eggs that year that would not have been consumed had the pre-1977 pattern continued. At a dollar a dozen, this increase amounted to additional retail sales of about $300,000,000 in 1979.

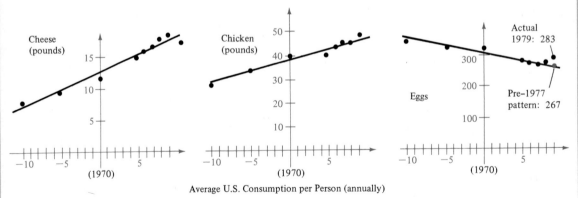

Average U.S. Consumption per Person (annually)

FIGURE 8.42

- **Least Squares Regression Line**
- **Least Squares Regression Quadratic**

We have already spent some time discussing the concept of a **mathematical model.** We pointed out that it is not always feasible to find an equation that *exactly* describes a particular phenomenon. In such cases we try to find a model that is both as *simple* and as *accurate* as possible. For instance, a simple linear model for the given points in Figure 8.43 is

$$y = 1.8566x - 5.0246$$

However, Figure 8.44 shows that by choosing the slightly more complicated quadratic model

$$y = 0.1996x^2 - 0.7281x + 1.3749$$

we can achieve significantly greater accuracy.

To measure how well the model $y = f(x)$ fits the collection of points

$$\{(x_1, y_1), (x_2, y_2), \ldots, (x_n, y_n)\}$$

we sum the squares of the differences between the actual $y$-values and the $y$-values given by the model. We call this value the **sum of the squared errors** and denote it by $S$.

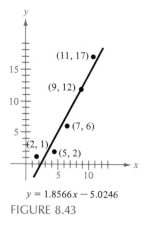

$y = 1.8566x - 5.0246$

FIGURE 8.43

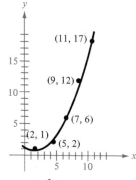

$y = 0.1996x^2 - 0.7281x + 1.3749$

FIGURE 8.44

**Definition of the sum of the squared errors**

The **sum of the squared errors** for the model $y = f(x)$ with respect to the points

$$\{(x_1, y_1), (x_2, y_2), \ldots, (x_n, y_n)\}$$

is given by

$$S = [f(x_1) - y_1]^2 + [f(x_2) - y_2]^2 + \cdots + [f(x_n) - y_n]^2$$

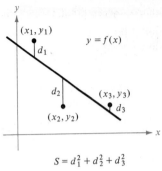

$$S = d_1^2 + d_2^2 + d_3^2$$

FIGURE 8.45

Graphically, $S$ is the sum of the squares of the vertical distances between the graph of $f$ and the given points in the plane, as shown in Figure 8.45.

**EXAMPLE 1**

**Finding the Sum of the Squared Errors**

Find the sum of the squared errors for the models

$$f(x) = 1.8566x - 5.0246$$

and

$$g(x) = 0.1996x^2 - 0.7281x + 1.3749$$

with respect to the points

$$\{(2, 1), (5, 2), (7, 6), (9, 12), (11, 17)\}$$

(See Figures 8.43 and 8.44.)

**SOLUTION**

We begin by evaluating each of the two models at the given values of $x$, as shown in Table 8.4.

TABLE 8.4

| $x$ | 2 | 5 | 7 | 9 | 11 |
|------|------|------|------|------|------|
| $y$ | 1 | 2 | 6 | 12 | 17 |
| $f(x)$ | $-1.3114$ | 4.2584 | 7.9716 | 11.6848 | 15.3980 |
| $g(x)$ | 0.7171 | 2.7244 | 6.0586 | 10.9896 | 17.5174 |

Now, for the linear model $f$, the sum of the squared errors is

$$S = (-1.3114 - 1)^2 + (4.2584 - 2)^2 + (7.9716 - 6)^2$$
$$+ (11.6848 - 12)^2 + (15.3980 - 17)^2$$

$$\approx 16.9959$$

Similarly, the sum of the squared errors for the quadratic model is

$$S = (0.7171 - 1)^2 + (2.7244 - 2)^2 + (6.0586 - 6)^2$$
$$+ (10.9896 - 12)^2 + (17.5174 - 17)^2$$
$$\approx 1.8968$$

■■■ **Remark:** Note in Example 1 that the sum of the squared errors for the quadratic model is less than the sum of the squared errors for the linear model, which confirms our initial claim that the quadratic model provides a better fit.

At this point you might be asking "What does all of this have to do with functions of several variables?" The answer is that we can use the optimization techniques of this chapter to find models with minimum least squared errors. Before looking at a general outline of this procedure, we look at a simple example. Study this example carefully.

**EXAMPLE 2**

**Using Partial Derivatives to Find the Best Linear Model**

Find the values of $a$ and $b$ such that the linear model

$$f(x) = ax + b$$

has a minimum sum of the squared errors for the points

$$\{(-3, 0), (-1, 1), (0, 2), (2, 3)\}$$

**SOLUTION**

Table 8.5 lists the $x$-values, the actual $y$-values, and the estimated $y$-values for these four points.

TABLE 8.5

| $x$ | $-3$ | $-1$ | $0$ | $2$ |
|---|---|---|---|---|
| $y$ | $0$ | $1$ | $2$ | $3$ |
| $f(x)$ | $-3a + b$ | $-a + b$ | $b$ | $2a + b$ |

The sum of the squared errors is

$$S = (-3a + b)^2 + (-a + b - 1)^2 + (b - 2)^2 + (2a + b - 3)^2$$

To minimize this error, we set the partial derivatives of $S$ (with respect to $a$ and $b$) equal to zero as follows:

$$\frac{\partial S}{\partial a} = 2(-3)(-3a + b) + 2(-1)(-a + b - 1) + 2(2)(2a + b - 3) = 0$$

$$18a - 6b + 2a - 2b + 2 + 8a + 4b - 12 = 0$$

$$28a - 4b - 10 = 0$$

$$14a - 2b - 5 = 0$$

$$\frac{\partial S}{\partial b} = 2(-3a + b) + 2(-a + b - 1) + 2(b - 2) + 2(2a + b - 3) = 0$$

$$-6a + 2b - 2a + 2b - 2 + 2b - 4 + 4a + 2b - 6 = 0$$

$$-4a + 8b - 12 = 0$$

$$-a + 2b - 3 = 0$$

Now we can solve for $a$ and $b$ as follows:

$$14a - 2b - 5 = 0$$

$$\underline{-a + 2b - 3 = 0}$$

$$13a \qquad - 8 = 0 \quad \Longrightarrow \quad a = \frac{8}{13}$$

Since $-a + 2b - 3 = 0$, we have

$$b = \frac{1}{2}(a + 3) = \frac{1}{2}\left(\frac{8}{13} + 3\right) = \frac{47}{26}$$

Finally, we conclude that the line that best fits the four given points is

$$f(x) = \frac{8}{13}x + \frac{47}{26}$$

(See Figure 8.46.)

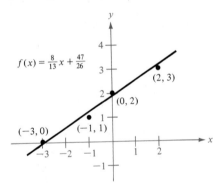

FIGURE 8.46

We call the line in Example 2 the **least squares regression line** for the given points. Before discussing a general procedure for finding such lines, we introduce the following notation for sums:

$$\sum_{i=1}^{n} x_i = x_1 + x_2 + \cdots + x_n$$

($\Sigma$ is the uppercase Greek letter sigma.) Now, using this shorthand notation for the linear model

$$f(x) = ax + b$$

and the points

$$\{(x_1, y_1), (x_2, y_2), \ldots, (x_n, y_n)\}$$

we have

$$S = \sum_{i=1}^{n} [f(x_i) - y_i]^2 = \sum_{i=1}^{n} (ax_i + b - y_i)^2$$

To minimize $S$, we set its partial derivatives (with respect to $a$ and $b$) equal to zero and solve for $a$ and $b$. We omit the details of this procedure (they are similar to the steps in Example 2) and simply summarize the result as follows.

**The least squares regression line**

The **least squares regression line** for the points

$$\{(x_1, y_1), (x_2, y_2), \ldots, (x_n, y_n)\}$$

is

$$y = ax + b$$

where

$$a = \frac{n \sum_{i=1}^{n} x_i y_i - \sum_{i=1}^{n} x_i \sum_{i=1}^{n} y_i}{n \sum_{i=1}^{n} x_i^2 - \left(\sum_{i=1}^{n} x_i\right)^2} \quad \text{and} \quad b = \frac{1}{n}\left(\sum_{i=1}^{n} y_i - a \sum_{i=1}^{n} x_i\right)$$

■■■ **Remark:** In the formula for the least squares regression line, note that if the $x$-values are symmetrically spaced about zero, then

$$\sum_{i=1}^{n} x_i = 0$$

and the formulas for $a$ and $b$ simplify to

$$a = \frac{\sum_{i=1}^{n} x_i y_i}{\sum_{i=1}^{n} x_i^2} \quad \text{and} \quad b = \frac{1}{n} \sum_{i=1}^{n} y_i$$

Note also that only the *development* of the least squares regression line involves partial derivatives. The *application* of this formula is simply a matter of computing the values of $a$ and $b$. This is demonstrated in the next example.

EXAMPLE 3

**Finding the Least Squares Regression Line by Formula**

The total number of degrees (bachelor's, master's, and doctoral) conferred each year in the United States from 1971 to 1975 is given in the following table.

| Year | 1971 | 1972 | 1973 | 1974 | 1975 |
|------|------|------|------|------|------|
| Number of degrees (in millions) | 1.148 | 1.224 | 1.280 | 1.321 | 1.316 |

Find the least squares regression line for these data and use the result to predict the number of degrees conferred in 1976.

**SOLUTION**

To simplify the calculations in this problem, we let $x = -2$ represent 1971, $x = -1$ represent 1972, and so on. Table 8.6 summarizes the calculations involved in applying the least squares regression line.

TABLE 8.6

| $x$ | $y$ | $xy$ | $x^2$ |
|-----|-----|------|-------|
| $-2$ | 1.148 | $-2.296$ | 4 |
| $-1$ | 1.224 | $-1.224$ | 1 |
| 0 | 1.280 | 0 | 0 |
| 1 | 1.321 | 1.321 | 1 |
| 2 | 1.316 | 2.632 | 4 |
| $\displaystyle\sum_{i=1}^{n} x_i = 0$ | $\displaystyle\sum_{i=1}^{n} y_i = 6.289$ | $\displaystyle\sum_{i=1}^{n} x_i y_i = 0.433$ | $\displaystyle\sum_{i=1}^{n} x_i^2 = 10$ |

Now, since the sum of the $x$-values is zero, we apply the simplified version of the formula for the least squares regression line to obtain

$$a = \frac{\displaystyle\sum_{i=1}^{n} x_i y_i}{\displaystyle\sum_{i=1}^{n} x_i^2} = \frac{0.433}{10} = 0.0433$$

and

$$b = \frac{1}{n} \sum_{i=1}^{n} y_i = \frac{1}{5}(6.289) = 1.2578$$

Thus, the least squares regression line is

$$y = 0.0433x + 1.2578$$

Finally, for 1976 we have $x = 3$, and we estimate the number of degrees conferred in that year to be

$$y = 0.0433(3) + 1.2578$$

$$= 1.3877 \text{ million}$$

(See Figure 8.47.)

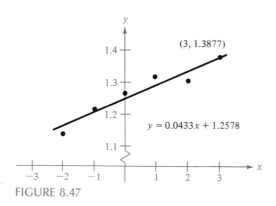

FIGURE 8.47

## Least Squares Regression Quadratic

Before concluding this section, we must emphasize that the least squares regression line provides the best *linear* model for a given set of points. It does not, however, necessarily provide the best possible model. For example, it seems from Figure 8.47 that in 1975 ($x = 2$) a downward trend in the number of degrees began to occur. For a more exact prediction for 1976 we might do better to use a quadratic model of the form

$$y = ax^2 + bx + c$$

As with the formula for the least squares regression line, the derivation of the formulas for the least squares regression quadratic involves the partial derivatives of the sum of the squared errors

$$S = \sum_{i=1}^{n} (ax_i^2 + bx_i + c - y_i)^2$$

We omit the details of the rather lengthy proof and simply state the theorem as follows.

**Least squares regression quadratic**

The **least squares regression quadratic** for the points

$$\{(x_1, y_1), (x_2, y_2), \ldots, (x_n, y_n)\}$$

is

$$y = ax^2 + bx + c$$

where $a$, $b$, and $c$ are the solutions to the system

$$a \sum_{i=1}^{n} x_i^4 + b \sum_{i=1}^{n} x_i^3 + c \sum_{i=1}^{n} x_i^2 = \sum_{i=1}^{n} x_i^2 y_i$$

$$a \sum_{i=1}^{n} x_i^3 + b \sum_{i=1}^{n} x_i^2 + c \sum_{i=1}^{n} x_i = \sum_{i=1}^{n} x_i y_i$$

$$a \sum_{i=1}^{n} x_i^2 + b \sum_{i=1}^{n} x_i + cn = \sum_{i=1}^{n} y_i$$

**EXAMPLE 4**

**Finding the Best Quadratic Fit for a Set of Points**

Find the least squares regression quadratic for the data in Example 3, and use the result to estimate the number of degrees conferred in the United States in 1976.

**SOLUTION**

As in Example 3, we let $x = -2$ represent 1971. Then, applying the least squares regression quadratic formula, we have

$$\sum_{i=1}^{n} x_i = 0 \qquad \text{From Example 3}$$

$$\sum_{i=1}^{n} x_i^2 = 10 \qquad \text{From Example 3}$$

$$\sum_{i=1}^{n} x_i^3 = (-2)^3 + (-1)^3 + (0)^3 + (1)^3 + (2)^3 = 0$$

$$\sum_{i=1}^{n} x_i^4 = (-2)^4 + (-1)^4 + (0)^4 + (1)^4 + (2)^4 = 34$$

$$\sum_{i=1}^{n} y_i = 6.289 \qquad \text{From Example 3}$$

$$\sum_{i=1}^{n} x_i y_i = 0.433 \qquad \text{From Example 3}$$

$$\sum_{i=1}^{n} x_i^2 y_i = (-2)^2(1.148) + (-1)^2(1.224) + (0)^2(1.280)$$

$$+ (1)^2(1.321) + (2)^2(1.316)$$

$$= 12.401$$

Thus, we have

$$34a + 10c = 12.401$$
$$10b = 0.433$$
$$10a + 5c = 6.289$$

Solving this system, we have

$$a = -0.01264, \qquad b = 0.0433, \qquad c = 1.2831$$

and the best quadratic fit is given by

$$y = -0.01264x^2 + 0.0433x + 1.2831$$

For 1976, we have $x = 3$ and

$$y = -0.01264(3^2) + 0.0433(3) + 1.2831$$

$$\approx 1.2992 \text{ million}$$

(See Figure 8.48.)

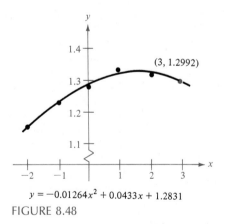

$$y = -0.01264x^2 + 0.0433x + 1.2831$$

FIGURE 8.48

**Remark:** The quadratic model in Example 4 has a sum of the squared errors of $S \approx 0.00009$, and the linear model in Example 3 has a sum of the squared errors of $S \approx 0.00232$.

In Exercises 1–4, (a) use the method of least squares to find the least squares regression line, and (b) calculate the sum of the squared errors.

**1.**

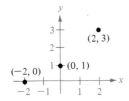

**2.**

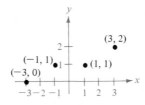

**3.**

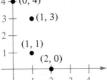

**4.**

In Exercises 5–14, find the least squares regression line for the given points.

**5.** $(-2, 0), (-1, 1), (0, 1), (1, 2), (2, 3)$
**6.** $(-4, -1), (-2, 0), (2, 4), (4, 5)$
**7.** $(-3, 0), (1, 4), (2, 6)$
**8.** $(-5, 1), (1, 3), (2, 3), (2, 5)$
**9.** $(-3, 4), (-1, 2), (1, 1), (3, 0)$
**10.** $(-10, 10), (-5, 8), (3, 6), (7, 4), (5, 0)$
**11.** $(0, 0), (1, 1), (3, 4), (4, 2), (5, 5)$
**12.** $(1, 0), (3, 3), (5, 6)$
**13.** $(0, 6), (4, 3), (5, 0), (8, -4), (10, -5)$
**14.** $(5, 2), (0, 0), (2, 1), (7, 4), (10, 6), (12, 6)$

In Exercises 15–18, find the values of $a$ and $b$ such that the linear model

$$f(x) = ax + b$$

has a minimum sum of the squared errors for the given points. (Use partial derivatives, following the pattern of Example 2 in this section.)

**15.** $(-2, 0), (0, 1), (2, 3)$ (See Exercise 1.)
**16.** $(-3, 0), (-1, 1), (1, 1), (3, 2)$ (See Exercise 2.)
**17.** $(0, 4), (1, 1), (1, 3), (2, 0)$ (See Exercise 3.)
**18.** $(1, 0), (2, 0), (3, 0), (3, 1), (4, 1), (4, 2), (5, 2),$ $(6, 2)$ (See Exercise 4.)

In Exercises 19–22, find the least squares regression quadratic for the given points. Then plot the points and sketch the graph of the least squares quadratic.

**19.** $(-2, 0), (-1, 0), (0, 1), (1, 2), (2, 5)$
**20.** $(-4, 5), (-2, 6), (2, 6), (4, 2)$
**21.** $(0, 0), (2, 2), (3, 6), (4, 12)$
**22.** $(0, 10), (1, 9), (2, 6), (3, 0)$

**23.** A store manager wants to know the demand for a certain product as a function of price. The daily sales for three different prices of the product are given in the following table.

| Price $(x)$ | $\$1.00$ | $\$1.25$ | $\$1.50$ |
| --- | --- | --- | --- |
| Demand $(y)$ | 450 | 375 | 330 |

(a) Find the least squares regression line for these data.

(b) Estimate the demand when the price is $1.40.

**24.** A hardware retailer wants to know the demand for a certain tool as a function of the price. The monthly sales for four different prices of the tool are listed in the following table.

| Price $(x)$ | $\$25$ | $\$30$ | $\$35$ | $\$40$ |
| --- | --- | --- | --- | --- |
| Demand $(y)$ | 82 | 75 | 67 | 55 |

(a) Find the least squares regression line for these data.

(b) Estimate the demand when the price is $32.95.

**25.** A farmer used four test plots to determine the relationship between wheat yield in bushels per acre and the amount of fertilizer in hundreds of pounds per acre. The results are given in the following table.

| Fertilizer $(x)$ | 1.0 | 1.5 | 2.0 | 2.5 |
| --- | --- | --- | --- | --- |
| Yield $(y)$ | 32 | 41 | 48 | 53 |

(a) Find the least squares regression line for these data.

(b) Estimate the yield for a fertilizer application of 160 pounds per acre.

**26.** The Consumer Price Index (CPI) for all items for five years is given in the following table.

| Year ($x$) | 1980 | 1981 | 1982 | 1983 | 1984 |
|---|---|---|---|---|---|
| CPI ($y$) | 247.0 | 272.3 | 288.6 | 297.4 | 310.7 |

Let $x = 0$ represent the year 1980.
(a) Find the least squares regression line for these data.
(b) Estimate the CPI for the year 1990. (This may be a poor estimate because the CPI is dependent on both domestic and international political decisions.)

**27.** The number of imported cars sold in the United States has increased dramatically in the past several years, as indicated in the following table (sales in millions).

| Year ($x$) | 1960 | 1965 | 1970 | 1975 | 1980 | 1983 |
|---|---|---|---|---|---|---|
| Sales ($y$) | 0.50 | 0.57 | 1.28 | 1.59 | 2.40 | 2.37 |

Let $x = 0$ represent the year 1960.
(a) Find the least squares regression line for these data.
(b) Estimate the sales of imports for the year 1990.

**28.** In order to analyze the decrease in the number of infant deaths per 1000 live births in the United States, a medical researcher obtains the following data.

| Year ($x$) | 1940 | 1950 | 1960 | 1970 | 1980 |
|---|---|---|---|---|---|
| Deaths ($y$) | 47.0 | 29.2 | 26.0 | 20.0 | 12.6 |

Let $x = 0$ correspond to the year 1960.
(a) Find the least squares regression line for these data.
(b) Estimate the number of infant deaths per 1000 live births for the year 1990.

**29.** The following table gives the world population in billions for five different years.

| Year ($x$) | 1960 | 1970 | 1975 | 1980 | 1985 |
|---|---|---|---|---|---|
| Population ($y$) | 3.0 | 3.7 | 4.1 | 4.5 | 4.8 |

Let $x = 0$ represent the year 1975.
(a) Find the least squares regression quadratic for these data.
(b) Use this quadratic to estimate the world population for the year 1990.

**30.** Repeat Exercise 28 using a least squares regression quadratic.

# Double Integrals and Area in the Plane

## INTRODUCTORY EXAMPLE
### Population Density of a City

Demographers have discovered that most cities have population densities that are greatest at the city's center. As one moves farther and farther from the city center, the population becomes less and less dense. The model most commonly used to represent this changing density (in number of people per square mile) is density $= Ce^{-ar^2}$, where $a$ and $C$ are constants and $r$ is the distance in miles from the city's center. To determine the total population of a city with a given density function, we assume that the center of the city lies at the origin, as shown in Figure 8.49. Since the distance from the origin to the point $(x, y)$ is $r = \sqrt{x^2 + y^2}$, we can rewrite the density function as a function of $x$ and $y$ as follows:

$$\text{Density} = f(x, y) = Ce^{-a(x^2+y^2)}$$

Now the total population of the city can be determined by integrating the density function over the plane region representing the city's limits. We do this by setting up a **double integral** to represent the area of the plane region. For example, a circular city with a radius of 5 miles has an area represented by the double integral

$$\text{Area} = \int_{-5}^{5} \int_{-\sqrt{25-x^2}}^{\sqrt{25-x^2}} dy\, dx$$

Thus, if the density function for this circular city is

$$f(x, y) = 100{,}000e^{-(x^2+y^2)}$$

then the total population for the city would be given by the double integral

$$\text{Population} = \int_{-5}^{5} \int_{-\sqrt{25-x^2}}^{\sqrt{25-x^2}} 100{,}000e^{-(x^2+y^2)}\, dy\, dx$$

Although evaluation of this *particular* double integral is beyond the scope of this text (the population for this model turns out to be 314,159), you will be able to evaluate a wide variety of double integrals after studying the techniques presented in this section.

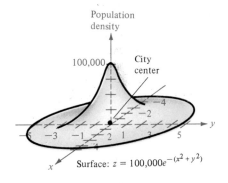

FIGURE 8.49

■ **Double Integrals**
■ **Evaluation of Double Integrals**
■ **Representation of Area by a Double Integral**

In Section 8.4, we showed that it is meaningful to differentiate functions of more than one variable by differentiating with respect to one variable at a time while holding the other variables constant. It should not be surprising to learn that we can *integrate* functions of two or more variables using a similar procedure. For example, if we are given the partial derivative $f_x(x, y) = 2xy$, then by holding $y$ constant we can integrate with respect to $x$ to obtain

$$f(x, y) = x^2y + C(y)$$

This procedure is sometimes referred to as **partial integration with respect to $x$.** Note that the constant of integration, $C(y)$, is assumed to be a function of $y$, since $y$ is fixed during integration with respect to $x$.

To evaluate the definite integral of a function of two or more variables, we can use the Fundamental Theorem of Calculus (Section 4.3) with one variable while holding the others constant. For instance,

$$\int_1^{2y} 2xy\,dx = x^2y \Big]_1^{2y} = (2y)^2y - (1)^2y = 4y^3 - y$$

$x$ is the variable
of integration and
$y$ is fixed

Replace $x$ by
the limits of
integration

The result is a
function of $y$

Note that we omit the constant of integration just as we do for the definite integral of a function of one variable.

A partial integral with respect to $y$ can be evaluated in a similar manner, as is demonstrated in Example 1.

**EXAMPLE 1**

**Partial Integration**

Evaluate the integrals

$$\int_1^x (2x^2y^{-2} + 2y)\,dy \qquad \text{and} \qquad \int_y^{5y} \sqrt{x - y}\,dx$$

**SOLUTION**

Considering $x$ to be constant while integrating with respect to $y$, we obtain

$$\int_1^x (2x^2y^{-2} + 2y)\,dy = \left[\frac{-2x^2}{y} + y^2\right]_1^x$$

$$= \left(\frac{-2x^2}{x} + x^2\right) - \left(\frac{-2x^2}{1} + 1\right)$$

$$= 3x^2 - 2x - 1$$

Similarly, considering $y$ to be constant, we have

$$\int_y^{5y} \sqrt{x-y}\ dx = \frac{2}{3}(x-y)^{3/2}\bigg]_y^{5y} = \frac{2}{3}[(5y-y)^{3/2} - (y-y)^{3/2}]$$

$$= \frac{2}{3}(4y)^{3/2} = \frac{16}{3}y^{3/2}$$

Note that in Example 1 the integral

$$\int_1^x (2x^2y^{-2} + 2y)\ dy = 3x^2 - 2x - 1$$

defines a function of one variable $x$ and, as such, can *itself* be integrated, as shown in the next example.

**EXAMPLE 2**

**Evaluating a Double Integral**

Evaluate the integral

$$\int_0^1 \left[\int_1^x (2x^2y^{-2} + 2y)\ dy\right] dx$$

**SOLUTION**

Using the results from Example 1, we have

$$\int_0^1 \left[\int_1^x (2x^2y^{-2} + 2y)\ dy\right] dx = \int_0^1 (3x^2 - 2x - 1)\ dx$$

$$= \left[x^3 - x^2 - x\right]_0^1 = -1$$

We call integrals of the form

$$\int_a^b \left[\int_{g_1(x)}^{g_2(x)} f(x,\ y)\ dy\right] dx \qquad \text{or} \qquad \int_c^d \left[\int_{h_1(y)}^{h_2(y)} f(x,\ y)\ dx\right] dy$$

**double integrals,** and we normally shorten the notation by omitting the brackets. Thus, by definition,

$$\int_a^b \int_{g_1(x)}^{g_2(x)} f(x,\ y)\ dy\ dx = \int_a^b \left[\int_{g_1(x)}^{g_2(x)} f(x,\ y)\ dy\right] dx$$

**Area of a Region in the Plane**

One of the simplest applications of a double integral is in finding the area of a plane region. For instance, if $R$ is the region bounded by

$$a \le x \le b \qquad \text{and} \qquad g_1(x) \le y \le g_2(x)$$

then by the methods of Section 4.4 we know that the area of $R$ is given by

$$\int_a^b [g_2(x) - g_1(x)] \, dx$$

(See Figure 8.50.) This same area is also given by the double integral

$$\int_a^b \int_{g_1(x)}^{g_2(x)} dy \, dx$$

because

$$\int_a^b \int_{g_1(x)}^{g_2(x)} dy \, dx = \int_a^b y \Big]_{g_1(x)}^{g_2(x)} dx = \int_a^b [g_2(x) - g_1(x)] \, dx$$

Figure 8.50 shows the two basic types of plane regions whose areas can be determined by a double integral.

In Figure 8.50, note that the position (vertical or horizontal) of the narrow rectangle indicates the order of integration. The "outer" variable of integration always corresponds to the width (thickness) of the rectangle.

**Determining area in the plane by double integrals**

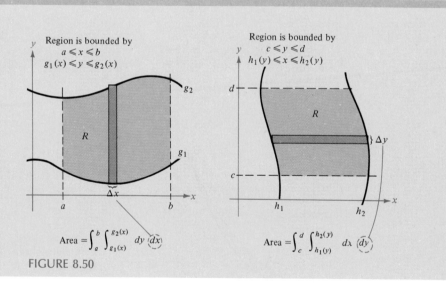

FIGURE 8.50

(Compare this procedure with the procedures of Section 4.4 for determining the area between two curves.) Note also that the outer limits of integration for a double integral are constant, whereas the inner limits may be functions of the outer variable.

**EXAMPLE 3**

**Finding Area by a Double Integral**

Set up a double integral and evaluate it to find the area of the region bounded by the graphs of $y = x^2$ and $y = x^3$.

**SOLUTION**    Using Figure 8.51, we see that the graphs of $y = x^2$ and $y = x^3$ intersect when $x = 0$ and $x = 1$. Therefore, the limits of integration for $x$ are

$$0 \le x \le 1$$

Using these constant limits, we choose $x$ to be the outer variable and $y$ to be the inner variable. The variable limits for $y$ are then

$$x^3 \le y \le x^2$$

Therefore, the area of the region is given by

$$\text{Area} = \int_0^1 \int_{x^3}^{x^2} dy\, dx$$

$$= \int_0^1 y \Big]_{x^3}^{x^2} dx$$

$$= \int_0^1 (x^2 - x^3)\, dx$$

$$= \left[ \frac{x^3}{3} - \frac{x^4}{4} \right]_0^1$$

$$= \frac{1}{3} - \frac{1}{4}$$

$$= \frac{1}{12}$$

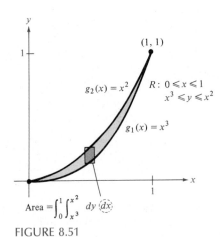

$$\text{Area} = \int_0^1 \int_{x^3}^{x^2} dy\, dx$$

FIGURE 8.51

In setting up double integrals, the most difficult task is likely to be the determination of the correct limits for the specified order of integration. This task can often be simplified by making a sketch of the region $R$ and identifying the appropriate bounds for $x$ and $y$. The next example illustrates this procedure.

**EXAMPLE 4**

**Changing the Order of Integration**

Given the double integral

$$\int_0^2 \int_{y^2}^4 dx\, dy$$

(a) sketch the region $R$ whose area is given by this integral.
(b) rewrite the integral so that $x$ is the outer variable.
(c) show that both orders of integration yield the same value.

**SOLUTION**

(a) From the limits of integration we know that

$$y^2 \le x \le 4$$

which means that the region $R$ is bounded on the left by the parabola $y^2 = x$ and on the right by the line $x = 4$. Furthermore, since

$$0 \le y \le 2$$

we have the region shown in Figure 8.52.

(b) If we interchange the order of integration so that $x$ is the outer variable, we see that $x$ has the constant bounds $0 \le x \le 4$. Solving for $y$ in the equation $y^2 = x$, we conclude that the bounds for $y$ are $0 \le y \le \sqrt{x}$ (see Figure 8.53). Therefore, with $x$ as the outer variable, we obtain the integral

$$\int_0^4 \int_0^{\sqrt{x}} dy\, dx$$

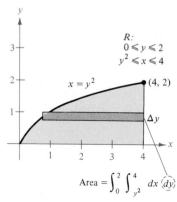

FIGURE 8.52

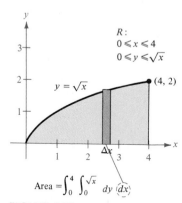

FIGURE 8.53

(c) Evaluating both integrals, we obtain

$$\int_0^2 \int_{y^2}^4 dx\, dy = \int_0^2 x \Big]_{y^2}^4 dy = \int_0^2 (4 - y^2)\, dy = \left[ 4y - \frac{y^3}{3} \right]_0^2 = \frac{16}{3}$$

and

$$\int_0^4 \int_0^{\sqrt{x}} dy\, dx = \int_0^4 y \Big]_0^{\sqrt{x}} dx = \int_0^4 \sqrt{x}\, dx = \frac{2}{3} x^{3/2} \Big]_0^4 = \frac{16}{3}$$

To designate a double integral or an area of a region without specifying a particular order of integration, we use the symbol

$$\int\int_R dA$$

where $dA = dx\,dy$ or $dA = dy\,dx$.

**EXAMPLE 5**

**Finding Area by a Double Integral**

Use a double integral to calculate the area denoted by

$$\int\int_R dA$$

where $R$ is the region bounded by $y = x$ and $y = x^2 - x$.

**SOLUTION**

From a sketch of region $R$ (Figure 8.54) we can see that vertical rectangles of width $dx$ are more convenient than horizontal ones. Therefore, $x$ is the outer variable of integration, and its constant bounds are $0 \le x \le 2$.
Thus, we have

$$\int\int_R dA = \int_0^2 \int_{x^2-x}^x dy\,dx$$

$$= \int_0^2 y\Big]_{x^2-x}^x dx$$

$$= \int_0^2 [x - (x^2 - x)]\,dx = \int_0^2 (2x - x^2)\,dx$$

$$= \left[x^2 - \frac{x^3}{3}\right]_0^2 = 4 - \frac{8}{3} = \frac{4}{3}$$

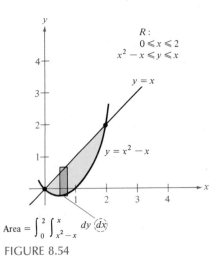

$$\text{Area} = \int_0^2 \int_{x^2-x}^x dy\,\widehat{dx}$$

FIGURE 8.54

At this point you may be wondering why we need double integrals, since we can already find the area between curves in the plane by single integrals. The real need for double integrals will be demonstrated in Section 8.9, where they will be used to find volumes and average functional values. When working the exercises in this section, remember that we have chosen to introduce double integrals by way of areas in the plane so that we can give primary attention to the procedures for finding the limits of integration. As the examples of this section show, this task is greatly simplified by making a sketch of the region under consideration.

## SECTION EXERCISES 8.8

In Exercises 1–10, evaluate the given integral.

**1.** $\displaystyle\int_0^x (2x - y)\, dy$

**2.** $\displaystyle\int_x^{x^2} \frac{y}{x}\, dy$

**3.** $\displaystyle\int_1^{2y} \frac{y}{x}\, dx$

**4.** $\displaystyle\int_0^{e^y} y\, dx$

**5.** $\displaystyle\int_0^{\sqrt{4-x^2}} x^2 y\, dy$

**6.** $\displaystyle\int_{x^2}^{\sqrt{x}} (x^2 + y^2)\, dy$

**7.** $\displaystyle\int_{e^y}^{y} \frac{y \ln x}{x}\, dx$

**8.** $\displaystyle\int_{-\sqrt{1-y^2}}^{\sqrt{1-y^2}} (x^2 + y^2)\, dx$

**9.** $\displaystyle\int_0^{x^3} ye^{-y/x}\, dy$

**10.** $\displaystyle\int_y^3 \frac{xy}{\sqrt{x^2 + 1}}\, dx$

In Exercises 11–24, evaluate the given double integral.

**11.** $\displaystyle\int_0^1 \int_0^2 (x + y)\, dy\, dx$

**12.** $\displaystyle\int_0^2 \int_0^2 (6 - x^2)\, dy\, dx$

**13.** $\displaystyle\int_0^4 \int_0^3 xy\, dy\, dx$

**14.** $\displaystyle\int_0^1 \int_0^x \sqrt{1 - x^2}\, dy\, dx$

**15.** $\displaystyle\int_1^2 \int_0^4 (x^2 - 2y^2 + 1)\, dx\, dy$

**16.** $\displaystyle\int_0^1 \int_y^{2y} (1 + 2x^2 + 2y^2)\, dx\, dy$

**17.** $\displaystyle\int_0^1 \int_0^{\sqrt{1-y^2}} (x + y)\, dx\, dy$

**18.** $\displaystyle\int_0^2 \int_{3y^2-6y}^{2y-y^2} 3y\, dx\, dy$

**19.** $\displaystyle\int_0^2 \int_0^{\sqrt{4-y^2}} \frac{2}{\sqrt{4 - y^2}}\, dx\, dy$

**20.** $\displaystyle\int_0^4 \int_0^x \frac{2}{(x + 1)(y + 1)}\, dy\, dx$

**21.** $\displaystyle\int_0^2 \int_0^{4-x^2} x^3\, dy\, dx$

**22.** $\displaystyle\int_0^a \int_0^{a-x} (x^2 + y^2)\, dy\, dx$

**23.** $\displaystyle\int_0^\infty \int_0^\infty e^{-(x+y)/2}\, dy\, dx$

**24.** $\displaystyle\int_0^\infty \int_0^\infty xye^{-(x^2+y^2)}\, dx\, dy$

In Exercises 25–32, sketch the region $R$ whose area is given by the double integral. Then switch the order of integration and show that both orders yield the same area.

**25.** $\displaystyle\int_0^1 \int_0^2 dy\, dx$

**26.** $\displaystyle\int_1^2 \int_2^4 dx\, dy$

**27.** $\displaystyle\int_0^1 \int_{2y}^2 dx\, dy$

**28.** $\displaystyle\int_0^4 \int_0^{\sqrt{x}} dy\, dx$

**29.** $\displaystyle\int_0^2 \int_{x/2}^1 dy\, dx$

**30.** $\displaystyle\int_0^4 \int_{\sqrt{x}}^2 dy\, dx$

**31.** $\displaystyle\int_0^1 \int_{y^2}^{\sqrt[3]{y}} dx\, dy$

**32.** $\displaystyle\int_{-2}^2 \int_0^{4-y^2} dx\, dy$

In Exercises 33–38, use a double integral to find the area of the specified region.

**33.**

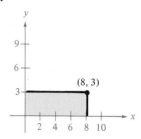

**34.**

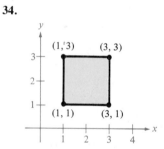

**35.**

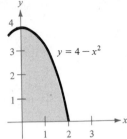

$y = 4 - x^2$

**36.**

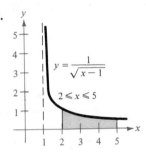

$y = \dfrac{1}{\sqrt{x-1}}$

$2 \leqslant x \leqslant 5$

**37.**

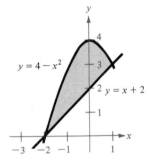

$y = 4 - x^2$

$y = x + 2$

**38.**

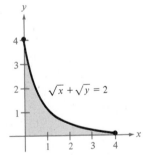

$\sqrt{x} + \sqrt{y} = 2$

In Exercises 39–44, use a double integral to find the area of the region bounded by the graphs of the given equations.

**39.** $y = 25 - x^2,\ y = 0$

**40.** $y = x^{3/2},\ y = x$

**41.** $2x - 3y = 0,\ x + y = 5,\ y = 0$

**42.** $xy = 9,\ y = x,\ y = 0,\ x = 9$

**43.** $y = x,\ y = 2x,\ x = 2$

**44.** $y = x^2 + 2x + 1,\ y = 3(x + 1)$

# Applications of Double Integrals

## INTRODUCTORY EXAMPLE
## Average Elevation

The highest point in the United States is Mt. McKinley in Alaska, with an elevation of 20,320 feet above sea level. The lowest point is in Death Valley, California, where the elevation drops to −282 feet (below sea level). Does this mean that the average elevation in the United States is $(20,320 − 282)/2 = 10,019$ feet? Quite clearly the answer to this question is no! Actually the average elevation of the United States is estimated to be about 2500 feet. The reason the average elevation is so much less than the mean of the two extremes is that considerably more land lies at lower elevations than at higher ones.

To see how the average elevation can be found, let us consider a simple example. Suppose that we want to find the average elevation of a triangular coastal county. (See Figure 8.55.) The elevation in miles above sea level at the point $(x, y)$ is given by

$$\text{Elevation} = f(x, y) = 0.25 − 0.025x − 0.01y$$

where $x$ and $y$ are measured in miles. Now, to find the average elevation of this county, we consider the first-octant solid bounded above by the plane $z = f(x, y)$ and below by the $xy$-plane. The average elevation is given by the double integral

$$\text{Average elevation} = \frac{\text{volume of solid}}{\text{area of base}}$$

$$= \frac{1}{A} \iint_R f(x, y) \, dA$$

where $A$ is the area of the base. Since the base is triangular, its area is given by

$$\text{Area of base} = \left(\frac{1}{2}\right)(10)(25) = 125 \text{ square miles}$$

Thus, the average elevation is given by

Average elevation

$$= \frac{1}{125} \int_0^{10} \int_0^{25-2.5x} (0.25 − 0.025x − 0.01y) \, dy \, dx$$

$$= \frac{1}{125} \int_0^{10} \left[ 0.25y − 0.025xy − 0.005y^2 \right]_0^{25-2.5x} dx$$

$$= \frac{1}{125} \int_0^{10} (0.03125x^2 − 0.625x + 3.125) \, dx$$

$$= \frac{1}{125} \left[ \frac{0.03125x^3}{3} − \frac{0.625x^2}{2} + 3.125x \right]_0^{10}$$

$$\approx 0.0833 \text{ miles} = 440 \text{ feet above sea level}$$

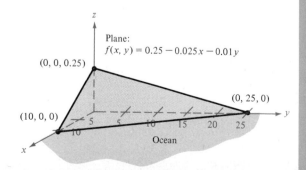

FIGURE 8.55

■ **Volume of a Solid Region**
■ **Average Value of a Function Over a Region**

In the preceding section, we demonstrated the use of a double integral as an alternative way to find the area of a plane region. In this section, we discuss more imperative uses of double integrals—namely, to find the volume of a solid region and to find the average value of a function.

Specifically, if $f$ is continuous and nonnegative over a region $R$, then the volume between $z = f(x, y)$ and the region $R$ in the $xy$-plane is given by the double integral

$$V = \int\int_R f(x, y)\, dA$$

where $dA = dy\, dx$ or $dA = dx\, dy$.

| | |
|---|---|
| **Determining volume by double integrals** | If $R$ is a bounded region in the $xy$-plane and $f$ is continuous and nonnegative over $R$, then the **volume of the solid** between the surface $z = f(x, y)$ and $R$ is given by $$V = \int\int_R f(x, y)\, dA$$ |

**EXAMPLE 1**

**Finding Volume by a Double Integral**

Find the volume of the solid bounded in the first octant by the plane

$$z = 2 - x - 2y$$

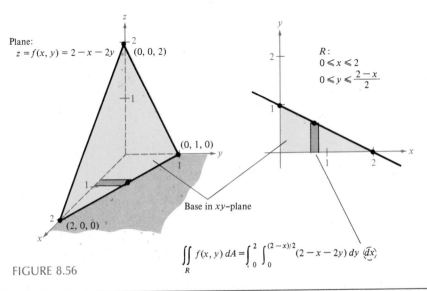

Plane:
$z = f(x, y) = 2 - x - 2y$    (0, 0, 2)

(0, 1, 0)

(2, 0, 0)

Base in $xy$-plane

$R:$
$0 \leqslant x \leqslant 2$
$0 \leqslant y \leqslant \dfrac{2-x}{2}$

$$\int\int_R f(x, y)\, dA = \int_0^2 \int_0^{(2-x)/2} (2 - x - 2y)\, dy\, dx$$

FIGURE 8.56

**SOLUTION**

To set up the double integral for the volume, it is helpful to sketch both the solid and the region $R$ in the $xy$-plane. From Figure 8.56 we can see that if $z = 0$, then $R$ is bounded in the $xy$-plane by the lines

$$x = 0, \qquad y = 0, \qquad y = \frac{2 - x}{2}$$

For a rectangle of width $dx$, we obtain

$$\text{Constant bounds for } x\colon\ 0 \le x \le 2$$

$$\text{Variable bounds for } y\colon\ 0 \le y \le \frac{2 - x}{2}$$

Thus, the volume of the solid region is

$$V = \int_0^2 \int_0^{(2-x)/2} (2 - x - 2y)\, dy\, dx = \int_0^2 \left[ (2 - x)y - y^2 \right]_0^{(2-x)/2} dx$$

$$= \int_0^2 \left[ (2 - x)\left( \frac{2 - x}{2} \right) - \left( \frac{2 - x}{2} \right)^2 \right] dx$$

$$= \frac{1}{4} \int_0^2 (2 - x)^2\, dx$$

$$= \frac{-1}{12}(2 - x)^3 \Big]_0^2$$

$$= \frac{8}{12} = \frac{2}{3}$$

Following the pattern of Example 1, we outline a procedure for determining the volume of a solid region:

1. Write the equation of the surface in the form $z = f(x, y)$ and (if possible) sketch the solid region.
2. Sketch region $R$ in the $xy$-plane and determine the order and limits of integration.
3. Evaluate the double integral

$$V = \int \int_R f(x, y)\, dA$$

using the order and limits determined in the second step.

In Example 1, the order of integration was arbitrary. We could have used $y$ as the outer variable, as shown in Figure 8.57. There are, however, some occasions on which one order of integration is much more convenient than the other. Example 2 shows such a case.

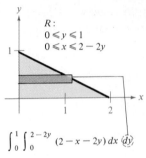

$$\int_0^1 \int_0^{2-2y} (2 - x - 2y)\, dx\, dy$$

Reversing the Order of Integration
(Compare with Figure 8.53.)

FIGURE 8.57

## EXAMPLE 2

**Comparing Different Orders of Integration**

Find the volume under the surface given by

$$f(x, y) = e^{-x^2}$$

bounded by the $xz$-plane and the planes $y = x$ and $x = 1$.

**SOLUTION**

In the $xy$-plane, the bounds of region $R$ are the lines $y = 0$, $x = 1$, and $y = x$ (see Figure 8.58). The two possible orders of integration are given in Figure 8.59. In attempting to evaluate these two double integrals, we discover that the one on the right requires the antiderivative

$$\int e^{-x^2}\, dx$$

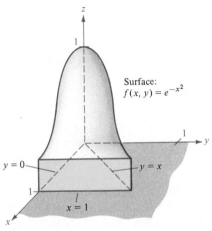

Surface:
$f(x, y) = e^{-x^2}$

$y = 0$

$y = x$

$x = 1$

FIGURE 8.58

which we know is not an elementary function. However, we can evaluate the integral on the left in the following manner:

$$V = \int_0^1 \int_0^x e^{-x^2} \, dy \, dx$$

$$= \int_0^1 e^{-x^2} y \Big]_0^x \, dx$$

$$= \int_0^1 x e^{-x^2} \, dx$$

$$= -\frac{1}{2} e^{-x^2} \Big]_0^1$$

$$= -\frac{1}{2}\left(\frac{1}{e} - 1\right)$$

$$= \frac{e - 1}{2e}$$

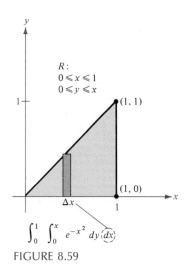

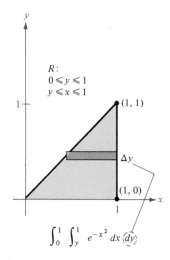

FIGURE 8.59

In Examples 1 and 2, we provided the sketches of both the solid whose volume we were finding and the plane region that formed the base of the solid. However, keep in mind that you do not have to be an artist to be able to find the volume of three-dimensional solids. If you are not able to sketch a particular solid, you can still find its volume by referring to a sketch of its base in the $xy$-plane. We demonstrate this in Example 3.

EXAMPLE 3

**Finding Volume by a Double Integral**

Find the volume of the solid bounded above by the surface

$$f(x, y) = 6x^2 - 2xy$$

and below by the plane region $R$ shown in Figure 8.60.

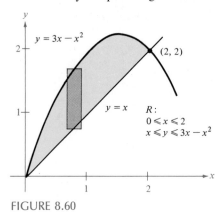

FIGURE 8.60

SOLUTION

Since the region $R$ is bounded by the parabola $y = 3x - x^2$ and the line $y = x$, $y$ has the limits

$$x \le y \le 3x - x^2$$

as shown in Figure 8.60. The volume of the solid is given by

$$V = \int_0^2 \int_x^{3x-x^2} (6x^2 - 2xy)\, dy\, dx$$

$$= \int_0^2 \left[ 6x^2 y - xy^2 \right]_x^{3x-x^2} dx$$

$$= \int_0^2 [(18x^3 - 6x^4 - 9x^3 + 6x^4 - x^5) - (6x^3 - x^3)]\, dx$$

$$= \int_0^2 (4x^3 - x^5)\, dx = \left[ x^4 - \frac{x^6}{6} \right]_0^2 = \frac{16}{3}$$

In the Introductory Example of Section 8.8, we saw that the population of a city whose population density is $f(x, y)$ is given by

$$\text{Population} = \int \int_R f(x, y)\, dA$$

where the region $R$ represents the city's limits. We demonstrate an application of this formula in the next example.

**EXAMPLE 4**

**Finding the Population of a City**

Find the population of a city whose density function is

$$f(x, y) = \frac{50,000}{x + |y| + 1}$$

and whose boundary is shown in Figure 8.61.

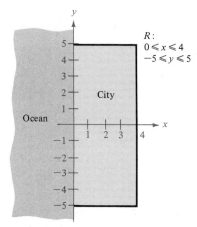

FIGURE 8.61

**SOLUTION**

To begin with, we note (from the absolute value of $y$ in the density function) that the population density at the point $(x, y)$ is the same as that at the point $(x, -y)$. Thus, we can conclude that the population in the first quadrant equals the population in the fourth quadrant. In other words, we can find the total population by doubling the population in the first quadrant, as follows:

Population

$$= 2 \int_0^4 \int_0^5 \frac{50,000}{x + y + 1} \, dy \, dx$$

$$= 100,000 \int_0^4 \left[ \ln (x + y + 1) \right]_0^5 dx$$

$$= 100,000 \int_0^4 \left[ \ln (x + 6) - \ln (x + 1) \right] dx$$

$$= 100,000 \left[ (x + 6) \ln (x + 6) - (x + 6) - (x + 1) \ln (x + 1) + (x + 1) \right]_0^4$$

$$= 100,000 \left[ (x + 6) \ln (x + 6) - (x + 1) \ln (x + 1) - 5 \right]_0^4$$

$$= 100,000[10 \ln (10) - 5 \ln (5) - 5 - 6 \ln (6) + 5]$$

$$\approx 422,810 \text{ people}$$

## The Average Value of a Function Over a Region

Recall from Chapter 4 that one important application of single integrals is in finding the **average value** of the function $f(x)$ over the interval $[a, b]$. We can generalize this concept to functions of two variables as follows.

**Average value of $f(x, y)$ over the region $R$**

If $f$ is integrable over the plane region $R$, then its **average value** over $R$ is given by

$$\text{Average value of } f(x, y) \text{ over } R = \frac{1}{A} \int \int_R f(x, y) \, dA$$

where $A$ is the area of the region $R$.

**EXAMPLE 5**

**Finding the Average Profit**

A manufacturer determines that the profit for selling $x$ units of one product and $y$ units of a second product is

$$P = -(x - 200)^2 - (y - 100)^2 + 5000$$

The weekly sales for product 1 vary between 150 and 200 units, and the weekly sales for product 2 vary between 80 and 100 units. Estimate the weekly profit for these two products.

**SOLUTION**

Since we are given $150 \le x \le 200$ and $80 \le y \le 100$, we estimate the weekly profit to be the average of the profit function over the rectangular region shown in Figure 8.62. Since the area of this region is given by

$$\text{Area} = (50)(20) = 1000$$

we have

$$\text{Average profit} = \frac{1}{1000} \int_{150}^{200} \int_{80}^{100} [-(x - 200)^2 - (y - 100)^2 + 5000] \, dy \, dx$$

$$= \frac{1}{1000} \int_{150}^{200} \left[ -(x - 200)^2 y - \frac{(y - 100)^3}{3} + 5000y \right]_{80}^{100} dx$$

$$= \frac{1}{1000} \int_{150}^{200} \left[ -20(x - 200)^2 + \frac{292{,}000}{3} \right] dx$$

$$= \frac{1}{3000} \left[ -20(x - 200)^3 + 292{,}000x \right]_{150}^{200}$$

$$= \frac{12{,}100{,}000}{3000} \approx \$4{,}033$$

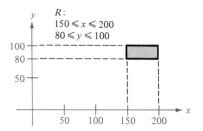

$$R:$$
$$150 \leqslant x \leqslant 200$$
$$80 \leqslant y \leqslant 100$$

FIGURE 8.62

## SECTION EXERCISES 8.9

In Exercises 1–6, sketch the region $R$ and evaluate the double integral

$$\int\int_R f(x, y) \, dA$$

**1.** $\displaystyle\int_0^2 \int_0^1 (1 + 2x + 2y) \, dy \, dx$

**2.** $\displaystyle\int_0^6 \int_{y/2}^3 (x + y) \, dx \, dy$

**3.** $\displaystyle\int_0^1 \int_y^{\sqrt{y}} x^2 y^2 \, dx \, dy$

**4.** $\displaystyle\int_{-a}^a \int_{-\sqrt{a^2-x^2}}^{\sqrt{a^2-x^2}} (x + y) \, dy \, dx$

**5.** $\displaystyle\int_0^1 \int_0^{\sqrt{1-x^2}} y \, dy \, dx$

**6.** $\displaystyle\int_0^2 \int_0^{4-x^2} xy^2 \, dy \, dx$

In Exercises 7–10, set up the integral for both orders of integration and use the more convenient order to evaluate the integral over the region $R$.

**7.** $\displaystyle\int\int_R xy \, dA$,  $R$: rectangle with vertices at $(0, 0)$, $(0, 5)$, $(3, 5)$, $(3, 0)$

**8.** $\displaystyle\int\int_R \frac{y}{x^2 + y^2} \, dA$,  $R$: triangle bounded by $y = x$, $y = 2x$, $x = 2$

**9.** $\displaystyle\int\int_R \frac{y}{1 + x^2} \, dA$,  $R$: region bounded by $y = 0$, $y = \sqrt{x}$, $x = 4$

**10.** $\displaystyle\int\int_R x \, dA$,  $R$: semicircle bounded by $y = \sqrt{25 - x^2}$ and $y = 0$

In Exercises 11–22, use a double integral to find the volume of the specified solid.

**11.**

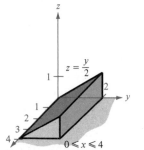

$$0 \leqslant x \leqslant 4$$
$$0 \leqslant y \leqslant 2$$

**12.**

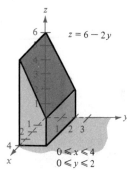

$$z = 6 - 2y$$
$$0 \leqslant x \leqslant 4$$
$$0 \leqslant y \leqslant 2$$

**13.**

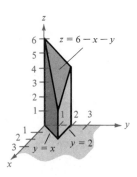

$$z = 6 - x - y$$
$$y = x$$
$$y = 2$$

**14.**

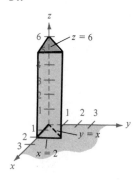

$$z = 6$$
$$y = x$$
$$x = 2$$

**15.**

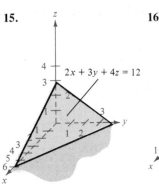

$2x + 3y + 4z = 12$

**16.**

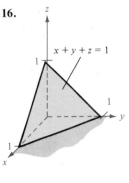

$x + y + z = 1$

In Exercises 23–26, use a double integral to find the volume of the solid bounded by the graphs of the equations.

**23.** $z = xy$, $z = 0$, $y = 0$, $y = 4$, $x = 0$, $x = 1$

**24.** $z = x$, $z = 0$, $y = x$, $y = 0$, $x = 0$, $x = 4$

**25.** $z = x^2$, $z = 0$, $x = 0$, $x = 2$, $y = 0$, $y = 4$

**26.** $z = x + y$, $x^2 + y^2 = 4$ (first octant)

In Exercises 27–30, find the average value of $f(x, y)$ over the region $R$.

**27.** $f(x, y) = x$, $R$: rectangle with vertices $(0, 0)$, $(4, 0)$, $(4, 2)$, $(0, 2)$

**28.** $f(x, y) = xy$, $R$: rectangle with vertices $(0, 0)$, $(4, 0)$, $(4, 2)$, $(0, 2)$

**29.** $f(x, y) = x^2 + y^2$, $R$: square with vertices $(0, 0)$, $(2, 0)$, $(2, 2)$, $(0, 2)$

**30.** $f(x, y) = e^{x+y}$, $R$: triangle with vertices $(0, 0)$, $(0, 1)$, $(1, 1)$

**17.**

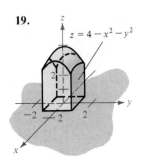

$z = 1 - xy$

$y = x$    $y = 1$

**18.**

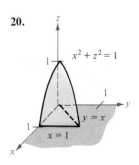

$z = 4 - y^2$

$y = x$    $y = 2$

**19.**

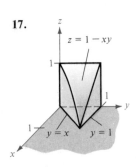

$z = 4 - x^2 - y^2$

**20.**

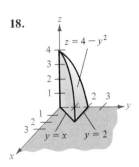

$x^2 + z^2 = 1$

$y = x$

$x = 1$

**31.** A company sells two products whose demand functions are

$$x_1 = 500 - 3p_1$$

and

$$x_2 = 750 - 2.4p_2$$

Therefore, the total revenue is given by

$$R = x_1p_1 + x_2p_2$$

Estimate the average revenue if the price $p_1$ varies between \$50 and \$75 and the price $p_2$ varies between \$100 and \$150.

**32.** A firm's weekly profit in marketing two products is given by

$$P = 192x_1 + 576x_2 - x_1^2 - 5x_2^2 - 2x_1x_2 - 5000$$

where $x_1$ and $x_2$ represent the number of units of each product sold weekly. Estimate the average weekly profit if $x_1$ varies between 40 and 50 units and $x_2$ varies between 45 and 50 units.

**33.** For a particular company, the Cobb-Douglas annual production function is

$$f(x, y) = 100x^{0.6}y^{0.4}$$

Estimate the average annual production level if the number of units of labor varies between 200 and 250 units and the number of units of capital varies between 300 and 325 units.

**21.**

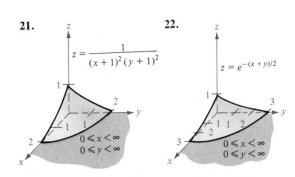

$z = \dfrac{1}{(x + 1)^2 (y + 1)^2}$

$0 \leqslant x < \infty$
$0 \leqslant y < \infty$

**22.**

$z = e^{-(x + y)/2}$

$0 \leqslant x < \infty$
$0 \leqslant y < \infty$

**34.** A corporation produces a product at two locations. Cost functions at plants 1 and 2 are

$$C_1 = 0.03x_1^2 + 8x_1 + 5000$$

and

$$C_2 = 0.06x_2^2 + 6x_2 + 8000$$

Therefore, the total cost for producing the product is

$$C = C_1 + C_2$$

Estimate the average annual cost if the annual production level at plant 1 varies between 200 and 225 units and the annual production level at plant 2 varies between 100 and 150 units.

## Important Terms

Solid analytic geometry
Three-dimensional coordinate system
Right-handed system
Coordinate planes
xy-plane, xz-plane, yx-plane
Octants
Distance formula
Midpoint formula
Standard equation of a sphere
Trace
Plane in space
Surface in space
Quadric surface
Elliptic cone
Elliptic paraboloid
Hyperbolic paraboloid
Ellipsoid
Hyperboloid of one sheet
Hyperboloid of two sheets
Function of two variables
Function of several variables
Domain
Range

Level curve
Contour map
Partial derivative
Partial differentiation
First partial derivative
Second partial derivative
Higher-order partial derivative
Mixed partial derivatives
Relative minimum of a function of two variables
Relative maximum of a function of two variables
Saddle point
Optimization problem
Objective function
Constraint
Lagrange multiplier
Sum of the squared errors
Least squares regression line
Least squares regression quadratic
Partial integration
Double integral
Limits of integration
Volume of a solid
Average value of a function

## Important Techniques

Plotting points in space
Finding the trace of a surface in a coordinate plane
Sketching the graph of a plane in space
Classifying quadric surfaces
Finding the domain and range of a function of two variables
Sketching the graph of a function of two variables
Sketching a contour map
Finding partial derivatives
Finding relative extrema and saddle points of a function of two variables

Using the method of Lagrange multipliers to solve optimization problems
Using a double integral to find the area of region in the plane
Finding the limits of integration of a double integral
Changing the order of integration of a double integral
Using a double integral to find the volume of a solid region
Using a double integral to find the average value of a function of two variables over a region.

## Important Formulas

Distance Formula:

$$d = \sqrt{(x_2 - x_1)^2 + (y_2 - y_1)^2 + (z_2 - z_1)^2}$$

Midpoint Formula:

$$\text{Midpoint} = \left( \frac{x_1 + x_2}{2}, \frac{y_1 + y_2}{2}, \frac{z_1 + z_2}{2} \right)$$

Standard equation of a sphere:

$$(x - h)^2 + (y - k)^2 + (z - 1)^2 = r^2$$

General equation of a plane:

$$ax + by + cz = d$$

General equation of a quadric surface:

$$Ax^2 + By^2 + Cz^2 + Dx + Ey + Fz + G = 0$$

First partial derivatives of $z = f(x, y)$ with respect to $x$ and to $y$:

$$\frac{\partial z}{\partial x} = \lim_{\Delta x \to 0} \frac{f(x + \Delta x, y) - f(x, y)}{\Delta x} \quad (y \text{ is held constant})$$

$$\frac{\partial z}{\partial y} = \lim_{\Delta y \to 0} \frac{f(x, y + \Delta y) - f(x, y)}{\Delta y} \quad (x \text{ is held constant})$$

**First-Partials Test:** If $f(x_0, y_0)$ is a relative extremum of $f$ on an open region $R$, and the first partial derivatives of $f$ exist in $R$, then

$$f_x(x_0, y_0) = 0 = f_y(x_0, y_0)$$

**Second-Partials Test:** If $f$ has continuous first and second partial derivatives on an open region and there exists a point $(a, b)$ in the region such that $f_x(a, b) = 0$ and $f_y(a, b) = 0$, then the quantity

$$d = f_{xx}(a, b)f_{yy}(a, b) - [f_{xy}(a, b)]^2$$

can be used as follows:

1. $f(a, b)$ is a relative minimum if $d > 0$ *and* $f_{xx}(a, b) > 0$.
2. $f(a, b)$ is a relative maximum if $d > 0$ *and* $f_{xx}(a, b) < 0$.
3. $(a, b, f(a, b))$ is a saddle point if $d < 0$.
4. The test gives no information if $d = 0$.

**Lagrange multipliers:** If $f(x, y)$ has a minimum or maximum subject to the constraint $g(x, y) = 0$, then it will occur at one of the critical numbers of the function $F$ defined by

$$F(x, y, \lambda) = f(x, y) - \lambda g(x, y)$$

Sum of squared errors for $y = f(x)$:

$$S = [f(x_1) - y_1]^2 + [f(x_2) - y_2]^2 + \cdots + [f(x_n) - y_n]^2$$

Least squares regression line for

$$\{(x_1, y_1), (x_2, y_2), \ldots, (x_n, y_n)\}$$

is $y = ax + b$, where

$$a = \frac{n \sum_{i=1}^{n} x_i y_i - \sum_{i=1}^{n} x_i \sum_{i=1}^{n} y_i}{n \sum_{i=1}^{n} x_i^2 - \left(\sum_{i=1}^{n} x_i\right)^2}$$

and

$$b = \frac{1}{n}\left(\sum_{i=1}^{n} y_i - a \sum_{i=1}^{n} x_i\right)$$

Least squares regression quadratic for $\{(x_1, y_1), (x_2, y_2), \ldots, (x_n, y_n)\}$ is $y = ax^2 + bx + c$, where $a$, $b$, and $c$ are the solutions to the system

$$a \sum_{i=1}^{n} x_i^4 + b \sum_{i=1}^{n} x_i^3 + c \sum_{i=1}^{n} x_i^2 = \sum_{i=1}^{n} x_i^2 y_i$$

$$a \sum_{i=1}^{n} x_i^3 + b \sum_{i=1}^{n} x_i^2 + c \sum_{i=1}^{n} x_i = \sum_{i=1}^{n} x_i y_i$$

$$a \sum_{i=1}^{n} x_i^2 + b \sum_{i=1}^{n} x_i + cn = \sum_{i=1}^{n} y_i$$

Volume of solid between the surface $z = f(x, y)$ and region $R$:

$$\text{Volume} = \int\int_R f(x, y) \, dA$$

Average value of $f(x, y)$ over region $R$ with area $A$:

$$\text{Average value} = \frac{1}{A} \int\int_R f(x, y) \, dA$$

In Exercises 1 and 2, (a) plot the points, (b) find the distance between the points, and (c) find the midpoint of the line segment joining the points.

**1.** $(-4, -2, 3)$, $(6, 2, -5)$

**2.** $(0, 4, -3)$, $(-6, 10, 8)$

In Exercises 3 and 4, find the standard form of the equation of the sphere.

**3.** Center: $(1, 0, -3)$; radius: 3

**4.** Center: $(4, -5, 3)$; radius: 10

In Exercises 5 and 6, find the center and radius of the sphere.

**5.** $x^2 + y^2 + z^2 + 8x - 4y + 2z + 5 = 0$

**6.** $16x^2 + 16y^2 + 16z^2 + 8x - 32y - 32z + 17 = 0$

In Exercises 7–10, sketch the graph of the plane.

**7.** $x + 2y + 3z = 6$    **8.** $2y + z = 4$

**9.** $4x - 3y + 6z = 12$    **10.** $x = 3$

In Exercises 11–20, identify the surface. (See Section 8.2.)

**11.** $x^2 + y^2 + z^2 - 2x + 4y - 6z + 5 = 0$

**12.** $16x^2 + 16y^2 - 9z^2 = 0$

**13.** $\dfrac{x^2}{16} + \dfrac{y^2}{9} + z^2 = 1$    **14.** $\dfrac{x^2}{16} + \dfrac{y^2}{9} - z^2 = 1$

**15.** $\dfrac{x^2}{16} - \dfrac{y^2}{9} + z^2 = -1$    **16.** $z = \dfrac{x^2}{9} + y^2$

**17.** $-4x^2 + y^2 + z^2 = 4$

**18.** $3x^2 + 4y^2 + z^2 + 6x - 4 = 0$

**19.** $z = \sqrt{x^2 + y^2}$    **20.** $z = 9x + 3y - 5$

In Exercises 21–24, describe the region $R$ in the $xy$-plane that corresponds to the domain of the function, and find the range of the function.

**21.** $f(x, y) = x^2 + y^2$

**22.** $f(x, y) = \sqrt{1 - x^2 - y^2}$

**23.** $f(x, y) = \ln(1 - x^2 - y^2)$

**24.** $f(x, y) = \dfrac{1}{x + y}$

In Exercises 25–34, find the first partial derivatives.

**25.** $f(x, y) = x\sqrt{y} + 3x - 2y$

**26.** $f(x, y) = 5x^2 + 4xy - 3y^2$

**27.** $g(x, y) = \ln\sqrt{2x + 3y}$    **28.** $g(x, y) = x^2 e^{-2y}$

**29.** $z = xe^y + ye^x$    **30.** $z = \ln(x^2 + y^2 + 1)$

**31.** $g(x, y) = \dfrac{xy}{x^2 + y^2}$    **32.** $f(x, y) = \dfrac{xy}{x + y}$

**33.** $w = 2\sqrt{xyz}$    **34.** $w = xyz^2$

In Exercises 35 and 36, find all second partial derivatives and verify that the second mixed partials are equal.

**35.** $f(x, y) = 3x^2 - xy + 2y^3$

**36.** $h(x, y) = \dfrac{x}{x + y}$

In Exercises 37 and 38, show that the function satisfies the Laplace equation

$$\frac{\partial^2 z}{\partial x^2} + \frac{\partial^2 z}{\partial y^2} = 0$$

**37.** $z = x^2 - y^2$

**38.** $z = \dfrac{y}{x^2 + y^2}$

In Exercises 39–42, locate and classify any extrema of the function.

**39.** $f(x, y) = x^3 - 3xy + y^2$

**40.** $f(x, y) = x^2 + 2xy + y^2$

**41.** $f(x, y) = 2x^2 + 6xy + 9y^2 + 8x + 14$

**42.** $z = 50(x + y) - (0.1x^3 + 20x + 150)$
$\quad\quad - (0.05y^3 + 20.6y + 125)$

In Exercises 43 and 44, locate any extrema of the function by using Lagrange multipliers.

**43.** $z = x^2 y$
$\quad$ Constraint: $x + 2y = 2$

**44.** $z = x^2 + y^2$
$\quad$ Constraint: $x + y = 4$

**45.** The production function for a manufacturer is $f(x, y) = 4x + xy + 2y$. Assume that the total amount available for labor $x$ and capital $y$ is \$2,000 and that units of labor and capital cost \$20 and \$4, respectively. Find the maximum production level for this manufacturer.

**46.** A manufacturer has an order for 1500 units that can be produced at two locations. Let $x_1$ and $x_2$ be the number of units produced at the two locations. Find the number that should be produced at each to meet the order and minimize cost if the cost function is

$$C = 0.20x_1^2 + 10x_1 + 0.15x_2^2 + 12x_2$$

In Exercises 47 and 48, find the least squares regression line for the given points. Plot the points and sketch the least squares regression line on the same coordinate axes.

**47.** $(1, 5)$, $(2, 4)$, $(3, 2)$, $(5, 1)$

**48.** $(2, 1)$, $(3, 3)$, $(4, 2)$, $(5, 4)$, $(6, 4)$

In Exercises 49 and 50, find the least squares regression quadratic for the given points. Plot the points and sketch the least squares quadratic on the same coordinate axes.

**49.** (1, 1), (3, 2), (4, 4), (5, 7)

**50.** (0, 10), (2, 9), (3, 7), (4, 4), (5, 0)

In Exercises 51–58, evaluate the double integral.

**51.** $\displaystyle\int_0^1 \int_0^{1+x} (3x + 2y) \, dy \, dx$

**52.** $\displaystyle\int_0^1 \int_0^{\sqrt{y}} xy \, dx \, dy$

**53.** $\displaystyle\int_1^2 \int_1^{2y} \frac{x}{y^2} \, dx \, dy$

**54.** $\displaystyle\int_0^2 \int_{x^2}^{2x} (x^2 + 2y) \, dy \, dx$

**55.** $\displaystyle\int_0^3 \int_0^{\sqrt{9-x^2}} 4x \, dy \, dx$

**56.** $\displaystyle\int_0^3 \int_{(x-1)^2-1}^{x} dy \, dx$

**57.** $\displaystyle\int_{-2}^4 \int_{y^2/4}^{(4+y)/2} (x - y) \, dx \, dy$

**58.** $\displaystyle\int_{-2}^2 \int_0^{4-y^2} (8x - 2y^2) \, dx \, dy$

In Exercises 59–64, write the limits to the double integral

$$\int \int_R f(x, y) \, dA$$

for both orders of integration. Compute the area of $R$ by letting $f(x, y) = 1$ and integrating.

**59.** Triangle: vertices (0, 0), (3, 0), (0, 1)

**60.** Triangle: vertices (0, 0), (3, 0), (2, 2)

**61.** Region bounded by the graphs of $y = 9 - x^2$ and $y = 5$

**62.** Region enclosed by the graphs of $xy = 4$, $y = 0$, $x = 1$, $x = 4$

**63.** Region bounded by the graphs of $x = y + 3$, $x = y^2 + 1$

**64.** Region bounded by the graphs of $x = -y$, $x = 2y - y^2$

In Exercises 65 and 66, use an appropriate double integral to find the volume of the solid.

**65.** Solid bounded by the graphs of $z = (xy)^2$, $z = 0$, $y = 0$, $y = 4$, $x = 0$, $x = 4$

**66.** Solid bounded by the graphs of $z = x + y$, $z = 0$, $x = 0$, $x = 3$, $y = x$, $y = 0$

# Series
# and Taylor
# Polynomials

# CHAPTER 9

# Sequences

We call a function that is defined only on the positive integers a **sequence,** and we usually denote its variable by $n$ rather than $x$.

A common example of a sequence is the approximation of an $x$-intercept of a function by Newton's Method.[*] For this method, the sequence

$$\begin{matrix} 1 & 2 & 3 & 4 & & n \\ \downarrow & \downarrow & \downarrow & \downarrow & & \downarrow \end{matrix}$$

$$x_1, \ x_2, \ x_3, \ x_4, \ \dots, \ x_n, \ \dots$$

consists of successively better approximations to a particular $x$-intercept of $y = f(x)$. To use this approximation method, we choose an initial estimate $x_1$ of the intercept and then calculate each successive term in the sequence by the rule

$$x_{n+1} = x_n - \frac{f(x_n)}{f'(x_n)}$$

For example, if we want to find the $x$-intercept of

$$f(x) = x^3 + x + 1$$

as shown in Figure 9.1, we have

$$x_{n+1} = x_n - \frac{x_n^{\ 3} + x_n + 1}{3x_n^{\ 2} + 1}$$

Now suppose that we guess that $f$ crosses the $x$-axis at $x_1 = 0$. This is *not* the actual $x$-intercept, and we apply Newton's Method to obtain a better estimate:

$$x_2 = 0 - \frac{0^3 + 0 + 1}{3(0^2) + 1} = -1$$

---

[*] We will study Newton's Method in detail in Section 9.6.

[Geometrically, we interpret $x_2$ to be the $x$-intercept of the tangent line to the graph of $f$ at the point $(x_1, f(x_1))$.] Next, using $x_2 = -1$, we can obtain an even better estimate:

$$x_3 = -1 - \frac{(-1)^3 - 1 + 1}{3(-1)^2 + 1} = -1 + \frac{1}{4} = -0.75$$

By continuing this process, we obtain the sequence shown in Table 9.1. If $a$ is the actual $x$-intercept of $f(x) = x^3 + x + 1$, then we say this sequence **converges** to $a$, and we write

$$\lim_{n \to \infty} x_n = a$$

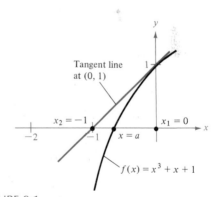

FIGURE 9.1

TABLE 9.1

| $n$ | 1 | 2 | 3 | 4 | 5 | 6 |
|---|---|---|---|---|---|---|
| $x_n$ | 0 | $-1$ | $-0.75$ | $-0.68605$ | $-0.68234$ | $-0.68233$ |

■ **Sequence**
■ **Limit of a Sequence**
■ **Pattern Recognition for Terms of a Sequence**

In mathematics the word ''sequence'' is used in much the same way as it is in ordinary English. When we say that a collection of objects or events is ''in sequence,'' we usually mean that the collection is ordered so that it has an identified first member, second member, third member, and so on. Mathematically, we define a sequence as a function whose domain is the set of positive integers. For instance, the equation

$$a(n) = \frac{1}{2^n}$$

defines the sequence

$$\frac{1}{2}, \frac{1}{4}, \frac{1}{8}, \frac{1}{16}, \ldots, \frac{1}{2^n}, \ldots$$

the terms of which correspond respectively to

$$a(1), a(2), a(3), a(4), \ldots, a(n), \ldots$$

We will generally write a sequence such as this one in the more convenient *subscript form,*

$$a_1, a_2, a_3, a_4, \ldots, a_n, \ldots$$

or we will denote it by the symbol $\{a_n\}$, where $a_n$ is the $n$th term of the sequence.

---

**Definition of a sequence**

A **sequence** $\{a_n\}$ is a function whose domain is the set of positive integers. The functional values $a_1, a_2, a_3, \ldots, a_n, \ldots$ are called the **terms** of the sequence.

---

■■ **Remark:** Occasionally it is convenient to begin subscripting a sequence with zero. In such cases we have

$$a_0, a_1, a_2, \ldots, a_n, \ldots$$

**EXAMPLE 1**

**Finding Terms in a Sequence**

List the first four terms of the sequences given by the following equations. (In each case assume $n$ begins with 1.)

(a) $a_n = 3 + (-1)^n$

(b) $b_n = \frac{2n}{1 + n}$

(c) $c_n = \frac{2^n}{2^n - 1}$

**SOLUTION**

(a) The first four terms of the sequence

$$\{a_n\} = \{3 + (-1)^n\}$$

are

$$3 + (-1)^1, \ 3 + (-1)^2, \ 3 + (-1)^3, \ 3 + (-1)^4$$

or

$$2, 4, 2, 4$$

(b) The first four terms of the sequence

$$\{b_n\} = \left\{\frac{2n}{1+n}\right\}$$

are

$$\frac{2(1)}{1+1}, \ \frac{2(2)}{1+2}, \ \frac{2(3)}{1+3}, \ \frac{2(4)}{1+4}$$

or

$$\frac{2}{2}, \frac{4}{3}, \frac{6}{4}, \frac{8}{5}$$

(c) The first four terms of the sequence

$$\{c_n\} = \left\{\frac{2^n}{2^n - 1}\right\}$$

are

$$\frac{2^1}{2^1 - 1}, \ \frac{2^2}{2^2 - 1}, \ \frac{2^3}{2^3 - 1}, \ \frac{2^4}{2^4 - 1}$$

or

$$\frac{2}{1}, \frac{4}{3}, \frac{8}{7}, \frac{16}{15}$$

In this chapter, our primary interest is in sequences whose terms approach a (unique) limiting value. Such sequences are said to **converge.** (Sequences having no limit are said to **diverge.**) For instance, the terms of the sequence

$$\frac{1}{2}, \frac{1}{4}, \frac{1}{8}, \frac{1}{16}, \ \cdots \ \frac{1}{2^n}, \ \cdots$$

approach 0 as $n$ increases, and we write

$$\lim_{n \to \infty} a_n = \lim_{n \to \infty} \frac{1}{2^n} = 0$$

Similarly, for the sequence $\{b_n\}$ in Example 1, we have

$$\lim_{n \to \infty} \frac{2n}{1+n} = 2\left(\lim_{n \to \infty} \frac{n}{1+n}\right) = 2(1) = 2$$

Although there are technical differences, for our purposes we can operate with limits of sequences in the same way as we operated with limits of continuous functions in Section 3.6. In other words, when we are asked to evaluate the limit of the sequence

$$\lim_{n \to \infty} \frac{2n}{1 + n}$$

we can imagine that $n$ is replaced by $x$ to obtain

$$\lim_{x \to \infty} \frac{2x}{1 + x} = \lim_{x \to \infty} \frac{2}{(1/x) + 1}$$

$$= \frac{2}{0 + 1} = 2$$

**EXAMPLE 2**

**Finding the Limit of a Sequence**

Determine the convergence or divergence of the sequences given by the following equations:

(a) $a_n = 3 + (-1)^n$   (b) $b_n = \dfrac{n}{1 - 2n}$   (c) $c_n = \dfrac{2^n}{2^n - 1}$

**SOLUTION**

(a) Since the sequence $\{a_n\} = \{3 + (-1)^n\}$ oscillates between 2 and 4,

$$2, 4, 2, 4, \ldots$$

we know that the limit of $a_n$ as $n \to \infty$ does not exist, and we conclude that $\{a_n\}$ diverges.

(b) For the sequence

$$\{b_n\} = \left\{ \frac{n}{1 - 2n} \right\}$$

we divide the numerator and denominator by $n$ to obtain

$$\lim_{n \to \infty} \frac{n}{1 - 2n} = \lim_{n \to \infty} \left[ \frac{1}{(1/n) - 2} \right] = -\frac{1}{2}$$

and we conclude that $\{b_n\}$ converges to $-\frac{1}{2}$.

(c) For the sequence

$$\{c_n\} = \left\{ \frac{2^n}{2^n - 1} \right\}$$

we divide the numerator and denominator by $2^n$ to obtain

$$\lim_{n \to \infty} \frac{2^n}{2^n - 1} = \lim_{n \to \infty} \frac{1}{1 - (1/2^n)} = 1$$

which implies that $\{c_n\}$ converges to 1.

Some of the most important sequences we will be dealing with involve factorials. If $n$ is a positive integer, then $n$ **factorial** is defined as

$$n! = 1 \cdot 2 \cdot 3 \cdot 4 \cdots (n-1) \cdot n$$

Moreover, for $n = 0$ we let $0! = 1$. Thus, we have

$$0! = 1$$

$$1! = 1$$

$$2! = 1 \cdot 2 = 2$$

$$3! = 1 \cdot 2 \cdot 3 = 6$$

$$4! = 1 \cdot 2 \cdot 3 \cdot 4 = 24$$

Factorials follow the same conventions for order of operation as exponents do. That is, just as $2x^3$ and $(2x)^3$ imply different orders of operations, $2n!$ and $(2n)!$ imply the following orders:

$$2n! = 2(n!) = 2(1 \cdot 2 \cdot 3 \cdot 4 \cdots n)$$

$$(2n)! = 1 \cdot 2 \cdot 3 \cdot 4 \cdots n \cdot (n+1) \cdots (2n)$$

**EXAMPLE 3**

**Finding the Limit of a Sequence**

Find the limit of the sequence

$$\left\{ \frac{(-1)^n}{n!} \right\}$$

**SOLUTION**

The first several terms of this sequence are listed in the following table.

| $n$ | 1 | 2 | 3 | 4 | 5 | 6 |
|-----|---|---|---|---|---|---|
| $\left\{ \dfrac{(-1)^n}{n!} \right\}$ | $-\dfrac{1}{1}$ | $\dfrac{1}{2}$ | $-\dfrac{1}{6}$ | $\dfrac{1}{24}$ | $-\dfrac{1}{120}$ | $\dfrac{1}{720}$ |

It is clear that the denominator is growing without bound while the numerator is bounded, and we have

$$\lim_{n \to \infty} \frac{(-1)^n}{n!} = 0$$

**Pattern Recognition for Sequences**

Occasionally, the terms of a sequence are generated by some rule that does not explicitly identify the $n$th term of the sequence. Under such circumstances we are required to discover a *pattern* in the sequence and come up with a description of the $n$th term. Once the $n$th term is specified, we can discuss the convergence or divergence of the sequence. This is demonstrated in the next example.

## EXAMPLE 4

### Finding a Pattern for a Sequence

Given $f(x) = e^{x/3}$, determine the convergence or divergence of the sequence given by

$$a_n = f^{(n-1)}(0)$$

where $f^{(0)} = f$ and $f^{(n)}$ is the $n$th derivative of $f$.

**SOLUTION**

We determine the first several terms of the sequence as shown in Table 9.2.

TABLE 9.2

| $n$ | 1 | 2 | 3 | 4 | 5 | $\cdots$ | $n$ |
|---|---|---|---|---|---|---|---|
| $f^{(n-1)}(x)$ | $e^{x/3}$ | $\dfrac{e^{x/3}}{3}$ | $\dfrac{e^{x/3}}{3^2}$ | $\dfrac{e^{x/3}}{3^3}$ | $\dfrac{e^{x/3}}{3^4}$ | $\cdots$ | $\dfrac{e^{x/3}}{3^{n-1}}$ |
| $f^{(n-1)}(0)$ | 1 | $\dfrac{1}{3}$ | $\dfrac{1}{3^2}$ | $\dfrac{1}{3^3}$ | $\dfrac{1}{3^4}$ | $\cdots$ | $\dfrac{1}{3^{n-1}}$ |

The pattern in the sequence $\{f^{(n-1)}(0)\}$ suggests that the $n$th term is

$$a_n = \frac{1}{3^{n-1}}$$

and therefore the sequence converges to 0, because

$$\lim_{n \to \infty} a_n = \lim_{n \to \infty} \left( \frac{1}{3^{n-1}} \right) = 0$$

Without a specific rule for generating the terms of a sequence it is not possible to determine the convergence or divergence of the sequence—knowing its first several terms is not enough. For instance, the first three terms of the four sequences given next are identical, yet from the description of their individual $n$th terms we see that the first two sequences converge to 0, the third sequence converges to $\frac{1}{9}$, and the fourth one diverges.

$$\{a_n\} = \left\{ \frac{1}{2}, \frac{1}{4}, \frac{1}{8}, \frac{1}{16}, \ldots, \frac{1}{2^n}, \ldots \right\}$$

$$\{b_n\} = \left\{ \frac{1}{2}, \frac{1}{4}, \frac{1}{8}, \frac{1}{15}, \ldots, \frac{6}{(n+1)(n^2-n+6)}, \ldots \right\}$$

$$\{c_n\} = \left\{ \frac{1}{2}, \frac{1}{4}, \frac{1}{8}, \frac{7}{62}, \ldots, \frac{n^2-3n+3}{9n^2-25n+18}, \ldots \right\}$$

$$\{d_n\} = \left\{ \frac{1}{2}, \frac{1}{4}, \frac{1}{8}, 0, \ldots, \frac{-n(n+1)(n-4)}{6(n^2+3n-2)}, \ldots \right\}$$

Thus, if only the first several terms of a sequence are given, there are many forms for an $n$th term of the sequence. In such a situation we can only determine the convergence or divergence of the sequence on the basis of our choice for the *n*th term.

**EXAMPLE 5**

**Finding a Pattern for a Sequence**

Determine an *n*th term for the sequence

$$\left\{ -\frac{1}{1}, \frac{3}{2}, -\frac{7}{6}, \frac{15}{24}, -\frac{31}{120}, \ldots \right\}$$

**SOLUTION**

Observe that the numerators are one less than $2^n$. Hence, we can generate the numerators by the rule

$$2^n - 1, \quad n = 1, 2, 3, 4, 5, \ldots$$

Now, factoring the denominators, we have

$$1, 1 \cdot 2, 1 \cdot 2 \cdot 3, 1 \cdot 2 \cdot 3 \cdot 4, 1 \cdot 2 \cdot 3 \cdot 4 \cdot 5, \ldots$$

This suggests that the denominators are represented by $n!$. Furthermore, since the signs alternate, we can write

$$a_n = (-1)^n \left( \frac{2^n - 1}{n!} \right)$$

as an *n*th term for the given sequence.

To aid you in finding patterns for sequences, we provide the following table summarizing the most common sequence patterns.

| Patterns for sequences | $n$ | 1 | 2 | 3 | 4 | 5 | 6 |
|---|---|---|---|---|---|---|---|
| Changes in sign | $(-1)^n$ | $-1$ | 1 | $-1$ | 1 | $-1$ | 1 |
| | $(-1)^{n+1}$ | 1 | $-1$ | 1 | $-1$ | 1 | $-1$ |
| Arithmetic sequences | $2n$ | 2 | 4 | 6 | 8 | 10 | 12 |
| | $2n - 1$ | 1 | 3 | 5 | 7 | 9 | 11 |
| | $an + b$ | $a + b$ | $2a + b$ | $3a + b$ | $4a + b$ | $5a + b$ | $6a + b$ |
| Binary sequence | $2^{n-1}$ | 1 | 2 | 4 | 8 | 16 | 32 |
| Geometric sequence | $ar^{n-1}$ | $a$ | $ar$ | $ar^2$ | $ar^3$ | $ar^4$ | $ar^5$ |
| Power sequences | $n^2$ | 1 | 4 | 9 | 16 | 25 | 36 |
| | $n^p$ | 1 | $2^p$ | $3^p$ | $4^p$ | $5^p$ | $6^p$ |
| Factorial sequence | $n!$ | 1 | 2 | 6 | 24 | 120 | 720 |
| Sequences of products | $2^n n!$ | 2 | $2 \cdot 4$ | $2 \cdot 4 \cdot 6$ | $2 \cdot 4 \cdot 6 \cdot 8$ | $2 \cdot 4 \cdot 6 \cdot 8 \cdot 10$ | $2 \cdot 4 \cdot 6 \cdot 8 \cdot 10 \cdot 12$ |
| | $\dfrac{(2n)!}{2^n n!}$ | 1 | $1 \cdot 3$ | $1 \cdot 3 \cdot 5$ | $1 \cdot 3 \cdot 5 \cdot 7$ | $1 \cdot 3 \cdot 5 \cdot 7 \cdot 9$ | $1 \cdot 3 \cdot 5 \cdot 7 \cdot 9 \cdot 11$ |

## EXAMPLE 6

**A Business Application**

A deposit of $1,000 is made in an account that earns 12% interest compounded monthly. Find a sequence to represent the balance in this account after $n$ months $(n = 1, 2, 3, \ldots)$.

## SOLUTION

Since an annual interest rate of 12% compounded monthly corresponds to a monthly rate of 1%, the balance after 1 month is

$$A_1 = 1000 + 1000(0.01) = 1000(1.01)$$

After 2 months the balance is

$$A_2 = 1000(1.01) + 1000(1.01)(0.01) = 1000(1.01)^2$$

Continuing this pattern, we find the $n$th term in the sequence representing the monthly balances to be

$$A_n = 1000(1.01)^n$$

and the first several terms in the sequence are

$$1000(1.01), \ 1000(1.01)^2, \ 1000(1.01)^3, \ 1000(1.01)^4, \ldots$$

or

$$\$1010.00, \ \$1020.10, \ \$1030.30, \ \$1040.60, \ldots$$

Note that this sequence is of the form $\{ar^n\}$ and is therefore a geometric sequence.

## SECTION EXERCISES 9.1

In Exercises 1–8, write out the first five terms of the specified sequence.

**1.** $a_n = 2^n$

**2.** $a_n = \dfrac{n}{n + 1}$

**3.** $a_n = \left(-\dfrac{1}{2}\right)^n$

**4.** $a_n = \dfrac{n - 1}{n^2 + 2}$

**5.** $a_n = \dfrac{3^n}{n!}$

**6.** $a_n = 5 - \dfrac{1}{n} + \dfrac{1}{n^2}$

**7.** $a_n = \dfrac{(-1)^n}{n^2}$

**8.** $a_n = \dfrac{3n!}{(n - 1)!}$

**17.** $2, \ 1 + \dfrac{1}{2}, \ 1 + \dfrac{1}{3}, \ 1 + \dfrac{1}{4}, \ldots$

**18.** $1 + \dfrac{1}{2}, \ 1 + \dfrac{1}{4}, \ 1 + \dfrac{1}{8}, \ 1 + \dfrac{1}{16}, \ldots$

**19.** $1, \ -1, \ 1, \ -1, \ 1, \ -1, \ldots$

**20.** $2, \ -4, \ 6, \ -8, \ 10, \ldots$

**21.** $1, \ \dfrac{1}{2}, \ \dfrac{1}{6}, \ \dfrac{1}{24}, \ \dfrac{1}{120}, \ldots$

**22.** $1, \ x, \ \dfrac{x^2}{2}, \ \dfrac{x^3}{6}, \ \dfrac{x^4}{24}, \ \dfrac{x^5}{120}, \ldots$

In Exercises 9–22, write an expression for the $n$th term of the given sequence.

**9.** $1, 4, 7, 10, \ldots$

**10.** $3, 7, 11, 15, \ldots$

**11.** $-1, 2, 7, 14, 23, \ldots$

**12.** $1, \dfrac{1}{4}, \dfrac{1}{9}, \dfrac{1}{16}, \ldots$

**13.** $\dfrac{2}{3}, \dfrac{3}{4}, \dfrac{4}{5}, \dfrac{5}{6}, \ldots$

**14.** $2, \dfrac{3}{3}, \dfrac{4}{5}, \dfrac{5}{7}, \dfrac{6}{9}, \ldots$

**15.** $2, -1, \dfrac{1}{2}, -\dfrac{1}{4}, \dfrac{1}{8}, \ldots$

**16.** $\dfrac{1}{3}, \dfrac{2}{9}, \dfrac{4}{27}, \dfrac{8}{81}, \ldots$

In Exercises 23–40, determine the convergence or divergence of the given sequence. If the sequence converges, find its limit.

**23.** $a_n = \dfrac{5}{n}$

**24.** $a_n = \dfrac{n}{2}$

**25.** $a_n = \dfrac{n + 1}{n}$

**26.** $a_n = \dfrac{1}{n^{3/2}}$

**27.** $a_n = (-1)^n \left(\dfrac{n}{n + 1}\right)$

**28.** $a_n = \dfrac{n-1}{n} - \dfrac{n}{n-1},\; n \geq 2$

**29.** $a_n = \dfrac{3n^2 - n + 4}{2n^2 + 1}$

**30.** $a_n = \dfrac{\sqrt{n}}{\sqrt{n+1}}$

**31.** $a_n = \dfrac{n^2 - 1}{n+1}$

**32.** $a_n = 1 + (-1)^n$

**33.** $a_n = \dfrac{1 + (-1)^n}{n}$

**34.** $a_n = \dfrac{n!}{n}$

**35.** $a_n = 3 - \dfrac{1}{2^n}$

**36.** $a_n = \dfrac{n}{\sqrt{n^2 + 1}}$

**37.** $a_n = \dfrac{3^n}{4^n}$

**38.** $a_n = (0.5)^n$

**39.** $a_n = \dfrac{(n+1)!}{n!}$

**40.** $a_n = \dfrac{(n-2)!}{n!}$

**41.** Consider the sequence $\{A_n\}$, whose $n$th term is given by $A_n = P[1 + (r/12)]^n$, where $P$ is the principal, $A_n$ is the amount at compound interest after $n$ months, and $r$ is the annual percentage rate.
(a) Is $\{A_n\}$ a convergent series?
(b) Find the first ten terms of the sequence if $P = \$9,000$ and $r = 0.115$.

**42.** A deposit of \$100 is made each month in an account that earns 12% interest compounded monthly. The balance in the account after $n$ months is given by

$$A_n = 100(101)[(1.01)^n - 1]$$

(a) Compute the first six terms of this sequence.
(b) Find the balance after 5 years by computing the 60th term of the sequence.
(c) Find the balance after 20 years by computing the 240th term of the sequence.

**43.** The sum of the first $n$ positive integers is given by

$$S_n = \frac{n(n+1)}{2}, \quad n = 1, 2, 3, \ldots$$

(a) Compute the first five terms of this sequence and verify that each term is the correct sum.
(b) Find the sum of the first fifty positive integers.

**44.** A ball is dropped from a height of 12 feet, and on each rebound it rises to $\frac{2}{3}$ its previous height.
(a) Write an expression for the height of the $n$th rebound.
(b) Determine the convergence or divergence of this sequence. If it converges, find the limit.

**45.** A government program that currently costs taxpayers \$2.5 billion per year is to be cut back by 20% per year.
(a) Write an expression for the amount budgeted for this program after $n$ years.
(b) Compute the budgets for the first four years.
(c) Determine the convergence or divergence of the sequence of reduced budgets. If the sequence converges, find its limit.

**46.** If the average price of a new car increases $5\frac{1}{2}\%$ per year and the average price is currently \$11,000, then the average price after $n$ years is

$$P_n = \$11,000(1.055)^n$$

Compute the average price for the first five years of increases.

**47.** Consider an idealized population with the characteristic that each population member produces 1 offspring at the end of every time period. If each population member has a lifespan of 3 time periods and the population begins with 10 newborn members, then the following table gives the population during the first five time periods.

| Age bracket | Time period | | | | |
| --- | --- | --- | --- | --- | --- |
| | 1 | 2 | 3 | 4 | 5 |
| 0–1 | 10 | 10 | 20 | 40 | 70 |
| 1–2 | | 10 | 10 | 20 | 40 |
| 2–3 | | | 10 | 10 | 20 |
| Total | 10 | 20 | 40 | 70 | 130 |

The sequence for the total population has the property that

$$S_n = S_{n-1} + S_{n-2} + S_{n-3}, \quad n > 3$$

Find the total population during the next five time periods.

# Series and Convergence

One of the many uses of sequences is in stabilization problems. To see how this works, let us consider a manufacturer who sells 10,000 units of a certain product each year. Suppose that in any given year each unit of this product (regardless of its age) has a 10% chance of breaking. In other words, after a year we would expect only 9000 of the previous year's 10,000 units to be still in use. During the next year this number would drop by an additional 10% to 8100, and so on. Assuming that 10,000 new units are sold every year, how many units would be in use after $n$ years?

To solve this problem, we set up a geometric sequence that represents the declining number of items (out of one year's sales of 10,000) in use after $n$ years:

| Year | 1 | 2 | $\cdots$ | $n$ |
|------|---|---|----------|-----|
| Number of items | 10,000 | 10,000(0.9) | $\cdots$ | $10,000(0.9)^{n-1}$ |

Now, since each year witnesses the introduction of 10,000 new units, we have the following totals in use:

1st year: 10,000

2nd year: $10,000 + 10,000(0.9)$

3rd year: $10,000 + 10,000(0.9) + 10,000(0.9)^2$

.

.

.

$n$th year: $10,000 + 10,000(0.9) + 10,000(0.9)^2$
$\qquad\qquad + \ldots + 10,000(0.9)^{n-1}$

These totals are represented graphically in Figure 9.2.

In this section, we will see that the sum of the **geometric series** representing the $n$th year's total is

$$\sum_{i=0}^{n-1} 10,000(0.9)^i = \frac{10,000[1 - (0.9)^n]}{1 - 0.9}$$

$$= 100,000[1 - (0.9)^n]$$

As $n$ increases, we have

$$\lim_{n \to \infty} (0.9)^n = 0$$

and we see that the stabilization point is 100,000 units.

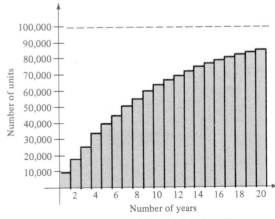

FIGURE 9.2

Total Number of Units in Use after $n$ Years

## SECTION TOPICS

- **Infinite Series**
- **Convergence and Divergence of a Series**
- **$n$th-Term Test for Divergence**
- **Geometric Series**

In this section, we investigate an important application of infinite sequences—namely, their use in representing infinite summations. As a simple illustration, suppose we write the decimal representation of $\frac{1}{3}$ as

$$\frac{1}{3} = 0.33333 \ldots = \frac{3}{10} + \frac{3}{10^2} + \frac{3}{10^3} + \frac{3}{10^4} + \frac{3}{10^5} + \cdots$$

We consider this representation to be an *infinite summation* whose value is $\frac{1}{3}$.

To facilitate the writing of sums involving many terms, we use the standard **sigma notation** for sums, as introduced in Section 8.7. Some illustrations of the use of sigma notation are given in the following example.

**EXAMPLE 1**

**Examples of Sigma Notation**

| *Sum* | | *Σ Notation* |
|---|---|---|

(a)  $1 + 2 + 3 + 4 + 5 + 6$ $\qquad\qquad\qquad\qquad\qquad\qquad\qquad\qquad \displaystyle\sum_{i=1}^{6} i$

(Note that for this particular sum the **index of summation** begins at $i = 1$. In general, we can use any variable as the index, and we can begin at any integer.)

(b)  $3^2 + 4^2 + 5^2 + 6^2 + 7^2$ $\qquad\qquad\qquad\qquad\qquad\qquad \displaystyle\sum_{n=3}^{7} n^2$

(c)  $3(1) + 3\left(\dfrac{1}{2}\right) + 3\left(\dfrac{1}{2}\right)^2 + \cdots + 3\left(\dfrac{1}{2}\right)^n$ $\qquad \displaystyle\sum_{k=0}^{n} 3\left(\dfrac{1}{2}\right)^k$

(d)  $a_1 + a_2 + a_3 + \cdots + a_n$ $\qquad\qquad\qquad\qquad\qquad \displaystyle\sum_{i=1}^{n} a_i$

Now, to obtain a better picture of what an infinite summation is, let $\{a_n\}$ be a sequence from which we form another sequence $\{S_N\}$ in the following manner:

$$S_0 = a_0$$
$$S_1 = a_0 + a_1$$
$$S_2 = a_0 + a_1 + a_2$$
$$\vdots$$
$$S_N = a_0 + a_1 + a_2 + \cdots + a_N = \sum_{n=0}^{N} a_n$$

As $n \rightarrow \infty$, the infinite summation

$$a_0 + a_1 + a_2 + a_3 + a_4 + \cdots$$

is denoted by the expression

$$\sum_{n=0}^{\infty} a_n$$

and is called an **infinite series.** The sequence $\{S_N\}$ is called a **sequence of partial sums,** and we have the following.

**Definition of convergence and divergence of an infinite series**

1. If $\displaystyle\lim_{N \rightarrow \infty} S_N = S$, the series $\displaystyle\sum_{n=0}^{\infty} a_n$ **converges.** $S$ is called the **sum of the series.**

2. If $\displaystyle\lim_{N \rightarrow \infty} S_N$ does not exist, the series $\displaystyle\sum_{n=0}^{\infty} a_n$ **diverges.**

The following properties are useful in determining the sum of an infinite series.

**Properties of an infinite series**

If $c$, $A$, and $B$ are real numbers such that

$$\sum_{n=0}^{\infty} a_n = A \qquad \text{and} \qquad \sum_{n=0}^{\infty} b_n = B$$

then

1. $\displaystyle\sum_{n=0}^{\infty} ca_n = c \sum_{n=0}^{\infty} a_n = cA$

2. $\displaystyle\sum_{n=0}^{\infty} (a_n \pm b_n) = \sum_{n=0}^{\infty} a_n \pm \sum_{n=0}^{\infty} b_n = A \pm B$

The two primary questions regarding an infinite series are as follows:

1. Does the series converge or does it diverge?
2. If the series converges, to what value does it converge?

We begin our pursuit of answers to these questions with a simple test for **divergence.**

| **nth-Term Test for divergence** | If |
| --- | --- |

$$\lim_{n \to \infty} a_n \neq 0$$

then the series $\displaystyle\sum_{n=0}^{\infty} a_n$ **diverges.**

**Remark:** Be sure you see that the *n*th-Term Test is a test for *divergence,* not for convergence. That is, if

$$\lim_{n \to \infty} a_n \neq 0$$

then we know that the series diverges. If the *n*th term of a series approaches zero as $n \to \infty$, then the series may diverge or it may converge. To determine which it does, we need additional information.

**EXAMPLE 2**

**Testing for Divergence**

Use the *n*th-Term Test to determine which, if any, of the following series diverge.

(a) $\displaystyle\sum_{n=0}^{\infty} 2^n$ \qquad (b) $\displaystyle\sum_{n=0}^{\infty} \frac{1}{2^n}$ \qquad (c) $\displaystyle\sum_{n=1}^{\infty} \frac{n!}{2n! + 1}$

**SOLUTION**

(a) The series

$$\sum_{n=0}^{\infty} 2^n = 1 + 2 + 4 + 8 + 16 + \cdots$$

diverges since

$$\lim_{n \to \infty} 2^n = \infty$$

(b) The *n*th-Term Test tells us nothing about the series

$$\sum_{n=0}^{\infty} \frac{1}{2^n} = 1 + \frac{1}{2} + \frac{1}{4} + \frac{1}{8} + \frac{1}{16} + \cdots$$

since

$$\lim_{n \to \infty} \frac{1}{2^n} = 0$$

(Later in this section we will see that this particular series happens to converge. The point here is that we cannot deduce this from the *n*th-Term Test.)

(c) The series

$$\sum_{n=1}^{\infty} \frac{n!}{2n! + 1} = \frac{1}{3} + \frac{2}{5} + \frac{6}{13} + \frac{24}{49} + \frac{120}{241} + \cdots$$

diverges since

$$\lim_{n \to \infty} \frac{n!}{2n! + 1} = \lim_{n \to \infty} \frac{1}{2 + (1/n!)} = \frac{1}{2 + 0} = \frac{1}{2}$$

## Geometric Series

We have already pointed out that the *n*th-Term Test is a test for divergence, not convergence. In the remaining portion of this section, we discuss a special type of series, called a **geometric series,** for which we will develop a test for convergence (as well as for divergence).

| **Definition of geometric series** | The series given by |
|---|---|

$$\sum_{n=0}^{\infty} ar^n = a + ar + ar^2 + \cdots + ar^n + \cdots, \quad a \neq 0$$

is called a **geometric series** with ratio *r*.

▬▬ **Remark:** Note that the first term in this geometric series is $ar^0 = a$. If the index had begun with $n = 1$, the first term would be $ar^1 = ar$. For a geometric series whose index begins at $n = 0$, we have the following formula for its *N*th partial sum.

| **Nth partial sum of a geometric series** | The *N*th partial sum of the geometric series |
|---|---|

$$\sum_{n=0}^{\infty} ar^n$$

is given by

$$S_N = \frac{a(1 - r^{N+1})}{1 - r}$$

Proof

Let

$$S_N = a + ar + ar^2 + \cdots + ar^N$$

Then multiplication by *r* yields

$$rS_N = ar + ar^2 + ar^3 + \cdots + ar^{N+1}$$

Subtracting these two equations, we obtain

$$S_N = a + ar + ar^2 + \cdots + ar^N$$
$$-rS_N = \quad - ar - ar^2 - \cdots - ar^N - ar^{N+1}$$
$$\overline{S_N - rS_N = a - ar^{N+1}}$$

Therefore,

$$S_N(1 - r) = a(1 - r^{N+1})$$

which implies that

$$S_N = \frac{a(1 - r^{N+1})}{1 - r}$$

▬▬

**EXAMPLE 3**

**Finding the Nth Partial Sum of a Geometric Series**

Find the Nth partial sum of the geometric series

$$\sum_{n=0}^{\infty} 3\left(\frac{1}{4}\right)^n$$

for $N = 3$, $N = 5$, and $N = 10$.

**SOLUTION**

For this geometric series, we have $a = 3$ and $r = \frac{1}{4}$. Therefore, the Nth partial sum is given by

$$S_N = \frac{a(1 - r^{N+1})}{1 - r}$$

$$= \frac{3[1 - (1/4)^{N+1}]}{1 - (1/4)}$$

$$= \frac{3[1 - (1/4)^{N+1}]}{3/4}$$

$$= 4\left[1 - \left(\frac{1}{4}\right)^{N+1}\right]$$

$$= 4 - \left(\frac{1}{4}\right)^N$$

When $N = 3$, the Nth partial sum is

$$S_3 = 4 - \left(\frac{1}{4}\right)^3 \approx 3.984$$

When $N = 5$, the Nth partial sum is

$$S_5 = 4 - \left(\frac{1}{4}\right)^5 \approx 3.999$$

When $N = 10$, the Nth partial sum is

$$S_{10} = 4 - \left(\frac{1}{4}\right)^{10} \approx 4.000$$

When applying the formula for the Nth partial sum of a geometric series, be sure you check to see if the index begins at $n = 0$. If it does not, you will have to adjust $S_N$ accordingly. For instance, comparing the two geometric series

$$\sum_{n=1}^{10} ar^n = ar + ar^2 + ar^3 + \cdots + ar^{10}$$

and

$$\sum_{n=0}^{10} ar^n = a + ar + ar^2 + ar^3 + \cdots + ar^{10}$$

shows that by subtracting $a$ from the second series we obtain the first. In other words,

$$\sum_{n=1}^{10} ar^n = \left(\sum_{n=0}^{10} ar^n\right) - a = \frac{a(1 - r^{11})}{1 - r} - a$$

We illustrate this procedure in the next example.

**EXAMPLE 4**

**Finding the Balance in an Increasing Annuity**

A deposit of $50 is made every month for 2 years in a savings account that pays 12% compounded monthly. What is the balance in the account at the end of the 2 years?

**SOLUTION**

The money that was deposited the first month will have become

$$A_{24} = 50\left(1 + \frac{0.12}{12}\right)^{24} = 50(1.01)^{24}$$

after 24 months. Similarly, the money deposited the second month will have become

$$A_{23} = 50(1.01)^{23}$$

after 23 months. Continuing this process, we see that the total balance resulting from the 24 deposits will be

$$A = A_1 + A_2 + \cdots + A_{24} = \sum_{n=1}^{24} A_n = \sum_{n=1}^{24} 50(1.01)^n$$

Now, noting that the index begins at $n = 1$, we apply the formula for the $N$th partial sum of a geometric series to determine

$$S_{24} = \sum_{n=0}^{24} 50(1.01)^n = \frac{50(1 - 1.01^{25})}{1 - 1.01}$$

$$= \frac{50(1 - 1.01^{25})}{-0.01}$$

Finally, accounting for the difference in the two indexes, we have a balance of

$$A = \frac{50(1 - 1.01^{25})}{-0.01} - 50 = \$1,362.16$$

Now that we have a formula for the $N$th partial sum of a geometric series, we have the necessary tool to determine which *infinite* geometric series converge and which ones diverge. The following result tells us that the only geometric series that converge are those in which the ratio $r$ lies between 1 and $-1$. The proof of this result hinges on the fact that the only numbers whose magnitudes decrease with repeated multiplication are numbers lying in the interval $-1 < r < 1$. For instance, if $r = -\frac{1}{2}$ or $r = \frac{1}{4}$, then the sequences

$$\left(-\frac{1}{2}\right)^0, \left(-\frac{1}{2}\right)^1, \left(-\frac{1}{2}\right)^2, \left(-\frac{1}{2}\right)^3, \left(-\frac{1}{2}\right)^4, \ldots$$

and

$$\left(\frac{1}{4}\right)^0, \left(\frac{1}{4}\right)^1, \left(\frac{1}{4}\right)^2, \left(\frac{1}{4}\right)^3, \left(\frac{1}{4}\right)^4, \ldots$$

both converge to 0. But if $r = 1$ or $r = -2$, then neither the sequence

$$(1)^0, (1)^1, (1)^2, (1)^3, (1)^4, \ldots$$

nor the sequence

$$(-2)^0, (-2)^1, (-2)^2, (-2)^3, (-2)^4, \ldots$$

converges to 0.

---

**Convergence of a geometric series**

The geometric series

$$\sum_{n=0}^{\infty} ar^n = a + ar + ar^2 + \cdots$$

has these properties:

1. If $|r| \geq 1$, it *diverges*.
2. If $|r| < 1$, it *converges* and the sum is given by

$$\sum_{n=0}^{\infty} ar^n = \frac{a}{1 - r}$$

---

**Proof**     If $|r| \geq 1$, then

$$\lim_{n \to \infty} ar^n \neq 0$$

and by the *n*th-Term Test the sequence diverges. On the other hand, if $|r| < 1$, then using the formula for the *N*th partial sum we have

$$\lim_{N \to \infty} S_N = \lim_{N \to \infty} \frac{a(1 - r^{N+1})}{1 - r} = \frac{a}{1 - r} \lim_{N \to \infty} (1 - r^{N+1}) = \frac{a}{1 - r}$$

and the series converges to

$$\frac{a}{1 - r}$$

**EXAMPLE 5**

**Finding the Sum of an Infinite Geometric Series**

Determine the convergence or divergence of the following geometric series:

(a) $\displaystyle\sum_{n=0}^{\infty} \left(-\frac{1}{2}\right)^n$  (b) $\displaystyle\sum_{n=0}^{\infty} \left(\frac{3}{2}\right)^n$  (c) $\displaystyle\sum_{n=1}^{\infty} \frac{4}{3^n}$

**SOLUTION**

(a) Since $a = 1$, $r = -\frac{1}{2}$, and $\left|-\frac{1}{2}\right| < 1$, the series converges to

$$\frac{a}{1-r} = \frac{1}{1-(-1/2)} = \frac{1}{3/2} = \frac{2}{3}$$

(b) Since $r = \frac{3}{2} > 1$, the series diverges.

(c) By rewriting the series as

$$\sum_{n=1}^{\infty} 4\left(\frac{1}{3}\right)^n$$

we see that $a = 4$ and $r = \frac{1}{3}$. Therefore, for the series beginning at $n = 0$, we have

$$\sum_{n=0}^{\infty} 4\left(\frac{1}{3}\right)^n = \frac{4}{1-(1/3)} = \frac{4}{2/3} = 6$$

Finally, for the given series (beginning at $n = 1$), we have

$$\sum_{n=1}^{\infty} 4\left(\frac{1}{3}\right)^n = \sum_{n=0}^{\infty} 4\left(\frac{1}{3}\right)^n - 4 = 6 - 4 = 2$$

Thus, the series converges to 2.

**EXAMPLE 6**

**Finding the Total Distance Traveled by a Bouncing Ball**

A ball is dropped from a height of 6 feet and begins bouncing. Suppose that the height of each bounce is $\frac{3}{4}$ that of the previous bounce, as shown in Figure 9.3. Find the total distance traveled by the ball.

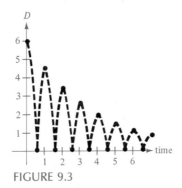

FIGURE 9.3

**SOLUTION**  When the ball hits the ground for the first time, it has traveled a distance of

$$D_1 = 6$$

Between the first and second times it hits the ground, it has traveled an additional distance of

$$D_2 = \underbrace{6\left(\frac{3}{4}\right)}_{\text{Up}} + \underbrace{6\left(\frac{3}{4}\right)}_{\text{Down}} = 12\left(\frac{3}{4}\right)$$

Between the second and third times it hits the ground, it has traveled an additional distance of

$$D_3 = \underbrace{6\left(\frac{3}{4}\right)\left(\frac{3}{4}\right)}_{\text{Up}} + \underbrace{6\left(\frac{3}{4}\right)\left(\frac{3}{4}\right)}_{\text{Down}} = 12\left(\frac{3}{4}\right)^2$$

Continuing this process, we have a total distance traveled of

$$D = 6 + 12\left(\frac{3}{4}\right) + 12\left(\frac{3}{4}\right)^2 + \cdots$$

$$= -6 + 12 + 12\left(\frac{3}{4}\right) + 12\left(\frac{3}{4}\right)^2 + \cdots$$

$$= -6 + \sum_{n=0}^{\infty} 12\left(\frac{3}{4}\right)^n$$

$$= -6 + \frac{12}{1 - (3/4)}$$

$$= -6 + 48 = 42 \text{ feet}$$

## SECTION EXERCISES 9.2

In Exercises 1–4, find the first five terms of the sequence of partial sums.

**1.** $\displaystyle\sum_{n=1}^{\infty} \frac{1}{n^2} = 1 + \frac{1}{4} + \frac{1}{9} + \frac{1}{16} + \frac{1}{25} + \cdots$

**2.** $\displaystyle\sum_{n=1}^{\infty} (-1)^{n+1}\frac{3^n}{2^{n-1}} = 3 - \frac{9}{2} + \frac{27}{4} - \frac{81}{8} + \frac{243}{16} - \cdots$

**3.** $\displaystyle\sum_{n=1}^{\infty} \frac{3}{2^{n-1}}$

**4.** $\displaystyle\sum_{n=1}^{\infty} \frac{(-1)^{n+1}}{n!}$

In Exercises 5–12, verify that the given infinite series diverges.

**5.** $\displaystyle\sum_{n=1}^{\infty} \frac{n}{n+1} = \frac{1}{2} + \frac{2}{3} + \frac{3}{4} + \frac{4}{5} + \cdots$

**6.** $\displaystyle\sum_{n=1}^{\infty} \frac{n}{2n+3} = \frac{1}{5} + \frac{2}{7} + \frac{3}{9} + \frac{4}{11} + \cdots$

**7.** $\displaystyle\sum_{n=1}^{\infty} \frac{n^2}{n^2+1} = \frac{1}{2} + \frac{4}{5} + \frac{9}{10} + \frac{16}{17} + \cdots$

**8.** $\displaystyle\sum_{n=1}^{\infty} \frac{n}{\sqrt{n^2+1}} = \frac{1}{\sqrt{2}} + \frac{2}{\sqrt{5}} + \frac{3}{\sqrt{10}} + \frac{4}{\sqrt{17}} + \cdots$

**9.** $\displaystyle\sum_{n=0}^{\infty} 3\left(\frac{3}{2}\right)^n = 3 + \frac{9}{2} + \frac{27}{4} + \frac{81}{8} + \cdots$

**10.** $\displaystyle\sum_{n=0}^{\infty} 5\left(-\frac{3}{2}\right)^n = 5 - \frac{15}{2} + \frac{45}{4} - \frac{135}{8} + \cdots$

**11.** $\displaystyle\sum_{n=0}^{\infty} 1000(1.055)^n = 1000 + 1055 + 1113.025 + \cdots$

**12.** $\displaystyle\sum_{n=0}^{\infty} 2(-1.03)^n = 2 - 2.06 + 2.1218 - \cdots$

In Exercises 13–16, verify that the given geometric series converges.

**13.** $\displaystyle\sum_{n=0}^{\infty} 2\left(\frac{3}{4}\right)^n = 2 + \frac{3}{2} + \frac{9}{8} + \frac{27}{32} + \frac{81}{128} + \cdots$

**14.** $\displaystyle\sum_{n=0}^{\infty} 2\left(-\frac{1}{2}\right)^n = 2 - 1 + \frac{1}{2} - \frac{1}{4} + \frac{1}{8} + \cdots$

**15.** $\displaystyle\sum_{n=0}^{\infty} (0.9)^n = 1 + 0.9 + 0.81 + 0.729 + \cdots$

**16.** $\displaystyle\sum_{n=0}^{\infty} (-0.6)^n = 1 - 0.6 + 0.36 - 0.216 + \cdots$

In Exercises 17–30, find the sum of the given convergent series.

**17.** $\displaystyle\sum_{n=0}^{\infty} \left(\frac{1}{2}\right)^n = 1 + \frac{1}{2} + \frac{1}{4} + \frac{1}{8} + \cdots$

**18.** $\displaystyle\sum_{n=0}^{\infty} 2\left(\frac{2}{3}\right)^n = 2 + \frac{4}{3} + \frac{8}{9} + \frac{16}{27} + \cdots$

**19.** $\displaystyle\sum_{n=0}^{\infty} \left(-\frac{1}{2}\right)^n = 1 - \frac{1}{2} + \frac{1}{4} - \frac{1}{8} + \cdots$

**20.** $\displaystyle\sum_{n=0}^{\infty} 2\left(-\frac{2}{3}\right)^n = 2 - \frac{4}{3} + \frac{8}{9} - \frac{16}{27} + \cdots$

**21.** $\displaystyle\sum_{n=0}^{\infty} 2\left(\frac{1}{\sqrt{2}}\right)^n = 2 + \sqrt{2} + 1 + \frac{1}{\sqrt{2}} + \cdots$

**22.** $\displaystyle\sum_{n=0}^{\infty} 4\left(\frac{1}{4}\right)^n = 4 + 1 + \frac{1}{4} + \frac{1}{16} + \cdots$

**23.** $1 + 0.1 + 0.01 + 0.001 + \cdots$

**24.** $8 + 6 + \frac{9}{2} + \frac{27}{8} + \cdots$

**25.** $3 - 1 + \frac{1}{3} - \frac{1}{9} + \cdots$   **26.** $4 - 2 + 1 - \frac{1}{2} + \cdots$

**27.** $\displaystyle\sum_{n=0}^{\infty} \left(\frac{1}{2^n} - \frac{1}{3^n}\right)$   **28.** $\displaystyle\sum_{n=0}^{\infty} [(0.7)^n + (0.9)^n]$

**29.** $\displaystyle\sum_{n=0}^{\infty} \left(\frac{1}{3^n} + \frac{1}{4^n}\right)$   **30.** $\displaystyle\sum_{n=0}^{\infty} [(0.4)^n - (0.8)^n]$

In Exercises 31–40, determine the convergence or divergence of the given series.

**31.** $\displaystyle\sum_{n=1}^{\infty} \frac{n + 10}{10n + 1}$   **32.** $\displaystyle\sum_{n=0}^{\infty} \frac{4}{2^n}$

**33.** $\displaystyle\sum_{n=1}^{\infty} \frac{n + 1}{n}$   **34.** $\displaystyle\sum_{n=1}^{\infty} \frac{n + 1}{2n - 1}$

**35.** $\displaystyle\sum_{n=1}^{\infty} \frac{3n - 1}{2n + 1}$   **36.** $\displaystyle\sum_{n=0}^{\infty} \frac{1}{4^n}$

**37.** $\displaystyle\sum_{n=0}^{\infty} (1.075)^n$   **38.** $\displaystyle\sum_{n=1}^{\infty} \frac{2^n}{100}$

**39.** $\displaystyle\sum_{n=0}^{\infty} (0.075)^n$   **40.** $\displaystyle\sum_{n=0}^{\infty} n!$

In Exercises 41–44, the repeated decimal is expressed as a geometric series. Find the sum of the geometric series and write the decimal as the ratio of two integers.

**41.** $0.6\overline{6} = 0.6 + 0.06 + 0.006 + 0.0006 + \cdots$

**42.** $0.23\overline{23} = 0.23 + 0.0023 + 0.000023 + \cdots$

**43.** $0.36\overline{36} = 0.36 + 0.0036 + 0.000036 + \cdots$

**44.** $0.21\overline{21} = 0.21 + 0.0021 + 0.000021 + \cdots$

**45.** A company produces a new product for which it estimates the annual sales to be 8000 units. Suppose that in any given year 10% of the units (regardless of age) will become inoperative.

(a) How many units will be in use after $n$ years?
(b) Find the market stabilization level of the product.

**46.** Repeat Exercise 45 with the assumption that 25% of the units will become inoperative each year.

**47.** A ball is dropped from a height of 16 feet. Each time it drops $h$ feet, it rebounds $0.81h$ feet. Find the total distance traveled by the ball.

**48.** The ball in Exercise 47 takes the following times for each fall. ($t$ is measured in seconds.)

$s_1 = -16t^2 + 16,$          $s_1 = 0$ if $t = 1$

$s_2 = -16t^2 + 16(0.81),$     $s_2 = 0$ if $t = 0.9$

$s_3 = -16t^2 + 16(0.81)^2,$   $s_3 = 0$ if $t = (0.9)^2$

$s_4 = -16t^2 + 16(0.81)^3,$   $s_4 = 0$ if $t = (0.9)^3$

.                               .

.                               .

.                               .

$s_n = -16t^2 + 16(0.81)^{n-1},$  $s_n = 0$ if $t = (0.9)^{n-1}$

Beginning with $s_2$, the ball takes the same amount of time to bounce up as to fall, and thus the total elapsed time before the ball comes to rest is given by

$$t = 1 + 2 \sum_{n=1}^{\infty} (0.9)^n$$

Find the total time it takes for the ball to come to rest.

**49.** Find the fraction of the total area of the square that is eventually shaded if the pattern of shading shown in Figure 9.4 is continued. (Note that the sides of the shaded corner squares are one-fourth those of the squares in which they are placed.)

FIGURE 9.4

**50.** A deposit of $100 is made at the beginning of each month for 5 years in an account that pays 10% interest compounded monthly. What is the balance $A$ in the account at the end of the 5 years?

$$A = 100\left(1 + \frac{0.10}{12}\right) + 100\left(1 + \frac{0.10}{12}\right)^2 + \cdots$$
$$+ 100\left(1 + \frac{0.10}{12}\right)^{60}$$

**51.** A deposit of $50 is made at the beginning of each month in an account that pays 12% interest compounded monthly. What is the balance in the account at the end of 10 years?

$$A = 50\left(1 + \frac{0.12}{12}\right) + 50\left(1 + \frac{0.12}{12}\right)^2 + \cdots$$
$$+ 50\left(1 + \frac{0.12}{12}\right)^{120}$$

**52.** A deposit of $P$ dollars is made every month for $t$ years in an account that pays an annual interest rate of $r$ compounded monthly. Let $N = 12t$ be the total number of deposits. Show that the balance in the account after $t$ years is

$$A = P\left[\left(1 + \frac{r}{12}\right)^N - 1\right]\left(1 + \frac{12}{r}\right)$$

**53.** Use the formula in Exercise 52 to find the amount in an account earning 9% interest compounded monthly after deposits of $50 have been made monthly for 40 years.

**54.** The number of direct ancestors a person has had is given by

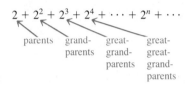

$$2 + 2^2 + 2^3 + 2^4 + \cdots + 2^n + \cdots$$

parents    grand-    great-    great-
parents    grand-    great-
parents    grand-
parents

This formula is valid *provided* the person has no common ancestors. (A common ancestor is one to whom you are related in more than one way. For example, one of your great-grandmothers on your father's side might also be one of your great-grandmothers on your mother's side, as shown in Figure 9.5.) How many direct ancestors have you had who have lived since the year A.D. 1? Assume that the average time between generations was 30 years (resulting in 66 generations) so that the total is given by

$$2 + 2^2 + 2^3 + 2^4 + \cdots + 2^{66}$$

Considering your total, is it reasonable to assume that you have had no common ancestors in the past 2000 years?

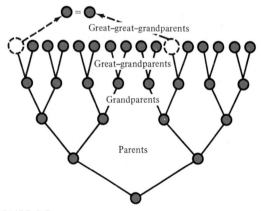

Have you had common ancestors?

Great–great-grandparents

Great–grandparents

Grandparents

Parents

FIGURE 9.5

**55.** Suppose that an employer offered to pay you 1 cent the first day, and then double your wages each day thereafter. Find your total wages for working 20 days.

# *p*-Series and the Ratio Test

INTRODUCTORY EXAMPLE
## Divergence of the Harmonic Series

In Section 9.2, we saw that the *n*th-Term Test for divergence tells us that a sufficient condition for the divergence of the series

$$\sum_{n=1}^{\infty} a_n$$

is that

$$\lim_{n \to \infty} a_n \neq 0$$

Moreover, we stated that knowing that the limit of $a_n$ is zero is not sufficient for convergence, and we now see a case in point. Let's consider the **harmonic series** given by

$$\sum_{n=1}^{\infty} \frac{1}{n} = \frac{1}{1} + \frac{1}{2} + \frac{1}{3} + \frac{1}{4} + \cdots$$

The *n*th-Term Test tells us nothing about the convergence or divergence of this series, since

$$\lim_{n \to \infty} \frac{1}{n} = 0$$

To see that this series actually diverges, we consider each of the terms in the series to represent the area of a rectangle. Then, as indicated in Figure 9.6, we can obtain the inequality

$$\int_{1}^{\infty} \frac{1}{x}\, dx \leq \sum_{n=1}^{\infty} \frac{1}{n}$$

However, from Section 6.6 we know that the improper integral

$$\int_{1}^{\infty} \frac{1}{x}\, dx$$

diverges. This implies that the harmonic series also diverges.

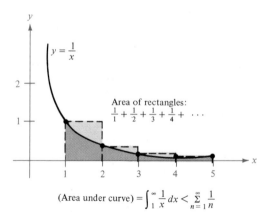

(Area under curve) $= \int_{1}^{\infty} \frac{1}{x}\, dx < \sum_{n=1}^{\infty} \frac{1}{n}$

FIGURE 9.6

■ **p-Series**
■ **The Harmonic Series**
■ **The Ratio Test**

In Section 9.2, we looked at a test to determine the convergence or divergence of a geometric series. We begin this section by looking at another common type of series called a **p-series.** Like geometric series, p-series have a simple arithmetic test for convergence or divergence.

| | |
|---|---|
| **Definition of p-series** | A series of the form $$\sum_{n=1}^{\infty} \frac{1}{n^p} = \frac{1}{1^p} + \frac{1}{2^p} + \frac{1}{3^p} + \cdots$$ is called a **p-series,** where $p$ is a positive constant. The series for which $p = 1$, $$\sum_{n=1}^{\infty} \frac{1}{n} = 1 + \frac{1}{2} + \frac{1}{3} + \cdots$$ is called the **harmonic series.** |

**EXAMPLE 1**

**Classifying p-Series**

Determine which of the following series are p-series. For those that are, find the value of $p$.

(a) $\displaystyle\sum_{n=1}^{\infty} \frac{1}{n^3}$      (b) $\displaystyle\sum_{n=1}^{\infty} \frac{1}{\sqrt{n}}$      (c) $\displaystyle\sum_{n=1}^{\infty} \frac{1}{3^n}$

**SOLUTION**

(a) The series

$$\sum_{n=1}^{\infty} \frac{1}{n^3} = \frac{1}{1^3} + \frac{1}{2^3} + \frac{1}{3^3} + \cdots$$

is a p-series with $p = 3$.

(b) The series

$$\sum_{n=1}^{\infty} \frac{1}{\sqrt{n}} = \frac{1}{1^{1/2}} + \frac{1}{2^{1/2}} + \frac{1}{3^{1/2}} + \cdots$$

is a p-series with $p = \frac{1}{2}$.

(c) The series

$$\sum_{n=1}^{\infty} \frac{1}{3^n} = \frac{1}{3^1} + \frac{1}{3^2} + \frac{1}{3^3} + \cdots$$

is *not* a p-series. (This series is geometric.)

To determine whether a *p*-series converges or diverges, we use the following test.

---

**Test for
convergence of
*p*-series**

The *p*-series

$$\sum_{n=1}^{\infty} \frac{1}{n^p} = \frac{1}{1^p} + \frac{1}{2^p} + \frac{1}{3^p} + \cdots$$

1. diverges if $0 < p \le 1$,
2. converges if $p > 1$.

---

It follows from this test that the *harmonic series*

$$\sum_{n=1}^{\infty} \frac{1}{n} = \frac{1}{1} + \frac{1}{2} + \frac{1}{3} + \cdots \qquad p = 1$$

diverges, and the series

$$\sum_{n=1}^{\infty} \frac{1}{n^{1.1}} = \frac{1}{1^{1.1}} + \frac{1}{2^{1.1}} + \frac{1}{3^{1.1}} + \cdots \qquad p = 1.1$$

whose terms are only *slightly* smaller, converges.

Notice that the test for convergence of a *p*-series tells us only whether the series converges or diverges. Unlike the test for convergence of a geometric series, this test does not give us the sum of a convergent *p*-series. We do not have a simple formula for the sum of a convergent *p*-series. Instead we usually must rely on the following approximation technique.

---

**Maximum error in
approximating the
sum of a
convergent *p*-series**

If the *p*-series

$$\sum_{n=1}^{\infty} \frac{1}{n^p} = \frac{1}{1^p} + \frac{1}{2^p} + \frac{1}{3^p} + \cdots$$

converges, then its sum $S$ differs from the $N$th partial sum $S_N$ by no more than

$$\frac{N^{1-p}}{p - 1}$$

---

**EXAMPLE 2**

**Approximation of the Sum of a *p*-Series**

Approximate the sum of the series

$$\frac{1}{1} + \frac{1}{4} + \frac{1}{9} + \frac{1}{16} + \cdots$$

using its first ten terms. Use the $p$-series error estimate to determine the maximum error of the approximation.

**SOLUTION**     The $n$th term of this series is $a_n = 1/n^2$. Since $p = 2$, we know that the series converges and we represent its sum as

$$S = \sum_{n=1}^{\infty} \frac{1}{n^2}$$

The tenth partial sum of this series is

$$S_{10} = \frac{1}{1} + \frac{1}{2^2} + \frac{1}{3^2} + \cdots + \frac{1}{10^2} \approx 1.55$$

Finally, using the $p$-series error estimate (with $p = 2$), we have a maximum error of

$$\frac{10^{1-2}}{2-1} = 0.1$$

and we conclude that $S_{10} < S < S_{10} + 0.1$. That is, the sum $S$ lies between the following bounds:

$$1.55 < S < 1.65$$

## The Ratio Test

At this point we have looked at two different convergence tests for series—one for a geometric series and one for a $p$-series. We conclude this section with a convenient test that can be applied to many series that do not happen to be geometric series or $p$-series. The test is called the **Ratio Test.**

**Ratio Test**     Let

$$\sum_{n=1}^{\infty} a_n$$

be a series with nonzero terms.

1. The series converges if $\displaystyle\lim_{n \to \infty} \left| \frac{a_{n+1}}{a_n} \right| < 1.$

2. The series diverges if $\displaystyle\lim_{n \to \infty} \left| \frac{a_{n+1}}{a_n} \right| > 1.$

3. The test is inconclusive if $\displaystyle\lim_{n \to \infty} \left| \frac{a_{n+1}}{a_n} \right| = 1.$

**Remark:** Although we list the Ratio Test with the index of summation beginning with $n = 1$, the index could begin with $n = 0$ (or any other integer).

Though the Ratio Test is not a cure for all ills associated with tests for convergence, it is particularly useful for series that *converge rapidly*. Series involving factorials or exponentials are frequently of this type.

**EXAMPLE 3**

**Using the Ratio Test**

Determine the convergence or divergence of the series

$$\sum_{n=0}^{\infty} \frac{2^n}{n!} = \frac{1}{1} + \frac{2}{1} + \frac{4}{2} + \frac{8}{6} + \frac{16}{24} + \cdots$$

**SOLUTION**

Since $a_n = 2^n/n!$, we have

$$\lim_{n \to \infty} \left| \frac{a_{n+1}}{a_n} \right| = \lim_{n \to \infty} \left[ \frac{2^{n+1}}{(n+1)!} \div \frac{2^n}{n!} \right]$$

$$= \lim_{n \to \infty} \left[ \frac{2^{n+1}}{(n+1)!} \cdot \frac{n!}{2^n} \right]$$

$$= \lim_{n \to \infty} \frac{2}{n+1}$$

$$= 0$$

Therefore, the Ratio Test shows that the series converges.

Example 3 tells us something about the rates at which the sequences $\{2^n\}$ and $\{n!\}$ increase as $n$ approaches infinity. For example, in Table 9.3 we can see that although the factorial sequence $\{n!\}$ has a slow start, it quickly overpowers the exponential sequence $\{2^n\}$.

TABLE 9.3

| $n$ | 0 | 1 | 2 | 3 | 4 | 5 | 6 | 7 | 8 | 9 |
|-----|---|---|---|---|----|----|-----|-------|--------|---------|
| $2^n$ | 1 | 2 | 4 | 8 | 16 | 32 | 64 | 128 | 256 | 512 |
| $n!$ | 1 | 1 | 2 | 6 | 24 | 120 | 720 | 5,040 | 40,320 | 362,880 |

From Table 9.3 we can also see that the sequence $\{n\}$ approaches infinity more slowly than the sequence $\{2^n\}$. This is further demonstrated in the next example, which shows that the series

$$\sum_{n=1}^{\infty} \frac{n}{2^n}$$

converges.

**EXAMPLE 4**

**Using the Ratio Test**

Determine the convergence or divergence of the series

$$\sum_{n=1}^{\infty} \frac{n}{2^n} = \frac{1}{2} + \frac{2}{4} + \frac{3}{8} + \frac{4}{16} + \cdots$$

**SOLUTION**

Since $a_n = n/2^n$, we have

$$\lim_{n \to \infty} \left| \frac{a_{n+1}}{a_n} \right| = \lim_{n \to \infty} \left[ \frac{n+1}{2^{n+1}} \div \frac{n}{2^n} \right]$$

$$= \lim_{n \to \infty} \left[ \frac{n+1}{2^{n+1}} \cdot \frac{2^n}{n} \right]$$

$$= \lim_{n \to \infty} \frac{n+1}{2n}$$

$$= \frac{1}{2}$$

and the Ratio Test shows that the series converges.

**EXAMPLE 5**

**Using the Ratio Test**

Determine the convergence or divergence of the series

$$\sum_{n=1}^{\infty} \frac{2^n}{n^2} = \frac{2}{1} + \frac{4}{4} + \frac{8}{9} + \frac{16}{16} + \frac{32}{25} + \cdots$$

**SOLUTION**

Since $a_n = 2^n/n^2$, we have

$$\lim_{n \to \infty} \left| \frac{a_{n+1}}{a_n} \right| = \lim_{n \to \infty} \left[ \frac{2^{n+1}}{(n+1)^2} \div \frac{2^n}{n^2} \right]$$

$$= \lim_{n \to \infty} \left[ \frac{2^{n+1}}{(n+1)^2} \cdot \frac{n^2}{2^n} \right]$$

$$= \lim_{n \to \infty} 2 \left( \frac{n}{n+1} \right)^2$$

$$= 2$$

and the Ratio Test shows that the series diverges.

We conclude this section with a summary of the various tests for convergence or divergence of a series that we have presented in this and the previous section.

## SUMMARY OF TESTS FOR SERIES

| Test | Series | Converges | Sum | Diverges |
|------|--------|-----------|-----|----------|
| $n$th-Term | $\sum\limits_{n=0}^{\infty} a_n$ | | | $\lim\limits_{n\to\infty} a_n \neq 0$ |
| Geometric | $\sum\limits_{n=0}^{\infty} ar^n$ | $\|r\| < 1$ | $S = \dfrac{a}{1-r}$ | $\|r\| \geq 1$ |
| $p$-Series | $\sum\limits_{n=1}^{\infty} \dfrac{1}{n^p}$ | $p > 1$ | $S \approx S_N + \dfrac{N^{1-p}}{p-1}$ | $0 < p \leq 1$ |
| Ratio | $\sum\limits_{n=1}^{\infty} a_n$ | $\lim\limits_{n\to\infty} \left\|\dfrac{a_{n+1}}{a_n}\right\| < 1$ | | $\lim\limits_{n\to\infty} \left\|\dfrac{a_{n+1}}{a_n}\right\| > 1$ |

## SECTION EXERCISES 9.3

In Exercises 1–10, determine the convergence or divergence of the given $p$-series.

**1.** $\sum\limits_{n=1}^{\infty} \dfrac{1}{n^3}$

**2.** $\sum\limits_{n=1}^{\infty} \dfrac{1}{n^{1/3}}$

**3.** $\sum\limits_{n=1}^{\infty} \dfrac{1}{\sqrt[3]{n}}$

**4.** $\sum\limits_{n=1}^{\infty} \dfrac{1}{n^{4/3}}$

**5.** $\sum\limits_{n=1}^{\infty} \dfrac{1}{n^{1.04}}$

**6.** $\sum\limits_{n=1}^{\infty} \dfrac{1}{n^{\pi}}$

**7.** $1 + \dfrac{1}{\sqrt{2}} + \dfrac{1}{\sqrt{3}} + \dfrac{1}{\sqrt{4}} + \cdots$

**8.** $1 + \dfrac{1}{4} + \dfrac{1}{9} + \dfrac{1}{16} + \dfrac{1}{25} + \cdots$

**9.** $1 + \dfrac{1}{2\sqrt{2}} + \dfrac{1}{3\sqrt{3}} + \dfrac{1}{4\sqrt{4}} + \cdots$

**10.** $1 + \dfrac{1}{2} + \dfrac{1}{3} + \dfrac{1}{4} + \dfrac{1}{5} + \cdots$

In Exercises 11–20, use the Ratio Test to determine the convergence or divergence of the given series.

**11.** $\sum\limits_{n=0}^{\infty} \dfrac{3^n}{n!}$

**12.** $\sum\limits_{n=1}^{\infty} n\left(\dfrac{2}{3}\right)^n$

**13.** $\sum\limits_{n=0}^{\infty} \dfrac{n!}{3^n}$

**14.** $\sum\limits_{n=1}^{\infty} n\left(\dfrac{3}{2}\right)^n$

**15.** $\sum\limits_{n=1}^{\infty} \dfrac{n}{4^n}$

**16.** $\sum\limits_{n=1}^{\infty} \dfrac{n^2}{2^n}$

**17.** $\sum\limits_{n=1}^{\infty} \dfrac{2^n}{n^3}$

**18.** $\sum\limits_{n=0}^{\infty} (-1)^n e^{-n}$

**19.** $\sum\limits_{n=0}^{\infty} \dfrac{(-1)^n 2^n}{n!}$

**20.** $\sum\limits_{n=0}^{\infty} \dfrac{4^n}{n!}$

In Exercises 21–24, approximate the sum of the given convergent series using the indicated number of terms. Include an estimate of the maximum error for your approximation.

**21.** $\sum\limits_{n=1}^{\infty} \dfrac{1}{n^5}$, four terms

**22.** $\sum\limits_{n=1}^{\infty} \dfrac{1}{n^4}$, four terms

**23.** $\sum\limits_{n=1}^{\infty} \dfrac{1}{n^{3/2}}$, ten terms

**24.** $\sum\limits_{n=1}^{\infty} \dfrac{1}{n^{10}}$, three terms

In Exercises 25–36, test for convergence or divergence using any appropriate test from this chapter. Identify the test used.

**25.** $\displaystyle\sum_{n=1}^{\infty} \frac{2n}{n+1}$

**26.** $\displaystyle\sum_{n=1}^{\infty} \frac{10}{3\sqrt[3]{n^2}}$

**27.** $\displaystyle\sum_{n=1}^{\infty} \frac{1}{n\sqrt{n}}$

**28.** $\displaystyle\sum_{n=0}^{\infty} \left(\frac{5}{6}\right)^n$

**29.** $\displaystyle\sum_{n=0}^{\infty} \frac{(-1)^n 2^n}{3^n}$

**30.** $\displaystyle\sum_{n=2}^{\infty} \ln n$

**31.** $\displaystyle\sum_{n=1}^{\infty} \left(\frac{1}{n^2} - \frac{1}{n^3}\right)$

**32.** $\displaystyle\sum_{n=1}^{\infty} \frac{n\, 3^n}{n!}$

**33.** $\displaystyle\sum_{n=0}^{\infty} \left(\frac{4}{3}\right)^n$

**34.** $\displaystyle\sum_{n=1}^{\infty} n(0.4)^n$

**35.** $\displaystyle\sum_{n=1}^{\infty} \frac{n!}{3^{n-1}}$

**36.** $\displaystyle\sum_{n=1}^{\infty} \frac{1}{n^{0.95}}$

# Power Series and Taylor's Theorem

How does a calculator (or a computer) compute values for the exponential function? Does it simply retrieve the values from a long list stored in memory, or does it actually compute values such as the following?

$$e^{0.25} \approx 1.2840254 \qquad \text{or} \qquad e^{-3.5} \approx 0.03019738$$

The answer is that a calculator actually computes these values each time the exponential key is activated. (You don't have to think very long about the problems of storing such a list to realize that this would not be feasible.)

When you enter $e^x$ in a calculator, there is usually a slight pause before the result is displayed. During this pause the calculator is summing up a series representation for $e^x$. In this section, we will see that the **power series** for the exponential function is

$$e^x = 1 + x + \frac{x^2}{2} + \frac{x^3}{3!} + \frac{x^4}{4!} + \frac{x^5}{5!} + \cdots$$

Although this series is infinite, a calculator only displays eight (or ten) digits, and thus it only sums enough terms to produce the necessary degree of accuracy.

For example, suppose we wanted to use this series to evaluate the first eight digits of $e^{0.25}$. We would have

$$e^{1/4} \approx 1 + \frac{1}{4} + \frac{1}{2(4^2)} + \frac{1}{3!(4^3)} + \frac{1}{4!(4^4)} + \frac{1}{5!(4^5)} + \cdots$$

The following values indicate that once we get past the eighth term in this series we are no longer adding enough to change the first eight digits. Thus, the first eight terms are sufficient to give us an eight-digit representation of $e^{0.25}$.

$$1.00000000 = 1$$

$$0.25000000 = \frac{1}{4}$$

$$0.03125000 = \frac{1}{2(4^2)}$$

$$0.00260417 = \frac{1}{3!(4^3)}$$

$$0.00016276 = \frac{1}{4!(4^4)}$$

$$0.00000814 = \frac{1}{5!(4^5)}$$

$$0.00000034 = \frac{1}{6!(4^6)}$$

$$+\ 0.00000001 = \frac{1}{7!(4^7)}$$

$$\overline{\phantom{1.28402542}}$$

$$1.28402542 \approx e^{0.25}$$

- **Power Series**
- **Interval of Convergence of a Power Series**
- **Taylor's Theorem**

Up to this point we have been dealing with series whose terms are constants. Now we want to consider series whose terms are variable. In particular, if $x$ is a variable, then a series of the form

$$\sum_{n=0}^{\infty} a_n x^n = a_0 + a_1 x + a_2 x^2 + a_3 x^3 + \cdots + a_n x^n + \cdots$$

is called a **power series.** More generally, we call a series of the form

$$\sum_{n=0}^{\infty} a_n (x - c)^n = a_0 + a_1 (x - c) + a_2 (x - c)^2 + \cdots + a_n (x - c)^n + \cdots$$

a **power series centered at** $c$. Note that we usually begin a power series at $n = 0$, and to simplify the $n$th term we agree that $(x - c)^0 = 1$ even if $x = c$.

Since a power series has variable terms, it can be viewed as a function of $x$,

$$f(x) = \sum_{n=0}^{\infty} a_n (x - c)^n$$

where the *domain of f* is the set of all $x$ for which the power series converges. The determination of this *domain of convergence* is one of the primary problems associated with power series.

Every power series converges at its center, since for $x = c$

$$f(c) = \sum_{n=0}^{\infty} a_n (c - c)^n = a_0 (1) + 0 + 0 + \cdots + 0 + \cdots = a_0$$

Thus, $x = c$ always lies in the domain of $f$. It may happen that the domain of $f$ consists *only* of the single point $x = c$. In fact, this is one of the three basic types of domains for power series. All three types are described in the following result.

**Convergence of a power series**

For a power series centered at $c$, precisely one of the following is true:

1. The series converges only for $x = c$.
2. The series converges for all $x$.
3. There exists an $R > 0$ such that the series converges for $|x - c| < R$ and diverges for $|x - c| > R$.

In the third part of this result, we call $R$ the **radius of convergence** of the power series. In parts 1 and 2, we consider the radius of convergence to be 0 and $\infty$, respectively. Furthermore, since this result states that the domain of convergence of a power series is always an interval, we call this domain the **interval of convergence,** as shown in Figure 9.7.

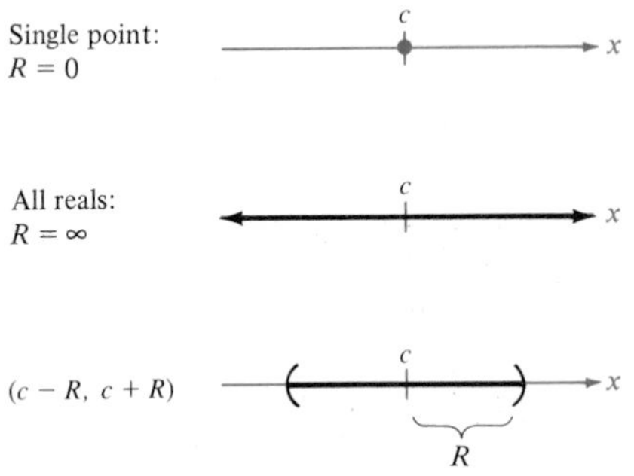

Single point:
$R = 0$

All reals:
$R = \infty$

$(c - R, c + R)$

$R$

Three Types of Domains

FIGURE 9.7

Note that the theorem about convergence of a power series says nothing about the convergence of the series at the endpoints of its interval of convergence. Determining the convergence or divergence at the endpoints can be a very difficult problem. In our presentation of power series, we will focus primarily on determination of the radius of convergence and (except for simple cases) leave the endpoint question open. To find the radius of convergence of a power series, we use the following test, which is called the Power Series Ratio Test.

| **Power Series Ratio Test** | The **radius of convergence** of the power series |
|---|---|

$$f(x) = \sum_{n=0}^{\infty} a_n(x - c)^n$$

is given by

$$\lim_{n \to \infty} \left| \frac{a_n}{a_{n+1}} \right| = R, \quad 0 \le R \le \infty$$

The Power Series Ratio Test is related to the Ratio Test discussed in the previous section in the following way. Using the Ratio Test, we know that for a given value of $x$, the power series

$$\sum_{n=0}^{\infty} a_n(x - c)^n$$

will converge if

$$\lim_{x \to \infty} \left| \frac{a_{n+1}(x - c)^{n+1}}{a_n(x - c)^n} \right| = \lim_{n \to \infty} \left| \frac{a_{n+1}}{a_n} \right| |x - c| < 1$$

This implies that the power series will converge if

$$|x - c| < \lim_{n \to \infty} \left| \frac{a_n}{a_{n+1}} \right| = R$$

In other words, the power series converges in the interval

$$c - R < x < c + R$$

**EXAMPLE 1**

**Finding the Radius of Convergence**

Find the radius of convergence for the series

$$\sum_{n=0}^{\infty} \frac{x^n}{n!}$$

**SOLUTION**

Since $a_n = 1/n!$, we have

$$\lim_{n \to \infty} \left| \frac{a_n}{a_{n+1}} \right| = \lim_{n \to \infty} \left| \frac{1/n!}{1/(n + 1)!} \right|$$

$$= \lim_{n \to \infty} \left| \frac{(n + 1)!}{n!} \right|$$

$$= \lim_{n \to \infty} (n + 1) = \infty$$

and the Power Series Ratio Test shows that the series converges for all $x$ ($R = \infty$).

**EXAMPLE 2**

**Finding the Radius of Convergence**

Find the radius of convergence for the series

$$\sum_{n=0}^{\infty} \frac{(-1)^n(x + 1)^n}{2^n}$$

**SOLUTION**

Since

$$\lim_{n \to \infty} \left| \frac{a_n}{a_{n+1}} \right| = \lim_{n \to \infty} \left| \frac{(-1)^n/2^n}{(-1)^{n+1}/2^{n+1}} \right|$$

$$= \lim_{n \to \infty} \left| \frac{2^{n+1}}{2^n} \right| = 2$$

the radius of convergence is $R = 2$. Since the series is centered at $x = -1$, it will converge in the interval $(-3, 1)$.

## Taylor's Theorem

Why are we interested in power series? One reason is that power series share many of the desirable properties of polynomials. For example, like polynomials, power series can be easily integrated. From our previous work we are well aware that there are many functions (such as $y = e^{x^2}$) that are not easily integrated. Thus, if we could find a power series to represent one of these difficult functions, we could integrate the series to obtain a power series representation of the antiderivative. The point is that we are interested in power series because the power series representation of a particular function may enable us to perform some otherwise difficult operation on the function.

This brings us to our next question. How can we find a power series for a given function? The answer to this question is given by **Taylor's Theorem,** named after the English mathematician Brook Taylor (1685–1731). This theorem gives the power series for a function $f$ in terms of the first (and higher-order) derivatives of $f$.

**Taylor's Theorem** — If $f$ is represented by a power series centered at $c$, then the power series is given by

$$f(x) = f(c) + f'(c)(x - c) + \frac{f''(c)}{2!}(x - c)^2 + \frac{f'''(c)}{3!}(x - c)^3 + \cdots$$

$$= \sum_{n=0}^{\infty} \frac{f^{(n)}(c)}{n!}(x - c)^n$$

**Proof** — Suppose that the function $f$ is represented by a power series centered at $x = c$,

$$f(x) = \sum_{n=0}^{\infty} a_n(x - c)^n$$

Thus, by successive differentiation, we have

$$f^{(0)}(x) = a_0 + a_1(x - c) + a_2(x - c)^2 + a_3(x - c)^3 + \cdots$$

$$f^{(1)}(x) = a_1 + 2a_2(x - c) + 3a_3(x - c)^2 + 4a_4(x - c)^3 + \cdots$$

$$f^{(2)}(x) = 2a_2 + 3!a_3(x - c) + 4 \cdot 3a_4(x - c)^2 + 5 \cdot 4a_5(x - c)^3 + \cdots$$

$$f^{(3)}(x) = 3!a_3 + 4!a_4(x - c) + 5 \cdot 4 \cdot 3a_5(x - c)^2 + 6 \cdot 5 \cdot 4(x - c)^3 + \cdots$$

$$\vdots$$

$$f^{(n)}(x) = n!a_n + (n + 1)!a_{n+1}(x - c) + \cdots$$

Now, evaluating each of these derivatives at $x = c$ yields

$$f^{(0)}(c) = a_0 = 0!a_0$$

$$f^{(1)}(c) = a_1 = 1!a_1$$

$$f^{(2)}(c) = 2a_2 = 2!a_2$$

$$f^{(3)}(c) = 3!a_3$$

.
.
.

$$f^{(n)}(c) = n!a_n$$

By solving for $a_n$ in the equation, we find that the $n$th coefficient of the power series representation of $f(x)$ is

$$a_n = \frac{f^{(n)}(c)}{n!}$$

The series obtained in Taylor's Theorem is called the **Taylor series** for $f(x)$ at $x = c$. If the series is centered at $x = 0$, it is called a **Maclaurin series,** after another English mathematician, Colin Maclaurin (1698–1746).

**EXAMPLE 3**

**Finding a Taylor Series**

Find the power series centered at $x = 1$ for $f(x) = 1/x$.

**SOLUTION**

Successive differentiation of $f(x)$ yields

$$f(x) = (x)^{-1} \qquad\qquad f(1) = 1 = 0!$$

$$f'(x) = -(x)^{-2} \qquad\quad f'(1) = -1 = -(1!)$$

$$f''(x) = 2(x)^{-3} \qquad\quad f''(1) = 2 = 2!$$

$$f'''(x) = -6(x)^{-4} \qquad f'''(1) = -6 = -(3!)$$

$$f^{(4)}(x) = 24(x)^{-5} \qquad f^{(4)}(1) = 24 = 4!$$

$$f^{(5)}(x) = -120(x)^{-6} \qquad f^{(5)}(1) = -120 = -(5!)$$

Following this pattern, we see that

$$f^{(n)}(1) = (-1)^n(n!)$$

Therefore, by Taylor's Theorem we have

$$\frac{1}{x} = f(1) + f'(1)(x - 1) + \frac{f''(1)(x - 1)^2}{2!} + \frac{f'''(1)(x - 1)^3}{3!} + \cdots$$

$$= 1 - (x - 1) + \frac{2!(x - 1)^2}{2!} - \frac{3!(x - 1)^3}{3!} + \frac{4!(x - 1)^4}{4!} - \cdots$$

$$= 1 - (x - 1) + (x - 1)^2 - (x - 1)^3 + (x - 1)^4 - \cdots$$

$$= \sum_{n=0}^{\infty} (-1)^n(x - 1)^n$$

The radius of convergence of the series in Example 3 is $R = 1$. Thus, the series converges in the interval $(0, 2)$. Furthermore, for any fixed value of $x$ this series is geometric, and by the test for convergence of a geometric series we know it diverges at the endpoints $x = 0$ and $x = 2$. Figure 9.8 compares the graphs of $f(x) = 1/x$ and this particular series. Remember that this series is valid *only* within the interval $(0, 2)$.

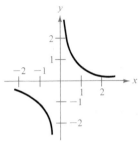

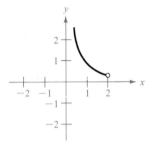

$f(x) = \dfrac{1}{x}$, Domain: all $x \neq 0$    $f(x) = \displaystyle\sum_{n=0}^{\infty} (-1)^n(x - 1)^n$, Domain: $0 < x < 2$

FIGURE 9.8

## EXAMPLE 4

### Finding a Maclaurin Series

Find a power series centered at $x = 0$ for $f(x) = e^{x^2}$.

## SOLUTION

To use Taylor's Theorem, we must calculate successive derivatives of $f(x) = e^{x^2}$. By calculating just the first two,

$$f'(x) = 2xe^{x^2} \quad \text{and} \quad f''(x) = (4x^2 + 2)e^{x^2}$$

we recognize this to be a rather cumbersome task. Fortunately, there is an alternative procedure that is simpler. Suppose we first consider the power series for $g(x) = e^x$. Then we have

$$g(x) = e^x \qquad g(0) = 1$$
$$g'(x) = e^x \qquad g'(0) = 1$$
$$g''(x) = e^x \qquad g''(0) = 1$$
$$\vdots \qquad\qquad \vdots$$
$$g^{(n)}(x) = e^x \qquad g^{(n)}(0) = 1$$

Therefore, the Taylor series for $e^x$ is

$$g(x) = e^x = \sum_{n=0}^{\infty} \frac{g^{(n)}(0)}{n!} x^n = \sum_{n=0}^{\infty} \frac{1}{n!} x^n$$

Now, since $e^{x^2} = g(x^2)$, we have

$$e^{x^2} = g(x^2) = \sum_{n=0}^{\infty} \frac{1}{n!}(x^2)^n$$

$$= 1 + x^2 + \frac{x^4}{2!} + \frac{x^6}{3!} + \frac{x^8}{4!} + \cdots$$

as the power series for $e^{x^2}$. The radius of convergence for this series is infinite, and hence this series converges for all $x$.

Example 4 illustrates an important point in determining power series representations of functions. Though Taylor's Theorem is applicable to a wide variety of functions, it is frequently tedious to use because of the complexity of finding derivatives. Therefore, the most practical use of Taylor's Theorem is in developing power series for a *basic list* of elementary functions. Then from this basic list we can determine power series for other functions by the operations of addition, subtraction, multiplication, division, differentiation, integration, or composition with known power series.

We provide the following list of power series for some basic elementary functions.

**Power series for elementary functions**

$$\frac{1}{x} = 1 - (x-1) + (x-1)^2 - (x-1)^3 + (x-1)^4 - \cdots + (-1)^n(x-1)^n + \cdots, \qquad 0 < x < 2$$

$$\frac{1}{1+x} = 1 - x + x^2 - x^3 + x^4 - x^5 + \cdots + (-1)^n x^n + \cdots, \qquad -1 < x < 1$$

$$\ln x = (x-1) - \frac{(x-1)^2}{2} + \frac{(x-1)^3}{3} - \frac{(x-1)^4}{4} + \cdots + \frac{(-1)^{n-1}(x-1)^n}{n} + \cdots, \qquad 0 < x \le 2$$

$$e^x = 1 + x + \frac{x^2}{2!} + \frac{x^3}{3!} + \frac{x^4}{4!} + \frac{x^5}{5!} + \cdots + \frac{x^n}{n!} + \cdots, \qquad -\infty < x < \infty$$

$$(1+x)^k = 1 + kx + \frac{k(k-1)x^2}{2!} + \frac{k(k-1)(k-2)x^3}{3!} + \frac{k(k-1)(k-2)(k-3)x^4}{4!} + \cdots, \qquad -1 < x < 1^*$$

$$(1+x)^{-k} = 1 - kx + \frac{k(k+1)x^2}{2!} - \frac{k(k+1)(k+2)x^3}{3!} + \frac{k(k+1)(k+2)(k+3)x^4}{4!} - \cdots, \qquad -1 < x < 1^*$$

The last two series in this basic list are called **binomial series.** In the final two examples in this section, we demonstrate how to use the basic list of power series to obtain series for functions that are related to those in the list.

---

*The convergence at $x = \pm 1$ depends on the value $k$.

**EXAMPLE 5**

**Using the Basic List of Power Series**

Determine the power series centered at $x = 0$ for $g(x) = e^{2x+1}$.

**SOLUTION**

First we consider $g(x)$ to be $e^{2x+1} = e^{2x}e$. Then, using the power series

$$f(x) = e^x = 1 + x + \frac{x^2}{2!} + \frac{x^3}{3!} + \frac{x^4}{4!} + \cdots$$

we write

$$ee^{2x} = e[f(2x)]$$

$$= e\left[ 1 + 2x + \frac{(2x)^2}{2!} + \frac{(2x)^3}{3!} + \frac{(2x)^4}{4!} + \cdots \right]$$

$$= e \sum_{n=0}^{\infty} \frac{(2x)^n}{n!}$$

$$= e \sum_{n=0}^{\infty} \frac{2^n}{n!}x^n, \quad -\infty < x < \infty$$

**EXAMPLE 6**

**Using the Basic List of Power Series**

Find the power series centered at $x = 0$ for $g(x) = \sqrt[3]{1 + x}$.

**SOLUTION**

Using the binomial series

$$(1 + x)^k = 1 + kx + \frac{k(k - 1)x^2}{2!} + \frac{k(k - 1)(k - 2)x^3}{3!} + \cdots$$

we let $k = \frac{1}{3}$ and write

$$(1 + x)^{1/3} = 1 + \frac{x}{3} - \frac{2x^2}{3^2 2!} + \frac{2 \cdot 5x^3}{3^3 3!} - \frac{2 \cdot 5 \cdot 8x^4}{3^4 4!} + \cdots$$

which converges for $-1 < x < 1$.

**SECTION EXERCISES 9.4**

In Exercises 1–20, find the radius of convergence for the given series.

**1.** $\displaystyle\sum_{n=0}^{\infty} \left(\frac{x}{2}\right)^n$

**2.** $\displaystyle\sum_{n=0}^{\infty} \left(\frac{x}{k}\right)^n$

**5.** $\displaystyle\sum_{n=0}^{\infty} \frac{x^n}{n!}$

**6.** $\displaystyle\sum_{n=0}^{\infty} \frac{(3x)^n}{n!}$

**3.** $\displaystyle\sum_{n=1}^{\infty} \frac{(-1)^n x^n}{n}$

**4.** $\displaystyle\sum_{n=1}^{\infty} (-1)^{n+1} n x^n$

**7.** $\displaystyle\sum_{n=0}^{\infty} n! \left(\frac{x}{2}\right)^n$

**8.** $\displaystyle\sum_{n=0}^{\infty} \frac{(-1)^n x^n}{(n + 1)(n + 2)}$

**9.** $\displaystyle\sum_{n=1}^{\infty} \frac{(-1)^{n+1}x^n}{4^n}$

**10.** $\displaystyle\sum_{n=0}^{\infty} \frac{(-1)^n n!(x-4)^n}{3^n}$

**11.** $\displaystyle\sum_{n=1}^{\infty} \frac{(-1)^{n+1}(x-5)^n}{n5^n}$

**12.** $\displaystyle\sum_{n=0}^{\infty} \frac{(x-2)^{n+1}}{(n+1)3^{n+1}}$

**13.** $\displaystyle\sum_{n=0}^{\infty} \frac{(-1)^{n+1}(x-1)^{n+1}}{n+1}$

**14.** $\displaystyle\sum_{n=1}^{\infty} \frac{(-1)^{n+1}(x-c)^n}{nc^n}$

**15.** $\displaystyle\sum_{n=1}^{\infty} \frac{(x-c)^{n-1}}{c^{n-1}}, 0 < c$

**16.** $\displaystyle\sum_{n=1}^{\infty} \frac{(-1)^{n+1}x^{2n-1}}{2n-1}$

**17.** $\displaystyle\sum_{n=1}^{\infty} \frac{n}{n+1}(-2x)^{n-1}$

**18.** $\displaystyle\sum_{n=0}^{\infty} \frac{(-1)^n x^{2n}}{n!}$

**19.** $\displaystyle\sum_{n=0}^{\infty} \frac{x^{2n+1}}{(2n+1)!}$

**20.** $\displaystyle\sum_{n=0}^{\infty} \frac{n!x^n}{(n+2)!}$

In Exercises 21–30, apply Taylor's Theorem to find the power series (centered at $c$) for the given function, and find the radius of convergence.

**21.** $f(x) = e^x$, $c = 0$
**22.** $f(x) = e^{-x}$, $c = 0$
**23.** $f(x) = e^{2x}$, $c = 0$
**24.** $f(x) = e^{-2x}$, $c = 0$
**25.** $f(x) = \dfrac{1}{x+1}$, $c = 0$
**26.** $f(x) = \dfrac{1}{2-x}$, $c = 0$
**27.** $f(x) = \sqrt{x}$, $c = 1$
**28.** $f(x) = \sqrt{x}$, $c = 4$
**29.** $f(x) = \ln x$, $c = 1$
**30.** $f(x) = \sqrt[3]{x}$, $c = 1$

In Exercises 31–34, apply Taylor's Theorem to find the binomial series (centered at $c = 0$) for the given function, and find the radius of convergence.

**31.** $f(x) = \dfrac{1}{(1+x)^2}$

**32.** $f(x) = \sqrt{1+x}$

**33.** $f(x) = \dfrac{1}{\sqrt{1+x}}$

**34.** $f(x) = \sqrt[3]{1+x}$

In Exercises 35–38, find the radius of convergence of (a) $f(x)$, (b) $f'(x)$, (c) $f''(x)$, and (d) $\int f(x)\,dx$.

**35.** $f(x) = \displaystyle\sum_{n=0}^{\infty} \left(\frac{x}{2}\right)^n$

**36.** $f(x) = \displaystyle\sum_{n=1}^{\infty} \frac{x^n}{n5^n}$

**37.** $f(x) = \displaystyle\sum_{n=0}^{\infty} \frac{(x-1)^{n+1}}{n+1}$

**38.** $f(x) = \displaystyle\sum_{n=0}^{\infty} \frac{(-1)^{n+1}(x-1)^n}{n}$

In Exercises 39–49, find the power series for the given function using the suggested method. Consult the basic list of power series for elementary functions provided in this section.

**39.** Use the power series for $e^x$:

$$f(x) = e^{x^2}$$

**40.** Use the series found in Exercise 39:

$$f(x) = e^{-x^2}$$

**41.** Differentiate the series found in Exercise 39:

$$f(x) = 2xe^{x^2}$$

**42.** Use the power series for $e^x$ and $e^{-x}$:

$$f(x) = \frac{e^x + e^{-x}}{2}$$

**43.** Use the power series for $1/(1+x)$:

$$f(x) = \frac{1}{1+x^2}$$

**44.** Use the series found in Exercise 43:

$$f(x) = \frac{2x}{1+x^2}$$

**45.** Integrate the series found in Exercise 44:

$$f(x) = \ln(1+x^2)$$

**46.** Integrate the series for $1/(1+x)$:

$$f(x) = \ln(1+x)$$

**47.** Integrate the series for $1/x$:

$$f(x) = \ln x$$

**48.** Differentiate the series found in Exercise 42:

$$f(x) = \frac{e^x - e^{-x}}{2}$$

**49.** Differentiate the series for $-1/(1+x)$:

$$f(x) = \frac{1}{(1+x)^2}$$

**50.** Differentiate the power series for $e^x$ term by term and use the resulting series to show that

$$\frac{d}{dx}[e^x] = e^x$$

# Taylor Polynomials

From elementary geometry we know that the area of a circle of radius $r$ is $A = \pi r^2$. Thus, the area of a circle whose radius is 1 is $\pi$. We can use this information to find a decimal approximation for $\pi$ as follows. (Recall that $\pi \approx 3.1416$.)

We know that the area of the quarter-circle pictured in Figure 9.9 is $\pi/4$, and thus we have

$$4 \int_0^1 \sqrt{1-x^2} \, dx = \pi$$

Using the Taylor series of the preceding section, we find that

$$(1-x^2)^{1/2} = 1 - \frac{x^2}{2} - \frac{x^4}{2^2 2!} - \frac{3x^6}{2^3 3!} - \frac{3 \cdot 5x^8}{2^4 4!} - \cdots$$

We call the $n$th partial sum of this series the $n$th-degree **Taylor polynomial** for $f(x) = \sqrt{1-x^2}$, and we can obtain a decimal approximation for $\pi$ by substituting one of these polynomials into the above integral. Moreover, the higher the degree of the polynomial we use, the better the approximation will be. To see this, let us begin with the second-degree Taylor polynomial for $\sqrt{1-x^2}$. Thus,

$$\pi \approx 4 \int_0^1 \left(1 - \frac{x^2}{2}\right) dx = 4\left[x - \frac{x^3}{6}\right]_0^1$$

$$= 4\left(1 - \frac{1}{6}\right) \approx 3.33$$

A better approximation is obtained by using the sixth-degree Taylor polynomial.

$$\pi \approx 4 \int_0^1 \left(1 - \frac{x^2}{2} - \frac{x^4}{8} - \frac{x^6}{16}\right) dx$$

$$= 4\left[x - \frac{x^3}{6} - \frac{x^5}{40} - \frac{x^7}{112}\right]_0^1$$

$$= 4\left(1 - \frac{1}{6} - \frac{1}{40} - \frac{1}{112}\right)$$

$$\approx 3.20$$

This process could be continued to obtain increasingly better approximations for $\pi$. [Actually, this particular sequence converges to $\pi$ quite slowly. That is, Taylor polynomials of large degree are needed to obtain close approximations. However, there are other Taylor series (involving functions we will not cover in this text) that converge to $\pi$ quite rapidly, and the principles demonstrated in this example can be applied to those series to obtain more accurate approximations of $\pi$.]

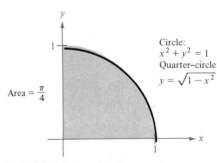

Circle:
$x^2 + y^2 = 1$
Quarter-circle:
$y = \sqrt{1-x^2}$

Area $= \dfrac{\pi}{4}$

FIGURE 9.9

■ **Taylor Polynomials and Approximation**
■ **Taylor's Theorem with Remainder**

In Section 9.4, we saw that it is sometimes possible to obtain an *exact* power series representation for a function. For example, the function $f(x) = e^{-x}$ can be represented exactly by the power series

$$e^{-x} = \sum_{n=0}^{\infty} \frac{(-1)^n}{n!} x^n$$

The problem with using this power series is that the exactness of its representation depends on the summation of an infinite number of terms. Since this is not feasible, in practice we must be content with a finite summation that approximates the given function rather than representing it exactly.

To obtain a better understanding of just how this particular series can be used to approximate $e^{-x}$, consider the following sequence of partial sums:

$$S_0(x) = 1$$

$$S_1(x) = 1 - x$$

$$S_2(x) = 1 - x + \frac{x^2}{2}$$

$$S_3(x) = 1 - x + \frac{x^2}{2} - \frac{x^3}{3!}$$

$$\cdot$$
$$\cdot$$
$$\cdot$$

$$S_n(x) = 1 - x + \frac{x^2}{2} - \frac{x^3}{3!} + \cdots + \frac{(-1)^n x^n}{n!}$$

The members of this sequence are called the **Taylor polynomials** for $e^{-x}$. As $n$ approaches infinity, the graphs of these Taylor polynomials become closer and closer approximations to the graph of $e^{-x}$. For example, the graphs of four of these polynomials are shown in Figure 9.10. Note that the graphs of $S_1$, $S_2$, $S_3$, and $S_4$ are successively better approximations to the graph of $e^{-x}$. Furthermore, from Figure 9.10 it appears that the closer $x$ is to the center of convergence ($x = 0$ in this case), the better the polynomial $S_n(x)$ approximates $e^{-x}$.

To reinforce this conclusion, consider the values shown in Table 9.4. Using the Power Series Ratio Test, we see that the power series for $e^{-x}$ converges for *all* $x$. However, from Table 9.4 and Figure 9.10 we see that the further $x$ is from 0, the more terms we need to obtain a good approximation.

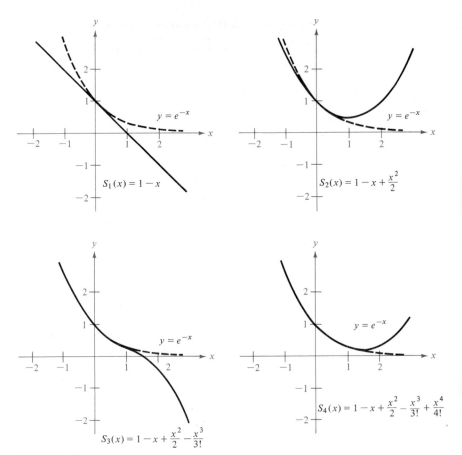

FIGURE 9.10

TABLE 9.4

| $x$ | $0$ | $0.5$ | $1.0$ | $1.5$ | $2.0$ |
|---|---|---|---|---|---|
| $S_1 = 1 - x$ | $1$ | $0.5000$ | $0$ | $-0.5000$ | $-1.0$ |
| $S_2 = 1 - x + \dfrac{x^2}{2}$ | $1$ | $0.6250$ | $0.5000$ | $0.6250$ | $1.0$ |
| $S_3 = 1 - x + \dfrac{x^2}{2} - \dfrac{x^3}{3!}$ | $1$ | $0.6042$ | $0.3333$ | $0.0625$ | $-0.3333$ |
| $S_4 = 1 - x + \dfrac{x^2}{2} - \dfrac{x^3}{3!} + \dfrac{x^4}{4!}$ | $1$ | $0.6068$ | $0.3750$ | $0.2734$ | $0.3333$ |
| $S_5 = 1 - x + \dfrac{x^2}{2} - \dfrac{x^3}{3!} + \dfrac{x^4}{4!} - \dfrac{x^5}{5!}$ | $1$ | $0.6065$ | $0.3667$ | $0.2102$ | $0.0667$ |
| $e^{-x}$ | $1$ | $0.6065$ | $0.3679$ | $0.2231$ | $0.1353$ |

A formal description of how closely a Taylor polynomial approximates a function $f$ is given in the following theorem.

**Taylor's Remainder Theorem**

Let $f$ have derivatives up through order $n + 1$ for every $x$ in an interval $I$ containing $c$. Then for all $x$ in $I$,

$$f(x) = f(c) + f'(c)(x - c) + \frac{f''(c)}{2!}(x - c)^2 + \cdots + \frac{f^{(n)}(c)}{n!}(x - c)^n + R_n$$

where

$$R_n = \frac{f^{(n+1)}(z)}{(n + 1)!}(x - c)^{n+1}$$

for some number $z$ between $c$ and $x$.

Although this theorem appears to give a formula for the exact remainder, note that the theorem does not specify which value of $z$ should be used to find $R_n$. In other words, the practical application of this theorem lies not in calculating $R_n$ but in finding bounds for $R_n$. Once we have found the bounds for $R_n$, we can tell how closely the Taylor polynomial of degree $n$ approximates the function $f(x)$. The following three examples demonstrate this concept.

**EXAMPLE 1**

**Finding Bounds for Taylor's Remainder**

Approximate $e^{-0.75}$ by a fourth-degree Taylor polynomial and determine bounds for the accuracy of this approximation.

**SOLUTION**

The fourth-degree Taylor polynomial for $e^{-x}$ is

$$e^{-x} \approx 1 - x + \frac{x^2}{2} - \frac{x^3}{3!} + \frac{x^4}{4!}$$

Now by Taylor's Remainder Theorem we know that the error in using this polynomial to approximate $e^{-0.75}$ is

$$R_4 = -\frac{e^{-z}}{5!}(0.75)^5, \quad 0 \leq z \leq 0.75$$

Furthermore, since we know that $e^{-z}$ has a maximum value of 1 in the interval $[0, 0.75]$, we can determine that the maximum possible error for this approximation is

$$|R_4| \leq \frac{1}{5!}(0.75)^5 = \frac{1}{5!}\left(\frac{3}{4}\right)^5 = \frac{243}{(120)(1024)} \approx 0.002$$

Finally, we have

$$e^{-0.75} \approx 1 - (0.75) + \frac{(0.75)^2}{2} - \frac{(0.75)^3}{3!} + \frac{(0.75)^4}{4!} \approx 0.474$$

This approximation is off by at most 0.002. Therefore, we know that the exact value of $e^{-0.75}$ lies in the interval

$$0.472 \le e^{-0.75} \le 0.476$$

**EXAMPLE 2**

**Using a Taylor Polynomial to Approximate a Function**

What degree Taylor polynomial must be used to approximate $f(x) = e^x$ in the interval $[-2, 2]$ to an accuracy of $\pm 0.001$?

**SOLUTION**

For $f(x) = e^x$, we have

$$f^{(n+1)}(x) = e^x$$

and thus the maximum value of $f^{(n+1)}(x)$ in the interval $[-2, 2]$ is

$$e^2 \approx (2.71828)^2 \approx 7.389 \le 7.4$$

Therefore, in the interval $[-2, 2]$ we know that the error in using the $n$th-degree Taylor polynomial to approximate $e^x$ is bounded by

$$|R_n| \le \left| \frac{7.4}{(n+1)!} x^{n+1} \right|, \quad -2 \le x \le 2$$

$$|R_n| \le \frac{7.4}{(n+1)!} 2^{n+1}$$

Since we want this error to be less than 0.001, we use trial and error to determine that

$$\frac{7.4}{10!}(2^{10}) = \frac{7.4(1024)}{3,628,800} \approx 0.002$$

$$\frac{7.4}{11!}(2^{11}) = \frac{7.4(2048)}{39,916,800} \approx 0.0004$$

Thus, since $n + 1 = 11$ we have $n = 10$, and we know that the tenth-degree Taylor polynomial for $e^x$ is sufficient to approximate $e^x$ (to the nearest 0.001) for *any* $x$ in the interval $[-2, 2]$.

**EXAMPLE 3**

**Using a Taylor Polynomial to Approximate a Function**

Use a Taylor polynomial to approximate $\ln \frac{3}{2}$. Choose the degree of the polynomial so that your approximation is accurate to within $\pm 0.001$.

**SOLUTION**    The $n$th derivative of $f(x) = \ln x$ is found as follows:

$$f(x) = \ln x$$

$$f'(x) = \frac{1}{x}$$

$$f''(x) = -\frac{1}{x^2}$$

$$f'''(x) = \frac{2}{x^3}$$

$$f^{(4)}(x) = -\frac{3!}{x^4}$$

$$\vdots$$

$$f^{(n)}(x) = \frac{(-1)^{n-1}(n-1)!}{x^n}$$

Now the Taylor series (centered at $x = 1$) for $\ln x$ is

$$\ln x = (x-1) - \frac{(x-1)^2}{2} + \frac{(x-1)^3}{3} - \frac{(x-1)^4}{4} + \cdots$$

Furthermore, the maximum value of $|f^{(n+1)}(x)|$ in the interval $[1, \frac{3}{2}]$ is

$$|f^{(n+1)}(1)| = \frac{n!}{(1)^{n+1}} = n!$$

Thus, we know that the remainder in approximating $\ln \frac{3}{2}$ by an $n$th-degree Taylor polynomial is bounded by

$$|R_n| \le \frac{n!}{(n+1)!}\left(\frac{3}{2} - 1\right)^{n+1} = \frac{1}{(n+1)(2^{n+1})}$$

Since we want this error to be less than 0.001, we use trial and error to determine that

$$\frac{1}{7(2^7)} = \frac{1}{7(128)} \approx 0.0011$$

$$\frac{1}{8(2^8)} = \frac{1}{8(256)} \approx 0.0005$$

Thus, since $n + 1 = 8$ we have $n = 7$, and we know that the seventh-degree Taylor polynomial for $\ln x$ is sufficient to approximate $\ln \frac{3}{2}$. Finally, we have

$$\ln \frac{3}{2} \approx \frac{1}{2} - \frac{(1/2)^2}{2} + \frac{(1/2)^3}{3} - \frac{(1/2)^4}{4} + \frac{(1/2)^5}{5} - \frac{(1/2)^6}{6} + \frac{(1/2)^7}{7} \approx 0.406$$

Using a calculator, we find that

$$\ln \frac{3}{2} \approx 0.4055$$

and thus our approximation is within 0.001 as specified.

In our last example, we demonstrate the use of a Taylor polynomial to evaluate an integral whose antiderivative we cannot find.

**EXAMPLE 4**

**Using a Taylor Polynomial to Approximate a Definite Integral**

Use the eighth-degree Taylor polynomial for $e^{-x^2}$ to approximate the value of

$$\int_0^1 e^{-x^2} \, dx$$

**SOLUTION**

Replacing $x$ with $-x^2$ in the series for $e^x$, we have

$$e^{-x^2} = 1 - x^2 + \frac{x^4}{2!} - \frac{x^6}{3!} + \frac{x^8}{4!} - \cdots.$$

Now, integrating the eighth-degree Taylor polynomial for $e^{-x^2}$, we have

$$\int_0^1 e^{-x^2} \, dx \approx \left[ x - \frac{x^3}{3} + \frac{x^5}{5 \cdot 2!} - \frac{x^7}{7 \cdot 3!} + \frac{x^9}{9 \cdot 4!} \right]_0^1$$

$$\approx 1 - \frac{1}{3} + \frac{1}{10} - \frac{1}{42} + \frac{1}{216}$$

Summing these terms, we have

$$\int_0^1 e^{-x^2} \, dx \approx 0.747$$

In Exercises 1–4, for the given function find the Taylor polynomial (centered at 0) of degree (a) 1, (b) 2, (c) 3, and (d) 4.

**1.** $f(x) = e^x$

**2.** $f(x) = \ln(x + 1)$

**3.** $f(x) = \sqrt{x + 1}$

**4.** $f(x) = \dfrac{1}{(x + 1)^2}$

In Exercises 5 and 6, for the given function find the Taylor polynomial (centered at 0) of degree (a) 2, (b) 4, (c) 6, and (d) 8.

**5.** $f(x) = \dfrac{1}{1 + x^2}$

**6.** $f(x) = e^{-x^2}$

In Exercises 7 and 8, complete the table using the given Taylor polynomials as approximations to the function $f$.

**7.** $f(x) = e^{x/2}$

| $x$ | | 0 | 0.25 | 0.50 | 0.75 | 1.0 |
|---|---|---|---|---|---|---|
| $e^{x/2}$ | | 1.0000· | 1.1331 | 1.2840 | 1.4550 | 1.6487 |
| $1 + \dfrac{x}{2}$ | | | | | | |
| $1 + \dfrac{x}{2} + \dfrac{x^2}{8}$ | | | | | | |
| $1 + \dfrac{x}{2} + \dfrac{x^2}{8} + \dfrac{x^3}{48}$ | | | | | | |
| $1 + \dfrac{x}{2} + \dfrac{x^2}{8} + \dfrac{x^3}{48} + \dfrac{x^4}{384}$ | | | | | | |

**8.** $f(x) = \ln(x^2 + 1)$

| $x$ | 0 | 0.25 | 0.50 | 0.75 |
|---|---|---|---|---|
| $\ln(x^2 + 1)$ | 0.00000 | 0.06062 | 0.22314 | 0.44629 |
| $x^2$ | | | | |
| $x^2 - \dfrac{x^4}{2}$ | | | | |
| $x^2 - \dfrac{x^4}{2} + \dfrac{x^6}{3}$ | | | | |
| $x^2 - \dfrac{x^4}{2} + \dfrac{x^6}{3} - \dfrac{x^8}{4}$ | | | | |

In Exercises 9–15, use a sixth-degree Taylor polynomial centered at $c$ for the function $f$ to obtain the required approximation.

| Function | Approximation |
|---|---|
| **9.** $f(x) = e^{-x}$, $c = 0$ | $f\left(\dfrac{1}{2}\right)$ |
| **10.** $f(x) = x^2 e^{-x}$, $c = 0$ | $f\left(\dfrac{1}{4}\right)$ |
| **11.** $f(x) = \ln x$, $c = 1$ | $f\left(\dfrac{3}{2}\right)$ |
| **12.** $f(x) = \sqrt{x}$, $c = 4$ | $f(5)$ |
| **13.** $f(x) = e^{-x^2}$, $c = 0$ | $\displaystyle\int_0^{1/2} e^{-x^2}\, dx$ |
| **14.**\* $f(x) = e^{-x^2/2}$, $c = 0$ | $\dfrac{1}{\sqrt{2\pi}}\displaystyle\int_{-1}^{1} e^{-x^2/2}\, dx$ |

| Function | Approximation |
|---|---|
| **15.** $f(x) = \dfrac{1}{\sqrt{1 + x^2}}$, $c = 0$ | $\displaystyle\int_0^{1/2} \dfrac{1}{\sqrt{1 + x^2}}\, dx$ |

**16.** It can be shown that

$$4 \int_0^1 \frac{1}{1 + x^2}\, dx = \pi$$

(a) Approximate $\pi$ by using the given integral and the Taylor polynomial for the integrand of degree (i) 6, (ii) 8, (iii) 10, and (iv) 12.

(b) Note that the approximations obtained in part (a) are alternately too large and too small. Average the last two approximations to obtain a better estimate of $\pi$.

---

\* This approximation yields the probability of being within one standard deviation of the mean in a normal distribution.

In Exercises 17 and 18, determine the degree of the Taylor polynomial centered at $c$ required to approximate $f$ in the given interval to an accuracy of $\pm 0.001$.

| Function | Interval |
|----------|----------|
| 17. $f(x) = e^x$, $c = 0$ | $[-1, 1]$ |
| 18. $f(x) = \dfrac{1}{x}$, $c = 1$ | $\left[1, \dfrac{3}{2}\right]$ |

In Exercises 19 and 20, determine the maximum error guaranteed by Taylor's Remainder Theorem when the given fifth-degree Taylor polynomial is used to approximate $f$ in the indicated interval.

19. $f(x) = e^{-x}$, $[0, 1]$

$$1 - x + \frac{x^2}{2!} - \frac{x^3}{3!} + \frac{x^4}{4!} - \frac{x^5}{5!}$$

20. $f(x) = \dfrac{1}{x}$, $\left[1, \dfrac{3}{2}\right]$

$$1 - (x - 1) + (x - 1)^2 - (x - 1)^3$$
$$+ (x - 1)^4 - (x - 1)^5$$

# Newton's Method

One difficulty in writing a mathematics text is choosing examples to illustrate a particular mathematical concept or technique. The problem lies basically in finding the appropriate level of difficulty for the example. If the example is too difficult, the point the author is attempting to illustrate becomes obscured. On the other hand, if the example is too simple, the new technique may seem unnecessary since the reader will be able to anticipate the solution without having to resort to the technique being illustrated.

As a case in point, consider the optimization problem discussed in Example 3 of Section 3.4. In that example, we found the minimum distance between the point $(0, 2)$ and the parabola $y = 4 - x^2$. To solve the problem, we set up the distance equation

$$d = \sqrt{(x - 0)^2 + (y - 2)^2}$$
$$= \sqrt{x^2 + (4 - x^2 - 2)^2}$$
$$= \sqrt{x^4 - 3x^2 + 4}$$

By differentiating the term inside the radical and setting the result equal to zero we obtained the equation

$$4x^3 - 6x = 2x(2x^2 - 3) = 0$$

and by factoring we concluded that the minimum distance occurred when $x = \sqrt{\frac{3}{2}}$.

Suppose we were to find the minimum distance between the point $(0, 5)$ and the parabola. We could go through the same steps, but it is clear from Figure 9.11 that the closest point on the parabola is its vertex and the minimum distance is 1. Thus, this example makes the optimization technique seem unnecessary.

Finally, suppose we are asked to find the minimum distance between the point $(1, 0)$ and the same parabola. In this case, we have

$$d = \sqrt{(x - 1)^2 + (y - 0)^2}$$
$$= \sqrt{(x - 1)^2 + (4 - x^2)^2}$$
$$= \sqrt{x^4 - 7x^2 - 2x + 17}$$

By differentiating and setting the result equal to zero, we obtain

$$4x^3 - 14x - 2 = 2(2x^3 - 7x - 1) = 0$$

This particular cubic cannot be solved by elementary factoring techniques, and thus this example is too difficult to be a good illustration of the optimization technique. In Section 9.6, we will introduce a technique called **Newton's Method** that can be used to solve this cubic.

FIGURE 9.11

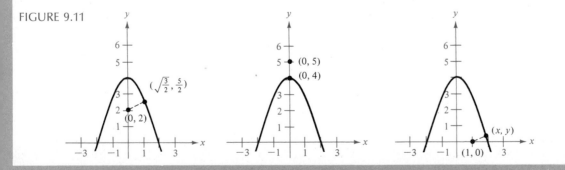

### ■ Newton's Method

In the preceding chapters of this text, we frequently needed to find the zeros of a function. [The zeros of $f$ are those values of $x$ for which $f(x) = 0$.] Until now our functions have been carefully chosen so that elementary algebraic techniques suffice for finding their zeros. For example, the zeros of

$$f(x) = x^2 - 6x + 8$$

$$g(x) = 2x^2 - 3x - 7$$

$$h(x) = x^3 - 2x^2 - x + 2$$

can all be found by factoring or by the quadratic formula. However, in practice we frequently encounter functions whose zeros are more difficult to find. For instance, the zeros of a function as simple as

$$f(x) = x^3 - x + 1$$

cannot be found by elementary algebraic methods. In such cases, the first-degree Taylor polynomial for the function may serve as a convenient tool for approximating the zeros. For example, the first-degree Taylor polynomial (centered at $x = x_1$) for $f(x)$ is

$$p(x) = f(x_1) + f'(x_1)(x - x_1)$$

Note that this linear polynomial is simply the equation of the tangent line to $f$ at the point $(x_1, f(x_1))$. (See Figure 9.12.) Newton's Method for approximating zeros is based on the assumption that $f$ and the tangent to $f$ at $(x_1, f(x_1))$ both cross the $x$-axis at about the same point. To illustrate this, suppose we wish to find the point in the interval $(a, b)$ at which the function $f$ crosses the $x$-axis. If we cannot find this point, we guess its value to be $x_1$. Now, since we can easily calculate the $x$-intercept for the tangent line, we use it as our second (and presumably better) estimate for the zero of

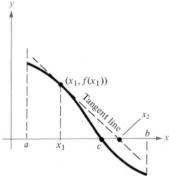

The $x$-intercept of the tangent line approximates the zero of $f$.

FIGURE 9.12

*f*. In other words, we use our initial guess, $x_1$, to find a better guess, $x_2$. To find $x_2$, we let $p(x) = 0$ and solve for $x$ as follows:

$$f(x_1) + f'(x_1)(x - x_1) = 0$$

$$x - x_1 = -\frac{f(x_1)}{f'(x_1)}$$

$$x = x_1 - \frac{f(x_1)}{f'(x_1)}$$

Thus, from our initial guess we arrive at a new estimate,

$$x_2 = x_1 - \frac{f(x_1)}{f'(x_1)}$$

At this point we may wish to improve on $x_2$ and calculate yet a third estimate:

$$x_3 = x_2 - \frac{f(x_2)}{f'(x_2)}$$

Repeated application of this process is called **Newton's Method.**

| | |
|---|---|
| **Newton's Method for approximating the zeros of a function** | Let $f$ be differentiable on $(a, b)$ and $f(c) = 0$, where $c$ is in $(a, b)$. To approximate $c$, we use the following steps: |

1. Make an initial estimate $x_1$ "close" to $c$.
2. Determine a new approximation using the formula

$$x_{n+1} = x_n - \frac{f(x_n)}{f'(x_n)}$$

3. If $|x_n - x_{n+1}|$ is less than the desired accuracy, let $x_{n+1}$ serve as a final approximation. Otherwise, return to step 2 and calculate a new approximation.

The process of calculating each successive approximation is called an **iteration.**

**EXAMPLE 1**

**Using Newton's Method**

Use three iterations of Newton's Method to approximate a zero of

$$f(x) = x^2 - 2$$

Use $x_1 = 1$ as the initial guess.

**SOLUTION**

Since

$$f(x) = x^2 - 2$$

we have

$$f'(x) = 2x$$

and the iterative process is given by the formula

$$x_{n+1} = x_n - \frac{f(x_n)}{f'(x_n)}$$

$$= x_n - \frac{x_n^2 - 2}{2x_n}$$

The calculations for three iterations are shown in Table 9.5, and the first iteration is graphically depicted in Figure 9.13.

TABLE 9.5

| $n$ | $x_n$ | $f(x_n)$ | $f'(x_n)$ | $\dfrac{f(x_n)}{f'(x_n)}$ | $x_n - \dfrac{f(x_n)}{f'(x_n)}$ |
|---|---|---|---|---|---|
| 1 | 1.000000 | −1.000000 | 2.000000 | −0.500000 | 1.500000 |
| 2 | 1.500000 | 0.250000 | 3.000000 | 0.083333 | 1.416667 |
| 3 | 1.416667 | 0.006945 | 2.833334 | 0.002451 | 1.414216 |
| 4 | 1.414216 | | | | |

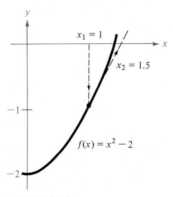

FIGURE 9.13

Of course, in this example we know that the two zeros of the function are $\pm\sqrt{2}$. To six decimal places,

$$\sqrt{2} = 1.414214$$

Thus, after only three iterations of Newton's Method, we have obtained an approximation (1.414216) that is within 0.000002 of the actual root.

As you might expect, Newton's Method has become more popular since the emergence of electronic calculators.

EXAMPLE 2

**Using Newton's Method**

Use Newton's Method to approximate the zeros of

$$f(x) = 2x^3 + x^2 - x + 1$$

Continue the iterations until two successive approximations differ by less than 0.0001.

**SOLUTION**

We begin by sketching a graph of $f$ and observing that it has only one zero, which occurs near $x = -1.2$ (see Figure 9.14). The calculations are shown in Table 9.6.

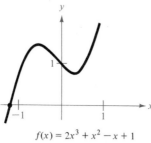

$f(x) = 2x^3 + x^2 - x + 1$

FIGURE 9.14

TABLE 9.6

| $n$ | $x_n$ | $f(x_n)$ | $f'(x_n)$ | $\dfrac{f(x_n)}{f'(x_n)}$ | $x_n - \dfrac{f(x_n)}{f'(x_n)}$ |
|---|---|---|---|---|---|
| 1 | $-1.20000$ | 0.18400 | 5.24000 | 0.03511 | $-1.23511$ |
| 2 | $-1.23511$ | $-0.00773$ | 5.68282 | $-0.00136$ | $-1.23375$ |
| 3 | $-1.23375$ | $-0.00001$ | 5.66539 | $-0.00000$ | $-1.23375$ |
| 4 | $-1.23375$ | | | | |

Thus, we estimate the zero of $f$ to be $-1.23375$, since two successive approximations differ by less than the required 0.0001.

When, as in Examples 1 and 2, the approximations approach a zero of the function, we say that the method **converges.** It is important to realize that Newton's Method does not always converge. Two situations in which it may not converge are as follows:

1. If $f'(x_n) = 0$ for some $n$ (see Figure 9.15).
2. If $\lim\limits_{n \to \infty} x_n$ does not exist (see Figure 9.16).

The type of problem illustrated in Figure 9.15 can usually be overcome with a better choice for $x_1$. However, the problem illustrated in Figure 9.16 is usually more serious and may be independent of the choice of $x_1$. For example, Newton's Method does not converge for any choice of $x_1$ (other than the actual zero) for the function $f(x) = x^{1/3}$.

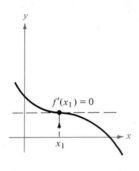

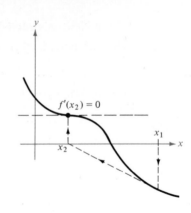

FIGURE 9.15

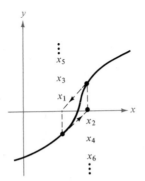

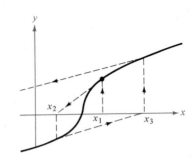

FIGURE 9.16

**EXAMPLE 3**

**An Example in Which Newton's Method Fails**

Refer to Figure 9.17. Using $x_1 = 0.1$, show that the limit of $x_n$ as $n \to \infty$ does not exist for the function $f(x) = x^{1/3}$.

**SOLUTION**

Table 9.7 and Figure 9.17 indicate that $x_n$ increases in magnitude as $n \to \infty$, and thus the limit does not exist.

TABLE 9.7

| $n$ | $x_n$ | $f(x_n)$ | $f'(x_n)$ | $\dfrac{f(x_n)}{f'(x_n)}$ | $x_n - \dfrac{f(x_n)}{f'(x_n)}$ |
|---|---|---|---|---|---|
| 1 | 0.10000 | 0.46416 | 1.54720 | 0.30000 | −0.20000 |
| 2 | −0.20000 | −0.58480 | 0.97467 | −0.60000 | 0.40000 |
| 3 | 0.40000 | 0.73681 | 0.61401 | 1.20000 | −0.80000 |
| 4 | −0.80000 | −0.92832 | 0.38680 | −2.40000 | 1.60000 |

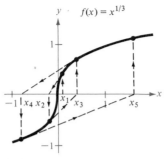

$f(x) = x^{1/3}$

FIGURE 9.17

## EXAMPLE 4

**Using Newton's Method to Find a Point of Intersection**

Estimate the point of intersection of the two curves given by

$$y = e^{-x^2} \qquad \text{and} \qquad y = x$$

as shown in Figure 9.18. Use Newton's Method and continue the iterations until two successive approximations differ by less than 0.0001.

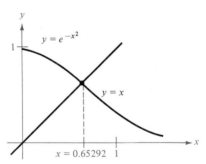

FIGURE 9.18

## SOLUTION

We seek the value of $x$ such that

$$e^{-x^2} = x$$

which implies that

$$0 = x - e^{-x^2}$$

To use Newton's Method, we let

$$f(x) = x - e^{-x^2}$$

and the iterative formula is

$$x_{n+1} = x_n - \frac{f(x_n)}{f'(x_n)} = x_n - \frac{x_n - e^{-x_n^2}}{1 + 2x_n e^{-x_n^2}}$$

The calculations are shown in Table 9.8, beginning with an initial guess of $x_1 = 0.5$.

TABLE 9.8

| $n$ | $x_n$ | $f(x_n)$ | $f'(x_n)$ | $\dfrac{f(x_n)}{f'(x_n)}$ | $x_n - \dfrac{f(x_n)}{f'(x_n)}$ |
|---|---|---|---|---|---|
| 1 | 0.50000 | −0.27880 | 1.77880 | −0.15674 | 0.65674 |
| 2 | 0.65674 | 0.00707 | 1.85331 | 0.00382 | 0.65292 |
| 3 | 0.65292 | 0.00000 | 1.85261 | 0.00000 | 0.65292 |
| 4 | 0.65292 | | | | |

Thus, the two curves intersect when $x \approx 0.65292$.

## SECTION EXERCISES 9.6

In Exercises 1 and 2, complete one iteration of Newton's Method for the given function using the indicated initial estimate.

| *Function* | *Initial estimate* |
|---|---|
| **1.** $f(x) = x^2 - 3$ | $x_1 = 1.7$ |
| **2.** $f(x) = 3x^2 - 2$ | $x_1 = 1$ |

In Exercises 3–12, approximate the indicated zero(s) of the function. Use Newton's Method, continuing the process until two successive approximations differ by less than 0.001.

**3.** $f(x) = x^3 + x - 1$

**4.** $f(x) = x^5 + x - 1$

**5.** $f(x) = 3\sqrt{x - 1} - x$

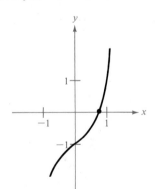

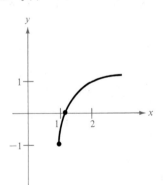

**6.** $f(x) = x^3 - 3.9x^2 + 4.79x - 1.881$

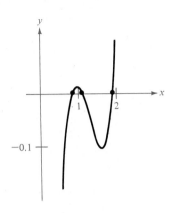

**7.** $f(x) = x^3 - 27x - 27$

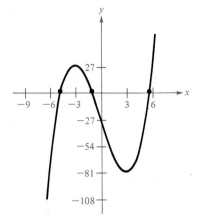

**8.** $f(x) = x^4 - 4x^3 + 6x^2 - 4x - 2$

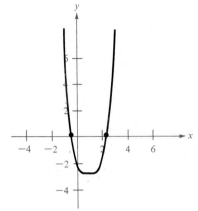

**9.** $f(x) = \ln x + x$

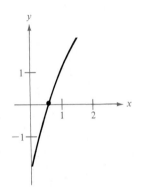

**10.** $f(x) = 2[e^{3x}(2 - x) - 1]$

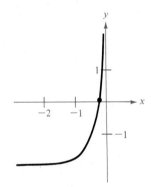

**11.** $f(x) = e^{-x^2} - x^2$

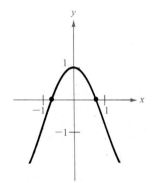

**12.** $f(x) = \ln x - \dfrac{1}{x}$

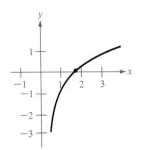

In Exercises 13–16, apply Newton's Method to approximate the $x$-value of the point of intersection of the two graphs. Continue the process until two successive approximations differ by less than 0.001.

**13.** $f(x) = 3 - x$, $g(x) = \ln x$

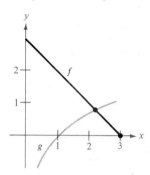

**14.** $f(x) = 2x + 1$, $g(x) = \sqrt{x + 4}$

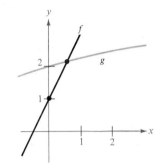

**15.** $f(x) = 3 - x$, $g(x) = \dfrac{1}{x^2 + 1}$

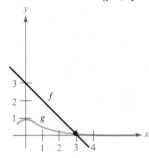

**16.** $f(x) = x$, $g(x) = e^{-x}$

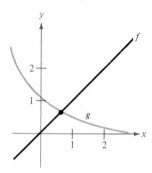

In Exercises 17–19, apply Newton's Method using the indicated initial estimate, and explain why the method fails.

**17.** $y = 2x^3 - 6x^2 + 6x - 1$, $x_1 = 1$

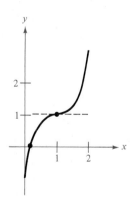

**18.** $y = 4x^3 - 12x^2 + 12x - 3$, $x_1 = \dfrac{3}{2}$

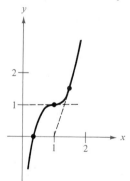

**19.** $y = -x^3 + 3x^2 - x + 1$, $x_1 = 1$

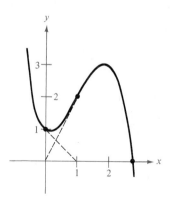

**20.** Use Newton's Method to obtain a general formula for approximating $\sqrt[n]{x}$. [Hint: Consider $f(x) = x^n - a$.]

In Exercises 21–24, use the result of Exercise 20 to approximate the given radical to three decimal places.

**21.** $\sqrt[3]{7}$

**22.** $\sqrt[4]{6}$

**23.** $\sqrt[5]{40}$

**24.** $\sqrt{5}$

**25.** In the Introductory Example of this section, the equation

$$2x^3 - 7x - 1 = 0$$

must be solved to find the coordinates of the point on the graph of $y = 4 - x^2$ that is closest to the point $(1, 0)$. Use Newton's Method to solve for $x$ (accurate to three decimal places). What is the $y$-coordinate of the point?

**26.** Find the coordinates (accurate to three decimal places) of the point on the graph of $y = x^2$ that is closest to the point $(4, -3)$.

**27.** A man is in a boat 2 miles from the nearest point on the coast, as shown in Figure 9.19. He is to go to a point $Q$, which is 3 miles down the coast and 1 mile inland. If he can row at 3 mi/hr and walk at 4 mi/hr, toward what point on the coast should he row in order to reach point $Q$ in the least time?

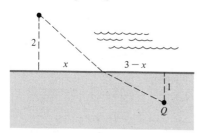

FIGURE 9.19

**28.** A company estimates that the cost in dollars of producing $x$ units of a certain product is given by the model

$$C = 0.0001x^3 + 0.02x^2 + 0.4x + 800$$

Find the production level that minimizes the average cost per unit.

**29.** The concentration $C$ of a certain chemical in the bloodstream $t$ hours after injection into muscle tissue is given by

$$C = \frac{3t^2 + t}{50 + t^3}$$

When is the concentration the greatest?

**30.** The ordering and transportation cost $C$ of the components used in manufacturing a certain product is given by

$$C = 100\left(\frac{200}{x^2} + \frac{x}{x + 30}\right), \quad 1 \le x$$

where $C$ is measured in thousands of dollars and $x$ is the order size in hundreds. Find the order size that minimizes the costs.

## Important Terms

Sequence
Terms of a sequence
Convergent sequence
Divergent sequence
$n$ factorial, $n!$
Index of summation
Sigma notation for sums
$N$th partial sum, $S_N$
Infinite series
Sum of a convergent series
Convergent series
Geometric series
Divergent series

Harmonic series
$p$-series
Remainder of a series, $R_N$
Power series
Power series centered at $c$
Radius of convergence
Interval of convergence
Taylor series
Maclaurin series
Binomial series
Taylor polynomial
Newton's method
Iteration
Convergence of Newton's method

## Important Techniques

Determining the convergence or divergence of a sequence
Finding a pattern for the $n$th term of a sequence
Determining the convergence or divergence of an infinite series
Finding the sum of convergent geometric series

Determining the convergence or divergence of a $p$-series
Approximating the sum of a $p$-series
Finding the radius of convergence of a power series
Finding the Taylor series for a function
Approximating a function with a Taylor polynomial
Using Newton's method to approximate the zeros of a function

## Important Formulas

$$0! = 1$$

$$1! = 1$$

$$n! = 1 \cdot 2 \cdot 3 \cdot 4 \cdots (n - 1) \cdot n$$

Properties of infinite series: If

$$\sum_{n=0}^{\infty} a_n = A \quad \text{and} \quad \sum_{n=0}^{\infty} b_n = B$$

then

1. $\displaystyle \sum_{n=0}^{\infty} ca_n = c \sum_{n=0}^{\infty} a_n = cA$

2. $\displaystyle \sum_{n=0}^{\infty} (a_n \pm b_n) = \sum_{n=0}^{\infty} a_n \pm \sum_{n=0}^{\infty} b_n = A \pm B$

$n$th-Term Test for divergence: If $\lim_{n \to \infty} a_n \neq 0$, then the

series $\displaystyle \sum_{n=0}^{\infty} a_n$ diverges.

$N$th partial sum of a geometric series:

$$S_N = \sum_{n=0}^{N} ar^n = \frac{a(1 - r^{N+1})}{1 - r}$$

The geometric series $\displaystyle \sum_{n=0}^{\infty} ar^n$ has these properties:

1. If $|r| \geq 1$, it diverges.

2. If $|r| < 1$, it converges and the sum is given by

$$\sum_{n=0}^{\infty} ar^n = \frac{a}{1 - r}$$

The $p$-series $\displaystyle \sum_{n=1}^{\infty} \frac{1}{n^p} = \frac{1}{1^p} + \frac{1}{2^p} + \frac{1}{3^p} + \cdots$

1. diverges if $0 < p \leq 1$.
2. converges if $p > 1$.

If the $p$-series $\displaystyle \sum_{n=1}^{\infty} \frac{1}{n^p}$ converges to $S$, then

$$S \approx S_N + \frac{N^{1-p}}{p - 1}$$

The Ratio Test: Let $\displaystyle\sum_{n=1}^{\infty} a_n$ be a series with nonzero terms.

1. The series converges if $\displaystyle\lim_{n\to\infty} \left|\frac{a_{n+1}}{a_n}\right| < 1$.

2. The series diverges if $\displaystyle\lim_{n\to\infty} \left|\frac{a_{n+1}}{a_n}\right| > 1$.

3. The test is inconclusive if $\displaystyle\lim_{n\to\infty} \left|\frac{a_{n+1}}{a_n}\right| = 1$.

For a power series centered at $c$, precisely one of the following is true:

1. The series converges only for $x = c$.
2. The series converges for all $x$.
3. There exists an $R > 0$ such that the series converges for $|x - c| < R$ and diverges for $|x - c| > R$.

Power Series Ratio Test: The radius of convergence of the power series $\displaystyle\sum_{n=0}^{\infty} a_n(x - c)^n$ is given by

$$\lim_{n\to\infty} \left|\frac{a_n}{a_{n+1}}\right| = R, \quad 0 \le R \le \infty$$

Taylor's Theorem: If $f$ is represented by a power series centered at $c$, then the power series is given by

$$f(x) = f(c) + f'(c)(x - c) + \frac{f''(c)}{2!}(x - c)^2 + \cdots$$

$$= \sum_{n=0}^{\infty} \frac{f^{(n)}(c)}{n!}(x - c)^n$$

Taylor's Theorem with Remainder: Let $f$ be a function such that $f^{(n+1)}(x)$ exists for every $x$ in an interval $I$ containing $c$. Then for all $x$ in $I$,

$$f(x) = f(c) + f'(c)(x - c) + \cdots + \frac{f^{(n)}(c)}{n!}(x - c)^n + R_n$$

where

$$R_n = \frac{f^{(n+1)}(z)}{(n+1)!}(x - c)^{n+1}$$

for some number $z$ between $c$ and $x$.

Newton's Method: Let $f$ be differentiable on $(a, b)$ and $f(c) = 0$, where $c$ is in $(a, b)$. To approximate $c$, we use the following steps:

1. Make an initial estimate $x_1$ "close" to $c$.
2. Determine a new approximation using the formula

$$x_{n+1} = x_n - \frac{f(x_n)}{f'(x_n)}$$

3. If $|x_n - x_{n+1}|$ is less than the desired accuracy, let $x_{n+1}$ serve as a final approximation. Otherwise, return to step 2 and calculate a new approximation.

In Exercises 1–4, find the general term of the given sequence.

**1.** $\dfrac{1}{3}, \dfrac{2}{5}, \dfrac{3}{7}, \dfrac{4}{9}, \cdots$

**2.** $\dfrac{1}{2}, \dfrac{2}{5}, \dfrac{3}{10}, \dfrac{4}{17}, \cdots$

**3.** $\dfrac{1}{3}, -\dfrac{2}{9}, \dfrac{4}{27}, -\dfrac{8}{81}, \cdots$

**4.** $1, \dfrac{2}{5}, \dfrac{4}{25}, \dfrac{8}{125}, \dfrac{16}{625}, \cdots$

In Exercises 5–10, determine the convergence or divergence of the sequence with the given general term.

**5.** $a_n = \dfrac{n+1}{n^2}$

**6.** $a_n = \dfrac{1}{\sqrt{n}}$

**7.** $a_n = \dfrac{n^3}{n^2+1}$

**8.** $a_n = 10e^{-n}$

**9.** $a_n = 5 + \dfrac{1}{3^n}$

**10.** $a_n = \dfrac{n}{\sqrt{n^2+1}}$

In Exercises 11–14, find the first five terms of the sequence of partial sums for the given series.

**11.** $\displaystyle\sum_{n=0}^{\infty} \left(\dfrac{3}{2}\right)^n$

**12.** $\displaystyle\sum_{n=1}^{\infty} \dfrac{(-1)^{n+1}}{2n}$

**13.** $\displaystyle\sum_{n=1}^{\infty} \dfrac{(-1)^{n+1}}{(2n)!}$

**14.** $\displaystyle\sum_{n=1}^{\infty} \dfrac{1}{n^2}$

In Exercises 15–18, find the sum of the given series.

**15.** $\displaystyle\sum_{n=0}^{\infty} \left(\dfrac{1}{5}\right)^n$

**16.** $\displaystyle\sum_{n=0}^{\infty} 3\left(\dfrac{1}{5}\right)^n$

**17.** $\displaystyle\sum_{n=0}^{\infty} \left(\dfrac{1}{4^n} - \dfrac{1}{3^n}\right)$

**18.** $\displaystyle\sum_{n=0}^{\infty} (0.8)^n$

In Exercises 19 and 20, express the repeating decimal as the ratio of two integers by finding the sum of the given geometric series.

**19.** $0.0\overline{909}$

**20.** $0.34\overline{2342}$

In Exercises 21–30, determine the convergence or divergence of the given series.

**21.** $\displaystyle\sum_{n=1}^{\infty} \dfrac{1}{n^{5/2}}$

**22.** $\displaystyle\sum_{n=0}^{\infty} \left(\dfrac{5}{2}\right)^n$

**23.** $\displaystyle\sum_{n=1}^{\infty} \dfrac{n^2+1}{n(n+1)}$

**24.** $\displaystyle\sum_{n=1}^{\infty} \dfrac{1}{n\sqrt[4]{n}}$

**25.** $\displaystyle\sum_{n=1}^{\infty} \dfrac{2^n}{n}$

**26.** $\displaystyle\sum_{n=0}^{\infty} \dfrac{n!}{4^n}$

**27.** $\displaystyle\sum_{n=1}^{\infty} \dfrac{n4^n}{n!}$

**28.** $\displaystyle\sum_{n=1}^{\infty} \dfrac{(-1)^{n+1}n3^{n-1}}{4^{n-1}}$

**29.** $\displaystyle\sum_{n=1}^{\infty} \dfrac{\sqrt[3]{n^2}}{n}$

**30.** $\displaystyle\sum_{n=1}^{\infty} \dfrac{1}{n^{2.5}}$

In Exercises 31–34, find the radius of convergence of the power series.

**31.** $\displaystyle\sum_{n=0}^{\infty} \dfrac{(-1)^n(x-2)^n}{(n+1)^2}$

**32.** $\displaystyle\sum_{n=0}^{\infty} (2x)^n$

**33.** $\displaystyle\sum_{n=0}^{\infty} n!(x-2)^n$

**34.** $\displaystyle\sum_{n=0}^{\infty} \dfrac{(x-2)^n}{2^n}$

In Exercises 35–38, apply Taylor's Theorem to find the power series for $f(x)$ centered at $c$.

**35.** $f(x) = e^{-2x}$, $c = 0$

**36.** $f(x) = \dfrac{1}{\sqrt{x}}$, $c = 1$

**37.** $f(x) = \dfrac{1}{x}$, $c = -1$

**38.** $f(x) = \sqrt{1+x}$, $c = 0$

In Exercises 39 and 40, find the series representation of $f(x)$.

**39.** $f(x) = \displaystyle\int_0^x \ln(t+1)\, dt$

**40.** $f(x) = \displaystyle\int_0^x (e^t - 1)\, dt$

In Exercises 41–46, use a Taylor polynomial of sixth degree to approximate the given value.

**41.** $\ln 1.75$

**42.** $\ln 1.5$

**43.** $e^{0.6}$

**44.** $e^{-0.25}$

**45.** $\displaystyle\int_0^{0.3} \sqrt{1+x^3}\, dx$

**46.** $\displaystyle\int_0^{0.5} e^{-x^2}\, dx$

**47.** Sketch the graph of $f(x) = e^x$ and its fifth-degree Taylor polynomial on the same axes.

**48.** A ball is dropped from a height of 8 feet. Each time it drops $h$ feet, it rebounds $0.7h$ feet. Find the total distance traveled by the ball.

In Exercises 49 and 50, use Newton's Method to approximate to three decimal places the zero(s) of the given function.

**49.** $f(x) = x^3 - 3x - 1$     **50.** $f(x) = x^3 + 2x + 1$

In Exercises 51 and 52, use Newton's Method to approximate to three decimal places the $x$-value of the points of intersection of the graphs of the given equations.

**51.** $f(x) = x^4$, $g(x) = x + 3$

**52.** $f(x) = \dfrac{1}{x}$, $g(x) = 5e^{-x}$

**53.** Show that the Ratio Test is inconclusive for a $p$-series.

# CHAPTER 10

# The Trigonometric Functions

# Radian Measure of Angles

## INTRODUCTORY EXAMPLE
### Winding a Pulley

One way to lift objects is by means of a simple pulley. As the cylindrical drum of the pulley is revolved, the pulley cable winds around the circumference of the drum. The length of cable wound around the drum corresponds precisely to the height the object is raised off the ground, as shown in Figure 10.1.

Suppose that we are asked to write a computer program to operate a pulley in an automated factory. If the radius of the pulley's drum is 1 foot, how many degrees should the drum be revolved in order to lift the object $s$ feet off the ground? The answer can be seen by observing that the entire circumference of this particular drum is given by

$$\text{Circumference} = (2\pi)(\text{radius}) = (2\pi)(1) = 2\pi$$

Thus, when the drum is turned one full revolution (an angle of $\theta = 360°$), the object is lifted $2\pi$ feet. (Since $\pi \approx 3.1416$, one revolution lifts the object about 6.2832 feet.) Now, since $360°$ corresponds to $2\pi$ feet, we have

$$360° \Longleftrightarrow 2\pi \text{ feet}$$

$$\frac{180°}{\pi} = \frac{360°}{2\pi} \Longleftrightarrow 1 \text{ foot}$$

and multiplying by $s$, we have

$$\left(\frac{180°}{\pi}\right)s \Longleftrightarrow s \text{ feet}$$

In other words, if we want to lift the object 2 feet off the ground, we should revolve the drum

$$\left(\frac{180°}{\pi}\right)(2) \approx 114.6°$$

Similarly, if we want to lift the object 7 feet off the ground, we should revolve the drum

$$\left(\frac{180°}{\pi}\right)(7) \approx 401.1°$$

Note that there is a one-to-one correspondence between the number of degrees revolved and the number of feet of cable wound around the drum. In this section, we make use of this correspondence to develop an alternative to the standard degree measure of an angle. This alternative measure is called **radian measure.** In this particular example (since the radius is 1), a revolution of $\theta$ degrees corresponds to a radian measure of $s$ radians. (See Figure 10.1.)

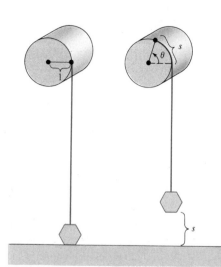

FIGURE 10.1

The concept of an angle is central to the study of trigonometry. We devote this section to a discussion of angles and their measure. As shown in Figure 10.2, an **angle** has three parts: an **initial ray,** a **terminal ray,** and a **vertex** (the point of intersection of these two rays). We say that an angle is in **standard position** if its initial ray coincides with the positive $x$-axis and its vertex is at the origin.

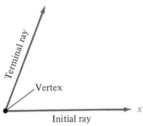

FIGURE 10.2

We assume that you are somewhat familiar with the measurement of angles in terms of degrees. Figure 10.3 shows several common angles with their corresponding degree measure. Note that we use $\theta$ (the lowercase Greek letter theta) to represent the measure of an angle. (Actually it is common practice to use $\theta$ to represent both the angle and its measure.)

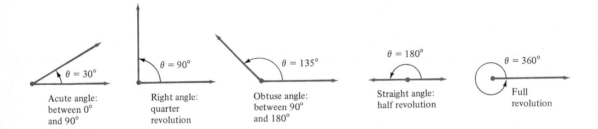

FIGURE 10.3

Also note in Figure 10.3 that we classify angles between $0°$ and $90°$ as **acute** and angles between $90°$ and $180°$ as **obtuse.** Positive angles are measured *counterclockwise* beginning with the initial ray. Negative angles are measured *clockwise*. For instance, Figure 10.4 shows an angle whose measure is $-45°$.

Be sure you understand from Figure 10.4 that merely knowing where an angle's initial and terminal rays are located does not allow us to assign a measure to an angle. To measure an angle we must also know how the terminal ray was revolved. For example, Figure 10.5 shows that the angle measuring $-45°$ has the same terminal ray as the angle measuring $315°$. We call such angles **coterminal.**

FIGURE 10.4     FIGURE 10.5     FIGURE 10.6

Although it may seem strange to consider angles that are larger than 360°, such angles have very useful applications in trigonometry. An angle that is larger than 360° is one whose terminal ray has revolved more than one full revolution counterclockwise. Figure 10.6 shows two angles measuring more than 360°. Similarly, we can generate angles whose measure is less than −360° by revolving a terminal ray more than one full revolution clockwise.

**EXAMPLE 1**

**Finding Coterminal Angles**

For each of the following angles, find a coterminal angle $\theta$ such that $0° \leq \theta < 360°$.

(a) 450°          (b) 750°          (c) −160°          (d) −390°

**SOLUTION**

To find the required coterminal angle, we must add (or subtract) the proper multiple of 360° to bring the result between 0° and 360°. (See Figure 10.7.)

(a) $450° - 360° = 90°$

(b) $750° - 2(360°) = 750° - 720° = 30°$

(c) $-160° + 360° = 200°$

(d) $-390° + 2(360°) = -390° + 720° = 330°$

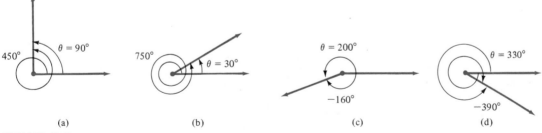

(a)          (b)          (c)          (d)

FIGURE 10.7

One of the most common applications of angle measure occurs in problems dealing with triangles. Many rules from geometry concern triangles. You need to be familiar with several of these, and so we summarize them next.

**Triangles: a summary of some rules**

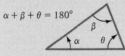

$\alpha + \beta + \theta = 180°$

1. The *sum of the angles* of a triangle is 180°.

$\alpha + \beta = 90°$

2. The *sum of the two acute angles* of a right triangle is 90°.

$a^2 + b^2 = c^2$

3. *Pythagorean Theorem:* The sum of the squares of the (perpendicular) sides of a right triangle is equal to the square of the hypotenuse.

$\dfrac{a}{b} = \dfrac{A}{B}$

4. *Similar Triangles:* If two triangles are similar (have the same angle measures), then the ratios of corresponding sides are equal.

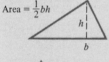

Area $= \dfrac{1}{2} bh$

5. The *area* of a triangle is equal to one-half the base times the height.

6. Each of the angles in an *equilateral triangle* measures 60°.

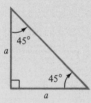

7. Each of the acute angles in an *isosceles right triangle* measures 45°.

## Radian Measure

A second way to measure angles is in terms of radians (rad). This second type of angle measure turns out to be very useful in calculus. (In Section 10.3, you will see why radian measure is preferable to degree measure in calculus.) To assign a radian measure to an angle $\theta$, we consider $\theta$ to be the central angle of a circular sector of radius 1, as shown in Figure 10.8. The radian measure of $\theta$ is then defined to be the

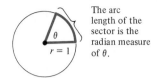

The arc length of the sector is the radian measure of θ.

FIGURE 10.8

length of the arc of this sector. Recall that the total circumference of a circle of radius 1 is given by

$$\text{Circumference} = (2\pi)(\text{radius})$$

Thus, the circumference of a circle of radius 1 is simply $2\pi$, and we may conclude that the radian measure of an angle measuring $360°$ is $2\pi$. In other words,

$$360° = 2\pi \text{ rad}$$

Figure 10.9 gives the radian measure for several common angles.

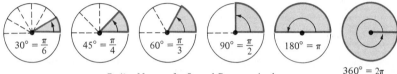

FIGURE 10.9

$30° = \dfrac{\pi}{6}$    $45° = \dfrac{\pi}{4}$    $60° = \dfrac{\pi}{3}$    $90° = \dfrac{\pi}{2}$    $180° = \pi$    $360° = 2\pi$

Radian Measure for Several Common Angles

It is important for you to be able to convert back and forth between the degree and radian measures of angles. We suggest that you memorize the common conversions (such as those pictured in Figure 10.9). For other conversions you can use one of the following conversion rules. (Many scientific calculators have keys programmed to perform these conversions.)

**Conversion rules: degrees ↔ radians**

$$180° = \pi \text{ rad} \qquad 1° = \frac{\pi}{180} \text{ rad} \qquad 1 \text{ rad} = \frac{180°}{\pi}$$

**EXAMPLE 2**

**Converting from Degrees to Radians**

Convert the following degree measures to radian measures.

(a) $135°$         (b) $40°$         (c) $540°$         (d) $-270°$

**SOLUTION**

(a) $135° = (135 \text{ deg})\left(\dfrac{\pi \text{ rad}}{180 \text{ deg}}\right) = \dfrac{3\pi}{4} \text{ rad}$

(b) $40° = (40 \text{ deg})\left(\dfrac{\pi \text{ rad}}{180 \text{ deg}}\right) = \dfrac{2\pi}{9} \text{ rad}$

(c) $540° = (540 \text{ deg})\left(\dfrac{\pi \text{ rad}}{180 \text{ deg}}\right) = 3\pi \text{ rad}$

(d) $-270° = (-270 \text{ deg})\left(\dfrac{\pi \text{ rad}}{180 \text{ deg}}\right) = -\dfrac{3\pi}{2} \text{ rad}$

## EXAMPLE 3

**Converting from Radians to Degrees**

Convert the following radian measures to degree measures.

(a) $-\dfrac{\pi}{2}$      (b) $\dfrac{7\pi}{4}$      (c) $\dfrac{11\pi}{6}$      (d) $\dfrac{9\pi}{2}$

**SOLUTION**

(a) $-\dfrac{\pi}{2}$ rad $= \left(-\dfrac{\pi}{2}\text{ rad}\right)\left(\dfrac{180\text{ deg}}{\pi\text{ rad}}\right) = -90°$

(b) $\dfrac{7\pi}{4}$ rad $= \left(\dfrac{7\pi}{4}\text{ rad}\right)\left(\dfrac{180\text{ deg}}{\pi\text{ rad}}\right) = 315°$

(c) $\dfrac{11\pi}{6}$ rad $= \left(\dfrac{11\pi}{6}\text{ rad}\right)\left(\dfrac{180\text{ deg}}{\pi\text{ rad}}\right) = 330°$

(d) $\dfrac{9\pi}{2}$ rad $= \left(\dfrac{9\pi}{2}\text{ rad}\right)\left(\dfrac{180\text{ deg}}{\pi\text{ rad}}\right) = 810°$

## SECTION EXERCISES 10.1

In Exercises 1–4, determine two coterminal angles (one positive and one negative) for the given angle. Give the answers in degrees.

**1.** (a)

$\theta = 36°$

(b)

$\theta = -45°$

**2.** (a)

$\theta = -120°$

(b)

$\theta = 390°$

**3.** (a)

$\theta = 300°$

(b)

$\theta = 740°$

**4.** (a)

$\theta = -420°$

(b)

$\theta = 230°$

In Exercises 5–8, determine two coterminal angles (one positive and one negative) for the given angle. Give the answers in radians.

**5.** (a)

$\theta = \dfrac{\pi}{9}$

(b)

$\theta = \dfrac{4\pi}{3}$

**6.** (a)

$\theta = \dfrac{11\pi}{6}$

(b)

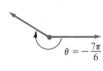
$\theta = -\dfrac{7\pi}{6}$

**7.** (a)

$\theta = -\dfrac{9\pi}{4}$

(b)

$\theta = -\dfrac{2\pi}{15}$

**8.** (a)

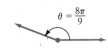
$\theta = \dfrac{8\pi}{9}$

(b)

$\theta = \dfrac{8\pi}{45}$

In Exercises 9–16, express the given angle in radian measure as a multiple of $\pi$.

9. 30°
10. 150°
11. 315°
12. 120°
13. −20°
14. −240°
15. −270°
16. 144°

In Exercises 17–24, express the given angle in degree measure.

17. $\dfrac{3\pi}{2}$
18. $\dfrac{7\pi}{6}$
19. $-\dfrac{7\pi}{12}$
20. $\dfrac{\pi}{9}$
21. $\dfrac{7\pi}{3}$
22. $-\dfrac{11\pi}{30}$
23. $\dfrac{11\pi}{6}$
24. $\dfrac{34\pi}{15}$

In Exercises 25–30, solve the triangle for the indicated side and/or angle.

25.

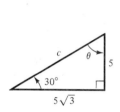

26.

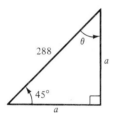

27.

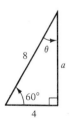

28.

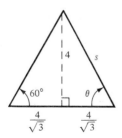

29.

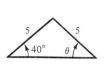

30.

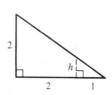

31. A person 6 feet tall standing 12 feet from a streetlight casts a shadow 8 feet long. (See Figure 10.10.) What is the height of the streetlight?

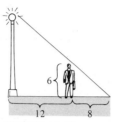

FIGURE 10.10

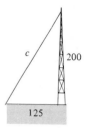

FIGURE 10.11

32. A guy wire is stretched from a broadcasting tower at a point 200 feet above the ground to an anchor 125 feet from the base. (See Figure 10.11.) How long is the wire?

33. Let $r$ represent the radius of a circle, $\theta$ the central angle (measured in radians), and $s$ the length of the arc subtended by the angle, as shown in Figure 10.12. Use the relationship $\theta = s/r$ to complete the following table.

| $r$ | 8 ft | 15 in | 85 cm | | |
|---|---|---|---|---|---|
| $s$ | 12 ft | | | 96 in | 8642 mi |
| $\theta$ | | 1.6 | $\dfrac{3\pi}{4}$ | 4 | $\dfrac{2\pi}{3}$ |

Circle of radius $r$

FIGURE 10.12

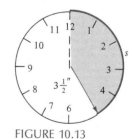

FIGURE 10.13

34. The minute hand on a clock is $3\frac{1}{2}$ inches long, as shown in Figure 10.13. Through what distance does the tip of the minute hand move in 25 minutes?

35. A man bends his elbow through 75°. The distance from his elbow to the tip of his index finger is $18\frac{3}{4}$ inches, as shown in Figure 10.14.
   (a) Find the radian measure of this angle.
   (b) Find the distance the tip of the index finger moves.

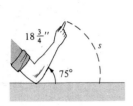

FIGURE 10.14

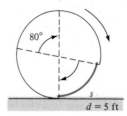

FIGURE 10.15

**37.** Assuming that the earth is a sphere of radius 4000 miles, what is the difference in latitude of two cities, one of which is 325 miles due north of the other? (Latitude lines on the earth run parallel to the equator and measure the angle from the equator to a point north or south of the equator.)

**38.** A car is moving at the rate of 45 mi/hr (66 ft/sec), and the diameter of each wheel is 2 feet. Find the number of revolutions per minute that the wheels make and the number of radians per second generated by a line segment from the center to the circumference of each wheel.

**36.** A tractor tire that is 5 feet in diameter is partially filled with a liquid ballast for additional traction. To check the air pressure, the tractor operator rotates the tire until the valve stem is at the top so that the liquid will not enter the gauge. On a given occasion the operator notes that the tire must be rotated 80° to have the stem in the proper position.
(a) Find the radian measure of this rotation.
(b) How far must the tractor be moved to get the valve stem in the proper position?

# The Trigonometric Functions

INTRODUCTORY EXAMPLE
## Peripheral Vision

When you look straight ahead, how great an arc can you see? How can we measure this angle of peripheral vision?

One way would be to stand at the center of a large eye-level circle that is marked off in degrees, as shown in Figure 10.16(a). While looking straight ahead, you could turn until your right eye could just barely see the 0° marker. At that point, the largest degree marker your left eye could read would represent the angle of your peripheral vision. This solution requires little mathematics, but unfortunately it does require the construction of a fairly elaborate structure.

A simpler solution (at least simpler physically) would be to stand facing the corner of a room and measure how far back along the wall you could see. Suppose that a person standing 1 foot from the corner can see an object located 2 feet down the wall, as shown in Figure 10.16(b).

Using the Pythagorean Theorem, we can determine the values of $x$ and $y$, as shown in Figure 10.17.

In this section, we will see that we can determine either of the acute angles of a right triangle if we know the ratio of any two sides of the triangle. The study of these ratios is called **trigonometry.** In this particular example, we call the ratio of $y$ to $x$ the **tangent** of the angle $\theta$ and write

$$\tan \theta = \frac{y}{x}$$

Since we know the values of $x$ and $y$, we can solve the problem by solving the equation

$$\tan \theta = \frac{\sqrt{2}}{\sqrt{2} - 1} \approx 3.414$$

The solution to this equation is $\theta \approx 73.7°$, which means that $\alpha/2 \approx 106.3°$ and $\alpha \approx 212.6°$. Thus, the person's angle of peripheral vision is about 212.6°.

(a)

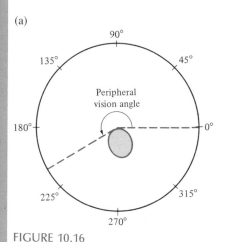

(b)

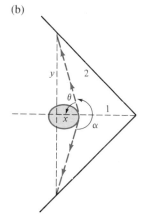

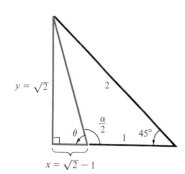

FIGURE 10.16

FIGURE 10.17

- **The Six Trigonometric Functions**
- **Trigonometric Identities**
- **Evaluation of Trigonometric Functions**
- **Solving Trigonometric Equations**

There are two common approaches to the study of trigonometry. In one case the trigonometric functions are defined as ratios of two sides of a right triangle. In the other case these functions are defined in terms of a point on the terminal side of an arbitrary angle. The first approach is the one generally used in surveying, navigation, and astronomy, where a typical problem involves a fixed triangle having three of its six parts (sides and angles) known and three to be determined. The second approach is the one normally used in physics, electronics, and biology, where the periodic nature of the trigonometric functions is emphasized. In the following definition we define the six trigonometric functions from both viewpoints.

**Definition of the six trigonometric functions**

*Right Triangle Definition:* $0 < \theta < \frac{\pi}{2}$. Refer to Figure 10.18.

$$\sin \theta = \frac{\text{opp.}}{\text{hyp.}} \qquad \csc \theta = \frac{\text{hyp.}}{\text{opp.}}$$

$$\cos \theta = \frac{\text{adj.}}{\text{hyp.}} \qquad \sec \theta = \frac{\text{hyp.}}{\text{adj.}}$$

$$\tan \theta = \frac{\text{opp.}}{\text{adj.}} \qquad \cot \theta = \frac{\text{adj.}}{\text{opp.}}$$

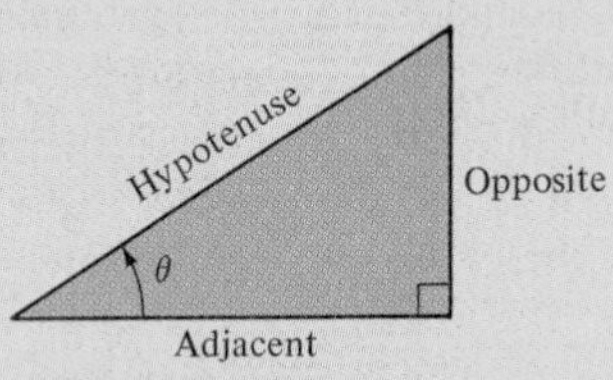

FIGURE 10.18

*Circular Function Definition:* $\theta$ is any angle in standard position and $(x, y)$ is a point on the terminal ray of the angle. Refer to Figure 10.19.

$$\sin \theta = \frac{y}{r} \qquad \csc \theta = \frac{r}{y}$$

$$\cos \theta = \frac{x}{r} \qquad \sec \theta = \frac{r}{x}$$

$$\tan \theta = \frac{y}{x} \qquad \cot \theta = \frac{x}{y}$$

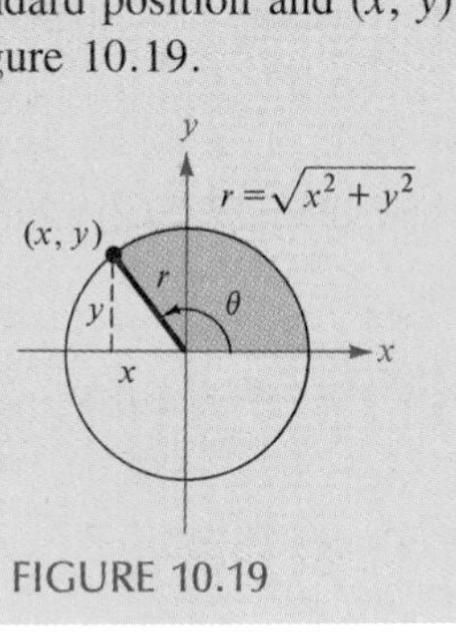

FIGURE 10.19

Note in Figure 10.19 that $r$ is always positive, and thus the quadrant signs of $x$ and $y$ determine the quadrant signs of the various trigonometric functions, as indi-

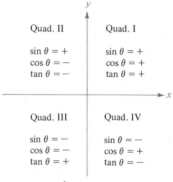

FIGURE 10.20

cated in Figure 10.20. Although the six trigonometric functions are usually represented by their three-letter abbreviations, their full names are actually

| | |
|---|---|
| sin: sine | csc: cosecant |
| cos: cosine | sec: secant |
| tan: tangent | cot: cotangent |

The following trigonometric identities are direct consequences of the definitions.

$$\csc \theta = \frac{1}{\sin \theta} \qquad \sec \theta = \frac{1}{\cos \theta} \qquad \cot \theta = \frac{1}{\tan \theta}$$

$$\tan \theta = \frac{\sin \theta}{\cos \theta} \qquad \cot \theta = \frac{\cos \theta}{\sin \theta}$$

Furthermore, since

$$\sin^2 \theta + \cos^2 \theta = \left(\frac{y}{r}\right)^2 + \left(\frac{x}{r}\right)^2$$

$$= \frac{x^2 + y^2}{r^2}$$

$$= \frac{r^2}{r^2} = 1$$

we can readily obtain the Pythagorean Identity

$$\sin^2 \theta + \cos^2 \theta = 1$$

[Note that we use $\sin^2 \theta$ to mean $(\sin \theta)^2$.] Additional trigonometric identities are
listed next, without proof.

*Pythagorean Identities:*

$$\sin^2 \theta + \cos^2 \theta = 1$$

$$\tan^2 \theta + 1 = \sec^2 \theta$$

$$\cot^2 \theta + 1 = \csc^2 \theta$$

*Sum or Difference of Two Angles:*[*]

$$\sin(\theta \pm \phi) = \sin \theta \cos \phi \pm \cos \theta \sin \phi$$

$$\cos(\theta \pm \phi) = \cos \theta \cos \phi \mp \sin \theta \sin \phi$$

$$\tan(\theta \pm \phi) = \frac{\tan \theta \pm \tan \phi}{1 \mp \tan \theta \tan \phi}$$

*Reduction Formulas:*

$$\sin(-\theta) = -\sin \theta$$

$$\cos(-\theta) = \cos \theta$$

$$\tan(-\theta) = -\tan \theta$$

$$\sin \theta = -\sin(\theta - \pi)$$

$$\cos \theta = -\cos(\theta - \pi)$$

$$\tan \theta = \tan(\theta - \pi)$$

*Double Angle:*

$$\sin 2\theta = 2 \sin \theta \cos \theta$$

$$\cos 2\theta = 2 \cos^2 \theta - 1 = 1 - 2 \sin^2 \theta$$

*Half Angle:*

$$\sin^2 \theta = \frac{1}{2}(1 - \cos 2\theta)$$

$$\cos^2 \theta = \frac{1}{2}(1 + \cos 2\theta)$$

▬▬ **Remark:** Although an angle can be measured in either degrees or radians, we have already mentioned that in calculus radian measure is preferred. Thus, all angles in the remainder of this chapter are assumed to be measured in radians *unless marked otherwise*. In other words, when we write sin 3 we mean the sine of three radians, and when we write sin 3° we mean the sine of three degrees.

### Evaluation of Trigonometric Functions

There are two common methods of evaluating trigonometric functions: decimal approximations using a calculator (or a table of values) and exact evaluations using trigonometric identities and formulas from geometry. The next three examples illustrate the second method. Study these examples carefully.

**EXAMPLE 1**

**Evaluating Trigonometric Functions**

Evaluate the sine, cosine, and tangent of $\pi/3$.

**SOLUTION**

We begin by drawing the angle $\theta = \pi/3$ in the standard position, as shown in Figure 10.21. Then, since $\pi/3$ rad $= 60°$, we imagine an equilateral triangle with sides of length 1 and with $\theta$ as one of its angles. Since the altitude of this triangle bisects its base, we know that

---

[*]Note that $\phi$ is the lowercase Greek letter phi.

$$x = \frac{1}{2}$$

Now, using the Pythagorean Theorem, we have

$$y = \sqrt{r^2 - x^2} = \sqrt{1 - \left(\frac{1}{2}\right)^2} = \sqrt{\frac{3}{4}} = \frac{\sqrt{3}}{2}$$

Finally, we have

$$\sin \frac{\pi}{3} = \frac{y}{r} = \frac{\sqrt{3}/2}{1} = \frac{\sqrt{3}}{2}$$

$$\cos \frac{\pi}{3} = \frac{x}{r} = \frac{1/2}{1} = \frac{1}{2}$$

$$\tan \frac{\pi}{3} = \frac{y}{x} = \frac{\sqrt{3}/2}{1/2} = \sqrt{3}$$

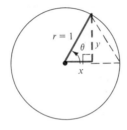 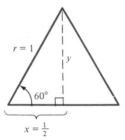

FIGURE 10.21

The degree and radian measures of several common angles are given in Table 10.1, along with the corresponding values of sine, cosine, and tangent.

TABLE 10.1

| Degrees | 0° | 30° | 45° | 60° | 90° |
|---------|-----|-----|-----|-----|-----|
| Radians | 0 | $\dfrac{\pi}{6}$ | $\dfrac{\pi}{4}$ | $\dfrac{\pi}{3}$ | $\dfrac{\pi}{2}$ |
| $\sin \theta$ | 0 | $\dfrac{1}{2}$ | $\dfrac{\sqrt{2}}{2}$ | $\dfrac{\sqrt{3}}{2}$ | 1 |
| $\cos \theta$ | 1 | $\dfrac{\sqrt{3}}{2}$ | $\dfrac{\sqrt{2}}{2}$ | $\dfrac{1}{2}$ | 0 |
| $\tan \theta$ | 0 | $\dfrac{1}{\sqrt{3}}$ | 1 | $\sqrt{3}$ | undefined |

As an aid to remembering the values in Table 10.1, note the pattern for sin $\theta$ and cos $\theta$ shown in the following table.

| $\theta$ | 0° | 30° | 45° | 60° | 90° |
|---|---|---|---|---|---|
| sin $\theta$ | $\dfrac{\sqrt{0}}{2}$ | $\dfrac{\sqrt{1}}{2}$ | $\dfrac{\sqrt{2}}{2}$ | $\dfrac{\sqrt{3}}{2}$ | $\dfrac{\sqrt{4}}{2}$ |
| cos $\theta$ | $\dfrac{\sqrt{4}}{2}$ | $\dfrac{\sqrt{3}}{2}$ | $\dfrac{\sqrt{2}}{2}$ | $\dfrac{\sqrt{1}}{2}$ | $\dfrac{\sqrt{0}}{2}$ |

Using the values in Table 10.1, you can determine the corresponding values for the tangent by using the identity

$$\tan \theta = \frac{\sin \theta}{\cos \theta}$$

To extend the use of the table to angles in quadrants other than the first quadrant, you can use the concept of a reference angle (as shown in Figure 10.22), together with the appropriate quadrant sign. The **reference angle** for an angle $\theta$ is the smallest positive angle between the terminal side of $\theta$ and the $x$-axis. For instance, the reference angle for 135° is 45° since

$$135° + 45° = 180°$$

and the reference angle for 210° is 30° since

$$210° - 30° = 180°$$

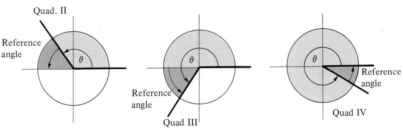

Reference Angle: $\pi - \theta$     Reference Angle: $\theta - \pi$     Reference Angle: $2\pi - \theta$

FIGURE 10.22

**EXAMPLE 2**

**Evaluating Trigonometric Functions**

Evaluate the following trigonometric functions:

(a) $\sin \dfrac{3\pi}{4}$          (b) $\tan 330°$          (c) $\cos \dfrac{7\pi}{6}$

**SOLUTION**

(a) Since the reference angle for $3\pi/4$ is $\pi/4$ and the sine is positive in the second quadrant (see Figure 10.23), we have

$$\sin \frac{3\pi}{4} = \sin \frac{\pi}{4} = \frac{\sqrt{2}}{2}$$

(b) Since the reference angle for 330° is 30° and the tangent is negative in the fourth quadrant, we have

$$\tan 330° = -\tan 30° = -\frac{1}{\sqrt{3}} = -\frac{\sqrt{3}}{3}$$

(c) Since the reference angle for $7\pi/6$ is $\pi/6$ and the cosine is negative in the third quadrant, we have

$$\cos \frac{7\pi}{6} = -\cos \frac{\pi}{6} = -\frac{\sqrt{3}}{2}$$

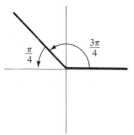
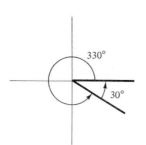
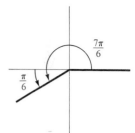

FIGURE 10.23

**EXAMPLE 3**

**Using Identities to Evaluate Trigonometric Functions**
Evaluate the following trigonometric functions:

(a) $\sin\left(-\frac{\pi}{3}\right)$      (b) $\sec 60°$      (c) $\cos 15°$

(d) $\sin 2\pi$      (e) $\cot 0°$      (f) $\tan \frac{9\pi}{4}$

**SOLUTION**

(a) Using the Reduction Formula,

$$\sin (-\theta) = -\sin \theta$$

we have

$$\sin\left(-\frac{\pi}{3}\right) = -\sin \frac{\pi}{3}$$

$$= -\frac{\sqrt{3}}{2}$$

(b) Using the Reciprocal Formula,

$$\sec \theta = \frac{1}{\cos \theta}$$

we have

$$\sec 60° = \frac{1}{\cos 60°}$$

$$= \frac{1}{1/2} = 2$$

(c) Using the Difference Formula,

$$\cos (\theta - \phi) = \cos \theta \cos \phi + \sin \theta \sin \phi$$

we have

$$\cos 15° = \cos (45° - 30°)$$

$$= (\cos 45°)(\cos 30°) + (\sin 45°)(\sin 30°)$$

$$= \left(\frac{\sqrt{2}}{2}\right)\left(\frac{\sqrt{3}}{2}\right) + \left(\frac{\sqrt{2}}{2}\right)\left(\frac{1}{2}\right)$$

$$= \frac{\sqrt{6} + \sqrt{2}}{4}$$

(d) Since the reference angle for $2\pi$ is 0, we have

$$\sin 2\pi = \sin 0 = 0$$

(e) Using the Reciprocal Formula,

$$\cot \theta = \frac{1}{\tan \theta}$$

we can see that $\cot 0°$ is *undefined* (since $\tan 0° = 0$).

(f) Since the reference angle for $9\pi/4$ is $\pi/4$ and the tangent is positive in the first quadrant, we have

$$\tan \frac{9\pi}{4} = \tan \frac{\pi}{4} = 1$$

---

Up to this point we have restricted our trigonometric evaluations to "standard" angles. How can we evaluate trigonometric functions of nonstandard angles? For example, what is the sine of $17°$ or the cosine of $\pi/11$? In such cases it is best to use decimal approximations obtained with a calculator. When doing this, be sure to remember to distinguish between degree and radian measure. Most calculators have a degree/radian key to distinguish between these two types of angle measure. Furthermore, most calculators have only three trigonometric function keys: sin, cos, and tan. To evaluate the other three functions, we combine one of these keys with the reciprocal key. For example, we can find the secant of 10° by pressing a sequence of four keys:

| Problem | Calculator steps | Display |
|---|---|---|
| To find sec 10° | $\boxed{1}$ $\boxed{0}$ $\boxed{\cos}$ $\boxed{1/x}$ | 1.015426611 |

We demonstrate an additional use of a calculator in the following example.

**EXAMPLE 4**

**Using Trigonometric Functions to Solve Triangles**

A surveyor is standing 50 feet away from the base of a large tree. The surveyor measures the angle of elevation to the top of the tree as 71.5°. How tall is the tree?

**SOLUTION**

Referring to Figure 10.24, we see that

$$\tan 71.5° = \frac{y}{x}$$

where $x = 50$ and $y$ is the height of the tree. Thus, we can determine the height as follows:

$$y = (x)(\tan 71.5°) \approx (50)(2.98868) \approx 149.4 \text{ feet}$$

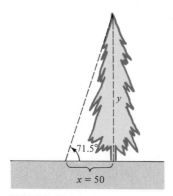

**FIGURE 10.24**

**Solving Trigonometric Equations**

To conclude this section, we look at some techniques for solving trigonometric equations. For example, consider the equation

$$\sin \theta = 0$$

We know $\theta = 0$ is one solution. Also, in Example 3 part (d), we saw that $\theta = 2\pi$ is another solution. But these are not the only solutions. Actually, there are infinitely many solutions to this equation! Any one of the following values will work:

$$\ldots, -3\pi, -2\pi, -\pi, 0, \pi, 2\pi, 3\pi, \ldots$$

To simplify the situation, we usually restrict our search for solutions to the interval $0 \le \theta \le 2\pi$, as shown in the following example.

**EXAMPLE 5**

**Solving Trigonometric Equations**

Solve for $\theta$ in the following equations. (In each case restrict the values of $\theta$ to the interval $0 \le \theta \le 2\pi$.)

(a) $\sin \theta = -\dfrac{\sqrt{3}}{2}$ 　　　　　(b) $\cos \theta = 1$ 　　　　　(c) $\tan \theta = 1$

**SOLUTION**

(a) To solve the equation

$$\sin \theta = -\frac{\sqrt{3}}{2}$$

we make two observations: the sine is negative in the third and fourth quadrants, and

$$\sin \frac{\pi}{3} = \frac{\sqrt{3}}{2}$$

Combining these two observations, we conclude that we are seeking values of $\theta$ in the third and fourth quadrants that have a reference angle of $\pi/3$, as shown in Figure 10.25. The two angles fitting these criteria are

$$\theta = \pi + \frac{\pi}{3} = \frac{4\pi}{3}$$

and

$$\theta = 2\pi - \frac{\pi}{3} = \frac{5\pi}{3}$$

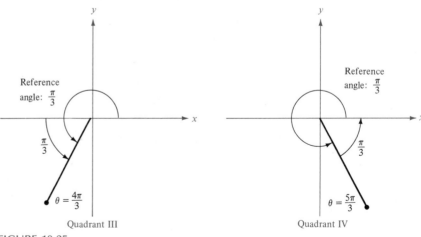

FIGURE 10.25

(b) To solve $\cos \theta = 1$, we observe that $\cos 0 = 1$ and note that in the interval $[0, 2\pi]$ the only angles whose reference angles are 0 are $0$, $\pi$, and $2\pi$. Since

$$\cos \pi = -1 \qquad \text{and} \qquad \cos 2\pi = 1$$

we conclude that there are two solutions:

$$\theta = 0 \qquad \text{or} \qquad \theta = 2\pi$$

(c) Since $\tan(\pi/4) = 1$ and the tangent is positive in the first and third quadrants, we have

$$\theta = \frac{\pi}{4} \qquad \text{or} \qquad \theta = \pi + \frac{\pi}{4} = \frac{5\pi}{4}$$

## EXAMPLE 6

**Solving Trigonometric Equations**

Solve the following equation for $\theta$:

$$\cos 2\theta = 2 - 3 \sin \theta, \quad 0 \le \theta \le 2\pi$$

## SOLUTION

Using the double angle identity

$$\cos 2\theta = 1 - 2 \sin^2 \theta$$

we have the following polynomial (in $\sin \theta$):

$$1 - 2 \sin^2 \theta = 2 - 3 \sin \theta$$

$$0 = 2 \sin^2 \theta - 3 \sin \theta + 1$$

$$0 = (2 \sin \theta - 1)(\sin \theta - 1) \qquad \text{See following Remark}$$

If $2 \sin \theta - 1 = 0$, we have $\sin \theta = \frac{1}{2}$ and

$$\theta = \frac{\pi}{6} \qquad \text{or} \qquad \theta = \frac{5\pi}{6}$$

If $\sin \theta - 1 = 0$, we have $\sin \theta = 1$ and

$$\theta = \frac{\pi}{2}$$

Thus, for $0 \le \theta \le 2\pi$, there are three solutions to the given equation:

$$\theta = \frac{\pi}{6}, \frac{\pi}{2}, \frac{5\pi}{6}$$

**Remark:** In Example 6, we factored the quadratic equation involving $\sin \theta$ just as if the sine of $\theta$ were a single unknown variable. For instance, if we let $\sin \theta = x$, then the factorization would have taken the standard algebraic form

$$2x^2 - 3x + 1 = (2x - 1)(x - 1)$$

## SECTION EXERCISES 10.2

In Exercises 1–6, determine all six trigonometric functions for the given angle $\theta$.

**1.**

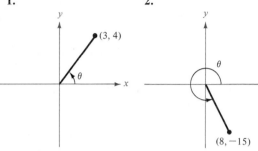

**2.**

**9.** Given $\cos \theta = \dfrac{4}{5}$;

find $\cot \theta$

**10.** Given $\sec \theta = \dfrac{13}{5}$;

find $\cot \theta$

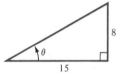

**11.** Given $\cot \theta = \dfrac{15}{8}$;

find $\sec \theta$

**12.** Given $\tan \theta = \dfrac{1}{2}$;

find $\sin \theta$

**3.**

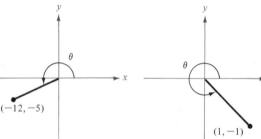

**4.**

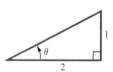

In Exercises 13–18, sketch a right triangle corresponding to the trigonometric function of the angle $\theta$ and find the other five trigonometric functions of $\theta$.

**13.** $\sin \theta = \dfrac{2}{3}$      **14.** $\cot \theta = 5$

**15.** $\sec \theta = 2$      **16.** $\cos \theta = \dfrac{5}{7}$

**17.** $\tan \theta = 3$      **18.** $\csc \theta = 4.25$

**5.**

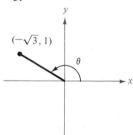

**6.**

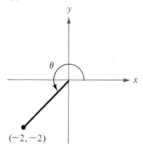

In Exercises 19–24, determine the quadrant in which $\theta$ lies.

**19.** $\sin \theta < 0$ and $\cos \theta < 0$
**20.** $\sin \theta > 0$ and $\cos \theta < 0$
**21.** $\sin \theta > 0$ and $\sec \theta > 0$
**22.** $\cot \theta < 0$ and $\cos \theta > 0$
**23.** $\csc \theta > 0$ and $\tan \theta < 0$
**24.** $\cos \theta > 0$ and $\tan \theta < 0$

In Exercises 7–12, find the indicated trigonometric function from the given one.

**7.** Given $\sin \theta = \dfrac{1}{2}$;

find $\csc \theta$

**8.** Given $\sin \theta = \dfrac{1}{3}$;

find $\tan \theta$

In Exercises 25–32, evaluate the sine, cosine, and tangent of the given angles *without* using a calculator.

**25.** (a) $60°$      (b) $\dfrac{2\pi}{3}$

**26.** (a) $\dfrac{\pi}{4}$      (b) $\dfrac{5\pi}{4}$

**27.** (a) $-\dfrac{\pi}{6}$      (b) $150°$

**28.** (a) $-\dfrac{\pi}{2}$      (b) $\dfrac{\pi}{2}$

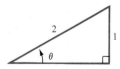

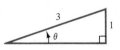

**29.** (a) 225°          (b) −225°

**30.** (a) 300°          (b) 330°

**31.** (a) 750°          (b) 510°

**32.** (a) $\dfrac{10\pi}{3}$          (b) $\dfrac{17\pi}{3}$

In Exercises 33–40, use a calculator to evaluate the given trigonometric functions to four decimal places.

**33.** (a) sin 10°          (b) csc 10°

**34.** (a) sec 225°          (b) sec 135°

**35.** (a) $\tan \dfrac{\pi}{9}$          (b) $\tan \dfrac{10\pi}{9}$

**36.** (a) cot 1.35          (b) tan 1.35

**37.** (a) cos (−110°)          (b) cos 250°

**38.** (a) tan 240°          (b) cot 210°

**39.** (a) csc 2.62          (b) csc 150°

**40.** (a) sin (−0.65)          (b) sin 5.63

In Exercises 41–46, find two values of $\theta$ corresponding to the given functions. List the measure of $\theta$ in radians ($0 \le \theta \le 2\pi$). Do not use a calculator.

**41.** (a) $\sin \theta = \dfrac{1}{2}$          (b) $\sin \theta = -\dfrac{1}{2}$

**42.** (a) $\cos \theta = \dfrac{\sqrt{2}}{2}$          (b) $\cos \theta = -\dfrac{\sqrt{2}}{2}$

**43.** (a) $\csc \theta = \dfrac{2\sqrt{3}}{3}$          (b) $\cot \theta = -1$

**44.** (a) $\sec \theta = 2$          (b) $\sec \theta = -2$

**45.** (a) $\tan \theta = 1$          (b) $\cot \theta = -\sqrt{3}$

**46.** (a) $\sin \theta = \dfrac{\sqrt{3}}{2}$          (b) $\sin \theta = -\dfrac{\sqrt{3}}{2}$

In Exercises 47–52, use the tables in the appendix to estimate two values of $\theta$ corresponding to the given functions. List the measure of $\theta$ in degrees ($0 \le \theta \le 360°$) and radians ($0 \le \theta \le 2\pi$).

**47.** (a) sin $\theta = 0.8191$          (b) sin $\theta = -0.2589$

**48.** (a) sin $\theta = 0.0175$          (b) sin $\theta = -0.6691$

**49.** (a) cos $\theta = 0.9848$          (b) cos $\theta = -0.5890$

**50.** (a) cos $\theta = 0.8746$          (b) cos $\theta = -0.2419$

**51.** (a) tan $\theta = 1.192$          (b) tan $\theta = -8.144$

**52.** (a) cot $\theta = 5.671$          (b) cot $\theta = -1.280$

In Exercises 53–62, solve the given equation for $\theta$ ($0 \le \theta \le 2\pi$). For some of the equations you should use the trigonometric identities listed in this section.

**53.** $2 \sin^2 \theta = 1$          **54.** $\tan^2 \theta = 3$

**55.** $\tan^2 \theta - \tan \theta = 0$          **56.** $2 \cos^2 \theta - \cos \theta = 1$

**57.** $\sin 2\theta - 2 \sin \theta = 0$          **58.** $\sec \theta \csc \theta = 2 \csc \theta$

**59.** $\sin \theta = \cos \theta$          **60.** $\cos 2\theta + 3 \cos \theta + 2 = 0$

**61.** $\cos^2 \theta + \sin \theta = 1$          **62.** $\cos \dfrac{\theta}{2} - \cos \theta = 1$

In Exercises 63–70, solve for $x$, $y$, or $r$ as indicated.

**63.** Solve for $y$.          **64.** Solve for $x$.

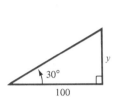
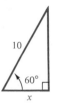

**65.** Solve for $x$.          **66.** Solve for $r$.

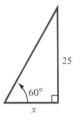
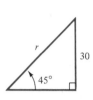

**67.** Solve for $r$.          **68.** Solve for $x$.

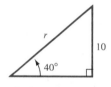
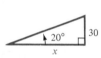

**69.** Solve for $y$.          **70.** Solve for $r$.

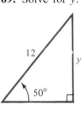
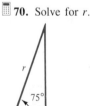

**71.** A 20-foot ladder leaning against the side of a house makes a 75° angle with the ground, as shown in Figure 10.26. How far up the side of the house does the ladder reach?

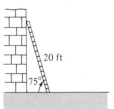

FIGURE 10.26

**72.** A biologist wants to know the width $w$ of a river in order to properly set instruments to study the pollutants in the water. From point $A$ the biologist walks downstream 100 feet and sights to point $C$. From this sighting it is determined that $\theta = 50°$, as shown in Figure 10.27. How wide is the river?

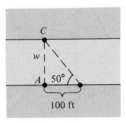

FIGURE 10.27

**73.** From a 150-foot observation tower on the coast, a Coast Guard officer sights a boat in difficulty. The angle of depression of the boat is 4°, as shown in Figure 10.28. How far is the boat from the shoreline?

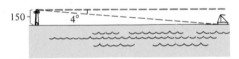

FIGURE 10.28

**74.** A ramp $17\frac{1}{2}$ feet in length rises to a loading platform that is $3\frac{1}{2}$ feet off the ground, as shown in Figure 10.29. Find the angle that the ramp makes with the ground.

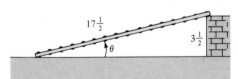

FIGURE 10.29

**75.** The average daily temperature (in degrees Fahrenheit) for a certain city is given by

$$T = 45 - 23 \cos\left[\frac{2\pi}{365}(t - 32)\right]$$

where $t$ is the time in days, with $t = 1$ corresponding to January 1. Find the average temperature on the following days:
(a) January 1
(b) July 4 ($t = 185$)
(c) October 18 ($t = 291$)

**76.** A company that produces a seasonal product forecasts monthly sales over the next 2 years to be

$$S = 23.1 + 0.442t + 4.3 \sin\frac{\pi t}{6}$$

where $S$ is measured in thousands of units and $t$ is the time in months, with $t = 1$ representing January 1987. Predict sales for the following months:
(a) February 1987    (b) February 1988
(c) September 1987    (d) September 1988

# Graphs of Trigonometric Functions

**INTRODUCTORY EXAMPLE**
**Biorhythms**

A popular theory that attempts to explain the ups and downs of everyday life states that each of us has three cycles, which begin at birth. These three *biorhythm cycles* have different periods.

The cycles are at their average energy levels at birth. Then the cycles begin increasing until they reach their peak energy levels during the 6th, 7th, and 8th days of life. Thereafter, the cycles begin dropping until they reach their lowest energy levels from the 16th through the 25th days of life. As the cycles become more out of phase, a person will have days in which one cycle is up and the other two are down, days in which two cycles are up and one is down, and so on.

The energy levels for each cycle can be represented as a sine function:

Physical (23 days):  $P(t) = \sin \dfrac{2\pi t}{23}$

Emotional (28 days):  $E(t) = \sin \dfrac{2\pi t}{28}$

Intellectual (33 days):  $I(t) = \sin \dfrac{2\pi t}{33}$

where $t$ represents the number of days since birth. For example, Figure 10.30 shows the September 1984 biorhythm graphs for a person born on 20 July 1964. Note that September had a dismal beginning for this person, but by the 18th through the 20th of the month the person's cycles were all at high energy levels. To compute the biorhythm energy levels for any particular day, we need to know that day's $t$-value. For instance, 1 September 1984 was the 7349th day of this person's life, and we could calculate the energy levels as follows:

$$P(7349) = \sin \frac{2\pi(7349)}{23} \approx -0.14$$

$$E(7349) = \sin \frac{2\pi(7349)}{28} \approx 0.22$$

$$I(7349) = \sin \frac{2\pi(7349)}{33} \approx -0.95$$

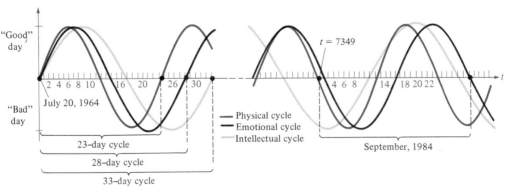

FIGURE 10.30

- **Graphs of Trigonometric Functions**
- **Period of a Trigonometric Function**
- **Amplitude of Sine and Cosine Function**

In the beginning of Section 10.2, we mentioned that there are two common approaches to the study of trigonometry: the right-triangle approach and the periodic-function approach. In Section 10.2, we stressed the right-triangle approach, and we followed the standard convention of using $\theta$ to represent one of the acute angles of a right triangle. In this section, we look at the second approach, which stresses the periodic nature of the six trigonometric functions. Since we are emphasizing the functional nature of trigonometry, we use $x$ (rather than $\theta$) to represent the angle and write

$$f(x) = \sin x$$

where $x$ is measured in radians.

We begin by examining the graph of the sine function. Table 10.2 gives the values for the sine for several values of $x$ between 0 and $2\pi$.

TABLE 10.2

| $x$ | 0 | $\frac{\pi}{6}$ | $\frac{\pi}{4}$ | $\frac{\pi}{3}$ | $\frac{\pi}{2}$ | $\frac{2\pi}{3}$ | $\frac{3\pi}{4}$ | $\frac{5\pi}{6}$ | $\pi$ | $\frac{7\pi}{6}$ | $\frac{5\pi}{4}$ | $\frac{4\pi}{3}$ | $\frac{3\pi}{2}$ | $\frac{5\pi}{3}$ | $\frac{7\pi}{4}$ | $\frac{11\pi}{6}$ | $2\pi$ |
|---|---|---|---|---|---|---|---|---|---|---|---|---|---|---|---|---|---|
| $\sin x$ | 0 | $\frac{1}{2}$ | $\frac{\sqrt{2}}{2}$ | $\frac{\sqrt{3}}{2}$ | 1 | $\frac{\sqrt{3}}{2}$ | $\frac{\sqrt{2}}{2}$ | $\frac{1}{2}$ | 0 | $-\frac{1}{2}$ | $-\frac{\sqrt{2}}{2}$ | $-\frac{\sqrt{3}}{2}$ | $-1$ | $-\frac{\sqrt{3}}{2}$ | $-\frac{\sqrt{2}}{2}$ | $-\frac{1}{2}$ | 0 |

Now, by plotting these points and connecting them with a smooth curve, we get the graph shown in Figure 10.31. Note in Figure 10.31 that the maximum value of $\sin x$ is 1 and the minimum value is $-1$. The **amplitude** of the sine function (or of the cosine function) is defined to be half of the difference between its maximum and

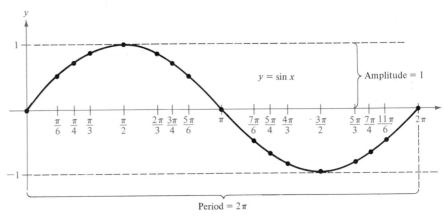

FIGURE 10.31

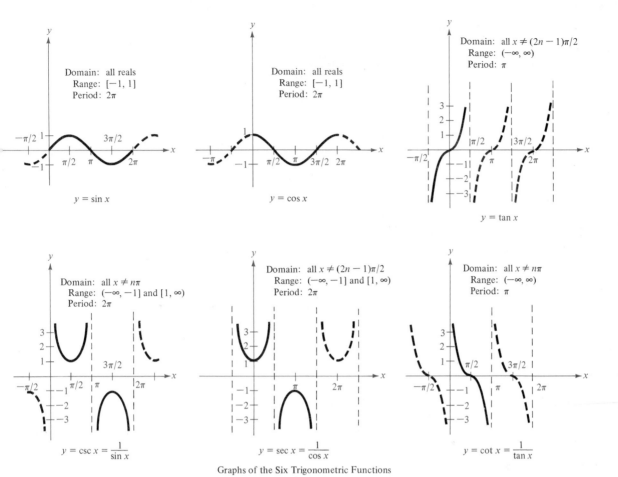

Graphs of the Six Trigonometric Functions

FIGURE 10.32

minimum values. Thus, the amplitude of $f(x) = \sin x$ is 1. The periodic nature of the sine function becomes evident when we observe that as $x$ increases beyond $2\pi$, the graph repeats itself over and over, continuously oscillating about the $x$-axis. The **period** of the function is the distance (measured on the $x$-axis) between successive cycles. Thus, the period of $f(x) = \sin x$ is $2\pi$.

Figure 10.32 gives the graphs of at least one cycle of all six trigonometric functions. Study these graphs carefully and try to memorize their basic features.

Familiarity with the graphs of the six basic trigonometric functions enables us to sketch graphs of more general functions, such as

$$y = a \sin bx \qquad \text{or} \qquad y = a \cos bx$$

Note that the function $y = a \sin bx$ oscillates between $-a$ and $a$ and hence has an amplitude of $|a|$. Furthermore, since $bx = 0$ when $x = 0$ and $bx = 2\pi$ when $x = 2\pi/b$, we may conclude that the function $y = a \sin bx$ has a period of $2\pi/|b|$. Table 10.3 summarizes the amplitudes and periods for some of the more general types of trigonometric functions.

**TABLE 10.3**

| Function | Period | Amplitude |
|---|---|---|
| $y = a \sin bx$ or $y = a \cos bx$ | $\dfrac{2\pi}{\lvert b \rvert}$ | $\lvert a \rvert$ |
| $y = a \tan bx$ or $y = a \cot bx$ | $\dfrac{\pi}{\lvert b \rvert}$ | —— |
| $y = a \sec bx$ or $y = a \csc bx$ | $\dfrac{2\pi}{\lvert b \rvert}$ | —— |

**EXAMPLE 1**

**Sketching the Graph of a Trigonometric Function**

Sketch the graph of

$$f(x) = 3 \cos 2x$$

**SOLUTION**

The graph of $f(x) = 3 \cos 2x$ has the following characteristics:

$$\text{Amplitude:} \quad 3$$

$$\text{Period:} \quad \frac{2\pi}{2} = \pi$$

Several cycles of the graph are shown in Figure 10.33, starting with the maximum point $(0, 3)$.

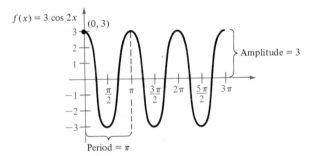

**FIGURE 10.33**

**EXAMPLE 2**

**Sketching the Graph of a Trigonometric Function**

Sketch the graph of

$$f(x) = -2 \tan 3x$$

**SOLUTION**

The graph of this function has a period of $\pi/3$. The vertical asymptotes of this particular tangent function occur at

$$x = \cdots - \frac{\pi}{6}, \frac{\pi}{6}, \frac{\pi}{2}, \frac{5\pi}{6}, \cdots$$

$$\text{Period} = \frac{\pi}{3}$$

Several cycles of the graph are shown in Figure 10.34, starting with the vertical asymptote $x = -\pi/6$.

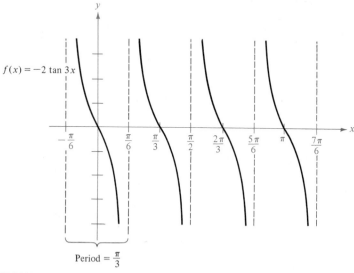

FIGURE 10.34

There are many examples of periodic phenomena in both business and biology. Many businesses have cyclical sales patterns, plant growth is affected by the day/night cycle, and human brain waves can be measured by the oscillating current produced by an electroencephalograph. The following example describes the cyclical pattern followed by many types of predator-prey populations, such as wolf-caribou, owl-mouse, and bird-insect.

**EXAMPLE 3**

**An Example of Predator-Prey Population Cycles**

Suppose that the population of a certain predator at time $t$ in months in a given region is estimated to be

$$P(t) = 10{,}000 + 3000 \sin \frac{2\pi t}{24}$$

and the population of its primary food source (its prey) is estimated to be

$$p(t) = 15{,}000 + 5000 \cos \frac{2\pi t}{24}$$

Sketch both of these functions on the same graph and explain the oscillations in the size of each population.

**SOLUTION**   Both of these functions have a period of

$$\frac{2\pi}{2\pi/24} = 24 \text{ months}$$

The predator population has an amplitude of 3000 and oscillates about the line $y = 10{,}000$, whereas the prey population has an amplitude of 5000 and oscillates about the line $y = 15{,}000$. The graphs of these two functions are given in Figure 10.35.

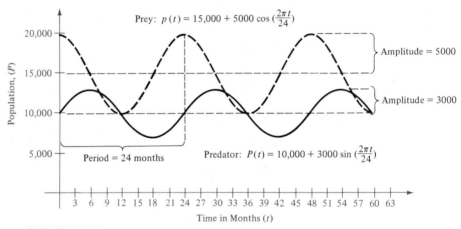

FIGURE 10.35

We can explain the cycles of this predator-prey population by noting the following cause and effect pattern:

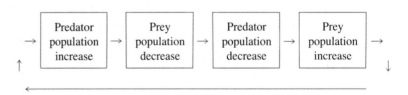

Our last example in this section examines the graph of a function that we will encounter again in Section 10.4.

**EXAMPLE 4**   **Using a Calculator to Graph a Trigonometric Function**

Use a calculator to evaluate the function

$$f(x) = \frac{\sin x}{x}$$

at several points in the interval $[-1, 1]$ and then use these points to sketch the function's graph. From your graph, estimate the limit

$$\lim_{x \to 0} \frac{\sin x}{x}$$

**SOLUTION**

Table 10.4 shows the values that were arrived at with a calculator set in radian mode for intervals of 0.1 from $-1$ to 1. Note that the function is undefined when $x = 0$.

TABLE 10.4

| $x$ | $-1.0$ | $-0.9$ | $-0.8$ | $-0.7$ | $-0.6$ | $-0.5$ | $-0.4$ | $-0.3$ | $-0.2$ | $-0.1$ |
|---|---|---|---|---|---|---|---|---|---|---|
| $\dfrac{\sin x}{x}$ | 0.841 | 0.870 | 0.897 | 0.920 | 0.941 | 0.959 | 0.974 | 0.985 | 0.993 | 0.998 |

| $x$ | 0.1 | 0.2 | 0.3 | 0.4 | 0.5 | 0.6 | 0.7 | 0.8 | 0.9 | 1.0 |
|---|---|---|---|---|---|---|---|---|---|---|
| $\dfrac{\sin x}{x}$ | 0.998 | 0.993 | 0.985 | 0.974 | 0.959 | 0.941 | 0.920 | 0.897 | 0.870 | 0.841 |

Figure 10.36 shows the result of plotting these points. It appears from the graph that the limit of this function as $x$ approaches 0 (from either side) is 1, and we have

$$\lim_{x \to 0} \frac{\sin x}{x} = 1$$

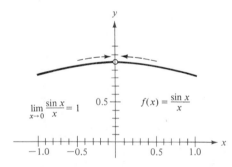

FIGURE 10.36

# SECTION EXERCISES 10.3

In Exercises 1–14, determine the period and amplitude of the given function.

**1.** $y = 2 \sin 2x$

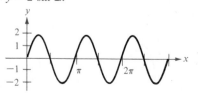

**2.** $y = 3 \cos 3x$

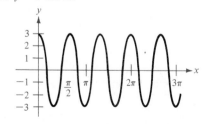

**3.** $y = \dfrac{3}{2} \cos \dfrac{x}{2}$

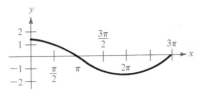

**8.** $y = -\cos \dfrac{2x}{3}$

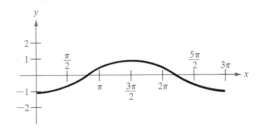

**4.** $y = -2 \sin \dfrac{x}{3}$

**9.** $y = -2 \sin 10x$      **10.** $y = \dfrac{1}{3} \sin 8x$

**11.** $y = \dfrac{1}{2} \cos \dfrac{2x}{3}$      **12.** $y = 5 \cos \dfrac{x}{4}$

**13.** $y = 3 \sin 4\pi x$      **14.** $y = \dfrac{2}{3} \cos \dfrac{\pi x}{10}$

**5.** $y = \dfrac{1}{2} \sin \pi x$

In Exercises 15–20, find the period of the given function.

**15.** $y = 5 \tan 2x$          **16.** $y = 7 \tan 2\pi x$
**17.** $y = 3 \sec 5x$          **18.** $y = \csc 4x$

**19.** $y = \cot \dfrac{\pi x}{3}$      **20.** $y = 5 \tan \dfrac{2\pi x}{3}$

**6.** $y = \dfrac{5}{2} \cos \dfrac{\pi x}{2}$

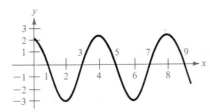

In Exercises 21–26, match the trigonometric function with the correct graph and give the period of the function. [Graphs are labeled (a), (b), (c), (d), (e), and (f).]

**21.** $y = \sec 2x$          **22.** $y = \dfrac{1}{2} \csc 2x$

**23.** $y = \cot \pi x$      **24.** $y = -\sec x$

**25.** $y = 2 \csc \dfrac{x}{2}$      **26.** $y = \tan \dfrac{x}{2}$

(a)                          (b)

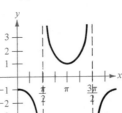

**7.** $y = -2 \sin x$

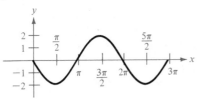

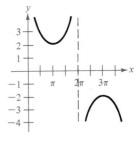

(c)

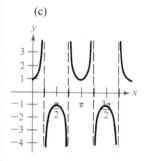

(d)

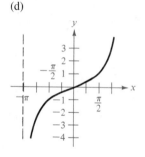

(e)

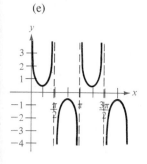

(f)

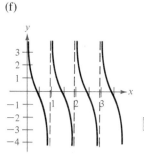

In Exercises 27–46, sketch the graph of the given function.

**27.** $y = \sin \dfrac{x}{2}$

**28.** $y = 4 \sin \dfrac{x}{3}$

**29.** $y = 2 \cos 2x$

**30.** $y = \dfrac{3}{2} \cos \dfrac{2x}{3}$

**31.** $y = -2 \sin 6x$

**32.** $y = -3 \cos 4x$

**33.** $y = \cos 2\pi x$

**34.** $y = \dfrac{3}{2} \sin \dfrac{\pi x}{4}$

**35.** $y = -\sin \dfrac{2\pi x}{3}$

**36.** $y = 10 \cos \dfrac{\pi x}{6}$

**37.** $y = 2 \tan x$

**38.** $y = 2 \cot x$

**39.** $y = \cot 2x$

**40.** $y = 3 \tan \pi x$

**41.** $y = \csc \dfrac{x}{2}$

**42.** $y = \csc \dfrac{x}{3}$

**43.** $y = 2 \sec 2x$

**44.** $y = \sec \pi x$

**45.** $y = \csc 2\pi x$

**46.** $y = -\tan x$

**47.** For a person at rest, the velocity $v$ in liters per second of air flow into and out of the lungs during a respiratory cycle is

$$v = 0.85 \sin \dfrac{\pi t}{3}$$

where $t$ is the time in seconds. Inhalation occurs when $v > 0$, and exhalation occurs when $v < 0$.

(a) Find the time for one full respiratory cycle.
(b) Find the number of cycles per minute.
(c) Sketch the graph of the velocity function.

**48.** After exercising for a few minutes, a person has a respiratory cycle for which the velocity of air flow is approximated by

$$v = 1.75 \sin \dfrac{\pi t}{2}$$

Use this model to repeat Exercise 47.

**49.** When tuning a piano, a technician strikes a tuning fork for the A above middle C and sets up wave motion that can be approximated by

$$y = 0.001 \sin 880\pi t$$

where $t$ is the time in seconds.
(a) What is the period $p$ of this function?
(b) What is the frequency $f$ of this note ($f = 1/p$)?
(c) Sketch the graph of this function.

**50.** The function

$$P = 100 - 20 \cos \dfrac{5\pi t}{3}$$

approximates the blood pressure $P$ in millimeters of mercury at time $t$ in seconds for a person at rest.
(a) Find the period of the function.
(b) Find the number of heartbeats per minute.
(c) Sketch the graph of the pressure function.

In Exercises 51 and 52, sketch the graph of the sales function over one year where $S$ is sales in thousands of units and $t$ is the time in months, with $t = 1$ corresponding to January.

**51.** $S = 22.3 - 3.4 \cos \dfrac{\pi t}{6}$

**52.** $S = 74.50 + 43.75 \sin \dfrac{\pi t}{6}$

In Exercises 53–56, complete the following table (using a calculator set in radian mode) to estimate $\lim\limits_{x \to 0} f(x)$.

| $x$ | −1.0 | −0.1 | −0.01 | −0.001 | 0.001 | 0.01 | 0.1 | 1.0 |
|---|---|---|---|---|---|---|---|---|
| $f(x)$ | | | | | | | | |

**53.** $f(x) = \dfrac{1 - \cos x}{x}$

**54.** $f(x) = \dfrac{\sin 2x}{\sin 3x}$

**55.** $f(x) = \dfrac{\sin x}{5x}$

**56.** $f(x) = \dfrac{3(1 - \cos x)}{x}$

# Derivatives of Trigonometric Functions

## INTRODUCTORY EXAMPLE
## The Construction of a Honeycomb

A honeycomb is constructed of two offset rows of hexagonal cells that fit together perfectly, with no spaces between the cells. Each cell has a hexagonal base and three rhombic upper faces which meet the altitude of the cell at an angle of $\theta$, as shown in Figures 10.37 and 10.38. There are 3 important geometrical properties of a honeycomb:

1. As long as each cell in the honeycomb has the *same angle* $\theta$, the cells will fit together so that each interior wall serves two cells.
2. The volume of each cell is given by

$$\text{Volume} = V = \frac{3\sqrt{3}}{2}s^2 h$$

3. The surface area of each cell is given by

$$\text{Surface area} = S = 6hs + \frac{3s^2}{2}\left(\frac{\sqrt{3} - \cos\theta}{\sin\theta}\right)$$

From these properties we note the value of $\theta$ does not affect the fit or the volume but does affect the surface area. Table 10.5 lists the surface area for several values of $\theta$. It appears that we can minimize the surface area and consequently minimize the amount of wax needed to build the honeycomb by choosing values of $\theta$ between 50° and

60°. Using the differentiation formulas in this section, we will see that we can minimize the surface area by taking its **derivative** with respect to $\theta$.

$$\frac{dS}{d\theta} = \frac{3s^2}{2}\left[\frac{\sin^2\theta - (\sqrt{3} - \cos\theta)\cos\theta}{\sin^2\theta}\right]$$

$$= \frac{3s^2}{2}\left(\frac{1 - \sqrt{3}\cos\theta}{\sin^2\theta}\right)$$

Finally, by setting $dS/d\theta = 0$, we have $\cos\theta = 1/\sqrt{3}$. The value of $\theta$ that satisfies this equation is $\theta \approx 54.74°$. It is not surprising that this value of $\theta$ is the one actually used by honeybees.

### TABLE 10.5

| $\theta$ | Surface area |
|---|---|
| 90° | $6hs + 2.598s^2$ |
| 80° | $6hs + 2.374s^2$ |
| 70° | $6hs + 2.219s^2$ |
| 60° | $6hs + 2.134s^2$ |
| 50° | $6hs + 2.133s^2$ |
| 40° | $6hs + 2.254s^2$ |
| 30° | $6hs + 2.598s^2$ |
| 20° | $6hs + 3.475s^2$ |
| 10° | $6hs + 6.455s^2$ |

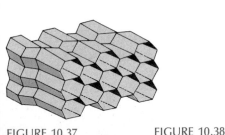

FIGURE 10.37

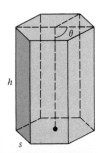
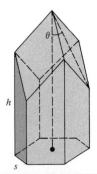
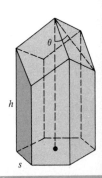

FIGURE 10.38

■ **Derivatives of Trigonometric Functions**
■ **Applications**

In Example 4 and Exercise 53 of the preceding section, we looked at the following two limits:

$$\lim_{\Delta x \to 0} \frac{\sin \Delta x}{\Delta x} = 1 \quad \text{and} \quad \lim_{\Delta x \to 0} \frac{1 - \cos \Delta x}{\Delta x} = 0$$

These two limits turn out to be crucial in the theoretical development of the derivative of the sine function.

$$\frac{d}{dx}[\sin x] = \lim_{\Delta x \to 0} \frac{\sin (x + \Delta x) - \sin x}{\Delta x}$$

$$= \lim_{\Delta x \to 0} \frac{\sin x \cos \Delta x + \cos x \sin \Delta x - \sin x}{\Delta x}$$

$$= \lim_{\Delta x \to 0} \frac{\cos x \sin \Delta x - \sin x\,(1 - \cos \Delta x)}{\Delta x}$$

$$= \lim_{\Delta x \to 0} \left[ \cos x \frac{\sin \Delta x}{\Delta x} - \sin x \frac{1 - \cos \Delta x}{\Delta x} \right]$$

$$= \cos x \left[ \lim_{\Delta x \to 0} \frac{\sin \Delta x}{\Delta x} \right] - \sin x \left[ \lim_{\Delta x \to 0} \frac{1 - \cos \Delta x}{\Delta x} \right]$$

$$= (\cos x)(1) - (\sin x)(0) = \cos x$$

This differentiation formula is shown graphically in Figure 10.39. Note that the *slope* of the sine curve determines the *value* of the cosine curve.

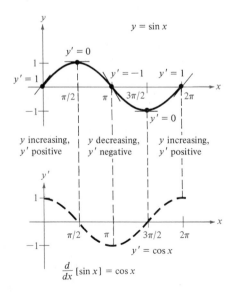

**FIGURE 10.39**

Recall from our previous work with derivatives that we usually listed two versions of each differentiation rule: a simple version and a general Chain Rule version. These two versions of the sine rule are as follows:

*Simple version*

$$\frac{d}{dx}[\sin x] = \cos x$$

*Chain Rule version*

$$\frac{d}{dx}[\sin u] = \cos u \, \frac{du}{dx}$$

We omit the development of the other five differentiation rules for trigonometric functions and simply summarize the results as follows.

**Derivatives of trigonometric functions**

$$\frac{d}{dx}[\sin u] = \cos u \, \frac{du}{dx} \qquad \frac{d}{dx}[\cos u] = -\sin u \, \frac{du}{dx}$$

$$\frac{d}{dx}[\tan u] = \sec^2 u \, \frac{du}{dx} \qquad \frac{d}{dx}[\cot u] = -\csc^2 u \, \frac{du}{dx}$$

$$\frac{d}{dx}[\sec u] = \sec u \tan u \, \frac{du}{dx} \qquad \frac{d}{dx}[\csc u] = -\csc u \cot u \, \frac{du}{dx}$$

■■■ **Remark:** As an aid to memorization, note that the cofunctions (cosine, cotangent, and cosecant) require a negative sign as part of their derivatives.

**EXAMPLE 1**

**Differentiating Trigonometric Functions**

Differentiate the following trigonometric functions:
(a) $y = \sin 2x$
(b) $y = \cos (x - 1)$
(c) $y = \tan 3x$

**SOLUTION**

(a) Considering $u = 2x$, we have

$$\frac{dy}{dx} = \cos u \, \frac{du}{dx}$$

$$= \cos 2x \, \frac{d}{dx}[2x]$$

$$= (\cos 2x)(2)$$

$$= 2 \cos 2x$$

(b) Considering $u = x - 1$, we see that $du/dx = 1$, and thus the derivative is simply

$$\frac{dy}{dx} = -\sin(x - 1)$$

(c) Considering $u = 3x$, we have $du/dx = 3$, and it follows that

$$\frac{dy}{dx} = 3 \sec^2 3x$$

**EXAMPLE 2**

**Differentiating Trigonometric Functions**

Differentiate

$$y = \cos 3x^2$$

**SOLUTION**

We consider $u = 3x^2$. Then

$$\frac{dy}{dx} = -\sin u \frac{du}{dx} = -\sin 3x^2 \frac{d}{dx}[3x^2]$$

$$= -(\sin 3x^2)(6x)$$

$$= -6x \sin 3x^2$$

**EXAMPLE 3**

**Differentiating Trigonometric Functions**

Differentiate

$$y = \tan^4 3x$$

**SOLUTION**

By the Power Rule we have

$$\frac{d}{dx}[(\tan 3x)^4] = 4(\tan 3x)^3 \frac{d}{dx}[\tan 3x]$$

$$= 4(\tan^3 3x)(3)(\sec^2 3x)$$

$$= 12 \tan^3 3x \sec^2 3x$$

**EXAMPLE 4**

**Differentiating Trigonometric Functions**

Differentiate

$$y = \csc \frac{x}{2}$$

**SOLUTION**

$$\frac{dy}{dx} = -\csc \frac{x}{2} \cot \frac{x}{2} \frac{d}{dx}\left[\frac{x}{2}\right] = -\frac{1}{2} \csc \frac{x}{2} \cot \frac{x}{2}$$

**EXAMPLE 5**

**Differentiating Trigonometric Functions**

Differentiate

$$f(t) = \sqrt{\sin 4t}$$

**SOLUTION**

We begin by rewriting the function as

$$f(t) = (\sin 4t)^{1/2}$$

Then by the Power Rule we obtain

$$f'(t) = \left(\frac{1}{2}\right)(\sin 4t)^{-1/2}\frac{d}{dt}[\sin 4t]$$

$$= \left(\frac{1}{2}\right)(\sin 4t)^{-1/2}(4\cos 4t)$$

$$= \frac{2\cos 4t}{\sqrt{\sin 4t}}$$

**EXAMPLE 6**

**Differentiating Trigonometric Functions**

Differentiate

$$y = x\sin x$$

**SOLUTION**

By the Product Rule we have

$$\frac{dy}{dx} = x\frac{d}{dx}[\sin x] + \sin x\frac{d}{dx}[x]$$

$$= x\cos x + \sin x$$

### Applications

**EXAMPLE 7**

**Finding Relative Extrema for Trigonometric Functions**

Determine the relative extrema of the graph of

$$y = \frac{x}{2} - \sin x$$

on the interval $(0, 2\pi)$.

**SOLUTION**

The derivative of $y$ is given by

$$\frac{dy}{dx} = \frac{1}{2} - \cos x$$

Now, setting this derivative equal to zero, we have

$$\cos x = \frac{1}{2}$$

Therefore, the critical numbers in the interval $(0, 2\pi)$ are $x = \pi/3$ and $x = 5\pi/3$. By either the First-Derivative Test or the Second-Derivative Test we determine that $\pi/3$ yields a relative minimum and $5\pi/3$ yields a relative maximum, as shown in Figure 10.40.

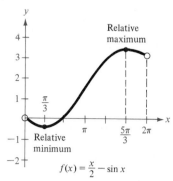

FIGURE 10.40

---

**EXAMPLE 8**

**Finding Relative Extrema for Trigonometric Functions**

Find the relative extrema and sketch the graph of

$$f(x) = 2 \sin x - \cos 2x$$

on the interval $[0, 2\pi]$.

**SOLUTION**

Setting $f'(x)$ equal to zero, we have

$$f'(x) = 2 \cos x + 2 \sin 2x = 0$$

Since $\sin 2x = 2 \cos x \sin x$, we have

$$2 \cos x + 4 \cos x \sin x = 0$$

$$2(\cos x)(1 + 2 \sin x) = 0$$

This equation has solutions when

$$\cos x = 0 \qquad \text{or} \qquad \sin x = -\frac{1}{2}$$

Thus, $f'(x) = 0$ if

$$x = \frac{\pi}{2}, \frac{3\pi}{2} \qquad \text{or} \qquad x = \frac{7\pi}{6}, \frac{11\pi}{6}$$

By the Second-Derivative Test we can determine that $(\pi/2, 3)$ and $(3\pi/2, -1)$ are relative maxima, and $(7\pi/6, -3/2)$ and $(11\pi/6, -3/2)$ are relative minima, as shown in Figure 10.41.

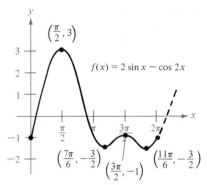

FIGURE 10.41

**EXAMPLE 9**

**A Business Application**

A fertilizer manufacturer finds that the national sales of fertilizer roughly follow the seasonal pattern

$$F = 100,000\left[1 + \sin\frac{2\pi(t - 60)}{365}\right]$$

where $F$ is measured in pounds and $t$ is measured in days, with $t = 1$ representing January 1.* (See Figure 10.42.) On which day of the year is the maximum amount of fertilizer sold?

**SOLUTION**

Taking the derivative, we have

$$\frac{dF}{dt} = 100,000\left[\frac{2\pi}{365}\cos\frac{2\pi(t - 60)}{365}\right]$$

Now, setting this derivative equal to zero, we have

$$\cos\frac{2\pi(t - 60)}{365} = 0$$

Since the cosine is zero at $\pi/2$ and $3\pi/2$, we have

$$\frac{2\pi(t - 60)}{365} = \frac{\pi}{2}$$

$$t - 60 = \frac{365}{4}$$

---

*For simplicity's sake we will follow the convention of saying January 1 is represented by $t = 1$ even though what we actually mean is that January 1 is represented by the *interval* [0, 1].

$$t = \frac{365}{4} + 60 \approx 151 \text{ (May 31)}$$

and
$$\frac{2\pi(t - 60)}{365} = \frac{3\pi}{2}$$

$$t - 60 = \frac{3(365)}{4}$$

$$t = \frac{3(365)}{4} + 60 \approx 334 \text{ (November 30)}$$

From Figure 10.42 we see that the maximum sales occur on May 31.

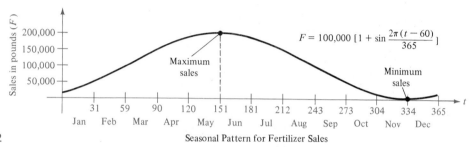

FIGURE 10.42                     Seasonal Pattern for Fertilizer Sales

## EXAMPLE 10

**An Application to Temperature Change**

The temperature during a given 24-hour period is approximated by the model

$$T = 70 + 15 \sin \frac{\pi(t - 8)}{12}$$

where $T$ is measured in degrees Fahrenheit and $t$ is measured in hours, with $t = 0$ representing midnight. (See Figure 10.43.) Find the rate at which the temperature is changing at 6 A.M.

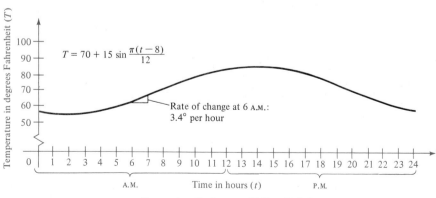

FIGURE 10.43                     Temperature Cycle over a 24–Hour Period

**SOLUTION** The rate of change of the temperature is given by the derivative

$$\frac{dT}{dt} = \frac{15\pi}{12} \cos \frac{\pi(t-8)}{12}$$

Since 6 A.M. corresponds to $t = 6$, the rate of change at 6 A.M. is

$$\frac{dT}{dt} = \frac{15\pi}{12} \cos \left( \frac{-2\pi}{12} \right)$$

$$= \frac{5\pi}{4} \cos \left( -\frac{\pi}{6} \right)$$

$$= \frac{5\pi}{4} \left( \frac{\sqrt{3}}{2} \right)$$

$$\approx 3.4° \text{ per hour}$$

## SECTION EXERCISES 10.4

In Exercises 1–24, find the derivative of the function.

**1.** $y = x^2 - \cos x$      **2.** $y = 5 + \sin x$

**3.** $y = \dfrac{1}{x} - 3 \sin x$      **4.** $g(t) = \pi \cos t$

**5.** $f(x) = 4\sqrt{x} + 3 \cos x$

**6.** $f(x) = 2 \sin x + 3 \cos x$

**7.** $f(t) = t^2 \sin t$      **8.** $f(x) = (x + 1) \cos x$

**9.** $g(t) = \dfrac{\cos t}{t}$      **10.** $f(x) = \dfrac{\sin x}{x}$

**11.** $y = \tan x + x^2$      **12.** $y = \ln |\sec x + \tan x|$

**13.** $y = 5x \csc x$      **14.** $y = e^{-x} \sin x$

**15.** $y = \sin 4x$      **16.** $y = \cos 3x$

**17.** $y = \sec x^2$      **18.** $y = \sin \pi x$

**19.** $y = \dfrac{1}{2} \csc 2x$      **20.** $y = \csc^2 x$

**21.** $y = x \sin \dfrac{1}{x}$      **22.** $y = x^2 \sin \dfrac{1}{x}$

**23.** $y = 3 \tan 4x$      **24.** $y = \tan e^x$

In Exercises 25–40, find the derivative of the function and simplify your answer by using the trigonometric identities listed in Section 10.2.

**25.** $y = \cos^2 x$      **26.** $y = \dfrac{1}{4} \sin^2 2x$

**27.** $y = \cos^2 x - \sin^2 x$      **28.** $y = \sin x \cos x$

**29.** $y = \dfrac{\cos x}{\sin x}$      **30.** $y = \dfrac{x}{2} + \dfrac{\sin 2x}{4}$

**31.** $y = \ln |\sin x|$      **32.** $y = -\ln |\cos x|$

**33.** $y = \ln |\csc x - \cot x|$      **34.** $y = x + \cot x$

**35.** $y = \tan x - x$      **36.** $y = \dfrac{\sec^7 x}{7} - \dfrac{\sec^5 x}{5}$

**37.** $y = \sqrt{\sin x}$      **38.** $y = \ln (\sin^2 x)$

**39.** $y = \dfrac{1}{2}(x \tan x - \sec x)$      **40.** $y = \ln |\cot x|$

In Exercises 41 and 42, use implicit differentiation to find $dy/dx$ and evaluate the derivative at the indicated point.

| Function | Point |
| --- | --- |
| **41.** $\sin x + \cos 2y = 1$ | $\left( \dfrac{\pi}{2}, \dfrac{\pi}{4} \right)$ |
| **42.** $\tan (x + y) = x$ | $(0, 0)$ |

In Exercises 43–46, show that the function satisfies the differential equation.

| Function | Differential equation |
| --- | --- |
| **43.** $y = 2 \sin x + 3 \cos x$ | $y'' + y = 0$ |
| **44.** $y = \dfrac{10 - \cos x}{x}$ | $xy' + y = \sin x$ |
| **45.** $y = \cos 2x + \sin 2x$ | $y'' + 4y = 0$ |
| **46.** $y = e^x(\cos \sqrt{2}x + \sin \sqrt{2}x)$ | $y'' - 2y' + 3y = 0$ |

In Exercises 47–52, find the slope of the tangent line to the given sine function at the origin. Compare this value to the number of complete cycles in the interval $[0, 2\pi]$.

**47.** $y = \sin 3x$

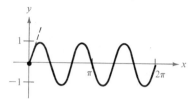

**48.** $y = \sin \dfrac{5x}{2}$

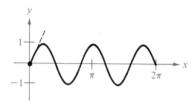

**49.** $y = \sin 2x$

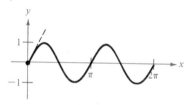

**50.** $y = \sin \dfrac{3x}{2}$

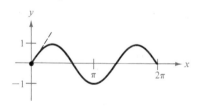

**51.** $y = \sin x$

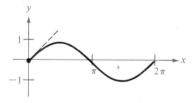

**52.** $y = \sin \dfrac{x}{2}$

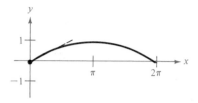

In Exercises 53 and 54, find an equation of the tangent line to the graph of the function at the indicated point.

| Function | Point |
|---|---|
| **53.** $f(x) = \tan x$ | $\left(-\dfrac{\pi}{4}, -1\right)$ |
| **54.** $f(x) = \sec x$ | $\left(\dfrac{\pi}{3}, 2\right)$ |

In Exercises 55–58, sketch the graph of each function on the indicated interval, making use of relative extrema and points of inflection.

| Function | Interval |
|---|---|
| **55.** $f(x) = 2 \sin x + \sin 2x$ | $[0, 2\pi]$ |
| **56.** $f(x) = 2 \sin x + \cos 2x$ | $[0, 2\pi]$ |
| **57.** $f(x) = x - 2 \sin x$ | $[0, 4\pi]$ |
| **58.** $f(x) = e^{-x} \sin x$ | $[0, 2\pi]$ |

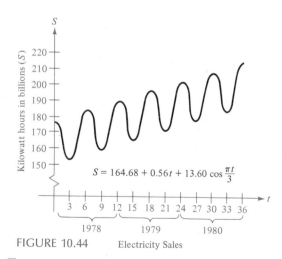

FIGURE 10.44    Electricity Sales

**59.** Electricity sales in the United States have had both an annual sales pattern and a monthly sales pattern. For the years 1978 and 1979, the sales pattern can be approximated by the model

$$S = 164.68 + 0.56t + 13.60 \cos \frac{\pi t}{3}$$

where $S$ is the sales per month in billions of kilowatt hours and $t$ is the time in months, with $t = 1$ corresponding to January 1978. (See Figure 10.44.)

(a) Find the relative extrema of this function for the years 1978 and 1979.

(b) Use this model to predict the sales in August 1980. (Use $t = 31.5$.)

**60.** Plants do not grow at constant rates during a normal 24-hour period, since their growth is affected by sunlight. Suppose that the growth of certain plant species in a controlled environment is given by the model

$$h = 0.20t + 0.03 \sin 2\pi t$$

where $h$ is the height of the plant in inches and $t$ is the time in days, with $t = 0$ corresponding to midnight. (See Figure 10.45.) During what time of day is the rate of growth of this plant

(a) the greatest     (b) the least

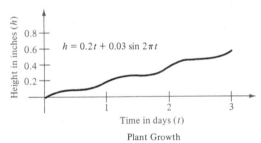

Plant Growth

FIGURE 10.45

**61.** The normal average daily temperature in degrees Fahrenheit for a certain city is given by

$$T = 45 - 23 \cos \frac{2\pi(t - 32)}{365}$$

where $t$ is the time in days, with $t = 1$ corresponding to January 1. Find the expected date of

(a) the warmest day     (b) the coldest day

**62.** For $f(x) = \sec^2 x$ and $g(x) = \tan^2 x$, show that $f'(x) = g'(x)$.

In Exercises 63 and 64, apply Taylor's Theorem to verify the power series centered at $c = 0$ for the given function, and find the radius of convergence.

**63.** $\sin x = x - \dfrac{x^3}{3!} + \dfrac{x^5}{5!}$

$$- \frac{x^7}{7!} + \cdots + \frac{(-1)^n x^{2n+1}}{(2n + 1)!} + \cdots$$

**64.** $\cos x = 1 - \dfrac{x^2}{2!} + \dfrac{x^4}{4!} - \dfrac{x^6}{6!} + \cdots + \dfrac{(-1)^n x^{2n}}{(2n)!} + \cdots$

In Exercises 65 and 66, use the power series for $\sin x$ to find the power series of the given function.

**65.** $f(x) = \sin x^2$

**66.** $g(x) = \dfrac{\sin x}{x}$

In Exercises 67 and 68, use the power series for $\cos x$ to find the power series of the given function.

**67.** $h(x) = \cos 2x$

**68.** $f(x) = \cos \sqrt{x}$

In Exercises 69 and 70, differentiate the appropriate power series to verify the given derivative formulas.

**69.** $\dfrac{d}{dx}[\sin x] = \cos x$

**70.** $\dfrac{d}{dx}[\cos x] = -\sin x$

# Integrals of Trigonometric Functions

If a 2-inch needle is tossed randomly onto a floor ruled with parallel lines that are 2 inches apart, what is the probability that the needle will touch one of the lines? At first glance this problem may seem to be very difficult. However, if we use trigonometry and integration, the solution becomes quite manageable.

We begin by letting $\theta$ represent the angle between the needle and the lines on the floor. Without loss of generality we can assume $0 \le \theta \le \pi/2$. Furthermore, we assume that each possible value of $\theta$ is equally likely. Now, for a given value of $\theta$, the needle will touch one of the lines if its center falls within $\sin \theta$ inches of one of the lines, as shown in Figure 10.46. In other words, there is a "touching region" of width $2 \sin \theta$ inches between each pair of lines on the floor. Since the distance between each pair of lines is 2 inches, the probability (for a given $\theta$) of touching one of the lines is given by

$$\text{Probability of touching} = \frac{\text{width of ``touching region''}}{\text{distance between lines}}$$

$$= \frac{2 \sin \theta}{2} = \sin \theta$$

(Note that for $\theta = 0$ the only way the needle could touch a line would be for it to fall exactly on one of the lines, and we consider this probability to be 0. Similarly, for $\theta = \pi/2$ we consider the probability of touching to be 1, since the only way a needle could not touch in that case would be for its center to fall exactly halfway between two of the lines.)

Finally, we define the total probability that the needle will touch the line to be the average of the probabilities as $\theta$ ranges between 0 and $\pi/2$. Using the **integration techniques** discussed in this section, we can determine the probability of touching as follows:

$$\text{Probability of touching} = \frac{1}{(\pi/2) - 0} \int_0^{\pi/2} \sin \theta \, d\theta$$

$$= \frac{2}{\pi} \left[ -\cos \theta \right]_0^{\pi/2} = \frac{2}{\pi} \approx 63.7\%$$

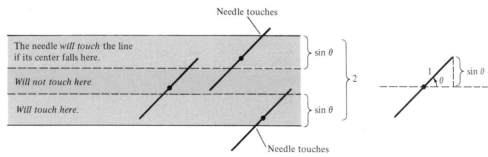

FIGURE 10.46

■ **Integrals Involving Trigonometric Functions**
■ **Integrals of Six Basic Trigonometric Functions**

Corresponding to each formula for differentiating a trigonometric function is an integration formula. For instance, corresponding to the differentiation formula

$$\frac{d}{dx}[\cos u] = -\sin u \frac{du}{dx}$$

is the integration formula

$$\int \sin u \, du = -\cos u + C$$

The following list contains the integration formulas corresponding to the derivatives of the six basic trigonometric functions.

| Trigonometric functions | *Integrals* | *Derivatives* |
|---|---|---|
| | $\int \cos u \, du = \sin u + C$ | $\frac{d}{dx}[\sin u] = \cos u \frac{du}{dx}$ |
| | $\int \sin u \, du = -\cos u + C$ | $\frac{d}{dx}[\cos u] = -\sin u \frac{du}{dx}$ |
| | $\int \sec^2 u \, du = \tan u + C$ | $\frac{d}{dx}[\tan u] = \sec^2 u \frac{du}{dx}$ |
| | $\int \sec u \tan u \, du = \sec u + C$ | $\frac{d}{dx}[\sec u] = \sec u \tan u \frac{du}{dx}$ |
| | $\int \csc^2 u \, du = -\cot u + C$ | $\frac{d}{dx}[\cot u] = -\csc^2 u \frac{du}{dx}$ |
| | $\int \csc u \cot u \, du = -\csc u + C$ | $\frac{d}{dx}[\csc u] = -\csc u \cot u \frac{du}{dx}$ |

**EXAMPLE 1**

**Integrating Trigonometric Functions**

Evaluate the integral

$$\int 2 \cos x \, dx$$

**SOLUTION**     Let $u = x$. Then $du = dx$. Now we have

$$\int 2 \cos x \, dx = 2 \int \cos x \, dx$$

$$= 2 \int \cos u \, du = 2 \sin u + C$$

$$= 2 \sin x + C$$

**EXAMPLE 2**     **Integrating Trigonometric Functions**

Evaluate the integral

$$\int 3x^2 \sin x^3 \, dx$$

**SOLUTION**     We let $u = x^3$. Then $du = 3x^2 \, dx$. Therefore, we can write

$$\int 3x^2 \sin x^3 \, dx = \int (\sin x^3) \, 3x^2 \, dx$$

$$= \int \sin u \, du$$

$$= -\cos u + C$$

$$= -\cos x^3 + C$$

**EXAMPLE 3**     **Integrating Trigonometric Functions**

Evaluate the integral

$$\int \sec 3x \tan 3x \, dx$$

**SOLUTION**     Let $u = 3x$. Then $du = 3 \, dx$, and we write

$$\int \sec 3x \tan 3x \, dx = \frac{1}{3} \int (\sec 3x \tan 3x)3 \, dx$$

$$= \frac{1}{3} \int \sec u \tan u \, du$$

$$= \frac{1}{3} \sec u + C$$

$$= \frac{1}{3} \sec 3x + C$$

**EXAMPLE 4**

**Integrating Trigonometric Functions**

Evaluate the integral

$$\int e^x \sec^2 e^x \, dx$$

**SOLUTION**

If we let $u = e^x$ it follows that $du = e^x \, dx$. This means that

$$\int e^x \sec^2 e^x \, dx = \int \sec^2 u \, du$$

$$= \tan u + C$$

$$= \tan e^x + C$$

**EXAMPLE 5**

**Finding Area by Integration**

Find the area of the region bounded by the $x$-axis and one arc of the sine curve $y = \sin x$.

**SOLUTION**

As indicated in Figure 10.47, this area is given by

$$\text{Area} = \int_0^\pi \sin x \, dx$$

$$= -\cos x \Big]_0^\pi$$

$$= -(-1) - (-1)$$

$$= 2$$

and we conclude that the area is 2.

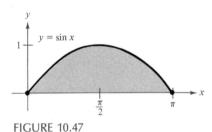

FIGURE 10.47

**EXAMPLE 6**

**Finding the Average Value of a Function Over an Interval**

The temperature during a 24-hour period is given by

$$T = 72 + 18 \sin \frac{\pi(t - 8)}{12}$$

where $T$ is measured in degrees Fahrenheit and $t$ is measured in hours, with $t = 0$ representing midnight. Find the average temperature during the 4-hour period from noon to 4 P.M.

**SOLUTION**

The average temperature during this 4-hour period is given by the following integral:

$$\text{Average temperature} = \frac{1}{4}\int_{12}^{16}\left[72 + 18\sin\frac{\pi(t-8)}{12}\right]dt$$

$$= \frac{1}{4}\left[72t + 18\left(\frac{12}{\pi}\right)\left(-\cos\frac{\pi(t-8)}{12}\right)\right]_{12}^{16}$$

$$= \frac{1}{4}\left[72(16) + 18\left(\frac{12}{\pi}\right)\left(\frac{1}{2}\right) - 72(12) + 18\left(\frac{12}{\pi}\right)\left(\frac{1}{2}\right)\right]$$

$$= \frac{1}{4}\left[288 + \frac{216}{\pi}\right]$$

$$= 72 + \frac{54}{\pi} \approx 89.2°$$

(See Figure 10.48.)

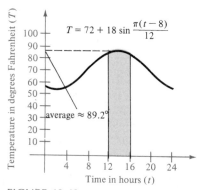

FIGURE 10.48

We frequently use the Power Rule and the Log Rule to evaluate integrals containing trigonometric functions. The key to their use lies in identifying as part of the integrand the derivative of one of the functions involved. The next three examples show how these rules are used. Recall that the Power and Log Rules for integration are

*Power Rule*

$$\int u^n \frac{du}{dx}\,dx = \frac{u^{n+1}}{n+1} + C, \ n \neq -1$$

*Log Rule*

$$\int \frac{du/dx}{u}\,dx = \ln|u| + C$$

**EXAMPLE 7**

Using the Power Rule on Trigonometric Integrals

Evaluate the integral

$$\int \sin^2 4x \cos 4x \, dx$$

**SOLUTION**

The integrand does not fit any of the six basic trigonometric integration formulas. However, letting $u = \sin 4x$, we have

$$\frac{du}{dx} = \frac{d}{dx}[\sin 4x] = 4 \cos 4x$$

Thus, using the Power Rule, we have

$$\int \sin^2 4x \cos 4x \, dx = \frac{1}{4} \int \overbrace{(\sin 4x)^2}^{u^2} \overbrace{(4 \cos 4x)}^{du/dx} \, dx$$

$$= \frac{1}{4} \int u^2 \, du$$

$$= \frac{1}{4} \frac{u^3}{3} + C$$

$$= \frac{1}{4} \frac{(\sin 4x)^3}{3} + C$$

$$= \frac{1}{12} \sin^3 4x + C$$

**EXAMPLE 8**

Using the Power Rule on Trigonometric Integrals

Evaluate the integral

$$\int \sec^2 x \tan^3 x \, dx$$

**SOLUTION**

Since

$$\frac{d}{dx}[\tan x] = \sec^2 x$$

we use the Power Rule with $u = \tan x$.

$$\int \sec^2 x \tan^3 x \, dx = \int \overbrace{(\tan x)^3}^{u^3} \overbrace{\sec^2 x}^{du/dx} \, dx$$

$$= \int u^3 \, du$$

$$= \frac{u^4}{4} + C$$

$$= \frac{(\tan x)^4}{4} + C$$

$$= \frac{1}{4} \tan^4 x + C$$

---

**EXAMPLE 9**

**Using the Log Rule on Trigonometric Integrals**

Evaluate the integral

$$\int \frac{\sin x}{\cos x}\, dx$$

**SOLUTION**

Knowing that

$$\frac{d}{dx}[\cos x] = -\sin x$$

we apply the Log Rule (with $u = \cos x$)

$$\int \frac{du/dx}{u}\, dx = \ln |u| + C$$

and write

$$\int \frac{\sin x}{\cos x}\, dx = -\int \frac{(-\sin x)}{\cos x}\, dx$$

$$= -\ln |\cos x| + C$$

We listed integration formulas for $\sin x$ and $\cos x$ in the beginning of this section. Now, using the result of Example 9, we also have an integration formula for $\tan x$. That is, since

$$\tan x = \frac{\sin x}{\cos x}$$

we have the following formula:

$$\int \tan x\, dx = -\ln |\cos x| + C$$

We omit the development of the integration formulas for the other three trigonometric functions and simply summarize all six formulas as follows.

---

**Integrals of the six basic trigonometric functions**

$$\int \sin u \ du = -\cos u + C \qquad\qquad \int \cos u \ du = \sin u + C$$

$$\int \tan u \ du = -\ln |\cos u| + C \qquad\qquad \int \cot u \ du = \ln |\sin u| + C$$

$$\int \sec u \ du = \ln |\sec u + \tan u| + C \qquad\qquad \int \csc u \ du = \ln |\csc u - \cot u| + C$$

## SECTION EXERCISES 10.5

In Exercises 1–34, evaluate the given integral.

**1.** $\displaystyle\int (2 \sin x + 3 \cos x) \ dx$   **2.** $\displaystyle\int (t^2 - \sin t) \ dt$

**3.** $\displaystyle\int (1 - \csc t \cot t) \ dt$   **4.** $\displaystyle\int (\theta^2 + \sec^2 \theta) \ d\theta$

**5.** $\displaystyle\int (\sec^2 \theta - \sin \theta) \ d\theta$

**6.** $\displaystyle\int (\sec y \tan y - \sec^2 y) \ dy$

**7.** $\displaystyle\int \sin 2x \ dx$   **8.** $\displaystyle\int \cos 6x \ dx$

**9.** $\displaystyle\int x \cos x^2 \ dx$   **10.** $\displaystyle\int x \sin x^2 \ dx$

**11.** $\displaystyle\int \sec^2 \frac{x}{2} \ dx$   **12.** $\displaystyle\int \csc^2 \frac{x}{2} \ dx$

**13.** $\displaystyle\int \tan 3x \ dx$   **14.** $\displaystyle\int \csc 2x \cot 2x \ dx$

**15.** $\displaystyle\int \tan^4 x \sec^2 x \ dx$   **16.** $\displaystyle\int \sqrt{\cot x} \csc^2 x \ dx$

**17.** $\displaystyle\int \cot \pi x \ dx$   **18.** $\displaystyle\int \tan 5x \ dx$

**19.** $\displaystyle\int \csc 2x \ dx$   **20.** $\displaystyle\int \sec \frac{x}{2} \ dx$

**21.** $\displaystyle\int \frac{\sec^2 x}{\tan x} \ dx$   **22.** $\displaystyle\int \frac{\sin x}{\cos^2 x} \ dx$

**23.** $\displaystyle\int \frac{\sec x \tan x}{\sec x - 1} \ dx$   **24.** $\displaystyle\int \frac{\cos t}{1 + \sin t} \ dt$

**25.** $\displaystyle\int \frac{\sin x}{1 + \cos x} \ dx$   **26.** $\displaystyle\int \frac{\sin \sqrt{x}}{\sqrt{x}} \ dx$

**27.** $\displaystyle\int \frac{\csc^2 x}{\cot^3 x} \ dx$   **28.** $\displaystyle\int \frac{1 - \cos \theta}{\theta - \sin \theta} \ d\theta$

**29.** $\displaystyle\int e^x \cos e^x \ dx$   **30.** $\displaystyle\int e^{\sin x} \cos x \ dx$

**31.** $\displaystyle\int e^{-x} \tan e^{-x} \ dx$

**32.** $\displaystyle\int e^{\sec x} \sec x \tan x \ dx$

**33.** $\displaystyle\int (\sin 2x + \cos 2x)^2 \ dx$

**34.** $\displaystyle\int (\csc 2\theta - \cot 2\theta)^2 \ d\theta$

In Exercises 35–38, evaluate the given integral using the integration by parts formula

$$\int u \ dv = uv - \int v \ du$$

**35.** $\displaystyle\int x \cos x \ dx$   **36.** $\displaystyle\int x \sin x \ dx$

**37.** $\displaystyle\int x \sec^2 x \ dx$   **38.** $\displaystyle\int \theta \sec \theta \tan \theta \ d\theta$

In Exercises 39–46, evaluate the definite integral.

**39.** $\displaystyle\int_0^{\pi/2} \cos \frac{2x}{3} \ dx$   **40.** $\displaystyle\int_0^{\pi/2} \sin 2x \ dx$

**41.** $\displaystyle\int_{\pi/2}^{2\pi/3} \sec^2 \frac{x}{2} \ dx$   **42.** $\displaystyle\int_0^{\pi/2} (x + \cos x) \ dx$

**43.** $\displaystyle\int_{\pi/12}^{\pi/4} \csc 2x \cot 2x \ dx$

**44.** $\displaystyle\int_0^{\pi/8} \sin 2x \cos 2x \ dx$

**45.** $\displaystyle\int_0^1 \sec (1 - x) \tan (1 - x) \ dx$

**46.** $\displaystyle\int_0^{\pi/4} (\sec x)^3 (\sec x \tan x) \ dx$

In Exercises 47–52, determine the area of the given region.

**47.** $y = \cos \dfrac{x}{2}$

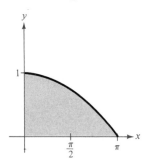

**48.** $y = x + \sin x$

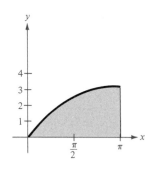

**49.** $y = \tan x$

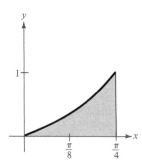

**50.** $y = 2 \sin x + \sin 2x$

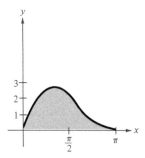

**51.** $y = \sin x + \cos 2x$

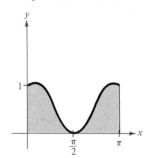

**52.** $y = x \sin x$

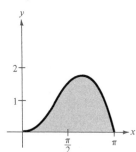

In Exercises 53 and 54, find the volume of the solid generated by revolving the region bounded by the graphs of the given equations about the x-axis.

**53.** $y = \sec x$, $y = 0$, $x = 0$, $x = \dfrac{\pi}{4}$

**54.** $y = \csc x$, $y = 0$, $x = \dfrac{\pi}{6}$, $x = \dfrac{5\pi}{6}$

In Exercises 55 and 56, find the general solution to the first-order linear differential equation.

**55.** $y' + 2y = \sin x$

**56.** $y' \cos^2 x + y = 1$

**57.** Use the Taylor polynomial

$$\frac{\sin x}{x} \approx 1 - \frac{x^2}{6} + \frac{x^4}{120} - \frac{x^6}{5040}$$

to approximate the integral

$$\int_0^{\pi/2} \frac{\sin x}{x}\, dx$$

**58.** Use the Taylor polynomial

$$\cos x^2 \approx 1 - \frac{x^4}{2} + \frac{x^8}{24} - \frac{x^{12}}{720}$$

to approximate the integral

$$\int_0^1 \cos x^2\, dx$$

**59.** Approximate

$$\int_0^{\pi/2} f(x)\, dx,\ f(x) = \begin{cases} \dfrac{\sin x}{x}, & x > 0 \\ 1, & x = 0 \end{cases}$$

by letting $n = 4$ and using
(a) the Trapezoidal Rule
(b) Simpson's Rule
Compare the result with that obtained in Exercise 57.

**60.** Approximate the integral of Exercise 58 by letting $n = 4$ and using

(a) the Trapezoidal Rule

(b) Simpson's Rule

**61.** The minimum stockpile level of gasoline in the United States can be approximated by the model

$$Q = 217 + 13 \cos \frac{\pi(t-3)}{6}$$

where $Q$ is measured in millions of barrels of gasoline and $t$ is the time in months, with $t = 1$ corresponding to January. Find the average minimum level given by this model during

(a) the first quarter $(0 \le t \le 3)$

(b) the second quarter $(3 \le t \le 6)$

(c) the entire year $(0 \le t \le 12)$

**62.** The sales of a seasonal product are given by the model

$$S = 74.50 + 43.75 \sin \frac{\pi t}{6}$$

where $S$ is measured in thousands of units and $t$ is the time in months, with $t = 1$ corresponding to January. Find the average sales during

(a) the first quarter $(0 \le t \le 3)$

(b) the second quarter $(3 \le t \le 6)$

(c) the entire year $(0 \le t \le 12)$

**63.** For a person at rest, the velocity $v$ in liters per second of air flow into and out of the lungs during a respiratory cycle is approximated by

$$v = 0.85 \sin \frac{\pi t}{3}$$

where $t$ is the time in seconds. Find the volume in liters of air inhaled during one cycle by integrating this function over the interval $[0, 3]$.

**64.** After exercising a few minutes, a person has a respiratory cycle for which the velocity of air flow is approximated by

$$v = 1.75 \sin \frac{\pi t}{2}$$

How much does the lung capacity of a person increase as a result of exercising? In other words, how much more air is inhaled during a cycle after exercising than is inhaled during a cycle at rest? (See Exercise 63.) (Note that the cycle is shorter and you must integrate over the interval $[0, 2]$.)

**65.** Suppose that the temperature in degrees Fahrenheit is given by

$$T = 72 + 12 \sin \frac{\pi(t-8)}{12}$$

where $t$ is the time in hours, with $t = 0$ representing midnight. Furthermore, suppose that it costs \$0.10 to cool a particular house 1° for 1 hour.

(a) Find the cost $C$ of cooling this house if the thermostat is set at 72° and the cost is given by

$$C = 0.1 \int_8^{20} \left[ 72 + 12 \sin \frac{\pi(t-8)}{12} - 72 \right] dt$$

(See Figure 10.49.)

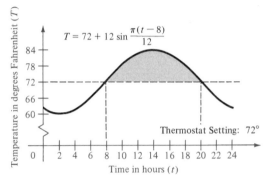

$$T = 72 + 12 \sin \frac{\pi(t-8)}{12}$$

Thermostat Setting: 72°

FIGURE 10.49

(b) Find the savings realized by resetting the thermostat to 78° by evaluating the integral

$$C = 0.1 \int_{10}^{18} \left[ 72 + 12 \sin \frac{\pi(t-8)}{12} - 78 \right] dt$$

(See Figure 10.50.)

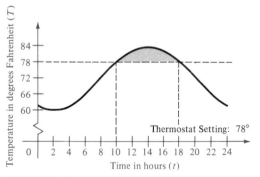

Thermostat Setting: 78°

FIGURE 10.50

**66.** In Example 9 of Section 10.4, the sales of a seasonal product were approximated by the model

$$F = 100{,}000 \left[ 1 + \sin \frac{2\pi(t - 60)}{365} \right]$$

where $F$ was measured in pounds and $t$ was the time in days, with $t = 1$ representing January 1. The manufacturer of this product wants to set up a manufacturing schedule to produce a uniform amount each day. What should this amount be? (Assume that there are 200 production days during the year.)

## Important Terms
Angle
Initial ray
Terminal ray
Standard position of an angle
Degree measure of an angle
Acute angle

Obtuse angle
Coterminal angles
Radian measure of an angle
Trigonometric function
Trigonometric identity
Amplitude of sine and cosine function
Period of trigonometric function

## Important Techniques
Converting between degree and radian measure of an angle
Applying trigonometric identities
Evaluating trigonometric functions

Solving trigonometric equations
Sketching the graph of a trigonometric function
Differentiating trigonometric functions
Finding relative extrema of a trigonometric function
Integrating trigonometric functions

## Important Formulas
The sum of the angles of a triangle is 180°.
The sum of the two acute angles of a right triangle is 90°.
Pythagorean Theorem: The sum of the squares of the (perpendicular) sides of a right triangle is equal to the square of the hypotenuse.
Similar Triangles: If two triangles are similar (have the same angle measures), then the ratios of corresponding sides are equal.
The area of a triangle is equal to one-half the base times the height.
Each of the angles in an equilateral triangle measures 60°.
Each of the acute angles in an isosceles right triangle measures 45°.
Conversion between degree and radian measure:

$$180° = \pi \text{ rad} \qquad 1° = \frac{\pi}{180} \text{ rad} \qquad 1 \text{ rad} = \frac{180°}{\pi}$$

Right-angle definition of trigonometric functions:

$$\sin \theta = \frac{\text{opp.}}{\text{hyp.}} \qquad \csc \theta = \frac{\text{hyp.}}{\text{opp.}}$$

$$\cos \theta = \frac{\text{adj.}}{\text{hyp.}} \qquad \sec \theta = \frac{\text{hyp.}}{\text{adj.}}$$

$$\tan \theta = \frac{\text{opp.}}{\text{adj.}} \qquad \cot \theta = \frac{\text{adj.}}{\text{opp.}}$$

Quadrant definition of trigonometric functions:

$$\sin \theta = \frac{y}{r} \qquad \csc \theta = \frac{r}{y}$$

$$\cos \theta = \frac{x}{r} \qquad \sec \theta = \frac{r}{x}$$

$$\tan \theta = \frac{y}{x} \qquad \cot \theta = \frac{x}{y}$$

Amplitude of $a \sin bx$ and $a \cos bx$ is $|a|$.
Period of $a \sin bx$, $a \cos bx$, $a \sec bx$, and $a \csc bx$ is $2\pi/|b|$.
Period of $a \tan bx$ and $a \cot bx$ is $\pi/|b|$.

Derivatives of trigonometric functions:

$$\frac{d}{dx}[\sin u] = \cos u \frac{du}{dx}$$

$$\frac{d}{dx}[\cos u] = -\sin u \frac{du}{dx}$$

$$\frac{d}{dx}[\tan u] = \sec^2 u \frac{du}{dx}$$

$$\frac{d}{dx}[\cot u] = -\csc^2 u \frac{du}{dx}$$

$$\frac{d}{dx}[\sec u] = \sec u \tan u \frac{du}{dx}$$

$$\frac{d}{dx}[\csc u] = -\csc u \cot u \frac{du}{dx}$$

Integrals of trigonometric functions:

$$\int \cos u \, du = \sin u + C$$

$$\int \sin u \, du = -\cos u + C$$

$$\int \sec^2 u \, du = \tan u + C$$

$$\int \sec u \tan u \, du = \sec u + C$$

$$\int \csc^2 u \, du = -\cot u + C$$

$$\int \csc u \cot u \, du = -\csc u + C$$

$$\int \tan u \, du = -\ln |\cos u| + C$$

$$\int \cot u \, du = \ln |\sin u| + C$$

$$\int \sec u \, du = \ln |\sec u + \tan u| + C$$

$$\int \csc u \, du = \ln |\csc u - \cot u| + C$$

In Exercises 1 and 2, sketch the angle in standard position and give a positive and a negative coterminal angle.

**1.** $\dfrac{11\pi}{4}$

**2.** $-405°$

In Exercises 3 and 4, convert the magnitude of the angle from radian to degree measure. Give the answer accurate to three decimal places.

**3.** $\dfrac{5\pi}{7}$

**4.** $-3.5$

In Exercises 5 and 6, convert the magnitude of the angle from degree to radian measure. Give the answer accurate to two decimal places.

**5.** $480°$

**6.** $-16.5°$

In Exercises 7 and 8, find the reference angle for the given angle.

**7.** $-\dfrac{6\pi}{5}$

**8.** $640°$

In Exercises 9 and 10, find the six trigonometric functions of the angle $\theta$ if it is in standard position and the terminal side passes through the given point.

**9.** $(-7, 2)$

**10.** $(4, -8)$

In Exercises 11 and 12, find the remaining five trigonometric functions of $\theta$ from the given information.

**11.** $\sec \theta = \dfrac{6}{5}$, $\tan \theta < 0$

**12.** $\tan \theta = -\dfrac{12}{5}$, $\sin \theta > 0$

In Exercises 13–20, find two values of $\theta$ in degrees ($0° \le \theta \le 360°$) and in radians ($0 \le \theta \le 2\pi$).

**13.** $\cos \theta = -\dfrac{\sqrt{2}}{2}$

**14.** $\sec \theta$ is undefined

**15.** $\csc \theta = -2$

**16.** $\tan \theta = \dfrac{\sqrt{3}}{3}$

**17.** $\sin \theta = 0.8387$

**18.** $\cot \theta = -1.5399$

**19.** $\sec \theta = -1.0353$

**20.** $\csc \theta = 11.4737$

In Exercises 21–24, sketch the graph of the given function.

**21.** $f(x) = 3 \sin \dfrac{2x}{5}$

**22.** $f(x) = 8 \cos \dfrac{x}{4}$

**23.** $f(x) = -\tan \dfrac{\pi x}{4}$

**24.** $f(x) = \sec 2\pi x$

**25.** An observer 2.5 miles from the launch pad of a space shuttle measures the angle of elevation to the base of the vehicle to be $28°$ soon after liftoff. How high is the shuttle at that instant if you assume that the shuttle is still moving vertically?

**26.** In traveling across relatively flat land, you notice a mountain directly in front of you. The angle of elevation to the peak is $3.5°$. After you drive 13 miles closer to the mountain, the angle of elevation is $9°$. Approximate the height of the mountain.

In Exercises 27–42, find $dy/dx$.

**27.** $y = \cos 5\pi x$

**28.** $y = \tan (4x - \pi)$

**29.** $y = -x \tan x$

**30.** $y = \csc 3x + \cot 3x$

**31.** $y = \dfrac{\sin x}{x^2}$

**32.** $y = \dfrac{\cos (x - 1)}{x - 1}$

**33.** $y = 3 \sin^2 4x + x$

**34.** $y = x \cos x - \sin x$

**35.** $y = 2 \csc^3 x$

**36.** $y = \sec^2 2x$

**37.** $y = e^x \tan x$

**38.** $y = \dfrac{1}{2} e^{\sin 2x}$

**39.** $x = 2 + \sin y$

**40.** $\sin (x + y) = x$

**41.** $\cos (x + y) = x$

**42.** $\cos x^2 = xe^y$

In Exercises 43–46, find the second derivative of the function.

**43.** $f(x) = \cot x$

**44.** $g(t) = \sin^2 t$

**45.** $h(x) = \dfrac{\cos x}{x}$

**46.** $f(x) = x \tan x$

In Exercises 47–56, evaluate the indefinite integral.

**47.** $\displaystyle\int \csc 2x \cot 2x \, dx$

**48.** $\displaystyle\int \sin^3 x \cos x \, dx$

**49.** $\displaystyle\int \tan \dfrac{\pi x}{4} \, dx$

**50.** $\displaystyle\int \dfrac{\cos x}{\sqrt{\sin x}} \, dx$

**51.** $\displaystyle\int \tan^n x \sec^2 x \, dx, \ n \ne -1$

**52.** $\displaystyle\int \dfrac{\sin \theta}{\sqrt{1 - \cos \theta}} \, d\theta$

**53.** $\displaystyle\int (3 - 5 \sin 5\pi x) \, dx$

**54.** $\displaystyle\int \dfrac{\tan (1/x)}{x^2} \, dx$

**55.** $\displaystyle\int x \cos x^2 \, dx$

**56.** $\displaystyle\int x \sin 3x^2 \, dx$

In Exercises 57 and 58, show that the given equation satisfies the differential equation.

**57.** $y = [a + \ln | \cos x |] \cos x + (b + x) \sin x$, $y'' + y = \sec x$

**58.** $y = a \cos 2x + b \sin 2x + 2x^2 - 1$,  $y'' + 4y = 8x^2$

In Exercises 59 and 60, find an equation of the tangent line to the graph of the equation at the specified point.

**59.** $y = x \cos x$,  $\left(\dfrac{\pi}{2}, 0\right)$      **60.** $y = \tan^2 \pi x$,  $\left(\dfrac{1}{4}, 1\right)$

**61.** Domestic energy consumption in the United States is seasonal. Suppose the consumption is approximated by the model

$$Q = 6.9 + \cos \frac{\pi(2t - 1)}{12}$$

where $Q$ is the total consumption in quads (quadrillion BTUs) and $t$ is the time in months, with $0 \le t \le 1$ corresponding to January, as shown in Figure 10.51. Find the dates for which this model predicts the greatest and the least consumption, respectively.

**62.** Find the average consumption rate of domestic energy for one year using the model of Exercise 61.

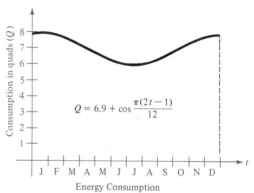

$$Q = 6.9 + \cos \frac{\pi(2t - 1)}{12}$$

Energy Consumption

FIGURE 10.51

**63.** Find the area of the largest rectangle that can be inscribed between one arc of the graph of $y = \cos x$ and the $x$-axis. Compare the area of the rectangle with the area of the region bounded by one arc of the cosine function and the $x$-axis.

# Reference Tables

## REFERENCE TABLE 1  The Greek Alphabet

| Letter | | Name | Letter | | Name | Letter | | Name |
|---|---|---|---|---|---|---|---|---|
| A | $\alpha$ | Alpha | I | $\iota$ | Iota | P | $\rho$ | Rho |
| B | $\beta$ | Beta | K | $\kappa$ | Kappa | $\Sigma$ | $\sigma$ | Sigma |
| $\Gamma$ | $\gamma$ | Gamma | $\Lambda$ | $\lambda$ | Lambda | T | $\tau$ | Tau |
| $\Delta$ | $\delta$ | Delta | M | $\mu$ | Mu | $\Upsilon$ | $\upsilon$ | Upsilon |
| E | $\epsilon$ | Epsilon | N | $\nu$ | Nu | $\Phi$ | $\phi$ | Phi |
| Z | $\zeta$ | Zeta | $\Xi$ | $\xi$ | Xi | X | $\chi$ | Chi |
| H | $\eta$ | Eta | O | $o$ | Omicron | $\Psi$ | $\psi$ | Psi |
| $\Theta$ | $\theta$ | Theta | $\Pi$ | $\pi$ | Pi | $\Omega$ | $\omega$ | Omega |

# REFERENCE TABLE 2    The Number of Each Day in the Year

| Day of Month | Jan. | Feb. | Mar. | Apr. | May | June | July | Aug. | Sept. | Oct. | Nov. | Dec. | Day of Month |
|---|---|---|---|---|---|---|---|---|---|---|---|---|---|
| 1 | 1 | 32 | 60 | 91 | 121 | 152 | 182 | 213 | 244 | 274 | 305 | 335 | 1 |
| 2 | 2 | 33 | 61 | 92 | 122 | 153 | 183 | 214 | 245 | 275 | 306 | 336 | 2 |
| 3 | 3 | 34 | 62 | 93 | 123 | 154 | 184 | 215 | 246 | 276 | 307 | 337 | 3 |
| 4 | 4 | 35 | 63 | 94 | 124 | 155 | 185 | 216 | 247 | 277 | 308 | 338 | 4 |
| 5 | 5 | 36 | 64 | 95 | 125 | 156 | 186 | 217 | 248 | 278 | 309 | 339 | 5 |
| 6 | 6 | 37 | 65 | 96 | 126 | 157 | 187 | 218 | 249 | 279 | 310 | 340 | 6 |
| 7 | 7 | 38 | 66 | 97 | 127 | 158 | 188 | 219 | 250 | 280 | 311 | 341 | 7 |
| 8 | 8 | 39 | 67 | 98 | 128 | 159 | 189 | 220 | 251 | 281 | 312 | 342 | 8 |
| 9 | 9 | 40 | 68 | 99 | 129 | 160 | 190 | 221 | 252 | 282 | 313 | 343 | 9 |
| 10 | 10 | 41 | 69 | 100 | 130 | 161 | 191 | 222 | 253 | 283 | 314 | 344 | 10 |
| 11 | 11 | 42 | 70 | 101 | 131 | 162 | 192 | 223 | 254 | 284 | 315 | 345 | 11 |
| 12 | 12 | 43 | 71 | 102 | 132 | 163 | 193 | 224 | 255 | 285 | 316 | 346 | 12 |
| 13 | 13 | 44 | 72 | 103 | 133 | 164 | 194 | 225 | 256 | 286 | 317 | 347 | 13 |
| 14 | 14 | 45 | 73 | 104 | 134 | 165 | 195 | 226 | 257 | 287 | 318 | 348 | 14 |
| 15 | 15 | 46 | 74 | 105 | 135 | 166 | 196 | 227 | 258 | 288 | 319 | 349 | 15 |
| 16 | 16 | 47 | 75 | 106 | 136 | 167 | 197 | 228 | 259 | 289 | 320 | 350 | 16 |
| 17 | 17 | 48 | 76 | 107 | 137 | 168 | 198 | 229 | 260 | 290 | 321 | 351 | 17 |
| 18 | 18 | 49 | 77 | 108 | 138 | 169 | 199 | 230 | 261 | 291 | 322 | 352 | 18 |
| 19 | 19 | 50 | 78 | 109 | 139 | 170 | 200 | 231 | 262 | 292 | 323 | 353 | 19 |
| 20 | 20 | 51 | 79 | 110 | 140 | 171 | 201 | 232 | 263 | 293 | 324 | 354 | 20 |
| 21 | 21 | 52 | 80 | 111 | 141 | 172 | 202 | 233 | 264 | 294 | 325 | 355 | 21 |
| 22 | 22 | 53 | 81 | 112 | 142 | 173 | 203 | 234 | 265 | 295 | 326 | 356 | 22 |
| 23 | 23 | 54 | 82 | 113 | 143 | 174 | 204 | 235 | 266 | 296 | 327 | 357 | 23 |
| 24 | 24 | 55 | 83 | 114 | 144 | 175 | 205 | 236 | 267 | 297 | 328 | 358 | 24 |
| 25 | 25 | 56 | 84 | 115 | 145 | 176 | 206 | 237 | 268 | 298 | 329 | 359 | 25 |
| 26 | 26 | 57 | 85 | 116 | 146 | 177 | 207 | 238 | 269 | 299 | 330 | 360 | 26 |
| 27 | 27 | 58 | 86 | 117 | 147 | 178 | 208 | 239 | 270 | 300 | 331 | 361 | 27 |
| 28 | 28 | 59 | 87 | 118 | 148 | 179 | 209 | 240 | 271 | 301 | 332 | 362 | 28 |
| 29 | 29 | * | 88 | 119 | 149 | 180 | 210 | 241 | 272 | 302 | 333 | 363 | 29 |
| 30 | 30 | — | 89 | 120 | 150 | 181 | 211 | 242 | 273 | 303 | 334 | 364 | 30 |
| 31 | 31 | — | 90 | — | 151 | — | 212 | 243 | — | 304 | — | 365 | 31 |

*On leap year, Feb 29 is day 60 and the number of each day after Feb 29 is increased by one.

## REFERENCE TABLE 3  Units of Measurement of Length

English System: Inch (in), Foot (ft), Yard (yd), Mile (mi)

1 mi = 5280 ft             1 mi = 1760 yd
1 yd = 3 ft                1 ft = 12 in

Metric System: Millimeter (mm), Centimeter (cm), Meter (m), Kilometer (km)

1 km = 1000 m              1 m = 100 cm
1 m = 1000 mm              1 cm = 10 mm

Conversion Factors (six significant figures)

| Metric to English | English to Metric |
|---|---|
| 1 mm = 0.0393701 in | 1 in = 25.4000 mm |
| 1 cm = 0.393701 in | 1 in = 2.54000 cm |
| 1 m = 3.28084 ft | 1 ft = 0.304800 m |
| 1 m = 1.09361 yd | 1 yd = 0.914400 m |
| 1 km = 0.621371 mi | 1 mi = 1.60934 km |

Miscellaneous

1 fathom = 6 ft
1 astronomical unit = 93,000,000 mi (average distance between earth and sun)
1 light year = 5,800,000,000,000 mi (distance traveled by light in one year)

## REFERENCE TABLE 4  Units of Measurement of Area

English System: Square Inch ($in^2$), Square Foot ($ft^2$), Square Yard ($yd^2$), Acre, Square Mile ($mi^2$)

$1\ mi^2 = 640$ acres         $1$ acre $= 43{,}560\ ft^2$
$1\ yd^2 = 9\ ft^2$           $1\ ft^2 = 144\ in^2$

Metric System: Square Centimeter ($cm^2$), Square Meter ($m^2$), Square Kilometer ($km^2$)

$1\ km^2 = 1{,}000{,}000\ m^2$         $1\ m^2 = 10{,}000\ cm^2$

## Conversion Factors (six significant figures)

<table>
<tr><td>

**Metric to English**

$1 \text{ cm}^2 = 0.155000 \text{ in}^2$
$1 \text{ m}^2 = 10.7640 \text{ ft}^2$
$1 \text{ m}^2 = 1.19599 \text{ yd}^2$
$1 \text{ km}^2 = 0.386102 \text{ mi}^2$

</td><td>

**English to Metric**

$1 \text{ in}^2 = 6.45160 \text{ cm}^2$
$1 \text{ ft}^2 = 0.0929030 \text{ m}^2$
$1 \text{ yd}^2 = 0.836127 \text{ m}^2$
$1 \text{ mi}^2 = 2.58999 \text{ km}^2$

</td></tr>
</table>

## Miscellaneous

1 square mile = 1 section = 4 quarters

## REFERENCE TABLE 5   Units of Measurement of Volume

**English System: Cubic Inch ($\text{in}^3$), Cubic Foot ($\text{ft}^3$), Cubic Yard ($\text{yd}^3$), Fluid Ounce (fl oz), Pint (pt), Quart (qt), Gallon (gal)**

<table>
<tr><td>

$1 \text{ yd}^3 = 27 \text{ ft}^3$
1 gal = 4 qt
1 qt = 32 fl oz
$1 \text{ gal} = 231 \text{ in}^3$

</td><td>

$1 \text{ ft}^3 = 1728 \text{ in}^3$
1 qt = 2 pt
1 pt = 16 fl oz
$1 \text{ ft}^3 = 7.48052 \text{ gal}$

</td></tr>
</table>

**Metric System: Cubic Centimeter ($\text{cm}^3$), Cubic Meter ($\text{m}^3$), Milliliter (cc), Liter**

<table>
<tr><td>

$1 \text{ m}^3 = 1000 \text{ liters}$
$1 \text{ cm}^3 = 1 \text{ cc}$

</td><td>

1 liter = 1000 cc

</td></tr>
</table>

## Conversion Factors (six significant figures)

<table>
<tr><td>

**Metric to English**

$1 \text{ cm}^3 = 0.0610237 \text{ in}^3$
$1 \text{ m}^3 = 35.3147 \text{ ft}^3$
$1 \text{ m}^3 = 1.30795 \text{ yd}^3$

</td><td>

**English to Metric**

$1 \text{ in}^3 = 16.3871 \text{ cm}^3$
$1 \text{ ft}^3 = 0.0283168 \text{ m}^3$
$1 \text{ yd}^3 = 0.764555 \text{ m}^3$

</td></tr>
</table>

## Miscellaneous

<table>
<tr><td>

1 gallon = 5 fifths
1 quart = 4 cups
1 tablespoon = 3 teaspoons (cooking)
1 tablespoon = 4 teaspoons (medical)
1 barrel (bl) = 42 gallons (petroleum)

</td><td>

1 fifth = 0.757 liters
1 cup = 16 tablespoons

</td></tr>
</table>

## REFERENCE TABLE 6   Units of Measurement of Mass and Force

English System (Force or Weight): Ounce (oz), Pound (lb), Ton

    1 ton = 2000 lb                  1 lb = 16 oz

Metric System (Mass): Gram (g), Kilogram (kg), Metric Ton

    1 metric ton = 1000 kg          1 kg = 1000 g

Conversion Factors (six significant figures at sea level)

| Metric to English | English to Metric |
|---|---|
| 1 g = 0.0352740 oz | 1 oz = 28.3495 g |
| 1 kg = 2.20462 lb | 1 lb = 0.453592 kg |
| 1 metric ton = 1.10231 ton | 1 ton = 0.907185 metric ton |

## REFERENCE TABLE 7   Units of Measurement of Time

Second (sec), Minute (min), Hour (hr)

    1 hr = 60 min                  1 min = 60 sec

Calendar Units: Day, Week, Month, Year (yr)

    1 yr = 365 days (366 days on leap year)
    1 yr = 365.256 mean solar days
    1 yr = 12 months
    1 week = 7 days
    1 day = 24 hr

## REFERENCE TABLE 8   Units of Measurement of Temperature

Fahrenheit (F), Celsius (C), Kelvin or Absolute (K)

| Celsius to Fahrenheit | Fahrenheit to Celsius |
|---|---|
| $°F = \dfrac{9}{5}°C + 32$ | $°C = \dfrac{5}{9}(°F - 32)$ |

| Celsius to Kelvin | Kelvin to Celsius |
|---|---|
| $°K = °C + 273.15$ | $°C = °K - 273.15$ |

Freezing temperature for water = 32° F = 0° C
Boiling temperature for water = 212° F = 100° C
Absolute zero temperature = 0° K = −273.15° C = −459.67° F
  (0° K is, by definition, the coldest possible temperature, at which there is no molecular activity.)


## REFERENCE TABLE 9   Miscellaneous Units and Number Constants

1 dozen (doz) = 12 units          1 gross = 12 dozen units
1 score = 20 units

$\pi \approx 3.1415926535$          $e \approx 2.7182818284$

Equatorial radius of the earth = 3963.34 mi = 6378.388 km
Polar radius of the earth = 3949.99 mi = 6356.912 km
Acceleration due to gravity at sea level = 32.1726 ft/sec$^2$
Speed of sound at sea level (standard atmosphere) = 1116.45 ft/sec
Speed of light in vacuum = 186,284 mi/sec
Density of water: 1 ft$^3$ = 62.425 lb


## REFERENCE TABLE 10   Algebra

## Operations with Exponents

1. $x^n x^m = x^{n+m}$          2. $\dfrac{x^n}{x^m} = x^{n-m}$          3. $(xy)^n = x^n y^n$

4. $\left(\dfrac{x}{y}\right)^n = \dfrac{x^n}{y^n}$          5. $(x^n)^m = x^{nm}$          6. $-x^n = -(x^n)$

7. $cx^n = c(x^n)$          8. $x^{n^m} = x^{(n^m)}$


## Exponents and Radicals (n and m are positive integers)

1. $x^n = \underbrace{x \cdot x \cdot x \cdots x}_{n \text{ factors}}$          2. $x^0 = 1,\ x \neq 0$

3. $x^{-n} = \dfrac{1}{x^n},\ x \neq 0$          *4. $\sqrt[n]{x} = a \implies x = a^n$

5. $x^{1/n} = \sqrt[n]{x}$          6. $x^{m/n} = (x^{1/n})^m = (\sqrt[n]{x})^m$
                                        $x^{m/n} = (x^m)^{1/n} = \sqrt[n]{x^m}$

7. $\sqrt[2]{x} = \sqrt{x}$

---

*If n is even, the principal nth root is defined to be positive.

---

## REFERENCE TABLE 10 (Continued)

### Operations with Fractions

1. $\dfrac{a}{b} + \dfrac{c}{d} = \dfrac{a}{b}\left(\dfrac{d}{d}\right) + \dfrac{c}{d}\left(\dfrac{b}{b}\right) = \dfrac{ad}{bd} + \dfrac{bc}{bd} = \dfrac{ad + bc}{bd}$

2. $\dfrac{a}{b} - \dfrac{c}{d} = \dfrac{a}{b}\left(\dfrac{d}{d}\right) - \dfrac{c}{d}\left(\dfrac{b}{b}\right) = \dfrac{ad}{bd} - \dfrac{bc}{bd} = \dfrac{ad - bc}{bd}$

3. $\left(\dfrac{a}{b}\right)\left(\dfrac{c}{d}\right) = \dfrac{ac}{bd}$

4. $\dfrac{a/b}{c/d} = \left(\dfrac{a}{b}\right)\left(\dfrac{d}{c}\right) = \dfrac{ad}{bc}$

$\dfrac{a/b}{c} = \dfrac{a/b}{c/1} = \left(\dfrac{a}{b}\right)\left(\dfrac{1}{c}\right) = \dfrac{a}{bc}$

5. $\dfrac{\cancel{a}b}{\cancel{a}c} = \dfrac{b}{c}$

$\dfrac{ab + ac}{ad} = \dfrac{\cancel{a}(b + c)}{\cancel{a}d} = \dfrac{b + c}{d}$

### Quadratic Formula

$$ax^2 + bx + c = 0 \quad \Rightarrow \quad x = \dfrac{-b \pm \sqrt{b^2 - 4ac}}{2a}$$

### Factors and Special Products

1. $x^2 - a^2 = (x - a)(x + a)$

2. $x^3 - a^3 = (x - a)(x^2 + ax + a^2)$

3. $x^3 + a^3 = (x + a)(x^2 - ax + a^2)$

4. $x^4 - a^4 = (x - a)(x + a)(x^2 + a^2)$

### Factoring by Grouping

$$acx^3 + adx^2 + bcx + bd = ax^2(cx + d) + b(cx + d) = (ax^2 + b)(cx + d)$$

### Binomial Theorem

1. $(x + a)^2 = x^2 + 2ax + a^2$

2. $(x - a)^2 = x^2 - 2ax + a^2$

3. $(x + a)^3 = x^3 + 3ax^2 + 3a^2x + a^3$

4. $(x - a)^3 = x^3 - 3ax^2 + 3a^2x - a^3$

5. $(x + a)^4 = x^4 + 4ax^3 + 6a^2x^2 + 4a^3x + a^4$

6. $(x - a)^4 = x^4 - 4ax^3 + 6a^2x^2 - 4a^3x + a^4$

7. $(x + a)^n = x^n + nax^{n-1} + \dfrac{n(n-1)}{2!}a^2x^{n-2} + \dfrac{n(n-1)(n-2)}{3!}a^3x^{n-3} + \cdots + na^{n-1}x + a^n$

8. $(x - a)^n = x^n - nax^{n-1} + \dfrac{n(n-1)}{2!}a^2x^{n-2} - \dfrac{n(n-1)(n-2)}{3!}a^3x^{n-3} + \cdots \pm na^{n-1}x \mp a^n$

## Miscellaneous

1. If $ab = 0$, then $a = 0$ or $b = 0$.

2. If $ac = bc$ and $c \neq 0$, then $a = b$.

3. Factorial: $0! = 1$, $1! = 1$, $2! = 2 \cdot 1$, $3! = 3 \cdot 2 \cdot 1$, $4! = 4 \cdot 3 \cdot 2 \cdot 1$, etc.

## Sequences

1. Arithmetic: $a, a + b, a + 2b, a + 3b, a + 4b, a + 5b, \ldots$

2. Geometric: $ar^0, ar^1, ar^2, ar^3, ar^4, ar^5, \ldots$

$$ar^0 + ar^1 + ar^2 + ar^3 + \cdots + ar^n = \frac{a(1 - r^{n+1})}{1 - r}$$

3. General Harmonic: $\dfrac{1}{a}, \dfrac{1}{a + b}, \dfrac{1}{a + 2b}, \dfrac{1}{a + 3b}, \dfrac{1}{a + 4b}, \dfrac{1}{a + 5b}, \ldots$

4. Harmonic: $\dfrac{1}{1}, \dfrac{1}{2}, \dfrac{1}{3}, \dfrac{1}{4}, \dfrac{1}{5}, \ldots$

5. $p$-Sequence: $\dfrac{1}{1^p}, \dfrac{1}{2^p}, \dfrac{1}{3^p}, \dfrac{1}{4^p}, \dfrac{1}{5^p}, \ldots$

## Series

$$\frac{1}{x} = 1 - (x - 1) + (x - 1)^2 - (x - 1)^3 + (x - 1)^4 - \cdots + (-1)^n(x - 1)^n + \cdots, \quad 0 < x < 2$$

$$\frac{1}{1 + x} = 1 - x + x^2 - x^3 + x^4 - x^5 + \cdots + (-1)^nx^n + \cdots, \quad -1 < x < 1$$

$$\ln x = (x - 1) - \frac{(x - 1)^2}{2} + \frac{(x - 1)^3}{3} - \frac{(x - 1)^4}{4} + \cdots + \frac{(-1)^{n-1}(x - 1)^n}{n} + \cdots, \quad 0 < x \leq 2$$

$$e^x = 1 + x + \frac{x^2}{2!} + \frac{x^3}{3!} + \frac{x^4}{4!} + \frac{x^5}{5!} + \cdots + \frac{x^n}{n!} + \cdots, \quad -\infty < x < \infty$$

$$\sin x = x - \frac{x^3}{3!} + \frac{x^5}{5!} - \frac{x^7}{7!} + \cdots, \quad -\infty < x < \infty$$

## REFERENCE TABLE 10 (Continued)

$$\cos x = 1 - \frac{x^2}{2!} + \frac{x^4}{4!} - \frac{x^6}{6!} + \cdots, \qquad -\infty < x < \infty$$

$$(1 + x)^k = 1 + kx + \frac{k(k-1)x^2}{2!} + \frac{k(k-1)(k-2)x^3}{3!} + \frac{k(k-1)(k-2)(k-3)x^4}{4!} + \cdots, \qquad -1 < x < 1*$$

$$(1 + x)^{-k} = 1 - kx + \frac{k(k+1)x^2}{2!} - \frac{k(k+1)(k+2)x^3}{3!} + \frac{k(k+1)(k+2)(k+3)x^4}{4!} - \cdots, \qquad -1 < x < 1*$$

---

*The convergence at $x = \pm 1$ depends on the value $k$.

## REFERENCE TABLE 11  Table of Square Roots and Cube Roots

| $n$ | $\sqrt{n}$ | $\sqrt[3]{n}$ | $n$ | $\sqrt{n}$ | $\sqrt[3]{n}$ | $n$ | $\sqrt{n}$ | $\sqrt[3]{n}$ |
|---|---|---|---|---|---|---|---|---|
| 1 | 1.00000 | 1.00000 | 31 | 5.56776 | 3.14138 | 61 | 7.81025 | 3.9365 |
| 2 | 1.41421 | 1.25992 | 32 | 5.65685 | 3.17480 | 62 | 7.87401 | 3.9578 |
| 3 | 1.73205 | 1.44225 | 33 | 5.74456 | 3.20753 | 63 | 7.93725 | 3.9790 |
| 4 | 2.00000 | 1.58740 | 34 | 5.83095 | 3.23961 | 64 | 8.00000 | 4.0000 |
| 5 | 2.23607 | 1.70998 | 35 | 5.91608 | 3.27107 | 65 | 8.06226 | 4.0207 |
| 6 | 2.44949 | 1.81712 | 36 | 6.00000 | 3.30193 | 66 | 8.12404 | 4.0412 |
| 7 | 2.64575 | 1.91293 | 37 | 6.08276 | 3.33222 | 67 | 8.18535 | 4.0615 |
| 8 | 2.82843 | 2.00000 | 38 | 6.16441 | 3.36198 | 68 | 8.24621 | 4.0816 |
| 9 | 3.00000 | 2.08008 | 39 | 6.24500 | 3.39121 | 69 | 8.30662 | 4.1015 |
| 10 | 3.16228 | 2.15443 | 40 | 6.32456 | 3.41995 | 70 | 8.36660 | 4.1212 |
| 11 | 3.31662 | 2.22398 | 41 | 6.40312 | 3.44822 | 71 | 8.42615 | 4.1408 |
| 12 | 3.46410 | 2.28943 | 42 | 6.48074 | 3.47603 | 72 | 8.48528 | 4.1601 |
| 13 | 3.60555 | 2.35133 | 43 | 6.55744 | 3.50340 | 73 | 8.54400 | 4.1793 |
| 14 | 3.74166 | 2.41014 | 44 | 6.63325 | 3.53035 | 74 | 8.60233 | 4.1983 |
| 15 | 3.87298 | 2.46621 | 45 | 6.70820 | 3.55689 | 75 | 8.66025 | 4.2171 |
| 16 | 4.00000 | 2.51984 | 46 | 6.78233 | 3.58305 | 76 | 8.71780 | 4.2358 |
| 17 | 4.12311 | 2.57128 | 47 | 6.85565 | 3.60883 | 77 | 8.77496 | 4.2543 |
| 18 | 4.24264 | 2.62074 | 48 | 6.92820 | 3.63424 | 78 | 8.83176 | 4.2726 |
| 19 | 4.35890 | 2.66840 | 49 | 7.00000 | 3.65931 | 79 | 8.88819 | 4.2908 |
| 20 | 4.47214 | 2.71442 | 50 | 7.07107 | 3.68403 | 80 | 8.94427 | 4.3088 |
| 21 | 4.58258 | 2.75892 | 51 | 7.14143 | 3.70843 | 81 | 9.00000 | 4.3267 |
| 22 | 4.69042 | 2.80204 | 52 | 7.21110 | 3.73251 | 82 | 9.05539 | 4.3444 |
| 23 | 4.79583 | 2.84387 | 53 | 7.28011 | 3.75629 | 83 | 9.11043 | 4.3620 |
| 24 | 4.89898 | 2.88450 | 54 | 7.34847 | 3.77976 | 84 | 9.16515 | 4.3795 |
| 25 | 5.00000 | 2.92402 | 55 | 7.41620 | 3.80295 | 85 | 9.21954 | 4.3968 |
| 26 | 5.09902 | 2.96250 | 56 | 7.48331 | 3.82586 | 86 | 9.27362 | 4.4140 |
| 27 | 5.19615 | 3.00000 | 57 | 7.54983 | 3.84850 | 87 | 9.32738 | 4.4310 |
| 28 | 5.29150 | 3.03659 | 58 | 7.61577 | 3.87088 | 88 | 9.38083 | 4.4479 |
| 29 | 5.38516 | 3.07232 | 59 | 7.68115 | 3.89300 | 89 | 9.43398 | 4.4647 |
| 30 | 5.47723 | 3.10723 | 60 | 7.74597 | 3.91487 | 90 | 9.48683 | 4.4814 |

| $n$ | $\sqrt{n}$ | $\sqrt[3]{n}$ | $n$ | $\sqrt{n}$ | $\sqrt[3]{n}$ | $n$ | $\sqrt{n}$ | $\sqrt[3]{n}$ |
|---|---|---|---|---|---|---|---|---|
| 91 | 9.53939 | 4.49794 | 128 | 11.3137 | 5.03968 | 165 | 12.8452 | 5.48481 |
| 92 | 9.59166 | 4.51436 | 129 | 11.3578 | 5.05277 | 166 | 12.8841 | 5.49586 |
| 93 | 9.64365 | 4.53065 | 130 | 11.4018 | 5.06580 | 167 | 12.9228 | 5.50688 |
| 94 | 9.69536 | 4.54684 | 131 | 11.4455 | 5.07875 | 168 | 12.9615 | 5.51785 |
| 95 | 9.74679 | 4.56290 | 132 | 11.4891 | 5.09164 | 169 | 13.0000 | 5.52877 |
| 96 | 9.79796 | 4.57886 | 133 | 11.5326 | 5.10447 | 170 | 13.0384 | 5.53966 |
| 97 | 9.84886 | 4.59470 | 134 | 11.5758 | 5.11723 | 171 | 13.0767 | 5.55050 |
| 98 | 9.89949 | 4.61044 | 135 | 11.6190 | 5.12993 | 172 | 13.1149 | 5.56130 |
| 99 | 9.94987 | 4.62606 | 136 | 11.6619 | 5.14256 | 173 | 13.1529 | 5.57205 |
| 100 | 10.0000 | 4.64159 | 137 | 11.7047 | 5.15514 | 174 | 13.1909 | 5.58277 |
| 101 | 10.0499 | 4.65701 | 138 | 11.7473 | 5.16765 | 175 | 13.2288 | 5.59344 |
| 102 | 10.0995 | 4.67233 | 139 | 11.7898 | 5.18010 | 176 | 13.2665 | 5.60408 |
| 103 | 10.1489 | 4.68755 | 140 | 11.8322 | 5.19249 | 177 | 13.3041 | 5.61467 |
| 104 | 10.1980 | 4.70267 | 141 | 11.8743 | 5.20483 | 178 | 13.3417 | 5.62523 |
| 105 | 10.2470 | 4.71769 | 142 | 11.9164 | 5.21710 | 179 | 13.3791 | 5.63574 |
| 106 | 10.2956 | 4.73262 | 143 | 11.9583 | 5.22932 | 180 | 13.4164 | 5.64622 |
| 107 | 10.3441 | 4.74746 | 144 | 12.0000 | 5.24148 | 181 | 13.4536 | 5.65665 |
| 108 | 10.3923 | 4.76220 | 145 | 12.0416 | 5.25359 | 182 | 13.4907 | 5.66705 |
| 109 | 10.4403 | 4.77686 | 146 | 12.0830 | 5.26564 | 183 | 13.5277 | 5.67741 |
| 110 | 10.4881 | 4.79142 | 147 | 12.1244 | 5.27763 | 184 | 13.5647 | 5.68773 |
| 111 | 10.5357 | 4.80590 | 148 | 12.1655 | 5.28957 | 185 | 13.6015 | 5.69802 |
| 112 | 10.5830 | 4.82028 | 149 | 12.2066 | 5.30146 | 186 | 13.6382 | 5.70827 |
| 113 | 10.6301 | 4.83459 | 150 | 12.2474 | 5.31329 | 187 | 13.6748 | 5.71848 |
| 114 | 10.6771 | 4.84881 | 151 | 12.2882 | 5.32507 | 188 | 13.7113 | 5.72865 |
| 115 | 10.7238 | 4.86294 | 152 | 12.3288 | 5.33680 | 189 | 13.7477 | 5.73879 |
| 116 | 10.7703 | 4.87700 | 153 | 12.3693 | 5.34848 | 190 | 13.7840 | 5.74890 |
| 117 | 10.8167 | 4.89097 | 154 | 12.4097 | 5.36011 | 191 | 13.8203 | 5.75897 |
| 118 | 10.8628 | 4.90487 | 155 | 12.4499 | 5.37169 | 192 | 13.8564 | 5.76900 |
| 119 | 10.9087 | 4.91868 | 156 | 12.4900 | 5.38321 | 193 | 13.8924 | 5.77900 |
| 120 | 10.9545 | 4.93242 | 157 | 12.5300 | 5.39469 | 194 | 13.9284 | 5.78896 |
| 121 | 11.0000 | 4.94609 | 158 | 12.5698 | 5.40612 | 195 | 13.9642 | 5.79889 |
| 122 | 11.0454 | 4.95968 | 159 | 12.6095 | 5.41750 | 196 | 14.0000 | 5.80879 |
| 123 | 11.0905 | 4.97319 | 160 | 12.6491 | 5.42884 | 197 | 14.0357 | 5.81865 |
| 124 | 11.1355 | 4.98663 | 161 | 12.6886 | 5.44012 | 198 | 14.0712 | 5.82848 |
| 125 | 11.1803 | 5.00000 | 162 | 12.7279 | 5.45136 | 199 | 14.1067 | 5.83827 |
| 126 | 11.2250 | 5.01330 | 163 | 12.7671 | 5.46256 | 200 | 14.1421 | 5.84804 |
| 127 | 11.2694 | 5.02653 | 164 | 12.8062 | 5.47370 | | | |

## REFERENCE TABLE 12    Exponential Tables

| $x$ | $e^x$ | $e^{-x}$ | $x$ | $e^x$ | $e^{-x}$ |
|-----|-------|----------|-----|-------|----------|
| 0.0 | 1.0000 | 1.0000 | 4.0 | 54.598 | 0.0183 |
| 0.1 | 1.1052 | 0.9048 | 4.1 | 60.340 | 0.0166 |
| 0.2 | 1.2214 | 0.8187 | 4.2 | 66.686 | 0.0150 |
| 0.3 | 1.3499 | 0.7408 | 4.3 | 73.700 | 0.0136 |
| 0.4 | 1.4918 | 0.6703 | 4.4 | 81.451 | 0.0123 |
| 0.5 | 1.6487 | 0.6065 | 4.5 | 90.017 | 0.0111 |
| 0.6 | 1.8221 | 0.5488 | 4.6 | 99.484 | 0.0101 |
| 0.7 | 2.0138 | 0.4966 | 4.7 | 109.95 | 0.0091 |
| 0.8 | 2.2255 | 0.4493 | 4.8 | 121.51 | 0.0082 |
| 0.9 | 2.4596 | 0.4066 | 4.9 | 134.29 | 0.0074 |
| 1.0 | 2.7183 | 0.3679 | 5.0 | 148.41 | 0.0067 |
| 1.1 | 3.0042 | 0.3329 | 5.1 | 164.02 | 0.0061 |
| 1.2 | 3.3201 | 0.3012 | 5.2 | 181.27 | 0.0055 |
| 1.3 | 3.6693 | 0.2725 | 5.3 | 200.34 | 0.0050 |
| 1.4 | 4.0552 | 0.2466 | 5.4 | 221.41 | 0.0045 |
| 1.5 | 4.4817 | 0.2231 | 5.5 | 244.69 | 0.0041 |
| 1.6 | 4.9530 | 0.2019 | 5.6 | 270.43 | 0.0037 |
| 1.7 | 5.4739 | 0.1827 | 5.7 | 298.87 | 0.0033 |
| 1.8 | 6.0496 | 0.1653 | 5.8 | 330.30 | 0.0030 |
| 1.9 | 6.6859 | 0.1496 | 5.9 | 365.04 | 0.0027 |
| 2.0 | 7.3891 | 0.1353 | 6.0 | 403.43 | 0.0025 |
| 2.1 | 8.1662 | 0.1225 | 6.1 | 445.86 | 0.0022 |
| 2.2 | 9.0250 | 0.1108 | 6.2 | 492.75 | 0.0020 |
| 2.3 | 9.9742 | 0.1003 | 6.3 | 544.57 | 0.0018 |
| 2.4 | 11.023 | 0.0907 | 6.4 | 601.85 | 0.0017 |
| 2.5 | 12.182 | 0.0821 | 6.5 | 665.14 | 0.0015 |
| 2.6 | 13.464 | 0.0743 | 6.6 | 735.10 | 0.0014 |
| 2.7 | 14.880 | 0.0672 | 6.7 | 812.41 | 0.0012 |
| 2.8 | 16.445 | 0.0608 | 6.8 | 897.85 | 0.0011 |
| 2.9 | 18.174 | 0.0550 | 6.9 | 992.27 | 0.0010 |
| 3.0 | 20.086 | 0.0498 | 7.0 | 1096.63 | 0.0009 |
| 3.1 | 22.198 | 0.0450 | 7.1 | 1211.97 | 0.0008 |
| 3.2 | 24.533 | 0.0408 | 7.2 | 1339.43 | 0.0007 |
| 3.3 | 27.113 | 0.0369 | 7.3 | 1480.30 | 0.0007 |
| 3.4 | 29.964 | 0.0334 | 7.4 | 1635.98 | 0.0006 |
| 3.5 | 33.115 | 0.0302 | 7.5 | 1808.04 | 0.0006 |
| 3.6 | 36.598 | 0.0273 | 7.6 | 1998.20 | 0.0005 |
| 3.7 | 40.447 | 0.0247 | 7.7 | 2208.35 | 0.0005 |
| 3.8 | 44.701 | 0.0224 | 7.8 | 2440.60 | 0.0004 |
| 3.9 | 49.402 | 0.0202 | 7.9 | 2697.28 | 0.0004 |

| $x$ | $e^x$ | $e^{-x}$ |
|-----|-------|----------|
| 8.0 | 2980.96 | 0.0003 |
| 8.1 | 3294.47 | 0.0003 |
| 8.2 | 3640.95 | 0.0003 |
| 8.3 | 4023.87 | 0.0002 |
| 8.4 | 4447.07 | 0.0002 |
| 8.5 | 4914.77 | 0.0002 |
| 8.6 | 5431.66 | 0.0002 |
| 8.7 | 6002.91 | 0.0002 |
| 8.8 | 6634.24 | 0.0002 |
| 8.9 | 7331.97 | 0.0001 |

| $x$ | $e^x$ | $e^{-x}$ |
|-----|-------|----------|
| 9.0 | 8103.08 | 0.0001 |
| 9.1 | 8955.29 | 0.0001 |
| 9.2 | 9897.13 | 0.0001 |
| 9.3 | 10938.02 | 0.0001 |
| 9.4 | 12088.38 | 0.0001 |
| 9.5 | 13359.73 | 0.0001 |
| 9.6 | 14764.78 | 0.0001 |
| 9.7 | 16317.61 | 0.0001 |
| 9.8 | 18033.74 | 0.0001 |
| 9.9 | 19930.37 | 0.0001 |
| 10.0 | 22026.47 | 0.0000 |

## REFERENCE TABLE 13   Natural Logarithmic Tables

|  | 0.00 | 0.01 | 0.02 | 0.03 | 0.04 | 0.05 | 0.06 | 0.07 | 0.08 | 0.09 |
|-----|------|------|------|------|------|------|------|------|------|------|
| 1.0 | 0.0000 | 0.0100 | 0.0198 | 0.0296 | 0.0392 | 0.0488 | 0.0583 | 0.0677 | 0.0770 | 0.0862 |
| 1.1 | 0.0953 | 0.1044 | 0.1133 | 0.1222 | 0.1310 | 0.1398 | 0.1484 | 0.1570 | 0.1655 | 0.1740 |
| 1.2 | 0.1823 | 0.1906 | 0.1989 | 0.2070 | 0.2151 | 0.2231 | 0.2311 | 0.2390 | 0.2469 | 0.2546 |
| 1.3 | 0.2624 | 0.2700 | 0.2776 | 0.2852 | 0.2927 | 0.3001 | 0.3075 | 0.3148 | 0.3221 | 0.3293 |
| 1.4 | 0.3365 | 0.3436 | 0.3507 | 0.3577 | 0.3646 | 0.3716 | 0.3784 | 0.3853 | 0.3920 | 0.3988 |
| 1.5 | 0.4055 | 0.4121 | 0.4187 | 0.4253 | 0.4318 | 0.4383 | 0.4447 | 0.4511 | 0.4574 | 0.4637 |
| 1.6 | 0.4700 | 0.4762 | 0.4824 | 0.4886 | 0.4947 | 0.5008 | 0.5068 | 0.5128 | 0.5188 | 0.5247 |
| 1.7 | 0.5306 | 0.5365 | 0.5423 | 0.5481 | 0.5539 | 0.5596 | 0.5653 | 0.5710 | 0.5766 | 0.5822 |
| 1.8 | 0.5878 | 0.5933 | 0.5988 | 0.6043 | 0.6098 | 0.6152 | 0.6206 | 0.6259 | 0.6313 | 0.6366 |
| 1.9 | 0.6419 | 0.6471 | 0.6523 | 0.6575 | 0.6627 | 0.6678 | 0.6729 | 0.6780 | 0.6831 | 0.6881 |
| 2.0 | 0.6931 | 0.6981 | 0.7031 | 0.7080 | 0.7129 | 0.7178 | 0.7227 | 0.7275 | 0.7324 | 0.7372 |
| 2.1 | 0.7419 | 0.7467 | 0.7514 | 0.7561 | 0.7608 | 0.7655 | 0.7701 | 0.7747 | 0.7793 | 0.7839 |
| 2.2 | 0.7885 | 0.7930 | 0.7975 | 0.8020 | 0.8065 | 0.8109 | 0.8154 | 0.8198 | 0.8242 | 0.8286 |
| 2.3 | 0.8329 | 0.8372 | 0.8416 | 0.8459 | 0.8502 | 0.8544 | 0.8587 | 0.8629 | 0.8671 | 0.8713 |
| 2.4 | 0.8755 | 0.8796 | 0.8838 | 0.8879 | 0.8920 | 0.8961 | 0.9002 | 0.9042 | 0.9083 | 0.9123 |
| 2.5 | 0.9163 | 0.9203 | 0.9243 | 0.9282 | 0.9322 | 0.9361 | 0.9400 | 0.9439 | 0.9478 | 0.9517 |
| 2.6 | 0.9555 | 0.9594 | 0.9632 | 0.9670 | 0.9708 | 0.9746 | 0.9783 | 0.9821 | 0.9858 | 0.9895 |
| 2.7 | 0.9933 | 0.9969 | 1.0006 | 1.0043 | 1.0080 | 1.0116 | 1.0152 | 1.0188 | 1.0225 | 1.0260 |
| 2.8 | 1.0296 | 1.0332 | 1.0367 | 1.0403 | 1.0438 | 1.0473 | 1.0508 | 1.0543 | 1.0578 | 1.0613 |
| 2.9 | 1.0647 | 1.0682 | 1.0716 | 1.0750 | 1.0784 | 1.0818 | 1.0852 | 1.0886 | 1.0919 | 1.0953 |
| 3.0 | 1.0986 | 1.1019 | 1.1053 | 1.1086 | 1.1119 | 1.1151 | 1.1184 | 1.1217 | 1.1249 | 1.1282 |
| 3.1 | 1.1314 | 1.1346 | 1.1378 | 1.1410 | 1.1442 | 1.1474 | 1.1506 | 1.1537 | 1.1569 | 1.1600 |
| 3.2 | 1.1632 | 1.1663 | 1.1694 | 1.1725 | 1.1756 | 1.1787 | 1.1817 | 1.1848 | 1.1878 | 1.1909 |
| 3.3 | 1.1939 | 1.1969 | 1.2000 | 1.2030 | 1.2060 | 1.2090 | 1.2119 | 1.2149 | 1.2179 | 1.2208 |
| 3.4 | 1.2238 | 1.2267 | 1.2296 | 1.2326 | 1.2355 | 1.2384 | 1.2413 | 1.2442 | 1.2470 | 1.2499 |

| | 0.00 | 0.01 | 0.02 | 0.03 | 0.04 | 0.05 | 0.06 | 0.07 | 0.08 | 0.09 |
|---|---|---|---|---|---|---|---|---|---|---|
| 3.5 | 1.2528 | 1.2556 | 1.2585 | 1.2613 | 1.2641 | 1.2669 | 1.2698 | 1.2726 | 1.2754 | 1.2782 |
| 3.6 | 1.2809 | 1.2837 | 1.2865 | 1.2892 | 1.2920 | 1.2947 | 1.2975 | 1.3002 | 1.3029 | 1.3056 |
| 3.7 | 1.3083 | 1.3110 | 1.3137 | 1.3164 | 1.3191 | 1.3218 | 1.3244 | 1.3271 | 1.3297 | 1.3324 |
| 3.8 | 1.3350 | 1.3376 | 1.3403 | 1.3429 | 1.3455 | 1.3481 | 1.3507 | 1.3533 | 1.3558 | 1.3584 |
| 3.9 | 1.3610 | 1.3635 | 1.3661 | 1.3686 | 1.3712 | 1.3737 | 1.3762 | 1.3788 | 1.3813 | 1.3838 |
| 4.0 | 1.3863 | 1.3888 | 1.3913 | 1.3938 | 1.3962 | 1.3987 | 1.4012 | 1.4036 | 1.4061 | 1.4085 |
| 4.1 | 1.4110 | 1.4134 | 1.4159 | 1.4183 | 1.4207 | 1.4231 | 1.4255 | 1.4279 | 1.4303 | 1.4327 |
| 4.2 | 1.4351 | 1.4375 | 1.4398 | 1.4422 | 1.4446 | 1.4469 | 1.4493 | 1.4516 | 1.4540 | 1.4563 |
| 4.3 | 1.4586 | 1.4609 | 1.4633 | 1.4656 | 1.4679 | 1.4702 | 1.4725 | 1.4748 | 1.4770 | 1.4793 |
| 4.4 | 1.4816 | 1.4839 | 1.4861 | 1.4884 | 1.4907 | 1.4929 | 1.4951 | 1.4974 | 1.4996 | 1.5019 |
| 4.5 | 1.5041 | 1.5063 | 1.5085 | 1.5107 | 1.5129 | 1.5151 | 1.5173 | 1.5195 | 1.5217 | 1.5239 |
| 4.6 | 1.5261 | 1.5282 | 1.5304 | 1.5326 | 1.5347 | 1.5369 | 1.5390 | 1.5412 | 1.5433 | 1.5454 |
| 4.7 | 1.5476 | 1.5497 | 1.5518 | 1.5539 | 1.5560 | 1.5581 | 1.5602 | 1.5623 | 1.5644 | 1.5665 |
| 4.8 | 1.5686 | 1.5707 | 1.5728 | 1.5748 | 1.5769 | 1.5790 | 1.5810 | 1.5831 | 1.5851 | 1.5872 |
| 4.9 | 1.5892 | 1.5913 | 1.5933 | 1.5953 | 1.5974 | 1.5994 | 1.6014 | 1.6034 | 1.6054 | 1.6074 |
| 5.0 | 1.6094 | 1.6114 | 1.6134 | 1.6154 | 1.6174 | 1.6194 | 1.6214 | 1.6233 | 1.6253 | 1.6273 |
| 5.1 | 1.6292 | 1.6312 | 1.6332 | 1.6351 | 1.6371 | 1.6390 | 1.6409 | 1.6429 | 1.6448 | 1.6467 |
| 5.2 | 1.6487 | 1.6506 | 1.6525 | 1.6544 | 1.6563 | 1.6582 | 1.6601 | 1.6620 | 1.6639 | 1.6658 |
| 5.3 | 1.6677 | 1.6696 | 1.6715 | 1.6734 | 1.6752 | 1.6771 | 1.6790 | 1.6808 | 1.6827 | 1.6845 |
| 5.4 | 1.6864 | 1.6882 | 1.6901 | 1.6919 | 1.6938 | 1.6956 | 1.6974 | 1.6993 | 1.7011 | 1.7029 |
| 5.5 | 1.7047 | 1.7066 | 1.7084 | 1.7102 | 1.7120 | 1.7138 | 1.7156 | 1.7174 | 1.7192 | 1.7210 |
| 5.6 | 1.7228 | 1.7246 | 1.7263 | 1.7281 | 1.7299 | 1.7317 | 1.7334 | 1.7352 | 1.7370 | 1.7387 |
| 5.7 | 1.7405 | 1.7422 | 1.7440 | 1.7457 | 1.7475 | 1.7492 | 1.7509 | 1.7527 | 1.7544 | 1.7561 |
| 5.8 | 1.7579 | 1.7596 | 1.7613 | 1.7630 | 1.7647 | 1.7664 | 1.7681 | 1.7699 | 1.7716 | 1.7733 |
| 5.9 | 1.7750 | 1.7766 | 1.7783 | 1.7800 | 1.7817 | 1.7834 | 1.7851 | 1.7867 | 1.7884 | 1.7901 |
| 6.0 | 1.7918 | 1.7934 | 1.7951 | 1.7967 | 1.7984 | 1.8001 | 1.8017 | 1.8034 | 1.8050 | 1.8066 |
| 6.1 | 1.8083 | 1.8099 | 1.8116 | 1.8132 | 1.8148 | 1.8165 | 1.8181 | 1.8197 | 1.8213 | 1.8229 |
| 6.2 | 1.8245 | 1.8262 | 1.8278 | 1.8294 | 1.8310 | 1.8326 | 1.8342 | 1.8358 | 1.8374 | 1.8390 |
| 6.3 | 1.8405 | 1.8421 | 1.8437 | 1.8453 | 1.8469 | 1.8485 | 1.8500 | 1.8516 | 1.8532 | 1.8547 |
| 6.4 | 1.8563 | 1.8579 | 1.8594 | 1.8610 | 1.8625 | 1.8641 | 1.8656 | 1.8672 | 1.8687 | 1.8703 |
| 6.5 | 1.8718 | 1.8733 | 1.8749 | 1.8764 | 1.8779 | 1.8795 | 1.8810 | 1.8825 | 1.8840 | 1.8856 |
| 6.6 | 1.8871 | 1.8886 | 1.8901 | 1.8916 | 1.8931 | 1.8946 | 1.8961 | 1.8976 | 1.8991 | 1.9006 |
| 6.7 | 1.9021 | 1.9036 | 1.9051 | 1.9066 | 1.9081 | 1.9095 | 1.9110 | 1.9125 | 1.9140 | 1.9155 |
| 6.8 | 1.9169 | 1.9184 | 1.9199 | 1.9213 | 1.9228 | 1.9242 | 1.9257 | 1.9272 | 1.9286 | 1.9301 |
| 6.9 | 1.9315 | 1.9330 | 1.9344 | 1.9359 | 1.9373 | 1.9387 | 1.9402 | 1.9416 | 1.9430 | 1.9445 |
| 7.0 | 1.9459 | 1.9473 | 1.9488 | 1.9502 | 1.9516 | 1.9530 | 1.9544 | 1.9559 | 1.9573 | 1.9587 |
| 7.1 | 1.9601 | 1.9615 | 1.9629 | 1.9643 | 1.9657 | 1.9671 | 1.9685 | 1.9699 | 1.9713 | 1.9727 |
| 7.2 | 1.9741 | 1.9755 | 1.9769 | 1.9782 | 1.9796 | 1.9810 | 1.9824 | 1.9838 | 1.9851 | 1.9865 |
| 7.3 | 1.9879 | 1.9892 | 1.9906 | 1.9920 | 1.9933 | 1.9947 | 1.9961 | 1.9974 | 1.9988 | 2.0001 |
| 7.4 | 2.0015 | 2.0028 | 2.0042 | 2.0055 | 2.0069 | 2.0082 | 2.0096 | 2.0109 | 2.0122 | 2.0136 |
| 7.5 | 2.0149 | 2.0162 | 2.0176 | 2.0189 | 2.0202 | 2.0215 | 2.0229 | 2.0242 | 2.0255 | 2.0268 |
| 7.6 | 2.0281 | 2.0295 | 2.0308 | 2.0321 | 2.0334 | 2.0347 | 2.0360 | 2.0373 | 2.0386 | 2.0399 |
| 7.7 | 2.0412 | 2.0425 | 2.0438 | 2.0451 | 2.0464 | 2.0477 | 2.0490 | 2.0503 | 2.0516 | 2.0528 |
| 7.8 | 2.0541 | 2.0554 | 2.0567 | 2.0580 | 2.0592 | 2.0605 | 2.0618 | 2.0631 | 2.0643 | 2.0656 |
| 7.9 | 2.0669 | 2.0681 | 2.0694 | 2.0707 | 2.0719 | 2.0732 | 2.0744 | 2.0757 | 2.0769 | 2.0782 |

| | 0.00 | 0.01 | 0.02 | 0.03 | 0.04 | 0.05 | 0.06 | 0.07 | 0.08 | 0.09 |
|---|---|---|---|---|---|---|---|---|---|---|
| 8.0 | 2.0794 | 2.0807 | 2.0819 | 2.0832 | 2.0844 | 2.0857 | 2.0869 | 2.0882 | 2.0894 | 2.0906 |
| 8.1 | 2.0919 | 2.0931 | 2.0943 | 2.0956 | 2.0968 | 2.0980 | 2.0992 | 2.1005 | 2.1017 | 2.1029 |
| 8.2 | 2.1041 | 2.1054 | 2.1066 | 2.1078 | 2.1090 | 2.1102 | 2.1114 | 2.1126 | 2.1138 | 2.1150 |
| 8.3 | 2.1163 | 2.1175 | 2.1187 | 2.1199 | 2.1211 | 2.1223 | 2.1235 | 2.1247 | 2.1258 | 2.1270 |
| 8.4 | 2.1282 | 2.1294 | 2.1306 | 2.1318 | 2.1330 | 2.1342 | 2.1353 | 2.1365 | 2.1377 | 2.1389 |
| 8.5 | 2.1401 | 2.1412 | 2.1424 | 2.1436 | 2.1448 | 2.1459 | 2.1471 | 2.1483 | 2.1494 | 2.1506 |
| 8.6 | 2.1518 | 2.1529 | 2.1541 | 2.1552 | 2.1564 | 2.1576 | 2.1587 | 2.1599 | 2.1610 | 2.1622 |
| 8.7 | 2.1633 | 2.1645 | 2.1656 | 2.1668 | 2.1679 | 2.1691 | 2.1702 | 2.1713 | 2.1725 | 2.1736 |
| 8.8 | 2.1748 | 2.1759 | 2.1770 | 2.1782 | 2.1793 | 2.1804 | 2.1815 | 2.1827 | 2.1838 | 2.1849 |
| 8.9 | 2.1861 | 2.1872 | 2.1883 | 2.1894 | 2.1905 | 2.1917 | 2.1928 | 2.1939 | 2.1950 | 2.1961 |
| 9.0 | 2.1972 | 2.1983 | 2.1994 | 2.2006 | 2.2017 | 2.2028 | 2.2039 | 2.2050 | 2.2061 | 2.2072 |
| 9.1 | 2.2083 | 2.2094 | 2.2105 | 2.2116 | 2.2127 | 2.2138 | 2.2148 | 2.2159 | 2.2170 | 2.2181 |
| 9.2 | 2.2192 | 2.2203 | 2.2214 | 2.2225 | 2.2235 | 2.2246 | 2.2257 | 2.2268 | 2.2279 | 2.2289 |
| 9.3 | 2.2300 | 2.2311 | 2.2322 | 2.2332 | 2.2343 | 2.2354 | 2.2364 | 2.2375 | 2.2386 | 2.2396 |
| 9.4 | 2.2407 | 2.2418 | 2.2428 | 2.2439 | 2.2450 | 2.2460 | 2.2471 | 2.2481 | 2.2492 | 2.2502 |
| 9.5 | 2.2513 | 2.2523 | 2.2534 | 2.2544 | 2.2555 | 2.2565 | 2.2576 | 2.2586 | 2.2597 | 2.2607 |
| 9.6 | 2.2618 | 2.2628 | 2.2638 | 2.2649 | 2.2659 | 2.2670 | 2.2680 | 2.2690 | 2.2701 | 2.2711 |
| 9.7 | 2.2721 | 2.2732 | 2.2742 | 2.2752 | 2.2762 | 2.2773 | 2.2783 | 2.2793 | 2.2803 | 2.2814 |
| 9.8 | 2.2824 | 2.2834 | 2.2844 | 2.2854 | 2.2865 | 2.2875 | 2.2885 | 2.2895 | 2.2905 | 2.2915 |
| 9.9 | 2.2925 | 2.2935 | 2.2946 | 2.2956 | 2.2966 | 2.2976 | 2.2986 | 2.2996 | 2.3006 | 2.3016 |

# REFERENCE TABLE 14  Properties of Logarithms

Inverse Properties

1. $\ln e^x = x$  2. $e^{\ln x} = x$

Properties of Logarithms

1. $\ln 1 = 0$  2. $\ln e = 1$

3. $\ln xy = \ln x + \ln y$  4. $\ln \dfrac{x}{y} = \ln x - \ln y$

5. $\ln x^y = y \ln x$  6. $\log_a x = \dfrac{\ln x}{\ln a}$

# REFERENCE TABLE 15    Geometry

## Triangles

### 1. General triangle

Sum of angles $\alpha + \beta + \theta = 180°$

$$\text{Area} = \frac{1}{2}(\text{base})(\text{height}) = \frac{1}{2}bh$$

### 2. Similar triangles

$$\frac{a}{b} = \frac{A}{B}$$

### 3. Right triangle

$c^2 = a^2 + b^2$ (Pythagorean Theorem)

Sum of acute angles $\alpha + \beta = 90°$

### 4. Equilateral triangle

$$\text{Height} = h = \frac{\sqrt{3}s}{2}$$

$$\text{Area} = \frac{\sqrt{3}s^2}{4}$$

### 5. Isosceles right triangle

$$\text{Area} = \frac{s^2}{2}$$

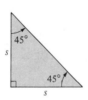

## Quadrilaterals (Four-Sided Figures)

### 1. Rectangle

$$\text{Area} = (\text{length})(\text{height}) = bh$$

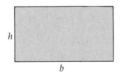

### 2. Square

$$\text{Area} = (\text{side})^2 = s^2$$

3. Parallelogram

$$\text{Area} = bh$$

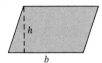

4. Trapezoid

$$\text{Area} = \frac{1}{2}h(a + b)$$

# Circles and Ellipses

1. Circle

$$\text{Area} = \pi r^2$$

$$\text{Circumference} = 2\pi r$$

2. Sector of circle ($\theta$ in radians)

$$\text{Area} = \frac{\theta r^2}{2}$$

$$s = r\theta$$

3. Circular ring

$$\text{Area} = \pi(R^2 - r^2)$$

4. Ellipse

$$\text{Area} = \pi ab$$

$$\text{Circumference} \approx 2\pi\sqrt{\frac{a^2 + b^2}{2}}$$

# Solid Figures

1. Cone ($A$ = area of base)

$$\text{Volume} = \frac{Ah}{3}$$

2. Right circular cone

$$\text{Volume} = \frac{\pi r^2 h}{3}$$

$$\text{Lateral surface area} = \pi r\sqrt{r^2 + h^2}$$

## REFERENCE TABLE 15 (Continued)

3. Frustum of right circular cone

$$\text{Volume} = \frac{\pi(r^2 + rR + R^2)h}{3}$$

Lateral surface area $= \pi s(R + r)$

4. Right circular cylinder

$$\text{Volume} = \pi r^2 h$$

Lateral surface area $= 2\pi rh$

5. Sphere

$$\text{Volume} = \frac{4}{3}\pi r^3$$

Surface area $= 4\pi r^2$

## REFERENCE TABLE 16   Trigonometry

## Definition of the Six Trigonometric Functions

Right triangle definition: $0 < \theta < \pi/2$

$$\sin \theta = \frac{\text{opp.}}{\text{hyp.}} \qquad \csc \theta = \frac{\text{hyp.}}{\text{opp.}}$$

$$\cos \theta = \frac{\text{adj.}}{\text{hyp.}} \qquad \sec \theta = \frac{\text{hyp.}}{\text{adj.}}$$

$$\tan \theta = \frac{\text{opp.}}{\text{adj.}} \qquad \cot \theta = \frac{\text{adj.}}{\text{opp.}}$$

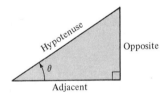

Circular function definition: $\theta$ is any angle and $(x, y)$ is a point on the terminal ray of the angle.

$$\sin \theta = \frac{y}{r} \qquad \csc \theta = \frac{r}{y}$$

$$\cos \theta = \frac{x}{r} \qquad \sec \theta = \frac{r}{x}$$

$$\tan \theta = \frac{y}{x} \qquad \cot \theta = \frac{x}{y}$$

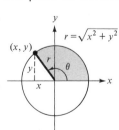

## Signs of the Trigonometric Functions by Quadrant

| Quadrant | sin | cos | tan | cot | sec | csc |
|----------|-----|-----|-----|-----|-----|-----|
| I | + | + | + | + | + | + |
| II | + | − | − | − | − | + |
| III | − | − | + | + | − | − |
| IV | − | + | − | − | + | − |

## Values of Trigonometric Functions at Common Angles

| Degrees | 0 | 30° | 45° | 60° | 90° |
|---------|---|-----|-----|-----|-----|
| Radians | 0 | $\dfrac{\pi}{6}$ | $\dfrac{\pi}{4}$ | $\dfrac{\pi}{3}$ | $\dfrac{\pi}{2}$ |
| $\sin \theta$ | 0 | $\dfrac{1}{2}$ | $\dfrac{\sqrt{2}}{2}$ | $\dfrac{\sqrt{3}}{2}$ | 1 |
| $\cos \theta$ | 1 | $\dfrac{\sqrt{3}}{2}$ | $\dfrac{\sqrt{2}}{2}$ | $\dfrac{1}{2}$ | 0 |
| $\tan \theta$ | 0 | $\dfrac{1}{\sqrt{3}}$ | 1 | $\sqrt{3}$ | undefined |

## Trigonometric Identities

### Reciprocal identities

$$\sin \theta = \frac{1}{\csc \theta} \qquad \cos \theta = \frac{1}{\sec \theta} \qquad \tan \theta = \frac{1}{\cot \theta}$$

$$\csc \theta = \frac{1}{\sin \theta} \qquad \sec \theta = \frac{1}{\cos \theta} \qquad \cot \theta = \frac{1}{\tan \theta}$$

$$\tan \theta = \frac{\sin \theta}{\cos \theta} \qquad \cot \theta = \frac{\cos \theta}{\sin \theta}$$

### Pythagorean identities

$$\sin^2 \theta + \cos^2 \theta = 1 \qquad \tan^2 \theta + 1 = \sec^2 \theta \qquad \cot^2 \theta + 1 = \csc^2 \theta$$

### Reduction formulas

$$\sin (-\theta) = -\sin \theta \qquad \cos (-\theta) = \cos \theta \qquad \tan (-\theta) = -\tan \theta$$
$$\sin \theta = -\sin (\theta - \pi) \qquad \cos \theta = -\cos (\theta - \pi) \qquad \tan \theta = \tan (\theta - \pi)$$

## REFERENCE TABLE 16 (Continued)

### Sum or difference of two angles

$\sin (\theta \pm \phi) = \sin \theta \cos \phi \pm \cos \theta \sin \phi$

$\cos (\theta \pm \phi) = \cos \theta \cos \phi \mp \sin \theta \sin \phi$

$\tan (\theta \pm \phi) = \dfrac{\tan \theta \pm \tan \phi}{1 \mp \tan \theta \tan \phi}$

$\sin (\theta + \phi) \sin (\theta - \phi) = \sin^2 \theta - \sin^2 \phi$

$\cos (\theta + \phi) \cos (\theta - \phi) = \cos^2 \theta - \sin^2 \phi$

### Double-angle identities

$\sin 2\theta = 2 \sin \theta \cos \theta$

$\tan 2\theta = \dfrac{2 \tan \theta}{1 - \tan^2 \theta}$

$\cos 2\theta = 2 \cos^2 \theta - 1 = 1 - 2 \sin^2 \theta$

### Multiple-angle identities

$\sin 3\theta = 3 \sin \theta - 4 \sin^3 \theta$

$\cos 3\theta = -3 \cos \theta + 4 \cos^3 \theta$

$\tan 3\theta = \dfrac{3 \tan \theta - \tan^3 \theta}{1 - 3 \tan^2 \theta}$

$\sin 4\theta = 4 \sin \theta \cos \theta - 8 \sin^3 \theta \cos \theta$

$\cos 4\theta = 8 \cos^4 \theta - 8 \cos^2 \theta + 1$

$\tan 4\theta = \dfrac{4 \tan \theta - 4 \tan^3 \theta}{1 - 6 \tan^2 \theta + \tan^4 \theta}$

### Half-angle identities

$\sin^2 \theta = \dfrac{1}{2}(1 - \cos 2\theta)$

$\cos^2 \theta = \dfrac{1}{2}(1 + \cos 2\theta)$

### Product identities

$\sin \theta \sin \phi = \dfrac{1}{2} \cos (\theta - \phi) - \dfrac{1}{2} \cos (\theta + \phi)$

$\cos \theta \cos \phi = \dfrac{1}{2} \cos (\theta - \phi) + \dfrac{1}{2} \cos (\theta + \phi)$

$\cos \theta \sin \phi = \dfrac{1}{2} \sin (\theta - \phi) + \dfrac{1}{2} \sin (\theta + \phi)$

# REFERENCE TABLE 17  Tables of Trigonometric Functions

1 degree ≈ 0.01745 radians
1 radian ≈ 57.29578 degrees

For $0 \leq \theta \leq 45$, read from upper left.
For $45 \leq \theta \leq 90$, read from lower right.
For $90 \leq \theta \leq 360$, use the identities:

| $\theta$ | Quadrant II | Quadrant III | Quadrant IV |
|---|---|---|---|
| $\sin \theta$ | $\sin(180-\theta)$ | $-\sin(\theta-180)$ | $-\sin(360-\theta)$ |
| $\cos \theta$ | $-\cos(180-\theta)$ | $-\cos(\theta-180)$ | $\cos(360-\theta)$ |
| $\tan \theta$ | $-\tan(180-\theta)$ | $\tan(\theta-180)$ | $-\tan(360-\theta)$ |
| $\cot \theta$ | $-\cot(180-\theta)$ | $\cot(\theta-180)$ | $-\cot(360-\theta)$ |

| Degrees | Radians | sin | cos | tan | cot | | |
|---|---|---|---|---|---|---|---|
| 0°00′ | .0000 | .0000 | 1.0000 | .0000 | – | 1.5708 | 90°00′ |
| 10 | .0029 | .0029 | 1.0000 | .0029 | 343.774 | 1.5679 | 50 |
| 20 | .0058 | .0058 | 1.0000 | .0058 | 171.885 | 1.5650 | 40 |
| 30 | .0087 | .0087 | 1.0000 | .0087 | 114.589 | 1.5621 | 30 |
| 40 | .0116 | .0116 | .9999 | .0116 | 85.940 | 1.5592 | 20 |
| 50 | .0145 | .0145 | .9999 | .0145 | 68.750 | 1.5563 | 10 |
| 1°00′ | .0175 | .0175 | .9998 | .0175 | 57.290 | 1.5533 | 89°00′ |
| 10 | .0204 | .0204 | .9998 | .0204 | 49.104 | 1.5504 | 50 |
| 20 | .0233 | .0233 | .9997 | .0233 | 42.964 | 1.5475 | 40 |
| 30 | .0262 | .0262 | .9997 | .0262 | 38.188 | 1.5446 | 30 |
| 40 | .0291 | .0291 | .9996 | .0291 | 34.368 | 1.5417 | 20 |
| 50 | .0320 | .0320 | .9995 | .0320 | 31.242 | 1.5388 | 10 |
| 2°00′ | .0349 | .0349 | .9994 | .0349 | 28.636 | 1.5359 | 88°00′ |
| 10 | .0378 | .0378 | .9993 | .0378 | 26.432 | 1.5330 | 50 |
| 20 | .0407 | .0407 | .9992 | .0407 | 24.542 | 1.5301 | 40 |
| 30 | .0436 | .0436 | .9990 | .0437 | 22.904 | 1.5272 | 30 |
| 40 | .0465 | .0465 | .9989 | .0466 | 21.470 | 1.5243 | 20 |
| 50 | .0495 | .0494 | .9988 | .0495 | 20.206 | 1.5213 | 10 |
| 3°00′ | .0524 | .0523 | .9986 | .0524 | 19.081 | 1.5184 | 87°00′ |
| 10 | .0553 | .0552 | .9985 | .0553 | 18.075 | 1.5155 | 50 |
| 20 | .0582 | .0581 | .9983 | .0582 | 17.169 | 1.5126 | 40 |
| 30 | .0611 | .0610 | .9981 | .0612 | 16.350 | 1.5097 | 30 |
| 40 | .0640 | .0640 | .9980 | .0641 | 15.605 | 1.5068 | 20 |
| 50 | .0669 | .0669 | .9978 | .0670 | 14.924 | 1.5039 | 10 |
| 4°00′ | .0698 | .0698 | .9976 | .0699 | 14.301 | 1.5010 | 86°00′ |
| 10 | .0727 | .0727 | .9974 | .0729 | 13.727 | 1.4981 | 50 |
| 20 | .0756 | .0756 | .9971 | .0758 | 13.197 | 1.4952 | 40 |
| 30 | .0785 | .0785 | .9969 | .0787 | 12.706 | 1.4923 | 30 |
| 40 | .0814 | .0814 | .9967 | .0816 | 12.251 | 1.4893 | 20 |
| 50 | .0844 | .0843 | .9964 | .0846 | 11.826 | 1.4864 | 10 |
| 5°00′ | .0873 | .0872 | .9962 | .0875 | 11.430 | 1.4835 | 85°00′ |
| 10 | .0902 | .0901 | .9959 | .0904 | 11.059 | 1.4806 | 50 |
| 20 | .0931 | .0929 | .9957 | .0934 | 10.712 | 1.4777 | 40 |
| 30 | .0960 | .0958 | .9954 | .0963 | 10.385 | 1.4748 | 30 |
| 40 | .0989 | .0987 | .9951 | .0992 | 10.078 | 1.4719 | 20 |
| 50 | .1018 | .1016 | .9948 | .1022 | 9.788 | 1.4690 | 10 |
| | | cos | sin | cot | tan | Radians | Degrees |

| Degrees | Radians | sin | cos | tan | cot | | |
|---|---|---|---|---|---|---|---|
| 6°00' | .1047 | .1045 | .9945 | .1051 | 9.514 | 1.4661 | 84°00' |
| 10 | .1076 | .1074 | .9942 | .1080 | 9.255 | 1.4632 | 50 |
| 20 | .1105 | .1103 | .9939 | .1110 | 9.010 | 1.4603 | 40 |
| 30 | .1134 | .1132 | .9936 | .1139 | 8.777 | 1.4573 | 30 |
| 40 | .1164 | .1161 | .9932 | .1169 | 8.556 | 1.4544 | 20 |
| 50 | .1193 | .1190 | .9929 | .1198 | 8.345 | 1.4515 | 10 |
| 7°00' | .1222 | .1219 | .9925 | .1228 | 8.144 | 1.4486 | 83°00' |
| 10 | .1251 | .1248 | .9922 | .1257 | 7.953 | 1.4457 | 50 |
| 20 | .1280 | .1276 | .9918 | .1287 | 7.770 | 1.4428 | 40 |
| 30 | .1309 | .1305 | .9914 | .1317 | 7.596 | 1.4399 | 30 |
| 40 | .1338 | .1334 | .9911 | .1346 | 7.429 | 1.4370 | 20 |
| 50 | .1367 | .1363 | .9907 | .1376 | 7.269 | 1.4341 | 10 |
| 8°00' | .1396 | .1392 | .9903 | .1405 | 7.115 | 1.4312 | 82°00' |
| 10 | .1425 | .1421 | .9899 | .1435 | 6.968 | 1.4283 | 50 |
| 20 | .1454 | .1449 | .9894 | .1465 | 6.827 | 1.4254 | 40 |
| 30 | .1484 | .1478 | .9890 | .1495 | 6.691 | 1.4224 | 30 |
| 40 | .1513 | .1507 | .9886 | .1524 | 6.561 | 1.4195 | 20 |
| 50 | .1542 | .1536 | .9881 | .1554 | 6.435 | 1.4166 | 10 |
| 9°00' | .1571 | .1564 | .9877 | .1584 | 6.314 | 1.4137 | 81°00' |
| 10 | .1600 | .1593 | .9872 | .1614 | 6.197 | 1.4108 | 50 |
| 20 | .1629 | .1622 | .9868 | .1644 | 6.084 | 1.4079 | 40 |
| 30 | .1658 | .1650 | .9863 | .1673 | 5.976 | 1.4050 | 30 |
| 40 | .1687 | .1679 | .9858 | .1703 | 5.871 | 1.4021 | 20 |
| 50 | .1716 | .1708 | .9853 | .1733 | 5.769 | 1.3992 | 10 |
| 10°00' | .1745 | .1736 | .9848 | .1763 | 5.671 | 1.3963 | 80°00' |
| 10 | .1774 | .1765 | .9843 | .1793 | 5.576 | 1.3934 | 50 |
| 20 | .1804 | .1794 | .9838 | .1823 | 5.485 | 1.3904 | 40 |
| 30 | .1833 | .1822 | .9833 | .1853 | 5.396 | 1.3875 | 30 |
| 40 | .1862 | .1851 | .9827 | .1883 | 5.309 | 1.3846 | 20 |
| 50 | .1891 | .1880 | .9822 | .1914 | 5.226 | 1.3817 | 10 |
| 11°00' | .1920 | .1908 | .9816 | .1944 | 5.145 | 1.3788 | 79°00' |
| 10 | .1949 | .1937 | .9811 | .1974 | 5.066 | 1.3759 | 50 |
| 20 | .1978 | .1965 | .9805 | .2004 | 4.989 | 1.3730 | 40 |
| 30 | .2007 | .1994 | .9799 | .2035 | 4.915 | 1.3701 | 30 |
| 40 | .2036 | .2022 | .9793 | .2065 | 4.843 | 1.3672 | 20 |
| 50 | .2065 | .2051 | .9787 | .2095 | 4.773 | 1.3643 | 10 |
| 12°00' | .2094 | .2079 | .9781 | .2126 | 4.705 | 1.3614 | 78°00' |
| 10 | .2123 | .2108 | .9775 | .2156 | 4.638 | 1.3584 | 50 |
| 20 | .2153 | .2136 | .9769 | .2186 | 4.574 | 1.3555 | 40 |
| 30 | .2182 | .2164 | .9763 | .2217 | 4.511 | 1.3526 | 30 |
| 40 | .2211 | .2193 | .9757 | .2247 | 4.449 | 1.3497 | 20 |
| 50 | .2240 | .2221 | .9750 | .2278 | 4.390 | 1.3468 | 10 |
| 13°00' | .2269 | .2250 | .9744 | .2309 | 4.331 | 1.3439 | 77°00' |
| 10 | .2298 | .2278 | .9737 | .2339 | 4.275 | 1.3410 | 50 |
| 20 | .2327 | .2306 | .9730 | .2370 | 4.219 | 1.3381 | 40 |
| 30 | .2356 | .2334 | .9724 | .2401 | 4.165 | 1.3352 | 30 |
| 40 | .2385 | .2363 | .9717 | .2432 | 4.113 | 1.3323 | 20 |
| 50 | .2414 | .2391 | .9710 | .2462 | 4.061 | 1.3294 | 10 |
| 14°00' | .2443 | .2419 | .9703 | .2493 | 4.011 | 1.3265 | 76°00' |
| 10 | .2473 | .2447 | .9696 | .2524 | 3.962 | 1.3235 | 50 |
| 20 | .2502 | .2476 | .9689 | .2555 | 3.914 | 1.3206 | 40 |
| 30 | .2531 | .2504 | .9681 | .2586 | 3.867 | 1.3177 | 30 |
| 40 | .2560 | .2532 | .9674 | .2617 | 3.821 | 1.3148 | 20 |
| 50 | .2589 | .2560 | .9667 | .2648 | 3.776 | 1.3119 | 10 |
| 15°00' | .2618 | .2588 | .9659 | .2679 | 3.732 | 1.3090 | 75°00' |
| 10 | .2647 | .2616 | .9652 | .2711 | 3.689 | 1.3061 | 50 |
| 20 | .2676 | .2644 | .9644 | .2742 | 3.647 | 1.3032 | 40 |
| 30 | .2705 | .2672 | .9636 | .2773 | 3.606 | 1.3003 | 30 |
| 40 | .2734 | .2700 | .9628 | .2805 | 3.566 | 1.2974 | 20 |
| 50 | .2763 | .2728 | .9621 | .2836 | 3.526 | 1.2945 | 10 |
| | | cos | sin | cot | tan | Radians | Degrees |

| Degrees | Radians | sin | cos | tan | cot | | |
|---------|---------|------|------|------|-------|--------|--------|
| 16°00' | .2793 | .2756 | .9613 | .2867 | 3.487 | 1.2915 | 74°00' |
| 10 | .2822 | .2784 | .9605 | .2899 | 3.450 | 1.2886 | 50 |
| 20 | .2851 | .2812 | .9596 | .2931 | 3.412 | 1.2857 | 40 |
| 30 | .2880 | .2840 | .9588 | .2962 | 3.376 | 1.2828 | 30 |
| 40 | .2909 | .2868 | .9580 | .2994 | 3.340 | 1.2799 | 20 |
| 50 | .2938 | .2896 | .9572 | .3026 | 3.305 | 1.2770 | 10 |
| 17°00' | .2967 | .2924 | .9563 | .3057 | 3.271 | 1.2741 | 73°00' |
| 10 | .2996 | .2952 | .9555 | .3089 | 3.237 | 1.2712 | 50 |
| 20 | .3025 | .2979 | .9546 | .3121 | 3.204 | 1.2683 | 40 |
| 30 | .3054 | .3007 | .9537 | .3153 | 3.172 | 1.2654 | 30 |
| 40 | .3083 | .3035 | .9528 | .3185 | 3.140 | 1.2625 | 20 |
| 50 | .3113 | .3062 | .9520 | .3217 | 3.108 | 1.2595 | 10 |
| 18°00' | .3142 | .3090 | .9511 | .3249 | 3.078 | 1.2566 | 72°00' |
| 10 | .3171 | .3118 | .9502 | .3281 | 3.047 | 1.2537 | 50 |
| 20 | .3200 | .3145 | .9492 | .3314 | 3.018 | 1.2508 | 40 |
| 30 | .3229 | .3173 | .9483 | .3346 | 2.989 | 1.2479 | 30 |
| 40 | .3258 | .3201 | .9474 | .3378 | 2.960 | 1.2450 | 20 |
| 50 | .3287 | .3228 | .9465 | .3411 | 2.932 | 1.2421 | 10 |
| 19°00' | .3316 | .3256 | .9455 | .3443 | 2.904 | 1.2392 | 71°00' |
| 10 | .3345 | .3283 | .9446 | .3476 | 2.877 | 1.2363 | 50 |
| 20 | .3374 | .3311 | .9436 | .3508 | 2.850 | 1.2334 | 40 |
| 30 | .3403 | .3338 | .9426 | .3541 | 2.824 | 1.2305 | 30 |
| 40 | .3432 | .3365 | .9417 | .3574 | 2.798 | 1.2275 | 20 |
| 50 | .3462 | .3393 | .9407 | .3607 | 2.773 | 1.2246 | 10 |
| 20°00' | .3491 | .3420 | .9397 | .3640 | 2.747 | 1.2217 | 70°00' |
| 10 | .3520 | .3448 | .9387 | .3673 | 2.723 | 1.2188 | 50 |
| 20 | .3549 | .3475 | .9377 | .3706 | 2.699 | 1.2159 | 40 |
| 30 | .3578 | .3502 | .9367 | .3739 | 2.675 | 1.2130 | 30 |
| 40 | .3607 | .3529 | .9356 | .3772 | 2.651 | 1.2101 | 20 |
| 50 | .3636 | .3557 | .9346 | .3805 | 2.628 | 1.2072 | 10 |
| 21°00' | .3665 | .3584 | .9336 | .3839 | 2.605 | 1.2043 | 69°00' |
| 10 | .3694 | .3611 | .9325 | .3872 | 2.583 | 1.2014 | 50 |
| 20 | .3723 | .3638 | .9315 | .3906 | 2.560 | 1.1985 | 40 |
| 30 | .3752 | .3665 | .9304 | .3939 | 2.539 | 1.1956 | 30 |
| 40 | .3782 | .3692 | .9293 | .3973 | 2.517 | 1.1926 | 20 |
| 50 | .3811 | .3719 | .9283 | .4006 | 2.496 | 1.1897 | 10 |
| 22°00' | .3840 | .3746 | .9272 | .4040 | 2.475 | 1.1868 | 68°00' |
| 10 | .3869 | .3773 | .9261 | .4074 | 2.455 | 1.1839 | 50 |
| 20 | .3898 | .3800 | .9250 | .4108 | 2.434 | 1.1810 | 40 |
| 30 | .3927 | .3827 | .9239 | .4142 | 2.414 | 1.1781 | 30 |
| 40 | .3956 | .3854 | .9228 | .4176 | 2.394 | 1.1752 | 20 |
| 50 | .3985 | .3881 | .9216 | .4210 | 2.375 | 1.1723 | 10 |
| 23°00' | .4014 | .3907 | .9205 | .4245 | 2.356 | 1.1694 | 67°00' |
| 10 | .4043 | .3934 | .9194 | .4279 | 2.337 | 1.1665 | 50 |
| 20 | .4072 | .3961 | .9182 | .4314 | 2.318 | 1.1636 | 40 |
| 30 | .4102 | .3987 | .9171 | .4348 | 2.300 | 1.1606 | 30 |
| 40 | .4131 | .4014 | .9159 | .4383 | 2.282 | 1.1577 | 20 |
| 50 | .4160 | .4041 | .9147 | .4417 | 2.264 | 1.1548 | 10 |
| 24°00' | .4189 | .4067 | .9135 | .4452 | 2.246 | 1.1519 | 66°00' |
| 10 | .4218 | .4094 | .9124 | .4487 | 2.229 | 1.1490 | 50 |
| 20 | .4247 | .4120 | .9112 | .4522 | 2.211 | 1.1461 | 40 |
| 30 | .4276 | .4147 | .9100 | .4557 | 2.194 | 1.1432 | 30 |
| 40 | .4305 | .4173 | .9088 | .4592 | 2.177 | 1.1403 | 20 |
| 50 | .4334 | .4200 | .9075 | .4628 | 2.161 | 1.1374 | 10 |
| 25°00' | .4363 | .4226 | .9063 | .4663 | 2.145 | 1.1345 | 65°00' |
| 10 | .4392 | .4253 | .9051 | .4699 | 2.128 | 1.1316 | 50 |
| 20 | .4422 | .4279 | .9038 | .4734 | 2.112 | 1.1286 | 40 |
| 30 | .4451 | .4305 | .9026 | .4770 | 2.097 | 1.1257 | 30 |
| 40 | .4480 | .4331 | .9013 | .4806 | 2.081 | 1.1228 | 20 |
| 50 | .4509 | .4358 | .9001 | .4841 | 2.066 | 1.1199 | 10 |
| | | cos | sin | cot | tan | Radians | Degrees |

| Degrees | Radians | sin | cos | tan | cot | | |
|---|---|---|---|---|---|---|---|
| 26°00' | .4538 | .4384 | .8988 | .4877 | 2.050 | 1.1170 | 64°00' |
| 10 | .4567 | .4410 | .8975 | .4913 | 2.035 | 1.1141 | 50 |
| 20 | .4596 | .4436 | .8962 | .4950 | 2.020 | 1.1112 | 40 |
| 30 | .4625 | .4462 | .8949 | .4986 | 2.006 | 1.1083 | 30 |
| 40 | .4654 | .4488 | .8936 | .5022 | 1.991 | 1.1054 | 20 |
| 50 | .4683 | .4514 | .8923 | .5059 | 1.977 | 1.1025 | 10 |
| 27°00' | .4712 | .4540 | .8910 | .5095 | 1.963 | 1.0996 | 63°00' |
| 10 | .4741 | .4566 | .8897 | .5132 | 1.949 | 1.0966 | 50 |
| 20 | .4771 | .4592 | .8884 | .5169 | 1.935 | 1.0937 | 40 |
| 30 | .4800 | .4617 | .8870 | .5206 | 1.921 | 1.0908 | 30 |
| 40 | .4829 | .4643 | .8857 | .5243 | 1.907 | 1.0879 | 20 |
| 50 | .4858 | .4669 | .8843 | .5280 | 1.894 | 1.0850 | 10 |
| 28°00' | .4887 | .4695 | .8829 | .5317 | 1.881 | 1.0821 | 62°00' |
| 10 | .4916 | .4720 | .8816 | .5354 | 1.868 | 1.0792 | 50 |
| 20 | .4945 | .4746 | .8802 | .5392 | 1.855 | 1.0763 | 40 |
| 30 | .4974 | .4772 | .8788 | .5430 | 1.842 | 1.0734 | 30 |
| 40 | .5003 | .4797 | .8774 | .5467 | 1.829 | 1.0705 | 20 |
| 50 | .5032 | .4823 | .8760 | .5505 | 1.816 | 1.0676 | 10 |
| 29°00' | .5061 | .4848 | .8746 | .5543 | 1.804 | 1.0647 | 61°00' |
| 10 | .5091 | .4874 | .8732 | .5581 | 1.792 | 1.0617 | 50 |
| 20 | .5120 | .4899 | .8718 | .5619 | 1.780 | 1.0588 | 40 |
| 30 | .5149 | .4924 | .8704 | .5658 | 1.767 | 1.0559 | 30 |
| 40 | .5178 | .4950 | .8689 | .5696 | 1.756 | 1.0530 | 20 |
| 50 | .5207 | .4975 | .8675 | .5735 | 1.744 | 1.0501 | 10 |
| 30°00' | .5236 | .5000 | .8660 | .5774 | 1.732 | 1.0472 | 60°00' |
| 10 | .5265 | .5025 | .8646 | .5812 | 1.720 | 1.0443 | 50 |
| 20 | .5294 | .5050 | .8631 | .5851 | 1.709 | 1.0414 | 40 |
| 30 | .5323 | .5075 | .8616 | .5890 | 1.698 | 1.0385 | 30 |
| 40 | .5352 | .5100 | .8601 | .5930 | 1.686 | 1.0356 | 20 |
| 50 | .5381 | .5125 | .8587 | .5969 | 1.675 | 1.0327 | 10 |
| 31°00' | .5411 | .5150 | .8572 | .6009 | 1.664 | 1.0297 | 59°00' |
| 10 | .5440 | .5175 | .8557 | .6048 | 1.653 | 1.0268 | 50 |
| 20 | .5469 | .5200 | .8542 | .6088 | 1.643 | 1.0239 | 40 |
| 30 | .5498 | .5225 | .8526 | .6128 | 1.632 | 1.0210 | 30 |
| 40 | .5527 | .5250 | .8511 | .6168 | 1.621 | 1.0181 | 20 |
| 50 | .5556 | .5275 | .8496 | .6208 | 1.611 | 1.0152 | 10 |
| 32°00' | .5585 | .5299 | .8480 | .6249 | 1.600 | 1.0123 | 58°00' |
| 10 | .5614 | .5324 | .8465 | .6289 | 1.590 | 1.0094 | 50 |
| 20 | .5643 | .5348 | .8450 | .6330 | 1.580 | 1.0065 | 40 |
| 30 | .5672 | .5373 | .8434 | .6371 | 1.570 | 1.0036 | 30 |
| 40 | .5701 | .5398 | .8418 | .6412 | 1.560 | 1.0007 | 20 |
| 50 | .5730 | .5422 | .8403 | .6453 | 1.550 | .9977 | 10 |
| 33°00' | .5760 | .5446 | .8387 | .6494 | 1.540 | .9948 | 57°00' |
| 10 | .5789 | .5471 | .8371 | .6536 | 1.530 | .9919 | 50 |
| 20 | .5818 | .5495 | .8355 | .6577 | 1.520 | .9890 | 40 |
| 30 | .5847 | .5519 | .8339 | .6619 | 1.511 | .9861 | 30 |
| 40 | .5876 | .5544 | .8323 | .6661 | 1.501 | .9832 | 20 |
| 50 | .5905 | .5568 | .8307 | .6703 | 1.492 | .9803 | 10 |
| 34°00' | .5934 | .5592 | .8290 | .6745 | 1.483 | .9774 | 56°00' |
| 10 | .5963 | .5616 | .8274 | .6787 | 1.473 | .9745 | 50 |
| 20 | .5992 | .5640 | .8258 | .6830 | 1.464 | .9716 | 40 |
| 30 | .6021 | .5664 | .8241 | .6873 | 1.455 | .9687 | 30 |
| 40 | .6050 | .5688 | .8225 | .6916 | 1.446 | .9657 | 20 |
| 50 | .6080 | .5712 | .8208 | .6959 | 1.437 | .9628 | 10 |
| 35°00' | .6109 | .5736 | .8192 | .7002 | 1.428 | .9599 | 55°00' |
| 10 | .6138 | .5760 | .8175 | .7046 | 1.419 | .9570 | 50 |
| 20 | .6167 | .5783 | .8158 | .7089 | 1.411 | .9541 | 40 |
| 30 | .6196 | .5807 | .8141 | .7133 | 1.402 | .9512 | 30 |
| 40 | .6225 | .5831 | .8124 | .7177 | 1.393 | .9483 | 20 |
| 50 | .6254 | .5854 | .8107 | .7221 | 1.385 | .9454 | 10 |
| | | cos | sin | cot | tan | Radians | Degrees |

| Degrees | Radians | sin | cos | tan | cot | | |
|---|---|---|---|---|---|---|---|
| 36° 00′ | .6283 | .5878 | .8090 | .7265 | 1.376 | .9425 | 54° 00′ |
| 10 | .6312 | .5901 | .8073 | .7310 | 1.368 | .9396 | 50 |
| 20 | .6341 | .5925 | .8056 | .7355 | 1.360 | .9367 | 40 |
| 30 | .6370 | .5948 | .8039 | .7400 | 1.351 | .9338 | 30 |
| 40 | .6400 | .5972 | .8021 | .7445 | 1.343 | .9308 | 20 |
| 50 | .6429 | .5995 | .8004 | .7490 | 1.335 | .9279 | 10 |
| 37° 00′ | .6458 | .6018 | .7986 | .7536 | 1.327 | .9250 | 53° 00′ |
| 10 | .6487 | .6041 | .7969 | .7581 | 1.319 | .9221 | 50 |
| 20 | .6516 | .6065 | .7951 | .7627 | 1.311 | .9192 | 40 |
| 30 | .6545 | .6088 | .7934 | .7673 | 1.303 | .9163 | 30 |
| 40 | .6574 | .6111 | .7916 | .7720 | 1.295 | .9134 | 20 |
| 50 | .6603 | .6134 | .7898 | .7766 | 1.288 | .9105 | 10 |
| 38° 00′ | .6632 | .6157 | .7880 | .7813 | 1.280 | .9076 | 52° 00′ |
| 10 | .6661 | .6180 | .7862 | .7860 | 1.272 | .9047 | 50 |
| 20 | .6690 | .6202 | .7844 | .7907 | 1.265 | .9018 | 40 |
| 30 | .6720 | .6225 | .7826 | .7954 | 1.257 | .8988 | 30 |
| 40 | .6749 | .6248 | .7808 | .8002 | 1.250 | .8959 | 20 |
| 50 | .6778 | .6271 | .7790 | .8050 | 1.242 | .8930 | 10 |
| 39° 00′ | .6807 | .6293 | .7771 | .8098 | 1.235 | .8901 | 51° 00′ |
| 10 | .6836 | .6316 | .7753 | .8146 | 1.228 | .8872 | 50 |
| 20 | .6865 | .6338 | .7735 | .8195 | 1.220 | .8843 | 40 |
| 30 | .6894 | .6361 | .7716 | .8243 | 1.213 | .8814 | 30 |
| 40 | .6923 | .6383 | .7698 | .8292 | 1.206 | .8785 | 20 |
| 50 | .6952 | .6406 | .7679 | .8342 | 1.199 | .8756 | 10 |
| 40° 00′ | .6981 | .6428 | .7660 | .8391 | 1.192 | .8727 | 50° 00′ |
| 10 | .7010 | .6450 | .7642 | .8441 | 1.185 | .8698 | 50 |
| 20 | .7039 | .6472 | .7623 | .8491 | 1.178 | .8668 | 40 |
| 30 | .7069 | .6494 | .7604 | .8541 | 1.171 | .8639 | 30 |
| 40 | .7098 | .6517 | .7585 | .8591 | 1.164 | .8610 | 20 |
| 50 | .7127 | .6539 | .7566 | .8642 | 1.157 | .8581 | 10 |
| 41° 00′ | .7156 | .6561 | .7547 | .8693 | 1.150 | .8552 | 49° 00′ |
| 10 | .7185 | .6583 | .7528 | .8744 | 1.144 | .8523 | 50 |
| 20 | .7214 | .6604 | .7509 | .8796 | 1.137 | .8494 | 40 |
| 30 | .7243 | .6626 | .7490 | .8847 | 1.130 | .8465 | 30 |
| 40 | .7272 | .6648 | .7470 | .8899 | 1.124 | .8436 | 20 |
| 50 | .7301 | .6670 | .7451 | .8952 | 1.117 | .8407 | 10 |
| 42° 00′ | .7330 | .6691 | .7431 | .9004 | 1.111 | .8378 | 48° 00′ |
| 10 | .7359 | .6713 | .7412 | .9057 | 1.104 | .8348 | 50 |
| 20 | .7389 | .6734 | .7392 | .9110 | 1.098 | .8319 | 40 |
| 30 | .7418 | .6756 | .7373 | .9163 | 1.091 | .8290 | 30 |
| 40 | .7447 | .6777 | .7353 | .9217 | 1.085 | .8261 | 20 |
| 50 | .7476 | .6799 | .7333 | .9271 | 1.079 | .8232 | 10 |
| 43° 00′ | .7505 | .6820 | .7314 | .9325 | 1.072 | .8203 | 47° 00′ |
| 10 | .7534 | .6841 | .7294 | .9380 | 1.066 | .8174 | 50 |
| 20 | .7563 | .6862 | .7274 | .9435 | 1.060 | .8145 | 40 |
| 30 | .7592 | .6884 | .7254 | .9490 | 1.054 | .8116 | 30 |
| 40 | .7621 | .6905 | .7234 | .9545 | 1.048 | .8087 | 20 |
| 50 | .7650 | .6926 | .7214 | .9601 | 1.042 | .8058 | 10 |
| 44° 00′ | .7679 | .6947 | .7193 | .9657 | 1.036 | .8029 | 46° 00′ |
| 10 | .7709 | .6967 | .7173 | .9713 | 1.030 | .7999 | 50 |
| 10 | .7738 | .6988 | .7153 | .9770 | 1.024 | .7970 | 40 |
| 30 | .7767 | .7009 | .7133 | .9827 | 1.018 | .7941 | 30 |
| 40 | .7796 | .7030 | .7112 | .9884 | 1.012 | .7912 | 20 |
| 50 | .7825 | .7050 | .7092 | .9942 | 1.006 | .7883 | 10 |
| 45° 00′ | .7854 | .7071 | .7071 | 1.0000 | 1.000 | .7854 | 45° 00′ |
| | | cos | sin | cot | tan | Radians | Degrees |

# REFERENCE TABLE 18   Plane Analytic Geometry

Distance between $(x_1, y_1)$ and $(x_2, y_2)$

$$d = \sqrt{(x_2 - x_1)^2 + (y_2 - y_1)^2}$$

Midpoint between $(x_1, y_1)$ and $(x_2, y_2)$

$$\text{Midpoint} = \left(\frac{x_1 + x_2}{2}, \frac{y_1 + y_2}{2}\right)$$

Slope of line passing through $(x_1, y_1)$ and $(x_2, y_2)$

$$m = \frac{y_2 - y_1}{x_2 - x_1}$$

Slopes of parallel lines

$$m_1 = m_2$$

Slopes of perpendicular lines

$$m_1 = -\frac{1}{m_2}$$

Equations of Lines

Point-slope form: $y - y_1 = m(x - x_1)$         General form: $Ax + By + C = 0$
Vertical line: $x = a$                            Horizontal line: $y = b$

Equations of Circles [Center: $(h, k)$, Radius: $r$]

Standard form: $(x - h)^2 + (y - k)^2 = r^2$
General form: $Ax^2 + Ay^2 + Dx + Ey + F = 0$

Equations of Parabolas [Vertex: $(h, k)$]

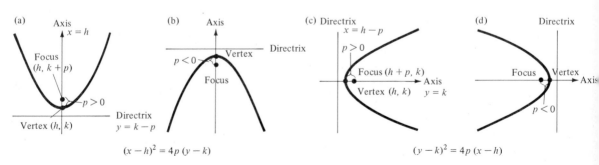

(a) Axis $x = h$; Focus $(h, k + p)$; $p > 0$; Vertex $(h, k)$; Directrix $y = k - p$; $(x - h)^2 = 4p(y - k)$; Vertical axis: $p > 0$

(b) Axis; Vertex; $p < 0$; Focus; Directrix; Vertical axis: $p < 0$

(c) Directrix $x = h - p$; $p > 0$; Focus $(h + p, k)$; Axis; Vertex $(h, k)$; $y = k$; $(y - k)^2 = 4p(x - h)$; Horizontal axis: $p > 0$

(d) Directrix; Focus; Vertex; Axis; $p < 0$; Horizontal axis: $p < 0$

## Equations of Ellipses [Center: $(h, k)$]

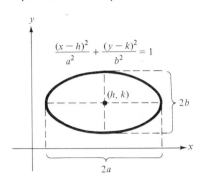

$$\frac{(x-h)^2}{a^2} + \frac{(y-k)^2}{b^2} = 1$$

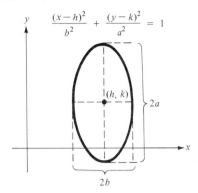

$$\frac{(x-h)^2}{b^2} + \frac{(y-k)^2}{a^2} = 1$$

## Equations of Hyperbolas [Center: $(h, k)$]

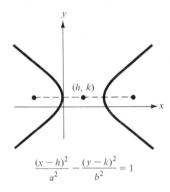

$$\frac{(x-h)^2}{a^2} - \frac{(y-k)^2}{b^2} = 1$$

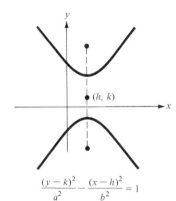

$$\frac{(y-k)^2}{a^2} - \frac{(x-h)^2}{b^2} = 1$$

## REFERENCE TABLE 19   Solid Analytic Geometry

Distance between $(x_1, y_1, z_1)$ and $(x_2, y_2, z_2)$

$$d = \sqrt{(x_2 - x_1)^2 + (y_2 - y_1)^2 + (z_2 - z_1)^2}$$

Midpoint between $(x_1, y_1, z_1)$ and $(x_2, y_2, z_2)$

$$\text{Midpoint} = \left(\frac{x_1 + x_2}{2}, \frac{y_1 + y_2}{2}, \frac{z_1 + z_2}{2}\right)$$

Equation of plane:

$$Ax + By + Cz + D = 0$$

Equation of sphere [Center: $(h, k, l)$, Radius: $r$]

$$(x - h)^2 + (y - k)^2 + (z - l)^2 = r^2$$

# REFERENCE TABLE 20 Differentiation Formulas

1. $\dfrac{d}{dx}[cu] = cu'$

2. $\dfrac{d}{dx}[u \pm v] = u' \pm v'$

3. $\dfrac{d}{dx}[uv] = uv' + vu'$

4. $\dfrac{d}{dx}\left[\dfrac{u}{v}\right] = \dfrac{vu' - uv'}{v^2}$

5. $\dfrac{d}{dx}[c] = 0$

6. $\dfrac{d}{dx}[u^n] = nu^{n-1}u'$

7. $\dfrac{d}{dx}[x] = 1$

8. $\dfrac{d}{dx}[|u|] = \dfrac{u}{|u|}(u'), \; u \neq 0$

9. $\dfrac{d}{dx}[\ln u] = \dfrac{u'}{u}$

10. $\dfrac{d}{dx}[e^u] = e^u u'$

11. $\dfrac{d}{dx}[\sin u] = (\cos u)u'$

12. $\dfrac{d}{dx}[\cos u] = -(\sin u)u'$

13. $\dfrac{d}{dx}[\tan u] = (\sec^2 u)u'$

14. $\dfrac{d}{dx}[\cot u] = -(\csc^2 u)u'$

15. $\dfrac{d}{dx}[\sec u] = (\sec u \tan u)u'$

16. $\dfrac{d}{dx}[\csc u] = -(\csc u \cot u)u'$


# REFERENCE TABLE 21 Integration Tables

## Forms Involving $u^n$

1. $\displaystyle \int u^n \, du = \dfrac{u^{n+1}}{n + 1} + C, \; n \neq -1$

2. $\displaystyle \int \dfrac{1}{u} \, du = \ln |u| + C$

## Forms Involving $a + bu$

3. $\displaystyle \int \dfrac{u}{a + bu} \, du = \dfrac{1}{b^2}[bu - a \ln |a + bu|] + C$

4. $\displaystyle \int \dfrac{u}{(a + bu)^2} \, du = \dfrac{1}{b^2}\left[\dfrac{a}{a + bu} + \ln |a + bu|\right] + C$

5. $\displaystyle \int \dfrac{u}{(a + bu)^n} \, du = \dfrac{1}{b^2}\left[\dfrac{-1}{(n - 2)(a + bu)^{n-2}} + \dfrac{a}{(n - 1)(a + bu)^{n-1}}\right] + C, \; n \neq 1, 2$

6. $\displaystyle \int \dfrac{u^2}{a + bu} \, du = \dfrac{1}{b^3}\left[-\dfrac{bu}{2}(2a - bu) + a^2 \ln |a + bu|\right] + C$

7. $\displaystyle \int \dfrac{u^2}{(a + bu)^2} \, du = \dfrac{1}{b^3}\left[bu - \dfrac{a^2}{a + bu} - 2a \ln |a + bu|\right] + C$

8. $\displaystyle \int \dfrac{u^2}{(a + bu)^3} \, du = \dfrac{1}{b^3}\left[\dfrac{2a}{a + bu} - \dfrac{a^2}{2(a + bu)^2} + \ln |a + bu|\right] + C$

9. $\displaystyle \int \dfrac{u^2}{(a + bu)^n} \, du = \dfrac{1}{b^3}\left[\dfrac{-1}{(n - 3)(a + bu)^{n-3}} + \dfrac{2a}{(n - 2)(a + bu)^{n-2}} - \dfrac{a^2}{(n - 1)(a + bu)^{n-1}}\right] + C, \; n \neq 1, 2, 3$

10. $\displaystyle\int \frac{1}{u(a + bu)}\, du = \frac{1}{a}\ln\left|\frac{u}{a + bu}\right| + C$

11. $\displaystyle\int \frac{1}{u(a + bu)^2}\, du = \frac{1}{a}\left[\frac{1}{a + bu} + \frac{1}{a}\ln\left|\frac{u}{a + bu}\right|\right] + C$

12. $\displaystyle\int \frac{1}{u^2(a + bu)}\, du = -\frac{1}{a}\left[\frac{1}{u} + \frac{b}{a}\ln\left|\frac{u}{a + bu}\right|\right] + C$

13. $\displaystyle\int \frac{1}{u^2(a + bu)^2}\, du = -\frac{1}{a^2}\left[\frac{a + 2bu}{u(a + bu)} + \frac{2b}{a}\ln\left|\frac{u}{a + bu}\right|\right] + C$

## Forms Involving $\sqrt{a + bu}$

14. $\displaystyle\int u^n\sqrt{a + bu}\, du = \frac{2}{b(2n + 3)}\left[u^n(a + bu)^{3/2} - na\int u^{n-1}\sqrt{a + bu}\, du\right]$

15. $\displaystyle\int \frac{1}{u\sqrt{a + bu}}\, du = \frac{1}{\sqrt{a}}\ln\left|\frac{\sqrt{a + bu} - \sqrt{a}}{\sqrt{a + bu} + \sqrt{a}}\right| + C,\ 0 < a$

16. $\displaystyle\int \frac{1}{u^n\sqrt{a + bu}}\, du = \frac{-1}{a(n - 1)}\left[\frac{\sqrt{a + bu}}{u^{n-1}} + \frac{(2n - 3)b}{2}\int \frac{1}{u^{n-1}\sqrt{a + bu}}\, du\right],\ n \neq 1$

17. $\displaystyle\int \frac{\sqrt{a + bu}}{u}\, du = 2\sqrt{a + bu} + a\int \frac{1}{u\sqrt{a + bu}}\, du$

18. $\displaystyle\int \frac{\sqrt{a + bu}}{u^n}\, du = \frac{-1}{a(n - 1)}\left[\frac{(a + bu)^{3/2}}{u^{n-1}} + \frac{(2n - 5)b}{2}\int \frac{\sqrt{a + bu}}{u^{n-1}}\, du\right],\ n \neq 1$

19. $\displaystyle\int \frac{u}{\sqrt{a + bu}}\, du = -\frac{2(2a - bu)}{3b^2}\sqrt{a + bu} + C$

20. $\displaystyle\int \frac{u^n}{\sqrt{a + bu}}\, du = \frac{2}{(2n + 1)b}\left[u^n\sqrt{a + bu} - na\int \frac{u^{n-1}}{\sqrt{a + bu}}\, du\right]$

## Forms Involving $u^2 - a^2,\ 0 < a$

21. $\displaystyle\int \frac{1}{u^2 - a^2}\, du = -\int \frac{1}{a^2 - u^2}\, du = \frac{1}{2a}\ln\left|\frac{u - a}{u + a}\right| + C$

22. $\displaystyle\int \frac{1}{(u^2 - a^2)^n}\, du = \frac{-1}{2a^2(n - 1)}\left[\frac{u}{(u^2 - a^2)^{n-1}} + (2n - 3)\int \frac{1}{(u^2 - a^2)^{n-1}}\, du\right],\ n \neq 1$

## Forms Involving $\sqrt{u^2 \pm a^2},\ 0 < a$

23. $\displaystyle\int \sqrt{u^2 \pm a^2}\, du = \frac{1}{2}[u\sqrt{u^2 \pm a^2} \pm a^2 \ln|u + \sqrt{u^2 \pm a^2}|] + C$

24. $\displaystyle\int u^2\sqrt{u^2 \pm a^2}\, du = \frac{1}{8}[u(2u^2 \pm a^2)\sqrt{u^2 \pm a^2} - a^4 \ln|u + \sqrt{u^2 \pm a^2}|] + C$

25. $\displaystyle\int \frac{\sqrt{u^2 + a^2}}{u}\, du = \sqrt{u^2 + a^2} - a \ln \left| \frac{a + \sqrt{u^2 + a^2}}{u} \right| + C$

26. $\displaystyle\int \frac{\sqrt{u^2 \pm a^2}}{u^2}\, du = \frac{-\sqrt{u^2 \pm a^2}}{u} + \ln \left| u + \sqrt{u^2 \pm a^2} \right| + C$

27. $\displaystyle\int \frac{1}{\sqrt{u^2 \pm a^2}}\, du = \ln \left| u + \sqrt{u^2 \pm a^2} \right| + C$

28. $\displaystyle\int \frac{1}{u\sqrt{u^2 + a^2}}\, du = \frac{-1}{a} \ln \left| \frac{a + \sqrt{u^2 + a^2}}{u} \right| + C$

29. $\displaystyle\int \frac{u^2}{\sqrt{u^2 \pm a^2}}\, du = \frac{1}{2}[u\sqrt{u^2 \pm a^2} \mp a^2 \ln \left| u + \sqrt{u^2 \pm a^2} \right|] + C$

30. $\displaystyle\int \frac{1}{u^2\sqrt{u^2 \pm a^2}}\, du = \mp \frac{\sqrt{u^2 \pm a^2}}{a^2 u} + C$

31. $\displaystyle\int \frac{1}{(u^2 \pm a^2)^{3/2}}\, du = \frac{\pm u}{a^2\sqrt{u^2 \pm a^2}} + C$

Forms Involving $\sqrt{a^2 - u^2},\ 0 < a$

32. $\displaystyle\int \frac{\sqrt{a^2 - u^2}}{u}\, du = \sqrt{a^2 - u^2} - a \ln \left| \frac{a + \sqrt{a^2 - u^2}}{u} \right| + C$

33. $\displaystyle\int \frac{1}{u\sqrt{a^2 - u^2}}\, du = \frac{-1}{a} \ln \left| \frac{a + \sqrt{a^2 - u^2}}{u} \right| + C$

34. $\displaystyle\int \frac{1}{u^2\sqrt{a^2 - u^2}}\, du = \frac{-\sqrt{a^2 - u^2}}{a^2 u} + C$

35. $\displaystyle\int \frac{1}{(a^2 - u^2)^{3/2}}\, du = \frac{u}{a^2\sqrt{a^2 - u^2}} + C$

Forms Involving $e^u$

36. $\displaystyle\int e^u\, du = e^u + C$

37. $\displaystyle\int ue^u\, du = (u - 1)e^u + C$

38. $\displaystyle\int u^n e^u\, du = u^n e^u - n \int u^{n-1} e^u\, du$

39. $\displaystyle\int \frac{1}{1 + e^u}\, du = u - \ln (1 + e^u) + C$

40. $\displaystyle\int \frac{1}{1 + e^{nu}}\, du = u - \frac{1}{n} \ln (1 + e^{nu}) + C$

## Forms Involving ln $u$

41. $\displaystyle\int \ln u\, du = u[-1 + \ln u] + C$

42. $\displaystyle\int u \ln u\, du = \frac{u^2}{4}[-1 + 2 \ln u] + C$

43. $\displaystyle\int u^n \ln u\, du = \frac{u^{n+1}}{(n+1)^2}[-1 + (n+1) \ln u] + C,\ n \neq -1$

44. $\displaystyle\int (\ln u)^2\, du = u[2 - 2 \ln u + (\ln u)^2] + C$

45. $\displaystyle\int (\ln u)^n\, du = u(\ln u)^n - n\int (\ln u)^{n-1}\, du$

## Forms Involving sin $u$ or cos $u$

46. $\displaystyle\int \sin u\, du = -\cos u + C$

47. $\displaystyle\int \cos u\, du = \sin u + C$

48. $\displaystyle\int \sin^2 u\, du = \frac{1}{2}(u - \sin u \cos u) + C$

49. $\displaystyle\int \cos^2 u\, du = \frac{1}{2}(u + \sin u \cos u) + C$

50. $\displaystyle\int \sin^n u\, du = -\frac{\sin^{n-1} u \cos u}{n} + \frac{n-1}{n}\int \sin^{n-2} u\, du$

51. $\displaystyle\int \cos^n u\, du = \frac{\cos^{n-1} u \sin u}{n} + \frac{n-1}{n}\int \cos^{n-2} u\, du$

52. $\displaystyle\int u \sin u\, du = \sin u - u \cos u + C$

53. $\displaystyle\int u \cos u\, du = \cos u + u \sin u + C$

54. $\displaystyle\int u^n \sin u\, du = -u^n \cos u + n\int u^{n-1} \cos u\, du$

55. $\displaystyle\int u^n \cos u\, du = u^n \sin u - n\int u^{n-1} \sin u\, du$

56. $\displaystyle\int \frac{1}{1 \pm \sin u}\, du = \tan u \mp \sec u + C$

57. $\displaystyle\int \frac{1}{1 \pm \cos u}\, du = -\cot u \pm \csc u + C$

58. $\displaystyle\int \frac{1}{\sin u \cos u}\, du = \ln |\tan u| + C$

### Forms Involving tan $u$, cot $u$, sec $u$, csc $u$

59. $\displaystyle\int \tan u\, du = -\ln |\cos u| + C$

60. $\displaystyle\int \cot u\, du = \ln |\sin u| + C$

61. $\displaystyle\int \sec u\, du = \ln |\sec u + \tan u| + C$

62. $\displaystyle\int \csc u\, du = \ln |\csc u - \cot u| + C$

63. $\displaystyle\int \tan^2 u\, du = -u + \tan u + C$

64. $\displaystyle\int \cot^2 u\, du = -u - \cot u + C$

65. $\displaystyle\int \sec^2 u\, du = \tan u + C$

66. $\displaystyle\int \csc^2 u\, du = -\cot u + C$

67. $\displaystyle\int \tan^n u\, du = \frac{\tan^{n-1} u}{n-1} - \int \tan^{n-2} u\, du,\ n \neq 1$

68. $\displaystyle\int \cot^n u\, du = -\frac{\cot^{n-1} u}{n-1} - \int (\cot^{n-2} u)\, du,\ n \neq 1$

69. $\displaystyle\int \sec^n u\, du = \frac{\sec^{n-2} u \tan u}{n-1} + \frac{n-2}{n-1} \int \sec^{n-2} u\, du,\ n \neq 1$

70. $\displaystyle\int \csc^n u\, du = -\frac{\csc^{n-2} u \cot u}{n-1} + \frac{n-2}{n-1} \int \csc^{n-2} u\, du,\ n \neq 1$

71. $\displaystyle\int \frac{1}{1 \pm \tan u}\, du = \frac{1}{2}(u \pm \ln |\cos u \pm \sin u|) + C$

72. $\displaystyle\int \frac{1}{1 \pm \cot u}\, du = \frac{1}{2}(u \mp \ln |\sin u \pm \cos u|) + C$

73. $\displaystyle\int \frac{1}{1 \pm \sec u}\, du = u + \cot u \mp \csc u + C$

74. $\displaystyle\int \frac{1}{1 \pm \csc u}\, du = u - \tan u \pm \sec u + C$

# REFERENCE TABLE 22  Formulas from Business

## Basic Terms

$x$ = number of units produced (or sold)
$p$ = price per unit
$R$ = total revenue from selling $x$ units
$C$ = total cost of producing $x$ units
$\overline{C}$ = average cost per unit
$P$ = total profit from selling $x$ units

## Basic Equations

$$R = xp \qquad \overline{C} = \frac{C}{x} \qquad P = R - C$$

## Demand function: $p = f(x)$ = price required to sell $x$ units

$$\eta = \frac{p/x}{dp/dx} = \text{price elasticity of demand (If } |\,\eta\,| < 1, \text{ the demand is inelastic. If } |\,\eta\,| > 1, \text{ the demand is elastic.)}$$

## Typical Graphs of Revenue, Cost, and Profit Functions

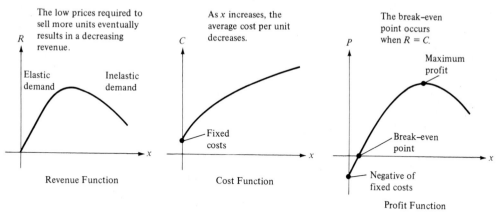

The low prices required to sell more units eventually results in a decreasing revenue.

Elastic demand    Inelastic demand

Revenue Function

As $x$ increases, the average cost per unit decreases.

Fixed costs

Cost Function

The break–even point occurs when $R = C$.

Maximum profit

Break–even point

Negative of fixed costs

Profit Function

## Marginals

$\dfrac{dR}{dx}$ = marginal revenue ≈ the *extra* revenue for selling one additional unit

$\dfrac{dC}{dx}$ = marginal cost ≈ the *extra* cost of producing one additional unit

$\dfrac{dP}{dx}$ = marginal profit ≈ the *extra* profit for selling one additional unit

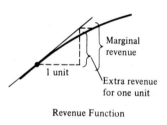

Marginal revenue

1 unit

Extra revenue for one unit

Revenue Function

## REFERENCE TABLE 23   Formulas from Finance

### Basic Terms

$P$ = amount of deposit
$r$ = interest rate
$n$ = number of times interest is compounded per year
$t$ = number of years
$A$ = balance after $t$ years

### Compound Interest Formulas

1. Balance when interest is compounded $n$ times per year

$$A = P\left(1 + \frac{r}{n}\right)^{nt}$$

2. Balance when interest is compounded continuously

$$A = Pe^{rt}$$

### Balance of an increasing annuity after $n$ deposits per year of $P$ for $t$ years

$$A = P\left[\left(1 + \frac{r}{n}\right)^{nt} - 1\right]\left(1 + \frac{n}{r}\right)$$

### Initial deposit for a decreasing annuity with $n$ withdrawals per year of $W$ for $t$ years

$$P = W\left(\frac{n}{r}\right)\left[1 - \left(\frac{1}{1 + [r/n]}\right)^{nt}\right]$$

### Monthly installment $M$ for a loan of $P$ dollars over $t$ years at $r\%$ interest

$$M = P\left[\frac{r/12}{1 - \left(\frac{1}{1 + [r/12]}\right)^{12t}}\right]$$

# Answers to Odd-Numbered Exercises

## CHAPTER 0

### Section 0.1

1. Rational

3. Irrational

5. Rational

7. Rational

9. Irrational

11. (a) Yes
    (b) No
    (c) Yes
    (d) No

13. (a) Yes
    (b) No
    (c) No
    (d) Yes

15.

| Interval Notation | Inequality | Graph |
|---|---|---|
| $[-2, 0)$ | $-2 \leq x < 0$ | 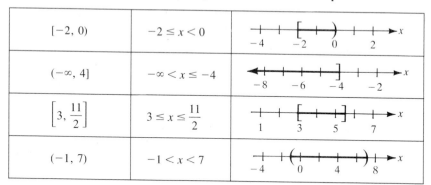 |
| $(-\infty, 4]$ | $-\infty < x \leq -4$ | |
| $\left[3, \dfrac{11}{2}\right]$ | $3 \leq x \leq \dfrac{11}{2}$ | |
| $(-1, 7)$ | $-1 < x < 7$ | |

17. $x \geq 12$

19. $x < -\dfrac{1}{2}$

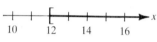

21. $x \geq \dfrac{1}{2}$

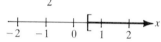

23. $x > 1$

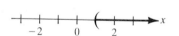

25. $-\dfrac{1}{2} < x < \dfrac{7}{2}$

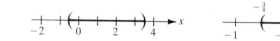

27. $-\dfrac{3}{4} < x < -\dfrac{1}{4}$

29. $x > 6$

31. $-3 \le x \le 1$

33. $-3 \le x \le 2$

35. $r > 12.5\%$

37. $x \ge 36$ units

39. Length of side $\ge 10\sqrt{5}$

41. (a) $\dfrac{355}{113} > \pi$

43. (a) False
    (b) True
    (c) True
    (d) True if $ab < 0$
        false if $ab > 0$

(b) $\dfrac{22}{7} > \pi$

## Section 0.2

1. (a) 4

3. (a) $\dfrac{23}{4}$

5. (a) $-51$

7. (a) $-14.99$

(b) $-4$

(b) $-\dfrac{23}{4}$

(b) 51

(b) 14.99

(c) 4

(c) $\dfrac{23}{4}$

(c) 51

(c) 14.99

9. 1

11. $-\dfrac{13}{4}$

13. 14

15. 1.25

17. $-5 < x < 5$

19. $x < -6$ or $x > 6$

21. $-7 < x < 3$

23. $x \le -7$ or $x \ge 13$

25. $x < 6$ or $x > 14$

27. $4 < x < 5$

29. $a - b \le x \le a + b$

31. $|x| \le 2$

33. $|x| > 2$

35. $|x - 4| \le 2$

37. $|x - 2| > 2$

39. $|x - 4| < 2$

41. $|y - a| \le 2$

43. $65.8 \le h \le 71.2$

45. $2{,}125{,}000 < x < 2{,}375{,}000$

# Section 0.3

1. $-24$

3. $\dfrac{1}{2}$

5. $3$

7. $44$

9. $5$

11. $9$

13. $\dfrac{1}{2}$

15. $\dfrac{1}{4}$

17. $908.3483$

19. $-5.3601$

21. $5x^6$

23. $24y^{10}$

25. $10x^4$

27. $7x^5$

29. $\dfrac{4}{3}(x+y)^2$

31. $3x$

33. $4x^4$

35. (a) $2\sqrt{2}$

(b) $3\sqrt{2}$

37. (a) $2x\sqrt[3]{2x^2}$

(b) $2|x|z\sqrt[4]{2z}$

39. (a) $\dfrac{5\sqrt{3}|x|}{y^2}$

(b) $(x-y)\sqrt{5(x-y)}$

41. $y^2(y+2)(y-2)$

43. $\dfrac{1}{4}(3x+2)$

45. $\sqrt{x}(1+x)$

47. $\dfrac{1}{3}(x^3-1)^4(3x^2)$

49. $\dfrac{5}{2}\sqrt[3]{1+x^2}(2x)$

51. $x^{1/2}(3+4x)$

53. $x^{-1/2}(3+4x^2)$

55. $\dfrac{1}{2}(x+1)^{-1/2}(3x+2)$

57. $x\geq 1$

59. $(-\infty, \infty)$

61. $(-\infty, 1)$ and $(1, \infty)$

63. $x>3$

65. $1\leq x\leq 5$

# Section 0.4

1. $\dfrac{1}{2}, -\dfrac{1}{3}$

3. $\dfrac{3}{2}$

5. $-2\pm\sqrt{3}$

7. $\dfrac{1\pm\sqrt{7}}{3}$

9. $\dfrac{7\pm\sqrt{17}}{4}$

11. No real roots

13. $\dfrac{-1\pm\sqrt{13}}{2}$

15. $(x-2)^2$

17. $(2x+1)^2$

19. $(x+2)(x-1)$

21. $(3x-2)(x-1)$

23. $(x-2y)^2$

25. $(2y-3z)(y+12z)$

27. $(4y+3)(4y-3)$

29. $(x+1)(x-3)$

31. $(3+y)(3-y)(9+y^2)$

33. $(x-2)(x^2+2x+4)$

35. $(y+4)(y^2-4y+16)$

37. $(x-3)(x^2+3x+9)$

39. $(x-4)(x-1)(x+1)$

41. $(2x-3)(x^2+2)$

43. $(x-2)(2x^2-1)$

45. $0, 5$

47. $\pm 3$

49. $\pm\sqrt{3}$

51. $0, 6$

53. $-2, 1$

55. $2, 3$

57. $-\dfrac{1}{2}, 1$

59. $-6, 8$

61. $-4$

63. $\pm 2$

65. $1, \pm 2$

67. $-2, -\dfrac{1}{2}, 1$

69. $x \le 3$ or $x \ge 4$

71. $-2 \le x \le 2$

73. $-4 \le x \le 3$

75. $-\infty < x < \infty$

77. $(x + 2)(x^2 - 2x + 4)$

79. $(x - 1)(2x^2 + x - 1)$

81. $(x - 3)(x^3 + 5x^2 + 9x + 9)$

83. $\pm 1$

85. $1, 2, 3$

87. $1, \pm\dfrac{1}{2}$

89. $4$

91. $-2, -3 \pm \sqrt{10}$

93. $\pm 2, \pm 3$

95. $(x - 2)^5$

## Section 0.5

1. $\dfrac{x + 5}{x - 1}$

3. $\dfrac{5x - 1}{x^2 + 2}$

5. $\dfrac{4x - 3}{x^2}$

7. $\dfrac{x - 6}{x^2 - 4}$

9. $\dfrac{2}{x - 3}$

11. $-\dfrac{x^2 + 3}{(x + 1)(x - 2)(x - 3)}$

13. $\dfrac{(A + B)x + 3(A - 2B)}{(x - 6)(x + 3)}$

15. $\dfrac{(A + B)x^2 + (10A + C)x + 5(5A - 5B - C)}{(x - 5)(x + 5)^2}$

17. $-\dfrac{(x - 1)^2}{x(x^2 + 1)}$

19. $\dfrac{x + 2}{(x + 1)^{3/2}}$

21. $-\dfrac{3t}{2\sqrt{1 + t}}$

23. $\dfrac{x(x^2 + 2)}{(x + 1)^{3/2}}$

25. $\dfrac{-2x - \Delta x}{x^2(x + \Delta x)^2}$

27. $\dfrac{2}{x^2\sqrt{x^2 + 2}}$

29. $\dfrac{1}{2\sqrt{x}(x + 1)^{3/2}}$

31. $\dfrac{\sqrt{3}}{3}$

33. $\dfrac{2}{3\sqrt{2}}$

35. $\dfrac{x\sqrt{x - 4}}{x - 4}$

37. $\dfrac{y}{6\sqrt{y}}$

39. $\dfrac{49\sqrt{x^2 - 9}}{x + 3}$

41. $\dfrac{\sqrt{14} + 2}{2}$

43. $\dfrac{x(5 + \sqrt{3})}{11}$

45. $\sqrt{6} - \sqrt{5}$

47. $\dfrac{1}{x(\sqrt{3} + \sqrt{2})}$

49. $\dfrac{2x - 1}{2x + \sqrt{4x - 1}}$

51. $\dfrac{\sqrt{x^2 - 2} + \sqrt{x}}{x - 2}$

53. $\dfrac{8x(\sqrt{17x - 1})}{17x - 1}$

## Review Exercises for Chapter 0

1. (a) $\dfrac{3}{2} < 7$

3. (a) $-\dfrac{3}{7} > -\dfrac{8}{7}$

5. $\dfrac{3}{2} \le x$

(b) $\pi > -6$

(b) $\dfrac{3}{4} > \dfrac{11}{16}$

**7.** $-1 \le x \le 5$

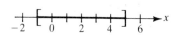

**9.** $x < -5$ or $x > -1$

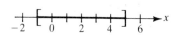

**11.** $\dfrac{27}{16}$

**13.** $x \ge 0$

**15.** $0 \le A \le 12$

**17.** $|x - 5| \le 3$

**19.** $|y - a| \ge 2$

**21.** $300x^2y^5$

**23.** $\dfrac{4}{81y}$

**25.** $\dfrac{a^6}{64b^9}$

**27.** $(x - y)^4$

**29.** $|xy|\sqrt{3}|y|$

**31.** $2x(x - 2)(x + 1)$

**33.** $7r(3s + r)(3s - r)$

**35.** $xy(2x - y)^2$

**37.** $6x(x - 2y)(x^2 + 2xy + 4y^2)$

**39.** $(x + y)(x - y)(x + 2y)(x - 2y)$

**41.** $2x(2x - 1)(4x - 1)$

**43.** $(x + 2y)(y + 2z)$

**45.** $\dfrac{1}{12}(9x^2 - 10x + 48)$

**47.** $(x - 1)(x^2 + x + 1)$

**49.** $(x + y)^2(x - y)^2$

**51.** $-\dfrac{3}{4}, \dfrac{1}{4}$

**53.** $-4 \pm 2\sqrt{5}$

**55.** No real roots

**57.** $\dfrac{17 + \sqrt{6817}}{102} \approx 0.976$

$\dfrac{17 - \sqrt{6817}}{102} \approx -0.643$

**59.** $3x^2 - 2x + 5$

**61.** $-x^2 + 3x - 6 + \dfrac{11}{x + 1}$

**63.** $-1, \dfrac{1}{2}, 5$

**65.** $0, 2, 1 \pm \sqrt{2}$

**67.** $\dfrac{x^3 - x + 3}{(x + 2)(x - 1)}$

**69.** $\dfrac{1}{\sqrt{x + \Delta x} + \sqrt{x}}$

**71.** 2000 units

# CHAPTER 1

## Section 1.1

**1.** $a = 4, b = 3, c = 5$
$4^2 + 3^2 = 5^2$

**3.** $a = 10, b = 3, c = \sqrt{109}$
$10^2 + 3^2 = (\sqrt{109})^2$

**5.** $d = 2\sqrt{5}$,

midpoint $= (3, 3)$

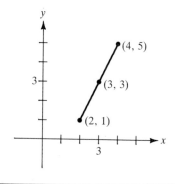

**7.** $d = 2\sqrt{10}$,

midpoint $= \left(-\dfrac{1}{2}, -2\right)$

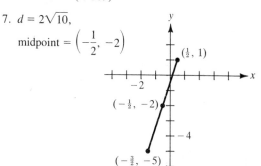

9. $d = 2\sqrt{37}$
   midpoint $= (3, 8)$

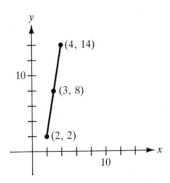

11. $d = \sqrt{8 - 2\sqrt{3}}$
    midpoint $= \left(0, \dfrac{\sqrt{3} + 1}{2}\right)$

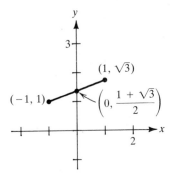

13. $d_1 = \sqrt{45}$, $d_2 = \sqrt{5}$, $d_3 = \sqrt{50}$

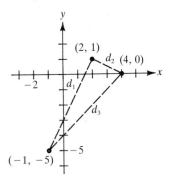

15. $d_1 = d_2 = d_3 = d_4 = \sqrt{5}$

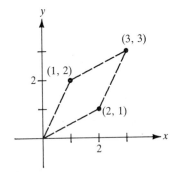

17. Collinear since $d_1 + d_2 = d_3$
    $2\sqrt{5} + \sqrt{5} = 3\sqrt{5}$

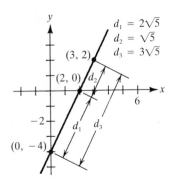

19. Not collinear since $d_1 + d_2 > d_3$
    $d_1 = \sqrt{2}$, $d_2 = \sqrt{13}$, $d_3 = 5$

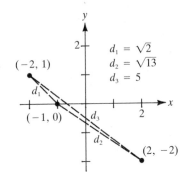

21. $x = \pm 3$

23. $y = \pm\sqrt{55}$

25. $3x - 2y - 1 = 0$

27. $\left(\dfrac{3x_1 + x_2}{4}, \dfrac{3y_1 + y_2}{4}\right), \left(\dfrac{x_1 + x_2}{2}, \dfrac{y_1 + y_2}{2}\right), \left(\dfrac{x_1 + 3x_2}{4}, \dfrac{y_1 + 3y_2}{4}\right)$

29. (a) $\left(\dfrac{7}{4}, -\dfrac{7}{4}\right), \left(\dfrac{5}{2}, -\dfrac{3}{2}\right), \left(\dfrac{13}{4}, -\dfrac{5}{4}\right)$  (b) $\left(-\dfrac{3}{2}, -\dfrac{9}{4}\right), \left(-1, -\dfrac{3}{2}\right), \left(-\dfrac{1}{2}, -\dfrac{3}{4}\right)$

31.

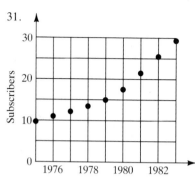

33. (a) 16.76 feet
   (b) 1341.04 square feet

35. \$630,000.00

37. (a) \$ 99 billion
   (b) \$103 billion
   (c) \$110 billion
   (d) \$113 billion

39. (a) 510,000
   (b) 410,000
   (c) 350,000
   (d) 380,000

41. $(2x_m - x_1, 2y_m - y_1)$

## Section 1.2

1. (a) Not a solution point
   (b) Solution point
   (c) Solution point

3. (a) Solution point
   (b) Solution point
   (c) Not a solution point

5. $(0, -3), \left(\dfrac{3}{2}, 0\right)$

7. $(0, -2), (-2, 0), (1, 0)$

9. $(0, 0), (-3, 0), (3, 0)$

11. $\left(0, \dfrac{1}{2}\right), (1, 0)$

13. $(0, 0)$          15. c

16. d          17. b

18. f          19. a

20. e

21.

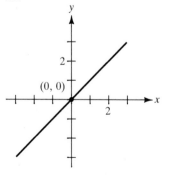

23.

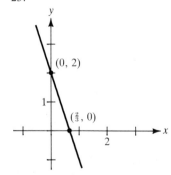

25.

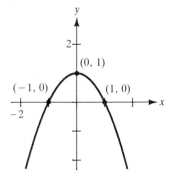

27.

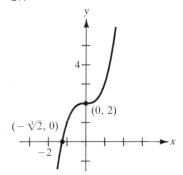

29.

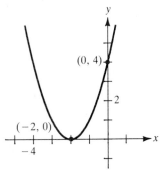

31.

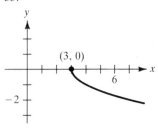

33.

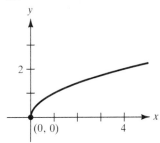

35.

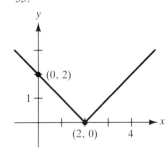

37.

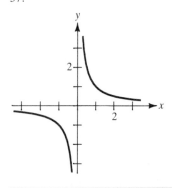

39.

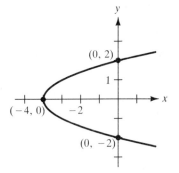

41. $x^2 + y^2 - 9 = 0$

43. $x^2 + y^2 - 4x + 2y - 11 = 0$

45. $x^2 + y^2 + 2x - 4y = 0$

47. $x^2 + y^2 - 6x - 8y = 0$

49. $(x - 1)^2 + (y + 3)^2 = 4$

51. $(x - 1)^2 + (y + 3)^2 = 0$

53. $\left(x - \dfrac{1}{2}\right)^2 + \left(y - \dfrac{1}{2}\right)^2 = 2$

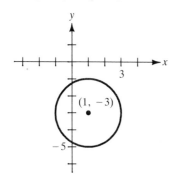

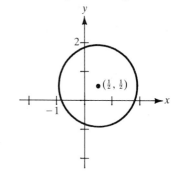

55. $\left(x + \dfrac{1}{2}\right)^2 + \left(y + \dfrac{5}{4}\right)^2 = \dfrac{9}{4}$

57. $(1, 1)$

59. $(5, 2)$

61. $(-1, -2), (2, 1)$

63. $(-1, -1), (0, 0), (1, 1)$

65. $(-1, 0), (0, 1), (1, 0)$

67. (a) $C = 21.6x + 5000$
$R = 34.1x$
(b) 400 units

69. 193 units

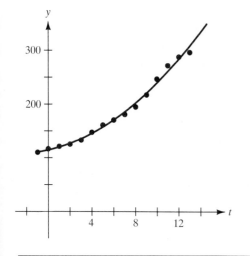

71. (a)

(b) 362.6

73. (a)

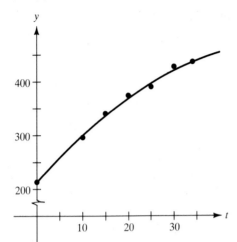

(b) 454.5 acres

## Section 1.3

1. 1    3. 0    5. −3

7. $m = 3$

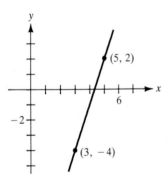

9. $m = 0$

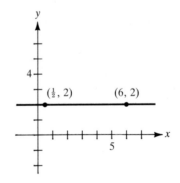

11. $m$ is undefined

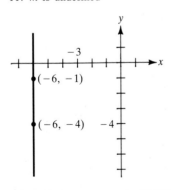

13. $m = 0$

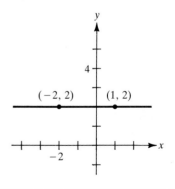

15. $(0, 1), (1, 1), (3, 1)$

17. $(6, -5), (7, -4), (8, -3)$

19. $(0, 10), (2, 4), (3, 1)$

21. $(-8, 0), (-8, 2), (-8, 3)$

23. $m = -\dfrac{1}{5}, (0, 4)$

25. $m = \dfrac{4}{3}, (0, -6)$

27. $m$ is undefined
no $y$-intercept

29. $2x - y - 3 = 0$

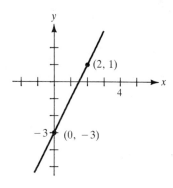

31. $3x + y = 0$

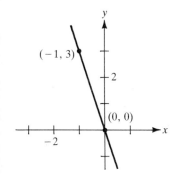

33. $x - 2 = 0$

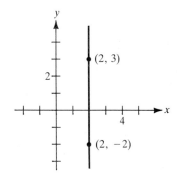

35. $y + 2 = 0$

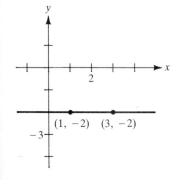

37. $3x - 4y + 12 = 0$

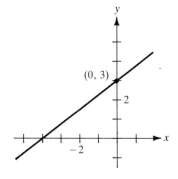

39. $2x - 3y = 0$

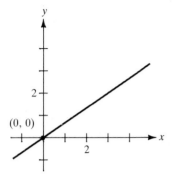

41. $2x + y - 5 = 0$

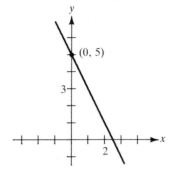

43. $4x - y + 2 = 0$

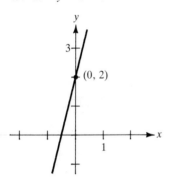

45. $9x - 12y + 8 = 0$

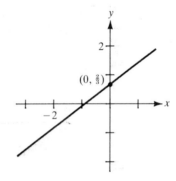

47. $x - 3 = 0$

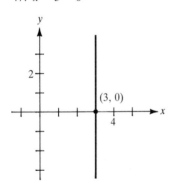

49. (a) $x + y + 1 = 0$
    (b) $x - y + 5 = 0$
51. (a) $3x + 4y + 2 = 0$
    (b) $4x - 3y + 36 = 0$
53. (a) $y = 0$
    (b) $x + 1 = 0$
55. $F = \dfrac{9}{5} C + 32$
57. $C = 0.25x + 95$
59. $y = 875 - 175t, \ 0 \le t \le 5$
61. (a) $x = \dfrac{1}{15}(1130 - p)$     (b) 45 units
                                          (c) 49 units
63. (a) $C = 14.75t + 26{,}500$
    (b) $R = 25t$
    (c) $P = 10.25t - 26{,}500$
    (d) 2585.4 hours

## Section 1.4

1. (a) $-3$
   (b) $-9$
   (c) $2x - 5$
   (d) $2\Delta x - 1$

3. (a) 4
   (b) 36
   (c) $c^2$
   (d) $x^2 + 2x\Delta x + (\Delta x)^2$

5. (a) 1
   (b) $-1$
   (c) 1
   (d) $\dfrac{|x - 1|}{(x - 1)} = \begin{cases} -1, & x < 1 \\ 1, & x > 1 \end{cases}$

7. 3

11. $x(x + 1)$

9. $3 + \Delta x$

13. Domain: $(-\infty, \infty)$
    range: $(-\infty, \infty)$

15. Domain: $(-\infty, \infty)$
    range: $[0, \infty)$

17. Domain: $[1, \infty)$
    range: $[0, \infty)$

19. Domain: $[-3, 3]$
    range: $[0, 3]$

21. Domain: $(-\infty, \infty)$
    range: $[0, \infty)$

23. $y$ is a function of $x$

25. $y$ is not a function of $x$

27. $y$ is a function of $x$

29. $y$ is a function of $x$

31. $y$ is not a function of $x$

33. $y$ is not a function of $x$

35. $y$ is a function of $x$

37. $y$ is a function of $x$

39. $y$ is not a function of $x$

41. (a) $2x$
    (b) $(x + 1)(x - 1) = x^2 - 1$
    (c) $\dfrac{x + 1}{x - 1}$
    (d) $x$
    (e) $x$

43. (a) $x^2 - x + 1$
    (b) $x^2(1 - x) = x^2 - x^3$
    (c) $\dfrac{x^2}{1 - x}$
    (d) $(1 - x)^2$
    (e) $1 - x^2$

45. (a) $x^2 + 5 + \sqrt{1 - x}$
    (b) $(x^2 + 5)\sqrt{1 - x}$
    (c) $\dfrac{x^2 + 5}{\sqrt{1 - x}}$
    (d) $|1 - x| + 5 = \begin{cases} 6 - x, & x \le 1 \\ 4 + x, & x > 1 \end{cases}$
    (e) Not defined

47. (a) $\dfrac{1}{x} + \dfrac{1}{x^2} = \dfrac{x + 1}{x^2}$
    (b) $\dfrac{1}{x^3}$
    (c) $x$
    (d) $x^2$
    (e) $x^2$

49. (a) $f(g(x)) = (\sqrt[3]{x})^3 = x$
    $g(f(x)) = \sqrt[3]{x^3} = x$
    (b)

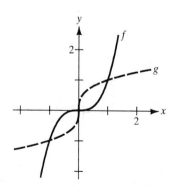

51. (a) $f(g(x)) = 5\left(\dfrac{x - 1}{5}\right) + 1 = x$
    $g(f(x)) = \dfrac{5x + 1 - 1}{5} = x$
    (b)

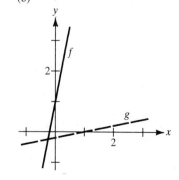

53. (a) $f(g(x)) = \sqrt{x^2 + 4} - 4 = x, \; x \geq 0$

$\qquad g(f(x)) = (\sqrt{x-4})^2 + 4 = x, \; x \geq 4$

(b)

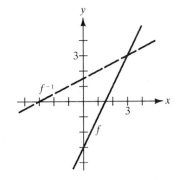

55. $f(x) = 2x - 3, \; f^{-1}(x) = \dfrac{x+3}{2}$

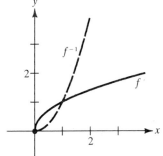

57. $f(x) = x^5, \; f^{-1}(x) = \sqrt[5]{x}$

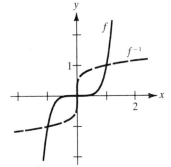

59. $f(x) = \sqrt{x}, \; f^{-1}(x) = x^2, \; x \geq 0$

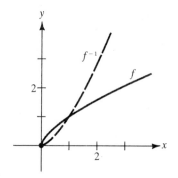

61. $f(x) = \sqrt{4 - x^2}$

$\qquad f^{-1}(x) = \sqrt{4 - x^2}, \; 0 \leq x \leq 2$

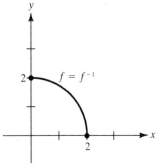

63. $f(x) = x^{2/3}, \; f^{-1}(x) = x^{3/2}, \; x \geq 0$

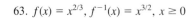

65. Yes, if $a \neq 0$, $f^{-1}(x) = \dfrac{x - b}{a}$

67. No

69. No

71. (a)

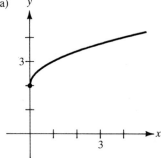

(b)

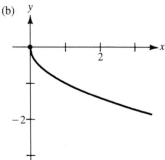

(c)

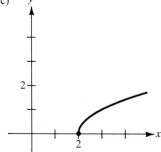

(d)

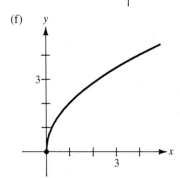

(e)

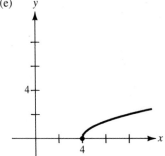

(f)

73. (a) $y = (x + 1)\sqrt{x + 4}$
    (b) $y = x\sqrt{x + 3} + 2$
    (c) $y = -x\sqrt{x + 3}$
    (d) $y = (1 - x)\sqrt{x + 2}$

75. $A = 50x - x^2$

77. $V = 1750x + 500{,}000$, $x \geq 0$

79. (a) $x = \dfrac{1475}{p} - 100$   (b) 47.5 units

81. $C = 15\sqrt{x^2 + \dfrac{1}{4}} + 10(3 - x)$

83. $P = kr^2(0.18 - r)$

## Section 1.5

1. (a) 1
   (b) 3

3. (a) 1
   (b) 3

5. (a) 1   (b) 1
   (c) 1

7. (a) 0   (b) 0
   (c) 0

9. (a) 3
(b) −3
(c) Limit does not exist.

11. 4

13. −1

15. −4

17. 2

19. $\dfrac{1}{2}$

21. (a) 5 (b) 6 (c) $\dfrac{2}{3}$

23. −2

25. $\dfrac{1}{6}$

27. $\dfrac{1}{10}$

29. 12

31. −∞

33. −∞

35. ∞

37. Limit does not exist.

39. Limit does not exist.

41. 2

43. 2

45. $2t$

47. $2x - 2$

49. $\dfrac{\sqrt{3}}{6}$

51. $-\dfrac{1}{4}$

53. $\dfrac{1}{4}$

55.

| $x$ | 1.9 | 1.99 | 1.999 | 2 | 2.001 | 2.01 | 2.1 |
|---|---|---|---|---|---|---|---|
| $f(x)$ | 13.5 | 13.95 | 13.995 | 14 | 14.005 | 14.05 | 14.5 |

$\lim\limits_{x\to 2} (5x + 4) = 14$

57.

| $x$ | 1.9 | 1.99 | 1.999 | 2 | 2.001 | 2.01 | 2.1 |
|---|---|---|---|---|---|---|---|
| $f(x)$ | 0.2564 | 0.2506 | 0.2501 | undefined | 0.2499 | 0.2494 | 0.2439 |

$\lim\limits_{x\to 2} \dfrac{x - 2}{x^2 - 4} = \dfrac{1}{4}$

59.

| $x$ | −0.1 | −0.01 | −0.001 | 0 | 0.001 | 0.01 | 0.1 |
|---|---|---|---|---|---|---|---|
| $f(x)$ | 0.3581 | 0.3540 | 0.3536 | undefined | 0.3535 | 0.3531 | 0.3492 |

$\lim\limits_{x\to 0} \dfrac{\sqrt{x + 2} - \sqrt{2}}{x} = \dfrac{1}{2\sqrt{2}} \approx 0.3536$

61.

| $x$ | −0.1 | −0.01 | −0.001 | 0 | 0.001 | 0.01 | 0.1 |
|---|---|---|---|---|---|---|---|
| $f(x)$ | −0.1316 | −0.1256 | −0.1251 | undefined | −0.1249 | −0.1244 | −0.1190 |

$\lim\limits_{x\to 0} \dfrac{[1/(2 + x)] - (1/2)}{2x} = -\dfrac{1}{8} = -0.1250$

63. 4

## Section 1.6

1. Continuous

3. Removable discontinuity at $x = -1$

5. Nonremovable discontinuity at $x = 1$

7. Continuous

9. Nonremovable discontinuity at $x = 1$

11. Continuous

13. Removable discontinuity at $x = -2$
nonremovable discontinuity at $x = 5$

15. Continuous

17. Nonremovable discontinuity at $x = 2$

19. Continuous

21. Nonremovable discontinuity at $x = -2$

23. Nonremovable discontinuity at every integer

25. Continuous in the domain $(1, \infty)$

27. Removable discontinuity at $x = 4$

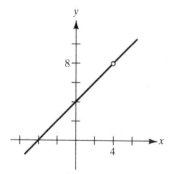

29. Removable discontinuity at $x = 0$

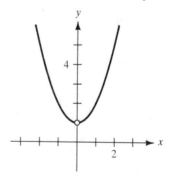

31. $a = 2$

33. Discontinuous at $t = 1, 2, 3, 4, 5$

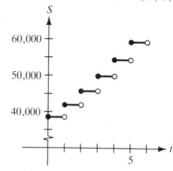

35. Discontinuous at $t = 2, 4, 6, 8$
    Replenish the inventory every two months.

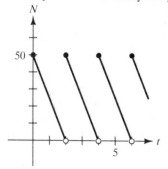

37. Continuous in the domain $[0, 100)$

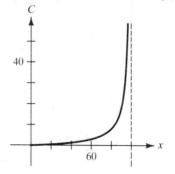

## Review Exercises for Chapter 1

1. (a) 6   (b) (3, 0)
   (c) $y = 0$

3. (a) 5   (b) $\left(0, \dfrac{1}{2}\right)$
   (c) $3x - 4y + 2 = 0$

5. (a) 13   (b) $\left(8, \dfrac{7}{2}\right)$
   (c) $5x - 12y + 2 = 0$

7. (a) $\sqrt{53}$   (b) $\left(\dfrac{5}{2}, 1\right)$
   (c) $2x - 7y + 2 = 0$

9. (a) $\dfrac{1}{6}\sqrt{53}$   (b) $\left(\dfrac{1}{2}, \dfrac{3}{4}\right)$
   (c) $21x + 6y - 15 = 0$

11. $t = \dfrac{7}{3}$

13. (a) $7x - 16y + 78 = 0$
    (b) $5x - 3y + 22 = 0$
    (c) $2x + y = 0$
    (d) $x + 2 = 0$

15.

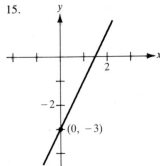

17.

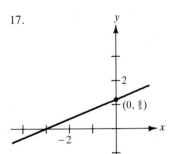

19. Center: $(-3, 1)$
    radius $= 3$

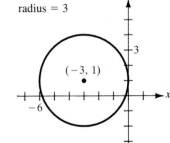

21. Single point: $(-3, 1)$

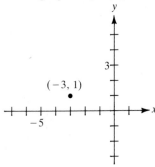

23. $x^2 + y^2 - 2x - 4y - 4 = 0$
    (a) On the circle
    (b) Inside the circle
    (c) Outside the circle
    (d) Inside the circle

25. $(4, 1)$

27. $y$ is a function of $x$

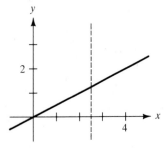

29. $y$ is not a function of $x$

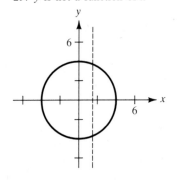

31. (a) $-x^2 + 2x + 2$
    (b) $-x^2 - 2x$
    (c) $-2x^3 - x^2 + 2x + 1$
    (d) $\dfrac{1 - x^2}{2x + 1}$
    (e) $-4x^2 - 4x$
    (f) $3 - 2x^2$

33. $P = x(500 - x)$

35. $x = 5(185 - p)$

37. 7

39. 91

41. $\dfrac{10}{3}$

43. Limit does not exist

45. $-\dfrac{1}{4}$

47. $-\infty$

49. $-1$

51. $3x^2 - 1$

53.

| $x$ | 1.1 | 1.01 | 1.001 | 1.0001 |
|---|---|---|---|---|
| $f(x)$ | 0.5680 | 0.5764 | 0.5773 | 0.5773 |

$$\lim_{x \to 1^+} \frac{\sqrt{2x + 1} - \sqrt{3}}{x - 1} = \frac{1}{\sqrt{3}} \approx 0.5774$$

55. False

57. False

59. False

61. Nonremovable discontinuity at each integer

63. Removable discontinuity at $x = 1$

65. Nonremovable discontinuity at $x = 2$

67. Nonremovable discontinuity at $x = -1$

69. $c = -\dfrac{1}{2}$

# CHAPTER 2

## Section 2.1

1.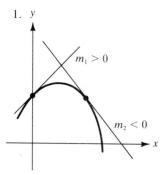

3.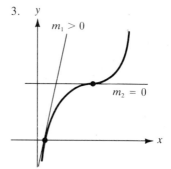

5. $m = 1$

7. $m = 0$

9. $m = -\dfrac{1}{3}$

11. $f(x) = 3$
   1. $f(x + \Delta x) = 3$
   2. $f(x + \Delta x) - f(x) = 0$
   3. $\dfrac{f(x + \Delta x) - f(x)}{\Delta x} = 0$
   4. $\lim\limits_{\Delta x \to 0} \dfrac{f(x + \Delta x) - f(x)}{\Delta x} = 0$

13. $f(x) = -5x + 3$
   1. $f(x + \Delta x) = -5x - 5\Delta x + 3$
   2. $f(x + \Delta x) - f(x) = -5\Delta x$
   3. $\dfrac{f(x + \Delta x) - f(x)}{\Delta x} = -5$
   4. $\lim\limits_{\Delta x \to 0} \dfrac{f(x + \Delta x) - f(x)}{\Delta x} = -5$

15. $f(x) = x^2$
   1. $f(x + \Delta x) = x^2 + 2x\Delta x + (\Delta x)^2$
   2. $f(x + \Delta x) - f(x) = 2x\Delta x + (\Delta x)^2$
   3. $\dfrac{f(x + \Delta x) - f(x)}{\Delta x} = 2x + \Delta x$
   4. $\lim\limits_{\Delta x \to 0} \dfrac{f(x + \Delta x) - f(x)}{\Delta x} = 2x$

17. $f(x) = 2x^2 + x - 1$
   1. $f(x + \Delta x) = 2x^2 + 4x\Delta x + 2(\Delta x)^2 + x + \Delta x - 1$
   2. $f(x + \Delta x) - f(x) = 4x\Delta x + 2(\Delta x)^2 + \Delta x$
      $= \Delta x(4x + 2\Delta x + 1)$
   3. $\dfrac{f(x + \Delta x) - f(x)}{\Delta x} = 4x + 2\Delta x + 1$
   4. $\lim\limits_{\Delta x \to 0} \dfrac{f(x + \Delta x) - f(x)}{\Delta x} = 4x + 1$

19. $h(t) = \sqrt{t - 1}$
   1. $h(t + \Delta t) = \sqrt{t + \Delta t - 1}$
   2. $h(t + \Delta t) - h(t) = \sqrt{t + \Delta t - 1} - \sqrt{t - 1}$
   3. $\dfrac{h(t + \Delta t) - h(t)}{\Delta t} = \dfrac{1}{\sqrt{t + \Delta t - 1} + \sqrt{t - 1}}$
   4. $\lim\limits_{\Delta t \to 0} \dfrac{h(t + \Delta t) - h(t)}{\Delta t} = \dfrac{1}{2\sqrt{t - 1}}$

21. $f(t) = t^3 - 12t$
   1. $f(t + \Delta t) = t^3 + 3t^2\Delta t + 3t(\Delta t)^2 + (\Delta t)^3 - 12t - 12\Delta t$
   2. $f(t + \Delta t) - f(t) = 3t^2\Delta t + 3t(\Delta t)^2 + (\Delta t)^3 - 12\Delta t$
   3. $\dfrac{f(t + \Delta t) - f(t)}{\Delta t} = 3t^2 + 3t\Delta t + (\Delta t)^2 - 12$
   4. $\lim\limits_{\Delta t \to 0} \dfrac{f(t + \Delta t) - f(t)}{\Delta t} = 3t^2 - 12$

23. $f'(x) = -2$
   $y = -2x + 6$

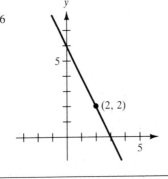

25. $f'(x) = 2x$
   $y = 4x - 6$

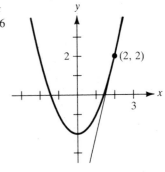

27. $f'(x) = 3x^2$
   $y = 12x - 16$

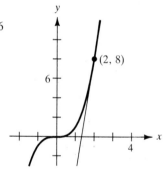

(2, 8)

6

4

29. $f'(x) = \dfrac{1}{2\sqrt{x} + 1}$
   $y = \dfrac{1}{4}(x + 5)$

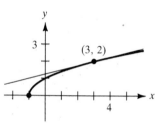

3

(3, 2)

4

31. $3x - y - 2 = 0$
   $3x - y + 2 = 0$

33. $2x - y + 1 = 0$
   $2x + y - 9 = 0$

35. $x = -3$

37. $x = 0$

## Section 2.2

1. (a) 2
   (b) $\dfrac{1}{2}$

3. (a) $-1$
   (b) $-\dfrac{3}{2}$

5. 0

7. 1

9. $2x$

11. $-4t + 3$

13. $3t^2 - 2$

15. $\dfrac{3}{t^{1/4}}$

17. $\dfrac{2}{\sqrt{x}}$

19. $-\dfrac{8}{x^3} + 4x$

| Function | Rewrite | Derivative | Simplify |
|---|---|---|---|
| 21. $y = \dfrac{1}{3x^3}$ | $y = \dfrac{1}{3}x^{-3}$ | $y' = -x^{-4}$ | $y' = -\dfrac{1}{x^4}$ |
| 23. $y = \dfrac{1}{(3x)^3}$ | $y = \dfrac{1}{27}x^{-3}$ | $y' = -\dfrac{1}{9}x^{-4}$ | $y' = -\dfrac{1}{9x^4}$ |
| 25. $y = \dfrac{\sqrt{x}}{x}$ | $y = x^{-1/2}$ | $y' = -\dfrac{1}{2}x^{-3/2}$ | $y' = -\dfrac{1}{2x^{3/2}}$ |

27. $-\dfrac{1}{x^2}, -1$

29. $\dfrac{3}{5t^2}, \dfrac{5}{3}$

31. $8x + 4, 4$

33. $\dfrac{2(x^3 + 2)}{x^2}$

35. $3x^2 - 3 + \dfrac{8}{x^5}$

37. $\dfrac{x^3 - 8}{x^3}$

39. $3x^2 + 1$

41. $\dfrac{4}{5x^{1/5}}$

43. $\dfrac{1}{3x^{2/3}} + \dfrac{1}{5x^{4/5}}$

45. $y = -2x + 2$

47. $(-1, 1), (0, 2), (1, 1)$

49. No horizontal tangents

51. $h'(x) = \dfrac{d}{dx}[f(x) + C]$

$\qquad = \dfrac{d}{dx}[f(x)] + \dfrac{d}{dx}[C]$

$\qquad = f'(x)$

53. (a) 3
 (b) 6
 (c) −3
 (d) −3

## Section 2.3

1. (a) 0.26
 (b) 0.426
 (c) 0.51
 (d) 0.34

3. Average rate: 2
 instantaneous rates:
 $f'(1) = 2, f'(2) = 2$

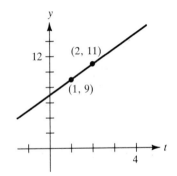

5. Average rate: −1
 instantaneous rates:
 $h'(0) = 0, h'(1) = -2$

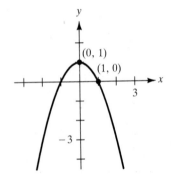

7. Average rate: 4.1
 instantaneous rates:
 $f'(2) = 4, f'(2.1) = 4.2$

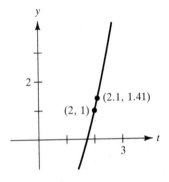

9. Average rate: $-\dfrac{1}{4}$

instantaneous rates:

$f'(1) = -1,\ f'(4) = -\dfrac{1}{16}$

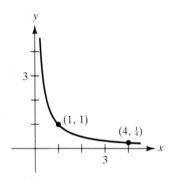

11. Average rate: 1

instantaneous rates:

$g'(1) = 2,\ g'(9) = \dfrac{2}{3}$

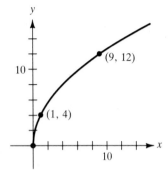

13. (a) Average rate: $\dfrac{11}{27}$

instantaneous rates:

$E'(0) = \dfrac{1}{3},\ E'(1) = \dfrac{4}{9}$

(b) Average rate: $\dfrac{11}{27}$

instantaneous rates:

$E'(1) = \dfrac{4}{9},\ E'(2) = \dfrac{1}{3}$

(c) Average rate: $\dfrac{5}{27}$

instantaneous rates:

$E'(2) = \dfrac{1}{3},\ E'(3) = 0$

(d) Average rate: $-\dfrac{7}{27}$

instantaneous rates:

$E'(3) = 0,\ E'(4) = -\dfrac{5}{9}$

15. (a) $-48$ ft/sec
    (b) $s'(1) = -32$ ft/sec
    $s'(2) = -64$ ft/sec
    (c) $\dfrac{15\sqrt{6}}{4} \approx 9.2$ seconds
    (d) $-120\sqrt{6} \approx -293.9$ ft/sec

17. (a) $s'(0) = 0$ ft/sec
    (b) $s'(1) = 15$ ft/sec
    (c) $s'(4) = 30$ ft/sec
    (d) $s'(9) = 45$ ft/sec

19. 1.47

21. $470 - \dfrac{1}{2}x$

23. $50 - x$

25. $-18x^2 + 16x + 200$

27. $-4x + 72$

29. $-\dfrac{1}{2000}x + 12.2$

31. (a) 1.999
    (b) 2

33. (a)

| $x$ | 10 | 15 | 20 | 25 | 30 | 35 | 40 |
|-----|-----|-----|-----|-----|-----|-----|-----|
| $C$ | 2025 | 1350 | 1012.50 | 810 | 675 | 578.57 | 506.25 |

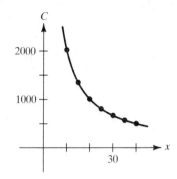

(b)

| x | 10 | 15 | 20 | 25 | 30 | 35 | 40 |
|---|---|---|---|---|---|---|---|
| $\dfrac{dC}{dx}$ | $-202.50$ | $-90.00$ | $-50.63$ | $-32.40$ | $-22.50$ | $-16.53$ | $-12.66$ |

(c) $C(11) - C(10) = -184.09$, $C'(10) = -202.50$

(d) $C(31) - C(30) = -21.77$, $C'(30) = -22.50$

35. (a) 37.40
    (b) 79.80
    (c) 14.83
    (d) $-48.40$

39. (a) 4.77 mi/yr
    (b) 3 mi/yr

37. (a) $R = \dfrac{1}{400}(-x^2 + 1100x)$

(b) $P = -\dfrac{1}{400}x^2 + \dfrac{3}{2}x - 65$

(c)

| x | 200 | 250 | 300 | 350 | 400 |
|---|---|---|---|---|---|
| $\dfrac{dR}{dx}$ | 1.75 | 1.50 | 1.25 | 1.00 | 0.75 |
| $\dfrac{dP}{dx}$ | 0.50 | 0.25 | 0 | $-0.25$ | $-0.50$ |
| P | 135.00 | 153.75 | 160.00 | 153.75 | 135.00 |

## Section 2.4

1. $f'(x) = 15x^4 - 2x$
   $f'(1) = 13$

3. $f'(x) = 2x^2$
   $f'(0) = 0$

5. $g'(x) = 3x^2 + 12x + 8$
   $g'(-2) = -4$

7. $h'(x) = -\dfrac{5}{(x-5)^2}$
   $h'(6) = -5$

9. $f'(t) = \dfrac{t^2 - 1}{4t^2}$
   $f'(1) = 0$

| Function | Rewrite | Derivative | Simplify |
|---|---|---|---|
| 11. $y = \dfrac{x^2 + 2x}{x}$ | $y = x + 2$ | $y' = 1$ | $y' = 1$ |

13. $y = \dfrac{7}{3x^3}$ $\qquad$ $y = \dfrac{7}{3}x^{-3}$ $\qquad$ $y' = -7x^{-4}$ $\qquad$ $y' = -\dfrac{7}{x^4}$

15. $y = \dfrac{3x^2 - 5}{7}$ $\qquad$ $y = \dfrac{1}{7}(3x^2 - 5)$ $\qquad$ $y' = \dfrac{1}{7}(6x)$ $\qquad$ $y' = \dfrac{6x}{7}$

17. $10x^4 + 12x^3 - 3x^2 - 18x - 15$ $\qquad\qquad$ 19. $15t^4 + 16t^3 - 45t^2 - 40t + 6$

21. $6s^5 - 12s^2$ $\qquad$ 23. $\dfrac{5}{6x^{1/6}} + \dfrac{1}{x^{2/3}}$ $\qquad$ 25. $-\dfrac{5}{(2x - 3)^2}$ $\qquad$ 27. $\dfrac{2}{(x + 1)^2}$

29. $\dfrac{3(x^4 + 1)}{x^2}$ $\qquad\qquad$ 31. $-\dfrac{t^2 + 2t}{(t^2 + 2t + 2)^2}$

33. $\dfrac{x - 1}{2x^{3/2}}$ $\qquad\qquad$ 35. $\dfrac{2x^2 + 8x - 1}{(x + 2)^2}$

37. $15x^4 - 48x^3 - 33x^2 - 32x - 20$ $\qquad$ 39. $-\dfrac{4c^2 x}{(x^2 - c^2)^2}$

41. $y = -x + 4$ $\qquad\qquad$ 43. $y = -x - 2$

45. $(0, 0)$ and $(2, 4)$ $\qquad\qquad$ 47. $-6$

49. (a) $-0.480$
$\quad$ (b) $0.120$
$\quad$ (c) $0.015$ $\qquad\qquad$ 51. 31.55 bacteria/hour

## Section 2.5

| $y = f(g(x))$ | $u = g(x)$ | $y = f(u)$ |
|---|---|---|
| 1. $y = (6x - 5)^4$ | $u = 6x - 5$ | $y = u^4$ |
| 3. $y = (4 - x^2)^{-1}$ | $u = 4 - x^2$ | $y = u^{-1}$ |
| 5. $y = \sqrt{x^2 - 1}$ | $u = x^2 - 1$ | $y = \sqrt{u}$ |
| 7. $y = \dfrac{1}{3x + 1}$ | $u = 3x + 1$ | $y = u^{-1}$ |

9. $6(2x - 7)^2$ $\qquad\qquad$ 11. $12x(x^2 - 1)^2$

13. $-6(4 - 2x)^2$ $\qquad\qquad$ 15. $6x(6 - x^2)(2 - x^2)$

17. $\dfrac{4x}{3(x^2 - 9)^{1/3}}$ $\qquad$ 19. $\dfrac{1}{2\sqrt{t + 1}}$ $\qquad$ 21. $\dfrac{t + 1}{\sqrt{t^2 + 2t - 1}}$

23. $\dfrac{6x}{(9x^2 + 4)^{2/3}}$ $\qquad$ 25. $-\dfrac{2x}{\sqrt{4 - x^2}}$ $\qquad$ 27. $-\dfrac{x}{(25 + x^2)^{3/2}}$

29. $\dfrac{4x^2}{(4 - x^3)^{7/3}}$ $\qquad$ 31. $-\dfrac{1}{(x - 1)^2}$ $\qquad$ 33. $-\dfrac{2}{(t - 3)^3}$

35. $-\dfrac{9x^2}{(x^3 - 4)^2}$

37. $-\dfrac{1}{2(x + 2)^{3/2}}$

39. $-\dfrac{3x^2}{(x^3 - 1)^{4/3}}$

41. $\dfrac{3(x + 1)}{\sqrt{2x + 3}}$

43. $\dfrac{t(5t - 8)}{2\sqrt{t - 2}}$

45. $2x(x - 2)^3(3x - 2)$

47. $\dfrac{5t^2 + 8t - 9}{2\sqrt{t + 2}}$

49. $-\dfrac{1}{2x^{3/2}\sqrt{x + 1}}$

51. $\dfrac{3t(t^2 + 3t - 2)}{(t^2 + 2t - 1)^{3/2}}$

53. $-\dfrac{1}{x^2\sqrt{x^2 + 1}}$

55. $y = \dfrac{9}{5}x - \dfrac{2}{5}$

57. (a) \$74.00
    (b) \$81.59
    (c) \$89.94

59. $\dfrac{100}{\sqrt{101}} \approx 9.95$

61.

| $t$ | 1 | 2 | 3 | 4 | 5 |
|---|---|---|---|---|---|
| $\dfrac{dC}{dt}$ | \$30.95 | \$32.93 | \$34.80 | \$36.57 | \$38.26 |

## Section 2.6

1. 0

3. 2

5. $2t - 8$

7. $\dfrac{4}{9t^{7/3}}$

9. $48x^2 - 16$

11. $\dfrac{4}{9x^{2/3}}$

13. $\dfrac{4}{(x - 1)^3}$

15. $6x - 8$

17. $60x^2 - 72x$

19. 12

21. $-\dfrac{9}{2x^5}$

23. $-\dfrac{3}{8(4 - x)^{5/2}}$

25. $4x$

27. $\dfrac{2}{x^2}$

29. 0

31. $f''(x) = 6(x - 3) = 0$
    when $x = 3$

33. $f''(x) = 6(x - 2) = 0$
    when $x = 2$

35. $f''(x) = 12(x - 1)(x - 3) = 0$
    when $x = 1$ or $x = 3$

37. $f''(x) = \dfrac{2x(x + 3)(x - 3)}{(x^2 + 3)^3} = 0$
    when $x = 0$, $x = -3$, or $x = 3$

39. (a) $v(t) = -32t + 48$
    $a(t) = -32$
    (b) 1.5 seconds
    (c) 36 feet

41.

| $t$ | 0 | 10 | 20 | 30 | 40 | 50 | 60 |
|---|---|---|---|---|---|---|---|
| $\dfrac{ds}{dt}$ | 0 | 45 | 60 | 67.5 | 72 | 75 | 77.14 |
| $\dfrac{d^2s}{dt^2}$ | 9.00 | 2.25 | 1.00 | 0.56 | 0.36 | 0.25 | 0.18 |

## Section 2.7

1. $3x$, $3$

3. $\dfrac{1}{2y}$, $-\dfrac{1}{4}$

5. $-\dfrac{y}{x}$, $-4$

7. $-\dfrac{x}{y}$, $-\dfrac{3}{4}$

9. $-\dfrac{y}{x+1}$, $-\dfrac{1}{4}$

11. $\dfrac{2x}{3y^2}$, $\dfrac{4}{3}$

13. $\dfrac{y-3x^2}{2y-x}$, $\dfrac{1}{2}$

15. $\dfrac{1-3x^2y^3}{3x^3y^2-1}$, $-1$

17. $-\sqrt{\dfrac{y}{x}}$, $-\dfrac{5}{4}$

19. $-\sqrt[3]{\dfrac{y}{x}}$, $-\dfrac{1}{2}$

21. $\dfrac{4xy-3x^2-3y^2}{2x(3y-x)}$, $-\dfrac{15}{28}$

23. $-\dfrac{x}{y}$, $\dfrac{4}{3}$

25. $-\dfrac{9x}{16y}$, $-\dfrac{\sqrt{3}}{4}$

27. At $(5, 12)$:  $5x + 12y - 169 = 0$
    at $(-12, 5)$: $12x - 5y + 169 = 0$

29. $-\dfrac{x^2}{100}$

31. Points of intersection: $(1, 2)$ and $(1, -2)$
    at $(1, 2)$, slope of $2x^2 + y^2 = 6$ is $-1$.
    at $(1, 2)$, slope of $y^2 = 4x$ is $1$.
    at $(1, -2)$, slope of $2x^2 + y^2 = 6$ is $1$.
    at $(1, -2)$, slope of $y^2 = 4x$ is $-1$.

33. $0.04224$


## Section 2.8

1. (a) $\dfrac{3}{4}$
   (b) $20$

3. (a) $-\dfrac{5}{8}$
   (b) $\dfrac{3}{2}$

5. (a) $24\pi$ in²/min
   (b) $96\pi$ in²/min

7. If $\dfrac{dr}{dt}$ is constant, $\dfrac{dA}{dt} = 2\pi r \dfrac{dr}{dt}$ and thus is proportional to $r$.

9. (a) $\dfrac{5}{\pi}$ ft/min
   (b) $\dfrac{5}{4\pi}$ ft/min

11. $\dfrac{8}{405\pi}$ ft/min

13. (a) $9$ cm³/sec
    (b) $900$ cm³/sec

15. (a) $-12$ cm/min
    (b) $0$ cm/min
    (c) $4$ cm/min
    (d) $12$ cm/min

17. (a) $-\dfrac{7}{12}$ ft/sec
    (b) $-\dfrac{3}{2}$ ft/sec
    (c) $-\dfrac{48}{7}$ ft/sec

19. (a) $-750$ mi/hr
    (b) $20$ min

21. $-\dfrac{28}{\sqrt{10}} \approx -8.85$ ft/sec

23. $650$/week

25. $60\pi \approx 188.5$ ft³/min

# Review Exercises for Chapter 2

1. $f(x) = 7x + 3$
   1. $f(x + \Delta x) = 7x + 7\Delta x + 3$
   2. $f(x + \Delta x) - f(x) = 7\Delta x$
   3. $\dfrac{f(x + \Delta x) - f(x)}{\Delta x} = 7$
   4. $\displaystyle\lim_{\Delta x \to 0} \dfrac{f(x + \Delta x) - f(x)}{\Delta x} = 7$

3. $h(t) = \sqrt{t + 9}$
   1. $h(t + \Delta t) = \sqrt{t + \Delta t + 9}$
   2. $h(t + \Delta t) - h(t) = \sqrt{t + \Delta t + 9} - \sqrt{t + 9}$
   3. $\dfrac{h(t + \Delta t) - h(t)}{\Delta t} = \dfrac{1}{\sqrt{t + \Delta t + 9} + \sqrt{t + 9}}$
   4. $\displaystyle\lim_{\Delta t \to 0} \dfrac{h(t + \Delta t) - h(t)}{\Delta t} = \dfrac{1}{2\sqrt{t + 9}}$

5. $3x^2 - 6x$

7. $3\left(x^2 - \dfrac{3}{x^4}\right)$

9. $\dfrac{x + 1}{2x^{3/2}}$

11. $2(6x^3 - 9x^2 + 16x - 7)$

13. $-\dfrac{4}{3t^3}$

15. $-\dfrac{x^2 + 1}{(x^2 - 1)^2}$

17. $\dfrac{6x}{(4 - 3x^2)^2}$

19. $-\dfrac{1}{(x + 1)^{3/2}}$

21. $\dfrac{3x^2}{2\sqrt{x^3 + 1}}$

23. $\dfrac{2x^2 + 1}{\sqrt{x^2 + 1}}$

25. $\dfrac{4}{3}\sqrt[3]{t + 1}$

27. $32x(1 - 4x^2)$

29. $18x^5(x + 1)(2x + 3)^2$

31. $2$

33. $\dfrac{30}{(1 - t)^4}$

35. $4(9x^2 - 9x + 8)$

37. $-8(1 - x^2)^2(1 - 7x^2)$

39. $-\dfrac{4}{x^{5/3}}$

41. $-\dfrac{2x + 3y}{3(x + y^2)}$

43. $\dfrac{x}{y}$

45. $y = 3x + 7$

47. $y = -\dfrac{1}{2}x + 5$

49. $y = \dfrac{2}{3}x - 1$

51. $650$

53. $\dfrac{25(x - 4)}{(x - 2)^{3/2}}$

55. $-0.0015x^2 + 10x - 1$

57. (a) $(0, -1)$, $\left(-2, \dfrac{7}{3}\right)$

    (b) $(-3, 2)$, $\left(1, -\dfrac{2}{3}\right)$

    (c) $\left(-1 + \sqrt{2}, \dfrac{2 - 4\sqrt{2}}{3}\right)$,

       $\left(-1 - \sqrt{2}, \dfrac{2 + 4\sqrt{2}}{3}\right)$

59. $v(t) = 1 - \dfrac{2}{(t + 1)^2}$, $a(t) = \dfrac{4}{(t + 1)^3}$

61. (a) $-18.667°/\text{hr}$
    (b) $-7.284°/\text{hr}$
    (c) $-3.240°/\text{hr}$
    (d) $-0.747°/\text{hr}$

63. (a) 6 board ft/in
    (b) 18 board ft/in
    (c) 30 board ft/in
    (d) 48 board ft/in

# CHAPTER 3

## Section 3.1

1. $f'(-1) = -\dfrac{8}{25}$, $f$ is decreasing

   $f'(0) = 0$, $f$ has a critical number

   $f'(1) = \dfrac{8}{25}$, $f$ is increasing

3. $f'(-3) = -\dfrac{2}{3}$, $f$ is decreasing

   $f'(-2)$ undefined, $f$ has a critical number

   $f'(-1) = \dfrac{2}{3}$, $f$ is increasing

5. Decreasing on $(-\infty, 3)$
   increasing on $(3, \infty)$

7. Increasing on $(-\infty, -2)$ and $(2, \infty)$
   decreasing on $(-2, 2)$

9. Increasing on $(-\infty, 0)$
   decreasing on $(0, \infty)$

11. No critical numbers
    increasing on $(-\infty, \infty)$

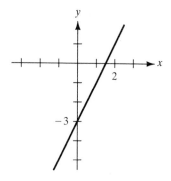

13. Critical number: $x = 1$
    increasing on $(-\infty, 1)$
    decreasing on $(1, \infty)$

15. Critical number: $x = 1$
    decreasing on $(-\infty, 1)$
    increasing on $(1, \infty)$

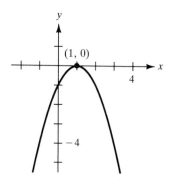

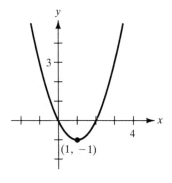

17. Critical numbers: $x = 0$, $x = 4$
    increasing on $(-\infty, 0)$ and $(4, \infty)$
    decreasing on $(0, 4)$

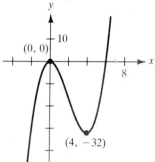

19. Critical number: $x = -1$
    decreasing on $(-\infty, \infty)$

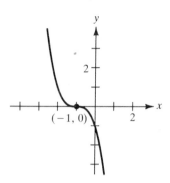

21. Critical number: $x = 1$
    increasing on $(-\infty, 1)$
    decreasing on $(1, \infty)$

23. Critical numbers: $x = -2$, $x = 1$
    increasing on $(-\infty, -2)$ and $(1, \infty)$
    decreasing on $(-2, 1)$

25. Critical number: $x = 0$
    decreasing on $(-\infty, 0)$
    increasing on $(0, \infty)$

27. Critical numbers: $x = 0$, $x = \dfrac{3}{2}$

    decreasing on $\left(-\infty, \dfrac{3}{2}\right)$

    increasing on $\left(\dfrac{3}{2}, \infty\right)$

29. Domain: $(-\infty, 3]$
    critical number: $x = 2$
    increasing on $(-\infty, 2)$
    decreasing on $(2, 3)$

31. Critical numbers: $x = 2$, $x = -2$
    decreasing on $(-\infty, -2)$ and $(2, \infty)$
    increasing on $(-2, 2)$

33. Critical numbers: $x = -1$, $x = 1$
    discontinuity: $x = 0$
    increasing on $(-\infty, -1)$ and $(1, \infty)$
    decreasing on $(-1, 0)$ and $(0, 1)$

35. Critical number: $x = 0$
    discontinuities: $x = \pm 3$
    increasing on $(-\infty, -3)$ and $(-3, 0)$
    decreasing on $(0, 3)$ and $(3, \infty)$

37. Discontinuity: $x = 0$
    increasing on $(-\infty, 0)$
    decreasing on $(0, \infty)$

39. Moving upward on $(0, 3)$
    moving downward on $(3, 6)$

41. Increasing on $[0, 84.34)$
    decreasing on $(84.34, 120]$

43. Decreasing on $[0, 6.0)$
    increasing on $(6.0, 14]$

45. $g'(0) < 0$

47. $g'(-6) < 0$

49. $g'(0) > 0$

## Section 3.2

1. Relative maximum: $(1, 5)$

3. Relative minimum: $(3, -9)$

5. Relative maximum: $(-2, 20)$
relative minimum: $(1, -7)$

7. No relative extrema

9. Relative maximum: $(0, 15)$
relative minimum: $(4, -17)$

11. Relative minimum: $\left(\dfrac{3}{2}, -\dfrac{27}{16}\right)$

13. No relative extrema

15. Relative minimum: $(0, 0)$

17. Relative maximum: $(-1, -2)$
relative minimum: $(1, 2)$

19. Relative maximum: $(0, 4)$

21. Yes　　　　　　　　23. No

25. Yes

27. (a) Maximum: $(1, 4)$
minimum: $(4, 1)$
(b) Maximum: $(1, 4)$
(c) Minimum: $(4, 1)$
(d) No extrema

29. (a) Maximum: $(0, 2)$
minima: $(-2, 0)$, $(2, 0)$
(b) Minimum: $(-2, 0)$
(c) Maximum: $(0, 2)$
(d) Maximum: $(1, \sqrt{3})$

31. Minimum: $(2, 2)$
maximum: $(-1, 8)$

33. Minimum: $(0, 0)$
maximum: $(2, 4)$

35. Minima: $(-1, -4)$ and $(2, -4)$
maxima: $(0, 0)$ and $(3, 0)$

37. Minimum: $(0, 0)$
maximum: $(-1, 5)$

39. Minimum: $(1, -1)$
maximum: $\left(0, -\dfrac{1}{2}\right)$

41. $\left| f''\left(\dfrac{\sqrt{3}}{3}\right) \right| = \dfrac{40\sqrt{3}}{3}$

43. $\left| f^{(4)}\left(\dfrac{1}{2}\right) \right| = 360$

45. 300 units

47. $\dfrac{2R}{3}$

## Section 3.3

1. Concave upward on $(-\infty, \infty)$

3. Concave upward on $(-\infty, -2)$ and $(2, \infty)$
concave downward on $(-2, 2)$

5. Concave upward on $(-\infty, -1)$ and $(1, \infty)$
concave downward on $(-1, 1)$

7. Relative maximum: $(3, 9)$

9. Relative minimum: $(5, 0)$

11. Relative maximum: $(0, 3)$
relative minimum: $(2, -1)$

13. Relative minimum: $(3, -25)$

15. Relative minimum: $(0, -3)$

17. Relative maximum: $(-2, -4)$
relative minimum: $(2, 4)$

19. Relative maximum: $(-2, 16)$
relative minimum: $(2, -16)$
point of inflection: $(0, 0)$

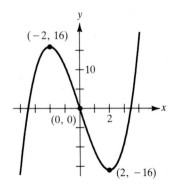

21. Point of inflection: (2, 0)

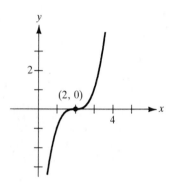

23. Relative maximum: (0, 0)
    relative minima: ($\pm 2$, $-4$)
    points of inflection: $\left( \pm \dfrac{2}{\sqrt{3}}, -\dfrac{20}{9} \right)$

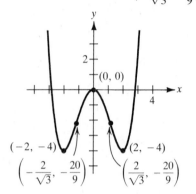

25. Relative maximum: ($-2$, 0)
    relative minimum: (0, $-4$)
    point of inflection: ($-1$, $-2$)

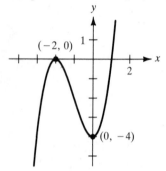

27. Relative minimum: ($-2$, $-2$)

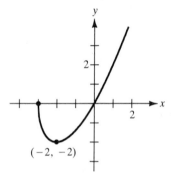

29. Relative maximum: (0, 4)
    points of inflection: $\left( \pm \dfrac{\sqrt{3}}{3}, 3 \right)$

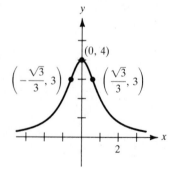

31. (3, 36)

33. 100 units

35.

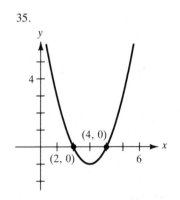

37. $f'(x)$
   (a) Positive on $(-\infty, 0)$
   (b) Negative on $(0, \infty)$
   (c) Not increasing
   (d) Decreasing on $(-\infty, \infty)$

$f(x)$
Increasing on $(-\infty, 0)$
Decreasing on $(0, \infty)$
Not concave upward
Concave downward on $(-\infty, \infty)$

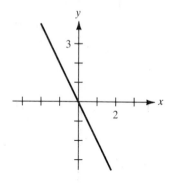

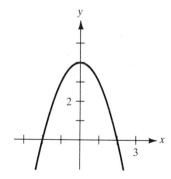

39. Relative minimum: $(6, 0)$
   relative maximum: $(2, 32)$
   point of inflection: $(4, 16)$

## Section 3.4

1. 55 and 55

3. 12 and 6

5. $\sqrt{192}$ and $\sqrt{192}$

7. 1

9. Length = width = 25 feet

11. Length = width = 8 feet

13. $x = 25$, $y = \dfrac{100}{3}$

15. Volume = 128 inches$^3$ when $x = 2$

17. 0.392 foot $\times$ 1.215 feet $\times$ 2.215 feet

19. 50 meters $\times \dfrac{100}{\pi}$ meters

21. $x = 3$, $y = \dfrac{3}{2}$

23. $(0, 0)$, $(1 + \sqrt[3]{4}, 0)$, $\left(0, 2 + \dfrac{2}{\sqrt[3]{4}}\right)$

25. $x = y = \dfrac{5}{\sqrt{2}} \approx 3.54$

27. $\left(\dfrac{7}{2}, \dfrac{\sqrt{7}}{2}\right)$

29. $r \approx 1.51$ inches, $h = 2r \approx 3.02$ inches

31. 18 inches $\times$ 18 inches $\times$ 36 inches

33. Radius: $\dfrac{8}{\pi + 4}$

    side of square: $\dfrac{16}{\pi + 4}$

35. $x = 1$ mile

37. $w = 8\sqrt{3}$, $h = 8\sqrt{6}$

## Section 3.5

1. 4500 units      3. 300 units      5. 200 units      7. 200 units

9. $60.00      11. $35.00      13. 3 units      15. (a) $80.00
         (b) $45.93

17. Maximum profit: $10,000;
point of diminishing returns: $5,833.33

19. 200 units      21. $92.50

23. Line should run from the power station to a point across the river $\dfrac{3}{2\sqrt{7}} \approx 0.57$ mile downstream

25. $10\sqrt[3]{165} \approx 54.8$ mi/hr

27. $-\dfrac{17}{3}$, elastic      29. $-\dfrac{1}{2}$, inelastic

31. $-\dfrac{3}{2}$, elastic

33. (a) $-6.83\%$
(b) $-1.37$
(c) $-\dfrac{4}{3}$
(d) $x = \dfrac{40}{3}, \; p = \sqrt{\dfrac{10}{3}} \approx \$1.83$

35. (a) $-\dfrac{14}{9}$
(b) $x = \dfrac{32}{3}, \; p = \dfrac{4\sqrt{3}}{3}$

## Section 3.6

1. Vertical: $x = 0$
horizontal: $y = 0$

3. Vertical: $x = -1, \; x = 2$
horizontal: $y = 1$

5. Vertical: $x = -1, \; x = 1$
horizontal: none

7. f      8. b      9. c      10. a      11. e

12. d      13. $\infty$      15. $-\infty$      17. $-\infty$      19. $-\infty$

21. $\dfrac{2}{3}$      23. 0      25. $-\infty$      27. $\infty$      29. 5

31.

| $x$ | $10^0$ | $10^1$ | $10^2$ | $10^3$ | $10^4$ | $10^5$ | $10^6$ |
|---|---|---|---|---|---|---|---|
| $f(x)$ | 2.000 | 0.348 | 0.101 | 0.032 | 0.010 | 0.003 | 0.001 |

$\lim\limits_{x \to \infty} \dfrac{x+1}{x\sqrt{x}} = 0$

33.

| $x$ | $10^0$ | $10^1$ | $10^2$ | $10^3$ | $10^4$ | $10^5$ | $10^6$ |
|---|---|---|---|---|---|---|---|
| $f(x)$ | 1.0 | 5.1 | 50.1 | 500.1 | 5,000.1 | 50,000.1 | 500,000.2 |

$\lim\limits_{x \to \infty} [x^2 - x\sqrt{x(x-1)}] = \infty$

35.

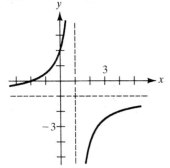

37.

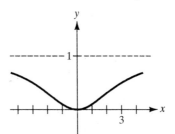

39.

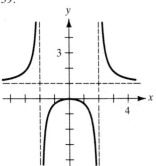

41.

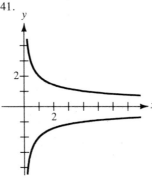

43.

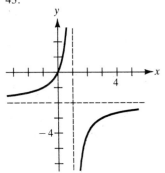

45.

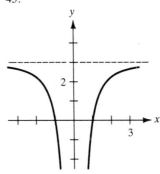

47.

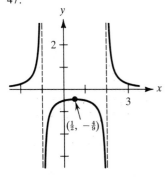

$(\frac{1}{2}, -\frac{4}{9})$

49.

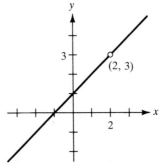

$(2, 3)$

51. (a) $47.05, $5.92
    (b) $1.35

53. (a) $176 million
    (b) $528 million
    (c) $1,584 million
    (d) ∞

55. *a*

57. (a) 167, 250, 400
    (b) 750

## Section 3.7

1. $a < 0$

3. $a > 0$

5.

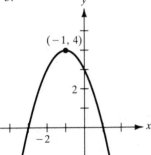

7.

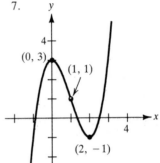

9.

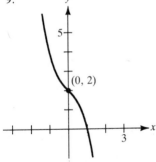

11.

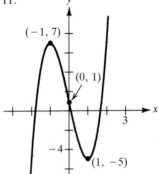

13.

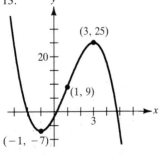

15.

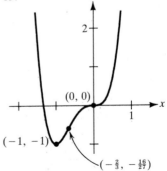

17.

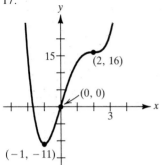

19.

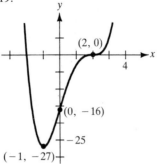

21.

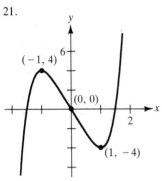

23.

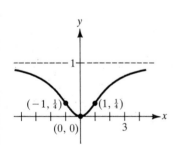

25.

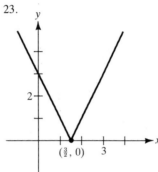

27.

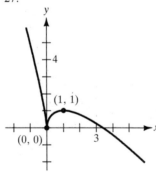

29.

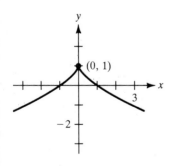

31. Domain: $(-\infty, 2)$, $(2, \infty)$

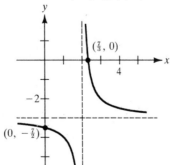

33. Domain: $(-\infty, -1)$, $(-1, 1)$, $(1, \infty)$

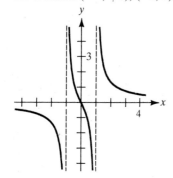

35. Domain: $(-\infty, 4]$

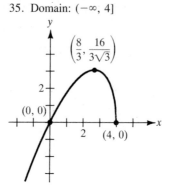

37. Domain: $(-\infty, 0)$, $(0, \infty)$

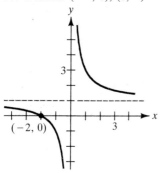

39.

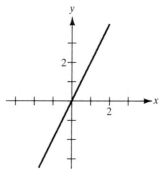

41.

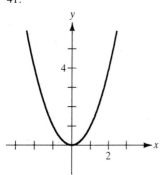

## Section 3.8

1. $6x \, dx$

3. $6(2x + 5)^2 \, dx$

5. $\dfrac{x}{\sqrt{x^2 + 1}} \, dx$

7. $\dfrac{1 - 2x^2}{\sqrt{1 - x^2}} \, dx$

9. $-\dfrac{3}{(2x - 1)^2} \, dx$

11.

| $dx = \Delta x$ | $dy$ | $\Delta y$ | $\Delta y - dy$ | $\dfrac{dy}{\Delta y}$ |
|---|---|---|---|---|
| 1.000 | 4.000 | 5.000 | 1.000 | 0.800 |
| 0.500 | 2.000 | 2.250 | 0.250 | 0.889 |
| 0.100 | 0.400 | 0.410 | 0.010 | 0.976 |
| 0.010 | 0.040 | 0.040 | 0.000 | 0.998 |
| 0.001 | 0.004 | 0.004 | 0.000 | 1.000 |

13.

| $dx = \Delta x$ | $dy$ | $\Delta y$ | $\Delta y - dy$ | $\dfrac{dy}{\Delta y}$ |
|---|---|---|---|---|
| 1.000 | 80.000 | 211.000 | 131.000 | 0.379 |
| 0.500 | 40.000 | 65.656 | 25.656 | 0.609 |
| 0.100 | 8.000 | 8.841 | 0.841 | 0.905 |
| 0.010 | 0.800 | 0.808 | 0.008 | 0.990 |
| 0.001 | 0.080 | 0.080 | 0.000 | 0.999 |

15. $3 + \dfrac{1}{54} \approx 3.0185$

(Using calculator: 3.0183)

17. $\dfrac{1}{3} + \dfrac{2}{243} \approx 0.3416$

(Using calculator: 0.3420)

19. (a) $dA = 2x\Delta x$, $\Delta A = 2x\Delta x + (\Delta x)^2$

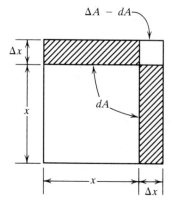

21. $\pm 7\pi$ square inches, 3.57%

23. $\pm 2.88\pi$ cubic inches, 0.01

25. $1,160.00, 2.67%

27. 19.00 deer

## Review Exercises for Chapter 3

1. b      2. c      3. f      4. g      5. h      6. d

7. e      8. a      9. $-\infty$      11. $\dfrac{1}{3}$      13. $\dfrac{5}{2}$

**15.**

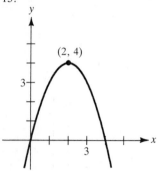

(2, 4)

**17.**

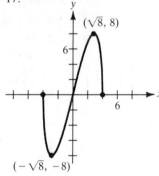

$(\sqrt{8}, 8)$

$(-\sqrt{8}, -8)$

**19.**

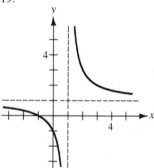

**21.**

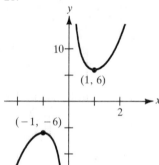

(1, 6)

(−1, −6)

**23.**

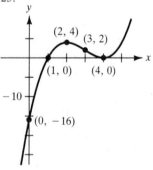

(2, 4)  (3, 2)

(1, 0)  (4, 0)

−10

(0, −16)

**25.**

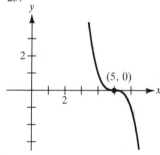

(5, 0)

**27.**

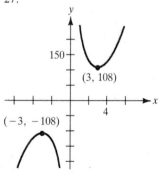

150

(3, 108)

(−3, −108)

**29.**

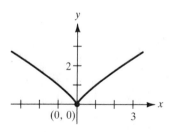

(0, 0)

31. $(0, 0)$, $(5, 0)$, $(0, 10)$           33. $12\sqrt{2} \times 4\sqrt{2}$

35. $48.00        37. 120        39. $x = \sqrt{\dfrac{2Qs}{r}}$

41. 2%, 3%        43. $\Delta p = dp = -\dfrac{1}{4}$

# CHAPTER 4

## Section 4.1

1. $6x + C$

$\dfrac{d}{dx}[6x + C] = 6$

3. $t^3 + C$

$\dfrac{d}{dt}[t^3 + C] = 3t^2$

5. $-\dfrac{1}{x^3} + C$

$\dfrac{d}{dx}\left[-\dfrac{1}{x^3} + C\right] = 3x^{-4}$

7. $u + C$

$\dfrac{d}{du}[u + C] = 1$

9. $\dfrac{2}{5}x^{5/2} + C$

$\dfrac{d}{dx}\left[\dfrac{2}{5}x^{5/2} + C\right] = x^{3/2}$

| | Given | Rewrite | Integrate | Simplify |
|---|---|---|---|---|
| 11. | $\displaystyle\int \sqrt[3]{x}\, dx$ | $\displaystyle\int x^{1/3}\, dx$ | $\dfrac{x^{4/3}}{4/3} + C$ | $\dfrac{3}{4}x^{4/3} + C$ |
| 13. | $\displaystyle\int \dfrac{1}{x\sqrt{x}}\, dx$ | $\displaystyle\int x^{-3/2}\, dx$ | $\dfrac{x^{-1/2}}{-1/2} + C$ | $-\dfrac{2}{\sqrt{x}} + C$ |
| 15. | $\displaystyle\int \dfrac{1}{2x^3}\, dx$ | $\dfrac{1}{2}\displaystyle\int x^{-3}\, dx$ | $\dfrac{1}{2}\left(\dfrac{x^{-2}}{-2}\right) + C$ | $-\dfrac{1}{4x^2} + C$ |

17. $\dfrac{x^4}{4} + 2x + C$

19. $\dfrac{2}{5}x^{5/2} + x^2 + x + C$

21. $\dfrac{3}{5}x^{5/3} + C$

23. $-\dfrac{1}{2x^2} + C$

25. $-\dfrac{1}{4x} + C$

27. $t - \dfrac{2}{t} + C$

29. $\dfrac{3}{4}u^4 + \dfrac{1}{2}u^2 + C$

31. $x^3 + \dfrac{x^2}{2} - 2x + C$

33. $\dfrac{2}{7}y^{7/2} + C$

35. $y = x^2 - x + 1$

37. $f(x) = x^2 + x + 4$

39. $f(x) = -4\sqrt{x} + 3x$

41. $R = 500x - \dfrac{5}{2}x^2$

$p = 500 - \dfrac{5}{2}x$

43. (a) $C = x^2 - 12x + 125$

$\overline{C} = x - 12 + \dfrac{125}{x}$

(b) $2,025

45. (a) $h(t) = \dfrac{t^2}{4} + 2t + 5$

(b) 26 inches

47. 56.25 feet

49. $v_0 = 40\sqrt{22} \approx 187.617$ ft/sec

## Section 4.2

$$\int u^n \frac{du}{dx}\, dx \qquad\qquad u \qquad\qquad \frac{du}{dx}$$

1. $\displaystyle\int (5x^2 + 1)^2 (10x)\, dx$ $\qquad\qquad 5x^2 + 1 \qquad\qquad 10x$

3. $\displaystyle\int \sqrt{1 - x^2}(-2x)\, dx$ $\qquad\qquad 1 - x^2 \qquad\qquad -2x$

5. $\displaystyle\int \left(4 + \frac{1}{x^2}\right)\left(\frac{-2}{x^3}\right) dx$ $\qquad\qquad 4 + \frac{1}{x^2} \qquad\qquad -\frac{2}{x^3}$

7. $\dfrac{1}{5}(1 + 2x)^5 + C$

9. $\dfrac{2}{3}(3x^2 + 4)^{3/2} + C$

11. $\dfrac{1}{15}(x^3 - 1)^5 + C$

13. $\dfrac{1}{16}(x^2 - 1)^8 + C$

15. $-\dfrac{1}{3(1 + x^3)} + C$

17. $-\dfrac{1}{2(x^2 + 2x - 3)} + C$

19. $\sqrt{x^2 - 8x + 1} + C$

21. $\dfrac{15}{8}(1 + x^2)^{4/3} + C$

23. $4\sqrt{1 + x^2} + C$

25. $-3\sqrt{2x + 3} + C$

27. $\dfrac{1}{2}\sqrt{1 + x^4} + C$

29. $\sqrt{x} + C$

31. $\sqrt{2x} + C$

33. $\dfrac{1}{4}t^4 - t^2 + C$

35. $6y^{3/2} - \dfrac{2}{5}y^{5/2} + C$

37. $\dfrac{1}{6}(2x - 1)^3 + C_1 = \dfrac{4}{3}x^3 - 2x^2 + x + C_2$

$\left(\text{Answers differ by a constant: } C_2 = C_1 - \dfrac{1}{6}\right)$

39. $\dfrac{1}{24}(3x^2 - 5)^4 + C$

41. $-\dfrac{2}{45}(2 - 3x^3)^{5/2} + C$

43. $\sqrt{x^2 + 25} + C$

45. $\dfrac{2}{3}\sqrt{x^3 + 3x + 4} + C$

47. $\dfrac{1}{3}[5 - (1 - x^2)^{3/2}]$

49. (a) $8\sqrt{x + 1} + 18$

    (b)   C

51. $x = \dfrac{1}{3}(p^2 - 16)^{3/2} + 41$

53. $x = \dfrac{6000}{\sqrt{p^2 - 16}} + 3000$

55. (a) $W = \dfrac{3}{2}(\sqrt{16t + 9} - 3)$

    (b) 55.67 pounds

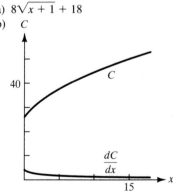

# Section 4.3

1. 1

3. $-\dfrac{5}{2}$

5. $-\dfrac{10}{3}$

7. $\dfrac{1}{3}$

9. $\dfrac{1}{2}$

11. 36

13. $-4$

15. $\dfrac{2}{3}$

17. $-\dfrac{1}{18}$

19. $-\dfrac{27}{20}$

21. 2

23. 0

25. 0

27. 1

29. $\dfrac{1}{6}$

31. $\dfrac{27}{4}$

33. $\dfrac{3}{8}(50\sqrt[3]{50} - 1) \approx 68.7$

35. $\dfrac{1}{6}$

37. $\dfrac{8}{5}$

39. 6

41. Area = 6

43. Area = $\dfrac{10}{3}$

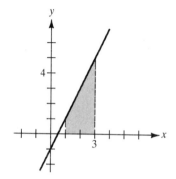

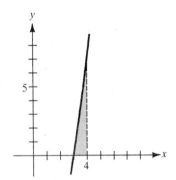

45. Area = $\dfrac{1}{4}$

47. 10

49. 6

51. Average = $\dfrac{8}{3}$

$$x = \dfrac{\pm 2\sqrt{3}}{3} \approx \pm 1.155$$

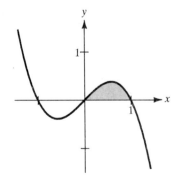

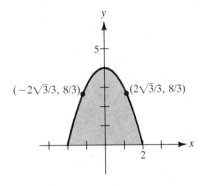

$(-2\sqrt{3}/3,\ 8/3)$  $(2\sqrt{3}/3,\ 8/3)$

53. Average $= \dfrac{4}{3}$

$x = \sqrt{2 + \dfrac{2\sqrt{5}}{3}} \approx 1.868$

$x = \sqrt{2 - \dfrac{2\sqrt{5}}{3}} \approx 0.714$

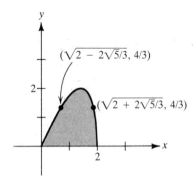

$(\sqrt{2 - 2\sqrt{5}/3},\ 4/3)$

$(\sqrt{2 + 2\sqrt{5}/3},\ 4/3)$

55. Average $= -\dfrac{2}{3}$

$x = \dfrac{4 + 2\sqrt{3}}{3} \approx 2.488$

$x = \dfrac{4 - 2\sqrt{3}}{3} \approx 0.179$

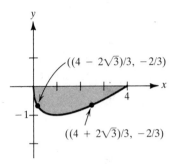

$((4 - 2\sqrt{3})/3,\ -2/3)$

$((4 + 2\sqrt{3})/3,\ -2/3)$

57. (a) $\dfrac{8}{3}$

   (b) $\dfrac{16}{3}$

   (c) $-\dfrac{8}{3}$

   (d) 8

59. (a) \$137,000.00
   (b) \$214,720.93
   (c) \$338,393.53

61. $V = \displaystyle\int_{0}^{3} 10{,}000(t - 6)\,dt = -\$135{,}000.00$

63. (a) $64.4°$
   (b) $68.6°$

## Section 4.4

1.

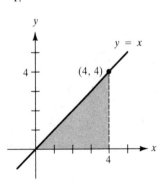

$y = x$

$(4, 4)$

3.

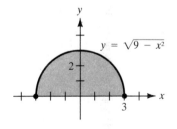

$y = \sqrt{9 - x^2}$

**5.**

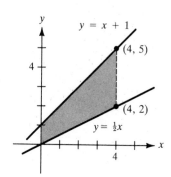

**7.** 36

**9.** 9

**11.** $\dfrac{3}{2}$

**13.** Area $= \dfrac{32}{3}$

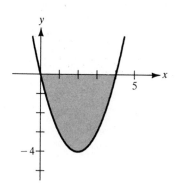

**15.** Area $= \dfrac{9}{2}$

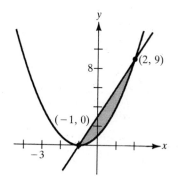

**17.** Area $= 1$

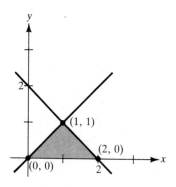

**19.** Area $= \dfrac{125}{6}$

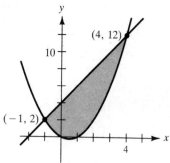

**21.** Area $= 2$

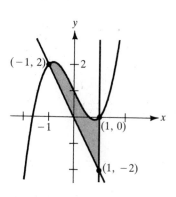

**23.** Area $= \dfrac{3}{2}$

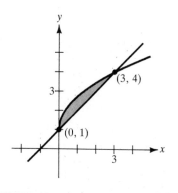

25. Area $= \dfrac{64}{3}$

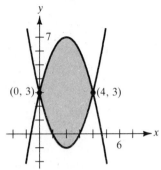

27. Area $= \dfrac{9}{2}$

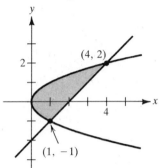

29. Area $= 6$

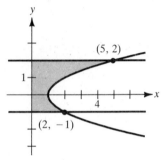

31. 8

33. $1.625 billion

35. 41.263 billion pounds

37. Point of equilibrium: (80, 10)
    consumer surplus $= 1600$
    producer surplus $= 400$

39. Point of equilibrium: (100, 200)
    consumer surplus $= 5000$
    producer surplus $= 5000$

41. Point of equilibrium: (100, 200)
    consumer surplus $\approx 6667$
    producer surplus $= 5000$

43. Point of equilibrium: (200, 25)
    consumer surplus $= 2500$
    producer surplus $\approx 1833$

45. Point of equilibrium: (300, 500)
    consumer surplus $= 50,000$
    producer surplus $\approx 25,497$

## Section 4.5

1. Midpoint Rule: 2

   exact area: 2

3. Midpoint Rule: 0.6730

   exact area: $\dfrac{2}{3} \approx 0.6667$

5. Midpoint Rule: 4.6250

   exact area: $\dfrac{14}{3} \approx 4.6667$

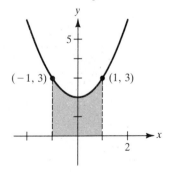

7. Midpoint Rule: 17.2500

   exact area: $\dfrac{52}{3} \approx 17.3333$

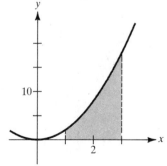

9. Midpoint Rule: 0.7578

    exact area: $\dfrac{3}{4} = 0.7500$

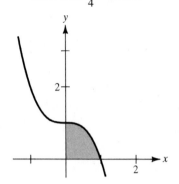

11. Midpoint Rule: 0.5703

    exact area: $\dfrac{7}{12} \approx 0.5833$

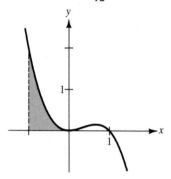

13. Midpoint Rule: 0.0859

    exact area: $\dfrac{1}{12} \approx 0.0833$

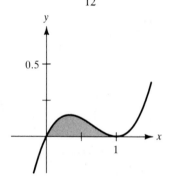

15. Midpoint Rule: 6.0000

    exact area: 6.0000

17.

| $n$ | 2 | 4 | 8 |
|---|---|---|---|
| **Approximation** | 13.0375 | 12.7357 | 12.6267 |

19.

| $n$ | 2 | 4 | 8 |
|---|---|---|---|
| **Approximation** | 3.1523 | 3.2202 | 3.2361 |

## Section 4.6

1. $\dfrac{\pi}{3}$      3. $\dfrac{16\pi}{3}$      5. $\dfrac{15\pi}{2}$      7. $\dfrac{2\pi}{35}$      9. $\dfrac{64\pi}{3}$

11. $\dfrac{128\pi}{5}$      13. $\dfrac{2\pi}{3}$      15. $\dfrac{243\pi}{5}$      17. $8\pi$      19. $\dfrac{\pi}{4}$

21. $\dfrac{\pi}{3}$      23. $\dfrac{256\pi}{15}$      25. $18\pi$      27. $V = \pi \displaystyle\int_{-r}^{r} (r^2 - x^2)\, dx = \dfrac{4\pi r^3}{3}$

29. $60\pi$                    31. $168,750\pi$ cubic feet, 1060 fish

## Review Exercises for Chapter 4

1. $\frac{2}{3}x^3 + \frac{1}{2}x^2 - x + C$

3. $x - \frac{3}{x} + C$

5. $-\frac{1}{3(x^3 - 1)} + C$

7. $\frac{1}{2}x^2 - 2x - \frac{1}{x} + C$

9. $x^{2/3} + C$

11. $\frac{2\sqrt{x}}{15}(3x^2 + 10x + 15) + C$

13. $\frac{2}{3}\sqrt{x^3 + 3} + C$

15. $-\frac{3}{2}\sqrt{1 - 2x^2} + C$

17. $\frac{x^7}{7} + \frac{3x^5}{5} + x^3 + x + C$

19. $\frac{2x^{3/2}}{5}(x + 5) + C$

21. 16

23. 0

25. 2

27. $\frac{422}{5}$

29. $\frac{\pi}{15}$

31. $\int_0^5 3\,dx$

33. $\int_0^2 x^2\,dx$

35. $\int_{-2}^2 (4 - x^2)\,dx = 2\int_0^2 (4 - x^2)\,dx$

37. $\int_0^2 y^3\,dy$

39. Area = 2

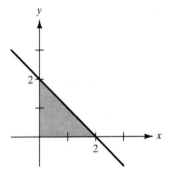

41. Area = 2

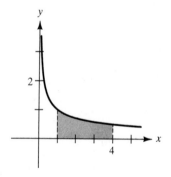

43. Area = 4

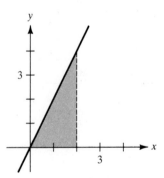

45. (a) 8
    (b) $-12$
    (c) 26
    (d) 30

47. 3.875

49. 0.7960

51. Area $= \dfrac{4}{5}$

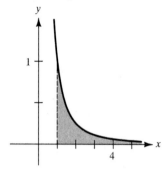

53. Area $= \dfrac{9}{2}$

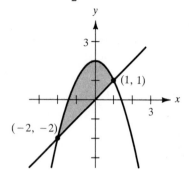

55. Area $= \dfrac{4}{3}$

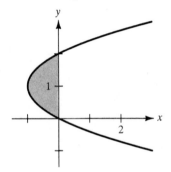

57. Area $= \dfrac{1}{6}$

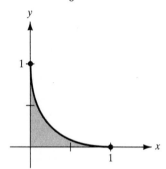

59. Area $= \dfrac{4}{15}$

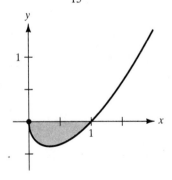

61. Average $= \dfrac{2}{5}$, $x = \dfrac{29}{4}$

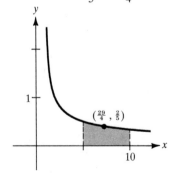

63. (a) $128\pi$

(b) $\dfrac{8192\pi}{15}$

67. 240 ft/sec

65. $f(x) = 2 - x^2$

69. (a) $\dfrac{20,650}{M}$

(b) $\dfrac{32,350}{M}$

# CHAPTER 5

## Section 5.1

1. (a) 625
   (b) 9
   (c) $16\sqrt{2}$
   (d) 9
   (e) 125
   (f) 4

3. (a) 3125
   (b) $\dfrac{1}{5}$
   (c) 625
   (d) $\dfrac{1}{125}$

5. (a) $\dfrac{1}{5}$
   (b) 27
   (c) 5
   (d) 4096

7. (a) $e^6$
   (b) $e^{12}$
   (c) $\dfrac{1}{e^6}$
   (d) $e^2$

9. 4

11. $-2$

13. 2

15. 16

17. $-\dfrac{5}{2}$

19. e

20. c

21. a

22. f

23. d

24. b

25. $f(x) = 5^x$

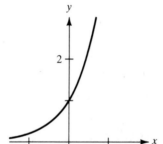

27. $f(x) = 5^{-x}$

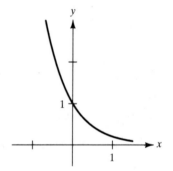

29. $y = 3^{-x^2}$

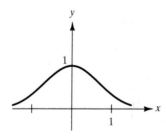

31. $y = 3^{-|x|}$

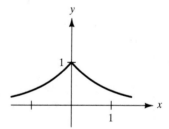

33. $s(t) = \dfrac{3^{-t}}{4}$

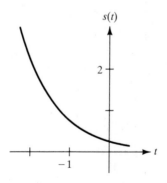

35. $h(x) = e^{x-2}$

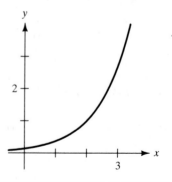

**37.** $N(t) = 1000e^{-0.2t}$

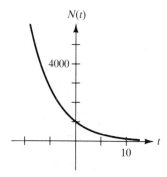

**39.** $g(x) = \dfrac{2}{1 + e^{x^2}}$

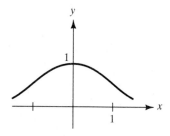

**41.** (a) \$2,593.74
(b) \$2,653.30
(c) \$2,685.06
(d) \$2,707.04
(e) \$2,717.91
(f) \$2,718.28

**43.** (a) \$88,692.04
(b) \$30,119.42
(c) \$9,071.80
(d) \$247.88

**45.** (a) \$849.53
(b) \$421.12

**47.** (a) $0.1535 = 15.35\%$
(b) $0.4866 = 48.66\%$
(c) $0.8111 = 81.11\%$

**49.** (a) 850
(b)

**51.** (a) 0.731
(b) 0.83

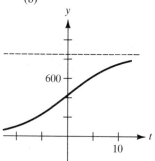

## Section 5.2

**1.** 3

**3.** 1

**5.** $-2$

**7.** $2e^{2x}$

**9.** $2(x - 1)e^{-2x+x^2}$

**11.** $-\dfrac{e^{1/x}}{x^2}$

**13.** $\dfrac{e^{\sqrt{x}}}{2\sqrt{x}}$

**15.** $(3x + 4)e^{3x}$

**17.** $3(e^x - e^{-x})(e^{-x} + e^x)^2$

**19.** $-\dfrac{2(e^x - e^{-x})}{(e^x + e^{-x})^2}$

**21.** $xe^x$

**23.** $6(3e^{3x} + 2e^{-2x})$

25. $32(x + 1)e^{4x}$

27. No relative extrema
    point of inflection: $(0, 1)$

29. Relative minimum: $(0, 0)$
    relative maximum: $(2, 4e^{-2})$
    points of inflection:
    $(2 - \sqrt{2}, 0.191), (2 + \sqrt{2}, 0.384)$

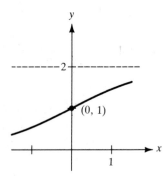

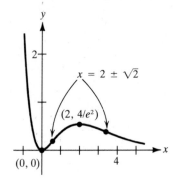

31. $y = -x + 1$

33. (a) $-5028.84$
    (b) $-406.89$

35. $113.51, 403.20, 5 \ln 19 \approx 14.7$ months

37. (a) $0.058$ million ft$^3$/yr
    (b) $0.073$ million ft$^3$/yr
    (c) $0.040$ million ft$^3$/yr

39. (a) $\$433.31$/year
    (b) $\$890.22$/year
    (c) $\$21,839.26$/year

41. $e^{5x} + C$

43. $e^{-x^4} + C$

45. $\dfrac{1}{2}(1 - e^{-2}) \approx 0.432$

47. $\dfrac{e^3 - e}{3} \approx 5.789$

49. $\dfrac{1}{2a}e^{ax^2} + C$

51. $\dfrac{e^3 - e}{3} \approx 5.789$

53. $e^x + 2x - e^{-x} + C$

55. $-\dfrac{2}{3}(1 - e^x)^{3/2} + C$

57. $e^5 - 1 \approx 147.413$

59. $\dfrac{1}{2}(1 - e^{-16}) \approx 0.500$

61. $\dfrac{\pi}{2}(e^2 - 1) \approx 10.036$

63. $\dfrac{12,500}{3}(e^{0.6} - 1) \approx \$3,425.50$

## Section 5.3

1. $e^{0.6931\ldots} = 2$

3. $e^{-0.6931\ldots} = 0.5$

5. $\ln 1 = 0$

7. $\ln (0.1353 \ldots ) = -2$

9. c                10. d

11. b                12. a

**13.**

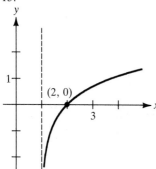

(2, 0)

**15.**

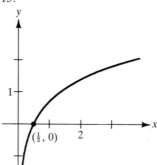

$(\frac{1}{2}, 0)$

**17.**

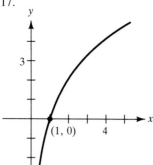

(1, 0)

**19.**

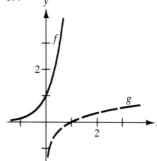

$f$   $g$

**21.**

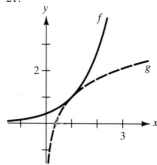

$f$   $g$

**23.** $x^2$

**25.** $5x + 2$

**27.** $\sqrt{x}$

**29.** (a) 1.7917
  (b) $-0.4055$
  (c) 4.3944
  (d) 0.5493

**31.** $\ln 2 - \ln 3$

**33.** $\ln x + \ln y + \ln z$

**35.** $\dfrac{1}{2} \ln (a - 1)$

**37.** $\ln 2 + \ln x - \dfrac{1}{2} \ln (x + 1) - \dfrac{1}{2} \ln (x - 1)$

**39.** $3[\ln (x + 1) + \ln (x - 1) - 3 \ln x]$

**41.** $\ln \dfrac{x - 2}{x + 2}$

**43.** $\ln \dfrac{x^3 y^2}{z^4}$

**45.** $\ln \left( \dfrac{x}{x^2 - 1} \right)^2$

**47.** $\ln \left[ \dfrac{x(x^2 + 1)}{x + 1} \right]^{3/2}$

49. $x = 4$

51. $x = 1$

53. $x = \ln 4 - 1 \approx 0.3863$

55. $t = \dfrac{\ln 5 - \ln 6}{0.11} \approx -1.6575$

57. $x = \dfrac{\ln 15}{2 \ln 5} \approx 0.8413$

59. $t = \dfrac{\ln 2}{\ln 1.07} \approx 10.2448$

61. (a) 6.64 years
    (b) 6.33 years
    (c) 6.30 years
    (d) 6.30 years

63.

| $r$ | 2% | 4% | 6% | 8% | 10% | 12% |
|---|---|---|---|---|---|---|
| $t$ | 54.93 | 27.47 | 18.31 | 13.73 | 10.99 | 9.16 |

65. (a) $x = \dfrac{\ln 300}{0.004} \approx 1426$

    (b) $x = \dfrac{\ln 400}{0.004} \approx 1498$

67.

| $x$ | $y$ | $\dfrac{\ln x}{\ln y}$ | $\ln \dfrac{x}{y}$ | $\ln x - \ln y$ |
|---|---|---|---|---|
| 1 | 2 | 0 | $-0.6931$ | $-0.6931$ |
| 3 | 4 | 0.7925 | $-0.2877$ | $-0.2877$ |
| 10 | 5 | 1.4307 | 0.6931 | 0.6931 |
| 4 | 0.5 | $-2.0000$ | 2.0794 | 2.0794 |

## Section 5.4

1. 3

3. 2

5. $\dfrac{3}{2}$

7. $\dfrac{2}{x}$

9. $\dfrac{1}{x}$

11. $\dfrac{2(x^3 - 1)}{x(x^3 - 4)}$

13. $\dfrac{4(\ln x)^3}{x}$

15. $1 + \ln x$

17. $\dfrac{2x^2 - 1}{x(x^2 - 1)}$

19. $\dfrac{1 - x^2}{x(x^2 + 1)}$

21. $\dfrac{2}{1 - x^2}$

23. $\dfrac{1}{1 - x^2}$

25. $\dfrac{1 - 2 \ln x}{x^3}$

27. $-\dfrac{4}{x(4 + x^2)}$

29. $\dfrac{x}{x^2 - 4}$

31. $e^{-x}\left(\dfrac{1}{x} - \ln x\right)$

33. $2x$

35. $-\dfrac{1}{p}$

37. $\dfrac{500(1 - p^2)}{p(p^2 + 1)}$

39. $\dfrac{1}{x}$

**41.** Relative minimum: (1, 1)

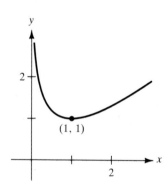

(1, 1)

**43.** Relative maximum: $\left(e, \dfrac{1}{e}\right)$;

point of inflection: $\left(e^{3/2}, \dfrac{3}{2e^{3/2}}\right)$

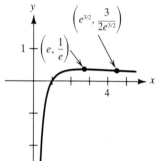

$\left(e^{3/2}, \dfrac{3}{2e^{3/2}}\right)$

$\left(e, \dfrac{1}{e}\right)$

**45.** Relative minimum: $\left(\dfrac{1}{\sqrt{e}}, -\dfrac{1}{2e}\right)$

point of inflection: $\left(\dfrac{1}{e^{3/2}}, -\dfrac{3}{2e^3}\right)$

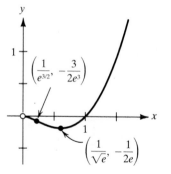

$\left(\dfrac{1}{e^{3/2}}, -\dfrac{3}{2e^3}\right)$

$\left(\dfrac{1}{\sqrt{e}}, -\dfrac{1}{2e}\right)$

**47.** $\ln|x+1| + C$

**49.** $-\dfrac{1}{2}\ln|3 - 2x| + C$

**51.** $\ln\sqrt{x^2 + 1} + C$

**53.** $\dfrac{1}{3}\ln 2 \approx 0.2310$

**55.** $\dfrac{x^2}{2} - 4\ln|x| + C$

**57.** $\dfrac{7}{3}$

**59.** $-\ln 3 \approx -1.099$

**61.** $2\sqrt{x + 1} + C$

**63.** $\dfrac{1}{3}\ln|x^3 + 3x^2 + 9x + 1| + C$

**65.** $\ln|e^x - e^{-x}| + C$

**67.** $\dfrac{15}{2} + 8\ln 2 \approx 13.045$

**69.** $P(t) = 1000[1 + \ln(1 + 0.25t)^{12}]$
$P(3) \approx 7715$

## Section 5.5

**1.** $y = 2e^{0.1014t}$

**3.** $y = 4e^{-0.4159t}$

**5.** $y = 0.6687e^{0.4024t}$

| Isotope | Half-life | Initial quantity | Amount after 1000 years | Amount after 10,000 years |
|---|---|---|---|---|
| **7.** Radium | 1620 | 10 grams | 6.52 grams | 0.14 gram |

9. Carbon   5730   6.70 grams   5.94 grams   2.00 grams

11. Plutonium   24,360   2.16 grams   2.10 grams   1.63 grams

13. 95.8%     15. $100e^{2.197} \approx 900$, 3.15 hours

| Initial investment | Annual percentage rate | Time to double investment | Amount after 10 years | Amount after 25 years |
|---|---|---|---|---|
| 17. $1,000 | 12% | 5.78 years | $3,320.12 | $20,085.54 |
| 19. $ 750 | 8.94% | 7.75 years | $1,833.67 | $ 7,009.86 |
| 21. $ 500 | 9.50% | 7.30 years | $1,292.85 | $ 5,375.51 |

23. $583,275.41

25. (a) $S(t) = 30e^{-1.7918/t}$
  (b) $30e^{-0.35836} \approx 20.9646$ or 20,965 units
  (c)

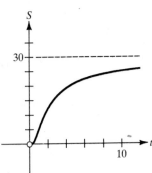

27. (a) $N(t) = 30(1 - e^{-0.0502t})$
  (b) $t = \dfrac{\ln 6}{0.0502} \approx 36$ days

29. (a) $p = 45\left(\dfrac{9}{8}\right)^5 e^{[x \ln (8/9)]/200}$
   $\approx 81.0915e^{-0.0005889x}$
  (b) $x = \dfrac{200}{\ln (9/8)} \approx 1698$ units
   $p = 45\left(\dfrac{9}{8}\right)^5 \dfrac{1}{e} \approx \$29.83$

## Review Exercises for Chapter 5

1.

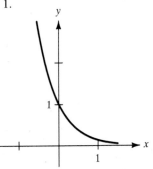

3.

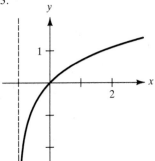

5. 3      7. $\dfrac{1}{3}$      9. $-2x$

11. $xe^x(x + 2)$    13. $\dfrac{e^{2x} - e^{-2x}}{\sqrt{e^{2x} + e^{-2x}}}$    15. $\dfrac{e^{2x}(2x - 1)}{x^2}$

17. $\dfrac{1 - e^x}{(e^x - x)^2}$

19. $\dfrac{1}{2x}$

21. $\dfrac{1 + 2 \ln x}{2\sqrt{\ln x}}$

23. $-\dfrac{y}{x(2y + \ln x)}$

25. $y(1 + \ln x)$

27. $\dfrac{x}{(a + bx)^2}$

29. $\dfrac{1}{x(a + bx)}$

31. $-\dfrac{1}{6}e^{-3x^2} + C$

33. $\dfrac{e^{4x} - 3e^{2x} - 3}{3e^x} + C$

35. $\ln |e^x - 1| + C$

37. $-\ln \sqrt{1 + e^{-2x}} + C$

39. $-\dfrac{1}{2} \ln |1 - x^2| + C$

41. $-\sqrt{1 - x^2} + C$

43. $\dfrac{1}{7} \ln |7x - 2| + C$

45. $-\dfrac{1}{14(7x - 2)^2} + C$

47. $(\ln x)^2 + C$

49. $\dfrac{1}{2}x^2 + 3 \ln |x| + C$

51. $\dfrac{1}{3} \ln |x^3 - 1| + C$

53. $x - \ln |x + 1| + C$

55. $e - e^{1/2} \approx 1.070$

57. $3 + \ln 4 \approx 4.386$

59. $\ln 5$

61. (a) \$525.64
    (b) \$824.36
    (c) \$74,206.58

63. \$3,499.38

65. (a) 27.73 years
    (b) 43.94 years

67. $A(t) = 500e^{[t \ln (3/5)]/40} \approx 500e^{-0.013t}$

69.

| Time elapsed between calls in minutes | 0–2 | 2–4 | 4–6 | 6–8 | 8–10 |
|---|---|---|---|---|---|
| Probability | 0.4866 | 0.2498 | 0.1283 | 0.0659 | 0.0338 |

# CHAPTER 6

## Section 6.1

1. $\dfrac{1}{15}(3x - 2)^5 + C$

3. $\dfrac{2}{9 - t} + C$

5. $\dfrac{2}{3}(1 + x^2)^{3/2} + C$

7. $\ln (3x^2 + x)^2 + C$

9. $-\dfrac{1}{10(5x + 1)^2} + C$

11. $\dfrac{1}{4}(\ln 2x^2)^2 + C$

13. $-\dfrac{1}{3} \ln |1 - e^{3x}| + C$

15. $\dfrac{1}{2}x^2 + x + \ln |x - 1| + C$

17. $-\dfrac{1}{6}(4 - 2x^2)^{3/2} + C$

19. $\dfrac{1}{5}e^{5x} + C$

21. $-\dfrac{2x + 1}{2(x + 1)^2} + C$

23. $\dfrac{1}{9}\left[ \ln |3x - 1| - \dfrac{1}{3x - 1} \right] + C$

25. $-\dfrac{(1-x)^5}{30}(1+5x)+C$

27. $\dfrac{1}{2}\ln|x^2-2x|+C$

29. $\dfrac{2}{5}(x-3)^{3/2}(x+2)+C$

31. $-\dfrac{2}{105}(1-x)^{3/2}(15x^2+12x+8)+C$

33. $\dfrac{\sqrt{2x-1}}{15}(3x^2+2x-13)+C$

35. $2\sqrt{t}-2\ln(\sqrt{t}+1)+C$

37. $4\sqrt{t}+\ln t+C$

39. $\dfrac{1}{3}\ln(3\sqrt{2x}+1)+C$

41. $\dfrac{26}{3}$

43. $\dfrac{3}{2}(e-1)\approx2.577$

45. $\ln 2-\dfrac{1}{2}\approx0.193$

47. $\dfrac{13}{320}$

49. $\dfrac{144}{5}$

51. $\dfrac{1209}{28}$

53. $\dfrac{4}{15}$

55. $\dfrac{4}{3}$

57. $\dfrac{4\pi}{15}\approx0.838$

59. $\dfrac{1}{2}$

61. 21.56%

63. (a) 0.353
    (b) 0.586

65. $24,520.95

## Section 6.2

1. $\dfrac{1}{2}e^{2x}+C$

3. $\dfrac{e^{2x}}{4}(2x-1)+C$

5. $\dfrac{1}{2}e^{x^2}+C$

7. $-\dfrac{e^{-2x}}{4}(2x^2+2x+1)+C$

9. $e^x(x^3-3x^2+6x-6)+C$

11. $\dfrac{x^4}{16}(4\ln x-1)+C$

13. $\dfrac{1}{4}[2(t^2-1)\ln(t+1)+t(2-t)]+C$

15. $x[(\ln x)^2-2\ln x+2]+C$

17. $\dfrac{1}{3}(\ln x)^3+C$

19. $\dfrac{2}{15}(x-1)^{3/2}(3x+2)+C$

21. $(x-1)^2e^x+C$

23. $\dfrac{e^{2x}}{4(2x+1)}+C$

25. $e-2\approx0.718$

27. $\dfrac{1}{25}(4e^5+1)\approx23.786$

29. $\dfrac{2}{5}(2x-3)^{3/2}(x+1)+C$

31. $\dfrac{2}{75}\sqrt{4+5x}(5x-8)+C$

33. $\displaystyle\int x^n\ln x\,dx=\dfrac{x^{n+1}}{n+1}\ln x-\int\dfrac{x^n}{n+1}\,dx$

35. $\dfrac{e^{5x}}{125}(25x^2-10x+2)+C$

$=\dfrac{x^{n+1}}{n+1}\ln x-\dfrac{x^{n+1}}{(n+1)^2}+C$

37. $\dfrac{x^6}{36}(6\ln x-1)+C$

$=\dfrac{x^{n+1}}{(n+1)^2}[-1+(n+1)\ln x]+C$

**39.** $1 - 5e^{-4} \approx 0.908$

**41.** (a) 1

(b) $\pi(e - 2) \approx 2.257$

**43.** (a) $3.2 \ln 2 - 0.2 \approx 2.018$

(b) $12.8 \ln 4 - 7.2 \ln 3 - 1.8 \approx 8.035$

**45.** $16,180.16

**47.** $737,817.01

**49.** $4,103.07

## Section 6.3

**1.** $\dfrac{5}{x - 5} - \dfrac{3}{x + 5}$

**3.** $\dfrac{9}{x - 3} - \dfrac{1}{x}$

**5.** $\dfrac{1}{x - 5} + \dfrac{3}{x + 2}$

**7.** $\dfrac{1}{x} - \dfrac{3}{x^2} + \dfrac{1}{x + 1}$

**9.** $\dfrac{1}{3(x - 2)} + \dfrac{1}{(x - 2)^2}$

**11.** $\dfrac{1}{2} \ln \left| \dfrac{x - 1}{x + 1} \right| + C$

**13.** $\dfrac{1}{4} \ln \left| \dfrac{4 + x}{4 - x} \right| + C$

**15.** $\ln \left| \dfrac{x}{x + 1} \right| + C$

**17.** $\ln \left| \dfrac{x}{2x + 1} \right| + C$

**19.** $\ln \left| \dfrac{x - 1}{x + 2} \right| + C$

**21.** $\dfrac{3}{2} \ln | 2x - 1 | - 2 \ln | x + 1 | + C$

**23.** $5 \ln | x - 2 | - \ln | x + 2 | - 3 \ln | x | + C$

**25.** $\dfrac{1}{2}[3 \ln | x - 4 | - \ln | x |] + C$

**27.** $2 \ln | x - 1 | + \dfrac{1}{x - 1} + C$

**29.** $\dfrac{1}{2}\left[ 5 \ln | x + 1 | - \ln | x | + \dfrac{3}{x + 1} \right] + C$

**31.** $\dfrac{1}{4} \ln \dfrac{5}{3} \approx 0.128$

**33.** $-\dfrac{4}{5} + 2 \ln \dfrac{5}{3} \approx 0.222$

**35.** $\dfrac{1}{5} \ln \left| \dfrac{e^x - 1}{e^x + 4} \right| + C$

**37.** $\dfrac{1}{4} \ln \left| \dfrac{\sqrt{4 + x^2} - 2}{\sqrt{4 + x^2} + 2} \right| + C$

**39.** $-\dfrac{2}{\sqrt{x} + 1} + C$

**41.** $\dfrac{1}{2a}\left( \dfrac{1}{a + x} + \dfrac{1}{a - x} \right)$

**43.** $6 - \dfrac{7}{4} \ln 7 \approx 2.595$

**45.** $y = \dfrac{10}{1 + 9e^{-10kt}}$

**47.** $x = \dfrac{n[e^{(n+1)kt} - 1]}{n + e^{(n+1)kt}}$

## Section 6.4

**1.** $\dfrac{1}{9}\left( \dfrac{2}{2 + 3x} + \ln | 2 + 3x | \right) + C$

**3.** $\dfrac{2(3x - 4)}{27}\sqrt{2 + 3x} + C$

5. $\ln (x^2 + \sqrt{x^4 - 9}) + C$

7. $\frac{1}{2}(x^2 - 1)e^{x^2} + C$

9. $\frac{1}{2}(x^2 + 1)[-1 + \ln (x^2 + 1)] + C$

11. $\ln \left| \dfrac{x}{1 + x} \right| + C$

13. $-\ln \left| \dfrac{1 + \sqrt{x^2 + 1}}{x} \right| + C$

15. $-\dfrac{1}{2} \ln \left| \dfrac{2 + \sqrt{4 - x^2}}{x} \right| + C$

17. $\frac{1}{4}x^2(-1 + 2 \ln x) + C$

19. $-e^{-x} - \ln \left( \dfrac{e^x}{1 + e^x} \right) + C$

21. $\frac{1}{4}(x^2\sqrt{x^4 - 9} - 9 \ln | x^2 + \sqrt{x^4 - 9} |) + C$

23. $\dfrac{1}{27}\left[ \dfrac{4}{2 + 3t} - \dfrac{4}{2(2 + 3t)^2} + \ln | 2 + 3t | \right] + C$

25. $\dfrac{1}{\sqrt{3}} \ln \left| \dfrac{\sqrt{3 + s} - \sqrt{3}}{\sqrt{3 + s} + \sqrt{3}} \right| + C$

27. $\frac{1}{2}x(x - 2) + \ln | x + 1 | + C$

29. $-\dfrac{\sqrt{1 - x^2}}{x} + C$

31. $\frac{1}{9}x^3(-1 + 3 \ln x) + C$

33. $\dfrac{1}{27}\left[ 3x - \dfrac{25}{3x - 5} + 10 \ln | 3x - 5 | \right] + C$

35. $\frac{1}{4}[x(x^2 + 2)\sqrt{x^2 + 4} - 8 \ln | x + \sqrt{x^2 + 4} |] + C$

37. $x - \dfrac{1}{2} \ln (1 + e^{2x}) + C$

39. $\frac{1}{4}[2 \ln x - 3 \ln | 3 + 2 \ln x |] + C$

41. (a) $(x + 3)^2 - 9$
   (b) $(x - 4)^2 - 7$
   (c) $(x^2 + 1)^2 - 6$
   (d) $4 - (x + 1)^2$

43. $\dfrac{1}{4} \ln \left| \dfrac{x - 3}{x + 1} \right| + C$

45. $-\ln \left| \dfrac{1 + \sqrt{x^2 - 2x + 2}}{x - 1} \right| + C$

47. $\dfrac{1}{8} \ln \left| \dfrac{x - 3}{x + 1} \right| + C$

49. $\frac{1}{2} \ln | x^2 + 1 + \sqrt{x^4 + 2x^2 + 2} | + C$

51. $\dfrac{40}{3}$

53. 42.58

55. Point of equilibrium: (12, 4)
   consumer surplus = 17.92
   producer surplus = 24

# Section 6.5

| Exact value | Trapezoidal Rule | Simpson's Rule |
|---|---|---|
| 1. 2.6667 | 2.7500 | 2.6667 |
| 3. 4.0000 | 4.2500 | 4.0000 |
| 5. 4.0000 | 4.0625 | 4.0000 |
| 7. 0.5000 | 0.5090 | 0.5004 |
| 9. 0.6931 | 0.6970 | 0.6933 |
| 11. | 3.41 | 3.22 |
| 13. | 0.342 | 0.372 |
| 15. | 0.749 | 0.771 |
| 17. | 0.772 | 0.780 |
| 19. | 0.286 | 0.274 |

21. $21,831.20

23. $0.3413 = 34.13\%$

25. 89,500 feet$^2$

27. (a) 0.5
 (b) 0

29. (a) $\dfrac{5e}{64} \approx 0.212$

 (b) $\dfrac{13e}{1024} \approx 0.035$

31. (a) $n = 130$
 (b) $n = 10$

33. Exact value: $\displaystyle\int_0^1 x^3\,dx = \dfrac{x^4}{4}\Big]_0^1 = \dfrac{1}{4}$

 Simpson's Rule: $\displaystyle\int_0^1 x^3\,dx = \dfrac{1}{6}\left[0^3 + 4\left(\dfrac{1}{2}\right)^3 + 1^3\right] = \dfrac{1}{4}$

# Section 6.6

1. 4

3. 6

5. 1

7. Diverges

9. 6

11. Diverges

13. 0

15. $\ln(2 + \sqrt{3}) \approx 1.317$

17. 1

19. Diverges

21. Diverges

23. Diverges

25. 2

27. $\dfrac{1}{4}$

29. (a) 1

 (b) $\dfrac{\pi}{3}$

31. (a) $748,367.34
 (b) $808,030.14
 (c) $900,000.00

33. (a)

| $x$ | 1 | 10 | 25 | 50 |
|---|---|---|---|---|
| $xe^{-x}$ | 0.3679 | 0.0005 | 0 | 0 |

(b)

| $x$ | 1 | 10 | 25 | 50 |
|---|---|---|---|---|
| $x^2 e^{-x}$ | 0.3679 | 0.0045 | 0 | 0 |

(c)

| x | 1 | 10 | 25 | 50 |
|---|---|---|---|---|
| $x^5e^{-x}$ | 0.3679 | 4.5400 | 0.0001 | 0 |

## Section 6.7

1. (a) $S = \{gggg, gggb, ggbg, gbgg, bggg, ggbb, gbgb, gbbg, bgbg, bbgg, bggb, gbbb, bgbb, bbgb, bbbg, bbbb\}$

(b)

| Random variable, x | 0 | 1 | 2 | 3 | 4 |
|---|---|---|---|---|---|
| Probability of x, P(x) | $\dfrac{1}{16}$ | $\dfrac{4}{16}$ | $\dfrac{6}{16}$ | $\dfrac{4}{16}$ | $\dfrac{1}{16}$ |

(c)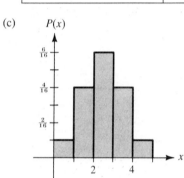

(d) $\dfrac{15}{16}$

3. $\displaystyle\int_0^1 6x(1-x)\,dx = \left[3x^2 - 2x^3\right]_0^1 = 1$

5. $\displaystyle\int_0^3 \frac{4}{27}x^2(3-x)\,dx = \frac{4}{27}\left[x^3 - \frac{x^4}{4}\right]_0^3 = 1$

7. $\displaystyle\int_0^\infty \frac{1}{3}e^{-x/3}\,dx = \lim_{b\to\infty}\left[-e^{-x/3}\right]_0^b = 1$

9. $\dfrac{1}{12}$

11. $\dfrac{3}{32}$

13. $\dfrac{1}{2}$

15. (a) $\dfrac{3}{5}$  (b) $\dfrac{1}{5}$

   (c) $\dfrac{1}{5}$  (d) $\dfrac{4}{5}$

17. (a) $\dfrac{\sqrt{2}}{4} \approx 0.354$

   (b) $1 - \dfrac{\sqrt{2}}{4} \approx 0.646$

   (c) $\dfrac{1}{8}(3\sqrt{3} - 1) \approx 0.525$

   (d) $\dfrac{3\sqrt{3}}{8} \approx 0.650$

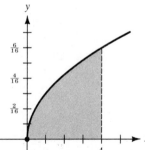

19. (a) $1 - e^{-1} \approx 0.632$

   (b) $e^{-1} \approx 0.368$

   (c) $e^{-1/2} - e^{-2} \approx 0.471$

   (d) 0

21. (a) $\dfrac{1}{4}$

   (b) $\dfrac{2}{5}$

23. (a) $1 - e^{-2/3} \approx 0.487$
   (b) $e^{-2/3} - e^{-4/3} \approx 0.250$
   (c) $e^{-2/3} \approx 0.513$

25. $1 - e^{-4/3} \approx 0.736$

27. (a) $1 - 2e^{-1} \approx 0.264$
   (b) $2e^{-1} - 3e^{-2} \approx 0.330$
   (c) $3e^{-2} \approx 0.406$

29. (a) 0.107
   (b) 0.353

## Section 6.8

1. (a) 4

   (b) $\dfrac{16}{3}$

   (c) 4

3. (a) $\dfrac{16}{3}$

   (b) $\dfrac{32}{9}$

   (c) $4\sqrt{2} \approx 5.66$

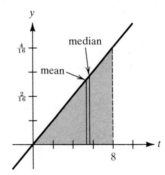

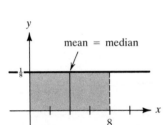

5. (a) $\dfrac{1}{2}$

   (b) $\dfrac{1}{20}$

   (c) $\dfrac{1}{2}$

7. (a) $\dfrac{5}{7}$

   (b) $\dfrac{20}{441}$

   (c) $2^{-2/5} \approx 0.758$

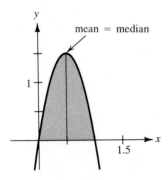

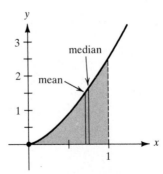

9. (a) $-1 + \dfrac{4}{3} \ln 4 \approx 0.848$

(b) $4 - \dfrac{16}{9}(\ln 4)^2 \approx 0.583$

(c) $\dfrac{3}{5}$

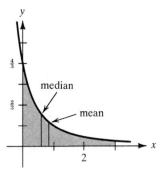

11. $7 \ln 2 \approx 4.852$

13. $\mu = \lambda$, $\sigma = \lambda$

15. $\mu = 8$, $\sigma^2 = 64$, $\sigma = 8$

17. (a) $f(t) = \dfrac{1}{4}e^{-t/4}$

(b) $0.528 = 52.8\%$

19. (a) $f(t) = \dfrac{1}{5}e^{-t/5}$

(b) $0.865 = 86.5\%$

21. (a) 12

(b) $0.203 = 20.3\%$

23. (a) $\mu = 3$, $\sigma = \dfrac{3\sqrt{5}}{5} \approx 1.342$

(b) 3

(c) $0.626 = 62.6\%$

25. $\mu = \dfrac{4}{7}$, $V(x) = \dfrac{8}{147}$

27. Relative maximum: $\left(0, \dfrac{1}{\sqrt{2\pi}}\right)$

points of inflection: $\left(\pm 1, \dfrac{1}{\sqrt{2\pi e}}\right)$

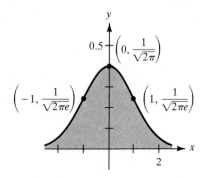

29. $0.3413 = 34.13\%$

## Review Exercises for Chapter 6

1. $\dfrac{x^2}{10}(-2x^3 + 15x^2 - 40x + 40) + C$

3. $\dfrac{2}{35}(x + 1)^{5/2}(5x - 2) + C$

5. $\dfrac{2}{5}(x + 3)^{3/2}(x - 2) + C$

7. $-\dfrac{2(1 - x)^{3/2}}{105}(15x^2 + 12x + 8) + C$

9. $x - 2\sqrt{x} + 2 \ln (\sqrt{x} + 1) + C$

11. $2\sqrt{x}(-2 + \ln 2x) + C$

13. $(x - 2)e^x + C$

15. $\dfrac{2}{9}x^{3/2}(-2 + 3 \ln x) + C$

17. $\dfrac{3}{2} \ln \left| \dfrac{x-3}{x+3} \right| + C$

19. $\dfrac{1}{5} \ln \left| \dfrac{x}{x+5} \right| + C$

21. $\dfrac{2}{3}\left[ 2 \ln |x-1| - \dfrac{1}{x-1} \right] + C$

23. $\sqrt{x^2+9} - 3 \ln \left| \dfrac{3+\sqrt{x^2+9}}{x} \right| + C$

25. Trapezoidal Rule: 0.305
    Simpson's Rule: 0.289

27. Trapezoidal Rule: 0.741
    Simpson's Rule: 0.737

29. $\dfrac{32}{3}$

31. Diverges

33. 4

35. $\dfrac{128}{15}$

37. 1

39. $\dfrac{\pi}{30}$

41. $\dfrac{1}{8}$

43. $\dfrac{1}{4}$

45. $\mu = \dfrac{10}{3}$

median $= 5(2 - \sqrt{2}) \approx 2.929$

$\sigma = \dfrac{5\sqrt{2}}{3} \approx 2.357$

(a) $\dfrac{9}{25}$  (b) $\dfrac{1}{25}$

47. $\mu = -1 + \ln 4 \approx 0.386$

median $= \dfrac{1}{3}$

$\sigma = 0.280$

(a) $\dfrac{2}{3}$  (b) $\dfrac{1}{3}$

49. $12 \ln 2 \approx 8.318$

51. (a) $2 - \sqrt{2} \approx 0.586$

(b) $\dfrac{13}{3}$ days

# CHAPTER 7

## Section 7.1

1. $\dfrac{dy}{dx} = -\dfrac{1}{x^2}$

3. $\dfrac{dy}{dx} = 4Ce^{4x} = 4y$

5. $2\dfrac{dy}{dt} + y - 5 = 2\left( -\dfrac{C}{2}e^{-t/2} \right) + (Ce^{-t/2} + 5) - 5 = 0$

7. $xy' - 3x - 2y = x(2Cx - 3) - 3x - 2(Cx^2 - 3x) = 0$

9. $x(y' - 1) - (y + 2) = x[(1 + \ln x + C) - 1] - [(x \ln x + Cx - 2) + 2] = 0$

11. $xy' + y = x\left( 2x + 2 - \dfrac{C}{x^2} \right) + \left( x^2 + 2x + \dfrac{C}{x} \right) = x(3x + 4)$

13. $2y'' + 3y' - 2y = 2\left( \dfrac{1}{4}C_1e^{x/2} + 4C_2e^{-2x} \right) + 3\left( \dfrac{1}{2}C_1e^{x/2} - 2C_2e^{-2x} \right) - 2(C_1e^{x/2} + C_2e^{-2x}) = 0$

15. $y' - \dfrac{ay}{x} = \left( \dfrac{4bx^3}{4-a} + aCx^{a-1} \right) - \dfrac{a}{x}\left( \dfrac{bx^4}{4-a} + Cx^a \right) = bx^3$

17. Implicit differentiation: $2x + 2yy' = Cy'$, $y' = \dfrac{2x}{C - 2y} = \dfrac{2xy}{Cy - 2y^2} = \dfrac{2xy}{(x^2 + y^2) - 2y^2} = \dfrac{2xy}{x^2 - y^2}$

19. $y' + 2xy = -\dfrac{4Cxe^{x^2}}{(1 + Ce^{x^2})^2} + 2x\left(\dfrac{2}{1 + Ce^{x^2}}\right) = xy^2$

21. Solution       23. Not a solution       25. $y = 3e^{-2x}$

27. $y = 5 + \ln\sqrt{|x|}$       29. $y = -\dfrac{4}{11}e^{6x} + \dfrac{4}{11}e^{-5x}$

31. $y = \dfrac{4}{3}(3 - x)e^{2x/3}$       33. $y^2 = \dfrac{1}{4}x^3$       35. $y = 3e^x$

37. $x^2 + y^2 = C$              39. $y = C(x + 2)^2$

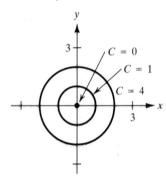      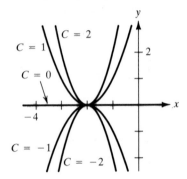

41. $y = x^3 + C$       43. $y = x - 2 \ln |x| + C$       45. $y = \dfrac{2}{5}(x - 3)^{3/2}(x + 2) + C$

47. (a) $N(t) = 750 - 650e^{-0.0484t}$
  (b)

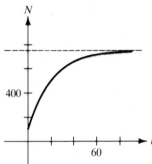

## Section 7.2

1. $y = x^2 + C$       3. $y = \sqrt[3]{x + C}$       5. $y = Ce^{x^2/2}$

7. $y = C(1 + x)^2$       9. $y = \ln |t^3 + t + C|$       11. $x^2 - y^2 = C$

13. $y = C(2 + x)^2$       15. $y = Ce^{(\ln x)^2/2}$       17. $y = Ce^{x(x+2)/2} - 1$

19. $x^2(1 - y^2) = C$       21. $y^2 = 2e^x + 14$       23. $y = e^{-x(x+2)/2}$

25. $P = P_0 e^{kt}$

27. $9x^2 + 16y^2 = 25$

29. $v = 34.56(1 - e^{-0.1t})$

31. $t = 119.7$ minutes

## Section 7.3

1. $y = 2 + Ce^{-3x}$

3. $y = e^{-x}(x + C)$

5. $y = x^2 + 2x + \dfrac{C}{x}$

7. $y = Ce^{-x^2} + 1$

9. $y = \dfrac{x^3 - 3x + C}{3(x - 1)}$

11. $y = \dfrac{x}{2} \ln x - \dfrac{x}{4} + \dfrac{C}{x}$

13. $y = \dfrac{1}{2}(e^x + 3e^{-x})$

15. $xy = 4$

17. $y = 5e^{-x} + x - 1$

19. $y = x^2(5 - \ln x)$

21. $A = \dfrac{P}{r^2}(e^{rt} - rt - 1)$

23. $S = t + 95(1 - e^{-t/5})$

| $t$ | 0 | 1 | 2 | 3 | 4 | 5 | 6 | 7 | 8 | 9 | 10 |
|---|---|---|---|---|---|---|---|---|---|---|---|
| $S$ | 0 | 18.22 | 33.32 | 45.86 | 56.31 | 65.05 | 72.39 | 78.57 | 83.82 | 88.30 | 92.14 |

## Section 7.4

1. $y = e^{(x \ln 2)/3} \approx e^{0.2310x}$

3. $y = 4e^{-(x \ln 4)/4} \approx 4e^{-0.3466x}$

5. $y = \dfrac{1}{2}e^{(\ln 2)x} \approx \dfrac{1}{2}e^{0.6931x}$

7. $y = \dfrac{20}{1 + 19e^{-0.5889x}}$

9. $y = \dfrac{5000}{1 + 19e^{-0.10156x}}$

11. $y = \dfrac{2000}{17t + 20}$

13. $y = 500e^{-1.6094e^{-0.1451t}}$

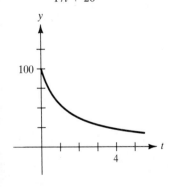

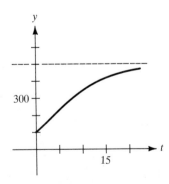

15. $\dfrac{dP}{dn} = kP(L - P)$, $P = \dfrac{CL}{e^{-Lkn} + C}$

17. $s = 25 - \dfrac{13 \ln (h/2)}{\ln 3}$, $2 \le h \le 15$

19. $y = Cx$

21. $A = \dfrac{P}{r}(e^{rt} - 1)$

23. \$7,305,295.15

25. (a) $C = C_0 e^{-Rt/V}$

    (b) 0

27. (a) $C(t) = \dfrac{Q}{R}(1 - e^{-Rt/V})$

    (b) $\dfrac{Q}{R}$

29. (a) $\dfrac{dQ}{dt} = q - kQ$

    (b) $Q = \dfrac{q}{k} + \left(Q_0 - \dfrac{q}{k}\right)e^{-kt}$

    (c) $\dfrac{q}{k}$

## Review Exercises for Chapter 7

1. $xy' - y = x(2 + \ln x^2 + 3x^{1/2} + C) - (x \ln x^2 + 2x^{3/2} + Cx) = x(2 + \sqrt{x})$

3. $\dfrac{dy}{dx} - \dfrac{2y}{x} = 2C(x - 1) - \dfrac{2}{x}[C(x - 1)^2] = \dfrac{1}{x}[2C(x - 1)] = \dfrac{1}{x}\dfrac{dy}{dx}$

5. $y' - \dfrac{3y}{x^2} = \dfrac{3C}{x^2}e^{-3/x} - \dfrac{3}{x^2}\left(-\dfrac{1}{3} + Ce^{-3/x}\right) = \dfrac{1}{x^2}$

7. $y'' - 2y' + y = [y + 2(C_2 + x^2)e^x + 2xe^x] - 2[y + (C_2 + x^2)e^x] + y = 2xe^x$

9. $2xy' - y = 2x\left(\dfrac{3}{5}x^2 - 1 + \dfrac{C}{2\sqrt{x}}\right) - \left(\dfrac{1}{5}x^3 - x + C\sqrt{x}\right) = x^3 - x$

11. $y = x^2 + 2x$

13. $y = 1 - 2e^{-x^2}$

15. $2x^3 - y^2 = C$

17. $y = \dfrac{3(1 + Ce^{2x^3})}{1 - Ce^{2x^3}}$

19. $y = C(x + 1) - 1$

21. $y = x \ln x^2 + 2x^{3/2} + Cx$

23. $y = -\dfrac{1}{3} + Ce^{-3/x}$

25. $y = \dfrac{1}{5}x^3 - x + C\sqrt{x}$

27. $t = \dfrac{\ln 0.25}{\ln 0.8} \approx 6.2$ hours

29. $A = \dfrac{P}{r} + \left(A_0 - \dfrac{P}{r}\right)e^{rt}$

31. $A_0 = \$284,838.90$

33. (a) 233,333

    (b) 5th year

# CHAPTER 8

## Section 8.1

1.

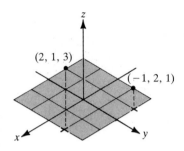

3.

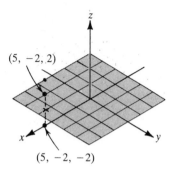

5. $\left(\dfrac{3}{2}, -3, 5\right)$

7. $(0.65, 1.1, -0.8)$

9. $(2, 3, 9)$

11. $3\sqrt{2}$

13. $\sqrt{206}$

15. $3, 3\sqrt{5}, 6$; right triangle

17. $6, 6, 2\sqrt{10}$; isosceles triangle

19. $x^2 + (y - 2)^2 + (z - 5)^2 = 4$

21. $(x - 1)^2 + (y - 3)^2 + z^2 = 10$

23. Center: $(1, -3, -4)$
    radius: 5

25. Center: $(0, 4, 0)$
    radius: 4

27. Center: $\left(\dfrac{1}{2}, -\dfrac{1}{2}, 1\right)$
    radius: 1

29. Center: $\left(\dfrac{1}{3}, -1, 0\right)$
    radius: 1

31.

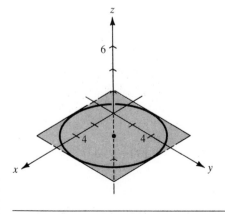

33.

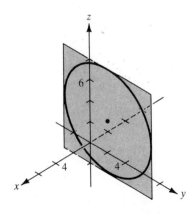

## Section 8.2

1.

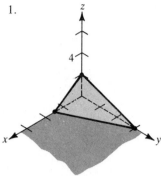

3.

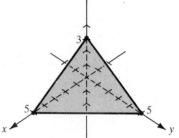

5.

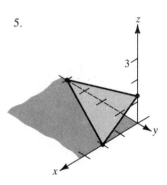

7.

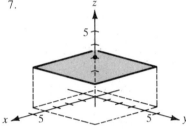

9.

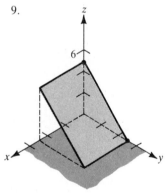

11.

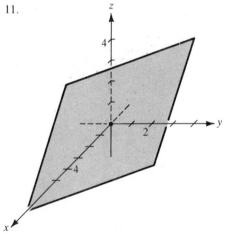

13. Perpendicular

15. Parallel

17. Parallel

19. Neither parallel nor perpendicular

21. $\dfrac{6\sqrt{14}}{7}$

23. $\dfrac{\sqrt{6}}{6}$

25. c          26. e

27. f          28. g

29. d    30. b    31. a    32. h

33. Ellipsoid    35. Hyperboloid of one sheet

37. Elliptic paraboloid    39. Hyperbolic paraboloid

41. Hyperboloid of two sheets    43. Elliptic cone

45. Hyperbolic paraboloid    47. Sphere

## Section 8.3

1. (a) $\dfrac{3}{2}$    (b) $-\dfrac{1}{4}$

   (c) 6    (d) $\dfrac{5}{y}$

   (e) $\dfrac{x}{2}$    (f) $\dfrac{5}{t}$

3. (a) 5
   (b) $3e^2$
   (c) $2e^{-1}$
   (d) $5e^y$
   (e) $xe^2$
   (f) $te^t$

5. (a) $\dfrac{2}{3}$
   (b) 0

7. (a) $90\pi$
   (b) $50\pi$

9. (a) $20,655.20
   (b) $1,397,672.67

11. (a) 4
   (b) 6

13. (a) $2x + \Delta x$
   (b) $-2$

15. Domain: all points inside and on the ellipse $x^2 + 4y^2 = 4$
   range: [0, 2]

17. Domain: all points in the $xy$-plane
   range: $(-\infty, 4]$

19. Domain: the half-plane below the line $y = -x + 4$
   range: $(-\infty, \infty)$

21. Domain: all points $(x, y)$ such that $x \neq 0$ and $y \neq 0$
   range: all real numbers except 0

23. Domain: all points $(x, y)$ such that $y \geq 0$
   range: $(-\infty, \infty)$

25. Circles centered at (0, 0) with radii $\leq 5$

27. Hyperbolas: $xy = c$

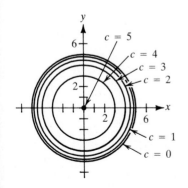

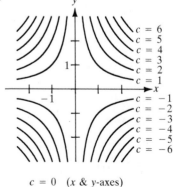

29. Lines with slope 1 passing through the fourth quadrant

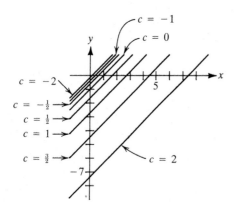

31. 135,540 units

33. $20,430.00

35. (a) 243 board feet
    (b) 507 board feet

37. (a) 12 minutes
    (b) 20 minutes
    (c) 10 minutes
    (d) 30 minutes

## Section 8.4

1. $f_x(x, y) = 2$
   $f_y(x, y) = -3$

3. $f_x(x, y) = \dfrac{5}{2\sqrt{x}}$
   $f_y(x, y) = -12y$

5. $f_x(x, y) = y$
   $f_y(x, y) = x$

7. $\dfrac{\partial z}{\partial x} = \sqrt{y}$

   $\dfrac{\partial z}{\partial y} = \dfrac{x}{2\sqrt{y}}$

9. $f_x(x, y) = \dfrac{x}{\sqrt{x^2 + y^2}}$

   $f_y(x, y) = \dfrac{y}{\sqrt{x^2 + y^2}}$

11. $\dfrac{\partial z}{\partial x} = 2xe^{2y}$

    $\dfrac{\partial z}{\partial y} = 2x^2 e^{2y}$

13. $h_x(x, y) = -2xe^{-(x^2+y^2)}$
    $h_y(x, y) = -2ye^{-(x^2+y^2)}$

15. $\dfrac{\partial z}{\partial x} = \dfrac{2x}{x^2 + y^2}$

    $\dfrac{\partial z}{\partial y} = \dfrac{2y}{x^2 + y^2}$

17. $\dfrac{\partial z}{\partial x} = -\dfrac{2y}{x^2 - y^2}$

    $\dfrac{\partial z}{\partial y} = \dfrac{2x}{x^2 - y^2}$

19. $f_x(x, y) = 1 - x^2$
    $f_y(x, y) = y^2 - 1$

21. $f_x(x, y) = 6x + y, \ 13$
    $f_y(x, y) = x - 2y, \ 0$

23. $f_x(x, y) = -\dfrac{y^2}{(x - y)^2}, \ -\dfrac{1}{4}$

    $f_y(x, y) = \dfrac{x^2}{(x - y)^2}, \ \dfrac{1}{4}$

25. $\dfrac{\partial w}{\partial x} = \dfrac{x}{\sqrt{x^2 + y^2 + z^2}}, \dfrac{2}{3}$

$\dfrac{\partial w}{\partial y} = \dfrac{y}{\sqrt{x^2 + y^2 + z^2}}, -\dfrac{1}{3}$

$\dfrac{\partial w}{\partial z} = \dfrac{z}{\sqrt{x^2 + y^2 + z^2}}, \dfrac{2}{3}$

29. $f_x(x, y, z) = 6xy - 5yz, 4$
$f_y(x, y, z) = 3x^2 - 5xz + 10z^2, 33$
$f_z(x, y, z) = -5xy + 20yz, -35$

33. $\dfrac{\partial^2 z}{\partial x^2} = 6x + 6y, \dfrac{\partial^2 z}{\partial y^2} = -10$

$\dfrac{\partial^2 z}{\partial y \partial x} = \dfrac{\partial^2 z}{\partial x \partial y} = 6x$

37. $\dfrac{\partial^2 z}{\partial x^2} = \dfrac{2y^2}{(x - y)^3}$

$\dfrac{\partial^2 z}{\partial y^2} = \dfrac{2x^2}{(x - y)^3}$

$\dfrac{\partial^2 z}{\partial y \partial x} = \dfrac{\partial^2 z}{\partial x \partial y} = -\dfrac{2xy}{(x - y)^3}$

41. $\dfrac{\partial^2 z}{\partial x^2} = 0, \dfrac{\partial^2 z}{\partial y^2} = 2xe^{-y^2}(2y^2 - 1)$

$\dfrac{\partial^2 z}{\partial y \partial x} = \dfrac{\partial^2 z}{\partial x \partial y} = -2ye^{y^2}$

47. (a) 2
(b) $-3$

51. (a) $-\dfrac{3}{4}$
(b) 0

55. (a) $f_x(x, y) = 60\left(\dfrac{y}{x}\right)^{0.4}$

(b) $f_y(x, y) = 40\left(\dfrac{x}{y}\right)^{0.6}$

59. (a) Complementary
(b) Substitute
(c) Complementary

27. $F_x(x, y, z) = \dfrac{x}{x^2 + y^2 + z^2}, \dfrac{3}{25}$

$F_y(x, y, z) = \dfrac{y}{x^2 + y^2 + z^2}, 0$

$F_z(x, y, z) = \dfrac{z}{x^2 + y^2 + z^2}, \dfrac{4}{25}$

31. $\dfrac{\partial^2 z}{\partial x^2} = 6x, \dfrac{\partial^2 z}{\partial y^2} = -8$

$\dfrac{\partial^2 z}{\partial y \partial x} = \dfrac{\partial^2 z}{\partial x \partial y} = 0$

35. $\dfrac{\partial^2 z}{\partial x^2} = -2, \dfrac{\partial^2 z}{\partial y^2} = -2$

$\dfrac{\partial^2 z}{\partial y \partial x} = \dfrac{\partial^2 x}{\partial x \partial y} = 0$

39. $\dfrac{\partial^2 z}{\partial x^2} = \dfrac{y^2}{(x^2 + y^2)^{3/2}}$

$\dfrac{\partial^2 z}{\partial y^2} = \dfrac{x^2}{(x^2 + y^2)^{3/2}}$

$\dfrac{\partial^2 z}{\partial y \partial x} = \dfrac{\partial^2 z}{\partial x \partial y} = -\dfrac{xy}{(x^2 + y^2)^{3/2}}$

43. $\dfrac{\partial^2 z}{\partial x^2} + \dfrac{\partial^2 z}{\partial y^2} = 0 + 0 = 0$

45. $\dfrac{\partial^2 z}{\partial x^2} + \dfrac{\partial^2 z}{\partial y^2} = 6x - 6x = 0$

49. (a) 18
(b) $-6$

53. $C_x(80, 20) = 183$

$C_y(80, 20) = 237$

57. An increase in either price will cause a decrease in the number of applicants.

## Section 8.5

1. Relative minimum: $(1, 3, 0)$

3. Relative minimum: $(-1, 1, -4)$

5. Relative maximum: $(8, 16, 74)$

7. Relative minimum: $(1, 2, -1)$

9. Saddle point: $(1, -2, -1)$

11. Saddle point: $(0, 0, 0)$

13. Relative minimum: $(1, 1, -1)$
   Saddle point: $(0, 0, 0)$

15. Relative maximum: $(1, 0, 1)$
   Saddle point: $\left(0, 0, \dfrac{1}{2}\right)$

17. Relative maxima: $(0, \pm1, 4)$;
   relative minimum: $(0, 0, 0)$;
   saddle points: $(\pm1, 0, 1)$

19. Saddle point: $(0, 0, 1)$

21. $x_1 = 3$, $x_2 = 6$

23. $p_1 = 2500$, $p_2 = 3000$

25. $x_1 = 275$, $x_2 = 110$

27. $x_1 \approx 94$, $x_2 \approx 157$

29. $36 \times 18 \times 18$ inches

31. $10, 10, 10$

33. $10, 10, 10$

## Section 8.6

1. $f(5, 5) = 25$

3. $f(2, 2) = 8$

5. $f\left(\dfrac{\sqrt{2}}{2}, \dfrac{1}{2}\right) = \dfrac{1}{4}$

7. $f(25, 50) = 2600$

9. $f(1, 1) = 2$

11. $f(2, 2) = e^4$

13. $f(2, 2, 2) = 12$

15. $f(8, 16, 8) = 1024$

17. $f\left(\sqrt{\dfrac{10}{3}}, \dfrac{1}{2}\sqrt{\dfrac{10}{3}}, \sqrt{\dfrac{5}{3}}\right) = \dfrac{5\sqrt{15}}{9}$

19. $f\left(\dfrac{1}{3}, \dfrac{1}{3}, \dfrac{1}{3}\right) = \dfrac{1}{3}$

21. $36 \times 18 \times 18$ inches

23. $x_1 = 377.5$ units, $x_2 = 622.5$ units

25. $\dfrac{\sqrt{13}}{13}$

27. $\sqrt{3}$

29. $x = 50\sqrt{2} \approx 71$
   $y = 200\sqrt{2} \approx 283$

31. (a) $f\left(\dfrac{3125}{6}, \dfrac{6250}{3}\right) \approx 147{,}314$
   (b) $1.473$
   (c) $184{,}139$

## Section 8.7

1. (a) $y = \dfrac{3}{4}x + \dfrac{4}{3}$
   (b) $\dfrac{1}{6}$

3. (a) $y = -2x + 4$
   (b) $2$

5. $y = \dfrac{7}{10}x + \dfrac{7}{5}$

7. $y = \dfrac{8}{7}x + \dfrac{10}{3}$

9. $y = -\dfrac{13}{20}x + \dfrac{7}{4}$

11. $y = \dfrac{37}{43}x + \dfrac{7}{43}$

13. $y = -\dfrac{175}{148}x + \dfrac{945}{148}$

15. $y = \dfrac{3}{4}x + \dfrac{4}{3}$

17. $y = -2x + 4$

19. $y = \dfrac{3}{7}x^2 + \dfrac{6}{5}x + \dfrac{26}{35}$

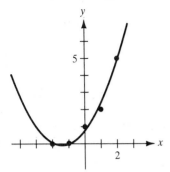

21. $y = x^2 - x$

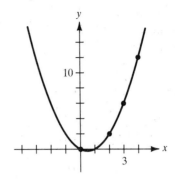

23. (a) $y = -240x + 685$
    (b) 349

25. (a) $y = 14x + 19$
    (b) 41.4 bushels per acre

27. (a) $y = \dfrac{1}{33,500}(3,089x + 11,048)$
    (b) 3.10 million

29. (a) $y = \dfrac{1}{67,900}(-3x^2 + 4957x + 278,140)$
    (b) 5.2 billion

## Section 8.8

1. $\dfrac{3x^2}{2}$

3. $y \ln |2y|$

5. $\dfrac{x^2}{2}(4 - x^2)$

7. $\dfrac{1}{2}y[(\ln y)^2 - y^2]$

9. $x^2(1 - e^{-x^2} - x^2 e^{-x^2})$

11. 3

13. 36

15. $\dfrac{20}{3}$

17. $\dfrac{2}{3}$

19. 4

21. $\dfrac{16}{3}$

23. 4

25. $\displaystyle\int_0^1 \int_0^2 dy\, dx = \int_0^2 \int_0^1 dx\, dy = 2$

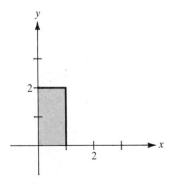

27. $\displaystyle\int_0^1 \int_{2y}^2 dx\, dy = \int_0^2 \int_0^{x/2} dy\, dx = 1$

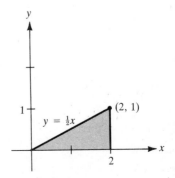

29. $\displaystyle\int_{0}^{2}\int_{x/2}^{1} dy\,dx = \int_{0}^{1}\int_{0}^{2y} dx\,dy = 1$

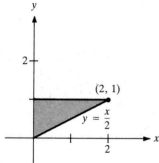

31. $\displaystyle\int_{0}^{1}\int_{y^2}^{\sqrt[3]{y}} dx\,dy = \int_{0}^{1}\int_{x^3}^{\sqrt{x}} dy\,dx \doteq \frac{5}{12}$

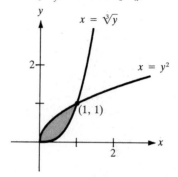

33. 24

35. $\dfrac{16}{3}$

37. $\dfrac{9}{2}$

39. $\dfrac{500}{3}$

41. 5

43. 2

## Section 8.9

1. 8

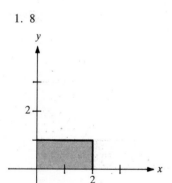

3. $\dfrac{1}{54}$

5. $\dfrac{1}{3}$

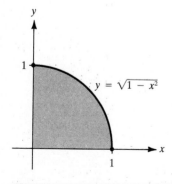

7. $\displaystyle\int_{0}^{3}\int_{0}^{5} xy\,dy\,dx = \int_{0}^{5}\int_{0}^{3} xy\,dx\,dy = \frac{225}{4}$

9. $\int_0^4 \int_0^{\sqrt{x}} \dfrac{y}{1+x^2}\, dy\, dx = \int_0^2 \int_{y^2}^4 \dfrac{y}{1+x^2}\, dx\, dy = \dfrac{1}{4}\ln 17 \approx 0.708$

11. 4

13. 8

15. 12

17. $\dfrac{3}{8}$

19. $\dfrac{40}{3}$

21. 1

23. 4

25. $\dfrac{32}{3}$

27. 2

29. $\dfrac{8}{3}$

31. \$75,125.00

33. 25,645 units

## Review Exercises for Chapter 8

1. (a)

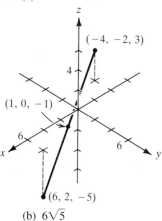

(b) $6\sqrt{5}$
(c) $(1, 0, -1)$

3. $(x-1)^2 + y^2 + (z+3)^2 = 9$

5. Center: $(-4, 2, -1)$
   radius: 4

7.

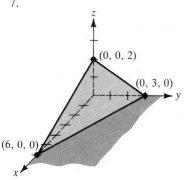

9.

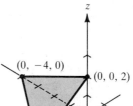

11. Sphere

13. Ellipsoid

15. Hyperboloid of two sheets

17. Hyperboloid of one sheet

19. Top half of a circular cone

21. Domain: all points in the $xy$-plane
    range: $[0, \infty)$

23. Domain: all points inside the circle $x^2 + y^2 = 1$
    range: $(-\infty, \infty)$

25. $f_x(x, y) = \sqrt{y} + 3$

$$f_y(x, y) = \frac{x}{2\sqrt{y}} - 2$$

27. $g_x(x, y) = \dfrac{1}{2x + 3y}$

$$g_y(x, y) = \frac{3}{2(2x + 3y)}$$

29. $\dfrac{\partial z}{\partial x} = e^y + ye^x$

$$\frac{\partial z}{\partial y} = xe^y + e^x$$

31. $g_x(x, y) = \dfrac{y(y^2 - x^2)}{(x^2 + y^2)^2}$

$$g_y(x, y) = \frac{x(x^2 - y^2)}{(x^2 + y^2)^2}$$

33. $\dfrac{\partial w}{\partial x} = \dfrac{yz}{\sqrt{xyz}}$

$$\frac{\partial w}{\partial y} = \frac{xz}{\sqrt{xyz}}$$

$$\frac{\partial w}{\partial z} = \frac{xy}{\sqrt{xyz}}$$

35. $f_{xx}(x, y) = 6$
    $f_{yy}(x, y) = 12y$
    $f_{xy}(x, y) = f_{yx}(x, y) = -1$

37. $\dfrac{\partial^2 z}{\partial x^2} + \dfrac{\partial^2 z}{\partial y^2} = 2 - 2 = 0$

39. Relative minimum: $\left(\dfrac{3}{2}, \dfrac{9}{4}, -\dfrac{27}{16}\right)$
    saddle point: $(0, 0, 0)$

41. Relative minimum: $\left(-4, \dfrac{4}{3}, -2\right)$

43. $\left(\dfrac{4}{3}, \dfrac{1}{3}, \dfrac{16}{27}\right)$, $(0, 1, 0)$

45. $f(49.4, 253) \approx 13{,}202$

47. $y = \dfrac{12}{35}(-3x + 17)$

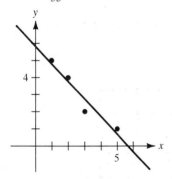

49. $y = \dfrac{1}{2}x^2 - \dfrac{3}{2}x + 2$

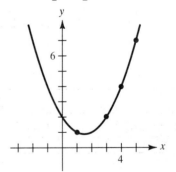

51. $\dfrac{29}{6}$

53. $\dfrac{7}{4}$

55. 36

57. $\dfrac{27}{5}$

59. $\displaystyle\int_0^3 \int_0^{(3-x)/3} dy\, dx = \int_0^1 \int_0^{3-3y} dx\, dy = \dfrac{3}{2}$

61. $\displaystyle\int_{-2}^2 \int_5^{9-x^2} dy\, dx = \int_5^9 \int_{-\sqrt{9-y}}^{\sqrt{9-y}} dx\, dy = \dfrac{32}{3}$

63. $\displaystyle\int_1^2 \int_{-\sqrt{x-1}}^{\sqrt{x-1}} dy\ dx + \int_2^5 \int_{x-3}^{\sqrt{x-1}} dy\ dx = \int_{-1}^2 \int_{y^2+1}^{y+3} dx\ dy = \frac{9}{2}$

65. $\dfrac{4096}{9}$

# CHAPTER 9

## Section 9.1

1. 2, 4, 8, 16, 32

3. $-\dfrac{1}{2}, \dfrac{1}{4}, -\dfrac{1}{8}, \dfrac{1}{16}, -\dfrac{1}{32}$

5. $3, \dfrac{9}{2}, \dfrac{27}{6}, \dfrac{81}{24}, \dfrac{243}{120}$

7. $-1, \dfrac{1}{4}, -\dfrac{1}{9}, \dfrac{1}{16}, -\dfrac{1}{25}$

9. $3n - 2$

11. $n^2 - 2$

13. $\dfrac{n + 1}{n + 2}$

15. $\dfrac{(-1)^{n-1}}{2^{n-2}}$

17. $\dfrac{n + 1}{n}$

19. $(-1)^{n-1}$

21. $\dfrac{1}{n!}$

23. Converges to 0

25. Converges to 1

27. Diverges

29. Converges to $\dfrac{3}{2}$

31. Diverges

33. Converges to 0

35. Converges to 3

37. Converges to 0

39. Diverges

41. (a) No

(b)

| Month | 1 | 2 | 3 | 4 | 5 |
|---|---|---|---|---|---|
| Amount | $9,086.25 | $9,173.33 | $9,261.24 | $9,349.99 | $9,439.60 |

| Month | 6 | 7 | 8 | 9 | 10 |
|---|---|---|---|---|---|
| Amount | $9,530.06 | $9,621.39 | $9,713.59 | $9,806.68 | $9,900.66 |

43. (a) $S_1 = 1$
$S_2 = 3$
$S_3 = 6$
$S_4 = 10$
$S_5 = 15$
(b) $S_{50} = 1275$

45. (a) $2.5(0.8)^n$ billion
(b) 1st year: $2 billion
2nd year: $1.6 billion
3rd year: $1.28 billion
4th year: $1.024 billion
(c) Converges to 0

47. $S_6 = 240$
    $S_7 = 440$
    $S_8 = 810$
    $S_9 = 1490$
    $S_{10} = 2740$

## Section 9.2

1. $S_1 = 1$
  $S_2 = \dfrac{5}{4} = 1.25$
  $S_3 = \dfrac{49}{36} \approx 1.361$
  $S_4 = \dfrac{205}{144} \approx 1.424$
  $S_5 = \dfrac{5269}{3600} \approx 1.464$

3. $S_1 = 3$
  $S_2 = \dfrac{9}{2} = 4.5$
  $S_3 = \dfrac{21}{4} = 5.25$
  $S_4 = \dfrac{45}{8} = 5.625$
  $S_5 = \dfrac{93}{16} = 5.8125$

5. $n$th-Term Test: $\lim\limits_{n \to \infty} \dfrac{n}{n+1} = 1 \neq 0$

7. $n$th-Term Test: $\lim\limits_{n \to \infty} \dfrac{n^2}{n^2+1} = 1 \neq 0$

9. Geometric series: $r = \dfrac{3}{2} > 1$

11. Geometric series: $r = 1.055 > 1$

13. $r = \dfrac{3}{4} < 1$

15. $r = 0.9 < 1$

17. 2

19. $\dfrac{2}{3}$

21. $4 + 2\sqrt{2} \approx 6.828$

23. $\dfrac{10}{9}$

25. $\dfrac{9}{4}$

27. $\dfrac{1}{2}$

29. $\dfrac{17}{6}$

31. $\lim\limits_{n \to \infty} \dfrac{n+10}{10n+1} = \dfrac{1}{10} \neq 0$; diverges

33. $\lim\limits_{n \to \infty} \dfrac{n+1}{n} = 1 \neq 0$; diverges

35. $\lim\limits_{n \to \infty} \dfrac{3n-1}{2n+1} = \dfrac{3}{2} \neq 0$; diverges

37. Geometric series: $r = 1.075 > 1$
    diverges

39. Geometric series: $r = 0.075 < 1$
    converges

41. $\dfrac{2}{3}$

43. $\dfrac{4}{11}$

45. (a) $80{,}000(1 - 0.9^n)$
    (b) $80{,}000$

47. $\dfrac{2896}{19} \approx 152.42$ feet

49. $\dfrac{1}{3}$ of area of square

51. $11{,}616.95

53. $235{,}821.51

55. $10{,}485.75

## Section 9.3

1. Converges

3. Diverges

5. Converges

7. Diverges

9. Converges

11. Converges

13. Diverges

15. Converges

17. Diverges

19. Converges

21. $1.036 < S < 1.037$

23. $1.995 < S < 2.628$

25. Diverges, $n$-th Term Test

27. Converges, $p$-series

29. Converges, geometric series

31. Converges, $p$-series

33. Diverges, geometric series

35. Diverges, Ratio Test

## Section 9.4

1. 2

3. 1

5. $\infty$

7. 0

9. 4

11. 5

13. 1

15. $c$

17. $\dfrac{1}{2}$

19. $\infty$

21. $\displaystyle\sum_{n=0}^{\infty} \frac{x^n}{n!}$, $R = \infty$

23. $\displaystyle\sum_{n=0}^{\infty} \frac{(2x)^n}{n!}$, $R = \infty$

25. $\displaystyle\sum_{n=0}^{\infty} (-1)^n x^n$, $R = 1$

27. $1 + \dfrac{1}{2}(x - 1) - \displaystyle\sum_{n=2}^{\infty} \frac{(-1)^n 1 \cdot 3 \cdot 5 \cdots (2n - 3)(x - 1)^n}{2^n n!}$, $R = 1$

29. $\displaystyle\sum_{n=1}^{\infty} \frac{(-1)^{n-1}}{n}(x - 1)^n$, $R = 1$

31. $\displaystyle\sum_{n=0}^{\infty} (-1)^n (n + 1) x^n$, $R = 1$

33. $1 + \displaystyle\sum_{n=1}^{\infty} \frac{(-1)^n 1 \cdot 3 \cdot 5 \cdots (2n - 1)}{2^n n!} x^n$, $R = 1$

35. $R = 2$ (all parts)

37. $R = 1$ (all parts)

39. $\displaystyle\sum_{n=0}^{\infty} \frac{x^{2n}}{n!}$

41. $2 \displaystyle\sum_{n=0}^{\infty} \frac{x^{2n+1}}{n!}$

43. $\displaystyle\sum_{n=0}^{\infty} (-1)^n x^{2n}$

45. $\displaystyle\sum_{n=0}^{\infty} \frac{(-1)^n x^{2n+2}}{n + 1}$

47. $\displaystyle\sum_{n=0}^{\infty} \frac{(-1)^n (x - 1)^{n+1}}{n + 1}$

49. $\displaystyle\sum_{n=1}^{\infty} (-1)^{n-1} n x^{n-1}$

## Section 9.5

1. (a) $S_1(x) = 1 + x$

(b) $S_2(x) = 1 + x + \dfrac{x^2}{2}$

(c) $S_3(x) = 1 + x + \dfrac{x^2}{2} + \dfrac{x^3}{6}$

(d) $S_4(x) = 1 + x + \dfrac{x^2}{2} + \dfrac{x^3}{6} + \dfrac{x^4}{24}$

3. (a) $S_1(x) = 1 + \dfrac{x}{2}$

(b) $S_2(x) = 1 + \dfrac{x}{2} - \dfrac{x^2}{8}$

(c) $S_3(x) = 1 + \dfrac{x}{2} - \dfrac{x^2}{8} + \dfrac{3x^3}{48}$

(d) $S_4(x) = 1 + \dfrac{x}{2} - \dfrac{x^2}{8} + \dfrac{3x^3}{48} - \dfrac{15x^4}{384}$

5. (a) $S_2(x) = 1 - x^2$
   (b) $S_4(x) = 1 - x^2 + x^4$
   (c) $S_6(x) = 1 - x^2 + x^4 - x^6$
   (d) $S_8(x) = 1 - x^2 + x^4 - x^6 + x^8$

7.

| $x$ | 0 | 0.25 | 0.50 | 0.75 | 1.00 |
|---|---|---|---|---|---|
| $f(x)$ | 1.0000 | 1.1331 | 1.2840 | 1.4550 | 1.6487 |
| $S_1(x)$ | 1.0000 | 1.1250 | 1.2500 | 1.3750 | 1.5000 |
| $S_2(x)$ | 1.0000 | 1.1328 | 1.2813 | 1.4453 | 1.6250 |
| $S_3(x)$ | 1.0000 | 1.1331 | 1.2839 | 1.4541 | 1.6458 |
| $S_4(x)$ | 1.0000 | 1.1331 | 1.2840 | 1.4549 | 1.6484 |

9. 0.607  11. 0.405  13. 0.461  15. 0.481

17. $n = 6$  19. $\dfrac{1}{6!} \approx 0.0083$

## Section 9.6

1.

| $n$ | $x_n$ | $f(x_n)$ | $f'(x_n)$ | $f(x_n)/f'(x_n)$ | $x_n - [f(x_n)/f'(x_n)]$ |
|---|---|---|---|---|---|
| 1 | 1.700 | $-0.110$ | 3.400 | $-0.032$ | 1.732 |

3. 0.682  5. 1.146  7. $-4.596, -1.042, 5.638$

9. 0.567  11. $\pm 0.753$  13. 2.208

15. 2.893  17. $f'(x_1) = 0$  19. $1 = x_1 = x_3 = \cdots$
                                    $0 = x_2 = x_4 = \cdots$

21. 1.913  23. 2.091  25. $(1.939, 0.240)$

27. $x \approx 1.563$ miles  29. $t \approx 4.486$ hours

## Review Exercises for Chapter 9

1. $\dfrac{n}{2n + 1}$, $n = 1, 2, 3, \ldots$

3. $(-1)^n \dfrac{2^n}{3^{n+1}}$, $n = 0, 1, 2, \ldots$

5. Converges to 0  7. Diverges  9. Converges to 5

11. $S_0 = 1$

$S_1 = \dfrac{5}{2} = 2.5$

$S_2 = \dfrac{19}{4} = 4.75$

$S_3 = \dfrac{65}{8} = 8.125$

$S_4 = \dfrac{211}{16} = 13.1875$

13. $S_1 = \dfrac{1}{2} = 0.5$

$S_2 = \dfrac{11}{24} \approx 0.4583$

$S_3 = \dfrac{331}{720} \approx 0.4597$

$S_4 = \dfrac{18,535}{40,320} \approx 0.4597$

$S_5 = \dfrac{1,668,151}{3,628,800} \approx 0.4597$

15. $\dfrac{5}{4}$

17. $-\dfrac{1}{6}$

19. $\dfrac{1}{11}$

21. Converges

23. Diverges

25. Diverges

27. Converges

29. Diverges

31. $R = 1$

33. $R = 0$

35. $\displaystyle\sum_{n=0}^{\infty} \dfrac{(-2)^n x^n}{n!}$

37. $-\displaystyle\sum_{n=0}^{\infty} (x + 1)^n$

39. $\displaystyle\sum_{n=1}^{\infty} \dfrac{(-1)^{n+1} x^{n+1}}{n(n + 1)}$

41. 0.548

43. 1.822

45. 0.301

47.

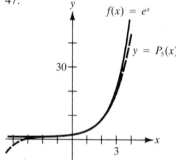

49. $-1.532,\ -0.347,\ 1.879$

51. $-1.164,\ 1.453$

# CHAPTER 10

## Section 10.1

1. (a) $396°,\ -324°$
   (b) $315°,\ -405°$

3. (a) $660°,\ -60°$
   (b) $20°,\ -340°$

5. (a) $\dfrac{19\pi}{9},\ -\dfrac{17\pi}{9}$
   (b) $\dfrac{10\pi}{3},\ -\dfrac{2\pi}{3}$

7. (a) $\dfrac{7\pi}{4},\ -\dfrac{\pi}{4}$
   (b) $\dfrac{28\pi}{15},\ -\dfrac{32\pi}{15}$

9. $\dfrac{\pi}{6}$  11. $\dfrac{7\pi}{4}$  13. $-\dfrac{\pi}{9}$  15. $-\dfrac{3\pi}{2}$

17. 270°  19. −105°  21. 420°  23. 330°

25. $c = 10$, $\theta = 60°$  27. $a = 4\sqrt{3}$, $\theta = 30°$

29. $\theta = 40°$  31. $h = 15$ feet

33.

| $r$ | 8 ft | 15 in | 85 cm | 24 in | $\dfrac{12{,}963}{\pi}$ mi |
|---|---|---|---|---|---|
| $s$ | 12 ft | 24 in | 200.28 cm | 96 in | 8642 mi |
| $\theta$ | 1.5 | 1.6 | $\dfrac{3\pi}{4}$ | 4 | $\dfrac{2\pi}{3}$ |

35. (a) $\dfrac{5\pi}{12}$  37. 4.655°

(b) $7.8125\pi$ inches

## Section 10.2

1. $\sin \theta = \dfrac{4}{5}$,  $\csc \theta = \dfrac{5}{4}$

$\cos \theta = \dfrac{3}{5}$,  $\sec \theta = \dfrac{5}{3}$

$\tan \theta = \dfrac{4}{3}$,  $\cot \theta = \dfrac{3}{4}$

3. $\sin \theta = -\dfrac{5}{13}$,  $\csc \theta = -\dfrac{13}{5}$

$\cos \theta = -\dfrac{12}{13}$,  $\sec \theta = -\dfrac{13}{12}$

$\tan \theta = \dfrac{5}{12}$,  $\cot \theta = \dfrac{12}{5}$

5. $\sin \theta = \dfrac{1}{2}$,  $\csc \theta = 2$

$\cos \theta = -\dfrac{\sqrt{3}}{2}$,  $\sec \theta = -\dfrac{2\sqrt{3}}{3}$

$\tan \theta = -\dfrac{\sqrt{3}}{3}$,  $\cot \theta = -\sqrt{3}$

7. $\csc \theta = 2$

9. $\cot \theta = \dfrac{4}{3}$

11. $\sec \theta = \dfrac{17}{8}$

13.

$\sin \theta = \dfrac{2}{3}$,  $\csc \theta = \dfrac{3}{2}$

$\cos \theta = \dfrac{\sqrt{5}}{3}$,  $\sec \theta = \dfrac{3\sqrt{5}}{5}$

$\tan \theta = \dfrac{2\sqrt{5}}{5}$,  $\cot \theta = \dfrac{\sqrt{5}}{2}$

15.

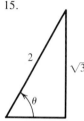

$$\sin \theta = \frac{\sqrt{3}}{2}, \qquad \csc \theta = \frac{2\sqrt{3}}{3}$$

$$\cos \theta = \frac{1}{2}, \qquad \sec \theta = 2$$

$$\tan \theta = \sqrt{3}, \qquad \cot \theta = \frac{\sqrt{3}}{3}$$

17.

$$\sin \theta = \frac{3\sqrt{10}}{10}, \qquad \csc \theta = \frac{\sqrt{10}}{3}$$

$$\cos \theta = \frac{\sqrt{10}}{10}, \qquad \sec \theta = \sqrt{10}$$

$$\tan \theta = 3, \qquad \cot \theta = \frac{1}{3}$$

19. Quadrant III 　　　　　21. Quadrant I 　　　　　23. Quadrant II

25. (a) $\sin 60° = \dfrac{\sqrt{3}}{2}$

　　　$\cos 60° = \dfrac{1}{2}$

　　　$\tan 60° = \sqrt{3}$

　(b) $\sin \dfrac{2\pi}{3} = \dfrac{\sqrt{3}}{2}$

　　　$\cos \dfrac{2\pi}{3} = -\dfrac{1}{2}$

　　　$\tan \dfrac{2\pi}{3} = -\sqrt{3}$

27. (a) $\sin \left(-\dfrac{\pi}{6}\right) = -\dfrac{1}{2}$

　　　$\cos \left(-\dfrac{\pi}{6}\right) = \dfrac{\sqrt{3}}{2}$

　　　$\tan \left(-\dfrac{\pi}{6}\right) = -\dfrac{\sqrt{3}}{3}$

　(b) $\sin 150° = \dfrac{1}{2}$

　　　$\cos 150° = -\dfrac{\sqrt{3}}{2}$

　　　$\tan 150° = -\dfrac{\sqrt{3}}{3}$

29. (a) $\sin 225° = -\dfrac{\sqrt{2}}{2}$

　　　$\cos 225° = -\dfrac{\sqrt{2}}{2}$

　　　$\tan 225° = 1$

　(b) $\sin(-225°) = \dfrac{\sqrt{2}}{2}$

　　　$\cos(-225°) = -\dfrac{\sqrt{2}}{2}$

　　　$\tan(-225°) = -1$

31. (a) $\sin 750° = \dfrac{1}{2}$

　　　$\cos 750° = \dfrac{\sqrt{3}}{2}$

　　　$\tan 750° = \dfrac{\sqrt{3}}{3}$

　(b) $\sin 510° = \dfrac{1}{2}$

　　　$\cos 510° = -\dfrac{\sqrt{3}}{2}$

　　　$\tan 510° = -\dfrac{\sqrt{3}}{3}$

33. (a) 0.1736 　　35. (a) 0.3640 　　37. (a) −0.3420 　　39. (a) 2.0070
　　(b) 5.7588 　　　　(b) 0.3640 　　　　(b) −0.3420 　　　　(b) 2.0000

41. (a) $\dfrac{\pi}{6}, \dfrac{5\pi}{6}$

   (b) $\dfrac{7\pi}{6}, \dfrac{11\pi}{6}$

43. (a) $\dfrac{\pi}{3}, \dfrac{2\pi}{3}$

   (b) $\dfrac{3\pi}{4}, \dfrac{7\pi}{4}$

45. (a) $\dfrac{\pi}{4}, \dfrac{5\pi}{4}$

   (b) $\dfrac{5\pi}{6}, \dfrac{11\pi}{6}$

| Degrees | Radians |
|---|---|

47. (a) 55°, 125°    $\dfrac{11\pi}{36}, \dfrac{25\pi}{36}$

   (b) 195°, 345°    $\dfrac{13\pi}{12}, \dfrac{23\pi}{12}$

49. (a) 10°, 350°    $\dfrac{\pi}{18}, \dfrac{35\pi}{18}$

   (b) 126°, 234°    $\dfrac{7\pi}{10}, \dfrac{13\pi}{10}$

51. (a) 50°, 230°    $\dfrac{5\pi}{18}, \dfrac{23\pi}{18}$

   (b) 97°, 277°    $\dfrac{97\pi}{180}, \dfrac{277\pi}{180}$

53. $\dfrac{\pi}{4}, \dfrac{3\pi}{4}, \dfrac{5\pi}{4}, \dfrac{7\pi}{4}$

55. $0, \dfrac{\pi}{4}, \pi, \dfrac{5\pi}{4}, 2\pi$

57. $0, \pi, 2\pi$

59. $\dfrac{\pi}{4}, \dfrac{5\pi}{4}$

61. $0, \dfrac{\pi}{2}, \pi, 2\pi$

63. $\dfrac{100\sqrt{3}}{3}$

65. $\dfrac{25\sqrt{3}}{3}$

67. 15.5572

69. 9.1925

71. $20 \sin 75° \approx 19.32$ feet

73. $150 \cot 4° \approx 2145.1$ feet

75. (a) 25.2°F
   (b) 65.1°F
   (c) 50.8°F

## Section 10.3

1. Period: $\pi$
   amplitude: 2

3. Period: $4\pi$
   amplitude: $\dfrac{3}{2}$

5. Period: 2
   amplitude: $\dfrac{1}{2}$

7. Period: $2\pi$
   amplitude: 2

9. Period: $\dfrac{\pi}{5}$
   amplitude: 2

11. Period: $3\pi$
   amplitude: $\dfrac{1}{2}$

13. Period: $\dfrac{1}{2}$
   amplitude: 3

15. Period: $\dfrac{\pi}{2}$

17. Period: $\dfrac{2\pi}{5}$

19. Period: 3

21. c

22. e

23. f

24. a

25. b

26. d

27.

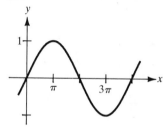

29.

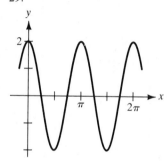

31.

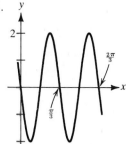

33.

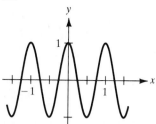

35.

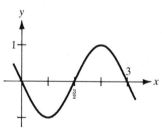

37.

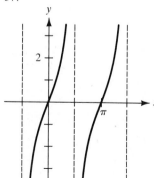

39.

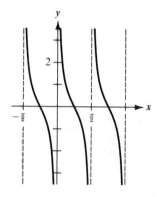

41.

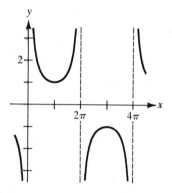

43.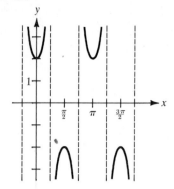

45.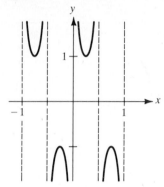

47. (a) 6 seconds
    (b) 10
    (c)

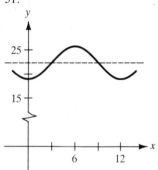

49. (a) $\dfrac{1}{440}$

    (b) 440

    (c) y

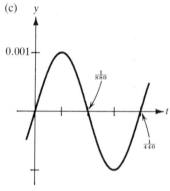

51.

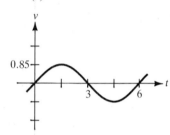

53.

| x | −1.0 | −0.1 | −0.01 | −0.001 | 0.001 | 0.01 | 0.1 | 1.0 |
|---|---|---|---|---|---|---|---|---|
| f(x) | −0.4597 | −0.0500 | −0.0050 | −0.0005 | 0.0005 | 0.0050 | 0.0500 | 0.4597 |

$$\lim_{x\to 0} \frac{1-\cos x}{x} = 0$$

**55.**

| x | $-1.0$ | $-0.1$ | $-0.01$ | $-0.001$ | 0.001 | 0.01 | 0.1 | 1.0 |
|---|---|---|---|---|---|---|---|---|
| **f(x)** | 0.1683 | 0.1997 | 0.2000 | 0.2000 | 0.2000 | 0.2000 | 0.1997 | 0.1683 |

$$\lim_{x \to 0} \frac{\sin x}{5x} = \frac{1}{5}$$

## Section 10.4

1. $2x + \sin x$

3. $-\dfrac{1}{x^2} - 3 \cos x$

5. $\dfrac{2}{\sqrt{x}} - 3 \sin x$

7. $t^2 \cos t + 2t \sin t$

9. $-\dfrac{t \sin t + \cos t}{t^2}$

11. $\sec^2 x + 2x$

13. $5 \csc x - 5x \csc x \cot x$

15. $4 \cos 4x$

17. $2x \sec x^2 \tan x^2$

19. $-\csc 2x \cot 2x$

21. $\sin \dfrac{1}{x} - \dfrac{1}{x} \cos \dfrac{1}{x}$

23. $12 \sec^2 4x$

25. $-2 \cos x \sin x = -\sin 2x$

27. $-4 \cos x \sin x = -2 \sin 2x$

29. $-\csc^2 x$

31. $\dfrac{1}{\sin x} \cos x = \cot x$

33. $\dfrac{-\csc x \cot x + \csc^2 x}{\csc x - \cot x} = \csc x$

35. $\sec^2 x - 1 = \tan^2 x$

37. $\dfrac{\cos x}{2\sqrt{\sin x}}$

39. $\dfrac{1}{2}(x \sec^2 x + \tan x - \sec x \tan x)$

41. $\dfrac{\cos x}{2 \sin 2y},\ 0$

43. $y'' + y = (-2 \sin x - 3 \cos x) + (2 \sin x + 3 \cos x) = 0$

45. $y'' + 4y = (-4 \cos 2x - 4 \sin 2x) + 4(\cos 2x + \sin 2x) = 0$

47. 3

49. 2

51. 1

53. $y = 2x + \left(\dfrac{\pi}{2} - 1\right)$

55. Relative maximum: $\left(\dfrac{\pi}{3}, \dfrac{3\sqrt{3}}{2}\right)$

relative minimum: $\left(\dfrac{5\pi}{3}, -\dfrac{3\sqrt{3}}{2}\right)$

point of inflection: $(\pi, 0)$

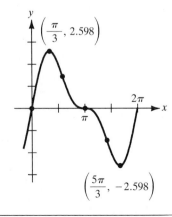

$\left(\dfrac{\pi}{3}, 2.598\right)$

$\left(\dfrac{5\pi}{3}, -2.598\right)$

57. Relative minima: $\left(\dfrac{\pi}{3}, \dfrac{\pi}{3} - \sqrt{3}\right)$,

$\left(\dfrac{7\pi}{3}, \dfrac{7\pi}{3} - \sqrt{3}\right)$

relative maxima: $\left(\dfrac{5\pi}{3}, \dfrac{5\pi}{3} + \sqrt{3}\right)$,

$\left(\dfrac{11\pi}{3}, \dfrac{11\pi}{3} + \sqrt{3}\right)$

points of inflection: $(0, 0)$, $(\pi, \pi)$, $(2\pi, 2\pi)$, $(3\pi, 3\pi)$, $(4\pi, 4\pi)$

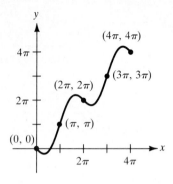

59. (a) Relative maxima: $(0.0376, 178.29)$, $(6.0376, 181.65)$, $(12.0376, 185.01)$, $(18.0376, 188.37)$
relative minima: $(3.0376, 152.79)$, $(9.0376, 156.15)$, $(15.0376, 159.51)$, $(21.0376, 162.87)$

(b) $182.32$ billion kilowatt hours

61. (a) $t = 214.5$ (August 2 and 3)

(b) $t = 32$ (February 1)

63. $\displaystyle\sum_{n=0}^{\infty} \dfrac{(-1)^n x^{2n+1}}{(2n+1)!}$, $R = \infty$

65. $\displaystyle\sum_{n=0}^{\infty} \dfrac{(-1)^n x^{4n+2}}{(2n+1)!}$

67. $\displaystyle\sum_{n=0}^{\infty} \dfrac{(-1)^n (2x)^{2n}}{(2n)!}$

## Section 10.5

1. $-2\cos x + 3\sin x + C$

3. $t + \csc t + C$

5. $\tan\theta + \cos\theta + C$

7. $-\dfrac{1}{2}\cos 2x + C$

9. $\dfrac{1}{2}\sin x^2 + C$

11. $2\tan\dfrac{x}{2} + C$

13. $-\dfrac{1}{3}\ln|\cos 3x| + C$

15. $\dfrac{1}{5}\tan^5 x + C$

17. $\dfrac{1}{\pi}\ln|\sin \pi x| + C$

19. $\dfrac{1}{2}\ln|\csc 2x - \cot 2x| + C$

21. $\ln|\tan x| + C$

23. $\ln|\sec x - 1| + C$

25. $-\ln|1 + \cos x| + C$

27. $\dfrac{1}{2}\tan^2 x + C$

29. $\sin e^x + C$

31. $\ln|\cos e^{-x}| + C$

33. $x - \dfrac{1}{4}\cos 4x + C$

35. $x\sin x + \cos x + C$

37. $x\tan x + \ln|\cos x| + C$

39. $\dfrac{3\sqrt{3}}{4}$

41. $2(\sqrt{3} - 1) \approx 1.4641$

43. $\dfrac{1}{2}$

45. $-1 + \sec 1 \approx 0.8508$

47. 2

49. $\ln \sqrt{2} \approx 0.3466$

51. 2

53. $\pi$

55. $y = \dfrac{1}{5}(2 \sin x - \cos x) + Ce^{-2x}$

57. 1.3707

59. (a) 1.3655
    (b) 1.3708

61. (a) 225.28 million barrels
    (b) 225.28 million barrels
    (c) 217 million barrels

63. 1.6234 liters

65. (a) $C \approx \$9.17$
    (b) $C \approx \$3.14$, Savings $\approx \$6.03$

## Review Exercises for Chapter 10

1. $\dfrac{3\pi}{4}, -\dfrac{5\pi}{4}$

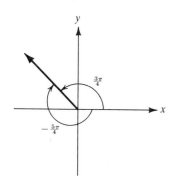

3. 128.571°

5. $\dfrac{8\pi}{3} \approx 8.38$

7. $\dfrac{\pi}{5}$

9. $\sin \theta = \dfrac{2\sqrt{53}}{53}$,   $\csc \theta = \dfrac{\sqrt{53}}{2}$

   $\cos \theta = -\dfrac{7\sqrt{53}}{53}$,   $\sec \theta = -\dfrac{\sqrt{53}}{7}$

   $\tan \theta = -\dfrac{2}{7}$   $\cot \theta = -\dfrac{7}{2}$

11. $\sin \theta = -\dfrac{\sqrt{11}}{6}$,   $\csc \theta = -\dfrac{6\sqrt{11}}{11}$

    $\cos \theta = \dfrac{5}{6}$,   $\sec \theta = \dfrac{6}{5}$

    $\tan \theta = -\dfrac{\sqrt{11}}{5}$,   $\cot \theta = -\dfrac{5\sqrt{11}}{11}$

| Degrees | Radians |
|---|---|
| 13. 135°, 225° | $\dfrac{3\pi}{4}, \dfrac{5\pi}{4}$ |
| 15. 210°, 330° | $\dfrac{7\pi}{6}, \dfrac{11\pi}{6}$ |
| 17. 57°, 123° | $\dfrac{19\pi}{60}, \dfrac{41\pi}{60}$ |
| 19. 165°, 195° | $\dfrac{11\pi}{12}, \dfrac{13\pi}{12}$ |

21.

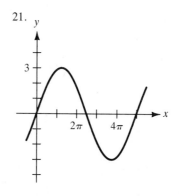

23.

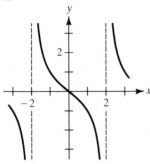

25. $13{,}200 \tan 28° \approx 7018$ feet $\approx 1.329$ miles

27. $-5\pi \sin 5\pi x$

29. $-x \sec^2 x - \tan x$

31. $\dfrac{x \cos x - 2 \sin x}{x^3}$

33. $24 \sin 4x \cos 4x + 1 = 12 \sin 8x + 1$

35. $-6 \csc^3 x \cot x$

37. $e^x(\sec^2 x + \tan x)$

39. $\sec y$

41. $-\dfrac{\sin(x + y) + 1}{\sin(x + y)} = -1 - \csc(x + y)$

43. $2 \csc^2 x \cot x$

45. $\dfrac{-x^2 \cos x + 2x \sin x + 2 \cos x}{x^3}$

47. $-\dfrac{1}{2} \csc 2x + C$

49. $-\dfrac{4}{\pi} \ln \left| \cos \dfrac{\pi x}{4} \right| + C$

51. $\dfrac{\tan^{n+1} x}{n + 1} + C$

53. $3x + \dfrac{1}{\pi} \cos 5\pi x + C$

55. $\dfrac{1}{2} \sin x^2 + C$

59. $y = -\dfrac{\pi}{2}\left(x - \dfrac{\pi}{2}\right)$

61. January 15, July 15

63. Area of rectangle: 1.12
area of region: 2

# Index

Second-Partials Test, 596
Separation of variables, 527
Sequence, 652
  convergence of, 653
  divergence of, 653
  limit of, 654
  patterns for, 657
  term of, 652
Sequence of partial sums, 662
Series
  binomial, 687
  convergence of, 662
  divergence of, 662
  geometric, 664
  harmonic, 673
  infinite, 662
  power, 681
  properties of, 662
  $p$-series, 673
  sum of, 662
  summary of tests for convergence or divergence, 678
Sigma notation, 661
Similar triangles, 718
Simple Power Rule
  for differentiation, 146
  for integration, 320
Simpson's Rule, 475
  error, 476
Sine function, 724
Slope
  of a curve, 133, 135
  of a line, 78
Slope-intercept form of the equation of a line, 75
Solid analytic geometry, 554
Solid region, volume of, 636
Solid of revolution, 366
Solution curve, 521
Solution of a differential equation, 520
Solution of an inequality, 5
  test intervals for, 7
Solution point of an equation, 61
Solution set of the inequality, 6
Speed, 196
Sphere, 557

Standard deviation, 506
Standard form
  of the equation of a circle, 65
  of the equation of a sphere, 557
Standard position of an angle, 716
Straight-line depreciation, 74
Substitution, 335, 430
Subtraction of fractions, 37
Sum Rule
  for differentiation, 149
  for integration, 320
Sum of a series, 662
Sum of the squared errors, 615
Sum of two functions, 93
Summary of differentiation rules, 187
Summation process, 359
Surface in space, 558
  quadric, 564
Synthetic division, 33

Tables of integrals, 460
Tangent function, 724
Tangent line, 132
Tangent line approximation, 306
Taylor polynomial, 691
Taylor series, 684
Taylor's Remainder Theorem, 693
Taylor's Theorem, 684
Techniques for evaluating limits, 105
Term of a sequence, 652
Terminal ray, 716
Test for concavity, 246
Test for increasing or decreasing functions, 224
Third derivative, 191
Three-dimensional coordinate system, 554
Total cost, 69, 161
Total profit, 161
Total revenue, 69, 161
Trace of a surface, 559
Transcendental function, 294
Trapezoidal Rule, 471
  error, 476

Triangle(s), 718
  equilateral, 718
  isosceles, 718
  right, 718
  similar, 718
Trigonometric function, 724
  differentiation of, 748
  graph of, 739
  integration of, 758, 764
Trigonometric identities, 725, 726
Truncation, 123
Two-point form of the equation of a line, 81

Unbounded function, 113
Upper limit of integration, 339

Variable
  dependent, 87
  discrete, 161
  independent, 87
Variance, 506, 507
Velocity, 159
Vertex of an angle, 716
Vertical asymptote, 279
Vertical line, 82
Vertical line test, 89
Volume by a double integral, 636
Volume of a solid, 636

$x$-axis, 2, 49, 554
$x$-coordinate, 49
$x$-intercept, 63
$xy$-plane, 554
$xz$-plane, 554

$y$-axis, 49, 554
$y$-coordinate, 49
$y$-intercept, 63
$yz$-plane, 554

$z$-axis, 554
Zero of a polynomial, 28

# Index of Applications (Continued)